# International Conference on Forest Products (ICFP) 2020: 12th International Symposium of IWoRS



IOP Conference Series: Materials Science and Engineering
Volume 935

Online
1 September 2020

ISBN: 978-1-7138-4771-7
ISSN: 1757-8981

**Printed from e-media with permission by:**

Curran Associates, Inc.
57 Morehouse Lane
Red Hook, NY 12571

**Some format issues inherent in the e-media version may also appear in this print version.**



Institute of Physics
Dirac House, Temple Back
Bristol BS1 6BE UK

Phone: 44 1 17 929 7481
Fax: 44 1 17 920 0979

techtracking@iop.org

**Additional copies of this publication are available from:**

Curran Associates, Inc.
57 Morehouse Lane
Red Hook, NY 12571 USA
Phone: 845-758-0400
Fax: 845-758-2633
Email: curran@proceedings.com
Web: www.proceedings.com

# TABLE OF CONTENTS

IOP Conf. Series: Materials Science and Engineering **935** (2020) 011001 doi:10.1088/1757-899X/935/1/011001

The International Conference on Forest Products (ICFP) 2020, held on September 1st, 2020, is the third international conference organized by Forest Products Research and Development Center, Ministry of Environment and Forestry, Republic of Indonesia. The forum this time is exceptional because it is held virtually during the COVID 19 pandemic, which has not ended. Besides, it also raises the thematic series of the 12th International Symposium of IWoRS (Indonesian Wood Research Society). It is a strategic step to develop a research network in the field of forest product processing with various areas for the management of forest resources and the environment. This conference is a forum for researchers, scientists, and all stakeholders from multiple research institutions and universities to exchange information and experiences and develop research and development collaborations and synergies. Therefore, we still held this conference even though it is virtually through the zoom application.

The theme of the **ICFP 2020 thematic series 12th International Symposium of IWoRS** is Forest Products Processing Innovation for Communities and Sustainable Forest and Environment Management, which is a strategic theme to encourage the role of science and technology in commercialization and to increase the complete application of science and technology.

More than 600 participants and invitees attended the conference. The conference was held with plenary sessions and parallel sessions based on topics. In this forum, we discussed and listened to information about policies and developments in science and technology in the area of forest product processing, with keynote speakers from Indonesia, Japan, and Korea. There are 75 local and international presenters from various institutions in Indonesia and other countries. They presented outstanding science and technology information on timber and non-timber forest products, biomass and renewable energy, bio-composite and nanotechnology, wood engineering, policy, and sustainable forest management.

Information on research results at this conference can be a benchmark for future research and development and become a modality for developing research and development collaborations.

Thank you.

Dr. Wening Sri Wulandari
Chairman of Organizing Committee

IOP Conf. Series: Materials Science and Engineering **935** (2020) 011002 doi:10.1088/1757-899X/935/1/011002

# Peer review declaration

All papers published in this volume of *IOP Conference Series: Materials Science and Engineering* have been peer reviewed through processes administered by the Editors. Reviews were conducted by expert referees to the professional and scientific standards expected of a proceedings journal published by IOP Publishing.

- **Type of peer review:** Single-blind
- **Conference submission management system:** https://conference.forpro.org
- **Number of submissions received:** 76
- **Number of submissions sent for review:** 76
- **Number of submissions accepted:** 75
- **Acceptance Rate (Number of Submissions Accepted / Number of Submissions Received X 100):** 98.68%
- **Average number of reviews per paper:** 4 reviews
- **Total number of reviewers involved:** 8 reviewers
- **Any additional info on review process:** -
- **Contact person for queries:** Dr. Wening Sri Wulandari (weningwulandari@yahoo.com)

IOP Conf. Series: Materials Science and Engineering **935** (2020) 012001 doi:10.1088/1757-899X/935/1/012001

# Waste extracts from forest plants and their application as natural coloring for fabrics

**E Basri[1*] and Saefudin[2]**

[1]Forest Products Research and Development Center, Ministry of Environment and Forestry, Bogor, 16610, Indonesia
[2]Biology Research Centre, Indonesian Institute of Science, Cibinong, Indonesia

*Email: denvig@yahoo.com

**Abstract**. Wood wastes and tree barks are rich in natural coloring-pigments, such as tannin, flavonoids, and anthocyanins. If extracted, those compounds can be beneficial as coloring agents for fabrics. The research aimed to obtain yields of dry extracts from bark wastes of secang (*Caesalpinia* sp.), bakau merah (*Rhizophora apiculata*), ketapang (*Terminalia* sp.), and mangium (*Acacia mangium*); as well as from wood wastes of secang and ketapang. Wet extracts were obtained through waste extraction using pure water solvent with waste/water ratio 1:4 and 1:6 (w/v), each implementing of two temperatures (60°C and 70°C) and three long immersion treatments before extracting (0/control, 12, and 24 hours). The obtained wet extracts were applied for coloring the fabrics, followed with 3 fixation treatment, i.e. tawas, kapur tohor, and tunjung; and then tested for color-fastness resistance of fabrics against, i.e. washing, ironing, and sun ray referred to the SNI-ISO standards. The results indicated that bakau minyak barks extract brought out the fabric colors to the best result, and the yield dry extract was the highest. The natural coloring extracts of bakau minyak barks produced the best coloring performance of the fabrics against consecutive washing and sun-ray exposure, that is moderate until good (score 3-4), and the ironing results in good until very good (4-5). The yield dry extract of the barks was 28% after implementing 24 hours immersion in water before extraction, weight ratio between the extracted barks and water at 1:4 (w/v), and extraction temperature at 70°C.

## 1. Introduction

Production of round timber (woods) from the Permit Endeavors in Utilization of Wood Forest Products at Natural Forests (PEUWFP-NF) averaged about 5,865,985.80 $m^3$/year [1]. The amount of wastes generated from activities of producing branch-free tree trunks reached 928,934.81 $m^3$/year [2]. This figure still did not include round timber production from the Permit Endeavors in Utilization of Wood Forest Products at Plantation Forests (PEUWFP-PF), whereby PF's timber production tended to increase each year. In 2018, production of round timber from plantation forests had already reached 40,945,378.90 $m^3$ [1]. Research results by [3] at Indonesia's paper mill company (PT Tanjung Enim Lestari Pulp & Paper) that used raw materials of round woods from mangium plantation forests disclosed that the amount of wastes especially in the forms of tree barks, which were generated throughout the pulp manufacture, could reach 216.49 ton/day or about 9% of the weight of round wood. Further, the wastes in the forms of slabs, wood sawdust, twigs, barks, etc. were not yet utilized optimally by wood-processing industries, as they focused only on utilization of wood portions. This

IOP Conf. Series: Materials Science and Engineering **935** (2020) 012001 doi:10.1088/1757-899X/935/1/012001

could create serious problems, because the presence of those generated and abandoned wastes became sooner or later accumulated enormously, thereby causing environment concerns in the forests as well as wood factories.

From several various research results [4,5,6], it revealed that wastes from particular tree portions (e.g. woods and barks) could serve as immense resources that were potentially rich in coloring compounds or pigments. Those pigments, when extracted, could be beneficially used as natural coloring-agents/chemicals for fabrics or weaving threads. Coloring chemicals or pigments encountered a lot in plant woods or barks are among others tannins. Tannins typify as inherently polyphenol compounds, which taste bitter in the mouth, soluble readily in water, and could coagulate proteins or other organic compounds such as amino acids and alkaloids [4]. The pigments from plant tannins could in color appear differently, depending of plant species and particular portions of the plants (e.g. woods, barks, twigs, leaves, etc.). For example, tannins from barks of nangka (*Artocarpus* sp.) trees exhibit red color, while those from nyirih (*Xylocarpus granatum*) wood trees emerge as brownish red color [7]. Natural coloring pigments could be obtained from their source origins through heating, storage, solvent treatment, or combination of any of those processes partially as well as entirely. The solvents which are used could be inorganic (water) or organic solvents. Organic solvents commonly employed are among others methanol and ethanol [8].

Plant pigments could be produced in liquid extracts or in dry extract forms. The advantage of liquid extracts is that they can be used directly for coloring the fabrics; but their disadvantage is that as organic substances the liquid extracts cannot be stored for a long time, because they are susceptible to deterioration and other degradation processes, getting slimy, or contamination by organisms. Accordingly, for long-term uses or for trades, liquid extracts should be processed into dry extract forms. The needs of natural color solutions for coloring the fabrics, depends on surface area of the fabrics and strength intensity of the desired colors at fabrics.

In coloring of fabrics, it involves two consecutive essential stages, which comprise mordanting and fixation. Mordanting intends to expel fat, oil, dirt that stick to the fabric fibers, in order facilitate the penetration of the fabric pores. Meanwhile, fixation aims to strengthen the bonds of colors to the fabrics that has been immersed in the dye solution [9]. In addition, fixation can keep the qualities of color-imparted fabrics from the wear off due to sweat, washing, ironing, and exposure to sun ray. To look into the performance of natural coloring which have been used for coloring the fabrics and then followed with fixative treatment should be tested for color-fading resistance against the washing, ironing, and sun ray exposure. The testing methods should refer to the recognized standards ISO (International Standard Organization) and SNI (Indonesia's National Standard).

This paper presents the results of the application of dye extracts from several types of forest plant waste on cloth and its color fastness tests on washing, rubbing, and exposure to the sun's heat, as well as the yield of dry extract with several treatments.

## 2. Materials and Methods

### *2.1. Materials*

The materials which were used in these research comprised wood waste and bark from secang (*Caesalpinia* sp.) and ketapang (*Terminalia* sp.) trees, which were obtained from Cilacap Regency (Central Java); bark wastes from bakau merah (*Rhizophora apiculata*) trees, obtained from Kubu Raya Regency (West Kalimantan); and bark wastes from mangium (*Acacia mangium*) trees, obtained from Sukabumi Regency (West Java). The materials serving as fixation agents were tawas ($Al_2[SO_4]_3$), kapur tohor ($CaCO_3$), and tunjung ($FeSO_4$). Meanwhile, the material as mordanting agent was tipol chemical. The fabrics that would be imparted with colors were mori or primissima fabrics. The tools which were incorporated consisted of gas stoves, pan for boiling forest plants' wastes, pails, jerry cans, graduated cylinders, stirrers, weighing scales, cloths for filtering, and oven.

IOP Conf. Series: Materials Science and Engineering **935** (2020) 012001 doi:10.1088/1757-899X/935/1/012001

### *2.2. Methods*

#### 2.2.1. Preparing of the wet extracts

Preparing of the wet extracts (in aqueous solution forms), application of wet extracts as natural-coloring at mori (white-colored) fabrics in the Forest Products Research and Development Center (FPRDC), Bogor; and testing on qualities of coloring performance of the extracts at the color-imparted fabrics in the Institute for Quality Testing on Goods, Ministry of Trades, Jakarta, as follows:

*2.2.1.1. Extraction of plant wastes*

Plants' wastes were cleaned. Afterwards, the cleaned barks and wood slabs were finely chopped off such that their dimension was reduced to about 1 cm size. The chopped barks and wood wastes were then boiled (extracted) in water for 2 hours, employing two levels of heating temperatures (60°C and 70°C); and concurrently implementing two ratios between the waste materials and water solvent (1:4 and 1:6, w/v). Before boiling, the chopped barks and wood wastes were immersed in water (regarded as pre-extraction), employing three immersion durations, comprising 0 hour (without immersion/control), 12 hours, and 24 hours. After that, the dye solution was filtered. The obtained liquid (wet extracts) was allowed for some time to cool down.

*2.2.1.2. Mordanting*

Mordanting was performed by immersing the mori fabrics (before fabric coloring with the extracts) in water, to which was already added 1% tipol (w/w), for 2 hours. Afterwards, the fabrics were cleansed by water several times without squeezing until the foams disappeared. Furthermore, the cloth is air-dried until dry.

*2.2.1.3. Coloring of fabrics*

Coloring was performed by immersing the fabric samples in the wet extracts (natural color) as many as 6-7 times intermittently, while turning over the samples repeatedly in the solution such that the immersed-fabrics became evenly colored. Each time of fabric immersion necessitated about 5-minute duration).

*2.2.1.4. Color fixation*

Fixation or locking intended to obtain the color-imparted fabrics with the colors suitable as desired by the users; and in order that the fabric colors were not easily faded away. The fixation process could be performed by immersing the color-imparted fabrics for 15 minutes in water, which was already added 2-3% fixation agents (w/v). In this research, three kinds of fixation agents were used separately, which comprised tawas, kapur tohor, and tunjung. Afterwards, the fabrics were cleansed with water, and then aerated.

*2.2.1.5. Testing on qualities of natural-coloring*

The testing methods for color-fade resistance of the fabrics consecutively against washing, referred to SNI ISO 105-C06: 2010; against ironing, referred to SNI ISO 105 x 12: 2012; and against sun-ray exposure, referred to SNI ISO 105-BO2: 2014. The stronger the color-fade resistance the greater would be the coloring qualities of the extracts at the fabrics. Qualities of the tested materials (only bark extracts) were categorized into five classes, i.e. class 1 (very poor), class 2 (poor), class 3 (moderate), class 4 (good), and class 5 (very good/best coloring performance).

#### 2.2.2. Dry coloring extracts

In this method, 20 ml of wet extract solution derived from various forest plant wastes was taken, then put into porcelain cups, and further heated in the oven (100°C) until it produces a dry extract. The yields of dry coloring extract were determined using the formula:

$$Y = ([Wi - Wf] / Wi) \times 100\% \qquad (1)$$

where, Y: Yield of the dry coloring extract (%)

Wi: Weight of coloring extract solution + weight of porcelain cup (g)

Wf: Weight of dry coloring extract + weight of porcelain cup (g)

*2.3 Data analysis*
The obtained yield data of dry coloring extracts was analyzed and assessed using a completely randomized design (CRD) with factorial pattern. The factors as incorporated were 6 kinds of forest plants' wastes (L), 3 immersing durations (S), 2 heating temperatures for extraction (H), and 2 weight ratios between the waste materials and water solvent (P). Each of the treatment combinations (LSHP) was replicated five times (R). When the effects of individual factors (L, S, H, P) or interaction (L x S x H x P) were significant on the yields, then the assessment continued with the Tukey tests. The data processing that implemented factorial CRD and Tukey tests was assisted by the SAS program of version 9.0.

## 3. Results and Discussion

*3.1. Application of natural coloring extracts at fabrics and quality testing*
Results regarding the application of natural coloring extracts, from six kinds of forest plants' wastes, at mori fabrics using pure water as solvent, before and after fixative treatments, were presented in Table 1.

Mori fabrics which were immersed in the aqueous solution of natural coloring extracts exhibited colors that differed from each other. Apparently this occurrence depended on the waste's species of forest plants as well as the waste's plant portion (woods or barks in the same species). In this research, extraction of secang woods using water solvent (pH≈7.0) brought out the wood extracts at mori fabrics with yellowish brown color. Meanwhile, extraction of secang barks using also water solvent resulted extracts with reddish brown color. The specific compound present in secang barks that imparted extracts with reddish color was the so-called brazilin or $C_{16}H_{14}O_5$ [10,11,12]. Conversely, the color of brazilin compound would change and become violet red under the base condition (pH>7.0). Likewise, brazilin compound in secang barks if extracted with organic solvent (e.g. methanol, ethanol), the color of obtained extracts would even become darker, i.e. strong brown. This occurred, because organic solvent could extract out almost all the compounds in secang barks, such as protein, carbohydrate, fat, besides extracting coloring pigments (i.e. brazilin and tannin). Meanwhile, the extracting ability of water solvent was limited only to the coloring-pigments and starches [13].

Further, extracts from ketapang woods exhibited grey color at mori fabrics, whereas correspondingly the color of extracts from ketapang barks was brown (Table 1). This visual appearance related to high content of tannin in tree barks, as inside the barks were present parenchyma tissues, which signified one that makes up tree bark tissues. The parenchyma tissues function to store reserve food and other compounds, including the tannins [14]. The thicker the tree barks, then the greater would be the tannin content inside. Barks of bakau and mangium trees contained a great amount of the condensed tannins or concentrated tannins, which was inherently the so-called flavonoid polymers [15,16]. Bark extracts from bakau minyak and mangium species brought out the tannin with brown color in the fabrics (Table 1).

At the color-imparted mori fabrics, except the extracts color from ketapang woods, the original color of the other extracts changes more immensely after the fixation or locking process. These color changes depended on kinds of fixation agents (Table 1). From those three kinds of fixation agents which were used, fixation using tunjung/ferro sulphate ($FeSO_4$) brought out the fabrics with aptly darker/black colors, compared to the fabric colors using two other fixation agents (kapur tohor/$CaCO_3$ and tawas/$Al_2(SO_4)_3$). Fixation using tawas brought out the fabric colors which were almost similar to or slightly weaker than the original colors of extracts at the fabrics (without fixatives); whereas, fixation using kapur tohor resulted in the fabric colors stronger than such extract's original colors at the fabrics. Meanwhile, fixation using tunjung exhibited fabric colors which contrasted with such also original extract colors at the fabrics (Table 1). Fixation using tunjung resulted in fabric colors which were darker (stronger black) compared with the original colors before fixation, because tannins in the color extracts reacted with $Fe^{+2}$ ions in tunjung, forming complex salts and thereby causing black colors of the fabrics [17]. From all the extract materials, the extracts from bakau minyak barks brought out the fabric colors which contrasted. The colors of mori fabrics after being immersed in bakau minyak extract solution became brown, but after kapur tohor fixation the fabric colors changed to marun red; whereas the treated with tunjung fixative became blackish brown (Table 1).

**Table 1.** Application of extracts from 6 kinds of forest plant wastes used as natural coloring pigments for mori fabrics, then treated with 3 kinds of fixation agents

| Kinds of wastes (as extract origins) | Colors of extract-immersed mori fabrics | | | |
|---|---|---|---|---|
| | Without fixation agents (control) | With fixation agents | | |
| | | Kapur tohor ($CaCO_3$) | Tawas ($Al_2[SO_4]_3$) | Tunjung ($FeSO_4$) |
| Secang barks | | | | |
| Secang woods | | | | |
| Ketapang barks | | | | |
| Ketapang woods | | | | |
| Bakau minyak barks | | | | |
| Mangium barks | | | | |

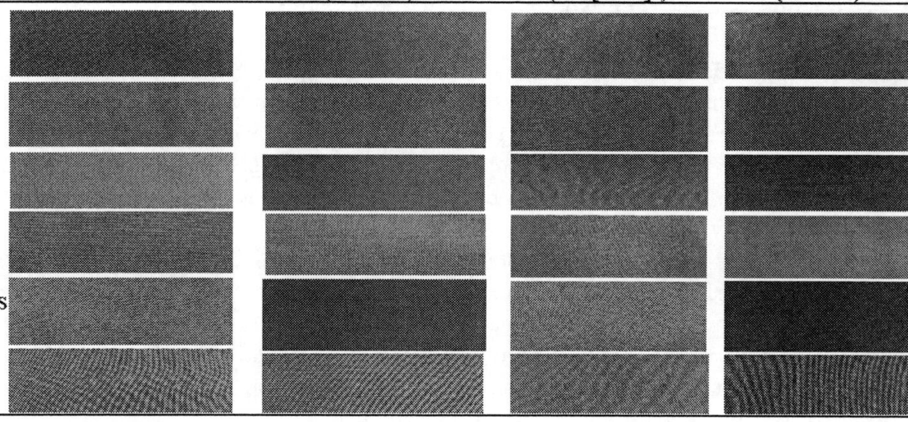

Qualities of bark extracts from wastes of forest plant, as natural-coloring pigments applied at mori fabrics (followed with kapur tohor fixative) were indicated by the color-fade resistance of washing, ironing, and sun-ray exposure. Relevantly, from the four kinds of bark extracts (Table 2), natural coloring extracts from bakau minyak barks afforded the highest qualities (best coloring performance of the fabrics) against consecutively the washing (moderate until good), the ironing (good until very good), and sun-ray exposure (moderate until good). Meanwhile at fabric dyed with natural dyes extracts from ketapang and secang barks, the qualities (except against the washing) were equally comparable to the coloring qualities of extracts from bakau minyak barks. Compared to the other three kinds of bark extracts (secang, ketapang, and bakau minyak) the qualities (coloring performance) of extracts from mangium barks at the fabrics were the lowest. The low color-fade resistance of mangium bark extracts at the fabrics in this research was presumably attributed to the mangium barks which were used in fact not the fresh ones. As such, the mangium barks had been stored for more than 3 months, before being extracted.

**Table 2**. The color-fade resistance of the bark extracts[*)] from four forest plant species

| No. | Color-leaching resistance against | Quality classes | | | |
|---|---|---|---|---|---|
| | | Secang | Ketapang | Bakau minyak | Mangium |
| 1. | Washing | 2 | 2 - 3 | 3 - 4 | 2 |
| 2 | Rubbing/ironing | 4 - 5 | 4 - 5 | 4 - 5 | 3 - 4 |
| 3. | Xenon ray (could representatively substitute for sun-ray exposure) | 3 - 4 | 3 - 4 | 3 - 4 | 2 - 3 |

Remarks: [*)] Mori fabric that has been dyed and fixed with kapur tohor; class 1 (very poor); class 2 (poor); class 3 (moderate); class 4 (good); and class 5 (very good/best coloring-performance)

### *3.2. Yields of dry extracts*

Natural coloring chemicals or pigments commonly used in industries are in the form of dry extracts, which are preferable to wet extract forms. This is because wet extracts cannot be stored for a long time. Wet extracts are usually used by the related crafters and home industries, whereby their locations are usually close to raw material (wet extracts) sources or places. Dry extracts, before being used, are added to the colorless materials, which could function as color fixing.

Data about yields of dry extracts, were presented in Table 3; and strengthened through statistical analysis (Appendix 1), which revealed that the acquirement of dry-extract yields varied significantly due to different kinds of forest plant origins (for the waste extracts), immersing durations in water, heating temperatures (during the waste extraction), and ratios between the waste materials and water.

Extraction of barks and woods waste, without undergoing water immersion (control) brought out the lowest yields of dry extracts (Table 1). Longer immersion durations (to 12 and 24 hours) tended to increase the yields of dry extracts (as confirmed through the Tukey test; Appendix 1). This is because during the immersion the diffusion of water into wood/bark tissues and cells aptly intensified with longer immersion duration, thereby softening the tissues, weakening the bondings between cell walls and coloring pigments, and hence causing them to easily move out during the further extraction stage. In this research, it turned out that 24 hours immersion imparted greater dry-extract yields significantly, compared to the yields with 12 hours immersion.

Weight ratios between the waste and water solvent, during the waste extraction significantly affected the dry-extract yields (Table 3, Appendix 1), whereby the greater the ratio (1:4) the higher would be the yields of dry extracts; and conversely the lower yields for the lower ratio (1:6). Apparently with such 1:6 lower ratio, the water might also intensify the hydrolysis on the removable matters forming more fragmented particles or simpler compounds with lower molecular weights in the filtrates (after passing through the cloths), which presumably prone to volatilization during the oven drying, thereby lowering the dry extract yields.

**Table 3**. Average yields of dry extracts from waste materials (woods and barks) of 4 forest plant species

| Immersion durations *) (before waste cooking), Hours | Weight ratio between waste materials and water | Cooking/ Extraction temperature (°C) | Forest plant species/Waste materials/Yields of dry extracts (%) | | | | | |
|---|---|---|---|---|---|---|---|---|
| | | | Secang woods | Secang barks | Ketapang woods | Ketapang barks | Bakau minyak barks | Mangium barks |
| 0 (control) | 1 : 4 | 60 | 13.51 | 14.30 | 15.27 | 13.72 | 17.72 | 11.83 |
| | | 70 | 14.14 | 15.10 | 16.27 | 16.42 | 19.83 | 15.43 |
| | 1 : 6 | 60 | 11.91 | 10.81 | 11.81 | 11.52 | 14.90 | 9.93 |
| | | 70 | 11.47 | 12.32 | 14.18 | 14.31 | 13.49 | 12.53 |
| 12 | 1 : 4 | 60 | 23.61 | 24.02 | 25.51 | 19.00 | 22.47 | 17.14 |
| | | 70 | 19.71 | 23.03 | 24.31 | 21.86 | 25.24 | 19.68 |
| | 1 : 6 | 60 | 19.07 | 19.11 | 20.35 | 16.59 | 20.01 | 14.93 |
| | | 70 | 18.02 | 19.11 | 20.35 | 19.66 | 19.16 | 17.71 |
| 24 | 1 : 4 | 60 | 20.60 | 21.51 | 22.67 | 20.83 | 24.96 | 18.75 |
| | | 70 | 21.35 | 22.50 | 23.84 | 24.92 | 28.04 | 22.87 |
| | 1 : 6 | 60 | 18.72 | 17.50 | 18.63 | 18.23 | 22.24 | 16.41 |
| | | 70 | 18.20 | 19.22 | 20.44 | 21.62 | 21.05 | 19.44 |

Remarks: Yield data average of five replications; *) Regarded as pre-extraction

IOP Conf. Series: Materials Science and Engineering **935** (2020) 012001 doi:10.1088/1757-899X/935/1/012001

Further, waste-cooking temperature (during the waste extraction) also significantly affected the dry-extract yields (Table 3, Appendix 1), but this depended on the species/portions of plant origins. For bark extracts (except for secang bark), highest yield was achieved at 70°C extraction temperature. This occurrence was commensurate with the research results by [15], whereby they performed the extraction on bakau (Rhizophora sp.) tree barks at also 70°C extraction temperature. It seemed that at 70°C temperature there occurred an increase in kinetic energy of the extractives as well as of the water solvent, thereby intensifying the movement of water molecules, enhancing the collisions between extractive molecules and water molecules which brought out the bark extracts in more fragmented shapes (smaller particle sizes) that could pass through the cloths even more easily into the filtrate, hence removing more intensively of bark extractives, and accordingly increasing the yields of dry extracts, compared to the yields at lower temperature (60°C). At 70°C extraction temperature, the yield of dry extracts (indicatively predominated by tannins) from bakau minyak barks was the highest, which achieved 28%, while at 60°C temperature the yield was about 25% (Table 3).

## 4. Conclusion

The research results of the 6 types of plant waste as a natural coloring agent showed:

1. The qualities (coloring performance) of bakau minyak barks extract that was applied for coloring mori fabrics (followed with kapur tohor fixation treatment), associated with color-fade/leaching resistance of the fabrics, were the best on rubbing/ironing process (good until very good quality), while the washing and sun-ray exposure shows moderate to good quality.
2. The highest dry extract yield was 28.04%. It is achieved by implementing 24 hours immersion of the extracts from bakau minyak (*Rhizophora apiculata*) barks in water (as pre-extraction or before extraction), followed with the extraction that employed weight ratio between the extracted barks and water at 1:4 and extraction temperature at 70°C.

## References


[1] [KLHK] (Kementerian Lingkungan Hidup dan Kehutanan) 2018 *Statistika Kementerian Lingkungan Hidup dan Kehutanan Tahun 2018* Jakarta

[2] Astana S, Soenarno, and Endom W 2015 Potensi penerimaan negara bukan pajak dari limbah kayu pemanenan di hutan alam dan hutan tanaman *Jurnal Penelitian Sosial dan Ekonomi Kehutanan* **12**(3) 227-43

[3] Suprapti S, Santoso E, Djarwanto, and Turjaman M 2012 Pemanfaatan kompos kulit kayu mangium untuk media pertumbuhan cendawan mikoriza arbuskula dan bibit *Acacia mangium* willd. *Jurnal Penelitian Hasil Hutan* **30**(2) 114-23

[4] Sjostrom E 2013 *Wood Chemistry: Fundamentals and Applications Second Edition* (California: Academic Press, Inc.)

[5] Rosyida A and Zulfiya A 2013 Pewarnaan bahan tekstil dengan menggunakan ekstrak kayu nangka dan teknik pewarnaannya untuk mendapatkan hasil yang optimal *Jurnal Rekayasa Proses* **7**(2) 52-8

[6] Puteri AYP 2012 Kadar tektokuinon pada ekstrak kayu dan kulit jati (*Tectona grandis* L.f) Jawa Barat dan Jawa Timur [Skripsi] Fakultas Kehutanan Institut Pertanian Bogor

[7] Saefudin, Basri E, and Santoso A 2018 *Bahan Alam Pewarna Batik* Eds. Roliadi H, Effendi R, and Setio P (Bogor: FORDA Press.) p. 88

[8] Boonsong P, Laohakunjit N, Kerdchoechuen O, and Matta FB 2011 Detection of pigments and natural colorants from Thai herbal plants for possible use as coloring dyes. *Hortscience* **46**(2) 265-72

[9] Sedarlah 2007 Pewarnaan kain: dahulu dan sekarang https://wol.jw.org/id/wol/d/r25/lp-in/102007129

[10] Putri US, Mukharomah AH, and Sulistyaningtyas AR 2018 Pengaruh konsentrasi pelarut etanol terhadap absorbansi brazilin pada simplisia kayu secang (*Caesalpinia sappan* L.) *Prosiding Seminar Nasional Mahasiswa Unimus 1* p. 283-88

[11] Fardhyanti DS and Riski R 2015 Pemungutan brazilin dari kayu secang (*Caesalpinia sappan* L) dengan metode maserasi dan aplikasinya untuk pewarnaan kain *Jurnal Bahan Alam Terbarukan* **4**(1) 6-13

[12] Nintasari R and Amaliyah DM 2016 Ektraksi zat warna alam dari kayu ulin (*Eusideroxylon zwageri*), kayu secang (*Caesalpinia* sp) dan kayu mengkudu (*Morinda citrifolia*) untuk bahan warna kain sasirangan *Jurnal Riset Industri Hasil Hutan* **8**(1) 25-32

[13] Hernani, Risfaheri, and Hidayat T 2017 Ekstraksi dan aplikasi pewarna alami kayu secang dan jambal dengan beberapa jenis pelarut *Dinamika Kerajinan dan Batik* **34**(2) 113-24

[14] Bowyer JL, Shmulsky R, and Haygreen JG 2007 *Forest Products & Wood Science: An Introduction 5th Edition* (Iowa State Press Ames)

[15] Kartikaningsih D, Bachroni MAA, and Danarto YC 2014 Pengambilan tanin dari kulit kayu bakau dan pemanfaatannya sebagai adsorben logam berat cuprum (Cu) dan timbal (Pb) *Ekuilibrium* **13**(1) 23-7

[16] Jessica 2018 Karakterisasi dan Pengembangan Tanin Ekstrak Kulit Mangium (*Acacia mangium* Willd.) sebagai Perekat Kayu Laminasi [Disertasi] Sekolah Pascasarjana Institut Pertanian Bogor

[17] Kristijanto AI and Soetjipto H 2013 Pengaruh jenis fiksatif terhadap ketuaan dan ketahanan luntur kain mori batik hasil pewarnaan limbah teh hijau *Prosiding Seminar Nasional* Sains dan *Pendidikan Sains VIII* **4**(1) 386-94

**Acknowledgement**
The authors in this occasion wish to express sincere gratitude to the Ministry of Research and Technology, through the Forest Products Research and Development Center, who has provided the funding partially for this research. Further appreciation is also extended to Ir. Fairus Mulya, owner of Kandelia Inc. in West Kalimantan, who has contributed bakau minyak barks for this research.

**Appendix 1**. Analysis of variance on yields of dry extracts, regarding the effect of kinds of forest plants' waste origins, waste-immersion durations, waste-boiling/extraction temperatures, and weight ratios between waste materials and water (w/v)

| Sources of variance | df | SS | MS | F-calculated | Pr > F |
|---|---|---|---|---|---|
| L (Kinds of waste origins) | 5 | 709.67073 | 141.93416 | 341.51** | 0.0001 |
| S (Immersion duration/treat-ment in water) | 2 | 3.743.05239 | 1.871.52619 | 4.503.13** | 0.0001 |
| H (Cooking/ Extraction temperature) | 1 | 143.56521 | 143.56521 | 345.44** | 0.0001 |
| P (Weight ratio between wastes and water solvent) | 1 | 931.61104 | 931.61104 | 2.241.58** | 0.0001 |
| Interactions: | | | | | |
| L*S | 10 | 176.66832 | 17.66683 | 42.51** | 0.0001 |
| L*H | 5 | 160.00539 | 32.00108 | 7..00 ** | 0.0001 |
| S*H | 2 | 25.47130 | 12.73565 | 30.64** | 0.0001 |
| L*S*H | 10 | 23.92751 | 2.39275 | 5.76** | 0.0001 |
| L*P | 5 | 49.18544 | 9.83709 | 23.67** | 0.0001 |
| S*P | 2 | 8.67834 | 4.33917 | 10.44** | 0.0001 |
| L*S*P | 10 | 10.81330 | 1.08133 | 2.60** | 0.0049 |
| H*P | 1 | 4.54276 | 4.54276 | 10.93** | 0.0011 |
| L*H*P | 5 | 62.61786 | 12.52357 | 30.13** | 0.0001 |
| S*H*P | 2 | 7.00585 | 3.50293 | 8.43** | 0.0003 |
| L*S*H*P | 10 | 9.96060 | 0.99606 | 2.40** | 0.0096 |

Remarks: df = degrees of freedom, SS = sums of squares, MS = mean squares; Pr = probability; ** = very significant effect (Pr < 0.01); CV (coeff of variation) of yield data = 3.502%; the Tukey tests consecutively for L factor, D0.05 (minimum significant difference) = 0.3777; for S factor, D0.05 = 0.1961; for H factor, D0.05 = 0.1338; for P factor, D0.05 = 0.1961; for highest level of interaction (L*S*H*P), D0.05 = 0.7164

IOP Conf. Series: Materials Science and Engineering **935** (2020) 012002 doi:10.1088/1757-899X/935/1/012002

# Works experience of chainsaw operators in peat swamp forest plantation in increasing timber productivity and efficiency

**S Suhartana*and Yuniawati**

Forest Products Research and Development Center, Ministry of Environment and forestry, Bogor 16610, Indonesia

*Email: sona.suhartana@gmail.com

**Abstract.** Felling is one of the logging activities that determines wood production. The appropriate and correct felling technique in plantation forests can increase the Efficiency of Timber Utilization (ETU). This research was conducted in one of the peat swamp forest plantation in Riau. The research objective is to determine the effect of work experience on the productivity and ETU in peat swamp forest plantations. The research method is by measuring the volume of wood harvested through local felling technique carried out by two chainsaw operators with different work experience. The results showed that: 1) The work experience of the chainsaw operator can effect felling productivity, production cost and ETU; 2) The average felling productivity produced by operator B is 10.151 $m^3$/hour higher than operator A which is 8.888 $m^3$/hour; 3) The average felling production cost for operator B is IDR 9,319.83/$m^3$ lower than operator A which is IDR 10,812.79/$m^3$; and 4) The lowest possible felling technique can increase ETU by 5% which is equivalent to an additional profit for the company of IDR 3,769,075,000/year.

## 1. Introduction

Timber harvesting in peat swamp plantation forest is a key point in a series of activities of forest utilization, and it should be well planned to avoid damage on dynamic stability of forest environment. However, timber harvesting in such area is still low in productivity and has relatively high cost, and create environmental damage [1]. Felling is an initial step in a series of timber harvesting activities, where one of the objectives is obtaining raw materials for wood industry. This harvesting activity should consider felling techniques, field topography, and equipments being used. Up to now felling still use conventional felling technique (felling technique $V_{6cm}$= felling technique which is commonly practiced by local loggers in accordance with company policy which utilizes log until diameter limit of 6 cm, with low ETU). On the other hand, if felling is practiced with technique resulting in stump with height as low as possible felling technique of $V_{5cm}$= which create stump with height as low as possible and utilized stem until diameter limit of 5 cm), this will create high level ulitization of wood [2,3]. Logging in peat swamp forest should consider land condition and characteristics of the majority of trees growing there, such as diameter for instance, that in the case where stem diameter is small or less than 40 cm, and with most of the stem ranging from soft to nearly hard, and therefore this require appropriate chainsaw [4,5]. Logging is an activity which affects very greatly the wood utilization efficiency, and thefore skill of chainsaw operator is one of the factors which determine the success of improving the ETU. Hiesl [6] reported that the skill of an operator requires at least three years of work experience to achieve 100% of the productivity potential. It is explained also that trained operators have 40% higher productivity as compared with those who have not received training. Although skill and work experience is one of the factors which affect work productivity, the ability of operator to adapt with change in working environment and with different situation should be given attention. Several things which should be considered in logging activities on peat swamp plantation forest are as follows: 1) Peat thickness which is more than 1 meter, with soil surface and foothold which always moving and is unstable. This condition makes it difficult for the chainsaw operator to arrange his foothold on the peat soil. Movement in peat soil can make the feet of the operator bog down in the peat soil. This is very dangerous for the safety of the operator during felling. Great thickness of peat affects the alertness of chainsaw operator, so that long duration of tree felling is needed; 2) Condition of fast growing trees in peat swamp plantation forest is dominated by species *Acacia crassicarpa*, where the characteristics of this species is desirable for pulp industry and is able to adapt to peat land. *Acacia crassicarpa* trees are harvested at average diameter of 19 cm and average tree height being utilized of 22 m. These tree characteristics which have small diameter and are not so tall, produce only smaller volume of wood as compared with those of pulpwood in dry land; and 3) Temperature at the surface of peat land is hotter than that in dry land [7,8]. Tono and Wawan [9] research results show that air temperature in peat forest area ranges between 30.04°C-34.45°C and soil temperature ranges between 25.9°C-30.21°C. Hot temperature can have impact on productivity of machine operator. High temperature on soil surface and air in the area of Peat Swamp Plantation Forest (PSPF) have consequences on rapid emergence of fatigue on the chainsaw operator. Based on such background, this study is intended to determine the effect of work experience on the productivity and ETU in peat swamp forest plantations.

## 2. Materials and Methods

### *2.1. Time and location*

Research was conducted in the beginning of year 2020 in the working area of IUPHHK-HT (Commercial Timber Forest Products Utilization Permit of Plantation Forest) of PT Satria Perkasa Agung (PT SPA), District of Simpang Kanan, Logging Block of RKT (Annual Working Plan) 2020, Logging Compartment No. SKNB 035301. This area belongs to the territory of Inderagiri Hilir regency, Riau province. Based on geographic position, this forest complex is situated between 102°39’10”-102° 58’50”E and 000°04’10”LS - 000°12’55”N.

### *2.2. Materials*

Equipments being used were chainsaw Stihl MS 382, Personal Protective Equipment (PPE), tally sheet, stationery/writing, apparatus and machette. Materials being used were trees, lubricants, and oil fuel.

### *2.3. Methods*

Determining (purposively), one logging compartment which will be immediately logged. Determining 2 personnel operators, one with working experience of less than 5 years, and the other with working experience of more than 5 years, as chainsaw operator. Each operator conducted logging with local technique (stem utilization until diameter of 6 cm in accordance with company policy = logging technique VØ 6cm) as many as 30 trees. Logging technique until diameter of 5 cm (logging technique VØ5cm) which was conducted through measurement of difference of stem lengths between those of Ø6cm and those of Ø5cm as many as 30 trees each.

Measurement of parameters, comprising productivity, ETU and felling production cost i.e. productivity: recording of working time (duration) and wood volume; production cost: recording all expenditures such as those for fuel, oil/lubricants, wage, productivity, depreciation, maintenance/repair, interest rate, tax and salary cost; and ETU: diameter of butt (lower end), diameter of top (upper end), stem length, and stump height.

### *2.4. Data analysis*
Productivity is calculated with the following formula:

$$P = V/T \qquad (1)$$

where, P: productivity ($m^3$/hour), V: volume = ¼ π $D^2L$ ($m^3$), W: felling time/duration (hour), π: 22/7, D: average diameter (m), L: stem length (m)

Efficiency of Timber Utilization (ETU):

$$ETU = \frac{V_p}{V_m} \times 100\% \qquad (2)$$

where, ETU: Efficiency of Timber Utilization (%), $V_p$: Volume being utilized ($m^3$), $V_m$: volume ought to be able to be utilized ($m^3$)

Logging production cost was calculated using formula from FAO [10]. For determining the appropriate felling technique which comprise the aspect of productivity, ETU, and aspect of cost, t-test was performed [11]. For learning the effect of chainsaw operator working experience on felling productivity increase and ETU, independent sample t test was performed using SPSS 25.0. Procedure for decision making was as follows :

- If the value of sig (2-tailed) > 0.05, then $H_0$ was accepted and $H_1$ was rejected, which implied that there was no difference in average productivity of felling and ETU between operator A and operator B.
- If the value of sig (2-tailed) < 0.05, then $H_0$ was rejected and $H_1$ was accepted, which implied that there is difference in average productivity of felling and ETU between operator A and operator B.

## 3. Results and Discussion

### *3.1 General condition of the company*
Most of the study area possess field slope between 0-15% , with elevation between 3-16 m asl. Soil types are podsolic, alluvial, organosol hemic and organosol sapric. Climate type based on Schmidt & Ferguson classification belongs to type A with monthly rainfall of 109 mm/month. Highest monthly rainfall is 195 mm/month and the lowest 29 mm/month and there is no dry month. Forest stand in the study area is akasia (*Acacia crassicarpa)* trees from family of *Leguminosae.* The akasia stand has density of around 1667 trees/ha (for trees with diameter of 10 cm or more). Most of the trees did not have buttress. On the average, most undergrowth vegetation had moderate density. In the Annual Working Plan (RKT) of year 2020, PT SPA of Simpang Kanan District harvests wood from area as large as 2354.9 ha with production target of 397,691 $m^3$ and average annual wood production of (average production realization in the past 5 years) 753,815 $m^3$, with maximum Annual Allowable Cutting (AAC) of 1,331,090 $m^3$ comprising wood species *Acacia crassicarpa* (data source: production realization of PT SPA District of Simpang Kanan, RKT 2015-2019). In the timber harvesting, the main equipments are as follows: chainsaw and its accessories for felling; excavator for skidding; excavator for loading and unloading, and hauling; logging truck for hauling in land route; and iron boat *(pompong*) or iron canoe for hauling in water route.

### *3.2 Productivity of felling by two chainsaw operators*
Chainsaw operators have roles in increasing felling productivity and lowering the production cost. Average felling productivity of the two chainsaw operators are presented in Table 1.

**Table 1** Average productivity of logging $V_{6cm}$, N = 30

| Aspect | $V_{6cm}(m^3)$ | Logging time (hour) | Productivity ($m^3$/hour) |
|---|---|---|---|
| Operator A* | | | |
| Min-Max | 0.231-0.664 | 0.032-0.057 | 6.841-12.823 |
| Average | 0.400 | 0.045 | 8.888 |
| SD | 0.110 | 0.006 | 1.667 |
| Operator B** | | | |
| Min-Max | 0.263-0.664 | 0.025-0.057 | 7.517-11.823 |
| Average | 0.423 | 0.042 | 10.151 |
| SD | 0.101 | 0.008 | 1.263 |

Remarks: $V_{6cm}$ = Wood volume until diameter limit of 6 cm; SD = standard deviation; N = Number of replications; * = Diameter range 15-23 cm, with average 18.5 cm; **= Diameter range 15-23 cm, with average 18.8 cm

Table 1 showed that average productivity of the two chainsaw operators was different from each other. Operator B showed average logging productivity which was higher than that of operator A, which was due to the fact that operator B possessed working experience as chainsaw operator as long as 9 years, as compared with that of operator A which was just 4.5 years. This difference in average productivity can be shown from average logging time of operator B which was faster than that of operator A, with average tree diameter at breast height of 18.8 cm greater than that in the location of operator A. Longer duration of working experience creates better working skill. These research results support the theory Robbins [12] which explains that "*We can say a positive relationship between tenure and job productivity*" which means that length of working experience (period) and skill is positively correlated. The longer the working period, the higher will be the experience and skill which will support their work, so that their productivity will increase. The results of this study are in agreement with empirical study from Adiati [13], Yudi and Wardana [14] which explained that working experience has positive and significant effect toward personnel productivity. Working experience as chainsaw operator is one of the factors which affect the average productivity of logging. This was in agreement with what was written by Muhdi [15 and Dulsalam *et al.* [16] which explains that skill of chainsaw operator affects average productivity of logging. Average productivity of logging by the two chainsaw operators is shown by results of t test, where the $t_{calculated} = 3.308^{**} > t_{table}$ (99%,58) = 2.682, so that it is suggested that chainsaw operator B was better than chainsaw operator A. Silayo [17] explained that the increasing of productivity average of felling could be achieved if the operators were provided with job training skill of logging. Fagbenle *et al.* [18] explained that one of the factors which affect employee's productivity is originated from the employee him/herself, namely the human factors which comprise among others the working experience. Working experience as explained by Manulang [19], is the process of establishment of knowledge and skill on method of a job for employee through involvement in the execution of the job. Working experience is a factor which has the greatest effect on the growth of a venture. The large amount of experience possessed by the employees will result in large growth of the venture. The longer the period of career that a worker is performing his job, the higher will be his/her skill. High skill will have positive impact upon his performance, such as less duration of time needed to complete the job and better quality of the job result [20].

Effect of working experience of *chainsaw* operator on increase of felling productivity is presented in Table 2. From Table 2 it is known that the value of sig 2 tailed 0.002 < 0.05 so that $H_0$ is rejected, which implies that there was significant difference of average logging productivity between chainsaw operator A and chainsaw operator B. Therefore, it can be suggested that average productivity of logging exhibited by the two operators with different working experience, is different. The column of *mean difference* shows that difference of average productivity of felling between operator A and that of operator B was -1,263.033, and because the value is negative, it means that operator A possessed average felling productivity which was lower than operator B. The longer experience of chainsaw operator, will be achieve the higher felling productivity avarage. These results were not different from research results of Suhartana and Yuniawati [21] in the area of industrial plantation forest (HTI) of PT SPA in 2013, which showed that logging activity in the study area used logging apparatus *Chainsaw* STIHL MS-381 which produced average volume of *Acacia crasicarpa* being felled as much as 0.595 $m^3$, with average duration of felling completion of 0.078 hour, so it was found that average felling productivity was 8 $m^3$/hour.

**Table 2** Effect of working experience of chainsaw operator on the increase of felling productivity

| | | Levene's Test for equality of variances | | T-test for equality of means | | | | | | |
|---|---|---|---|---|---|---|---|---|---|---|
| | | | | | | | | | 95% Confidence Interval of the Difference | |
| | | F | Sig. | t | df | Sig. (2-tailed) | Mean Difference | Std. Error Difference | Lower | Upper |
| Producti-vity | Equal variances assumed | 2.932 | .092 | -3.307 | 58 | .002 | -1,263.033 | 381.884 | -2,027.456 | -498.610 |
| | Equal variances not assumed | | | -3.307 | 54.044 | .002 | -1,263.033 | 381.884 | -2,028.650 | -497.417 |

Several research results in relation with average felling productivity in plantation forest are as follows: 1) Research results by Câmpu and Ciubotaru [22] showed that average felling productivity in casuarina tree in Rumania using *chainsaw* husqvarna 365 was 10.138 $m^3$/hour (4.55 trees/hour) in the first stand and 11.344 $m^3$/hour (4.33 trees/hour) in the second stand; 2) Sitohang *et al.* [23] reported that the highest average productivity of felling was obtained by using *chainsaw* with trade mark Stihl, namely as large as 77.91$m^3$/hour, whereas that which use chainsaw with trade mark *New West* was 70.39 $m^3$/hour. In the use of chainsaw with trademark *Stihl* and *chainsaw* with trade mark *New West*, capacity of the oil tank was 0.6 liters. It means that the chainsaw being used were of small sizes. From several samples being observed, it was known that the average time being used for felling of one tree was 12.15 seconds, it comprised the work of cutting one tree and walking to the next tree and felling down the tree. Therefore, one minute work was able to fell down 5 trees. One hour work would be able to fell 300 trees. If productive work per day was 8 hours, then the average productivity per day was 2,400 trees; and 3) Study by Sukadaryati *et al.* [24] showed that tree felling in people/privately owned forest in Ciamis used *chainsaw* with average felling productivity for gmelina trees in Kertabumi village as large as 4.880 $m^3$/hour, whereas logging productivity of mixed trees in Bojonggedang village was on the average 3.392 $m^3$/hour.

### *3.3 Felling production cost*

For learning the production cost of timber harvesting activities, there is a need for calculation of machine cost for each felling activity (Table 3). Machine cost divided by productivity in each activity will reveal production cost in each activity. Average production cost of felling is presented in Table 4. From the data of machine cost, the amount of felling production cost is known. From calculation of felling cost, the average cost of felling for each operator was IDR 10,812.79/$m^3$ (operator A) and IDR 9,319.83/$m^3$ (operator B). Such cost calculation showed that operator B exhibited felling production cost which was lower than that of operator A with difference of IDR 10,812.79/$m^3$-IDR 9,319.83/$m^3$ = IDR 1,492.96/$m^3$. If the production of the harvested timber is more than 100 $m^3$ in one year, then the difference is very large. Average cost of felling production for operator B was lower than that of operator A, and this was due to average productivity of felling by operator B which was higher than that of operator A.

**Table 3.** Components of felling cost

| Cost components | Amount (IDR/hour) |
|---|---|
| Depreciation cost | 7,200 |
| Insurance cost | 144 |
| Interest cost | 720 |
| Tax cost | 96 |
| Fuel cost | 5,724 |
| Oil / lubricant cost | 572.4 |
| Maintenance / repair cost | 7,200 |
| Wage cost | 71,428.6 |
| Total cost of machine | 93,085 |

Productivity of timber harvesting can affect production cost; the higher the productivity, the lower will be the production cost. Study by Stover [25], Suhartana and Yuniawati [26], explained that cost per unit volume of a course follow the level of productivity. Meanwhile, study by Germain *et al.* [27] explained that operator experience affects production cost. To prove the difference of logging production cost of the two operators, t test was performed. Result of t- test showed that $t_{calculated}$ =3,593$^{**}$ > $t_{table}$ (99%,58) = 2.682. This implies that from the aspect of production cost, operator B exhibited better performance than that of operator A, so that the average logging production cost of operator B was lower than that of operator A.

**Table 4.** Average felling production cost $V_{6cm}$, N = 30

| Aspect | Productivity ($m^3$/hour) | Felling production cost (IDR/$m^3$) |
|---|---|---|
| Operator A | | |
| Min-Max | 6.841-12.823 | 7,259.236-13,606.548 |
| Average | 8.888 | 10,812.79 |
| SD | 1.667 | 1,898.154 |
| Operator B | | |
| Min-Max | 7.517-11.823 | 7,872.965-12,383.001 |
| Average | 10.151 | 9,319.83 |
| SD | 1.263 | 1,255.923 |

Average felling production cost in this research was higher than that of study results from [22], for logging of A. *mangium* wood with chainsaw Stihl type, namely IDR 5,235/$m^3$ (RIL) and IDR 6,171/$m^3$ (local technique). The low felling cost was due to the phenomenon that all components of felling cost, particularly the costs for purchasing the consumable materials (oil fuel, oil lubricant and service) were cheaper than the purchase during this research in the year 2020.

*3.4. Efficiency of timber utilization (ETU)*

Average ETU is presented in Table 5 which shows that operator B exhibited average ETU $V_{5cm}$, volume of wood being utilized ($V_{5cm}$), volume difference ($V_{5\text{-}6cm}$) and stem length difference ($V_{5\text{-}6cm}$) which were higher than those of operator A. The larger wood volume being produced or collected, the greater will be the efficiency value of wood utilization. Also, the average height of stump exhibited by operator B was shorter than that of operator A. Average stump height being exhibited will also affect the average volume of wood being utilized. The lower the stump height, the higher will be the average volume of wood being utilized. Therefore, operator B in the application of logging technique $V_{5cm}$ produced average ETU, which was better than that of operator A.

**Table 5** Average ETU (%) by two chainsaw operators

| Aspect | $V_{5cm}$ ($m^3$) | $V_{6cm}$ ($m^3$) | Efficiency (%) | | $\Delta V_{5\text{-}6cm}$ ($m^3$) | $\Delta t_{5\text{-}6cm}$ (cm) | SH (cm) |
|---|---|---|---|---|---|---|---|
| | | | $V_{5cm}$ | $V_{6cm}$ | | | |
| Operator A | | | | | | | |
| Min-Max | 0.247-0.706 | 0.231-0.664 | 0.981-0.992 | 0.897-0.977 | 0.005-0.048 | 0.2-1.4 | 0.1-0.17 |
| Average | 0.422 | 0.400 | 0.987 | 0.935 | 0.022 | 0.810 | 0.132 |
| SD | 0.114 | 0.110 | 0.003 | 0.023 | 0.011 | 0.344 | 0.020 |
| Operator B | | | | | | | |
| Min-Max | 0.281-0.706 | 0.263-0.664 | 0.984-0.992 | 0.907-0.979 | 0.005-0.042 | 0.200-1.200 | 0.100-0.150 |
| Average | 0.447 | 0.423 | 0.990 | 0.940 | 0.023 | 0.830 | 0.116 |
| SD | 0.106 | 0.101 | 0.002 | 0.020 | 0.010 | 0.277 | 0.014 |

Remarks: $V_{5cm}$= Wood volume until diameter limit 5 cm; $V_{6cm}$ = Wood volume until diameter limit 6 cm; SD = Standard deviation; $\Delta V_{5\text{-}6cm}$ = Stem volume difference between those of 5 and 6 cm; $\Delta t_{5\text{-}6cm}$ = Stem height difference between those of diameter 5 and 6 cm; SH = stump height

From calculation results it was known that with felling technique of $V_{5cm}$, this could increase ETU by 0.99-0.94 = 5%. Based on field data and quotation from company office, it was known that average log production per year is 753,815 $m^3$, so that on the basis of increase of wood utilization by 5%, the company would get additional advantage in the form of production increase per year by 5% x 753,815 $m^3$ = 37,690.75 $m^3$ per year. Wood price was IDR 500,000/$m^3$, and the apropriate profit for the company was 20% or IDR 100,000/$m^3$, so that the company will get additional profit of 37,690.75 $m^3$ x IDR 100,000/$m^3$ = IDR 3,769,075,000/year. Considering the profit which will be obtained, then there is opportunity for the company to adopt the felling technique of $V_{5cm}$. Effect of working experience of chainsaw operator on the increase of ETU is presented in Table 6.

**Table 6** Effect of working experience of chainsaw operator on the increase of ETU

| | | Levene's Test for equality of variances | | T-test for equality of means | | | | | | |
|---|---|---|---|---|---|---|---|---|---|---|
| | | | | | | | | | 95% Confidence Interval of the Difference | |
| | | F | Sig. | t | df | Sig. (2-tailed) | Mean Difference | Std. Error Difference | Lower | Upper |
| Efficiency | Equal variances assumed | 43.407 | .000 | -4.646 | 58 | .000 | -.00633 | .00136 | -.00906 | -.00360 |
| | Equal variances not assumed | | | -4.646 | 36.444 | .000 | -.00633 | .00136 | -.00910 | -.00357 |

Table 6 shows that the value of sig 2 tailed 0,000 < 0,05, so that $H_0$ is rejected, which implies that there was significant difference in ETU between chainsaw operator A and chainsaw operator B. Therefore, it can be suggested that average ETU exhibited by the two operators with different working experience, was different. Column of mean difference shows that average ETU between operator A and operator B was -0.000633, and because the value is negative, it is said that operator A produced average ETU which was lower than operator B. The longer the experience of chainsaw operator, the higher will be the average ETU being exhibited. These research results showed average ETU which was higher than those of 1). Study by Suhartana *et al.* [28] in the felling of *Acacia mangium* in East Kalimantan with ETU by felling technique of $V_{5cm}$ as large as 98.7% and $V_{6cm}$ as large as 96.0%, so that the company will get additional profit as large as IDR 345,866,400/ year. Such difference was due to the phenomenon that in the East Kalimantan company, average volume of wood stem being utilized was lower, namely that for felling technique $V_{5cm}$ was as large as 0.298 $m^3$ and that for $V_{6cm}$ was as large as 0.348 $m^3$. Average volume of wood stem being produced affected the average ETU; and 2) Study by Suhartana and Yuniawati [2] which showed that felling of mixed forest wood in West Kalimantan exhibited average ETU with felling technique $V_{5cm}$ as large as 99.3% and that of $V_{6cm}$ as large as 93.3% with average volume of each felling technique were 0.817 $m^3$ and 1.004 $m^3$, with diference of 0.055 $m^3$. The company will get additional profit of IDR 74,400,000/year.

## 4. Conclusion

This study concluded that chainsaw operator with working experience of more than 5 years (B) exhibited productivity (10.151 $m^3$/hour as compared with 8.888 $m^3$/hour) and ETU (99% as compared with 94%) which were higher; and production cost (IDR 9,319.83/$m^3$ as compared with IDR 10,812.79/$m^3$) which was lower than those of operator with working experience of less than 5 years (A).

## References

[1] Suhartana S and Yuniawati 2010 *J. Penelitian Hasil Hutan* **28**(2) 119-29
[2] Suhartana S and Yuniawati 2011 *J. Penelitian Hasil Hutan* **29**(4) 369-384
[3] Suhartana S, Yuniawati, and Rahmat 2012 *Sepilok Bulletin* ***15*** *&* ***16*** 27-35
[4] Suhartana S and Yuniawati 2011 *J. Penelitian Hasil Hutan* **29**(4) 369-84
[5] Suhartana S and Yuniawati 2010. *J. Penelitian Hasil Hutan* **28**(2) 119-29
[6] Hiesl P 2013 Productivity standards for whole tree and cut to length harvesting systems in Maine [Disertation] The University of Maine
[7] Suhartana S and Yuniawati 2010 The effect of logging on peat land conditions: A case study at a peat swamp forest company in Riau *Proceedings the First International Symposium of Indonesian Wood Research Society* Pp. 300-6
[8] Suhartana S and Yuniawati 2016 *J. Hutan Tropis* **4**(3) 273-81
[9] Tono S and Wawan W 2016 Fluks $CO_2$ pada berbagai kondisi hutan rawa gambut di areal konsesi PT Diamond Raya Timber Kecamatan Bangko Kabupaten Rokan Hilir [Dissertation] Riau University
[10] FAO 1992 *Cost control in forest harvesting and road construction (FAO Forestry Paper No 99)* (Rome: FAO)
[11] Steel RGD and Torrie JH 1980 *Principles and Procedures Of Statistics* (New York: McGraw-Hill Book Co. Inc.) pp. 633
[12] Robbins S 2007 *Perilaku 0rganisasi* (New Jersey: Prentice Hall, Inc.)
[13] Adiati A 2013 *J. Ekonomi Pembangunan* **2**(1-10)
[14] Yudi AWP and Wardana G 2019 *J. Ekonomi Pembangunan Universitas Udayana* **8**(3) 669-97
[15] Muhdi 2016 *J. of Science and Research* **5** (10) 141-3
[16] Dulsalam, Sukadaryati, and Yuniawati 2018 *J. Penelitian Hasil Hutan* **36**(1) 1-12
[17] Silayo DSA 2015 *Tanzania American Journal of Operational Research* **5**(2) 29-38
[18] Fagbenle, Olabosipo I, Phillip LO, Omuh, and Igartius O 2012 *International Journal of Management Sciences and Business Research* **1**(7) 1-11
[19] Manulang M 2005 *Dasar-Dasar Manajemen* (Yogyakarta: Gajah Mada Univ. Press)
[20] Sulaeman A 2014 *Trikonomika* **13**(1) 91-100
[21] Suhartana S and Yuniawati 2016 *J. Hutan Tropis* **4**(3) 273-81
[22] Câmpu V R and Ciubotaru A 2017 *Silva Fennica* **51**(2) 1-19
[23] Sitohang WRM, Muhdi, and Afifuddin Y 2016 *Peronema Forestry Science Journal* **5**(2) 192-203
[24] Sukadaryati, Yuniawati, and Dulsalam 2018 *J. Ilmu Kehutanan* **12** 142-55
[25] Stover D 2018 *Comparative costs of harvesting irregular and regular mixed conifer stands at Stourhead (Western) Estate* (A dissertation submitted in partial fulfilment of the requirements for the degree of *master of science* (msc)) in forestry of Bangor University pp. 1-99
[26] Suhartana S and Yuniawati 2019 The economic and environmentally tree felling techniques in natural forest *IOP Conf.Ser.: Earth Environ.Sci.* **359** 01200724
[27] Germain R, Regula J, Bick S, and Zhang L 2019 *The Forestry Chronicle* **95**(1) 16-23
[28] Suhartana S, Yuniawati, and Rahmat 2014 *J. Wahana Foresta* **8**(2) 42-52

**Acknowledgement**
The authors express their gratitude to Ir. Joko Pramono (Planning Division) and Mr. Sumadi (Harvesting Division) who have helped in this study.

**Authors' Contribution**
S. Suhartana (main contributor) designed the study, developed the methodology, performed the experiment, analyzed the data and wrote the manuscript. Yuniawati (co-author) analyzed the data and wrote the manuscript.

IOP Conf. Series: Materials Science and Engineering **935** (2020) 012003 doi:10.1088/1757-899X/935/1/012003

# Distribution of permethrin on laminated bamboo after the preservative treatment using SC-$CO_2$ as a carrier solvent

**Z Uslinawaty[1*], M Muin[2], and Suhasman[2]**

[1]Department of Forestry, Haluoleo University, Kendari 93232, Indonesia.
[2]Department of Forestry, Hasanuddin University, Makassar, 90245, Indonesia

*Email: zakiahuslinawati@gmail.com

**Abstract**. This study was conducted to determine the presence of the insecticide permethrin and its retention gradients within the structure of laminated bamboo after treatment using supercritical carbon dioxide (SC-$CO_2$) as the carrier solvent. the presence of permetrin was based on functional group differences indicated by FT-IR analyses before and after the preservative treatment. Permethrin retentions in the treated bamboo materials ranged from 0.0020 to 0.0022%. This result indicated that the preservative treatment of laminated bamboo using SC-$CO_2$ as a carrier solvent produced evenly distributed chemical an effective level. There was no signicant distribution gradient of the preservative observed over the layers of the treated laminated bamboo arranged with outer-inner surface laminations. The current treatment condition is potential to produce laminated bamboo products completely resistant to biodegradation.

## 1. Introduction

Laminated bamboo products are economically and environmentally attractive considered since bamboo is categorized as fast growing species that has better flexural strength than other raw materials [1-4]. However, bamboo has also low durability against biological degradation and therefore, some efforts have been made to overcome this unfavorable characteristic mostly with the use of preservatives [5]. Unfortunately, conventional preservatives treatments of bamboo product require liquid preservatives that potentially decrease mechanical properties due to the need to re-drying after treatment [2,6,7]. Conventional treatment has been also cause detrimental effects on environment because of chemical wastes [8-10]. In addition, conventional treatment method has a limited enactment in distributing preservative throughout the treated materials. In this case, the preservative was found mostly on the surface parts with higher concentration compared to the center parts [11,12].

One alternative approach to preservative treatment of composite products is the use of supercritical carbon dioxide (SC-$CO_2$) as carrier solvent. This method is environmentally friendly and can effectively increase the durability of the wood-based composite products [11-15]. SC-$CO_2$ is liquefied gas and able to easily penetrate into the inside of wood structure, maintaining wood dimension and stability. A previous study has indicated the feasibility of the SC-$CO_2$ treatment in producing bamboo laminates with high strength and durability properties [16]. However, the effectiveness of SC-$CO_2$ to impregnating and distributed preservatives in to bamboo-based composite products has never been evaluated. The current study intended to assess the distribution of permethrin within the matrix of bamboo products impregnated using SC-$CO_2$ as carrier solvent. Permethrin is commercially available and has been found to be effective against white rot fungi *Pycnoporoussanguineus* and the termite *Coptotermes curvignathus* [5,17,18].

## 2.Materials and Methods

### *2.1 Materials*

Three representative bamboo culms (age 4 years, diameter 22-25 cm, and thickness 1-1.5 cm) of *Dendrocalamus asper,* grown at Anduna Wolasi village, South Konawe district, South east Sulawesi, Indonesia, were harvested. Samples were taken at the distance of about 1 m from bottom of the bamboo culms. The samples were then axially cut into 2.5 cm width and 15 cm length. The outermost and innermost layers were laminated. Prior to treatments, all samples were then conditioned at 60°C for 24 hours.

Permethrin 100 EC (Bioindustries) in a liquid form with an active ingredient concentration 100 g/L permethrin. A preservative concentration of 0.5% was prepared for SC-$CO_2$ impregnation by addition of methanol p.a. (Merck) as a co-solvent. The preservative concentration was applied based on the manufacturer's recommendation.

### *2.2. Methods*

#### 2.2.1. Bamboo lamination

Bamboo samples 1 (R) x 2.5 (T) x 15 (L) cm were assembled for laminations. Samples were arranged with different laminated surface areas of bamboo samples (outer-inner, outer-outer, and inner-inner surfaces). The laminated bamboo was produced by cold press method at the pressure of 10 kg/cm$^2$ for 3 hours after applying a glue type of isocyanates (polyurethanes) with the glue spread of 250 g/m$^2$.

#### 2.2.2. Preservative treatment

Preservative treatments were applied in two separate ways, i.e. to bamboo raw materials (prefabrication or before lamination) and to laminated bamboo products (post-fabrication or after lamination). SC-$CO_2$ with 5 mL of permethrin 0.5% was impregnated using the procedure of Multi-Phase Carbon Dioxide (MP$CO_2$) treatment developed by Muin [19]. The treatment procedure in this study consisted of introducing the samples of six bamboo strips or three laminated bamboo products into the treatment vessel, settingthe vessel temperature to 15°C, introducing $CO_2$ with the preservative at the pressure equal to the $CO_2$ source bomb approximately 6 MPa, gradually increasing the temperature and pressure to about 35°C and 8 MPa, and finally releasing the pressure to ambient atmosphere.

#### 2.2.3. Analysis of preservative distribution

The presence of preservative permethrin within the structure of the treated bamboo materials was determined using Fourier Transform Infra-Red (FTIR) Spectroscopy (Shimadzu IR Prestige 21 type with 8400 S). FTIR spectroscopy has been found appropriate to determine the intensity of specific bonds and functional groups within the polymeric structure. It was used in this study to indicate the changes of functional groups on bamboo products before and after permethrin impregnation using SC-$CO_2$ as carier solvent. For the analysis of the preservative distribution, the treated bamboo samples of either bamboo strip or bamboo laminations were sliced into four sections with a proportional thickness. As the use of conventionl method using weight gain in determining preservative retentions is thought to be unappropriate, the permethrin concentrations within the treated bamboo products were analyzed using spectrophotometer (Perkinelmer Type Lamda 25). The treated sample was milled to pass a 40-mesh screen. Two grams of the milled materials were dissolved in10 mL of Kl 5% and 10 mL of $AgNO_3$ 0.25% and then macerated for 30 min. The permethrin concentration was analyzed using UV-VIS spectrophotometer employed at the wave length of 339 nm.

IOP Conf. Series: Materials Science and Engineering **935** (2020) 012003 doi:10.1088/1757-899X/935/1/012003

## 3. Results and Discussion

Results of this study showed that the use of SC-$CO_2$ as a carrier solvent was found to have capability in introducing preservative pemethrin into the treated bamboo products.It is indicated by the results of FTIR analysis showing the changes of existing functional groups as shown in Figure 1. There was no different functional groups between bamboo samples with and without the SC-$CO_2$ treatments, indicating there was no chemical interaction between $CO_2$ and bamboo materials during the impregnating process. However, results of FTIR analysis on the bamboo samples before and after the preservative treatment indicated the changes of existing functional groups. Figure1 clearly demonstrates the presence of different compositions in the functional groupsof bamboo samples before and after impregnated with the preservative permethrin.The presence of the preservative compound after treatment is shown by C-Cl linkage functional groups of permethrin. This fact was associated with the solubility of permethrin in SC-$CO_2$ ranging from 0.1765 g/L (without co-solvent) to 0.6497 g/L (with

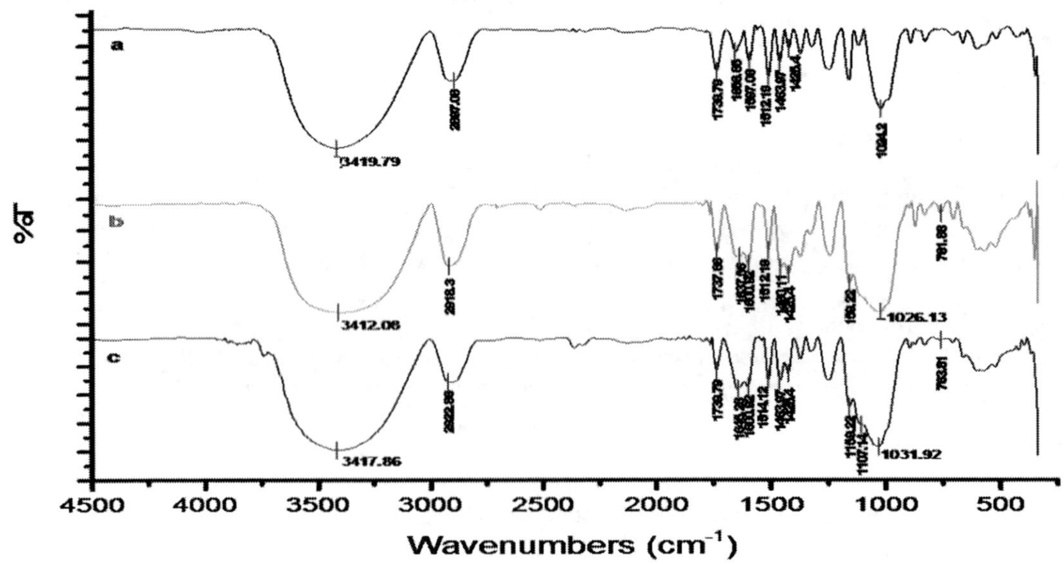


**Figure 1.** FTIR spectra of bamboo products: (a) bamboo strip without treatment; (b) laminated bamboo after SC-$CO_2$ impregnation with permethrin to bamboo strips (before lamination); (c) laminated bamboo after SC-$CO_2$impregnation with permethrin (after lamination)

Figure 1 shows that the FTIR spectra pattern is generally unchanged as indicated by the peak of 1512 cm$^{-1}$, 1514 cm$^{-1}$, and 1739 cm$^{-1}$ on the tested bamboo specimens. These conditions are related to the nature of lignin and hemicellulose, respectively, characterized by the presence of C=C groups around the wavelength of 1505-1512 cm$^{-1}$ [21] and the peak of 1739 cm$^{-1}$ [22,23]. However, new peaks of 763.81 cm$^{-1}$ and 761.88 cm$^{-1}$ in the FTIR spectrum were found to occur on laminated bamboo products after treatment, indicating the presence of permethrin in the specimen. The peak of 763.81 cm$^{-1}$ occurred at the specimens of laminated bamboo with permethrin treatment before lamination, while the peak of 761.88 cm$^{-1}$ occurred at the specimens of laminated bamboo with permethrin treatment after lamination.

Further anaysis on the presence of permethrin within the treated bamboo samples demonstrated that the preservative was distributed with different levels of significancy, depending on the treated bamboo materials (Table 1). The retentions of the preservative permethrinwere significantly different from outer to inner layers (layer 1 to layer 4) of the treated bamboo splits. The significant differences of gradient retentions were also found in the laminated bamboo with outer-outer and inner-inner laminations, but not in the outer-inner laminated bamboo. Overall, the retention values of the preservative in all treated samples were approximately 0.002% It was comparable to the findings of Cookson [20] showing the permethrin retention of 0.005% on *Eucalyptus oblique* after treatment using SC-$CO_2$ at the pressure of 15 MPa and temperature of 50°C. These results suggest that the SC-$CO_2$ treatment condition in the current study requires the application of higher amount of the preservative to be able to transport the same concentration of 0.005%. However, it is important to further test the effectiveness of the preservative retention found in this study since permethrin is becoming increasingly used as components in low retention wood preservative [24]. Results also showed that there was no significant difference in overall permethrin retention among the different surface areas of bamboo laminations as shown in Table 1.

**Table 1**. Retentions of permethrin bamboo products after SC-$CO_2$ impregnation

| | Retention (%) | | | | Average |
|---|---|---|---|---|---|
| Specimen | Layer 1 | Layer 2 | Layer 3 | Layer 4 | |
| Bamboo splits | 0.002 | 0.002 | 0.0028 | 0.0023 | 0.0022 |
| Outer-outer laminated bamboo | 0.0018 | 0.0021 | 0.0023 | 0.0021 | 0.002 |
| Outer-inner laminated bamboo | 0.0021 | 0.0021 | 0.0021 | 0.0021 | 0.0021 |
| Inner-inner laminated bamboo | 0.0021 | 0.0023 | 0.0022 | 0.0021 | 0.0021 |

## 4. Conclusion

Preservative treatment with SC-$CO_2$ as a carrier solvent is feasible with laminated bamboo products. The current condition of SC-$CO_2$ treatment proved effective in solubilizing and transporting a sufficient amount of permethrin into laminated bamboo prevent biological attacks. In this study, permethrin retentions on the treated bamboo products were approximately 0.002% without any significant difference among the surface laminated areas. The preservative permethrin within the samples of outer-inner laminated bamboo was more evenly distributed compared to those of outer-outer and inner-inner laminated bamboo products, indicating the importance of considering surface layer laminations in the manufacture of biological resistant laminated bamboo.

## Reference

[1] Kariuki J, Nyomboi T, and Mumenya S 2014 Effect of orentation and arrangement of bamboo strips on structural strenght of laminated bamboo beam *Int. J. Eng. Sci. Emerg. Technol.* **7** 55-67

[2] Li T, Cheng Dl, Wålinder MEP, and Zhou DG 2015 Wettability of oil heat-treated bamboo and bonding strength of laminated bamboo board *Industrial Crops and Products* **69** 15-30

[3] Sharma B, Gatóo A, Bock M, and Ramage M 2015 Engineered bamboo for structural applications *Construction Building Material* **81** 66-73

[4] Verma CS, Sharma NK, Chariar VM, Maheshwari S, and Hada MK 2014 Comparative study of mechanical properties of bamboo laminae and their laminates with woods and wood based composites *Compos Part B Eng.* **60** 523-30

[5] Hanim AZ, Zaidon A, Anwar UMK, and Rafidah S 2013 Effects Chemical treatment durability properties of gigantochloa scortechinii strips ply-bamboo *Asian Journal Biology Sci.* **6** 153-60

[6] Silviana, Kareth S, and Petermann M 2014 durability assessment and physical properties investigation of modified petung bamboo (*Dendrocalamus asper*) as resulted on acetylation, assisted by supercritical $CO_2$ *Procedia Chem.* **9** 273-83

[7] Sulaiman O, Hashim R, Wahab R, Ismail ZA, Samsi HW, and Mohamed A 2006 Evaluation of shear strength of oil treated laminated bamboo *Bioresour. Technol.* **97** 2466-69

[8] Lebow S 1996 Leaching of wood preservative components and their mobility in the environment General Technical Report FPL-GTR-93 (United States Department of Agriculture) pp. 47

[9] Lebow ST, Lebow PK, Foster DO, and Brooks KM 2000 *Environmental Impact of Preservative-Treated Wood in a Wetland Boardwalk* (Madison Wisconsin: USDA Forest Service Forest Products Laboratory)

[10] Hedley MD 1997 An assessment of risks associated with use of CCA-treated timber in sensitive environments and options for its substitution with alternative timber materials (Conservation Advisory Science Notes No. 154) Department of Conservation, Wellington

[11] Kang S, Levien KL, and Morrell JJ 2006 Effect of process variations during supercritical fluid impregnation on cyproconazole retention and distribution in Ponderosa Pine sapwood *Wood Fiber Sci.* **38** 64-73

[12] Tripathi S and Poonia PK 2015 Treatability of *Melia composita* using vacuum pressure impregnation *Maderas Ciencia y Tecnologia* **17** 373-84

[13] Muin M and Tsunoda K 2003 Preservative treatment of wood-based composites with 3-iodo-2-propynyl butylcarbamate using supercritical carbon dioxide impregnation *J. Wood Sci.* **49** 430-6

[14] Lucas S, González E, Calvo MP, Palencia C, Alonso E, and Cocero MJ 2007 Supercritical $CO_2$ impregnation of Radiata pine with organic fungicides: Effect of operating conditions and two-parameters modeling *J. Supercrit. Fluids* **40** 462-8

[15] Kjellow 2009 Supercritical wood impegnation [Disertassion] Faculty of Life Sciences University of Copenhagen

[16] Uslinawaty Z, Muin M, Putranto B, and Suhasman S 2017 Properties of laminated bamboo produced with the involvement of supercritical carbon dioxide treatment *J. Indian Acad. Wood Sci.* **14** 74-8

[17] Freeman MH, Shupe TF, Vlosky RP, and Barnes HM 2003 Past, present, and future of the wood preservation industry *Forest Product Journal* **5** 9-15

[18] Cao J and Jiang X 2014 Preservation of wood and other sustainable biomaterials in China *ACS Symposium Series American Chemical Society Washington* pp. 364-84

[19] Muin M 2011 Physical, mechanical, and biological properties of silafluofen- treated tropical wood using multi-phase carbon dioxide impregnation *Wood Research Journal* **2** 42-5

[20] Cookson LJ, Qader A, Creffield JW, and Scown DK 2009 Treatment of timber with permethrin in supercritical carbon dioxide to control termites *J. of Supercrit. Fluids* **49** 203-8

[21] Wijaya M and Arcana M 2009 Characterization and making polyurethane from mahoni wood of sawdust by biodegradable polymer *Jurnal Ilmu dan Teknologi Hasil Hutan* **2** 64-7

[22] Faix O 1991 Classifcation of lignins from different botanical origins by FT-IR spectroscopy *Holzforschung* **45** 21-7

[23] Pandey KK and Pitman AJ 2003 FTIR studies of the changes in wood chemistry following decay by brown-rot and white-rot fungi *International Biodeterioration& Biodegradation* **52** 151-60

[24] Tame NW, Dlugogorski BZ, and Keneddy EM 2007 Formation of dioxins and furans during combustion of treated wood *Progress in Energy and Combustion Science* **33** 384-408

# Screening of antioxidant and antidiabetic activity from fruit bark, stem bark and leaves of *Gyrinops versteegii* (Gilg.) Domke

**A Sukito[1*], S Darmawan[2], and A I Prihantini[1]**

[1]Non Timber Forest Product Tecnology Research and Development Institute, Ministry of Environment and Forestry, Mataram, 83111, Indonesia
[2]Forest Product Research and Development Center, Ministry of Environment and Forestry, Bogor, 16610, Indonesia

[*]Email: agus_su@yahoo.com

**Abstract.** Agarwood is a commodity that has many benefits for the community, one of which is as medicine. *Gyrinops versteegi* (Gilg.) Domke, as one of plant producing agarwood and an endemic plant in West Nusa Tenggara, has less information of biological activity. The present study aims to investigate the biological activities from *G. versteegii*, particularly for anti-oxidants and anti-diabetic activities and their cytotoxicity. The 1,1-diphenyl-2-picrylhydrazyl (DPPH)-radical scavenging activity assay was used to determine the anti-oxidant activity, whereas the α-glucosidase inhibition assay was conducted to determine the antidiabetic activity. The cytotoxicity test used brine shrimp lethality test (BSLT) and phytochemical analysis were also conducted. The highest anti-oxidant and antidiabetic activity was found in the methanol extract of leaves from infected *G. versteegii* with $IC_{50}$ values of 32.89±2.7 and 55.01±2.4 μg/mL, respectively. The cytotoxicity test showed high toxicity with the $LC_{50}$ value of 1.09 μg/mL. Therefore, their high activities showed that those extracts are potent for antioxidant and anti-diabetic agent.

## 1. Introduction

Agarwood is a commodity that has many benefits and high economic value. Recently, much attention has been paid to its resins and oil productions. Several technique on agarwood production have been developed. An inoculation technique using fungi such as *Fusarium spp.* are widely used nowadays. Some of them using drill or Agar-Wit to induce the fungi [1-3]. A modified inoculation techniques were also reported on fungal inoculation using a porous nail system and implant technique [4].
Many studies have successfully explored other benefits that can be derived from agarwood-producing plants. *Aquilaria sinensis* are also broadly used as a main ingredient in several health foods including tea, honey and flavour [5]. The phytochemical and antioxidant activity of *A. malaccensis* were studied [6]. The ethanolic extract of *A. sinensis* leaves was reported to have analgesic, anti-inflammatory, and nitrite scavenging activities [7,8]. The essential oil extract from *A. crassna* was a safe material and suggested as a potential anti-tumor candidate [9].

Moreover, *A. sinensis* have been widely explored for their chemical constituents. Isolation compounds from healthy stem of *A. sinensis* successfully confirmed Aquilarin A, a new benzenoid derivative, balanophonin and lariciresinol, which the last two exhibited cytotoxicity against SGC-7901 and SMMC-7721 cell lines [10]. Other study was also isolated flavonoids including apigenin 7,4 dimethylether, genkwain, quercetin, and kaempferol from wild *A. sinensis* [5]. Moreover, from methanolic extract of the leaves of *A. sinensis* were reported the isolation of five new benzophenone glycosides, aquilarinensides A-E [11]. However, the literatures mostly studied on *Aquilaria spp.*, meanwhile the pharmacological study of *Gyrinops versteegii*, another plant producing agarwood, are less informed.

*G. versteegii* (Gilg.) Domke is mostly distributed in the eastern part of Indonesia [12]. It is an endemic plant in West Nusa Tenggara. However, less reports studied on its biological activities compared with its resinous heartwood production. The antioxidant assays have been studied from *G. versteegii* extracts [13]. Furthermore, the antioxidant and cytotoxic activities of the chloroform extract of *G. versteegii* leaves against Hela cell lines have been reported [14]. Moreover, our previous research reported that agarwood from *G. versteegii* potent for antioxidant and antidiabetic agents [15]. Nevertheless, the antidiabetic activity and cytotoxicity of different parts of *G. versteegii* are less studied. Therefore, this study aims to investigate the biological activity from different parts of *G. versteegii* for their antioxidant and antidiabetic activities as well as their cytotoxicity.

## 2. Materials and Methods

### *2.1. Materials*

2.1.1. Plant materials
There are 2 types of *G. versteegii* trees used in this study: health trees and fungi-infected trees. The fruit barks, stem barks and leaves of *G. versteegii* were collected from each type of trees. *G. versteegii* were collected from Genggelang, Lombok Utara District, West Nusa Tenggara Barat Province. All samples were dried and then mashed into powder (40 mesh) before extraction. Each sample was extracted using methanol solvent. All specimens were deposited in Laboratory of Bioassay, Non-Timber Forest Product of Research and Development Agency, Mataram.

2.1.2. Chemicals
Quercetin dehydrate, dimethylsulfoxide (DMSO), p-nitrophenyl-D-glucopyranoside, α-glucosidase, 1,1–diphenyl-2-picrylhydrazl (DPPH), bovine serum albumin, and quercetin as a control were purchased from Wako Chemical Co. Ltd. (Osaka, Japan). All solvents were purchased from Wako Chemical Co. Ltd. (Osaka, Japan) and distilled prior to use.

### *2.2. Methods*

2.2.1. Extraction
The fruit barks, stem barks and leaves were dried by avoiding direct sunlight, mashed into powder and stored at room temperature before extracting. The dried material was extracted for 48 hours at room temperature with methanol solvent. The extraction was repeated in twice. The methanol extract was concentrated with a rotary evaporator under reduced pressure. The extracts were dried and prepared for the biological activity assays.

2.2.1. DPPH-radical scavenging assay
Antioxidant activity of the extracts was estimated using the DPPH radical method described by Sukito et al [15]. Approximately 0.1 mL methanolic solution containing 2 to 4 mg of extracts was mixed with 2 mL of methanol, and a methanolic solution of 1,1-diphenyl-2-picrylhydrazyl (DPPH) (1mM, 0.5 mL) was then added. The mixture was stirred for 15 seconds, and then left to stand at room temperature for 30 minutes. The absorbance of this solution was then read at 517 nm. Percentage of antioxidant activity was calculated using the following equation:

$$\% \text{ Antioxidant activity (AA)} = [(A0-A1)/A0] \times 100 \quad (1)$$

where, A0 is the absorbance of the control and A1 is the absorbance of the extract or standard. Assays were carried out in triplicate and the results were expressed as the mean ± standard deviation. Quercetin was used as standard.

2.2.3. α-Glucosidase inhibitory activity assay
Antidiabetic activity of the extracts was determined by α-glucosidase inhibition assay described by Sukito and Tachibana [16]. The reaction mixture consisted of 250 μL of 5 mM p-nitrophenyl-D-glucopyranoside and 1 mL of 100 mM phosphate buffer (pH 7.0) and was added to a flask containing 250 μL of the sample dissolved in DMSO at various concentrations. The reaction mixture was pre-incubated for 5 min at 37°C and the reaction was then initiated by adding 250 μL α-glucosidase, and

the incubation was continued for 15 minutes. The reaction was stopped by adding 1 mL of 200 mM $Na_2CO_3$. The activity of α-glucosidase was determined by measuring the release of p-nitrophenol at UV-VIS max of 400 nm. The percentage of inhibitory activity was calculated using the following equation:

$$\% \text{ inhibition} = [(A0-A1)/A0] \times 100 \quad (2)$$

where, A0 is the absorbance of the control and A1 is the absorbance of the extract or standard. The 50% inhibition concentration ($IC_{50}$) of each compound was determined from the graph of percentage α-glucosidase inhibitory activity against sample concentrations. Assays were performed in triplicate and the results were expressed as mean values ± standard deviations. Quercetin was used as the positive control.

2.2.4. Citotoxicity test
Cytotoxicity test was conducted using Brine Shrimp Lethality Test (BSLT) method described by Sukito *et al.* [16]. In this experiment, the cytotoxicity of each sample is calculated at various concentrations of 10, 100, 250, 500, 750 and 1000 µg/mL in seawater containing 1% DMSO (v/v). Three replications were used for each concentration. A parallel series of tests with the standard potassium dichromate solution ($LC_{50}$=20-40µg/mL) the blank control was always included. After 24 hours, survivors were counted, by dissection microscope, and percentage of mortality (% M) of each dose was calculated as compared with control. Cytotoxicity was considered the significant if the $LC_{50}$ value is less than 20-30 µg/mL.

2.2.5. Phytochemical Analisys
The extract was tested for the phytochemical characteristics according to standard methods described by Sukito et al [15] as below:

*Test for saponin*: About 5 mL of the extract was boiled in 20 mL of distilled water in a water bath and filtered. 10 mL of the filtrate was mixed with 5 mL of distilled water and shaken vigorously for a stable persistent froth. The frothing was mixed with 3 drops of olive oil and shaken vigorously, then observed for the formation of emulsion, which confirms a positive presence of saponin.

*Test for alkaloid*: Various methanol and acetone extracts of the samples were acidified by adding few drops of Dragendroff's reagent (Potassium bismuth iodide). Appearance of orange red precipitate indicated presence of alkaloids.

*Test for flavonoid*: 3 mL of 1% Aluminium chloride solution were added to 5 mL of each extract. A yellow coloration was observed indicating the presence of flavonoids. 5 mL of dilute ammonia solution were added to the above mixture followed by addition of concentrated $H_2SO_4$. A yellow coloration disappeared on standing. The yellow coloration, which disappeared on standing, indicates a positive test for flavonoids.

## 3. Results

### *3.1. DPPH radical scavenging assay*
The antioxidant activity of each methanol extract and quercetin as positive control were evaluated based on their DPPH-radical scavenging activities. To determine the $IC_{50}$ value of anti-oxidant activity for each extract, further testing was carried out at several concentrations and analysed using a regression equation to obtain the results as shown in Table 1 below.

**Table 1.** $IC_{50}$ value of DPPH radical scavenging assay on methanol extract of fruit barks, stem barks and leaves of *G. versteegii*

| Explant type | $IC_{50}$ (µg/mL) |
|---|---|
| Fruit barks A | 43.61±2.5 |
| Fruit barks B | 41.90±3.3 |
| Stem barks A | 57.71±2.2 |
| Stem barks B | 50.27±1.9 |
| Leaves A | 33.17±2.3 |
| Leaves B | 32.89±2.7 |
| Quercetin | 8.59±0.3 |

Note: A: tree without artificial inoculation using fungi; B: tree with artificial inoculation using fungi

Extracts with the highest scavenging activity were leaves extracts in both of trees with and without inoculating fungi, with the $IC_{50}$ value of 32.89±2.7 and 33.17±2.3 $\mu$g/mL, respectively. However, none of the extracts of *G. verstegii* were found to be more active than the standard (quercetin, $IC_{50}$ value of 8.59±0.3 $\mu$g/mL) for DPPH scavenging activity.

*3.2 α-Glucosidase inhibition assay*
The inhibition of α-glucosidase inhibitory activity was monitored on each methanol extract. The 50% inhibition concentration ($IC_{50}$) of each extract was determined from the graph of percentage α-glucosidase inhibitory activity against sample concentration as shown on Table 2.

**Table 2.** $IC_{50}$ value of inhibitory activity of the enzyme α-glucosidase methanol extract of fruit barks, stem barks and leaves of *G. versteegii*

| Explant type | $IC_{50}$ ($\mu$g/mL) |
|---|---|
| Fruit barks A | 97.03±3.3 |
| Fruit barks B | 95.14±2.8 |
| Stem barks A | 116.43±5.4 |
| Stem barks B | 112.67±4.3 |
| Leaves A | 55.63±3.1 |
| Leaves B | 55.01±2.4 |
| Quercetin | 5.34±0.2 |

Note: A: tree without artificial inoculation using fungi; B: tree with artificial inoculation using fungi

The leaves extracts of *G. versteegii* collected from tree with and without artificial inoculating fungi were shown the highest α-glucosidase inhibitory activity with the $IC_{50}$ value of 55.01±2.4 and 55.63±3.1 $\mu$g/mL, respectively.

*3.3 Phytochemical screening*
The quantitative phytochemical screening was carried out to determine the total phenolic content (TPC), total flavonoid content (TFC), and total saponin/triterpenoid content (TSC) as shown in Table 3.

**Table 3.** Quantitative results of phytochemical screening on methanol extracts of fruit barks, stem barks and leaves of *G. versteegii*

| Explant type | Content (% or mg/100 g dried extract) | | |
|---|---|---|---|
| | TPC | TFC | TSC |
| Fruit barks A | 8.28 | 13.12 | 0.99 |
| Fruit barks B | 8.56 | 13.15 | 0.91 |
| Stem barks A | 9.71 | 12.13 | 0.61 |
| Stem barks B | 9.57 | 12.35 | 0.50 |
| Leaves A | 14.93 | 18.75 | 7.81 |
| Leaves B | 15.18 | 18.90 | 7.95 |

Note: A: tree without artificial inoculation using fungi; B: tree with artificial inoculation using fungi

The phytochemical screening of methanol extract of different part of *G. versteegii* revealed the presence of some secondary metabolites such as alkaloid, flavonoid and saponin contents in all extract. Qualitative analysis showed that saponin and tannin were detected in stem barks however absent in leaves and fruit barks (data not displayed in this paper).

*3.3 Cytotoxicity test*
The cytotoxicity test in this study used the BSLT method with shrimp larvae (*Artemia salina* Leach.) as a bio-indicator. The result of BSLT on the crude fruit barks, stem barks and leaves methanol extract were shown on Table 4.

**Table 4.** Value $LC_{50}$ of Brine Shrimp Lethality Test of fruit bark, stem bark and leaves of *G. versteegii* methanol extract

| Explant type | $LC_{50}$ ($\mu$g/mL) |
|---|---|
| Fruit barks A | 11.01±0.7 |
| Fruit barks B | 10.35±0.4 |
| Stem barks A | 3.55±0.5 |
| Stem barks B | 2.63±0.3 |
| Leaves A | 1.71±0.3 |
| Leaves B | 1.53±0.4 |

Note: A: tree without artificial inoculation using fungi; B: tree with artificial inoculation using fungi

In the BSLT for $LC_{50}$ values, we found that leaves extract from both tree with and without inoculating fungi exhibited most active, which obtained $LC_{50}$ values of 1.53±0.4 and 1.71±0.3 $\mu$g/mL, respectively.

## 4. Discussion

The antioxidant (DPPH scavenging), α-glucosidase inhibition and cytotoxic activities are associated with phytochemical compounds. We found that the methanol extract of leaves was the most active as a radical scavenger and α-glucosidase inhibitor than other extracts of different *G. versteegii* explant. However, their activities showed lower activity than quercetin as positive control. For the cytotoxic activity according to Kowalska [17], when assessing the toxicity of plant extract with BSLT, a $LC_{50}$ value is less than 1000 $\mu$g/mL is considered to be bioactive. Therefore, we found that the samples analysed in this study were most active for biochemical and pharmacological activities. Table 4 showed that leaves extract (A and B) have the highest activity in the BSLT with $LC_{50}$ value of 1.71±0.3 and 1.53±0.4 $\mu$g/mL, respectively. The result showed that leaves extract had high both radical scavenging and α-glucosidase inhibition activities. Their activities is though to be related with the presence of phenolic, flavonoid and saponin contents. According to Table 3, the TPC, TFC and TSC of leaves extract were higher than fruit bark and stem bark. Several study revealed the antioxidant and α-glucosidase inhibitory activities related with the presence of phenolic, flavonoid and saponin contents [18-22]. Triterpenoid and saponin are reported to have antioxidant and cytotoxic activities [23,24]. The results of this study are also corroborated by our previous research that *G. versteegii* wood methanol extract has high DPPH radical scavenging and α-glucosidase inhibition activities [15].

## 5. Conclusions

Leaves extract from *G. vesteegii* has potential as an antioxidant and antidiabetic agent due to their high bioactivity ($IC_{50}$ value of 32.89±2.7 and 55.01±2.4 $\mu$g/mL, respectively). The highest antioxidant value found in both of trees with and without inoculating fungi leaves extract. The high antioxidant and α-glucosidase inhibitory activities might be due to their phenolic, flavonoid and triterpenoid/saponin containing in the extract. However, further isolation is needed to explain the bioactive compounds that have responsible to their activities.

## References

[1] Faizal A, Esyanti RR, Aulianisa EN, Iriawati, Santoso E, and Turjaman M 2017 Formation of agarwood from *Aquilaria malaccensis* in response to inoculation of local strains of *Fusarium solani Trees* **31**189-97

[2] Putri N, Karlinasari L, Turjaman M, Wahyudi I, and Nandika D 2017 Evaluation of incense-resinous wood formation (*Aquilaria malaccensis* Lam.) using sonic tomography *Agriculture and Natural Resources* **51**(2) 84-90

[3] Liu Y, Chen H, Yang Y, Zhang Z, Wei J, Meng H, Chen W, Feng J, Gan B, Chen X, Gao Z, Huang J, Chen B, and Chen H 2013 Whole-tree agarwood-inducing technique: an efficient novel technique for producing high-quality agarwood in cultivated *Aquilaria sinensis Trees Molecules* **18**(3) 3086-106

[4] Wahyuni R and Prihantini AI 2020 The best inoculation technique applied on gyrinops versteegii tree trunk at 4 to 5 meters' height *Biosaintifika: Journal of Biology & Biology Education* **12**(2)

[5] Yang MX, Liang YG, Chen HR, Huang YF, Gong HG, Zhang TY, and Ito Y 2018 Isotion of flavonoids from wild *Aquilaria sinensis* leaves by an improved preparative high-speed counter-current chromatography apparatus *Journal of Chromatographic Science* **56**(1) 18-24

[6] Huda AWN, Munira MAS, Fitrya SD, and Salmah M 2009 Antioxidant activity of *Aquilaria malaccensis* (thymelaeaceae) leaves *Pharmacognocy Research* **1**(5) 270-73

[7] Zhou MH, Wang HG, Suolangjiba, Kou JP, and Yu BY 2008 Antinociceptive and anti-inflammatory activities of *Aquilaria sinensis* (Lour.) Gilg. leaves extract *Journal of Ethnopharmacology* **117** 345-50

[8] Yang MX, Mao SS, and Chen HR 2012 Scavenging effect of different fractions extracted from leaves of *Aquilaria sinensis* on $NO_2$ *Journal of Beihua University* **13** 406-09

[9] Dahham SS, Hassan LEA, Ahamed MBK, Majid ASA, Majid AMSA, and Zulkepli NK 2016 In vivo toxicity and antitumor activity of essential oils extract from agarwood (*Aquilaria crassna*) *BMC Complementary Medicine and Therapies* **16**(1) 236

[10] Wang QH, Peng K, Tan LH, Dai HF 2010 Aquilarin A a new benzenoid derivative from the fresh stem of *Aquilaria sinensis Molecules* **15**(6) 4011-16

[11] Sun J, Wang S, Xia F, Wang KY, Chen JM, and Tu PF 2014 Five new benzophenone glycosides from the leaves of *Aquilaria sinensis* (Lour.) Gilg. *Chinese Chemical Letters* **25**(12) 1573-76

[12] Lee SY, Turjaman M, and Mohamed 2018 Phylogenetic Relatedness of Several Agarwood-Producing Taxa (Thymelaeceae) from Indonesia *Tropical Life Science Research* **29**(2) 13-28

[13] Prihantini AI and Rizqiani KD 2019 Various antioxidant assays of agarwood extracts (*Gyrinops versteegii*) from West Lombok, west Nusa Tenggara, Indonesia *Asian Journal of Agriculture* **3**(1) 1-5

[14] Nuringtyas TR, Isromarina R, Septia Y, Hidayati L, Wijayanti N, and Moeljopawiro S 2018 The antioxidant and cytotoxic activities of the chloroform extract of agarwood (*Gyrinops versteegii* (Gilg.) Domke) leaves on Hela cell lines *AIP Conference Proceedings* 2002 020067

[15] Sukito A, Darmawan S, and Turjaman M 2020 Anti-oxidant and anti-diabetes activities of agarwood extracts from Gyrinops versteegii (Gilg.) Domke and their cytotoxicity *IOP Conference Series: Earth and Environmental Science* **415**(1) 012001

[16] Sukito A and Tachibana S 2014 Potent α-Glucosidase Inhibitors Isolated from Ginkgo biloba Leaves *Pakistan Journal of Biological Sciences* **17** 1170-78

[17] Kowalska K 2011 Natural compounds involved in adipose tissue mass control in in vitro studies *Postepy higieny i medycyny doswiadczalnej* **65** 515-23

[18] Tofighi Z, Alipour F, Hadavinia H, Abdollahi M, Hadjiakhoondi A, and Yassa N 2014 Effective antidiabetic and antioxidant fractions of Otostegia persica extract and their constituents, *Pharmaceutical Biology* **52**(8) 961-66

[19] Chen D, Bi D, Song YL, and Tu PF 2012 Flavanoids from the stems of *Aquilaria sinensis Chin. J. Nat. Med.* **10** 287-91

[20] Sarian MN, Ahmed QU, Mat So'ad SZ, Alhassan AM, Murugesu S, Perumal V, Syed Mohamad SN, Khatib A, and Latip J 2017 Antioxidant and antidiabetic effects of flavonoids A structure-activity relationship based study *BioMed research international* 2017

[21] Pranakhon R, Aromdee C, and Pannangpetch P 2015 Effects of iriflophenone 3-C-β-glucoside on fasting blood glucose level and glucose uptake *Pharmacogn Mag.* **11**(82)
[22] Qi J, Lu JJ, Liu JH, and Yu BY 2009 Flavonoid and a rare benzophenone glycoside from the leaves of *Aquilaria sinensis Chemical & Pharmaceutical Bulletin* **57**(2) 134-37
[23] Tapondjou LA, Nyaa LBT, Tane P, Ricciutelli M, Quassinti L, Bramucci M, Lupidi G, Ponou BK, and Barboni L 2011 Cytotoxic and antioxidant triterpene saponins from Butyrospermum parkii (Sapotaceae) *Carbohydrate Research* **346** 2699-704
[24] Ponou BK, Teponno RB, Ricciutelli M, Quassinti L, Bramucci M, Lupidi G, Barboni L, and Tapondjou LA 2010 Dimeric antioxidant and cytotoxic triterpenoid saponins from Terminalia ivorensis *A. Chev. Phytochemistry* **71**(17-18) 2108-15

## Authors' Contribution

All authors contributed equally to this work.

# The effect of pre-freezing treatment on the color changes and wettability of teak wood (*Tectona grandis* Linn.F.)

**U Adzkia*, I S Rahayu, L Karlinasari**
Department of Forest Product, Faculty of Forestry, IPB University, Bogor, 16689

*Email: ulfa23adzkia@apps.ipb.ac.id

**Abstract.** Pre-freezing treatment is one of the methods used for increased the properties of wood. The pre-freezing treatment affects the physical properties of wood, extractive, and surface conditions, this will affect the wettability of wood and wood color. The purpose of this research was to evaluate the effect of pre-freezing treatment on the color changes and changes in contact angle and wettability of wood. The study was conducted on teak wood (*Tectona grandis*) with the temperature -16°C for 24 hours and 48 hours. The results of this study showed that the pre-freezing treatment made decreasing of brightness with a reduction of the k-value. Besides, the pre-freezing time affected the wettability of wood. The wettability of wood was declined with increased time of pre-freezing. Wood utilization as a raw material of composite needed more treatment to increase the wettability of wood.

## 1. Introduction

Teak wood (*Tectona grandis*) is one of the most important species of wood in the world [1]. Teak wood is an expensive species hardwood with physical properties, durability, and good esthetic [2,3]. The superior properties of teak wood increasing demand in the community cause decreasing in the availability of teak wood. To solve the problem, the community and Perhutani Corporation are maximizing wood production per unit area through genetic selection and silvicultural techniques to produce fast-growing teak [4]. Fast growing teak wood had weaknesses properties such as low dimensional stability, strength, and durability [5]. Therefore, treatment is required to improve the quality of teak wood. One of the alternatives treatment is pre-freezing treatment.

The pre-freezing treatment has been done in the some of wood species in the winter in America. Wood store at the negative temperature has advantages such as reduced the biodeterioration, drying defects, and, dimensional shrinkage, and rate of drying [6,7]. Pre-freezing treatment caused the water transfer from the cell wall to the lumen, so that the cell wall is thickening and cells become rigid [8]. Cell walls under pressure make the cell more rigid and tend to get steady. Pre-freezing treatment in a short time has a positive effect on the strength properties of wood. This was also stated by Szmutku *et al.* [7] that fast pre-freezing results in better quality than products with long pre-freezing.

The study of pre-freezing treatment has been carried out on the several species of wood such as spruce wood [7], pinewood [8], and fast-growing species like rose gum and Gympie messmate [5]. Based on the results mentioned that wood has lower surface energy with higher pH value, decreasing weight loss, and increasing the water content. Furthermore, the pre-freezing treatment caused decreasing of MOE value and compression parallel to a grain of spruce wood [5]. Alteration of wood properties due to pre-freezing treatment is thought to affect wettability and the color of the wood. Changes of the wettability and color was occured on the same species after heat treatment [9,10].

Based on the research that the high temperature affected the ability wetting the water and the color of wood. After heat treatment the color of wood became darker. In addition, wood is also keeped in low temperatures. The wood that is stored at low temperatures effect to changes the components of wood, so it is thought to also affect the wetting and color of the wood.Therefore, this study was conducted on the fast-growing teak wood with pre-freezing treatment to evaluated the color change and the wettability of teak wood. The research aimed were to evaluate the effect of the pre-freezing treatment on the color change and the wettability of the teak wood fast-growing wood species.

## 2. **Materials and Methods**

### *2.1. Materials*

The research was conducted at Wood Anatomy and Quality Improvement Laboratory, and Wood Design and Engineering Laboratory, Department of Forest Product, Faculty of Forestry, IPB University. The material used for this study was the teak wood with the size of the length, width, and thickness 10 x 10 x 2 cm. The color testing and wettability was perfomed three replication.

### 2.2. *Methods*

#### 2.2.1. Pre-freezing treatment

Samples were scanned before and after pre-freezing treatment. Samples were ready to put into the freezer at a temperature of -16°Cwith a variation of time for 24 hours and 48 hours. After that, samples were drying in an oven at 60°C for 4 days. Dried were occured until the moisture content of wood under 20%. Dried samples put into desiccator for 15 minutes.

#### 2.2.2. The color changes

The measurement of wood color was used the Cannon MP 145 Scanner which is connected to a laptop, then the data was processed used ImageJ which is produced using ImageJ which produces L*, a* and b* values. The measurement of the color change in the wood refers to the research of [11] using the Cannon MP 145 Scanner the data formed in a photograph of a wood surface. The color change was tested at nine different points as in Figure 1.

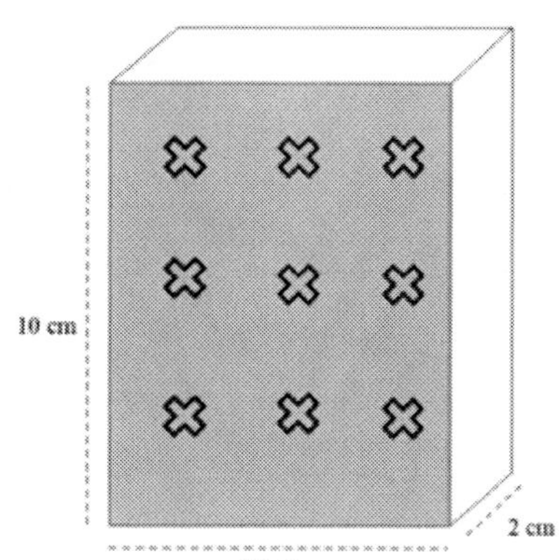


**Figure 1**. Sketch of the sample for the color changes teak wood

The color changes were measured before and after the pre-freezing treatment. The data obtained were processed using the ImageJ application. The color change was calculated using the L* a* b* formula [12]. L* was the change in the brightness from 0 (black) to 100 (white). a* showed the change in color red (positive) to green (negative). b* was indicated the change in color from yellow (positive) to blue (negative). The difference of color (ΔE) was calculated by the CIELab formula [13,14].

$$\Delta E = \sqrt{(\Delta L *)^2 + (\Delta a *)^2 + (\Delta b *)^2} \quad (1)$$

where, $\Delta$E : Different of color
$\Delta$L*: Different of light = L* treatment – L* control
$\Delta$a*: Different of color red or green
$\Delta$b*: Different of color yellow or blue

The value of changed or differenced in wood color before and after pre-freezing can be seen in Table 1.

**Table 1.** The differences in color ($\Delta E$) [15]

| Different of color ($\Delta E$) | The influenced |
|---|---|
| $0.2 < \Delta E*$ | Invisible difference |
| $0.2 < \Delta E* < 2$ | Small difference |
| $2 < \Delta E* < 3$ | Color change visible with a high-quality filter |
| $3 < \Delta E* < 6$ | Color change visible with medium-quality filter |
| $6 < \Delta E* < 12$ | High color changes |
| $\Delta E* > 12$ | Different color |

2.2.3. The measurement of the contact angle

Measurement of the contact angle between liquid and wood surface using the sessile drop method. The liquid used was distilled water. A lot of liquid that is used 0.02 mL was dripped on a wooden surface, then recorded using a portable microscope camera. The results obtained were analyzed using GOM player software to get photos every 9 seconds for 3 minutes thus obtained 19 photos for each point. The photos were measured its contact angle using Image-J (Figure 2). Each board is dripped at three different points (Figure 3). The measuring contact angel were conducted into the two sample. There sample were freeze treatment for 24 hours and 48 hours. Contact angles were tested on samples before and after pre-freezing treatment.

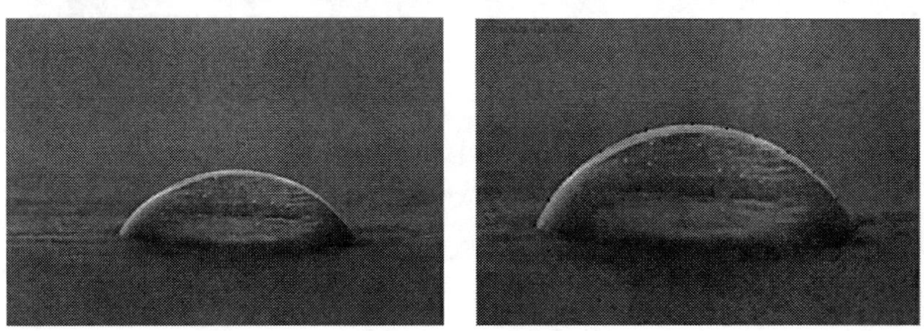

**Figure 2.** The measuring of the contact angle

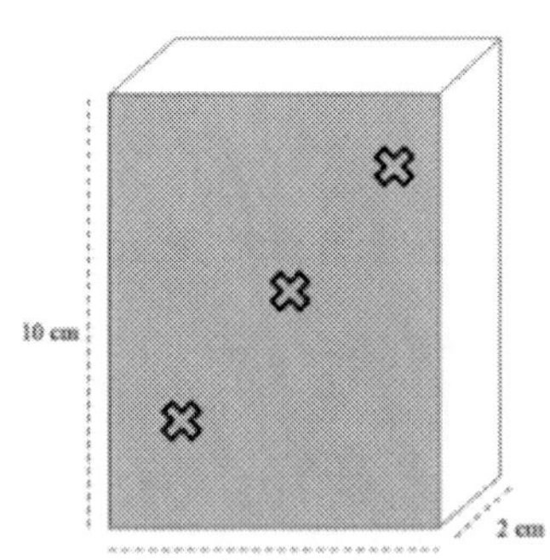


**Figure 3.** Sketch of the sample for wettability test

Over time, the contact angle is formed will also change. Changes of the contanct angle in the surface of wood was illustrated the ease of wetting. The Changed in contact angles was analyzed using the polynomial regression model based on Equation 1.

$$y = ax^2 + bx\ c \tag{2}$$

The measuring of the equilibrium contact angle used PROC NLIN software from SAS 9.1 with a *segmented regression model* based on Equation 2.

$$y = a + bX + cX^2 + E \tag{3}$$

where, y: contact angle; x: time; a, b, and c: constanta; E: error

2.2.4. The wettability based on the k-value
The contact angle value was determined based on a segmented regression equation between time (t) and contact angle (θ) using PROC NLIN SAS. Parameter of k-value is determined based on the S/G model (Shi and Gardener 2001). The S/G model uses the formula

$$\theta = \frac{\theta i.\theta e}{\theta i + (\theta e - \theta i)\exp[K\left(\frac{\theta e}{\theta e - \theta i}\right)t]} \tag{4}$$

where,
θ : Contact angle at the specific time
θi : Contact angle at the initial time
θe : Contact angle at the equilibrium
t : Time
k : The constant rate of change the contact angle
The k-value measuring from S/G model with the XLSTAT 2019 software.

## 3. Results and Discussion

### *3.1. The Color changes*
Color is one of the determinants of wood quality that will affect customer perceptions of the product [16]. Color is shown visual perception that is determined by spectra light makeup reflected off the surface. The L* value was indicated the level of brightness of the coloring, the more positive the brighter the wood. But on the contrary, if the value of L* decreased was the darker the coloring result. The value of a* represents the color produced in red or green, while the value of b* represents yellow or blue [17].

The color changes of teak wood after pre-freezing treatment can be seen in Figure 4. The L* value of teak wood for 24 hours treatment before and after pre-freezing treatment is 80.18, and 76.95, respectively. The average brightness decrease of 4.02%. Meanwhile, wood brightness before 48 hours pre-freezing treatment is 83.28 and decreased to 77.78. The average brightness reduction of 6.60%. The a* and b* values of teak wood due to pre-freezing treatment tended to increase. The value of b* after 24-hour pre-freezing treatment was increasing by 18.23%, while the 48-hour pre-freezing treatment higher increasing by 22.31%. The percentage change in the value of b* 48-hour pre-freezing treatment is greater than the 24 hours treatment. This relates to changes in the color of teak wood which becomes darker as increasing of the cooling time. The decreasing trend that occurs in teak wood due to pre-freezing treatment be expected by changes in chemical composition such as degradation of hemicellulose. Karlinasari *et al.* [18] stated that the freeze treatment can be presented the highest variation in holocelulose content, wherein the decrease was from 68.44% to 63.55%. Degradation on the holocelulose effected to the color wood. This was stated by [15,10,19] on the species of sengon, teak, mangium, mahagony, shorea, merbau, oak trees, scotch pine trees, and silver bitch tress that the color change becomes darker is more influenced by the degradation of hemicellulose.

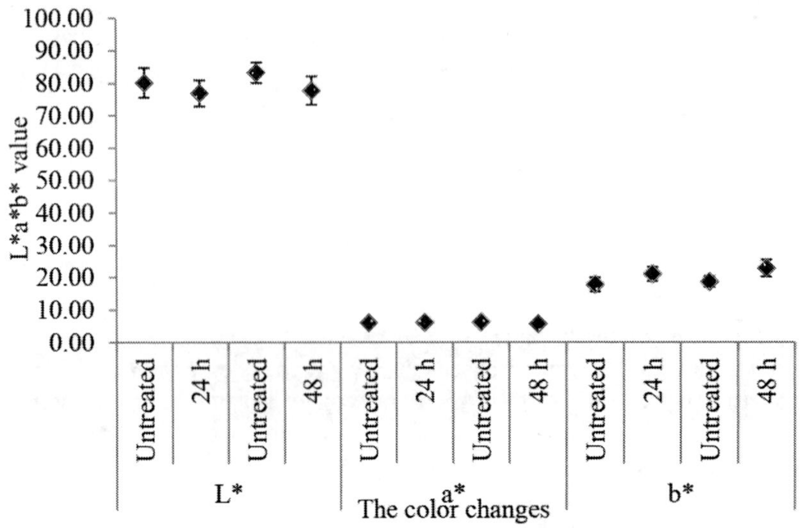


**Figure 4.** The color changes of teak wood pre-freezing treatment

The total color change was calculated by the value ΔE. ΔE is strongly influenced by temperature differences in the wood [12]. Figure 5 explains that the color changes occurred on the surface of the wood. Figure 6 explains that the amount of color changes quantitatively. The average color change of 24 hours treatment is (ΔE) 5.647, this color change is classified as a moderate color change. The 48 hours pre-freezing treatment had a color change (ΔE) of 7.157 and was classified as a high color change. Barcík *et al.* [15] stated that if a change of 3 <ΔE <6 that color changes with a medium quality filter, whereas if a change occurs in the range of 6 <ΔE <12 that a high color changed.

Color changes that occurred in wood due to pre-freezing treatment are caused by the degradation of the wood components. The decreasing trend of the bringhtness of wood that occurs in teak wood due to pre-freezing treatment is caused by changes in chemical composition such as degradation of hemicellulose [18]. Also, changes in the value of a* were affected by condensation, decomposition, and / or degradation of lignin content. Meanwhile, changes in the value of b* was affected by the lignin content in wood, lignin has a role in the presence of yellow in the wood [12]. This is following the research of Missio *et al.* [5] that the pre-freezing treatment effects to the chemical components of wood consisting of extractives, lignin, holocellulose. Particularly for teak wood, quinones are classified as colouring matters in plants and give yellow, orange, red and green or black. Based on the results of the study, Gympie messmate wood extractive substances have increased (1.16-1.23%), while the holocellulose content has decreased (64.15-62.83%). Therefore, the wood had a pre-freezing treatment will be darker.

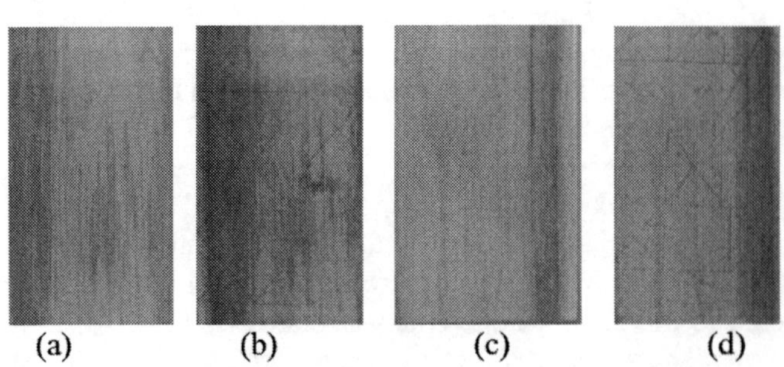


**Figure 5.** Changes of the color wood (a) Untreated, (b) after 24 hours, (c) Untreated, (d) after pre-freezing 48 hours

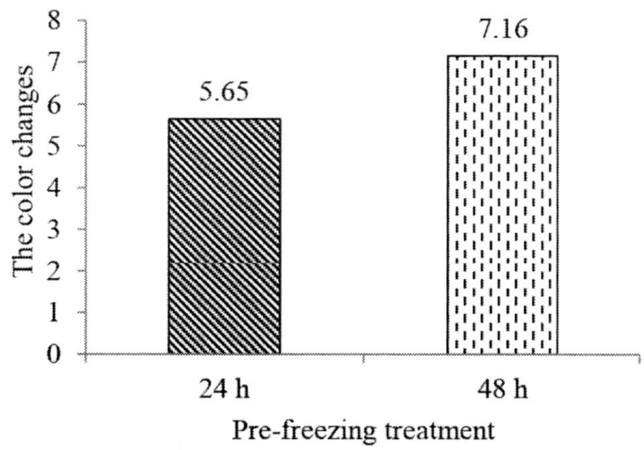


**Figure 6.** The changes of ΔE values on the pre-freezing treatment for 24 hours and 48 hours

### *3.2. The contact angle and wettability*

The wettability is the ability of a material to absorb the liquid [20]. Wettability was influenced by several factors related to adhesives (surface tension, temperature, and viscosity) and wood (density, porosity, extractives, and surface roughness) [18]. One of the criteria that can be used in determining the wetting of wood is the contact angle. Figure 7 shows the distillate water droplets on teak wood without pre-freezing treatment (untreated) and after pre-freezing treatment. If the contact angle value is more than 90° then wood wetting is not good. As a result, the surface of the wood is difficult to get wet by the liquid (hydrophobic) [21]. Changes in contact angle that occur based on changes in time indicate the wettability of a material [22]. The liquid that falls on the surface of the wood will spread and penetrate until reach a constant angle (θe).

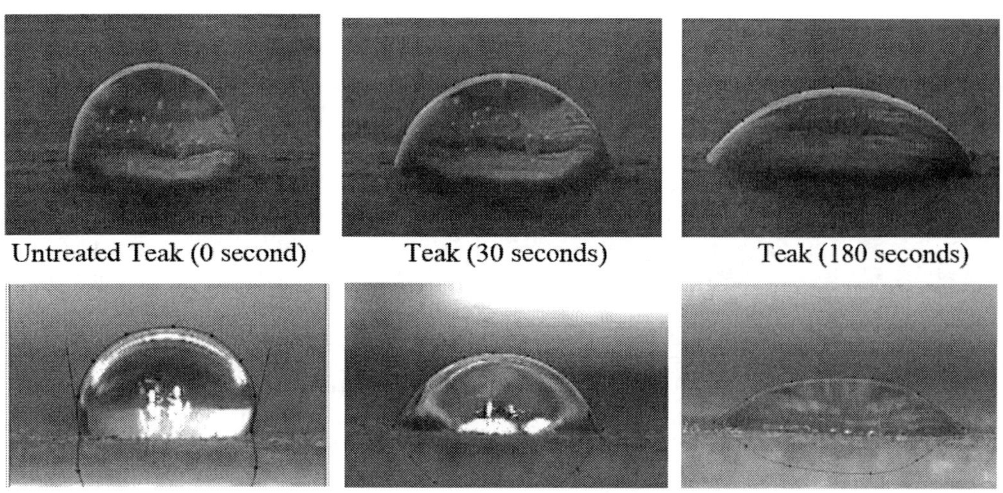

Untreated Teak (0 second) Teak (30 seconds) Teak (180 seconds)

Freeze treated teak (0 second) Teak Treatment (30 seconds) Teak Treatment (180 seconds)

**Figure 7.** The contact angle untreated and pre-freezing treatment for 24 hours

Figure 8 shows the change in the contact angle value of the teak wood treated 24 hours and 48 hours. Significant changes in contact angle occur in the initial 10 seconds. The contact angle without treatment is in the range between 80°-90°, while the contact angle after treatment is in the range between 114°-116°. The average contact angle untreated wood in the 0° second by 79° ± 7°, 30 second by 61° ± 6, and 180 second by 42° ± 4. The average contact angle treated wood in the 0 second by 102°

± 12°, 30 second by 65° ± 10, and 180 second by 44° ± 9.This shows that the contact angle of the pre-freezing treatment is higher than the contact angle without the pre-freezing treatment. Pre-freezing treatment can reduce wood wettability. This is influenced by wood factors such as species, wood anatomy (composition), wood chemicals (extractives, cellulose, hemicellulose, and lignin), and surface roughness [23]. Wood is a heterogeneous material that will have an impact on the results of varying wettability on each species [24]. Wettability of teak wood is quite poor, this is due to the high density of wood and some deposits can inhibit the process of water absorption. The type of red banana wood also has a deposit which caused a low wettability [24].

Wettability of teak wood is influenced by extractive substances in wood. As a result of pre-freezing treatment, it is assumed that the extractive moves to the surface so that the content of the extractive substance tends to be higher and cause lower wettability. This was also conveyed by Ilic [25] that pre-freezing treatment causes extractive substances to move to the surface of the wood so it causes the wood more rigid. The wood extractive substances make wood more hydrophobic [26]. Besides, the results of the research Karlinasari *et al.* [18] on wood species of rose gum and Gympie messmate explained that pre-freezing treatment affects the chemical content, especially wood extractive substances. The pre-freezing treatment affects the increase of extractive content of 1.16% to 1.23%. Some studies suggest that extractive substances affect wettability [23,27]. Extractive substances make the interaction between the adhesive and lignocellulose material less than optimal. As a result, the surface of the wood is difficult to pass by liquid.

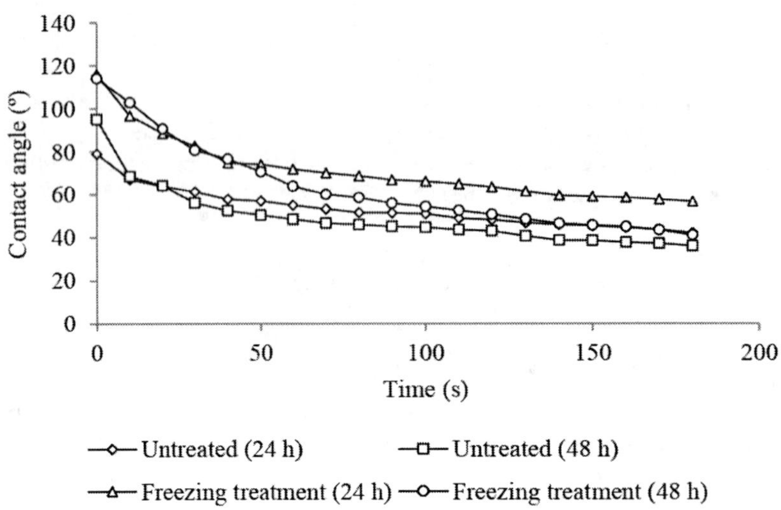


**Figure 8**. The transformation of the contact angle pre-freezing treatment 24 hours and 48 hours

The liquid that falls on the surface of the wood will spread and penetrate the wood until a constricted contact angle (θe) is reached. The value θe is used to determine the rate of change in the constant contact angle (k-value). The value of k is an indicator of the value of wettability. The higher the value of k, the better the wettability of material, this indicates that the liquid is more easily spread and absorbed in the material [28]. Figure 9 shows the k-value treatment after pre-freezing treatment for 24 hours and 48 hours tends to decrease compared to without pre-freezing treatment. This indicates that pre-freezing treatment for 24 hours and 48 hours reduces the wettability of teak wood. When associated with wood absorption values, 24 hours and 48 hours pre-freezing treatments are lower than without pre-freezing treatment. This causes the longer wood to be pre-freezing, the better dimensional stability. According to the research of Szmutku *et al.* [7] on spruce wood that the dimensional stability of wood is increasing with increasing pre-freezing time. The pre-freezing treatment is a method that makes the water turn into ice which causes the expansion of the lumen and constriction of the cell wall. This causes the wood absorption value to decrease and wood wettability decreases while the contact angle increased.

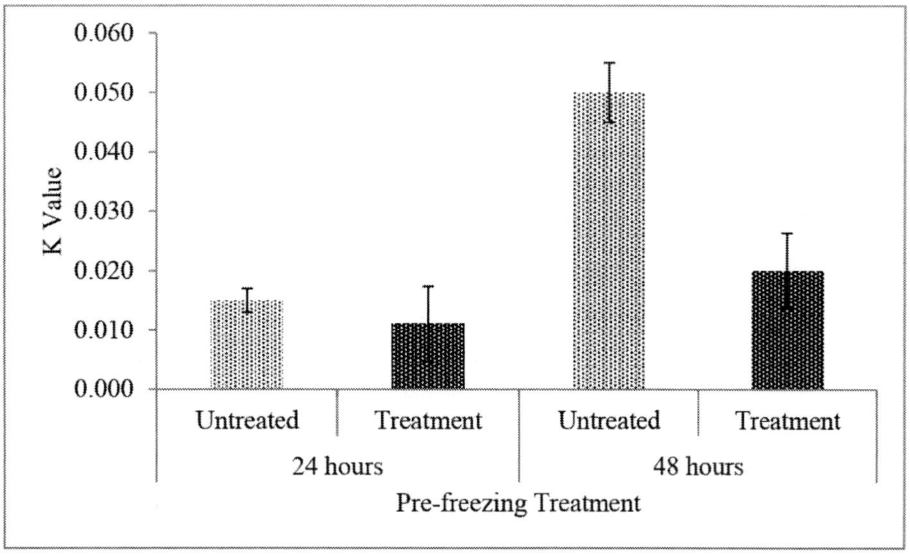


**Figure 9**. K-value of teak wood from pre-freezing treatment for 24 hours and 48 hours

## 4. Conclusion

The pre-freezing treatment occurred with the variation of time 24 hours and 48 hours. Based on the treatment, the color teak wood was darker with decreasing of L* value and $\Delta E$ value of teak wood is 5.65 for 24 h and 7.16 for 48 hours. The pre-freezing treatment caused decreasing the wettability with decreasing of the k-value while the contact angle was increased.

## References

[1] [FAO] Food and Agricultural Organization 2013 Natural teak forests decline, while planted teak forests increase (UN FAO) http://www.fao.org/news/story/en/item/129569/icode/ Accessed 04 Juli 2020
[2] Sanwo S 1987 The characteristics of the crown-formed and stem-formed wood in plantation grown teak (*Tectona grandis* L.f.) in Nigeria *J. Inst. Wood Sci.* **11(2)** 85-8
[3] Kjaer ED, Kajornsrichon S, and Lauridsen EB 1999 Heartwood, calcium and silica content in five provenances of Teak (*Tectona grandis*) *Silvae Genet* **48** 1-3
[4] Bhat KM and Ok Ma H 2004 Teak growers unite *Trop. for Update* **14** 3-5
[5] Missio AL, De Cademartori PHG, Mattos BD, and Santini EJ 2016 Physical and mechanical properties of fast-growing *European J. of Wood and Wood Product* **11** 10378-90
[6] Ilic J 1995 Advantages of prepre-freezing for reducing shrinkage-related degrade in eucalypts: General considerations and review of the literature *Wood Sci. Technol.* **29**(4) 277-85
[7] Szmutku MB, Campean M, and Porojan M 2013 Strength reduction of spruce wood through slow pre-freezing *Eur. J. Wood Wood Prod.* **71**(2) 205-10
[8] Awoyemi L, Femi-Ola TO, and Aderibigbe EY 2010 Pre-freezing as a pre-treatment for thermal modification of wood Part 2: Surface properties and termite resistance *J. Indian Acad. Wood Sci.* **7**(1-2) 19-24
[9] Karlinasari L, Lestari AT, and Priadi T 2018 Evaluation of surface roughness and wettability of heat-treated, fast-growing tropical wood species sengon (*Paraserianthes falcataria* (L.) I.C.Nielsen), jabon (*Anthocephalus cadamba* (Roxb.) Miq), and acacia (*Acacia mangium* Willd.) *Int. Wood Prod. J.* **9**(3) 142-8

[10] Karlinasari L, Azmi MI, and Priadi T 2018 The changes in color and dynamic modulus of elasticity of five important Indonesian tropical wood species after 10 months of outdoor exposure *J. Indian Acad. Wood Sci.* **15**(2) 149-157
[11] Kleeberger KS and Moser BC 2002 Flatbed scanners: An alternative tool for gathering horticultural data *Horttechnology* **12**(3) 444-6
[12] Sikora A, Kačík F, Gaff M, Vondrová V, Bubeníková T, and Kubovský I 2018 Impact of thermal modification on color and chemical changes of spruce and oak wood *J. Wood Sci.* **64**(4) 406-16
[13] ISO 11664-4 2008 Colorimetry - Part 4: CIE 1976 L* a* b* colour space (ISO 11664-4:2008)
[14] ISO 11664-6 2014 Colorimetry- Part 6: CIEDE2000 Colour-difference formula (ISO 11664-6: 2014)
[15] Barcík S, Gašparík M, and Razumov EY 2015 Effect of temperature on the color changes of wood during thermal modification *Cellul. Chem. Technol.* **49**(9-10) 789-98
[16] Torres SS, Jomaa W, and Marc F 2010 Causes of color changes in wood during drying *For. Stud. China* **12**(4) 167-75
[17] Muflihati 2013 The extraction of the bark samak woof (*Syzgium inophyllum DC.*) for the natural wood coloring [Thesis] Bogor: IPB
[18] Karlinasari L, Yoresta FS, and Priadi T 2018 Color canges and hardness properties of thermally modified wood at various temperatures and wood species ) *JITKT* **16**(1) 69-83
[19] Lestari AT, Darmawan IW, and Nandika D 2016 Effects of surface conditions on the adhession strength of surface coatings *JITKT* **14**(1) 18-22
[20] Missio AL, Mattos BD, De Cademartori PHG, Pertuzzatti A, Conte B, and Gatto DA, 2015 Thermochemical and physical properties of two fast-growing eucalypt kinds of wood subjected to two-step freeze-heat treatments *Thermochim. Acta* **615** 15-22
[21] Bracco G and Holst B 2013 *Surface science techniques* **51**(1)
[22] Shi SQ and Gardner DJ 2001 Dynamic adhesive wettability of wood *Wood Fiber Sci.* **33**(1) 58-68
[23] Kamke FA and Lee JN 2007 Adhesive penetration in wood - A review *Wood Fiber Sci.* **39**(2) 205-20
[24] Marbun SD 2018 Basic Characteristics and Skills of Some Underutilized Wood Species (Lower Species of Wood Used) from the Kalimantan Tropical Rain Forest [Tesis] Pascasarjana Institut Pertanian Bogor
[25] Ilic J 1999 Shrinkage-related degrade and its association with some physical properties in Eucalyptus regnans F. Muell *Wood Sci. Technol.* **33**(5) 425-37
[26] Qin ZQ. Zhang GS, and J. Li 2014 Surface free energy and dynamic wettability of differently machined poplar woods *BioResources* **9**(2) 3088-3103
[27] Nussbaum RM and Sterley M 2002 The effect of wood extractive content on glue adhesion and surface wettability of wood *Wood Fiber Sci.* **34**(1) 57-71
[28] Darmawan IW, Nandika D, Noviyanti E, Alipraja I, Gardner D, and Geraldin P 2017 Wettability and bonding quality of exterior coatings on jabon and sengon wood surfaces *J. Cost. Technol Res* **15**(1) 95-104

IOP Conf. Series: Materials Science and Engineering **935** (2020) 012006 doi:10.1088/1757-899X/935/1/012006

# The residual stand damages due to log skidding using crawler tractor at one logging company in Central Kalimantan

**Dulsalam[*], Sukadaryati and Soenarno**

Forest Products Research and Development Center, Ministry of Environment and Forestry, Bogor, 16610, Indonesia

[*]E-mail: dul.salam@gmail.com

**Abstract.** An investigation on residual stand damages due to log skidding using a skidding tractor was carried out at one logging company in Central Kalimantan in 2012. The objective is to find out information about residual stand damages on trees, poles, and sapling caused by Conventional Log Skidding (CLS system) and Low Impact Log Skidding (LILS system). Data on stand potency consists of tree, pole and sapling levels as well as the number of trees felled were collected. The investigation revealed that damage to the level of standing trees, poles, and sapling skidding caused by CLS system at PT Gunung Meranti was 11.40% (Coefficient of variation/CV) 7.95%; 9.53% (CV 14.74%) and 14.84% (CV 16.92%), respectively. The average damage of standing trees, poles and sapling caused by LILS system in PT Gunung Meranti respectively was 9.85% (CV 14.00%); 7.54% (CV 9.81%) and 12,55% (CV 16.60%). LILS system could reduce the damages to the standing trees.

## 1. Introduction

Log skidding is one of the elements of timber harvesting activities intended to move logs from the felling sites to the log landing placest on the edge of the forest roads. The tool used to skim wood in Indonesia's natural forests is usually a crawler tractor. The use of the tractor has many other advantages: (1) it can move freely between the core trees in the selective cutting system, (2) it can be used safely up to 40% slope, and (3) it can be used at a sufficiently long skid distance [1]. Steel chain tractors have a low-pressure point [2]. The Caterpillar D7G tractor has been widely used to extract wood from the felling area to the landing [3]. The use of this type of tractor requires a large investment [4]. Timber skidding activities can cause damage to residual stands, especially at the tree, pole, and sapling levels.

To reduce the damage of the residual stand, skidding activities need to be carried out in a controlled manner. This is intended to make optimal use of forest resources and minimal not only a waste of resources but also environmental disturbances. The optimal use of forest resources will increase forest productivity and harvest efficiency and will reduce the amount of harvesting waste that occurs. Increasing forest productivity and the efficiency of timber harvesting are highly demanded in the context of supplying wood industry raw materials that are adequate in quantity and quality on the one hand and minimizing environmental disruptions that occur on the other.

Logging of natural forests in Indonesia involves employs chainsaws for felling and crawler bulldozers for skidding [5]. Residual stand damage as a result of logging or forest harvesting is not only caused by felling but also by skidding. Further explained that before logging begins, the company has to design a logging plan that includes tree maps as well as topographic maps and skidding roadmaps. In practice, many companies conduct inaccurate inventories owing to reasons such as having prepared inappropriate skidding road plans. One of the ways to improve harvesting efficiency is to use efficient and effective skidding techniques and cause minimal damage to standing stands. The magnitude of damage to residual stands in intensive silviculture areas was 24.58% and 15.43%, which consisted of 5.86% due to felling and 9.59% due to skidding [6,7]. The closer the stand distance from the skid trail, the higher the residual stand damage occurs [8].

Damage to the residual stand mentioned above is damage to the stand at all levels of the residual stand with a certain diameter limit. Damage to the residual stand at the level of regeneration of the stand is still very limited. This paper tries to present the residual stand damage that happens at a forest company in Central Kalimantan. The aim is to provide information on damage to residual stands at various levels of residual stands. The information is expected to be useful for implementing timber skidding activities in the field and for policymakers in log skidding.

## 2. Materials and Methods

### *2.1. Materials*

The study was conducted in the concession areas at PT Gunung Meranti in Central Kalimantan. The geographical location of PT Gunung Meranti's IUPHHK-HA area is 113°46 '- 114°04' and 0°16'- 0°43'LS. The research plot included Lawang Tamang Village, Sungai Hanyu District, Kapuas Regency, Central Kalimantan Province. Topographic field topography (45%) is rather steep (15-25%). The research location has an altitude of 200-800 m above sea level. Research locations included climate A (Smith & Fergusson) with the highest rainfall per month of 356 mm/month and the lowest of 86 mm/month. This research was conducted in 2012.

The materials in this study are wood that is ready for skidding, diesel, wood paint, plastic mining, and steel cables. The tools used in this study was field equipment which included skidding tractor, cameras, and stationery.

### *2.2. Methods*

Primary data and secondary data were collected in this study. Primary data were collected through observing the potential of residual stands before skidding and observing the damage to residual stands after skidding. Observation of the potential of stands at tree level (plant that have a trunk diameter of 20 cm and up) was carried out a census in the observation plot. Observation of the potential of pole level stands (plants that have stem diameter between 10 cm to less than 20 cm) was carried out by taking samples measuring 10 m x 10 m in the sample plot. Observation of the potential of stock level saplings (plant that have stem diameter less than 10 cm and a heigt of 1.5 m) was carried out by taking samples of 5 x 5 m in the sample plot. Each plot is recorded by species according to the stand-level.

The procedures of the investigation were as follows:

a. Determine the location of the study purposively based on the ease of researching to achieve the stated goals and still consider population representation.
b. Make 3 observation plots for the skidding technique control and 3 plots for the conventional skidding technique at the selected location (Figure 1).
c. Observed the potential of residual stands at the level of trees, poles, and sapling before skidding which was stated by the number of stems per level of residual stands per plot. To estimate the degree of damage to trees, the Forestry Department's criteria [9] were used. Trees are considered damaged if they are one or more of the following conditions: (1) Tree canopy is damaged above 30% or a tree branch or large branch is broken and (2) Stem injury above ¼ around the trunk with a length of more than 1.5 m. The criteria for stand damage based on the percentage of standing stand damage are as follows [10]:
   - Standing stand is called mild damage if damage < 25%
   - Standing stand is called moderately damaged if damage between 25-50%
   - Standing stand is called heavily damaged if damage > 50%

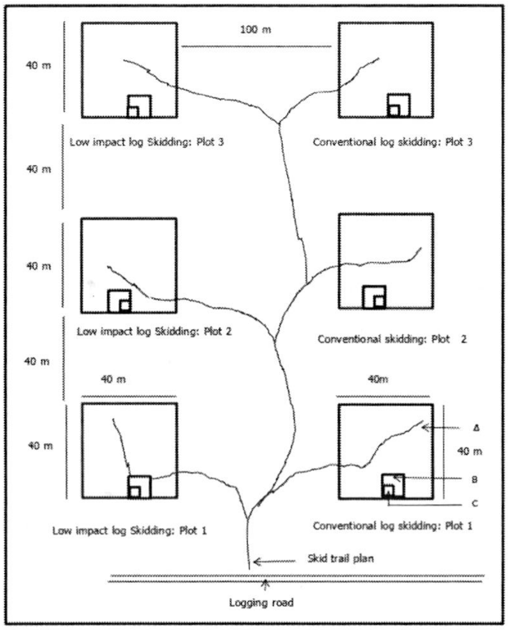


Remarks: A = The observation plots of the residual stands of tree level measuring 40 m x 40 m; B = The observation plots of the residual stands of pole level measuring 10 m x 10 m; C = The observation plots of the residual stands of sapling level measuring 5 m x 5 m

**Figure 1.** Sketch of the research plot placement

Observed residual stand damages due to CLS and LILS at the chosen location. CLS was done according to the skidding habits in the field. LILS was carried out as follows [11]: (1) skid trails were well planned; (2) good cooperation between the felling team and the skidding team; (3) improved tractor operator skills; (4) tractor operator made a planned skid trail; (5) the tractor only moved on the specified trail; (6) operators used the prepared winching path; (7) log skidding was sought not to cross rivers/grooves; (8) when skidding, the front of the skidding wood was lifted. Observations of damage to residual stands at tree level, poles and sapling were stated in the number of stems per residual stand-level per plot

### *2.3. Data analysis*

Damage to the residual stand is calculated according to the formula:

- tree level :

$$KTP = \frac{JRP}{JTP} \qquad (1)$$

where, KTP = damage to standing trees left due to skidding (%), JRP = the number of residual stands at tree level damaged by skidding (trees), JTP = the number of trees remaining before the skidding (trees)

- pole level:

$$KTT = \frac{JRT}{JTT} \qquad (2)$$

where, KTT = standing pole level damage due to skidding (%), JRT = number of residual stands at pole level damaged by skidding (poles), JTT = number of standing pole level stands before skidding (poles)

- Sapling level:

$$KTPc = \frac{JRPc}{JTPc} \quad (3)$$

where, KTPc = stand damage left to sapling due to skidding (%), JRPc = the number of residual stand damage at sapling level by skidding (sapling), JTPc = the number of stands left before the skidding (sapling)

## 3. Results and Discussion

### *3.1. Residual damages to tree level*

The results of observations of tree potential and tree level stand damage are presented in Appendix 1. From Appendix 1 it can be calculated that residual stand damages on tree are as presented in Table 1.

**Table 1**. Residual stand damages to tree level

| CLS system | | LILS system | |
|---|---|---|---|
| Plot | Damages (%) | Plot | Damages (%) |
| 1 | 9.42 | 1 | 9.17 |
| 2 | 12.22 | 2 | 10.71 |
| 3 | 13.46 | 3 | 10.10 |
| Total | 35.10 | Total | 29.98 |
| Mean | 11.70 | Mean | 9.99 |
| SD | 1.19 | SD | 0.45 |
| CV | 10.21 | CV | 4.48 |

SD = Standard of deviation; CV = Coefficient of variation

Damage to residual stands at tree level due to timber skidding also varies from one felling technique to another skidding technique (Table 1). The average stand damage caused by the CSL system tends to be higher when compared to the average stand damage caused by the LILS system i.e. sequentially 11.70% compared to 9.99% in PT Gunung Meranti. Meanwhile, the CV value of residual stand damage caused by the LILS system is lower than the CSL system, ie 4.48%, and 10.21%, respectively. This reflected that there were still many wood skidding work teams at PT Gunung Meranti that use the CLS system. In general, it can be said that the skidding of wood using LILS system causes the amount of damage relatively few stands compared to the CLS system. The reduced impact logging system in the tropical natural forest caused lower damage on a residual stand when compared with the conventional timber harvesting [12]. The residual stand damage caused by harvesting activity in hilly tropical forests ranged from 19.27 to 34.90% with an average of 24.37% [13]. Furthermore, the residual stand damage varied from 17 to 40% [14].

### *3.2. Residual damages to poles level*

The recapitulation of a pole stands residual damage measuerement based on attachment 2 is presented in Table 2 below.

**Table 2.** Residual damages to pole level

| CLS system | | LILS system | |
|---|---|---|---|
| Plot | Damages (%) | Plot | Damages (%) |
| 1 | 12.68 | 1 | 10.00 |
| 2 | 11.95 | 2 | 11.25 |
| 3 | 12,.8 | 3 | 9.48 |
| Total | 36.81 | Total | 30.73 |
| Mean | 12.27 | Mean | 10.24 |
| SD | 0.22 | SD | 0.52 |
| CV | 1.76 | CV | 5.13 |

SD = Standard of deviation; CV = Coefficient of variation

The average of pole level damage caused by the CSL system tends to be higher when compared to the average of pole level damage caused by the LILS system i.e. sequentially 12.27% compared to 10.24%. As well as damage to residual stands at the tree level, LILS system also causes damage to residual stands that occur at a lower level.

Damage to residual stands due to the selective logging system at tropical forests in Central Kalimantan was higher than in West Sumatera, which was 27.8% and 22.4%, respectively [5]. Further mentioned the percentage of damage to residual stands increased with increasing intensity of logging in both locations.

### *3.3. Residual damages to sapling level*

The recapitulation of a pole stands residual damage measuerement based on attachment is presented in Table 3 below. Based on Table 3, stand damage at the level of sapling caused by the activities of skidding wood using the LILS system was lower than the CSl system. The average sapling level damage caused by the CSL system and LILS system were 14.99% and 13.01%, respectively.

**Table 3.** Residual damages to sapling level

| CLS system | | LILS system | |
|---|---|---|---|
| Plot | Damages (%) | Plot | Damages (%) |
| 1 | 12.59 | 1 | 13.24 |
| 2 | 18.38 | 2 | 14.04 |
| 3 | 14.01 | 3 | 11.86 |
| Total | 44.99 | Total | 39.05 |
| Mean | 14.99 | Mean | 13.01 |
| SD | 1.74 | SD | 0.63 |
| CV | 11.62 | CV | 4.86 |

SD = Standard of deviation; CV = Coefficient of variation

The results of this study showed that the average value of percentage of standing damage in the tree level was lower than at the pole level, both in the CSL system and in the LILS system, i.e. 11.70% versus 12.27% for the CSL system and 9.9% compared 10.24% for the LILS system. Likewise, the pole stand level, where the average value of the residual stand damage was lower than the sapling level either due to the CSL system or the LILS system, which was 12.27% compared to 14.99% for the CLS system and 10.24% compared to 13.01% in the LILS system. Meanwhile the results of that study by others researchers were follow 17%-40% [14]; 24.3% [13]; 33.15% for a conventional system and 19.53% for RIL system [12]. The residual damage value of damage to the residual stand from the results of this study was lower when compared to the results by other reseacrchers. This can be understood because of the difference in calculation of damage to residual stand, where the value on the reference results was the calculation of damage to residual stands due to logging activities while in the results of this study was caused by the activity of skidding wood. Degradation of residual stand in the form of a sapling, poles, and trees that be caused by forest harvesting activity namely 14.07%; 14.73% and 11.04%, respectively [15]. That is included in the low degradation category because it implements part of reducing impact logging principles.

## 4. Conclusion

Damage to the level of standing trees, poles, and sapling in the CLS system at PT Gunung Meranti was 11.40%, 9.53%, and 14.84%. Meanwhile, the average damage of standing trees, poles, and sapling due to the LILS system was 9.85%, 7.54%, and 12.55%, respectively. LILS system could reduce the damages of residual standing trees. Reduced impact log skidding is recommended to be implemented in order to reduce the residual stand damaged.

## References

[1] Suparto RS 1979 *Eksploitasi Hutan Modern* (Bogor: Fakultas Kehutanan Institut Pertanian Bogor)

[2] Haryanto 1986 *Pemungutan Hasil Hutan* (Yogyakarya: Yayasan Pembina Fakultas Kehutanan Universias Gadjah Mada)

[3] Dulsalam and Sukanda 1989 Produktivitas traktor Catterpillar D7G di suatu perusahaan hutan di Jambi *Jurnal Penelitian Hasil Hutan* **6**(1) 6-11

[4] Muhdi 2015 Cost analysis and productivity of skidding in natural tropical rain forests of PT Inhutani II, North Kalimantan *Jurnal Ekosistem Dipterokarpa* **1**(2) 63-8

[5] Matangaran JR, Putra EI, Diatin I, Mujahid M, and Adlan Q 2019 Residual stand damage from selective logging of tropical forests: a comparative case study in Central Kalimantan and West Sumatra, Indonesia *Global Ecology and Conservation* **19**

[6] Idris MM and Soenarno 2012 Teknik tree length logging di hutan produksi lahan kering (*Laporan Hasil Penelitian*) (Bogor: Pusat Penelitian and Pengembangan Keteknikan Kehutanan and Pengolahan Hasil Hutan)

[7] Idris MM and Soenarno 2015 Penerapan tree length logging secara operasional di areal teknik silvikultur intensif (Studi kasus di PT Sarmiento Parakantja Timber Provinsi Kalimantan Timur)

[8] Susilawati and Jaya NS 2003 Evaluating logged over stand using landsat 7 ETM+ in Sri Buana Dumai Co. Ltd. concession area, Riau Province

[9] Departemen Kehutanan 1993 *Petunjuk Teknis Tebang Pilih Tanam Indonesia (TPTI) pada Hutan Alam Dataran Rendah* (Jakarta: Direktorat Jenderal Pengusahaan Hutan Departemen Kehutanan Republik Indonesia)

[10] Elias 2008 *Pembukaan Wilayah Hutan* (Bogor: IPB Press)

[11] Elias 1998 *Forest Harvesting Case Study: Reduced Impact Timber Harvesting in the Tropical Natural Forest in Indonesia* (Rome: FAO)

[12] Muhdi and Hanifah DS 2007 The effect of reduced impact logging to residual stand damage in the tropical natural forest (A case study in Forest Concession areas of PT Suka Jaya Makmur, West Kalimantan)

[13] Soenarno, Endom W, and Bustomi S 2018 Kerusakan tegakan tinggal akibat pemanenan kayu pada hutan tropis berbukit *Jurnal Penelitian Hasil Hutan* **35**(4) 273-88

[14] Elias 2016 Penerapan reduced impact loging dalam rangka reformasi eksploitasi hutan alam tropik Indonesia (Bogor: Fakulas Kehutanan IPB)

[15] Puspita R, Wahyudi, and Joni H 2014 Analisis kerusakan tegaakan tinggal akibat pemanenan kayu sistem TPTI di PT Dwimajaya Utama Provinsi Kalimantan tengah *Jurnal Hutan Tropika* **9**(2) 21-29

**Appendix 1.** Stand potency and residual stand damage on tree level caused by log skidding

| Skidding technique/Plot | Stand potency before skidding (trees/ha) | Stand damages (trees/ha) | Damage percentage /% |
|---|---|---|---|
| CSL system | | | |
| 1 | 276 | 26 | 9.42 |
| 2 | 180 | 22 | 12.22 |
| 3 | 156 | 21 | 13.46 |
| LILS system | | | |
| 1 | 240 | 22 | 9.17 |
| 2 | 168 | 18 | 10.71 |
| 3 | 198 | 20 | 10.10 |

**Appendix 2.** Stand potency and residual stand damage on poles level caused by log skidding

| Skidding technique/Plot | Stand potency before skidding (poles/ha) | Stand damages (poles/ha) | Damage percentage /% |
|---|---|---|---|
| CSL system | | | |
| 1 | 426 | 54 | 12.68 |
| 2 | 318 | 22 | 11.95 |
| 3 | 258 | 18 | 12.18 |
| LILS system | | | |
| 1 | 114 | 12 | 10.00 |
| 2 | 134 | 15 | 11.25 |
| 3 | 132 | 12 | 9.48 |

**Appendix 3**. Stand potency and residual stand damage on sapling level caused by log skidding

| Skidding technique/Plot | Stand potency before skidding (sapling/ha) | Stand damages (sapling/ha) | Damage percentage /% |
|---|---|---|---|
| CSL system | | | |
| 1 | 6190 | 800 | 12.50 |
| 2 | 6950 | 1200 | 18.38 |
| 3 | 5500 | 800 | 14.01 |
| LILS system | | | |
| 1 | 3200 | 424 | 13.24 |
| 2 | 4500 | 632 | 14.04 |
| 3 | 5500 | 652 | 11.86 |

IOP Conf. Series: Materials Science and Engineering **935** (2020) 012007 doi:10.1088/1757-899X/935/1/012007

# Phytochemical test and physical chemical properties of rubber honey from three types of bees (*Apis mellifera, Apis dorsata* and *Trigona itama)*

**Y Adalina[1*], E Kusmiati[2], and M Pudjiani[3]**

[1]Forest Research and Development Center, Ministry of Environment and Forestry, Bogor 16610, Indonesia
[2]Forest Product Research and Development Center, Ministry of Environment and Forestry, Bogor 16610, Indonesia

*Email: yelinadalina@yahoo.com

**Abstract.** The rubber plant is one type of plant as a source of nectar for honey bees. This study aims to determine the characteristics of rubber honey from *Apis melifera*, *Apis dorsata*, and *Trigona itama* bees based on phytochemical screening and honey's physicochemical properties. The observed variables consisted of active compound content, water content, honey color, pH value, acidity, Hydroxyl Methyl Furfural (HMF) and reducing sugars. Rubber honey samples were taken by purposive sampling from *A. mellifera* honey beekeepers in Jepara Regency, Central Java Province, *T. itama* honey beekeepers in Tapin District, South Kalimantan Province and *A. dorsata* honey beekeepers in West Bangka Regency, Bangka-Belitung Province. The results showed that rubber honey from the three types of bees met the honey quality standard based on SNI 8664-2018, except for the water content. There are differences in honey's Physico chemical characteristics from the three types of bees, which include differences in the value of color intensity, acidity, HMF levels, and reducing sugar levels. All honey samples contain active compounds saponins, flavonoids, and phenols but do not contain alkaloids, tannins, triterpenoids, and steroids. The need for post-harvest honey treatment reduces the water content of honey with a dehumidifier tool to comply with SNI standards.

## 1. Introduction

Honey contains many nutrients that are very beneficial for humans and are widely used in the food, pharmaceutical, and cosmetics industries [1]. Honey can be used as a traditional medicine because it does not cause harmful side effects and is widely used to cure various diseases. This is because of secondary metabolites found in honey, such as alkaloids, flavonoids, saponins, tannins, and essential oils. To find secondary metabolite compounds in honey through phytochemical screening [2].

Besides containing secondary metabolites, honey contains many compounds, including carbohydrates (primarily fructose and glucose), vitamins, minerals, enzymes, organic compounds, free amino acids, and volatile compounds that contribute to color, aroma, and taste [3]. The content of honey consists of 80-85% carbohydrates (glucose and fructose), 15-17% water, 0.1-0.4% protein, 0.2% ash, small amounts of amino acids, enzymes, vitamins, and other substances [4]. The composition and quality of honey are very diverse and depends on the source of honey bee feed as a source of nectar, climate differences, honey maturity and processing and storage [3,5,6]. The honey content significantly affects the physico chemical properties of honey [1]. Honey has physico chemical and microbiological characteristics that can be used as honey quality parameters [7]. In Indonesia the honey quality standard is regulated in the Indonesian National Standard (SNI) 8664-2018 [8].

Various studies have been conducted to determine the quality of honey, among others [9], about local honey from several regions in Temanggung Districts and Wulandari [10] concerning storage on honey quality. Several studies related to rubber honey characteristics, among others, Harjo *et al.* [11] stated that Malang rubber honey from *A. mellifera* bees has a water content of 21.59%, diastase enzyme activity of 11.5 DN and HMF 17.23 mg/kg. The results of research by Evahelda *et al.* [12] show that rubber honey from forest bees (*A. dorsata*) from Central Bangka Regency has a moisture content of 24.25%, HMF 0 mg/kg, a pH value of 3.92, the total sugar content of 74.77%. The study results by Ridoni *et al.* [13] that the rubber honey from the *Trigona* sp. originating from Banjar Regency has a moisture content of 25%, 54.13% reducing sugar, 146.79 mL of NaOH 1N/kg acidity, and 5% water-insoluble solids. However, research on honey from rubber tree nectar from various bees has not been done much. Therefore, it is necessary to test honey's physicochemical properties to determine the quality of rubber honey from three different types of bees.

This study aims to determine the secondary metabolite compounds contained in honey and honey's physicochemical properties with the source of rubber tree nectar (*Hevea brasiliensis* (Willd.ex A. Juss.) Müll. Arg) from three different types of honey bees. This study's results are expected to provide information about the quality of rubber honey produced from various types of honey bees and active compounds contained in rubber honey.

## 2. Materials and Methods

### *2.1. Materials*

The material used in this study was honey from rubber tree nectar (*H. brasiliensis*) from three types of bees, namely *A. mellifera* and *T. itama* bees obtained from honey beekeepers, while from *A. dorsata* bees were obtained from forest honey collectors. The chemicals used in this study include glacial acetic acid, sulfuric acid, aquades, ethanol, Mayer reagents, Dragendorf reagents, Wagner reagents, HCl, $H_2SO_4$, chloroform, ammonia, $FeCL_3$, Mg powder, NaOH, 1% PP indicator and Oxalic acid. The equipment used includes a spectrophotometer, shaker, analytical balance, hotplate, oven, vortex mixer refractometer, and honey color analyzer.

Three rubber honey samples were taken by purposive sampling and measurements were carried out 3 times. Rubber honey from *A. mellifera* bee species was obtained from Telogowungu District, Pati Regency, Central Java Province; *A. dorsata* bee honey from Coconut District, West Bangka Regency, Bangka Belitung Province, and *T. itama* bee honey from Lok Paikat District, Tapin Regency, South Kalimantan Province.

### *2.2. Methods*

This research approach uses a quantitative approach with the support of a qualitative approach. The research method was a survey and interview with *A. dorsata* forest honey collectors and *A. mellifera* and *T. itama* honey beekeepers.

Qualitative analysis of phytochemical constituents refers to Harborne [14]. Phytochemical testing parameters consist of alkaloids, tannin, flavonoid, saponin, phenol, quinone, steroid, and triterpenoid. The physico-chemical characteristics of the analyzed honey consist of color intensity, water content (SNI 8664: 2018) [8], referring to [15], pH [16], acidity [17], hydroxymethylfurfural (HMF) according to AOAC [15], and sugar was reducing methods DNS [18]. Phytochemical analysis was carried out at the Biopharmaca laboratory Bogor and honey physicochemical analyses were carried out at the Forest Products Research and Development Laboratory, Bogor. The results of the analysis are compared with the Indonesian National Standard (SNI) 8664-2018 [8] concerning honey. The results of data analysis are presented descriptively.

## 3. Results and Discussion

### *3.1. Phytochemical constituents*

#### 3.1.1. Analysis of alkaloids compounds
The analysis results showed that rubber honey from the three types of bees with Dragendorff, Mayer, and Wagner reagents did not form sediment. This is indicated by the formation of orange-red precipitates with Dragendorff reagents, not forming white deposits with Mayer reagents, and the formation of brown-red precipitates with Wagner reagents. The analysis showed that all negative honey samples contained alkaloids (Table 1). This study's results are in line with Adalina and Kuntadi [19] that various types of honey monofloral (rubber honey, kapok honey, and mango honey) from *A. mellifera* bees originating from Java islands negative contain alkaloids.

#### 3.1.2. Analysis of steroid compounds and triterpenoids
The steroid analysis results are not formed by greenish-blue or triterpenoid analysis with no brownish color, which indicates that all negative honey samples contain steroids and triterpenoids. Rubber honey is a monofloral honey, honey, with the dominant source of nectar from one plant type. This study's results are in line with Adalina and Kuntadi [19] that various types of monofloral honey from the *A. mellifera* species from Java Island are adverse containing steroids and triterpenoids. Likewise, the study results by Gunawan *et al.* [20], Sumarlin *et al.,* [21] and Emmasitah *et al.* [22] that *Trigona* honey was negative for steroids, while Adalina [23] forest honey (*A. dorsata*) from South Kalimantan which was negative for steroids. In this test, acetic acid and sulfuric acid will bind to terpenoids/steroids so that the color changes [24].

#### 3.1.3. Analysis of flavonoid compounds
The principle of the method analysis flavonoid compounds by reducing benzopyrene with the addition of HCl and Mg metal, the formation of dark red color on flavonoids [25]. The results of the flavonoid analysis showed that all samples of positive research honey contained flavonoid compounds (Table 1), which were marked by the formation of deep red color. Rubber honey from all three types of bees has flavonoid content with weak positive flavonoid results. This research is in line with Adalina and Kuntadi [19]. That honey monofloral (kapok, mango, and rubber) from the native Java Island *A. mellifera* types have the same flavonoid that weak positive results. Mahardika [26] suggested that flavonoid compounds can function as antioxidants.

#### 3.1.4. Analysis of tannin compounds
The results of the analysis showed that all negative honey samples contained tannins (Table 1). The 1% FeC13 reaction results with one of the hydroxyl groups present in tannin compounds showed no dark blue or greenish-black formation. Tannins function as secondary antioxidants because they can bind iron ions and slow down oxidation [27].

#### 3.1.5. Analysis of saponin compounds
The results obtained from the analysis of saponin compounds showed that all positive honey samples contained saponins characterized by the formation of foam or stable foam with moderate positive saponin results (Table 1). Ningsih *et al.* [28] suggested that the formation of foam in the saponin test was due to glycosides, which can form foam in water and hydrolyze into glucose and other compounds. Prasetyo *et al.* [29] stated that saponin is a secondary metabolic compound with antibacterial ability.

#### 3.1.6. Analysis of phenol compounds
The principle of the method of analysis of phenol compounds by adding 1% FeCL3 reagent in ethanol extract, if a green or blue-green color is formed, indicates a positive phenol compound. The analysis results that all rubber honey from the three types of positive bees contains phenolic compounds marked by the formation of green color (Table 1). Rubber honey from *A. mellifera* and *T. itama* bees with actual phenolic content is weak, whereas *A. dorsata* is moderately positive.

According to Alvárez-Suárez *et al.* [30] and Biluca *et al.* [31] reported the presence of individual phenolic compound in their stingless bee honey sample. Whereas plants pollinated by bees produce nectar that is used for the production of honey, it is possible that the bioactive compounds synthesized by the parent plants are transferred to the honey. Thus, the location of the apiaries, and or the geographical origin where the bees collect the nectar can influence the composition of the polyphenols of the honey, since these compounds act in the plant as protectors of environmental stress, in the change of temperature, light level, water content, UV exposure and mineral nutrient deficiency. These compounds, may therefore vary due to the exposure of the plant to the environment. According to Da Silva *et al.* [32] there are differences in honey phenol content from the same bee species because it influenced by geographical differences related to variations in the flowers found at each location, climate, and other factors. Likewise, there are differences in honey phenol content from different bee species at the same beekeeping location with the same harvest period.

**Table 1**. Phytochemical identification results of rubber honey

| Testing parameters | Type of bee | | |
|---|---|---|---|
| | *Apis mellifera* | *Apis dorsata* | *Trigona itama* |
| Alkaloid | | | |
| Wagner | - | - | - |
| Mayer | - | - | - |
| Dragendorf | - | - | - |
| Steroid | - | - | - |
| Triterpenoid | - | - | - |
| Tanin | - | - | - |
| Saponin | ++ | ++ | ++ |
| Phenol | + | ++ | + |
| Flavonoid | + | + | + |

Remarks: - negative; + weak positive; ++ positive medium; +++ strong positive; ++++ very positive

### 3.2. *Analysis of the physicochemical properties of honey*

#### 3.2.1. Honey color

Honey has a different color, aroma, and taste, depending on the type of plant as a nectar source. Honey color, aroma, and taste are essential indicators for consumers [33]. However, they cannot directly determine honey quality because honey color depends on the type of flower nectar [5]. Nevertheless, consumers often judge honey based on their color [12].

Honey classification can be distinguished by color. There are seven color calcifications of honey, from extra white to dark amber [12]. The study results show that rubber honey has extra light amber - light amber color criteria with a color intensity value of 36-73 mmP fund (Table 2). There is a difference in the color of honey from *A. mellifera* bees and the two other types of bees, namely *A. dorsata* and *T. itama*. The lowest color intensity values were obtained from *A. dorsata* bee species and the highest from *A. mellifera* bee species. This is because honey's color depends on the type of nectar that is dominant and the variety of plants as a source of nectar consumed by bees at each location where honey bees nest. Kek *et al.* [34] state that the color of honey is affected by mineral content and also geographical and botanical origin. Diverse condition may produce different honey color either white, colorless, pale yellow, golden or amber. Storage duration and light exposure are affecting the color of honey. By the opinion of Wibowo *et al.* [5], and Nayik and Nanda [33], that the type of plant influences the color of honey as the dominant source of nectar consumed by honey bees. Besides nectar as a source of honey, honey's color is also influenced by the length of storage and processing or heating [35].

#### 3.2.2. Water content

One of the parameters that determine the quality of honey is moisture content. The water content in honey determines the durability of honey. Honey with high water content is natural to ferment. In Indonesia, the quality of honey refers to SNI 8664-2018 [8] concerning honey, which limits the water content of honey of *A. dorsata* and *A. mellifera* bees to a maximum of 22% and maximum stingless bees

(*Trigona* sp.) 27.5% (Table 2). The water content of stingless bee honey depends on various factors such as fruit or flower season, the geography that affect the temperature, bee species and maturity of the honey [36].

The results showed that rubber honey's water content from the three types of bees was above the limit of the Indonesian National Standard (SNI). High levels of honey water are affected by the humidity of the surrounding air, ambient temperature, and early harvest by the opinion of Kahraman *et al.* [37]. The factors that influence the water content of honey include the humidity in the surrounding air, harvest season and the level of maturity of honey in the comb.

There are differences in the water content of rubber honey from the three types of bees. This is because of differences in harvest location. According to the statement of Sjamsiah [1], each honey from a different area has a different water content because the ambient temperature influences it. The study results showed that the water content of *T. itama* honey was higher than the honey content of *A. mellifera* and *A. dorsata*. This is in line with Tanuwidjaya [38] that the water content of *Trigona* honey is higher than *A. mellifera* and *A. dorsata* bees. The high-water content of *T. itama* honey is due to the cell storage in open pot-shaped contact with air, the water from the surrounding environment will be absorbed by honey while Apis bees store honey in wax-covered cells.

The water content of rubber honey bee type *A. mellifera* is 22.5%. However, compared to Savitri *et al.* [9], it shows that the water content of rubber honey is 20.9%, and that meets SNI standards, as well as Harjo *et al.* [11] with a moisture content of 21.59%. This difference is due to various factors that affect the water content of honey. Savitri *et al.* [9] state that the high and low levels of water content are caused by the influence of environmental temperature and way of production and time of harvest. Wulandari [10] states that honey has hygroscopic properties, which is easy to absorb water. The higher the environmental humidity, the higher the honey water content. Likewise, in Hilmanto [39] low water levels are caused by low humidity levels. Handling honey harvest that is too early, most honeycombs are still not covered by wax, also affects honey's water content [40]. Honey harvested with wax-covered nests has lower water content than honey that is harvested. Most honeycombs are still not covered by wax [10].

The water content in honey determines the quality of honey. The higher the water content of honey, the lower it is quality [41]. The water content in honey determines the durability of honey. Microbes cannot live in honey with low water content. Harjo *et al.* [11] explained that honey with high water content would be easily fermented by the yeast of the genus Zygosaccharomyces, which is resistant to high sugar concentrations to grow and develop in honey.

### 3.2.3. PH value

The study results that the rubber honey from the three types of bees has a low pH of 3.80 - 4.50 (Table 2). Although from the same dominant source of nectar, rubber honey from the three types of bees has a different honey pH. The difference in pH values is due to differences in mineral and acid content in honey [42]. However, the type of bee has no significant effect on honey pH [43].

The lowest pH value of *T. itama* bee is 3.80 and has a more acidic taste than rubber honey from *A. dorsata* and *A. mellifera*. All of the honey types are characterized as acidic. The results showed that although with different geographical and botanical source of the honey samples, their pH was almost within the same range. The finding may suggest that stingless bee honey has lower range of pH regardless their origin. By the opinion of Khalil *et al.* [44], the pH of honey can affect honey's flavor and aroma.

### 3.2.4. Acidity level

The acidity of honey is one of the parameters used to determine the quality of honey. Acidity shows the amount of free acid per kg of honey [10]. The analysis results all samples showed acidity levels, quality requirements were a maximum of 50 mL of 0.1N/kg NaOH for *A. mellifera* and *A. dorsata* bees and a maximum of 200 mL of 0.1N/kg NaOH for *Trigona* bees (Table 2). The acidity of honey from the three different bees is 2.94 mL NaOH 0.1N/kg *A. mellifera* bee; 5.92 mL NaOH 0.1N/kg *A. dorsata* bee and equal to 11.29 mL NaOH 0.1N/kg *T. itama* (Table 2). This difference is due to the soil's biophysical conditions and the environment of the different harvest locations so that it affects the acidity of the honey, by the opinion of Savitri *et al.* [9] that the acidity of honey significantly affects different regions. The highest acidity level of honey was obtained from *T. itama*. This shows that the type of bee influences the acidity. By Tanuwidjaya [38] statement, bee species have a very significant effect on acidity. SNI standard stipulates the maximum acidity of honey reaches only 50 mL NaOH/kg. Due to the higher acidity of *Trigona* bees, the SNI reference used is based on the type of bees. The acidity of *Trigona* honey is not caused by fermentation but due to the content of free acids, acidic minerals, and amino acids of honey.

The acidity of rubber honey in this study is different from [10], where the average acidity of rubber honey is 45 mL NaOH 0.1N/kg, while Savitri *et al.* [9] is 34.59 mL NaOH 0.1 N/kg. As stated by Savitri *et al.* [9], acidity levels significantly affected different regions. Some factors that influence honey's acidity are the biophysical conditions of the honey harvesting location, climate, honey maturity, and the way it is processed and stored.

3.2.5. Hidroksi Metil Furfural (HMF)

HMF content is widely recognised as a parameter of honey freshness. HMF is a breakdown product of simple sugars, especially fructose. Ultimately, fructose in the honey is decomposing and reacting with acid. HMF levels are one indicator of honey damage by overheating or due to counterfeiting with inverted sugar and duration of storage [12]. HMF in honey is produced from an overhaul of honey monosaccharides (glucose and fructose) in an acidic atmosphere and with the help of heat [11].

The results of the study show that rubber honey from the three types of bees has a low HMF content of 0.0897 - 0.4590 mg/kg and meets SNI 2018 [8] standards. This shows that the honey has no heating or counterfeiting treatment with invert sugar. HMF levels of *A. mellifera* bee species results of this study were lower than research by Harjo *et al.* [11] with HMF levels of rubber honey by 17.23 mg/kg. Following the statement of Harjo *et al.* [11] that storage temperatures caused an increase in HMF levels. Low HMF levels indicate that the honey has just been harvested. According to Al-Diab *et al.* [45], newly harvested honey contains HMF with HMF levels below 1 mg/kg.

**Table 2.** The results of the physicochemical analysis of rubber honey

| Parameters | Type of bee | | | |
|---|---|---|---|---|
| | *Apis mellifera* | *Apis dorsata* | *Trigona itama* | SNI 8664-2018 |
| Color criteria | light amber | extra light amber | extra light amber | - |
| Color intensity (mmP fund) | 73 | 36 | 49 | - |
| Water content (%) | 22.5 | 30.0 | 28.4 | *A. mellifera* dan *A. dorsata* maks 22%; *T. itama* maks 27.5% |
| pH | 4.50 | 4.00 | 3.80 | - |
| acidity (mL N NaOH/kg) | 2.94 | 5.92 | 11.29 | *A. mellifera* dan *A. dorsata* maks 50(mL N NaOH/kg); *T. itama* maks 200 (mL N NaOH/kg) |
| HMF (mg/kg) | 0.0897 | 0,2372 | 0.4590 | Maks 50 (mg/kg) |
| Reducing sugar (%) | 65.76 | 65.60 | 62.97 | *A. mellifera* dan *A. dorsata* Min 65%; *T. itama* min 55% |

3.2.6. Reducing sugar
The honey quality can be determined from the sugar content contained in it [5]. The results showed that the sugar reducing levels of rubber honey from bee species *A. mellifera*, *A. dorsata*, and *T. itama* met the SNI 8664-2018 [8] standard, namely 65.76%, 65.60%, and 62.97%. The three types of bees have different reducing sugar levels. Rubber honey from *T. itama* bee species has lower reducing sugar levels than *A. dorsata* and *A. mellifera* bees but meets SNI 2018 standards. According to Tanuwidjaya [38] statement that bee species have a very significant effect on reducing sugar levels.

Factors affecting sugar composition include bee food types, regional differences, and climatic conditions [46]. Temperature can affect the nectar production of feed source plants. The higher the ambient temperature, the less nectar produced by feed sources, but the nectar's sugar content is increasing and vice versa. The drier the air, the lower the nectar produced [47]. The difference in humidity affects honey glucose levels. High humidity causes more nectar production but low sugar content. When the air is dry, the nectar production is low, but the sugar content increases [49].

## 4. Conclusion
The phytochemical analysis of honey from the three positive bee species has active compounds of phenol and flavonoid saponins. Nevertheless, it does not contain alkaloid compounds, steroids, triterpenoids, tannins, and quinones. Rubber honey from *A. dorsata* bees has higher phenol active compounds compared to *A. mellifera* and *T. itama* bees.

All rubber honey samples from *A. mellifera*, *A. dorsata*, and *T. itama* species have physicochemical characteristics with honey quality in accordance with the SNI 8664-2018 based on hydroxymethylfurfural parameters, reducing sugar content and acidity but have water content above SNI standards. Rubber honey has extra light amber color criteria to light amber.

## References
[1] Sjamsiah, Sikanna R, Rifkah A, and Saleh A 2018 Penentuan sifat fisikokimia madu hutan (*Apis dorsata*) Sulawesi Selatan *Al-kimia* **6**(2) 185-93
[2] Noral A, Wilapanggal A, and Novianti T 2018 Antioxidant activity, antibacterial activity, water content, and ash content in Baduy honey *Bioscience* **2**(1) 38-44
[3] Pavlova T, Stamatovska V, Kalevska T, Dimov I, and Nakov G 2018 Quality characteristics of honey: a review *Proceedings of University of Ruse* **57**(10.2) 32-7
[4] Buba F, Gidado A, and Shugaba A 2013 Analysis of biochemical composition of honey samples from North-East Nigeria *Biochemistry and Analytical Biochemstry Journal* **2**(3) 1-7
[5] Wibowo BA, Rivai M, and Tasripan 2016 Alat uji madu menggunakan polarimeter dan sensor warna *Jurnal Teknik* **5**(1) 28-33
[6] Fitrianingsih SP, Khairat A, and Choesrina R 2017 Aktivitas antibakteri madu hitam pahit dan madu hitam manis terhadap *Escheria coli* dan *Staphylococcus aureus Jurnal Farmasi Galenika* **1**(2) 32-7
[7] Prabowo S. Yuliana, Prayitno YA, Lestari K, and Kusesvara A 2019 Penentuan karakteristik fisiko-kimia beberapa jenis madu menggunakan metode konvensional dan metode kimia *Journal of Tropical AgriFood* **1**(2) 66-73
[8] [BSN] Badan Standardisasi Nasional 2018 *Madu SNI 8664:2018* (Jakarta: Badan Standardisasi Nasional)
[9] Savitri NPTS, Hastuti ED, and Suedy SWA 2017 Kualitas madu lokal dari beberapa wilayah di Kabupaten Temanggung *Buletin Anatomi dan Fisiologi* **2**(1) 58-66
[10] Wulandari DD 2017 Kualitas madu (keasaman, kadar air dan kadar gula pereduksi) berdasarkan perbedaan suhu penyimpanan *Jurnal Kimia* **2**(1) 16-22
[11] Harjo SST, Radiati LE, and Rosyidi 2015 Perbandingan madu karet dan madu rambutan berdasarkan kadar air, aktivitas enzim diastase dan hidroksimetilfurfural *Jurnal Ilmu dan Teknologi Hasil Ternak* **10**(1) 18-21

[12] Evahelda E, Pratama F, Malahayati N, and Santoso B 2017 Sifat fisik dan kimia madu dari nektar pohon karet di Kabupaten Bangka Tengah, Indonesia *Agritech* **37**(4) 363-8

[13] Ridoni R, Radam R, and Fatriani 2020 Analisis kualitas madu kelulut (*Trigona* sp) dari Desa Mangkauk, Kecamatan Pengaron, Kabupaten Banjar *Jurnal Sylva Scienteae* **3**(2) 346-55

[14] Harborne JB 1996 *Metode Fitokimia Penuntun Cara Modern Menganalisis Tumbuhan* [Terjemahan] Kosasih P and Imama S Edition II pp. 4-7; 69-76 (Bandung: Publisher ITB)

[15] [AOAC] Association of Official Analytical Chemists 2005 *Official Methods of Analysis* (Washington DC: Association of Official Analytical Chemists)

[16] [AOAC] Association of Official Analytical Chemists 1990 *Official Methods of Analysis* (Washington DC: Association of Official Analytical Chemists)

[17] [BSN] Badan Standardisasi Nasional 2013 *Mutu Madu SNI 01-3545-2013* (Jakarta: Badan Standardisasi Nasional)

[18] Miller G 1959 Use of dinitrosalicylic acid reagen for determination of reducing sugar *Analysis Chemistry* **31**(3) 426-8

[19] Adalina Y, and Kuntadi 2019 Phytochemical indentification of honey from several regions in Java and Sumbawa *AIP Conference Proceedings* **2120**(1) 080024

[20] Gunawan R, Erwin, and Syafrizal 2018 Uji fitokimia dan penentuan aktivitas antioksidan dari madu *Trigona insica Jurnal Atomik* **3**(1) 18-21

[21] Sumarlin LO, Suprayogi A, Rahminiwati M, Tjachja A, and Sukandar D 2015 Bioaktivitas ekstrak methanol daun Namnam serta kombinasinya dengan madu Trigona *Jurnal Teknologi dan Industri Pangan* **26**(2) 144-54

[22] Emmasitah, Raya I, Usman AN, Mauludiyah I, and Prihatin AR 2020 Uji FTIR dan uji fitokimia dari madu Trigona sp untuk persiapan suplemen wanita prakonsepsi *Jurnal Keperawatan Muhammadiyah* 1-6

[23] Adalina Y 2018 Analisis habitat koloni lebah hutan Apis dorsata dan kualitas madu yang dihasilkan dari kawasan hutan dengan tujuan khusus (KHDTK) Rantau, Kalimantan Selatan *Jurnal Penelitian Hutan dan Konservasi* **15**(1) 25-40

[24] Minhatun N, Tukiran, Suyatno, and Hidayati N 2014 Uji skrining fitokimia pada ekstrak heksan, kloroform, dan metanol dari tanaman Patikan Kebo (*Euphorbiaehirtae*) *Prosiding Seminar Nasional Kimia* pp. 279-86

[25] Agustina W, Nurhamidah, and Handayani D 2017 Skrining fitokimia dan aktivitas antioksidan beberapa fraksi dari kulit batang jarak (*Ricinus communis* L.) *Jurnal Pendidikan dan Ilmu Kimia* **1**(2) 117-22

[26] Mahardhika C 2013 Fraksionasi ekstrak kulit petai berpotensi antioksidan [Skripsi] Institut Pertanian Bogor

[27] Amarowicz R 2007 Tannins: the new natural antioxidants *European Journal of Lipid Science and Technology* **109** 549-51

[28] Ningsih DR, Zusfahair, and Kartika D 2016 Indentifikasi senyawa metabolit sekunder serta uji aktivitas ekstrak daun sirsak sebagai antibakteri *Molekul* **11**(1) 101-11

[29] Prasetyo BF, Wientarsih I, and Priosoeryanto BP 2008 Aktivitas sediaan gel ekstrat batang pohon pisang ambon dalam proses penyembuhan luka pada mencit *Jurnal Veteriner* **11**(2) 70-3

[30] Alvárez-Suárez JM, Giampieri F, Brenciani A, Mazzoni L, Gasparrini M, Gonzalez-Paramas AM, Santos-Buelga C, Morroni G, Simoni S, Forbes-Hernandez TY, Afrin S, Givanetti E, and Battino M 2018 *Apis mellifera* vs *Melipona beecheii* cuban polifloral honeys: a comparison base on their physicochemical parameters, chemical composition and biology properties *LWT-Food Science and Technology* **87** 272-9

[31] Biluca, FC, de Gois JS, Schulz M, Braghini F, Gonzaga LV, Maltez HF, and Fett R 2017 Phenolic compounds, antioxidant capacity and bioaccessibility of minerals of stingless bee honey (*Meliponinae*) *Journal of Food Composition and Analysis* **63** 89-97

[32] Da Silva, PM, Gauche C, Gonzaga LV, Costa ACO, and Fett R 2016 Honey: Chemical composition: stability and authenticity *Food Chemistry* **196** 309-23

[33] Nayik GA and Nanda V 2015 Physico-chemical, enzymatic, mineral and colour characterization of three different varieties of honey from khasmir valley of India with a multivariate approach *Polish Journal of Food and Nutritions Sciences* **65**(2) 101-8
[34] Kek SP, Chin NL, Yusof YA, Tan SW, and Chua LS 2014 Total phenolic contents and colour intensity of Malaysian honeys from the *Apis spp.* and *Trigona spp.* bees *Agriculture and Agricultural Science Procedia* **2** 150-5
[35] Eleazu CO, Iroaganachi MA, Eleazu KC, and Okoronkwo JO 2013 Determination of the physicochemical composition microbial quality and free radical scavenging activities of some commercially sold honey samples in Aba Nigeria: The effect of varying colours *International Journal of Biomedical Research* **4**(1) 32-41
[36] de Sousa JMB, de Souza EL, and Marques G 2016 Sugar profile, physicochemical and sensory aspects of monofloral honeys produced by different stingless bee species in Brazilian semi-arid region *LWT- Food Science and Technology* **65** 645-51
[37] Kahraman T, Buyukunal SK, Vural A, and Altunatmaz SS 2010 Physicochemical properties in honey from different regions of Turkey *Food Chem.* **123** 41-4
[38] Tanuwidjaya SJ 2014 Karakteristik kimia dan organoleptik madu dari lebah *Apis mellifera*, *Apis cerana*, *Apis dorsata* dan *Trigona* sp. [Skripsi] Institut Pertanian Bogor
[39] Hilmanto R 2010 Analisis paket teknologi lokal dalam pengelolaan produksi madu organik untuk pasar global dan industri *Jurnal Ilmu Pertanian Indonesia* **15**(2) 88-95
[40] Ajeng P, Minarti S, and Junus M 2014 Perbandingan kadar air dan aktivitas enzim diastase madu lebah *Apis mellifera* di kawasan penggembalaan mangga (*Mangifera Indica*) dan kawasan penggembalaan karet (*Hevea Brasilliensis*) [Disertasi] Universitas Brawijaya Malang
[41] Adalina Y 2017 Kualitas madu putih asal Provinsi Nusa Tenggara Barat *Prosiding Seminar Nasional Biodiversitas Indonesia* **3**(2) 189-93
[42] Gulfraz M, Ifftikhar F, Asif S, Raja GK, Asad MJ, Abbasi K, and Zeenat A 2010 Quality assement and antimicrobial activity of various honey types of Pakistan *African Journal of Biotechnology* **9**(41) 6902-6
[43] Pramesti AG 2014 Karakteristik fisik madu dari lebah *Apis mellifera*, *Apis dorsata*, *Apis cerana* dan *Trigona spp.* [Skripsi] Institut Pertanian Bogor
[44] Khalil MI, Moniruzzaman M, Boukraa L, Benhanifa M, Islam MA, Islam MN, Sulaiman SA and Gan SH 2012 Physicochemical and antioxidant properties ofalgerian honey *Molecules* **17**(9) 11199-215
[45] Al-Diab D and Jarkas B 2015 Effect of storage and thermal treatment on the quality of some local brand of honey from Latakia markets *Journal of Entomology and Zoology Studies* **3**(3) 328-34
[46] Silvia PMD, Gauche C, Gonzaga LV, Costa ACO, and Fett R. 2015 Honey: Chemical composition, stability and authenticity *Brazil Food Chemistry* **196** 309-23
[47] Sihombing DTH 2005 *Ilmu Ternak Lebah Madu* (Yogyakarta: Gajah Mada University Press)
[48] Prasetyo BA 2014 Perbandingan mutu madu lebah *Apis mellifera* berdasarkan kandungan gula pereduksi dan non pereduksi di kawasan karet (*Hevea brasiliensis*) dan rambutan (*Nephelium lappaceum*) [Tesis] Fakultas Peternakan Universitas Brawijaya Malang

**Acknowledgments**
Acknowledgments are conveyed to the management of the Forest Research and Development Center, who has financed and facilitated the implementation of the research and thanked the authors also to all parties who have helped in the implementation of this research.

IOP Conf. Series: Materials Science and Engineering **935** (2020) 012008 doi:10.1088/1757-899X/935/1/012008

# Bamboo, farmer motivation on cultivation and its affecting factors

**E Fauziyah* and Sanudin**

Agroforestry Technology Research and Development Institute, Ciamis, 46201, Indonesia

*Email: fauziyah_eva@yahoo.com

**Abstract**. Motivation is an important factor to encourage community on bamboo cultivation or maintain existing bamboo plantation. This study aims to determine farmer motivation on bamboo cultivation and its affecting factors. This research was conducted in Sukaharja Village, Rajadesa Sub-district, Ciamis District, West Java on November 2016. Primary data collection is done through interviews and Focus Group Discussion (FGD) with 30 farmers involved in the development of bamboo agroforestry demonstration plots. The questionnaire consisted of socioeconomic characteristics of farmers and farmers motivations for bamboo cultivation. A survey was conducted to determine the motivation statements of farmers on bamboo cultivation using five-point Likert-type scale from 1 = strongly disagree to 5 = strongly agree. Data analysis was carried out using descriptive methods and Likert's Summated Rating (SLR). The result showed that farmer motivation on bamboo cultivation in Sukaharja Village was low. The number of family dependents and the number of information sources are affecting farmer motivation on bamboo cultivation. The farmer motivation on bamboo cultivation can be increased through providing a comprehensive information from stakeholders such as research institute, universities and others stakeholder and increasing the added value of bamboo such as post-harvest processing and expanding its marketing network.

## 1. Introduction

Bamboo is one of the most multi-use natural products available [1]. Bamboo can be used as a substitute for wood in construction, furniture, scaffolding, flooring [2], roofing, fabrics and cloth, pulp and paper, charcoal, food (bamboo shoots), and ornamental garden planting [1,3], and used in housing and for other construction purposes [4]. Bamboo is also known to be a valuable ecological resource for soil and water conservation and restoration of degraded lands [1]. The most widely known features of bamboo are its fast growth [1,5], adaptability, resilience and substantial biomass production [1].

About 1.5 billion people depend on bamboo for their daily lives [1]. In Indonesia, bamboo areas reported by the agricultural census belong to farmers and thus are classified as private and bamboo forest areas are owned by the government and are classified as public [3].

In spite of the fact that most bamboo resources grow naturally, more prominent consideration has been paid as of late to the foundation of planted bamboo. Bamboos have long been cultivated in villages and, historically, the rural poor have been the prime users of naturally regenerating bamboo [3]. In the Sukaharja Village Ciamis District, population growth and new bamboo processing opportunities have led to the over exploitation of bamboo resources and farmer replaces bamboo with other commodities because economic reason. It can threaten the existence of bamboo plantation.

Widyangsih *et al.* [6] stated that community perception on bamboo in Sukaharja Village was positive, but not many people cultivate bamboo. The community has assumed that bamboo is a legacy plant that does not need to be cultivated and its existence will not be exhausted. Limited land ownership is also one of the problems, if the land was cultivated with bamboo, they cannot cultivate other plants. Planting with agroforestry based bamboo have not yet understood by the community.

Motivation is an important factor to encourage community on bamboo cultivation or maintain the existing bamboo plantation. Hammond *et al.* [7] said that understanding the motivations of smallholder farmers to adapt their practices is essential in designing appropriate interventions. The motivations data can also be used to inform the design of mechanisms that encourage farmers to adapt their behaviors. This study aims to determine farmer motivation on bamboo cultivation and its affecting factors.

## 2. Materials and Methods

This research was conducted in Sukaharja Village, Rajadesa Sub District, Ciamis District, West Java on November 2016. The research location is presented in Figure 1.

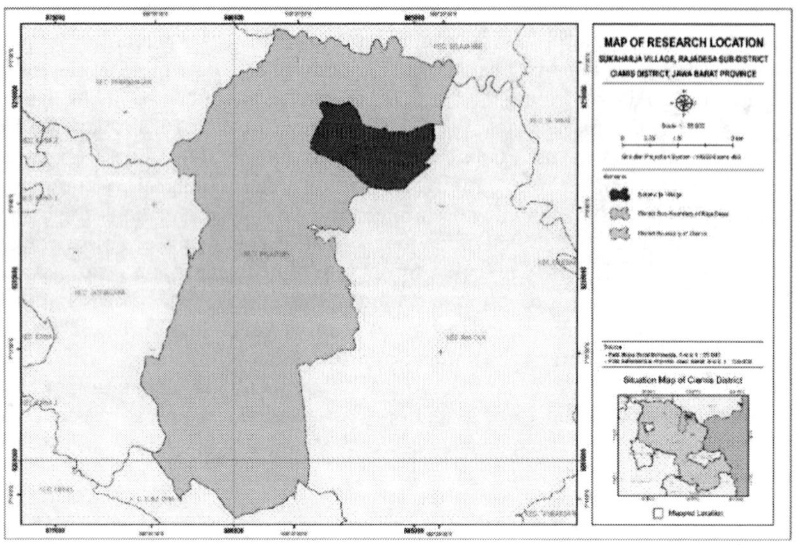


**Figure 1.** The research location in Sukaharja Village Ciamis District

Primary data collection was done through interviews and Focus Group Discussion (FGD). Interviews were conducted with 30 farmers involved in the development of bamboo agroforestry demonstration plots. FGDs were carried out by involving cultivating farmer's, farmer group and village officials. Secondary data were obtained from Sukaharja Village Office, Rajadesa Sub District, Ciamis District.

The questionnaire consisted of two major parts: the first part included socioeconomic characteristics of farmers, such as age, education level, number of family dependents, farming experience, land ownership, and income. The second part included statements of farmer's motivations on bamboo cultivation. For measuring farmer's motivations, the respondents were asked a set of questions addressing their motivation. A list of 10 motivation factors was finally created and considered, namely: internal motivation (needs, wants, personal expectations, pride, and job satisfaction) and external motivation (type and character of job, working group, workplace situation, workplace organization, and the current reward system).

A survey was conducted to determine the motivation statements of farmers motivations on bamboo cultivation using five-point Likert-type scale that captured farmers' responses to the tested set of motivations from 1 = emphatically oppose this idea to 5 = unequivocally concur. The independence of the farmer characteristics and the motivations was tested using a Spearman's test.

Data analysis was carried out using descriptive methods and Likert's Summated Rating (SLR). Descriptive method is a method of analysing and describing the existing research data and is related to theories that have to do with problems in order to draw conclusions. While the Likert scale (1-5) is used to measure the attitudes, opinions and perceptions of a person or group about social events or phenomena then a range of each category is made using the formula 1 and scores of farmer motivation for bamboo cultivation are presented in Table 1:

$$interval = \frac{max\ score - \min score}{number\ of\ category} - 0.01 \qquad (1)$$

**Table 1.** Scores of farmer motivation [8]

| Category | Scale | Score |
|---|---|---|
| Very high | 5 | 4.20 – 5.00 |
| High | 4 | 3.40 – 4.19 |
| Medium | 3 | 2.60 – 3.39 |
| Low | 2 | 1.80 – 2.59 |
| Very low | 1 | 1.00 – 1.79 |

## 3. Results and Discussion

### 3.1. *Bamboo in Sukaharja Village*

Ciamis District has the potential of bamboo with an area of 5,169 ha and the number of stands of 3,751,357 stems [9]. Bamboo in Sukaharja Village grows naturally and has a potential of 765 bamboo stems per hectare with 53% composed of young bamboo and 47 % old bamboo. The most common types of bamboo found in Sukaharja Village are apus bamboo (*Gigantochloa apus*), other types of bamboo are petung (*Dendrocalamus asper*), buluh (*Gigantochloa atter*), gombong (*Gigantochloa pseudoarundinaceae*), and haur (*Bambusa vulgaris* v. Vitata) [10].

This shows that Sukaharja Village has considerable bamboo's potential. The apus bamboo has been used for the construction of buildings/houses with inter-regency marketing areas. There are many other uses of bamboo as conveyed by Velez [11] that bamboo is a renewable resource in the areas of construction and infrastructure, which is used structurally in homes and other buildings. Bamboo can be utilized for furniture, carpets, paneling, flooring, allotments, plumbing, roof, structure and shapes, among numerous others.

Apus bamboo is the type of bamboo that is most widely traded by the community which is sold at IDR 2500-3000/stem (prices at locations) and IDR 5000-6000/stem (prices at roadside) [6]. Although this price is still considered very low for farmers, however, farmers still can feel the economic benefits of this type of bamboo. The apus bamboo can be harvested every year without having to do maintenance and can provide better results from young stems and old stems [10].

### 3.2. *Internal and external characteristics*

Farmer's internal characteristics are factors that come from within the farmer himself and affect the role and performance of the farmer. Farmer's internal characteristics in this study include age, education level, number of family dependents, farming experience, income and land ownership. According to Budiono [12], demographic characteristics that affect farmers in managing farming both shifting cultivation and dryland agriculture include age, education level, knowledge, number of family dependents, income, farming experience, land ownership and cosmopolitanism. Farmer's internal characteristics on bamboo cultivation in Sukaharja Village are presented in Table 2.

**Table 2.** Internal characteristics of farmer's

| No. | Description | Interval | Average | Score | Category |
|---|---|---|---|---|---|
| 1. | Age (years) | 20 - 70 | 48.8 | 3.4 | high |
| 2. | Education level (years) | 0 - 12 | 6.2 | 2.1 | low |
| 3. | Number of family dependents (person) | 0 - 5 | 2.2 | 3.2 | medium |
| 4. | Farming experience (years) | 0 - 50 | 21.3 | 2.2 | low |
| 5. | Income (IDR/year) | 0 – 98,000,000 | 1,282,824 | 2.4 | low |
| 6. | Land ownership (hectares) | 0 – 0.37 | 0.02 | 2.2 | low |
| | Average | | | 2.58 | low |

Table 2 shows that almost all internal characteristics are in the low category except age (high) and number of family dependents (medium) with an average score of 2.58 which is in the range of 1.80-2.59 and included in the low category. Age is one of the important factors in supporting someone who will increase work motivation and productivity. Maturity level signifies his mindset. The greater the number of family dependents will increase community motivation to work harder in meeting their needs.

External characteristics of farmers examined in this study are extension intensity, extension channel and number of information sources. External characteristics of respondents in Sukaharja Village on bamboo cultivation are presented in Table 3.

**Table 3.** External characteristics of farmer

| No. | Description | Score | Category |
|---|---|---|---|
| 1. | Extension intensity | 2.4 | low |
| 2. | Extension channel | 2.2 | low |
| 3. | Number of information sources | 2.8 | medium |
| | Average | 2.47 | low |

All external characteristics of farmer's are in the low category with an average score of 2.47 (1.80-2.59) except number of information sources (medium). At present, information technology can make it easier for community to obtain knowledge related to bamboo, so that it can increase communities motivation in bamboo cultivation. Armstrong and Taylor [13] said that knowledge is what people know about concepts, ideas, theories, methods, phenomena, activities and objects.

Farmer's knowledge of bamboo can be seen by asking a number of things starting from the aspects of cultivation, utilization, processing and marketing. Buckley and Shortle [14] said that farmers use and integrate knowledge from various sources in order to meet their diverse knowledge needs. In numerous cases the farmer's information was the foremost noticeable and trusted information source, because it is seen as locally significant and meaningful. In their daily operations, farmers rely primarily on their knowledge they have accumulated, often over extended periods of time, from their practical experience: doing, experimenting and observing.

The potential of existing bamboo plantation can be increased in value through the processing into handicrafts (glass/cups) to get added value and increase community income. The research team has facilitated the community through training activities to become bamboo craftsmen and succeeded in recruiting 1-2 people who until now have been able to make bamboo crafts as their main source of livelihood. However, these efforts have not succeeded in increasing the motivation of the community to cultivation bamboo.

The frequency of extension workers' visits to Sukaharja Village is a frequent category. Extension workers regularly visit farmer groups on a schedule once a month depending on needs, but of course extension workers have much information and knowledge limitations regarding bamboo in all its aspects (cultivation, utilization, processing and marketing). The people in the research location are people who depend their lives on vegetables/rops so that the knowledge of extension workers is limited to the commodity.

### *3.3. Internal and external motivation*

Motivation is the strength and heading of conduct and the components that impact individuals to act in certain ways. The term motivation can allude differently to the objectives that people have, the ways in which people chose their objectives and the ways in which others attempt to alter their behavior. Individuals are motivated when they anticipate that a course of activity is likely to lead to the fulfillment of an objective and an esteemed remunerate-one that satisfies their needs and wants [13].

Factors influencing farmer motivations are categorised into intrinsic and extrinsic factors [15] and [13]. Intrinsic motivation takes to put when people feel that their work is critical, curiously and challenging which it gives them with a sensible degree of independence (opportunity to act), openings to attain and progress, and scope to utilize and create their abilities. It can be depicted as inspiration from the work itself [13].

Extrinsic motivation happens when things are done to or for individuals in arrange to motivate them. These include rewards such as motivating forces, expanded pay, laud or advancement, and punishments such as disciplinary activity, with holding pay, or criticism. Extrinsic motivators can have an immediate and powerful effect, but it will not necessarily last long [13]. Extrinsic factors are demographic, economic, geographical, and intrinsic factors are related to knowledge, perceptions, attitudes; and found that intrinsic factors in particular are often overlooked [15]. An intrinsically motivated action is not reliant upon any outcome separable from the behavior itself. For example, a farmer may undertake an environmental activity, such as planting trees, for no other reason than because it is innately satisfying. Conversely, extrinsic motivation is instrumental in nature and so is performed to attain some other outcome. For instance, a farmer might undertake environmental activity as part of an Agri Environmental Services (AES) in order to receive a payment [16]. Intrinsic motivation refers to doing something because it is inherently interesting or enjoyable, and extrinsic motivation, which refers to doing something because it leads to a separable outcome [17].

Farmers internal motivation on bamboo cultivation (Table 4) is influenced by needs, wants, personal expectations, pride, and job satisfaction [13]. Table 4 shows that the internal motivational factors are in the low category with an average score of 2.57 (1.80-2.59), except for the needs and wants factors that are in the medium category. This is consistent with Amstrong and Taylor [13] that people are motivated when they expect that a course of action is likely to lead to the attainment of a goal and a valued reward one that satisfies their needs and wants.

The link between needs and behaviours is direct and automatic, rather than mediated by human consciousness, values and choice [18]. Intrinsic motivation alludes to accomplishing something since it is naturally intriguing or pleasant [17]. Bamboo has not fulfilled the farmers' expectations, especially in the economic aspects to be used as the main source of livelihood. The lack of diversification in the use of bamboo, which so far has only been limited to construction/buildings is thought to influence this assumption. The price of a bamboo stick measuring around 6 m in length is priced at IDR 6000/stick.

People certainly think that the selling price is still not meet with their expectations, especially if it is associated with the time needed to grow the bamboo plants (the average harvest age is 3-5 years).

**Table 4.** Internal motivation of farmers

| No. | Description | Score | Category |
|---|---|---|---|
| 1. | Needs | 2,80 | medium |
| 2. | Wants | 2,83 | medium |
| 3. | Personal expectations | 2,50 | low |
| 4. | Pride | 2,37 | low |
| 5. | Job satisfaction | 2,37 | low |
| | Average | 2,57 | low |

Economic dependence is diminishing farmer motivation to participate in conservation programs, but found that income was less important than other factors for responders [19]. In their review of about 160 studies on factors that affect farmer participation in biodiversity policies, Siebert *et al.* [20] found that many analysis corroborate that economic motivations play a key role-which is not surprising since farmers need to manage their farms in an economically viable way. Financial compensation and incentives function as a necessary, though clearly not sufficient, condition to explain farmer support for agri-environmental measures. Burton and Schawarz [21] and Home *et al.* [22] said that they concluded by pointing to a need for more conceptual and empirical attention for influencing social norms and expectations, which escape scientific attention when focussing principally on farmer's individual economic interests.

For decades the people in Sukaharja Village have made the agricultural sector, especially the cultivation of vegetables/crops as the main source of livelihood to meet the needs of daily life. The inherent habit is certainly difficult to let go of if the community has not seen other work alternatives that are more promising, especially agroforestry bamboo cultivation which so far has not attracted their attention. Farmers external motivation (Table 5) is influenced by type and character of job, working group, workplace situation, workplace organization, and current reward system [13].

**Table 5.** Farmers external motivation

| No. | Description | Score | Category |
|---|---|---|---|
| 1. | Type and character of job | 2.10 | low |
| 2. | Working group | 2.60 | medium |
| 3. | Environmental situation | 2.40 | low |
| 4. | Workplace organization | 3.53 | high |
| 5. | Current rewards system | 2.43 | low |
| | Average | 2.41 | low |

The most of farmers external motivation have a low score, except for working group (medium category) with an average score of 2.41 (1.80-2.59). Farmers feel comfortable with the existence of farmer groups where they share experiences/information/gatherings.

### *3.4. Factors affecting farmers motivation on bamboo cultivation*

Based on the calculation results (Appendix), relationship between characteristics (internal and external) and motivation (internal and external) are presented in Table 6. The correlation between number of family dependents (internal characteristic) and number of information sources (external characteristic) on workplace situations (external motivation) is quite strong, significant and unidirectional. This means that the greater number of family dependents will increase the motivation of the community to work harder to meet their needs, so that it requires a comfortable workplace situation. At present, the community is very easy to access data and information through technology. So the source of information is not only through counseling (face to face). This condition allows farmers to get comfortable in working. Comfortable work situations, one of which can be obtained from various things such as other farmers who have the same goals.

**Table 6**. Statistical test results using a Spearman's test

| No | Relationship between | Correlation coefficient | Significant |
|---|---|---|---|
| 1. | Farming experience (internal characteristic) with pride (internal motivation) | -0.411 | 0.024 |
| 2. | Number of family dependents (internal characteristic) with workplace situation (external motivation) | 0.438 | 0.016 |
| 3. | Income (internal characteristic) with current rewards system (external motivation) | -0.378 | 0.040 |
| 4. | Extension channel (external characteristic) with workplace situation (external motivation) | -0.400 | 0.028 |
| 5. | Number of information sources (external characteristic) with workplace situation (external motivation) | 0.365 | 0.047 |

A farmer's motivations could not, in this case, be reliably inferred from his livelihood and farm characteristics without having gathered separate, specific information regarding motivations. Such data are not usually collected, and socio-demographic proxies are usually used instead [23]. In this study, the usual proxies (age, ethnicity, education) showed no significant predictive power of farmer's motivation type. Our results here show that predicting how likely a farmer is to adapt his behaviour based on the usual farm typology data of farm structural characteristics, livelihoods or demographics [24] is not a very reliable strategy, and that consideration of intrinsic motivations [15] should be done separately. However, we acknowledge that further work is required to test the degree to which these motivation types accurately predict actual behaviour. A panel survey approach would be very valuable in this regard, particularly where farmer's responses to an intervention or set of interventions could be monitored [7].

Some people will be much more motivated by money than others. It cannot be assumed that money motivates everyone in the same way and to the same extent. Avoid the introduction of a performance-related pay scheme in the belief that it will miraculously transform everyone overnight into well-motivated, high-performing individuals [13].

### *3.5. Policy implication*

Motivation will be more successful in leading to the desired behaviour if incentives or directives are tailored to both complement existing or intrinsic motivations and to remove barriers [25]. Intrinsic motivation emerging from the work itself can be more impressive than extrinsic motivation.

The interplay between intrinsic and extrinsic motivations is more complex and cannot be straightforwardly separated [26]. In the initial stages an extrinsic motivation approach can be used to increase community motivation in bamboo cultivation. Although Ryan and Deci [17] said that some extrinsic motivations can undermine and suppress intrinsic motivations and Greiner and Gregg [27] said that it can even lead to crowding out of intrinsic motivations.

Shifting farmer's extrinsic motivations for undertaking environmental management activities to more intrinsic ones is necessary to ensure sustained and widespread environmental improvements. As environmental practices that are undertaken voluntarily, without coercion or incentives, have a greater potential for sustained and durable benefits. Farmer motivation on bamboo cultivation can be increased through the transfer of knowledge and technology from research institutes, universities and others stakeholders and increasing the capacity of extension workers.

## 4. Conclusion

The motivation of farmers on bamboo cultivation in Sukaharaja Village is low because bamboo is considered unable to provide promising economic value to farmers. Factors that can effect farmer motivation are the number of family members and the number of sources of information. The farmer motivation on bamboo cultivation can be increased through providing a comprehensive information from stakeholders such as research institute, universities and others stakeholder and increasing the added value of bamboo such as post-harvest processing and expanding its marketing network.

## References

[1] Flynn A, Chan KW, Zhu ZH, and Yu L 2017 Sustainability, space and supply chains: The role of bamboo in Anji *J. Rural Stud.* **49** 128-39

[2] Nirmala C and Bisht MS 2015 Bamboo: A prospective ingredient for functional food and nutraceuticals *Bamboo J. (Japan Bamboo Soc.)* **30** 82-99

[3] Lobovikov M, Paudel S, Piazza M, Ren H, and Wu J 2007 *World Bamboo Resources: A Thematic Study Prepared In The Framework Of The Global Forest Resources Assessment 2005* (Rome)

[4] De Flander K and Rovers R 2009 One laminated bamboo-frame house per hectare per year *Constr. Build. Mater.* **23** 210-8

[5] Esteve-Sendra C, Moreno-Cuesta R, Portalés-Mañanós A, and Magal-Royo T 2012 Bamboo, from traditional crafts to contemporary design and architecture *Procedia-Soc. Behav. Sci.* **51** 777-81

[6] Widyaningsih TS, Fauziyah E, and Kuswantoro P 2015 Persepsi masyarakat terhadap bambu *Prosiding Seminar Nasional Agroforestry* pp. 654-700

[7] Hammond J, Van Wijk MT, Smajgl A, Ward J, Pagella T, Xu J, Su Y, Yi Z, and Harrison RD 2017 Farm types and farmer motivations to adapt: Implications for design of sustainable agricultural interventions in the rubber plantations of South West China *Agric. Syst.* **154** 1-12

[8] Arifin Z, Cepriadi, and Muwardi D 2015 Analisis faktor-faktor yang mempengaruhi motivasi petani dalam meningkatkan produksi padi di Desa Bungaraya Kecamatan Bungaraya Kabupaten Siak *Jom Faperta* **2** 1-15

[9] Kustiawan E 2020 Revitalisasi peran dan fungsi hutan rakyat di Gunung Sawal Bagi Ekonomi Ekologi dan Sosial-Menegaskan pengarusutamaan kepedulian Ciamis terhadap Kelestarian Gunung Sawal (Balai Besar Konservasi Sumberdaya Alam dan Ekosistem Jawa Barat) pp. 1-15

[10] Hani A, Fauziyah E, Widyaningsih TS, and Kuswantoro P 2018 Agroforestry patterns that support bamboo sustainability in Sukaharja Village Ciamis District *Wasian* **5** 115-25

[11] Velez G 2001 El Humilde bambú "acero vegetal *In Congreso Virtual de Arquitectura* (Caracas)

[12] Budiono RP 2006 *Karakteristik Petani Tepi Hutan Dan Kompetensinya Dalam Melestarikan Hutan Lindung di 12 Desa di Provinsi Lampung* (Bogor: IPB)

[13] Amstrong M and Taylor S 2014 *Armstrong's Handbook Of Human Resource Management Practice* (United Kingdom: Ashford Colour press Ltd)

[14] Buckley C and Shortle G 2015 Farmer adoption of a new nutrient management technology, Republic of Ireland *RETHINK Case Study Report The Agriculture and Food Development Authority (TEAGASC)* (Wexford, Ireland: Johnstown Castle Research Centre)

[15] Meijer SS 2014 The role of knowledge, attitudes and perceptions in the uptake of agricultural and agroforestry innovations among smallholder farmers in sub-Saharan Africa *Int. J. Agric. Sustain.* 1-15

[16] Legault L 2017 Intrinsic and Extrinsic Motivation
[17] Ryan RM and Deci EL 2000 Self-determination theory and the facilitation of intrinsic motivation, social development, and well-being *Am. Psichol.* **55** 68-78
[18] Shields J and North-Samardzic A 2007 Employee benefits *Shields, John: Managing employee performance and reward: concepts, practices, strategies* **1** 318-39
[19] Cross JE, Keske CM, Lacy MG, Hoag DLK, and Bastian CT 2011 Landscape and urban planning adoption of conservation easements among agricultural landowners in Colorado and Wyoming : The role of economic dependence and sense of place *Landsc. Urban Plan.* **101** 75-83
[20] Siebert R, Toogood M, and Knierim A 2006 Factors affecting european farmers' participation in biodiversity policies *Eur. Soc. Rural Sociol.* **46** 319-40
[21] Burton RJF and Paragahahewa HU 2011 Creating culturally sustainable agri-environmental schemes *J. Rural Stud.* **27** 95-104
[22] Home R, Balmer O, Jahrl I, Stolze M, and Pfiffner L 2014 Motivations for implementation of ecological compensation areas on Swiss farms *J. Rural Stud.* **34** 26-36
[23] Subhrendu K, Pattanayak, Mercer E, Sills E, and Yang J C 2003 Taking stock of agroforestry adoption studies. *Agrofor. Syst.* **57**
[24] van der Ploeg JD, Laurent C, Blondeau F, and Bonnafous P 2009 Farm diversity, classification schemes and multifunctionality *J. Environ. Manag.* **90** S124-31
[25] Moon K and Coklin C 2011 Participation in biodiversity conservation: Motivations and barriers of Australian landholders *J. Rural Stud.* **27** 331-42
[26] Darragh H, and Emery S 2018 What can and can't crowding theories tell us about farmers' environmental intentions in post -agri-environment scheme contexts? *Sociol. Ruralis* ***58***(2) 370-391
[27] Greiner R and Gregg D 2011 Farmers' intrinsic motivations, barriers to the adoption of conservation practices and effectiveness of policy instruments: Empirical evidence from northern Australia Land Use Policy Farmers' intrinsic motivations, barriers to the adoption of conserv *Land use policy* **28** 257-65

**Acknowledgments**
Thank you to team of researcher, R Supriatna (apparatus of Sukaharja Village) who help with data collection in the field and Agroforestry Technology Research and Development Ciamis which facilitated the author in conducting research and development of bamboo in Sukaharja Village, Rajadesa Sub-District, Ciamis District.

**Apendix 1**. The relationship between internal characteristic and internal motivation

| No. | Internal characteristic | Internal motivation | | | | | | | | | |
|---|---|---|---|---|---|---|---|---|---|---|---|
| | | Needs | | Wants | | Personal expectations | | Pride | | Job satisfaction | |
| | | Corr. Coeff. | Sig. | Corr. Coeff. | Sig. | Corr. Coeff. | Sig. | Corr. Coeff. | Sig. | Corr. Coeff. | Sig. |
| 1. | Age | 0.001 | 0.996 | 0.087 | 0.647 | -0.038 | 0.842 | -0.091 | 0.633 | -0.151 | 0.425 |
| 2. | Education | -0.117 | 0.539 | -0.111 | 0.561 | 0.110 | 0.564 | -0.084 | 0.657 | 0.130 | 0.492 |
| 3. | Farming experience | -0.156 | 0.411 | -0.013 | 0.945 | -0.088 | 0.644 | -0.411 | 0.024 | -0.353 | 0.056 |
| 4. | Number of family dependents | 0.173 | 0.361 | 0.250 | 0.183 | 0.352 | 0.057 | 0.188 | 0.320 | 0.033 | 0.864 |
| 5. | Income | 0.092 | 0.629 | -0.031 | 0.872 | 0.261 | 0.163 | 0.216 | 0.252 | 0.087 | 0.647 |
| 6. | Land ownsership | -0.013 | 0.945 | -0.157 | 0.409 | 0.018 | 0.925 | 0.127 | 0.503 | 0.027 | 0.886 |

IOP Conf. Series: Materials Science and Engineering **935** (2020) 012008 doi:10.1088/1757-899X/935/1/012008

**Apendix 2**. The relationship between external characteristic and internal motivation

| No. | Exsternal characteristic | Internal motivation | | | | | | | | | |
|---|---|---|---|---|---|---|---|---|---|---|---|
| | | Needs | | Wants | | Personal expectations | | Pride | | Job satisfaction | |
| | | Corr. Coeff. | Sig. | Corr. Coeff. | Sig. | Corr. Coeff. | Sig. | Corr. Coeff. | Sig. | Corr. Coeff. | Sig. |
| 1. | Extension intensity | -0.162 | 0.391 | -0.003 | 0.987 | 0.110 | 0.562 | 0.027 | 0.889 | 0.204 | 0.279 |
| 2. | Extension channel | -0.028 | 0.884 | -0.033 | 0.861 | 0.285 | 0.127 | -0.340 | 0.066 | -0.155 | 0.414 |
| 3. | Number of information sources | 0.153 | 0.419 | 0.217 | 0.249 | 0.054 | 0.775 | 0.107 | 0.573 | -0.064 | 0.736 |

**Apendix 3**. The relationship between internal characteristic and external motivation

| No | Internal characteristic | External motivation | | | | | | | | | |
|---|---|---|---|---|---|---|---|---|---|---|---|
| | | Type and character of job | | Working group | | Workplace situation | | Organization of the place of work | | Current rewards system | |
| | | Corr. Coeff. | Sig. | Corr. Coeff. | Sig. | Corr. Coeff. | Sig. | Corr. Coeff. | Sig. | Corr. Coeff. | Sig. |
| 1. | Age | 0.045 | 0.812 | 0.221 | 0,240 | 0.030 | 0.876 | 0.232 | 0.217 | 0.160 | 0.398 |
| 2. | Education | -0.202 | 0.284 | 0.074 | 0,699 | -0.091 | 0.634 | -0.088 | 0.644 | -0.083 | 0.662 |
| 3. | Farming experience | 0.032 | 0.866 | 0.287 | 0,125 | 0.235 | 0.211 | 0.325 | 0.080 | 0.232 | 0.217 |
| 4. | Number of family dependents | 0.105 | 0.579 | 0.329 | 0,076 | 0.438 | 0.016 | -0.111 | 0.558 | 0.079 | 0.678 |
| 5. | Income | -0.066 | 0.729 | -0.080 | 0,673 | 0.196 | 0.299 | -0.304 | 0.103 | -0.378 | 0.04 |
| 6. | Land ownership | -0.048 | 0.801 | 0.024 | 0,901 | 0.000 | 1 | -0.027 | 0.885 | -0.080 | 0.675 |

**Apendix 4**. The relationship between external characteristic and internal motivation

| No | Exsternal characteristic | External motivation | | | | | | | | | |
|---|---|---|---|---|---|---|---|---|---|---|---|
| | | Type and character of job | | Working group | | Workplace situation | | Organization of the place of work | | Current rewards system | |
| | | Corr. Coeff. | Sig. | Corr. Coeff. | Sig. | Corr. Coeff. | Sig. | Corr. Coeff. | Sig. | Corr. Coeff. | Sig. |
| 1. | Extension intensity | 0.258 | 0.168 | 0.251 | 0.181 | 0.096 | 0.614 | 0.077 | 0.684 | 0.135 | 0.478 |
| 2. | Extension channel | -0.250 | 0.182 | 0.183 | 0.334 | 0.365 | 0.047 | 0.319 | 0.085 | 0.244 | 0.194 |
| 3. | Number of information sources | 0.025 | 0.894 | -0.434 | 0.017 | -0.400 | 0.028 | 0.178 | 0.346 | -0.045 | 0.815 |

# Physical properties of furfurylated mangium (*Acacia mangium* Willd.) and pine (*Pinus merkusii* Jungh et de Vriese) woods

**D Mulyosari[1], Y S Hadi[1*], E N Herliyana[2], G Pari[3], R Pari[3], and I B Abdillah[1]**

[1]Departement of Forest Products, Faculty of Forestry, IPB University, 16680, Bogor, Indonesia
[2]Departement of Silviculture, Faculty of Forestry, IPB University, 16680, Bogor, Indonesia
[3]Forest Products Research and Development Centre, Ministry of Environment and Forestry, Bogor, 16610 Indonesia

*Email: yshadi@indo.net.id

**Abstract.** Wood from plantation forest contains a lot of sapwood and juvenile wood which has inferior in physical properties. Furfurylation could improve the properties through a reaction between Furfuryl Alcohol (FA) with a hydroxyl group of wood components. The purpose of the study was to improve the physical properties of mangium (*Acacia mangium* Willd.) and pine (*Pinus merkusii* Jungh et de Vriese) woods. The wood specimens for physical properties was prepared according to the ASTM D 143-94. They were vacuumed at 600 mmHg during 30 minutes, when the vacuum has been released, the FA reagent was inserted to the tank and followed with pressure at 10 kg/cm$^2$ during 30 minutes. The specimens were wrapped with aluminium foil and heated in the oven at 100°C for 24 hours. After conditioned for three weeks, the specimens were tested for physical properties including Weight Percent Gain (WPG), density, Moisture Content (MC), water absorption, swelling, shrinkage, and anti-swelling efficiency (ASwE) according to the ASTM D 143-94. Untreated specimens were also prepared for comparison purpose with five replications for each treatment. The results showed that furfurylated mangium and pine woods reached WPG 11.34% and 17.74%, densities were increased 8.95% and 3.75%, MC were decreased 88.41% and 84.74%, water absorption were decreased 41.76% and 46.84%, volume swelling were decreased 55.18% and 61.57%, volume shrinkage were decreased 34.99% and 64.75%, respectively. ASwE of mangium and pine woods was 55.1% and 61.6%. It could be mentioned that the furfurylated wood had better physical properties compared to untreated woods.

## 1. Introduction

Forest plantation was developed to substitute natural forest, especially for timber supply purposes. Mostly the planted trees were fast-growing species, because the demand for wood was increased year by year, and the stand could be exploited after 6-10 years old. Some favourite species for industrial plantation forest are mangium (*Acacia mangium* Willd.) and pine (*Pinus merkusii* Jungh et de Vriese). Total log production in Indonesia in 2017 reached 49.13 million m$^3$, 63.36% of total production was mangium wood, and 0.39% of total production was pine wood [1]. The density of mangium wood was reported in the range of 0.59-0.61 [2]. The density of pine wood was 0.48-0.54 [3]. The wood from plantation forest was dominated with sapwood and containing juvenile wood which had lower physical-mechanical properties [4]. The low physical and mechanical properties of young mangium and pine wood causes the limited utilization of the wood, hence special treatment is needed to improve the wood quality.

Wood modification was an effective way to increase wood stability, water resistance, and wood life [5]. Several technique in wood modification are chemical, thermal, and surface modification [5]. Chemical modification processes for wood modification are acetylation and furfurylation. Furfurylation was based upon the reaction of wood with the bio-chemical furfuryl alcohol at the cell wall level. Furfurylation significantly improved wood properties, protected against decay, decreased hygroscopicity and improved dimensional stability. Furfurylated wood has high dimensional stability and surface hardness [5-7]. Furfurylated wood had better dimensional stability and mechanical properties, and was more resistant to organism attack. Modification of wood with furfurylation shows better results and less impact on the environment, so this process was in high demand, especially in Europe [8].

The use of catalysts further develops the furfurylation process. Furfurylation study using boric acid catalyst on *Crytomeria japonica* D. and *Pinus sylvestris* L. wood produced a higher weight percent gain (WPG) value than furfurylation without catalyst and produced wood with 70-86% ASwE value [9]. Tartaric acid was an organic compound derived from ascorbic acid, which is white, odourless, and stable in the air. Tartaric acid is very soluble in water and easily soluble in ethanol. The use of the tartaric acid catalyst in the furfurylation process of mangium and pine wood had not been studied. In this study, the furfurylation was conducted with the addition of tartaric acid as a catalyst. The purpose of this study was to improve the physical properties of mangium and pine woods through the furfurylation process.

## 2. Materials and Methods

### *2.1. Materials*
The mangium and pine woods were obtained from Bogor, West Java, Indonesia. The wood samples taken from 6-10 years old trees. The materials for furfurylation process were Furfuryl Alcohol (FA) and tartaric acid.

### *2.2. Methods*

#### 2.2.1 Sample preparation
For physical properties test, the woods were cut into specimens with a size of 10 cm by 2.5 cm by 2.5 cm according to ASTM D 143-94 [10]. Impregnation process was carried out to fill the cell voids with FA solution. To accelerate the polymerization process, the FA solution was added with tartaric acid at the amount of 5/100 (w/w). The wood specimens were placed in the chamber and then vacuumed at 600 mmHg for 30 minutes. During the released period of vacuum process, the FA solution was allowed to flow into the chamber and subsequently pressured at 10 kg/cm$^2$ for 30 minutes. After the vacuum-pressure process was finished, the wood specimens were wrapped with aluminium foil and then heated in an oven at 100°C for 24 hours. Afterwards, the aluminium sheet was removed, and the wood specimens were weighed for WPG calculation. For comparison purposes, untreated wood was included as the control, and five samples as replication were prepared for each treatment. The specimens were conditioned in a room condition for three weeks.

#### 2.2.2 Physical properties test
The physical properties of wood were density, moisture content, water absorption, volume swelling, volume shrinkage, and anti-swelling efficiency (ASwE) were tested according to ASTM D 143-94 [10].

#### 2.2.3 Functional groups test
Functional group testing was performed using Fourier Transform Infra-Red Spectroscopy (FTIR). This process was conducted to determine the functional groups in wood furfurylation. The process was carried out by mixing the powder from furfurylated wood with dry potassium bromide (KBr) (1: 100). The mixed powder was moulded and compacted to form a thin pellet, then the functional groups were

seen at wavelengths between 4000-500 $cm^{-1}$. Untreated specimens were also prepared for comparison purpose.

2.2.4 Data analysis
To analyze the effect of treatments upon all responses, i.e. wood density, a 2 × 2 factorial in completely randomized design was used for data analysis. The first factor was wood species (mangium and pine), and the second factor was treatments (untreated and furfurylated woods). The data were analyzed with analysis of variance (ANOVA) by Microsoft Excel.

## 3. Results and Discussion

### *3.1. Effectiveness of the impregnation process*
The effectiveness of the impregnation process could be observed from the amount of WPG. The WPG value for mangium and pine wood were 11.34%, 17.74%, respectively. The difference in WPG value was influenced by wood anatomical characteristics and also the chemical components of wood in the form of extractive substances whose proportions differ in the two types of wood. The higher WPG in furfurylated pine wood because pine belongs to conifer which had simpler anatomical characteristic, and mangium was hardwood which had more complex characteristics. The other factor that mangium wood contains more extractives which could inhibit chemical reactions between FA and wood [11,12].

### *3.2. Physical properties of furfurylated wood*
Density was one of the important physical properties related to the use of wood [13]. The higher density of wood means that the content in the wood cell wall was higher which means the thicker the cell wall [14]. The results of physical properties testing can be seen in Table 1. The density of wood increased after the furfurylation process. Analysis of variance showed that the FA treatment had a significant effect on increasing the wood density (Table 2). According to Dong's research [15], the physical properties of furfurylated wood were higher than untreated wood because FA filled the cell walls.

**Table 1** Physical properties of untreated and furfurylated mangium and pine woods

| Wood Species | Treatments | Wood Density ($g/cm^3$) | MC (%) | Water Absorption (%) | Volume Swelling (%) | Volume Shrinking (%) |
|---|---|---|---|---|---|---|
| Mangium | Untreated | 0.67 | 15.53 | 33.71 | 8.10 | 7.63 |
| | Furfurylated | 0.73 | 1.80 | 19.63 | 3.63 | 4.96 |
| Pine | Untreated | 0.80 | 14.09 | 25.98 | 9.37 | 10.13 |
| | Furfurylated | 0.83 | 2.15 | 13.81 | 3.60 | 3.57 |

Furfurylated wood has a lower moisture content compared to untreated wood. Analysis of variance showed that the FA treatment had significantly affected the moisture content of the wood. The moisture content of wood was reduced after the furfurylation process [16]. Furfurylated wood was more hydrophobic or difficult to absorb water or water vapour from the environment due to chemical bonds between the FA and OH groups in the wood, so that furfurylated wood has a lower water content than untreated wood.

Furfurylated wood has lower water absorption compared to the untreated wood. Analysis of variance showed that the FA treatment had significantly reduced the ability of water absorption both in mangium and pine wood. The low absorption of wood to water showed that the furfural groups impregnated into the wood reacted with the chemical compounds in the wood. Penetration of chemical material into wood can produce cross bonds between hydroxyl groups in the cell walls of wood. Cross-linking could reduce the hygroscopicity of wood by reducing the bonding place for water inside the cell wall [14]. The lower water absorbtion of wood affects its dimensional stability [17]. Wood tissue will maintain water content in balance with its environment, through absorption and release of water [18]. Analysis of variance showed that the FA treatment had high significantly affected the volume swelling and volume shrinkage of the wood.

Dimensional stability was the ability of wood not to expand and shrink when it is in environmental conditions with fluctuating temperature and humidity changes. Criteria that could be used to determine the dimensional stability of wood was the value of the Anti-Swelling Efficiency (AswE) [19]. The ASwE value for furfurylated mangium and pine woods in this study were 55.1% and 61.6% respectively. The ASwE value in this study had a highervalue than ASwE from another study; those indicated that the furfurylated wood was more stable in dimensions. Furfurylation process could stabilize the dimensions of wood due to reduced empty volume in wood [20].

**Table 2** Variance analysis of physical properties.

| Responses | Wood species | Treatment |
|---|---|---|
| Density | ** | * |
| Moisture Content | ns | ** |
| Water Absorption | * | ** |
| Volume Swelling | * | ** |
| Volume Shrinking | * | ** |

** Highly significantly different ($P \leq 0.01$); * Significantly different ($P \leq 0.05$)

### *3.3. Functional group of furfurylated wood*

Analysis using FTIR was aimed to investigate changes in the chemical structure of wood after being treated. The results of FTIR analysis on FA treated wood and non-treated wood showed that the wave peak position and spectrum shape was relatively same in both wood species, although in mangium wood there was a difference in intensity. In general, O-H strain absorption bands (about 3400 $cm^{-1}$) and C-H absorption bands (around 2927 $cm^{-1}$) were derived from all wood components. Functional clusters located in the fingerprint area (1800-650 $cm^{-1}$) can be used as a differentiator between wood components [21], but it is difficult to identify. The hydroxyl group intensity of furfurylated wood was lower than untreated wood. The decrease in intensity was presumably due to some OH groups have been linked to FA. The decrease in the intensity of the hydroxyl group makes the wood more hydrophobic, so it was difficult to absorb water, and the dimensional stability is increased. The FTIR spectra of mangium wood for untreated and furfurylated wood (Figure 1) shows that there was an aromatic skeletal vibration of lignin at wave number 1590 $cm^{-1}$, indicated that furfural compounds impregnated was binded with lignin compounds in wood. These results are consistent with research conducted by Dong's research [7].

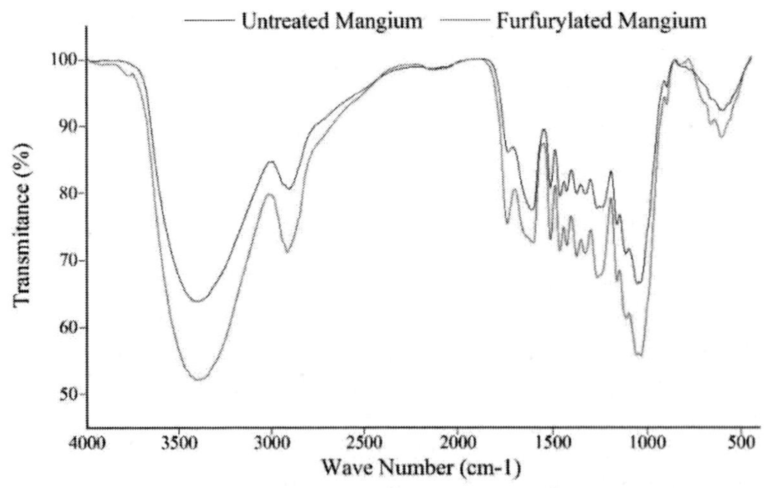


**Figure 1.** FTIR Spectra of mangium wood

IOP Conf. Series: Materials Science and Engineering **935** (2020) 012009 doi:10.1088/1757-899X/935/1/012009

## 4. Conclusion

Modification of the furfurylation process with the tartaric acid catalyst can improve the physical properties of mangium and pine wood. Furfurylation increases the density and decreases the water content of mangium and pine wood. The dimensional stability of furfurylated wood increases compared to untreated wood. Improved physical properties of furfurylated wood are caused by chemical bonds between FA and wood components as evidenced by the changes in the intensity of functional groups in furfurylated wood.

## References

[1] [BPS] Badan Pusat Statistik 2018 *Statistik Produksi Kehutanan* 2017 (Jakarta: Badan Pusat Statistik)
[2] Arsad E 2011 Sifat fisik dan mekanik kayu akasia mangium (*Acacia mangium* Willd.) dari hutan tanaman industri Kalimantan Selatan *J. Riset Indonesia Hasil Hutan* **3**(1) 20-3
[3] Pasaribu 2008 Perbedaan sifat fisis-mekanis dan anatomi kayu tusam (Pinus merkusii) strain Tapanuli dan strain Aceh *JPHH* **26**(2) 166-73
[4] Martawijaya A 1990 Sifat dasar berbagai jenis kayu yang berasal dari hutan alam dan hutan tanaman *Prosiding Diskusi Hutan Tanaman Industri*
[5] Hill CAS 2006 *Wood Modification: Chemical, Thermal and Other Processes* (Chichester, UK: John Wiley & Sons)
[6] Esteves BM, Nunes L, and Pereira H 2011 Properties of furfurilated wood (*Pinus pinaster*) *Eur. J. Wood Prod.* **69**(4) 521-5
[7] Dong Y, Yan Y, Zhang S, Li J, and Wang J 2015 Flammability and physical-mechanical properties assessment of wood treated with furfuryl alcohol and nano-$SiO_2$ *Eur. J. Wood Prod.* **73**(4) 457-64
[8] Lande S, Westin M, and Schneider M 2008 Development of modified wood products based on furan chemistry *Mol. Cryst. Liq. Cryst.* **484**(1) 367-78
[9] Baysal E, Ozaki SK, and Yalinkilic MK 2004 Dimensional stabilization of wood treated with furfuryl alcohol catalyzed by borates *Wood Sci. Technol.* **38** 405-15
[10] [ASTM] American Society for Testing Materials 2000 *Standard Test Methods for Small Clear Specimens of Timber ASTM D 143-94* (West Conshohocken, USA: ASTM International)
[11] Balfas J 1993 Stabilisasi dimensi pada kayu tanaman karri (*Eucalyptus diversicolor*) dan jarrah (*E. marginata*)-bagian II: furfurilasi. *JPHH* **11**(4) 134-6
[12] Balfas J and Sumarni G 1995 Keawetan kayu tusam (*Pinus merkusii* Jungh. et de Vr.) dan mangium (*Acacia mangium* Willd) setelah furfurilasi *JPHH* **13**(7) 259-65
[13] Pandit IKN and Ramdan H 2002 *Anatomi Kayu: Pengantar Sifat Kayu Sebagai Bahan Bangunan* (Bogor: Yayasan Penerbit Fakultas Kehutanan IPB)
[14] Bowyer JL, Shmulsky R, and Haygreen JG 2003 *Forest Products and Wood Science: An Introduction.* Fourt Edition (Iowa, US: Iowa State Press)
[15] Dong Y, Qin Y, Wang K, Yan Y, Zhang S, Li J, and Zhang S 2016 Assessment of the performance of furfurylated wood and acetylated wood: comparison among four fast-growing wood species *Bioresources* **11**(2) 3679-3690
[16] Epmeier H, Johansson M, Kliger R, and Westin M 2007 Material properties and their interrelation in chemically modified clear wood of scots pine *Holzforschung* **61** 34-42
[17] Shmulsky R and Jones PD. 2011. *Forest Products & Wood Science, an Introduction, sixth edition* (Iowa, USA: John Wiley & Sons)
[18] Nandika D, Soenaryo, and Saragih A 1996 *Kayu dan Pengawetan Kayu* (Jakarta: Dinas Kehutanan DKI)
[19] Pandey KK, Jayashree, and Nagaveni HC 2009 Study of dimensional stability, decay resistance, and light stability of phenylisothiocyanate modified rubberwood *Bioresources* **4**(1) 257-67
[20] Hill CAS 2011 Wood modification: An update *Bioresources* **6**(2) 918-9

[21] Zhang FD, Xu CH, Li MY, Huang AM, and Sun SQ 2014 Rapid identification of *Pterocarpus santalinus* and *Dalbergia louvelii* by FTIR and 2D correlation IR spectroscopy *J. Mol. Struct.* **1069** 89-95

**Acknowledgments**
The authors also thank to IPB University and Forest Products Research and Development Centre for supporting the research facilities.

IOP Conf. Series: Materials Science and Engineering **935** (2020) 012010 doi:10.1088/1757-899X/935/1/012010

# Effect of combined strip and zephyr laminated bamboo composite on physical and mechanical properties

**I Sumardi[1*], R Dungani[1], I M Sulastiningsih[2], and D Amalia[1]**

[1]Postharvest Technology, School of Life Sciences and Technology, Bandung Institute of Technology, Bandung, 40132, Indonesia
[2]Forest Product Research and Development Center, Ministry of Environment and Forestry, Bogor, 16110, Indonesia

*Email: ihak@sith.itb.ac.id

**Abstract**. Bamboo (*Gigantochloa apus)* is a potential material for wood substitution. However, the bamboo uses are limited according to the dimensions. Therefore, it needs modification technology such as transform bamboo into laminated bamboo boards. The combination of bamboo strips and zephyr as well as different fiber directions affected the physical and mechanical properties of laminated bamboo boards. The study was carried out on *Gigantochloa apus*, which were made into strip and zephyr manually. Each strip and zephyrs are dried and then formed into a laminated board, then glued using isocyanate adhesive in a double spread with a weight of 300 g/m$^2$. Arrangement of laminated bamboo boards perpendicularly between face/ back and core (Ply bamboo) and the parallel between face/back and core (Laminated Bamboo Lumber/LBL). The laminated bamboo board was cold pressed at a pressure of 22.22 kgf/cm$^2$ for 1 hour. Physical and mechanical properties have been evaluated. Laminated bamboo boards have an average moisture content and density that meet JAS standards. The face/back of the strip has better dimensional stability value compared to zephyr. The combination of raw materials on the face/back and variations in the direction of the fiber can improve the mechanical properties of the bamboo laminate board.

## 1. Introduction

The ability of forests as a provider of raw materials has been unable to meet the increasing needs of the wood manufacturing industry. Especially, the demand for wood as structural components, furniture, and household equipment continues to increase along with population growth [1]. The main factor causing the lack of availability of wood as raw materials is the slow growth of wood from natural forests. Therefore, alternative substitution of wood as raw material is needed to meet industrial needs. Bamboo is a potential plant to be used as an alternative material to replace wood. Besides fast-growing time, bamboo is also easy to find, easy to plant in a variety of soil conditions, relatively cheap, and has good mechanical properties. However, bamboo has several characteristics that are taken into consideration when used as raw material, such as having node and internode so that it tends not to be homogeneous in each bamboo culms, then cylindrical and hollow shape which can limit the use of bamboo as a raw material that should be modified. The modification process aims to make the shape of bamboo more flexible and can be used as a structural or non-structural material [2].

Research related to bamboo modification has been developed, such as the manufacture of wood composite bamboo boards, laminated bamboo boards, ply bamboo, bamboo chipboard, Laminated

Bamboo Lumber (LBL) and other bamboo products [3]. The manufacture of a LBL is an alternative modification of the shape of the bamboo that can increase the use of bamboo as a raw material. A previous study related to LBL made of zephyr, that is processed through the destruction of bamboo culmns which are given a hot press and glued using resorcinol adhesive is an alternative to improve the adhesion and mechanical properties of bamboo [4]. Composite bamboo boards from Andong bamboo strips glued using isocyanate with varying composition of fiber direction and combination of glued bamboo strips have good adhesion quality and good dimensional stability [5]. Combination of bamboo Andong in the form of strips and zephyr and solid wood as a core part with isocyanate adhesives can be used in making wood composite bamboo boards that increasing strength and other mechanical properties [6]. Then, based on the anatomical structure of bamboo, in one internode of bamboo changes from the inner layer to the outer layer due to the greater concentration of cellulose fibers in the outermost layer compared to the inner layer which is more dominated by lignin. Therefore, the arrangement of bamboo and its orientation will affect the mechanical properties of the laminated bamboo board. Besides, bamboo also has inter-section fibers (node) which also affect the physical and mechanical properties of the laminated bamboo board [7]. This study aimed to determine the effect of a combination of strips and zephyr on face/back and core ply bamboo and LBL difficult to catch on physical and mechanical properties.

## 2. Materials and Methods

### *2.1. Materials and equipment*

The materials used in this study were borax and 4 years old bamboo (*Gigantochloa apus*) culms obtained from Sumedang area, and isocyanate adhesive (P.I. Bond). Then equipment used in this study are calipers, axes, mixer, cold press machine, moisture meter, oven, planer machine, running saw machine, sander machine, table saw machine, digital scales, universal testing machine (UTM) Shimadzu AG -I, adhesive container, and water bath.

### *2.2. Methods*

#### 2.2.1. Bamboo culms preparation

Fresh bamboo culms are cut along ± 40 cm bamboo is preserved by soaking using borax solution ($Na_2B_4O_7.5H_2O$) for 7 days.

#### 2.2.2. Bamboo zephyr preparation

The preserved bamboo culmn is crushed first to form flat sheets with an axe then dried at room temperature for 7 days or until it reaches a moisture content of 25%. Then, bamboo zephyr is planned on the outside and inside with a planer machine. Furthermore, zephyr is dried to reach 20% of moisture content. The target thickness of the zephyr layer is 0.5 cm. Then, cut and arranged into zephyr bamboo sheets of length and width of 40 cm.

#### 2.2.3. Bamboo strips preparation

Preparation of making bamboo strips is carried out by drying at room temperature for 7 days or until it reaches a water content of 20%. Then, the bamboo culmn is removed from the outer and inner parts and cut into 40 cm long, 2 cm wide and 0.4 cm thick by 20 pieces, then arranged into sheets of bamboo strips.

#### 2.2.4. Producing laminated bamboo boards

Laminated bamboo boards with dimensions of 400 x 400 x 15 mm are arranged following the arrangement variations and material combination that prescribed and presented in Table 1 with 3 layers. Each layer of zephyr or strip is given an isocyanate adhesive using a brush with a double spread technique. The glue spread rate is 300 g/m$^2$. Then cold pressed at room temperature with a pressure of 22.22 kgf/cm$^2$ for 1 hour. After the cold pressing process, laminated bamboo boards were conditioned for 7 days. Then finishing laminated bamboo board by sanding and cutting with running saw machine or table saw machine according to the size needed for testing.

**Table 1.** Laminated bamboo board arrangement and combination

| Code | Name | Fiber direction | Material | |
|---|---|---|---|---|
| | | | Face/Back | Core |
| LBL-Z | LBL zephyr | Parallel | Bamboo zephyr | Bamboo strip |
| LBL-S | LBL strip | | Bamboo strip | Bamboo zephyr |
| Ply-Z | Ply bamboo zephyr | Perpendicular | Bamboo zephyr | Bamboo strip |
| Ply-S | Ply bamboo strip | | Bamboo strip | Bamboo zephyr |

2.2.5. Testing

*2.2.5.1. Physical Properties.*
The laminated bamboo board evaluated by water content, density, water absorption, and shrinkage. Determination of water content and density according to JAS for plywood and JAS for laminated veneer lumber [8,9]. Testing of shrinkage and water absorption according to modified ASTMD 1037 standard [10].

2.2.5.2. *Mechanical properties.*
The mechanical properties of ply bamboo and LBL include MOE, MOR, shear strength, and screw withdrawal strength. The determination of MOE and MOR ply bamboo board is according to JAS for plywood in the long and cross direction of the sample. Then, determination of MOE and MOR LBL according to JAS for laminated veneer lumber in flat and edge direction of the sample. The testing of shear strength is according to JAS for plywood standard with ply bamboo sample size (81 x 25 mm). Whereas, the LBL Horizontal shear strength was carried out according to the JAS standard for laminated veneer lumber in two directions of testing, flat direction (81 x 39 mm) and edge direction (99 mm x thickness) [8,9]. The standard used in determining screw withdrawal strength is SNI 03-2105-2006 with a sample size of 5 x 10 cm and a screw depth of 0.7 mm [11].

*2.3. Data analysis*
The design of the experiment was carried out using the Completely Randomized Design method with 2 factors, a combination of raw materials and variations in fiber direction. Each treatment used 4 replications. Data were analyzed using the Independent-sample T-test with $\alpha \leq 0.05$. Data analysis was performed using SPSS 16.0 software.

## 3. Result and Discussion

*3.1. Physical properties: moisture content and density*
The average result of moisture content, density, and thickness change are presented in Table 2. The thickness is measured in conditions before pressing (before gluing), after pressing, and after conditioning to determine the thickness changes that occur in producing laminated bamboo board. The value of moisture content after conditioning for 7 days is varied from 11.63 to13.46%. These results indicate that the moisture content of the laminated bamboo board is in the range of JAS [8-9] ≤14%. Then, the density of laminated bamboo boards after conditioning is in the range of 0.77-0.83 g/cm$^3$. The density of bamboo culm is 0.65 g/cm$^3$ so that the manufacture of laminated boards can increase the density of bamboo due to the adhesive and pressing process [12]. The previous study related to composite bamboo boards made of Andong bamboo has a density varied from 0.78 to 0.80 g/cm$^3$ [5].

According to Table 2, it can be seen that both ply bamboo and LBL boards have experienced a thickness change. It is suspected that compression can cause thickness shrinkage of laminated bamboo boards. Shrinkage that occurs after conditioning is due to environmental conditions and the hygroscopic nature of the board so that it releases water into the environment until it reaches an equilibrium condition. Laminated bamboo boards with strips on the face/back (LBL-S and Ply-S) have an average thickness change higher than bamboo laminate boards with zephyr. This probably due to a decrease in the water content of the bamboo laminate board with strips on the face/back greater than the water content in the laminated bamboo board with zephyr. Zephyr on the face/back causing greater shrinkage. The results of the Independent-sample T-test show that the combination of strip and zephyr and the variation of fiber direction are significantly different for the value of water content, while not significantly different in the value of density.

**Table 2.** Moisture content, density, and thickness changes of laminated bamboo boards

| Testing | Code | | | |
|---|---|---|---|---|
| | LBL-Z | Ply-Z | LBL-S | Ply-S |
| Moisture content after cold press (%) | 14.45 (1.52) | 13.90 (1.94) | 16.80 (0.70) | 17.53 (0.43) |
| Moisture content after conditioning (%) | 11.63 (0.70) | 12.40 (0.41) | 12.71 (0.43) | 13.46 (0.43) |
| Density (g/cm$^3$) | 0.77 (0.03) | 0.79 (0.05) | 0.83 (0.02) | 0.81 (0.04) |
| Thickness change (%)* | 1.30 (1.77) | 0.63 (0.58) | 1.901.58) | 2.47 (1.61) |

*Thickness after conditioning - Thickness after cold press

### *3.2. Physical properties: swelling, shrinkage, and water absorption*

Testing of water absorption, swelling, and shrinkage was carried out to determine the dimensional stability of laminated bamboo boards. Swelling and shrinkage are evaluated from changes in the dimensions of thickness and length of the sample, while water absorption is measured from changes in sample weights. Better discuss WA first, the others as effect of WA. Swell and shrink after soak in water.

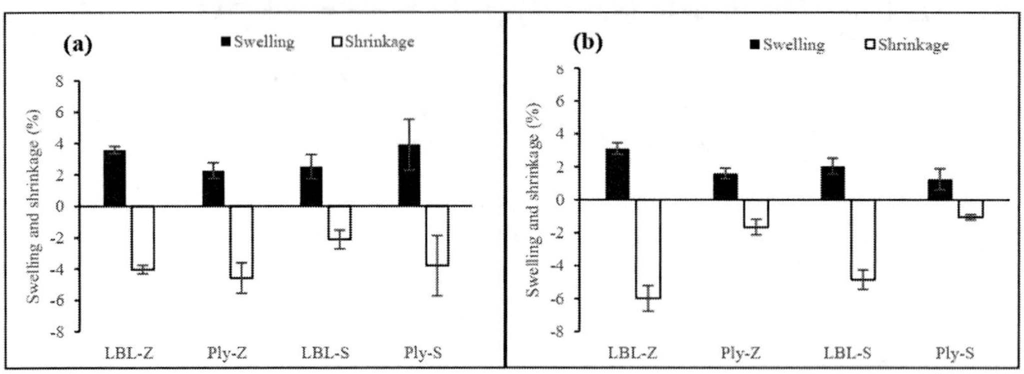


**Figure 1.** Swelling and shrinkage (a) Thickness (b) Length

According to Figure 1, it can be seen that several shrinkage values are greater than the swelling value. This can occur due to the condition of the final water content which is lower than the initial condition. Based on these results, it is known that the lowest value of thickness changes is on LBL-S with a range between swelling and shrinkage of 4.63%. Then, the lowest length changes value is on the Ply-S with a range between swelling and shrinkage of 2.30%. Both of these results show that laminated bamboo boards with strips on the face/back can reduce the value of shrinkage both in the thickness direction and in the length direction. This probably due to more gaps found on the laminated bamboo board with zephyr on the face/back which can decrease the stability of the board. According to Figure 1 (a), crossed layers in the composition of the laminated bamboo board can withstand or reduce the occurrence of swelling or shrinkage. Layer direction of the laminated bamboo board greatly affects the stability of dimensions [5]. The results of the Independent-sample T-test show that the variation in direction of parallel and perpendicular fiber is significantly different for the value of length shrinkage and thickness swelling of laminated bamboo boards. Whereas the combination of strip and zephyr material is only significantly different in length swelling.

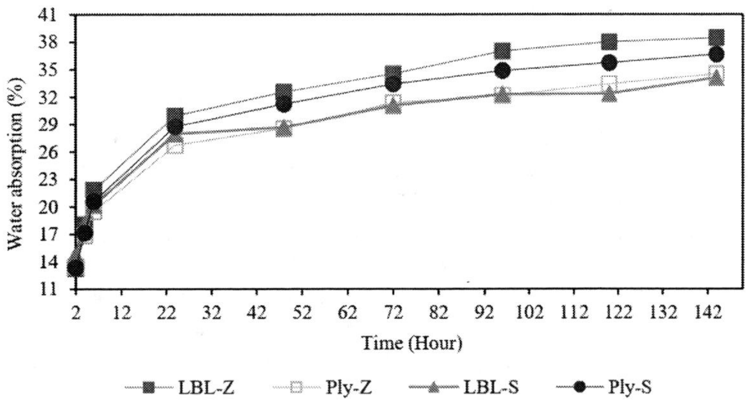


**Figure 2.** Water absorption of laminated bamboo

Water absorption on the laminated bamboo board is obtained by soaking for 6 days (144 hours). Water absorption increases exponentially at a time of 0 to 24 hours immersion, then starts to be constant at a time of 120 hours. Based on Figure 4. It is known that LBL-Z has the highest water absorption (29.95% during 24 hours immersion). This probably due to more gaps found on the laminated board with zephyr on the face/back. Significant changes in water absorption in the first 2 hours are supposedly due to stress release on the laminated bamboo board which causes thickness swelling. This indicates that the bamboo laminate board absorbs more water. Composite board stresses occur due to compression, which is then released during immersion and will cause thickness swelling [13].

*3.3. Mechanical properties: Modulus of Elasticity (MOE) and Modulus of Rupture (MOR)*

MOE values indicate the nature of the board stiffness while the MOR indicates the strength of the fiber that occurs at maximum load. The density of the sample can affect MOE and MOR values. The relationship between density values and ply bamboo MOE/MOR values are presented in Table 3.

**Table 3.** MOE and MOR values of ply bamboo

| Testing Direction | Code | Density ($g/cm^3$) | MOE(MPa) | MOR (MPa) |
|---|---|---|---|---|
| Long | Ply-Z | 0.80 (0.04) | 10,168.15 (611.97) | 55.19 (13.04) |
| | Ply-S | 0.80 (0.03) | 6,665.55 (1398.30) | 43.23 (1.08) |
| Cross | Ply-Z | 0.78 (0.03) | 644.75 (67.20) | 19.59 (10.56) |
| | Ply-S | 0.80 (0.06) | 783.94 (162.20) | 19.24 (2.09) |

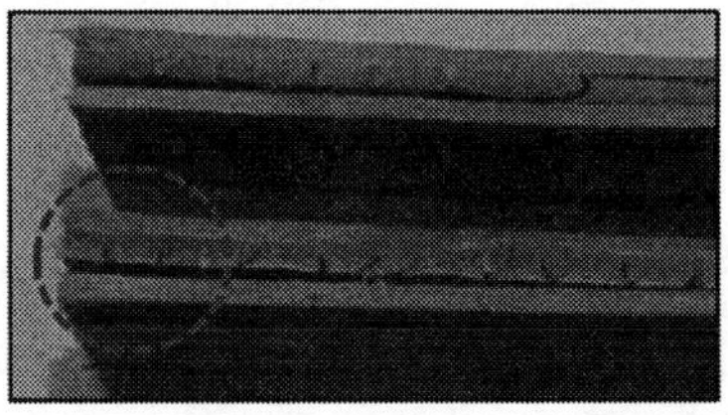

**Figure 3.** Bonding damage of ply bamboo strip sample

MOE/MOR values on ply bamboo are affected by the face and back of the laminated bamboo board because in this test the upper surface experiences compressive forces while the lower surface experiences tensile forces. Based on Table 3. It is known that the laminated bamboo board with perpendicular direction of fiber (ply bamboo) has a different MOE value in terms of the test direction (long and cross). In the long direction of testing (parallel to grain), the MOE value on the Ply-Z is greater than the Ply-S. This probably due to bonding damage during Ply-S testing which can be seen in Figure 3, thus reducing the MOE value. MOE values on laminated bamboo boards are affected by the nature of bamboo and the gluing of bamboo strips. Initial shear damage due to weak adhesion on the sample can reduce the MOE value [14]. In the cross direction of testing (perpendicular to the grain) the MOE value on Ply-S is greater than Ply-Z. According to JAS for plywood [8], the standard for MOE in long and cross direction is 5000 MPa and 4000 MPa, therefore MOE value in long direction meets the standard, while the MOE value in cross direction does not meet the standard. This was supposedly due to damage that occurred between the connections on the laminated bamboo board.

The MOR value of Ply-Z in both the long and cross directions of testing are greater than Ply-S. This probably due to bond damage on the laminated bamboo board with the strip on the face/back. According to JAS for plywood [8], the standard for MOR (long and cross direction) is 24 MPa and 20 MPa, so the results of MOR in long direction meet the standard, while the MOR in cross direction does not meet the standard.

**Table 4.** MOE and MOR values of LBL

| Testing Direction | Code | Density (g/cm$^3$) | MOE (MPa) | MOR (MPa) |
|---|---|---|---|---|
| Flat | LBL-Z | 0.80 (0.07) | 19747.53 (3638.18) | 76.04 (14.64) |
| | LBL-S | 0.77 (0.03) | 15326.30 (2855.22) | 75.73 (7.85) |
| Edge | LBL-Z | 0.82 (0.04) | 23917.63 (3455.13) | 84.44 (14.86) |
| | LBL-S | 0.81 (0.02) | 20173.03 (1804.74) | 108.62 (7.15) |

**Figure 4.** LBL zephyr sample damage

According to Table 4. It is known that LBL has different MOE values in different testing direction (flat and edge). In both the flat and edge testing direction, the LBL-Z MOE values were greater than LBL-S. Then, for the MOR value in the flat testing direction, LBL-Z has a greater value compared to LBL-S. While the MOR value on the edge testing direction LBL-S has a greater value compared to LBL-Z. According to JAS for LVL [9], the standard for structural type A MOE and MOR values for both LBL-S and LBL-Z are in the 180E class. Independent-sample T-test results proved that a combination of zephyr and strip raw materials was significantly different for the LBL MOE flat testing direction value and ply bamboo MOE long testing direction.

MOE/MOR values are greater on laminated bamboo boards with parallel fiber direction (LBL) compared to laminated boards with perpendicular fiber direction (ply bamboo). The tensile and flexural strength decrease with increasing lamina angle. On the laminated bamboo board, the aligned fiber direction shows the fibers are arranged in one direction with a continuous polymer [15]. Independent sample T-test results proved that the variation in the direction of parallel and perpendicular fibers was significantly different for the MOE and MOR values of the laminated bamboo board.

According to the Indonesian Timber Construction Regulations NI-5 [16], related to the strength of wood class in terms of the modulus of elasticity, Ply-Z is in the wood strength class II (10,000 MPa),

Ply-S is in the wood strength class IV (6000 MPa), and both LBL-Z and LBL-S are in the wood strength class I (12,500 MPa).

### *3.4. Mechanical properties: shear strength*

The value of the shear strength indicates the quality of adhesion on the laminated bamboo board. According to Figure 5. It is known that the shear strength value of Ply-Z is greater than Ply-S. This is supposedly due to the silica on the surface that glued together in a laminated bamboo board with the strip on the face/back. The waxy layer and the silica layer found on the surface of the bamboo can reduce the shear strength of the laminated bamboo board [17]. According to JAS for plywood [8], the standard for shear strength is 0.7 MPa, so the value of shear strength in ply bamboo meets these standards.

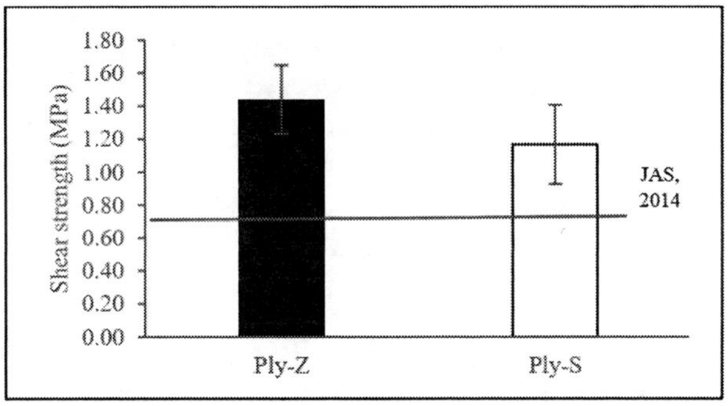


**Figure 5.** Shear strength value of ply bamboo

**Figure 6.** Damage sample of ply bamboo

According to Figure 7, it is known that in the direction of flat and edge testing the shear strength value of LBL-S is greater than LBL-Z. This is supposedly due to more gaps in the zephyr as face/back which can reduce the shear strength of the laminated bamboo board. Then, according to JAS for LVL [9] standard for structural type A, the value of shear strength of both LBL-S and LBL-Z are in the 65V-55H class. The quality of adhesion on bamboo lamination is affected by the cut in or wax layers and surface roughness. The presence of the cut in layer and a rough surface can reduce the value of shear strength or the adhesion of laminated bamboo boards [18]. Independent-sample T-test results proved that the combination of zephyr and strip is significantly different for the value of shear strength LBL in flat and edge testing direction.

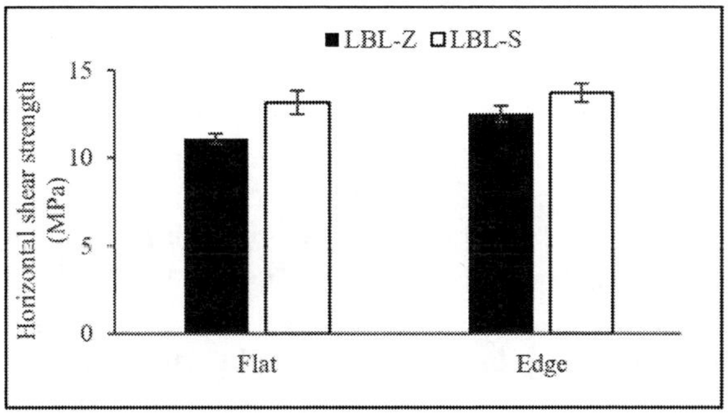


**Figure 7.** Horizontal shear strength value of LBL

a b

**Figure 8.** Damage Sample of (a) Edge direction (b) Flat direction

*3.5. Mechanical properties: screw withdrawal*

The screw withdrawal strength is carried out on the surface of the laminated bamboo board in a flat direction. Screw withdrawal strength value is different based on the arrangement of the fiber directions. Ply bamboo has a greater value than LBL. According to Figure 9, it is known that Ply-Z has the highest value of screw withdrawal strength. Zephyr on the face/back of the laminated bamboo board gives a higher value of screw pulling strength on both ply bamboo and LBL.

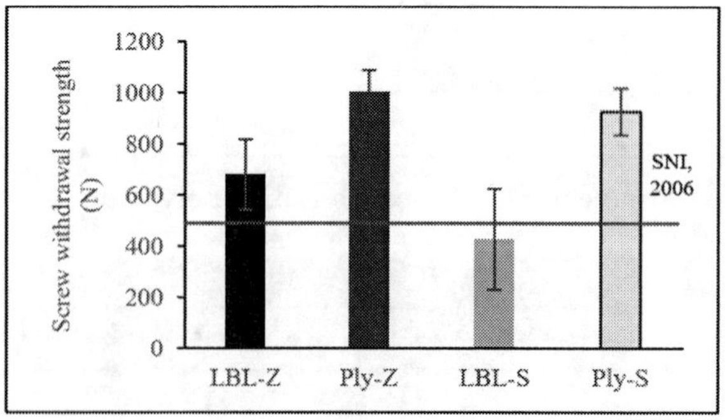


**Figure 9.** Screw withdrawal strength value

**Figure 10.** Splitting damage of LBL-S dample

The perpendicular direction of fiber (ply bamboo) gives a higher value than the parallel direction of fiber (LBL). The screw withdrawal strength value is affected by the moisture content, density, direction of the board fibers, the board layers, and the use of adhesive [19]. According to SNI 03-2105-2006 [11] for particleboard decorative and structural requirements, screw withdrawal strength minimum value is 500.14 N. so that the screw withdrawal strength value in this study meets the standard except for the LBL-S. This probably due to aligned orientation tends not to be able to withstand the load of the screw, so it caused splitting damage as can be seen in Figure 10. The screw withdrawal value on the laminated board can be affected by the orientation of the fiber, the thickness of the layer, and the quantity of adhesive on the laminated board [20]. Independent-sample T-test results proved that the combination of strip and zephyr and variation in the direction of parallel and perpendicular fibers is significantly different for the screw withdrawal strength value.

## 4. Conclusion

Based on this study it can be concluded that:

1) Laminated boards from bamboo strips have better dimensional stability and better shear strength than laminated bamboo boards from bamboo zephyr. Whereas laminated bamboo board from bamboo zephyr has better MOE and MOR values and better screw withdrawal strength.
2) The perpendicular fiber direction of laminated bamboo boards can increase dimensional stability, reduce water absorption, and increase the value of the screw withdrawal strength. The parallel fiber direction of laminated bamboo boards can increase the MOE/MOR values.

## References

[1] Sulastiningsih IM, Indrawan DA, Balfas J, Santoso A, and Iskandar MI 2017 *Jurnal Penelitian Hasil Hutan* **35**(3) 197-209
[2] Sharma B, Gatóo A, Bock M, and Ramage M 2015 *Construction and Building Materials* **81** 66-73
[3] Cheng R, and Zhang Q 2011 *Journal of Adhesion Science and Technology* **25**(14) 1619-28
[4] Nugroho N, and Ando N 2001 *Journal of Wood Science* **47**(3) 237-42
[5] Sulastiningsih IM, Ruhendi S Massijaya MY Darmawan W, and Santoso A 2014 *Jurnal Penelitian Hasil Hutan* **32**(3) 221-34
[6] Sulastiningsih IM, Damayanti R, and Supriadi A 2018 *Journal of Agricultural Science and Technology* **35**(8) 122-30
[7] Kariuki J, Nyomboi T, and Mumenya S 2015 *International Journal of Engineering Sciences and Emerging Technologies* **7**(2) 555-68
[8] [JAS] Japan Agricultural Standard 2014 *Japanese Agricultural Standard for Plywood* (Japan: Japan Plywood Inspection Corporation)
[9] [JAS] Japan Agricultural Standard 2013 *Japanese Agricultural Standard for Laminated Veneer Lumber* (Japan: Japan plywood inspection corporation)
[10] [ASTM] American Society for Testing and Materials 2012 *Standard methods of evaluating properties of wood-base fiber and particles materials ASTM D 103* (West Conshohocken, PA: ASTM International)
[11] [BSN] Badan Standardisasi Nasional 2006 *Particle Board Testing Method SNI 03-2105-2006* (Jakarta: Badan Standardisasi Nasional)
[12] Sulastiningsih IM and Nurwati 2009 *Journal of Tropical Forest Science* 246-51
[13] Geimer RL, Kwon JH, and Bolton J 2007 *Wood and fiber science* **30**(4) 326-38
[14] Aini N, Morisco M, and Anita A 2009 *Civil Engineering Forum Teknik Sipil* **19**(1) 979-86
[15] Verma CS and Chariar VM 2012 *Composites Part B: Engineering* **43**(3) 1063-9
[16] [BSN] Badan Standardisasi Nasional 1961 *Indonesian Timber Construction Regulations SNI-51961* (Jakarta: Badan Standardisasi Nasional)
[17] Nugroho N and Ando N 2000 *Journal of wood science* **46**(1) 68-74
[18] Jimenez JP and Natividad RA 2019 *Philippine Journal of Science* **148**(1) 21-31
[19] Rahman KS, Alam DM, and Islam MN 2012 *Journal of Biological Science* **1**(12) 61-4
[20] Suhaily SS, Gopakumar DA, Aprilia NS, Samsul R, Paridah MT, and Khalil HA 2019 *BioResources* **14**(4) 8376-90

## Acknowledgements

The authors would like to thank the Institut Teknologi Bandung (ITB), Indonesia for providing Research Grant of Research Program ITB 2020. The author would also like to thank to the Forest Product Research and Development Center, Bogor and PT Sumber Graha Sejahtera (SGS), Tangerang.

# A simple reducing water content technique for stingless bee honey (*Heterotrigona itama*) in South Kalimantan

**W Halwany*, S S Hakim, B Rahmanto, R S Wahyuningtyas, Siswadi, S Andriani, F Lestari**

Banjarbaru Environmental and Forestry Research Development Institute, Banjarbaru, Indonesia

*Email: wawanh73@gmail.com

**Abstract**. Water content is one of the parameters of the honey product quality. Unlike Apis honey product, honey from stingless bee (*Heterotrigona itama*) have higher water content. To meet the required standard, reducing water content of stingless bee honey need to be done. Therefore, in this study we aim to find out the effect of using honey water content reducer device to water content and quality of stingless bee honey. We modified simple drying cupboard with heat pad to reduce honey water content. Moreover, nutrient analysis was carried out to know whether the drying process affect its nutrient quality. Six sample of stingless bee honey which was placed in a container (thickness=0.5 cm) and the temperature was set at 40°C. According to the study, to reach the recommended honey water content (21-22%) it takes approximately 3 hours with a value of R=0.848 in addition, drying treatment (40°C) did not affect the nutritional quality of stingless bee honey.

## 1. Introduction

The demand for honey continues to increase along with the increase in population and public awareness of honey consuming [1]. The types of honey in Indonesia have a variety of products. The diversity of honey is influenced by differences in regional origin [2], climate [3], bee species [3,4], nectar source plant species [5], cultivated bees or wild bees, how to harvest and how to handle post-harvest [1,3]. The honey standard in Indonesia was developed into three categories namely forest honey, cultivated honey and stingless bee honey [1].

Stingless bee has great potential to be used in medicine and has a higher drug value than other types of honey [6,7]. One of the advantages of stingless bee honey is that there are antioxidants that can break the chain of free radicals [8]; antimicrobial properties can overcome bacterial contamination; abundant potassium, calcium, sodium, magnesium and manganese content [9], wound healing, anti-diabetes, anti-bacterial and anti-inflammatory [7,10].

However, the weakness of this stingless bee honey is high water content [11-13]. Honey with high water content tends to do fermentation and crystallization [12,14,15]. This will cause a decrease in the quality of honey [16,17]. Decreasing quality in terms of sour taste and undesirable appearance as the presence of osmophilic yeast and fungus in honey is inevitable [16]. Therefore, it is necessary to reduce the water content of honey in order extending the shelf life. The purpose of this research is to find out the effect of using modified honey water content reducer device to water content and quality of honey.

## 2. Materials and Methods

### *2.1. Materials*

Sampling was carried out in July 2019 (during the dry season). The honey sample that was tested came from honey bee farmer which located in Layuh Village, Batu Benawa Subdistrict, Hulu Sungai Tengah District (115°28'37.7" East; 2°36'30.1" South) Barabai, South Kalimantan with a height of 27 meters above sea level. The bee species on site is *Heterotrigona itama*. The temperature at the time of honey sampling ranged between 27.8 °C-28.5 °C. Harvested honey was put in a clean, sealed plastic bottle and then stored in the refrigerator at 4°C.

The activity of reducing water content was carrried out done less than a month after harvesting kelulut honey. This experiment was carried out at the Banjarbaru Environment and Forestry Research and Development Institute Laboratory.

### *2.2. Methods*

2.2.1. Designing a simple honey water content reducer device

The design of the simple honey water content reducer device (SHWCRD) was carried out by designing and making water content reducing equipment that was easy, affordable and applicable but did not damage the quality of honey. The honey water content reduction was a modified tool that has been developed by the community of South Kalimantan kelulut honey farmers. Materials and equipment needed were closed glass cabinets (used showcases) with size of 50 x 50 x 100 cm, mini dehumidifiers of 0.5 L of capacity, mini/exhaust fans, heat pads, digital thermo hygrometer, and reflectometer, a gauge for measuring moisture content (Figure 1). Exhaust fan mounted on the topside inside of the glass cabinet. The heat pad was mounted on the left and right side of the inside of the glass cabinet. Dehumidifier was placed at the bottom inside the cupboard. Honey samples were placed on cabinet shelves.

The working principle of the device made was evaporation technique to reduce water content by using dehumidifier. A closed glass cabinet used in this device was functioning to keep the honey from contamination with outside air. Heat pad was added to this tool to reduce air humidity inside the air-tight glass cabinet that resulted a faster water content reduction. Heat pad was set on temperature below 40°C to minimize honey quality decrease. Temperature and humidity inside the glass cabinet can be read from thermo hygrometer that installed in the device. Reduced water content process was focused on contact area between honeys with drying air inside air-tight glass cabinet. Mini fan was set to ensure drying air circulated evenly inside the glass cabinet, so each honey shelf exposed to drying air flow. The specification of mini fan is AC200V/240V 50/60Hz 0.14A21W 2P.

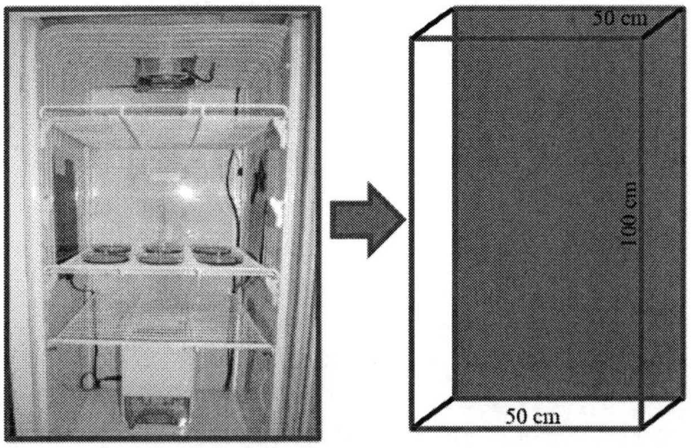


**Figure 1.** Water content reducer device of stingless bee honey

IOP Conf. Series: Materials Science and Engineering **935** (2020) 012011 doi:10.1088/1757-899X/935/1/012011

2.2.2. Test for reducing water content of honey

The honey sample was put into a container /petri dish with a thickness of honey liquid in the container was 0.5 cm. In this study, the sample used were 6 samples. The temperature set at 40°C.

Before honey was put into the glass cabinet, the honey's initial moisture content was measured. Observations were made every one hour of observation. The experiments were carried out until honey levels were as expected (21-22%).

2.2.3. Stingless bee honey quality

Laboratory analysis was performed to find out whether the stingless bee honey quality change before and after the treatment using simple honey water content reducer device. The honey quality parameter analysed were diatase enzyme, reducing sugar (glucose), acidity, and metal pollutant Lead (Pb).

*2.3. Data analysis*

Data were analysed using simple regression analysis with OLS (Ordinary Least Square) method. The analysis uses the Microsoft Excel 2010 program.

## 3. Results and Discussion

*3.1. Water Content*

The result of reducing the honey water content with a simple honey water content reducer device (SHWCRD) showed that the water content decreases with the length of drying duration on temperature of 40°C. Decreasing the water content required less than 3 hours (Figure 2). The results of the regression analysis showed that the water content of honey was closely related to the duration of drying.

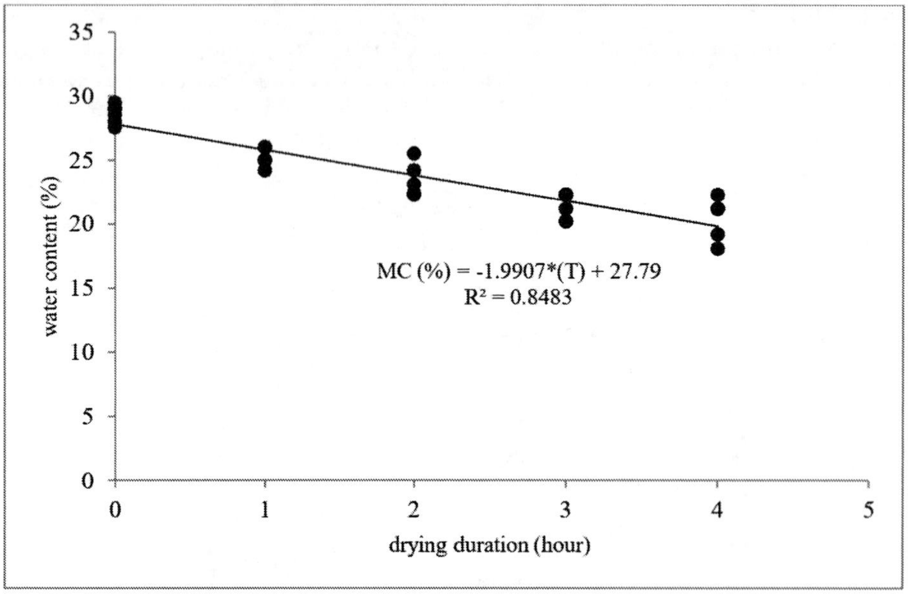


**Figure 2.** Decreased water content of stingless bee honey on some drying duration

Honey water content reducer device which used by some stingless bee honey farmers in South Kalimantan usually made of glass cabinet (used showcase), dehumidifier and fan. Water content of stingless bee honey of *Heterotrigona itama* in South Kalimantan ranged from 25% to 29%. Likewise, water content of stingless bee honey of *Heterotrigona itama* in Malaysia as 26.53% to 28.87% [18]. Honey with water content of more than 22% tends to be easily fermented and decrease honey quality [16,19].

Modification a simple honey water content reducer device commonly used by honey farmers by adding heat pad that attached inside of the device. Heat pad was functioning to increase temperature inside the device. Temperature was set on 40°C as stated by [14] that dehydrating stingless bee honey on 40°C and 55°C did not damage the nutrient content. Based on test results, it needed about 3 hours of treatment to get recommended honey water content (21-22%). Decreasing water content of stingless bee honey was expected to maintan its quality.

### *3.2 Honey quality*

The honey quality analyses results after shwcrd treatment on temperature of 40°C showed decreasing on the four of honey quality parameters measured (Table 1). In diastase enzyme, acidity and Lead (Pb) metal pollutant was still fit with the honey quality standard requirements based on SNI 8664: 2018. In reducing sugar parameter, its value was below the requirement standard whether before or after the treatment.

**Table 1**. Physicochemical test results of stingless bee honey

| parameter | IHC Standard | SNI 8664 : 2018 | Before treatment | After treatment |
|---|---|---|---|---|
| Diastase Activity (DN) | Min 8 | Min 1 | 3.411 | 3.141 |
| Sum of Fructose and Glucose (g/100 g) | Min 60 | Min 55 (Glucose) | 22.48 | 11.44 |
| Free Acidity (meq/100 g) | Max 50 | Max 200 | 203.93 | 107.27 |
| Lead (Pb) metal pollutant (mg/kg) | - | Max 1 | 0.897 | $< 0.001$ |

Heating on honey was done to reduce its water content. Based on some references, water content can accelerate phytochemical changes of honey. In general, heating process on temperature of 40°C does not degrade the honey quality. Reducing water content in stingless bee honey is needed to maintain the quality and extend shelf life so that its storage and marketing will be easier [15].

Diastase enzyme content is one of quality and originality validity parameter of honey [1]. Laboratory analyses showed that diastase enzyme DN value declined after the heating (reducing water content) but still fit the quality requirement standard.

Reducing sugar of stingless bee honey of *Heterotrigona itama* sample value was below the honey quality requirement standard. Low reducing sugar content in kelulut honey happened because the honey harvested when it was not matured enough (honey cell was not fully closed) so that inversion by amylase enzyme that break complex sugar to simple sugar (reduction) on honey was not complete [20].

Acidity of stingless bee honey was decreasing along with reducing honey water content. It was confirmed with honey reviewed by [4] that showed stingless bee honey that had a high water content also had a high acidity.

Lead (Pb) metal pollutant in stingless bee honey before and after heating was still meets the honey quality requirement standard. In addition, heating stingless bee honey can reduce Lead content in honey.

## 4. Conclusion

Using modified simple honey water content reducer device (SHWCRD) on temperature of 40°C in 3 hours can reduce water content of stingless bee honey (*Heterotrigona itama*) up to 6% of initial water content. Honey with lower water content can be stored longer that can help the farmer to distribute their honey product.

## References

[1] [BSN] Badan Standardisasi Nasional 2018 *Madu SNI 8664:2018 ICS 65.020.99* (Jakarta: Badan Standardisasi Nasional)

[2] Maringgal B, Hashim N, Tawakkal ISMA, Mohamed MTM, and Shukor NIA 2019 Phytochemical compositions and antioxidant activities of malaysian stingless bee honey *Pertanika J. Sci. Technol.* **27**(S1) 15-28

[3] Sant'ana RDS, de Carvalho CAL, OdaSouza M, Souza BDA, and Dias FDS 2020 Characterization of honey of stingless bees from the Brazilian semi-arid region *Food Chemistry* 127041

[4] Nordin A, Sainik NQAV, Chowdhury SR, Saim AB, and Idrus RBH 2018 Physicochemical properties of stingless bee honey from around the globe: A comprehensive review *Journal of Food Composition and Analysis* **73** 91-102

[5] Shamsudin NH, Wong CF, Rahman RARNZ, and Ali MSM 2017 Tight repression of elastase strain K overexpression by Pt7 (A1/O4/O3) shuttle expression system *Galeri Warisan Sains* **1**(1) 20-2

[6] Yaacob M, Rajab NF, Shahar S, and Sharif R 2018 Stingless bee honey and its potential value: a systematic review *Food Research* **2**(2) 124-33

[7] Issaro N, Weerakul T, Machana S, Ornnim P, Phanudulkitti C, Srijan T, and Pattarapanich C. 2013 Stingless bee honey II: Qualitative and quantitative studies on honey produced by three stingless bee species collected from a mangosteen garden in Chantaburi province Thailand. *Thai J. Pharm. Sci.* **38** 16-18

[8] Abd Jalil MA, Kasmuri AR, and Hadi H 2017 Stingless bee honey the natural wound healer: A review *Skin Pharmacology and Physiology* **30**(2) 66-75

[9] Biluca FC, Della Betta F, de Oliveira GP, Pereira LM, Gonzaga LV, Costa ACO, and Fett R 2014 5-HMF and carbohydrates content in stingless bee honey by CE before and after thermal treatment *Food Chemistry* **159** 244-49

[10] Ali H, Abu Bakar MF, Majid M, Muhammad N, and Lim SY 2020 In vitro anti-diabetic activity of stingless bee honey from different botanical origins *Food Research* **4**(5) 1421-26

[11] Chuttong B, Chanbang Y, Sringarm K, and Burgett M 2016 Physicochemical profiles of stingless bee (Apidae: Meliponini) honey from South east Asia (Thailand) *Food Chemistry* **192** 149-55

[12] Ramli AS, Basrawi F, Idris DMND, bin Yusof MH, Ibrahim TK, Mustafa Z, and Sulaiman SA 2017 A new dewatering technique for stingless bees honey *MATEC Web of Conferences* **131** 03014

[13] Ávila S, Hornung PS, Teixeira GL, Beux MR, Lazzarotto M, and Ribani RH 2018 A chemometric approach for moisture control in stingless bee honey using near infrared spectroscopy *Journal of Near Infrared Spectroscopy* **26**(6) 379-88

[14] Yap SK, Chin NL, Yusof YA, and Chong KY 2019 Quality characteristics of dehydrated raw Kelulut honey *Int. J. Food Prop.* **22**(1) 556-71

[15] Contrera FAL, Menezes, and Venturieri GC 2011 New horizons on stingless beekeeping (Apidae, Meliponini) *Rev. Bras. Zootec.* **40** 48-51

[16] Chong KY, Chin NL, and Yusof YA 2017 Thermosonication and optimization of stingless bee honey processing *Food Sci. Technol. Int.* **23**(7) 608-22

[17] Fuad AMA, Anwar NZR, Zakaria AJ, Shahidan N, and Zakaria Z 2017 Physicochemical Characteristics Of Malaysian Honeys Influenced By Storage Time And Temperature *J. Fundam. Appl. Sci.* **9**(1) 841-51

[18] Abu Bakar MF, Sanusi SB, Abu Bakar FI, Cong OJ, and Mian Z 2017 Physicochemical and antioxidant potential of raw unprocessed honey from malaysian stingless bees *Pakistan J. Nutr.* **16**(11) 888-94

[19] bin M. Yunus MA, Abid A, binti Sahlan S, bin M. Khairi MT, and binti B. Hisham AA 2020 Automated honey dehydrator based on optimized drying air heat *Conf. Pap.* pp. 145-150

[20] Küçük M, Kolayli S, Karaoğlu Ş, Ulusoy E, Baltaci C, and Candan F 2007 Biological activities and chemical composition of three honeys of different types from Anatolia *Food Chem.* **100**(2) 526-34

**Acknowledgment**
Thank you for the Head of Forestry Service of South Kalimantan Province for funding this research. We also thank to the Head of Hulu Sungai Forest Management Unit that had helped and facilitated the stingless bee honey research. A great thanks to stingless bee honey farmers, Mr. Fahrudin (CV. Hutan Mandiri Trigona), Mr. Yusuf, Mr. Warhamni and Ibnu kasir for the information about stingless bee cultivation; Mr. Syaifuddin for the information about stingless bee cultivation and information about honey water content reducer device; Mr. Abdul Qirom for data analyses advice; Hendra Ambo Basiang, Arief Susianto, Sukma Alamsyah and Dian Cahyo Buwono for a great teamwork.

IOP Conf. Series: Materials Science and Engineering **935** (2020) 012012 doi:10.1088/1757-899X/935/1/012012

# Field evaluation of subterranean termites palatability on treated pine wood in Alam Sinarsari Residence, West Java

**Arinana[1*], F E Hutapea[1], D Nandika[1], and N F Haneda[2]**

[1]Forest Product Department, Faculty of Forestry, IPB University, Bogor, 16680, Indonesia.

[2]Department of Silviculture, Faculty of Forestry, IPB University, Bogor, 16680, Indonesia

*Email: arinana@apps.ipb.ac.id

**Abstract**. Pine wood has been commonly utilized as bait wood in controlling termites by using a baiting system. This research aims to evaluate the palatability of subterranean termites against pine wood by boiled treatment, steamed, hot vapor pressured, and oven in Alam Sinarsari Residence, Bogor, West Java, Indonesia and analyze the distribution of termites and identity the termite species that attack bait wood. The size of bait wood refers to ASTM D 1758-06. There were four treatments of bait wood; steam treatment (100°C), boiling treatment (100°C), hot vapor pressure treatment (1 bar and 105°C) for five hours each and oven treatment (103 ± 2°C) for 48 hours. As a comparison, bait wood without treatment was prepared as a control. The result showed that bait wood using boiling treatment had more active subterranean termites than others. The identification result showed that there were 4 species of subterranean species (*Macrotermes* sp., *Odontotermes* sp., *Microtermes* sp., and *Schedorhinotermes* sp). The highest frequency of termites attack was in bait wood using oven treatment, and the highest damage intensity was in hot vapor pressure treatment. Based on the rating system the best treatment for bait wood treatment is hot vapor pressure treatment and boiling.

## 1. Introduction

Termites are wood destroying organisms such as furniture, frames, and building construction that cannot be ignored. The loss caused by termite attack is tremendous and even the amount continues to increase every year. The amount of economic loss in Indonesia caused by termite attacks according to Nandika *et al.* [1] reached 1.67 trillion in 1995, 2.79 trillion in 2000, 5.17 trillion in 2010, and 8.68 trillion in 2015. This significant economic loss cannot be overlooked; therefore, an efficient termite's management is necessary. One way to control termites is to provide bait in the form of material preferred by termites, one of which is pine wood.

Pine wood is the most preferred wood by termite. According to Subekti [2] pine wood is one of the wood favored by termites, and it is also supported by Arinana *et al.* [3] which showed that termites preference towards pine wood is higher than other species that are investigated in the study, namely acacia wood (*Acacia mangium*), rubber wood (*Hevea brasiliensis*), and falcata wood (*Falcataria molucana*). Pine is also an attractant for termites therefore it can be used as bait for termites. Attractant function can be obtained from various types of plants that have a distinctive aroma. Referring to the research of [4], pine wood is one of the food sources for termite thus, it is suitable to be used as bait for termite control.

Pine wood has been used commercially in Sentricon Colony Elimination System as termite bait [5]. Baiting system is an environmentally friendly method to termite control [6]. The feeding method is a method that applies slow action poisonous substances to bait wood and utilizes the termites trophallaxis. Trophallaxis is termite's behavior of feeding each other so that termites that have eaten the bait wood will spread the poisons contained in the bait wood into their colonies [1]. However, the utilization of pine wood as bait wood for subterranean termites apparently does not provide sufficient attraction for termites, therefore further efforts are needed to elevate termite preferability to pine wood.

Arinana *et al.* [7] has conducted research to increase *C. curvignathus* preference towards pine wood in laboratory scale by several treatments including steaming, boiling, and presto for 5 hours each. The results showed that the treatment has the potential to increase termite preference for pine wood with the highest percentage of weight loss increased by 20.32% in the presto treatment. However, further field scale is necessary to evaluate the effectiveness of this treatment. The suitable place to install the baiting stations is in the residence area. Therefore the research is conducted in Alam Sinarsari Residence, Bogor, West Java, Indonesia.

This study is expected to provide information about the termite species in Alam Sinarsari Residence, Bogor, West Java, and data regarding the preeminent treatment that can be established to the bait wood for subterranean termites to solidify the diversity of subterranean termites in the location.

## 2. Materials and Methods

### 2.1. Time and location

This research was conducted from January to July 2019 in Alam Sinarsari Residance, Bogor, West Java, Indonesia. The position of the northern part wood sample was at coordinates of 6°34'45.90"S 106°43'58.26"E, western part at coordinates 6°34'53.87"S 106°43'49.11"E, southern part at coordinates of 6°35'10.02"S 106°43'51.15"E, and eastern part at coordinates of 6°35'3.17"S 106°43'57.40"E. Location and installed wood samples as shown in Figure 1. Sample preparation and identification of termites was conducted at the Laboratory of Wood Quality Improvement Technology, Department of Forest Products, Faculty of Forestry, IPB University. Soil analysis was carried out at the Indonesian Center for Biodiversity and Biotechnology (ICBB), Bogor.

**Figure 1.** Location and installed wood samples in Alam Sinarsari Residance, Bogor, West Java

### 2.2. Materials and equipment

The material used in this study was pine (*Pinus merkusii*) logs from Ciampea, Bogor with a diameter of ±40 cm. The log was then sawed into boards and turned into wood samples (stakes) in accordance to the

American Standard Testing and Materials (ASTM) D 1758-06 [8], with the dimension of 2 x 2 x 46 cm. The wood part that was utilized in this research was sapwood. Another material that was employed in this study is water for the boiling and steaming process, alcohol 70%, and oil paint (red color). The equipment used in this research includes Global Positioning System (GPS), bottles for termite collection, brushes, a digital camera, sandpaper, a stereo microscope, using LCD Digital Temperature Humidity Data Recording Logger Meter, transparent plastic, and a soil driller.

### *2.3. Methods*

#### 2.3.1. Sample test preparation

The wood samples (stakes) were sanded to remove the residual fibers after the cutting process. Treatment for the wood sample consisted of boiling, hot vapor pressure treatment using an autoclave, and oven. The arrangement for boiled and steamed treatment were 5 hours at 100°C. Hot vapor pressure treatment was carried out for 5 hours with a pressure of 1 bar and a temperature of 105°C. Meanwhile, the oven treatment was carried out for 2 x 24 hours with a temperature of 100 ± 2°C. As a control, untreated pine wood was prepared. After treatment, the wood sample was air dried until the moisture content was < 20%. The dry wood samples were then sanded to remove resin that sticks to the surface. After that, the tip of each wood sample was marked by paint and markers so each wood sample can be easily distinguished. The marking of the wood sample was done based on the following manners: K (steamed treatment), R (boiled treatment), A (hot vapor pressure treatment). O (oven treatment), and C (control). Furthermore, the wood sample was conditioned to air dry and weighed to find out the initial weight ($W_1$) before being exposed at the research site.

#### 2.3.2. Field test and wood sample baiting

The wood sample was installed in Alam Sinarsari Residence, Bogor, West Java. The stakes were installed around the house and outdoor structures on the property in areas that were not covered by concrete or other artificial barriers. Some stakes were also installed in outdoor community areas such as playgrounds, gardens, *etc*. Stakes amounted to as many as 285 pieces, spread around the housing areas in Alam Sinarsari Residence. Stakes were installed vertically into the soil so that half was below the soil surface [8], as shown in Figure 2. The coordinates of all stakes were determined by using GPS. Stakes were being installed for three months and observation was performed every month. After three months, the stakes were removed, cleaned, conditioned until air dry weighed to find out the initial weight ($W_2$), then the attack frequency and the damage intensity of stakes was calculated.

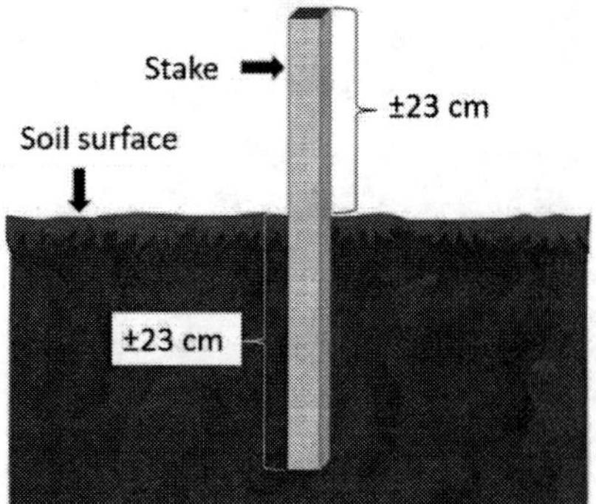


**Figure 2**. Installation of stakes

2.3.3. Soil characteristic analysis

Soil sampling was conducted at five points: four points in the cardinal directions (Eastern, Western, Southern, Northern part) and one point in the middle of the field test location. The soil sample was taken at least 1 kg at each point with 0-25 cm depth and then put in transparent plastic. All soil samples were analyzed at the ICBB Laboratory, Bogor (an ISO/IEC17025 certified laboratory) to measure the soil pH, C-organic content, and texture.

2.3.4. Temperature and humidity measurement

The temperature and humidity measurement were performed using LCD Digital Temperature Humidity Data Recording Logger Meter. The device was placed in one of the houses of the resident. Measurement of temperature and humidity was conducted for three months with data collection for every five minutes at the observed area.

2.3.5. Identification of termites species

Termites on the attacked samples were taken and put into specimen bottles that already contained alcohol 70%. The species of the soldier termites were than identified based on literature and key identification [1,10]. The collected termites were photographed and observed by microscope under 10x magnification for the entire termite body shape.

2.3.6. The attack frequency of the subterranean termites on the stakes

Attack frequency is the ratio between the number of attacked samples and the total wood samples in the Alam Sinarsari Residence, expressed in percent. According to Cookson and Trajstman [11], the attack frequency data are classified into six classes that then modified as shown in Table 1.

**Table 1**. The classification of attack frequency of subterranean termites [11]

| No | Frequency (%) | Category |
|---|---|---|
| 1 | 0 | None (Uninfested) |
| 2 | 01-10 | Very low |
| 3 | 11-20 | Low |
| 4 | 12-30 | Moderate |
| 5 | 31-40 | High |
| 6 | > 40 | Very high |

2.3.7. The damage intensity of the stakes

Damage intensity of the stakes consists of three data, namely weight loss, wood damage percentage of stakes, and wood damage grading system. Weight loss is the ratio between the weight of stakes (before and after baited in Alam Sinarsari Residence) and the initial weight of the stake, expressed in percent. The weight loss value is calculated using this following equation:

$$WL = \frac{(W_1 - W_2)}{W_1} x100 \tag{1}$$

where:
WL: Weight Loss (%)
$W_1$: Weight of air-dried stake before baiting(g)
$W_2$: Weight of air-dried stake after baiting(g)

The wood damage percentage of stakes is the ratio between the length of cross-sectional damage on the attacked sample with the initial width or thickness expressed in percent. The damage percentage is classified into seven classes: uninfested to more than 70% of cross-sectional damage [8].

The wood damage grading system are classified based on the percent loss of cross-section to attacking subterranean termite [8]. The grading system of stakes attacked by subterranean termite can be seen in Table 2.

**Table 2.** The grading system of stakes attacked by subterranean termite [8]

| Grade no. | Description of condition |
|---|---|
| 10 | No attack, 1to 2 small nibbles permitted |
| 9 | Nibbles to 3% cross section |
| 8 | Penetration 3 to 10% of cross-section |
| 7 | Penetration 10 to 30% cross-section |
| 6 | Penetration 30 to 50% cross-section |
| 4 | Penetration 50 to 75% cross-section |
| 0 | Failure |

### *2.4. Data analysis*
The qualitative data was presented in pictures or graphics and then was explained descriptively, while quantitative data were delivered using means values that were regenerated using Microsoft Excel 2016. The preeminent treatment was determined based on the highest total score (1-5) of the infestation frequency of subterranean termites and the damage intensity of the stakes.

## 3. Results and Discussion

### *3.1. Soil characteristic analysis*
The results showed that the C-organic content in Alam Sinarsari Residence ranged between 0.69-1.78%. Mahaney *et al.* [12] states that subterranean termites have a role in increasing the content of organic matter in the soil, which is also supported by the research of Lee and Wood [13] that the land in which there is termite activity has a high content of organic matter. The C-organic content at subterranean termites *Odontotermes* sp. tunnel was 0.576%, and the C-organic content of the surrounding soil was 0.536% [13]. Meanwhile, the research result of Arinana *et al.* [14] showed that the average C-organic content in residential areas in DKI Jakarta Province was 2.02%. The C-organic content in the study site is lower from the results of Arinana *et al.* [14] located in DKI Jakarta Province.

The results showed that the soil pH at the study site was 5.49-6.85. Soil pH values are included in the slightly acidic group. Kaschuk *et al.* [15] states that aside from living in the soil that has a neutral pH, subterranean termite are also able to live in the soil has a slightly acidic pH, which is between 4.23-5.46, so the soil pH is at the research location is still suitable to support termite life.

Meanwhile, the soil texture at the research location is classified as a clay type. In general, termites dislike sandy soils and chooses clay-rich soils [13]. Based on these results, the Alam Sinarsari Residence environment is a habitat that is suitable for supporting termite life. Soil characteristics in Alam Sinarsari Residance can be seen in Table 3.

**Table 3.** Soil characteristic in five points in Alam Sinarsari Residence, Bogor, West Java

| Soil characteristic | Measurement points | | | | |
|---|---|---|---|---|---|
| | Center | South | North | West | East |
| C-Organic (%) | 1.78 | 1.43 | 1.34 | 0.69 | 1.70 |
| pH | 6.40 | 6.29 | 6.28 | 6.85 | 5.49 |
| Sand (%) | 14 | 19 | 11 | 14 | 11 |
| Silt (%) | 15 | 20 | 14 | 9 | 14 |
| Clay (%) | 71 | 60 | 75 | 77 | 74 |
| Soil Texture | Clay | Clay | Clay | Clay | Clay |

### *3.2. Temperature and humidity*
The results showed that the temperature and humidity in Alam Sinarsari Residence are very suitable for the development of termites. The highest temperature recorded at the research location was 35.8°C and the lowest temperature was 23.2°C. Meanwhile the highest humidity is 94.6% and the lowest humidity is 56%. Some data of temperature and humidity fluctuations in Alam Sinarsari Residence can be seen in Figure 3.

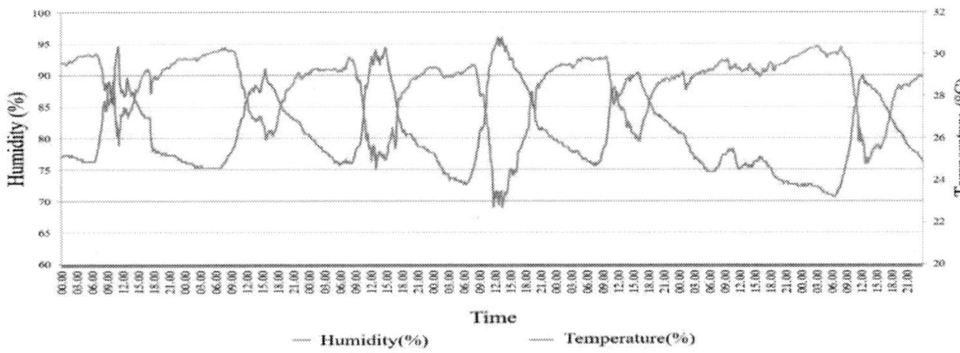


**Figure 3**. Data of temperature and humidity fluctuations in Alam Sinarsari Residance, Bogor

The optimal temperature for termites is between 28-32°C, and the optimal humidity is 75-90% [1,16]. Meanwhile, the research by Arinana *et al.* [16] showed that the average temperature in the tunnel of the *Coptotermes curvignathus* range is between 29.4-33.8°C. Environmental factors that can influence the development of subterranean termite population are temperature, humidity, rainfall, and food availability. These factors interacting as well as influencing each other. Temperature and humidity are strong factors that influencing termite's activity [17]. Areas with temperature and humidity that are relatively high are very suitable for the growth and development of wood destroying organisms like termites [18].

### *3.3. Termite identification*
The results showed that not all damaged stakes were found an active subterranean termite in the stakes. The number of stakes with boiled treatment is the highest amount of active subterranean termite found in wood bait that is 19 units followed by stakes with hot vapor pressure treatment and oven with the same amount, each with 14 units. Meanwhile, the number of control stakes was found in 11 units and 9 units of steamed treatment. Based on the identification of species of subterranean termite that attacked the stakes in Alam Sinarsari Residence there were four species namely *Macrotermes* sp., *Odontotermes* sp., *Microtermes* sp., and *Schedorhinotermes* sp.

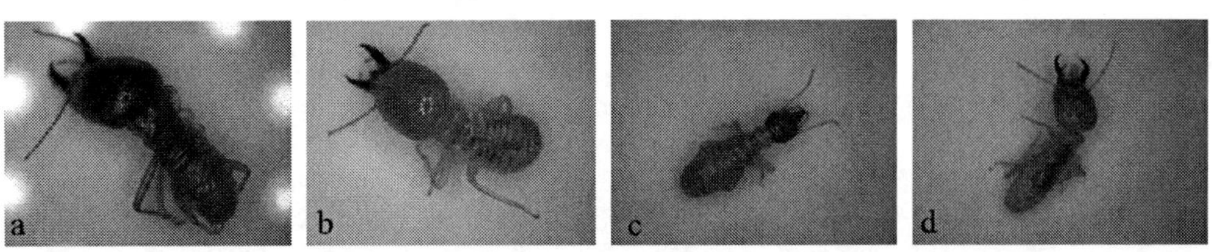


**Figure 4**. Subterranean termites found on the infested stakes in 10X magnification: (a) *Macrotermes* sp., (b) *Odontotermes* sp., (c) *Microtermes* sp., dan (d) *Schedorhinotermes* sp.

The body shape morphological of the subterranean termites are presented in Figure 4. Species of *Macrotermes* sp., *Odontotermes* sp., and *Microtermes* sp. members of Termitidae family. Species belonging to the Termitidae family are commonly found in Asia. Termites *Schedorhinotermes* sp. Members of Rhinotermitidae family commonly found in the tropics region.

All species (four species) of active subterranean termite were found to attack stakes with boiled treatment. This suggests that wood bait with boiled treatment is preferred by all subterranean termite at the research location. Meanwhile, stakes with oven treatment found three species of active subterranean termites. Not founded *Odontotermes* sp. attack the stakes with oven treatment. Hot vapor pressure treatment, steamed treatment, and control each found two active subterranean termite species namely *Microtermes sp.* and *Schedorhinotermes sp.* Both species of termite are found in all stakes. The number of stakes found by active subterranean termite and species of subterranean termite attacking each treatment can be seen in Table 4.

**Table 4**. Amount of stakes found that found active termites and termites species attacked stakes species in each treatment

| Treatments | Number of stakeswith active termites | Subterranean termites species |
|---|---|---|
| Control | 11 | *Schedorhinotermes* sp. (5)<br>*Microtermes* sp. (6) |
| Steamed | 9 | *Schedorhinotermes* sp. (4)<br>*Microtermes* sp. (5) |
| Oven | 14 | *Macrotermes* sp. (1)<br>*Schedorhinotermes* sp. (3)<br>*Microtermes* sp. (10) |
| Boiled | 19 | *Macrotermes* sp. (1)<br>*Odontotermes* sp. (1)<br>*Schedorhinotermes* sp. (4)<br>*Microtermes* sp. (13) |
| Hot vapor pressure | 14 | *Schedorhinotermes* sp. (5)<br>*Microtermes* sp. (9) |

Note: The number in parentheses is the amount of stakes that found active termites

The species of subterranean termite were found unevenly distributed at the research location. The northern part is dominated by *Microtermes sp.* and the southern part is dominated by *Schedorhinotermes* sp. *Macrotermes* sp. were found attacking two stakes while *Odontotermes* sp. found attacking one stake. In general, the most dominant termite species attacking stakes is *Microtermes* sp. Subsequently followed by *Schedorhinotermes* sp., *Macrotermes* sp., and *Odontotermes* sp. Distribution of subterranean termite species found attacking stakes in the research location can be seen in Figure 5. The research on the diversity of subterranean termite in residential locations has been done in DKI Jakarta Province [14], Taman Darmaga Permai I Ciampea, Bogor [19], and Bumi Bekasi Baru National Housing, Rawalumbu, Bekasi [20]. Arinana *et al.* [14] stated that found four species of subterranean termites were *Coptotermes curvignathus, Microtermes insperatus, Macrotermes gilvus*, and *Schedorhinotermes javanicus*. Meanwhile Arinana *et al.* [19] stated that found subterranean termite *of Macrotermes gilvus, Odontotermes javanicus, Coptotermes curvignathus, Schedorhinotermes javanicus,* and *Schedorhinotermes sarawakensis* at the research location, while Arinana *et al.* [20] found subterranean termite species of *Schedorhinotermes sp.* and *Coptotermes sp.* at the research location. Different from the three results of the research, in the three residence locations species found in the subterranean termite of *Coptotermes sp.* which was not found in this s research. However, it has similarities that in each research location *Schedorhinotermes* sp.

**Figure 5**. Distribution termite in Alam Sinarsari Residence, Bogor

### *3.4. The attack frequency of the subterranean termites on stakes*

The number of stakes installed at the study site was 285 units. After three months, the stakes were taken out. A total of 84 samples were missing and the number of stakes found in which frequency of attacks and damage intensity are observable and assessable was 201 samples. From the total of 201 stakes, 63 stakes were not infested by subterranean termites so that the frequency of attacks on the overall stakes was 68.5% and stakes damaged by subterranean termite attack but have no active subterranean termite inside of it are as many as 71 samples. According to the classification of Cookson and Trajstman [11] in Table 1, the frequency of termite attacks at the research site can be classified as very high. Compared with the research result of [14] in DKI Jakarta Province, the frequency of termite attacks on bait wood at this research site is greater. In DKI Jakarta Province the frequency of attacks is only 20.7%.

Meanwhile, the attack frequency of termites on stakes for each treatment ranged from 11.4% to 15.4%. The attack frequency of gthe termites on the stakes by oven treatment has the highest frequency of 15.4%, followed by boiled treatment with 14.4% which is then followed by control with 13.9%. The attack frequency that was found in stakes with hot vapor pressure and steam treatment is below the control sample, consecutively 13.4% and 11.4%. This shows that oven and steam treatment can increase the palatability of termites as it has a higher attack frequency than the control. Complete data on the attack frequency of subterranean termites on stakes are presented in Table 5.

**Table 5**. Attack frequency of subterranean termites on stakes

| Treatments | Amount of initial stakes | Amount of missing stakes | Amount of uninfested stakes | Amount of infested stakes | Attack frequency (%) |
|---|---|---|---|---|---|
| Control | 57 | 16 | 13 | 28 | 13.9 |
| Steamed | 57 | 19 | 15 | 23 | 11.4 |
| Oven | 57 | 16 | 10 | 31 | 15.4 |
| Boiled | 57 | 14 | 14 | 29 | 14.4 |
| Hot vapor pressure | 57 | 19 | 11 | 27 | 13.4 |
| Total | 285 | 84 | 63 | 138 | 68.5 |

IOP Conf. Series: Materials Science and Engineering **935** (2020) 012012 doi:10.1088/1757-899X/935/1/012012

### *3.5. The damage intensity of the stakes*

The results showed that the percentage of weight loss of stakes due to termite attack ranged between 18.58-24.64%. The highest weight loss was found in the control sample (24.64%), followed by hot vapor pressure, oven, boiled, and steam treatments, consecutively 22.15%, 20.10%, 19.85%, and 18.58%. The weight loss of all treatments was lower than the control sample. The complete data can be seen in Figure 6. This shows that all the treatments are not able to increase the palatability of subterranean termites towards stakes. The higher the value of wood weight loss, the higher the attack occurred in the bait wood. Subterranean termites have extraordinary destructive speed, depending on the size of their colonies [21].

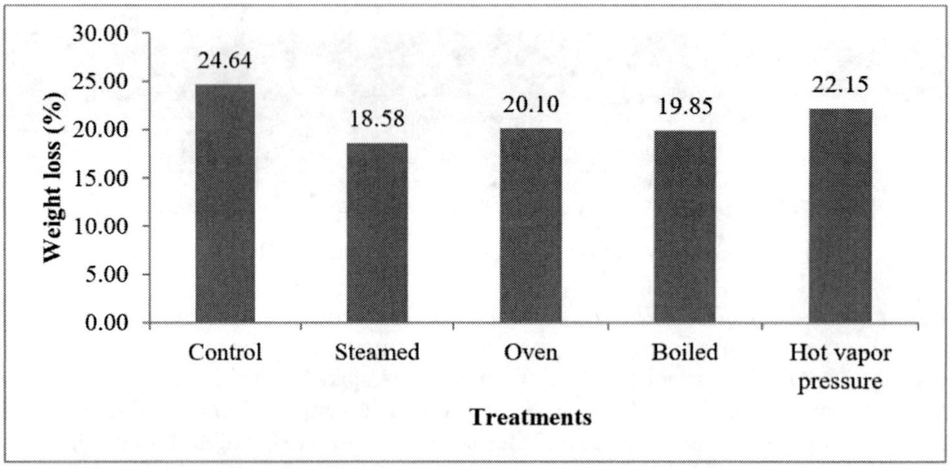


**Figure 6**.Weight loss of stakes with various treatments

The highest of wood damage percentage is the hot vapor pressure treatment about 65.77%, followed by boiled, control, and oven treatments sequentially about 65.67%, 65.03%, and 61.070%. The lowest wood damage percentage is steamed treatment which is 52.98%. The complete data can be seen in Figure 7. The wood damage percentage with hot vapor pressure and boiled treatments have greater values than control treatment. This shows the increment of hot vapor pressure and boiled treatments can increase the palatability of termites. Thus, it proves the subterranean termites prefer both of those stakes. Allegedly, the hot vapor pressure and boiled treatments make the resin production of wood higher and the wood becomes more malleable, so it rises the termite consumption rate. The hot vapor pressure and boiled treatments help wood releases the extractive substances excessively [22,23]. The termite consumption rate is supported by the mechanical characteristic of termite bite which prefer the malleable wood or plants [24].

The wood damage percentage with oven treatment is greater than the steamed treatment. In accordance with the research of Widyorini *et al.* [23], the oven treatment gives a greater response than steamed treatment. The malleable process is also done by boiling or cooking in hot vapor pressure treatment [25]. Based on the obtained results, the hot vapor pressure treatment for 5 hours gives greater result than the steamed, oven, and boiled treatments because of the pressure on the hot vapor pressure treatment will reduce the mechanical characteristic of wood, so the wood becomes more malleable. According to Sukatik [26], applying pressure on wood will reduce the mechanical characteristic of wood.

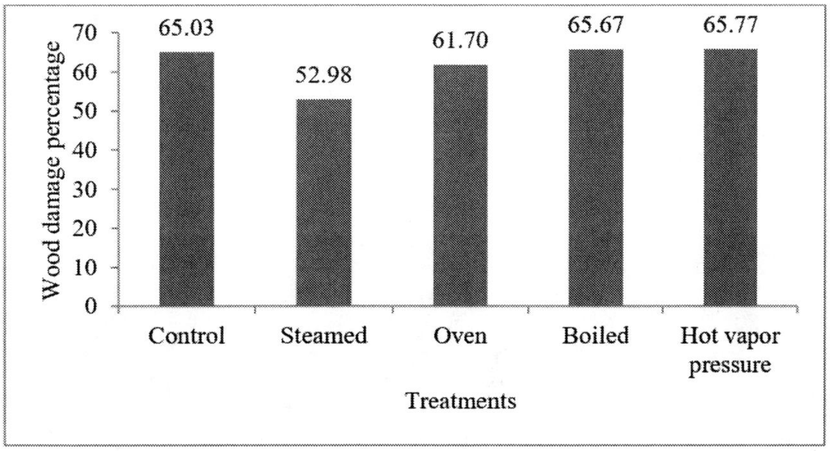


**Figure 7**. Wood damage percentage due to termite attack in Alam Sinarsari Residence, Bogor

The result shows the wood damage percentage of control wood is higher than the treatment woods by steamed and oven, but lower than the treatment woods by hot vapor pressure and boiled. This happens because the field test in a residential environment has a different environmental condition so the termite attack on wood bait is not certainty known. The influence of various environmental factors, such as predator, temperature, humidity, and the availability of litter on the land surface also influence the presence and growth of termites at the location [27]. The sample location is also being an attack factor for the wood bait. If the amount of termite consumption in the research location is already available in large quantities, that will be possibility termites will not attack while the supplies still exist [28].

### *3.6. Stakes grading score*

The results showed that the stakes grading score was classified as a value of 4. The complete grading score can be seen in Table 6. There was no difference in the grade score rate, both in the control and treatment of wood. The results of the grade score rate classification analysis differ from the percentage of the stakes damage. Classification into the grading score that no treatment can increase the termite palatability of the stakes.

**Table 6**. Percentage and grade number of stakes

| Treatments | Wood damage percentage | Grade Number |
|---|---|---|
| Control | 65.03 | 4 |
| Steamed | 52.98 | 4 |
| Oven | 61.70 | 4 |
| Boiled | 65.67 | 4 |
| Hot vapor pressure | 65.77 | 4 |

The form of a stakes grading system with various treatments after being installed for three months at the study site can be seen in Figure 8. Some stakes have lost their mass due to being eaten by subterranean termite. Not only does the termite make a chewed spot transversely, but they take up the entire transverse direction of stakes. Visual observation is known to be able to measure the level of wood damage by using calipers for the depth of attack. The higher the level of attack on stakes, also shows the higher level of termites' preference for wood.

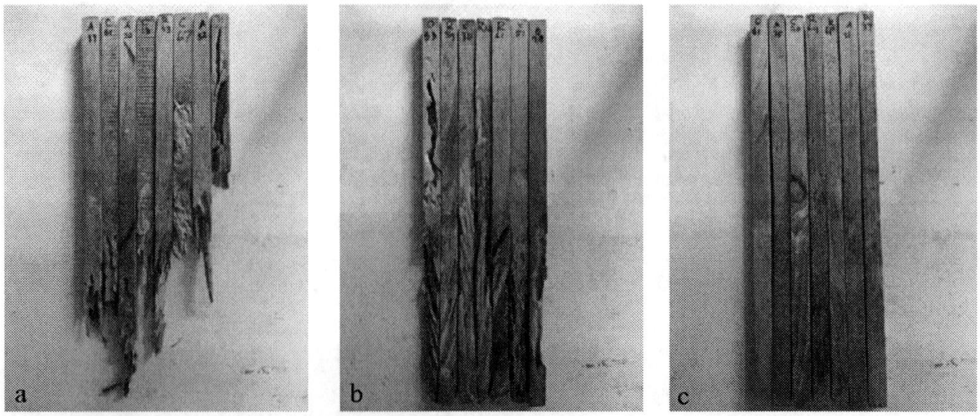

**Figure 8**. Form of various treated stakes damages after baiting for three months period in Alam Sinarsari Residance, Bogor: failure (a), medium damage (b), no attack

*3.7. Determination of the highest quality treatment*

Based on the highest total score of the total number of active termites found in stakes, the frequency of termite attacks, and the intensity of stakes damage (weight loss, percentage of wood bait damage, wood grading score), the best treatment is boiled and hot vapor pressure (each with a score of 16). The recapitulation of the best treatment score can be seen in Table 7. Boiled treatment and hot vapor pressure are thought to reduce the resin and make the wood softer without changing the chemical components in the wood. The chemical structure of wood does not change until the temperature of 150°C, instead the wood begins to experience chemical structure changes starting at a temperature of 180°C [29].

**Table 7**. Recapitulation of the best treatment scores from various treatments

| Parameter | Treatment Score | | | | |
|---|---|---|---|---|---|
| | Control | Steamed | Oven | Boiled | Hot vapor pressure |
| Amount of stakes found that found active termites | 3 | 2 | 4 | 5 | 4 |
| Attack frequency | 3 | 1 | 5 | 4 | 2 |
| Damage intensity: | | | | | |
| Weight loss | 5 | 1 | 3 | 2 | 4 |
| Damage Percentage | 3 | 1 | 2 | 4 | 5 |
| Grading score | 1 | 1 | 1 | 1 | 1 |
| Total skor | 15 | 6 | 15 | 16 | 16 |

Note: The score value is from 1 to 5. The higher the score shows that the treatment is the best treatment

## 4. Conclusions

Boiled treatment at 100°C for 5 hours and treatment with hot vapor pressure treatment (1 bar 105°C for 5 hours) using autoclave against pine wood has the potential to increase the palatability of subterranean termites. Both of these treatments are the best treatments in increasing the possibility of subterranean termite against pine wood. Subterranean termite are found in Alam Sinarsari Residence, Bogor, West Java are *Macrotermes* sp., *Odontotermes* sp., *Microtermes* sp., and *Schedorhinotermes* sp. Termite species are spread unevenly and dominated by *Microtermes* sp. (64.17%).

**References**

[1] Nandika D, Rismayadi Y, and Diba F 2015 *Rayap: Biologi dan Pengendaliannya Edisi ke-2* (Surakarta: Muhammadiyah University Press)

[2] Subekti N 2012 Biodeteriorasi kayu pinus (*Pinus merkusii*) oleh rayap tanah *Macrotermes gilvus* Hagen (Blattodea: Termitidae) *Bioteknologi* **9**(2) 57-65

[3] Arinana, Tsunoda K, Herliyana EN, and Hadi YS 2012 Termite susceptible spesies of wood for inclusion as reference in indonesian standardized laboratory testing *Insects* **3** 396-401

[4] Amran, Ahmad I, Putra RE, and Koeswanto E 2015 Aplikasi campuran serbuk kayu pinus dan fipronil sebagai umpan rayap tanah *Macrotermes gilvus* (Hagen) (Isoptera: Termitidae) di Bandung *Jurnal Entomologi Indonesia* **12**(2) 73-9

[5] Jones DT, Verkerk RHJ, and Eggleton P 2005 *Methods for sampling termites* in Leather S Insects sampling in forest ecosystems (Australia: Blackwell Science Ltd)

[6] Kadarsah A 2005 Studi keragaman rayap tanah dengan teknik pengumpanan pada tumpukan jerami padi dan ampas tebu di Perusahaan Jamur PT Zeta Agro Corporation Jawa Tengah *Jurnal Bioscientiae* **2**(2) 17-22

[7] Arinana, Simamora S, Hanindita FA, Metapara J, and Nandika D 2020 Palatability of subterranean termiter *Coptotermes curvignathus* Holmgren treated pine wood (*Pinus merkusii*) *Pakistan Journal of Biological Sciences* **23**(2) 181-9

[8] [ASTM] American Society for Testing and Material 2006 *Standard Tes Method of Evaluating Wood Preservatives by Field Test with Stakes ASTM D 1758-06* (United States: American Society for Testing and Material)

[9] Arinana A, Aldina R, Nandika D, Rauf A, Harahap IS, Sumertajaya IM, and Bahtiar ET 2016 Termite diversity in urban landscape, South Jakarta, Indonesia *Insects* **7**(2) 20

[10] Ahmad M 1958 *Key to the Indomalayan Termite Biologia Vol 4* (Lahore: Department of Zoology University of the Panjab)

[11] Cookson LJ and Trajstman AC 2002 Termite Survey and Hazard Mapping (CSIRO FFP Technical Report Nomor 137) (Clayton South,Victoria: CSIRO Forestry and Forest Products)

[12] Mahaney WC, Zippin J, Milner MW, Sanmugadas K, Kancock RGV, and Aufreiter S 1999 Chemistry, mineralogy and microbiology of termite mound soil eaten by champanzees of the Mahal mountains, Western Tanzania *J. Tropical Ecology* **15** 565-88

[13] Lee KE and Wood TG 1971 *Termite and soil* (London: London Academic Press)

[14] Arinana, Rauf A, Nandika D, Harahap IS, and Sumertajaya IM 2019 Model prediksi risiko kelas serangan rayap tanah di Provinsi DKI Jakarta berbasis spesies, tanah, dan iklim *Prosiding Seminar Nasional PEI Cabang Bandung* 170-8

[15] Kaschuk G, Santos JCP, Almeida JA, Sinhorati DC, and Berton-Junior JF 2006 Termite activity in relation to natural grassland soil attributes *Sci. Agric.* **63**(6) 583-8

[16] Arinana, Philippines I, Bahtiar ET, Koesmaryono Y, Nandika D, Rauf A, Harahap IS, and Sumertajaya IM 2016 *Coptotermes curvignathus* Holmgren (Isoptera: Rhinotermitidae) capability to maintain the suhue fluctuation inside its nests *J. Entomology* **13**(5-6) 199-202

[17] Subekti N 2010 Karakteristik populasi rayap tanah *Coptotermes spp* (Blattodea: Rhinotermitidae) dan dampak serangannya *Biosaintifika* **2**(2) 110-4

[18] Tafsir A, Wardenaar E, and Wahdina 2015 Uji aktivitas anti rayap ekstrak rimpang lempuyang gajah (*Zingiber zerumbet* Smith) terhadap rayap tanah (*Coptotermes curvignathus* Holmgren). *Jurnal Hutan Lestari* **3**(2) 293

[19] Arinana, Haneda NF, Nandika D, Lestaria SDW, Bahtiar ET, Harahap IS, Rauf A, and Sumertajaya IM 2016 Termite biodiversity and intensity of residential houses deterioration in Taman Darmaga Permai I Ciampea Bogor *Proceedings The 7th International Symposium of Indonesian Wood Research Society* 281-289

[20] Arinana, Haneda NF, Nandika, and Prawitasari WA 2015 Damage intensity of house building and termite diversity in Perumahan Nasional Bumi Bekasi Baru, Rawalumbu, Bekasi *Proceeding The 6th International Symposium of Indonesian Wood Research Society* 134-42

[21] Simbolon G, Hartono R, Sucipto T, and Iswanto AH 2014 Variation of particle pretreatment of subterranean termite attack on particle board from oil palm trunk waste with phenol formaldehyde adhesive *Peronema Forestry Science Journal* **3**(1) 5-9
[22] Amin Y and Dwianto W 2018 Pengaruh suhu dan tekanan uap air terhadap fiksasi kayu kompresi dengan menggunakan close system compression temperature and steam pressure dependency on the fixation of compressed wood by close system compression *Jurnal Ilmu dan Teknologi Kayu Tropis* **4**(2) 55-60
[23] Widyorini R, Khotimah K, and Prayitno TA 2014 Pengaruh suhu dan metode perlakuan panas terhadap sifat fisika dan kualitas finishing kayu mahoni *Jurnal Ilmu Kehutanan* **8**(2) 65-74
[24] Permana RD and Husni H 2017 Efektivitas bioatraktan dari bahan alami terhadap rayap tanah (*Coptotermes curvignathus* Holmgren) *Jurnal Hutan Lestari* **5**(3) 629-88
[25] Darmawan T, Jayadi, Sudijono, Wahyuni I, Amin Y, and Dwianto W 2006 Modifikasi alat pelengkung kayu skala pilot dengan menggunakan pemanas *J. Tropical Wood Science & Technology* **4**(1) 1-8
[26] Sukatik 2003 Sifat mekanis kayu kelapa sawit hasil impregnasi dengan poliropilena bekas hasil modifikasi dengan asam akrilat *Jurnal R & B* **3**(2) 14-8
[27] Syaukani 2013 Termites species richness and distribution at residential area in PT Arun lng *Jurnal Natural* **13**(1) 43-9
[28] Mariana E, Ariyanti, and Erniwati 2013 Uji retensi dan efektivitas tanaman kumis kucing (*Orthosiphon aristatus*) terhadap serangan rayap tanah (*Coptotermes sp*) pada kayu durian (*Durio zibethinus*) *Warta Rimba* **1**(1) 1-8
[29] Esteves BM and Pereira HM 2009 Wood modification by heat treatment: a review *BioResources* **4**(1) 370-404

**Acknowledgements**
The authors wish to express their appreciation to the Ministry of Research, Technology and Higher Education of the Republic Indonesia and Bogor Agricultural University for the supports thus this study could be conducted.

IOP Conf. Series: Materials Science and Engineering **935** (2020) 012013 doi:10.1088/1757-899X/935/1/012013

# The grading classification for *Styrax sumatrana* resins based on physico chemical characteristics using two-step cluster analysis

**A Aswandi* and C R Kholibrina**

Environment and Forestry Research Development Institute of Aek Nauli, Simalungun, 21174, Indonesia

*Email: andiasw@yahoo.com

**Abstract**. Non timber forest management of styrax resin has become the main community livelihood in the Tapanuli region, North Sumatra, Indonesia. However, thousands of tons of incense resin produced have not been graded simply and inaccurately. The resin grading has been determined visually based on shape, size, and color. This is a disadvantage regarding the important value of this commodity is its chemical content. The objectives of the study were (a) to investigate the physico chemical properties of styrax resin based on current local grading, and (b) to develop a new grading classification of Sumatran incense (*Styrax sumatrana*) resin. A completely random design was carried out to determine the physico chemical properties of styrax from six incense grade traded. The results showed that the current quality grading affects the physico chemical properties of the resin. The first grade contains the best physico chemical properties, conversely the sixth is the lowest among others. Isolated cinnamic acid from various grades produces different levels of purity. The highest balsamic acid content and purity were derived from the highest quality incense (current local Grade I) while the lowest was contained by Grade VI. A two-step cluster analysis was performed to identify an accurate classification of resin grade according to both numerical and categorical attributes. Clustering algorithms with the two-step method worked efficiently and determined the optimal number of clusters, four grade incense profiles. The most important grade contains higher balsamic content and purity, lower impurities, ash content, melting point, and moisture content. Ash content and impurities were the most important predictor for differentiating each segment. The information obtained is expected to facilitate more accurate and honest resin quality grading application for farmers and traders.

## 1. Introduction

Non-timber forest management of styrax resin has become the main community livelihood in the Tapanuli region, North Sumatra, Indonesia [1]. However, thousands of tons of incense resin produced have not been graded simply and accurately [2,3]. The mechanism for determining prices was practically fluctuated and determined unilaterally by big buyers or whole traders [1].

The resin grading in local trading has been determined visually based on shape, size, and color. This is a disadvantage because the important value of this commodity is its chemical content. Several studies were examined the physico chemical characteristics of various incense resins grade, especially the application of significant difference tests for resin quality classification [4]. However, most of these algorithms work restricted to numerical data. Nevertheless, physico chemical resin data contains both numerical and categorical attributes. The two-step cluster is an SPSS method that solves this problem. Data clustering is a method applied to distinguish classes of objects with similar characteristics [5,6,7]. This accurate method is an advantage for determining the proper number of clusters, so that resin grading is simpler and more applicable.

The objectives of the study were (a) to investigate the physico chemical properties of styrax resin based on current local grading, and (b) to develop a new grading classification of Sumatran incense (*Styrax sumatrana*) resin using two-step cluster method. The information obtained is expected to facilitate farmers and traders for more accurate resin quality grading applications. Simple and accurate postharvest processing technology that can be applied by farmers and traders is expected to increase added value in initial processing of incense resin.

## 2. Materials and Methods

### *2.1. Materials*

The main ingredient includes incense toba (*Styrax sumatrana*) resins that harvested in three periods in December 2019 to March 2020 from the community forest in Humbang Hasundutan District, North Sumatra, Indonesia. The analytical materials included some distilled water, xylene solvent, ethanol 96%, ethanol p.a., KOH-ethanol 0.5 N, $MgSO_4$, HCl, ether, $NaHCO_3$, chloroform, NaOH, oxalic acid, phenolphthalein indicator, and filter paper. While equipments applied include some glasses, analytical scales, erlenmeyer bottles, oven, thermometer, separating funnel, spiritus lamp, condenser and others.

### *2.2. Methods*

#### 2.2.1. Grading resin quality test

A completely random design and followed by Tukey HSD-test were carried out to determine both quality and quantity differences of styrax resin properties and the balsamic acid content of the six grading incense quality where traded in the local market. The resins with its bark were collected from tapping 20 incense trees for three harvest periods (first/main, second, and third). All resins were cleaned from impurities and sorted visually according to commonly traded grading (6 grading) and harvest periods. Visual observations made include resin color, size, and shape.

Some tests were applied to the physico chemical characteristics of each resin quality including water content, ash content, impurities, melting point, balsamic acid content, and purity. The analysis procedure is based on SNI 7940: 2013 and several previous related studies [4,8,9].

a. Moisture content, calculated by dissolving 5 grams of incense into a 300 mL erlenmeyer glass and added 100 to 150 mL xylene solvent. Then, erlenmeyer was connected to the *Aufhauser* equipment and heated until the contained water in resin came out completely. The moisture content was read on the scale on the *Aufhauser* device.

b. Ash content, analyzed by heating 3 grams resin into the kiln. The samples with porcelain containers were heated to a temperature ±625°C until the resin turns into ash form. Then, the sample and container was cooled and weighed to a constant weight. Ash content were calculated by:

$$Ash\ content = \frac{(W1-W)}{(W2-W)}\ x\ 100\% \tag{1}$$

where:
W: weight of empty container (cup)
W1: weight of ashes with container
W2: weight of incense sample with container

c. Level of impurities, investigated by dissolving 2 grams of incense into 100 mL erlenmeyer containing 25 mL of ethanol 90%. The solution was filtered with filter paper. Furthermore, the residue was washed with 90% warm ethanol and dried at 100°C for 6 hours to a constant weight. The level of impurities was calculated by the formula:

$$Level\ of\ impurities = \frac{W1-W2}{W}\ x\ 100\% \quad (2)$$

where
W: weight of incense sample
W1: weight of filter paper and insoluble ingredients at 90% ethanol
W2: weight of filter paper

d. Melting point, tested by melting 30 grams of incense fine powder into a ring and fills up to the entire surface. After the molten resin cools, a ring containing hardened resin was placed on a ring holder and a steel ball were placed above the ring. Furthermore, the ring, steel ball, and thermometer were inserted into a glass cup filled with 800 mL water, followed by heating so that the resin softens and the steel ball slowly drops and touches the bottom of the device. The melting point was determined based on the temperature read on thermometer when the ball falls on the bottom of device.

e. Cinnamic acid content, run based on SII 2044-1987. A total of 3 grams of incense powder was put into a closed erlenmeyer containing 50 mL KOH-ethanol 0.5 N and refluxed for one hour. Then, ethanol was evaporated to dry condition, and then residual evaporation was dissolved with 75 mL hot distilled water until it was homogeneous and cooled. After it cools down, the solution was added with 80 mL distilled water and 75 mL of $MgSO_4$ 3% solution then stirred thoroughly and left for 10 minutes, and then it was filtered. The residue was washed with distilled water and 96% ethanol. Furthermore, the filtrate was collected and acidified with HCl 4 N, then extracted with ether 3 times. The ether phases were collected and extracted gradually with 40 mL and 20 mL of $NaHCO_3$ solution 5%. The aquades phase was collected and washed with ether, while the ether phase is removed. Hereinafter, the aquades phase was acidified with HCl 4 N and then it was extracted gradually with 60, 40, and 20 mL of chloroform. The chloroform solution was collected in an erlenmeyer bottle and chloroform fraction was evaporated to dryness. The residue was added with hot water (heated) and filtered under hot conditions. Furthermore, the filter was cooled with ice. The resulting crystal was filtered and washed with cold water, then dried in an oven at ± 50°C for 3 hours. The results obtained are expressed as cinnamic acid crystal.

$$Cinnamic\ acid\ content = \frac{W1}{W}\ x\ 100\% \quad (3)$$

where:
W: weights of sample incense (grams);
W1: dry weight of cinnamic acid crystal (grams)

f. The purity of cinnamic acid, the test was carried out by dissolving 0.1 gram of cinnamic acid crystal into an erlenmeyer glass containing 20 mL of ethanol p.a. neutral. Then, sample was heated in oven at ± 50°C for one hour. Subsequently, the sample was cooled, followed by titration with NaOH 0.5 N solution standardized by oxalic acid 0.0318 N and phenolphthalein as an indicator. Color changes were observed from yellow to reddish orange. The purity was calculated by the formula

$$The\ purity\ of\ cinnamic\ acid\ (\%) = \frac{V\ x\ N\ x\ 14.2}{W}\ x\ 100\% \quad (4)$$

where:
V: volume of NaOH used (mL)
N: normality of NaOH (N)
W: weight of sample crystal (mg)
148.2 = BE of cinnamic acid (g/ek)

2.2.2. Compilation of new quality classification

A two-step cluster analysis was performed to identify an accurate classification of resin grade according to qualitative and quantitative parameters i.e. shapes, size, color, moisture content, ash content, impurity, melting point, and cinnamic acid content. Two-step cluster analysis was carried out using the SPSS statistical software package [5-7]. The process consisted of two major steps; the first step, initial clustering of observations into small sub-clusters performed, and further on these sub-clusters treated as separate observations. The decision of whether the observation was joined in an already formed cluster or a new cluster shall be formed was made on the bases of distance criteria. The grouping of these new observations was applied by hierarchical cluster method. The second step was groping, where the sub-clusters were bases for analysis, and it was grouped into required number of clusters [5-7].

In this analysis there were one or more categorical variable, the log-likelihood distance measure was applied. The log-likelihood distance measure represents the distance based on probability. The distance between two clusters in relation to decrease of value of log-likelihood distance measure, when two clusters were joined in one. While calculating the log-likelihood distance measure, normal distribution for continuous variables, and multinomial for categorical variables were assumed. Also, the independence of variables and independence of observations were also assumed. The distance between clusters $R$ and $S$ were defined as

$$d_{(R)(S)} = \xi_R + \xi_S - \xi_{(R,S)} \tag{5}$$

where:

$$\xi_v = -N_v\left(\left(\sum_{k=1}^{K^A}\frac{1}{2}\log(\hat{\sigma}_k^2 + \hat{\sigma}_{v-k}^2)\right) + \left(\sum_{k=1}^{K^B}\hat{E}_{v-k}\right)\right)$$

$$\hat{E}_{v.k} = -\sum_{l=1}^{L_k}\left(\frac{N_{v.k.l}}{N_v}\log\left(\frac{N_{v.k.l}}{N_v}\right)\right)$$

here $K^A$ is a total number of continuous variables in analysis; $K^B$ is a total number of categorical variables; $R_k$ is interval or range of $k$ continuous variable; $N$ is a number of observations; $N_k$ is a number of objects in $k$ cluster; $\hat{\sigma}_k^2$ is estimated variance of $k$ continuous variable; $\hat{\sigma}_{Rk}^2$ is estimated variance of $k$ continuous variable in $R$ cluster; $N_{Rkl}$ is number of objects in $R$ cluster, where $k$ categorical variable takes $l$ category; $d_{(R)(S)}$ is distance between $R$ and $S$ clusters; ), $(R,S)$ is index that represents cluster which is formed by joining of clusters $R$ and $S$ [5-7].

If the $\hat{\sigma}_k^2$ is ignored in equation, the distance between clusters $R$ and $S$ will be equal to the decreased value of log-likelihood distance measure when two clusters were joined. The expression $\hat{\sigma}_k^2$ given as a solution to rising problem, if $\hat{\sigma}_v^2 = 0$, by undefined values for natural logarithm were reached.

For automatic determination of clusters number, SPSS has developed the two-step procedure which is compatible with hierarchical cluster analysis. In the first step, the *BIC* or Bayes Information Criterion was calculated for each different cluster solution with different clusters number. In the second step, initial estimate was improved by finding the highest distance increase between two closest clusters during each stage in hierarchical clustering [5-7].

The statistics *BIC* for $R$ clusters were defined as

$$BIC_R = -2\sum_{i=1}^{R}\xi_R + m_R\log(N) \tag{6}$$

$$m_R = R\left\{2K^A + \sum_{k=1}^{K}(L_k - 1)\right\} \tag{7}$$

where $L_k$ is the number of groups in $k$ categorical variable

## 3. Results and Discussion

### 3.1. Physico chemical properties

Styrax resin is produced by inner epithelium layers of injured bark, therefore bark tapping is the best practice to harvest this valuable resin [2]. Styrax forest management has been carried out traditionally, including the harvesting techniques. Some simple tools are used include a *guris* (a bark cleaning tool), a *tuhil* (a tapping tool), a peeler-scraping, climbing ropes, baskets and plastic drying mats. The farmer ability to harvest increased from 6-8 trees initially to 8-12 trees in single day.

The tapping process begins by cleaning the stems and removing mosses that grow on the trunk surface. The number of tapping holes (8 to 12 holes each tree) were made by puncturing the bark surface and making a space between bark and woody parts (cambium). This tapping wound has a 5 to 10 cm in width. Therefore, the injured bark was closed again by knocking gently. The spaces created in the inner bark will be filled with resin during the following 3-4 months. After the tapping holes were filled with hardened resin, harvesting activities were carried out by prying all resins including the bark pieces. This technique causes some sores in form of sizable holes in bark surface.

Furthermore, harvested resins were air dried for two weeks until one month. Then resin sorted out according to size, color, and cleanliness or impurities of resin, qualitatively. There were no technological applications in this stage, chunks resins are sold by farmers. Based on the grading set by traders in Tapanuli region, there were six classes of incense resin quality namely Coarse *Mata* (Grade I), Fine *Mata* (Grade II), *Tahir (*Grade III), *Jurur/jarir* (Grade IV), *Barbar* (Grade V) and Ash-flakes (Grade VI) *(*Table 1).

**Table 1**. Grading of styrax resin at farmer and local trader

| Grading | Color | Shapes | Sizes | Remarks |
|---|---|---|---|---|
| Grade I (Coarse *Mata*) | White, pure white | Chunk, medium plates | width ≥ 3 cm lenght ≥ 5 cm | First harvest Main quality |
| Grade II (Fine *Mata*) | White - yellowish white | Small - medium plates | width 2-3 cm lenght 3-5 cm | First harvest |
| Grade III (*Tahir*) | Light yellow to reddish brown | Medium to small plates | width ≤ 2 cm lenght ≤ 3 cm | Second harvest, mixed with dirt |
| Grade IV (*Jurur/jarir*) | Reddish yellow | Coarse flakes | Diameter 1-2 cm | Third harvest, mixed with dirt |
| Grade V (*Barbar*) | Light reddish brown | flakes | Diameter < 1 cm | collected during bark cleansing |
| Grade VI (Ash - flakes) | Light reddish brown | Granules, flakes | Granules, fines | cleaning remnants |

The first quality is a large plate resin with 3 cm width, 5 cm length, and white or pure white color. The second class is a smaller white and yellowish white plate resin with 2-3 cm width and 3-5 cm length. These two highest grades were produced in the first harvest period (main harvest), generally 3 months after tapping were applied. Both of these were hardened resin between the inner bark and the woody part of the trunk (cambium).

Third grade is small light yellow to reddish-brown plates up to 2 cm wide and up to 3 cm long in. In this local grade trade, it called *Tahir*, produced by the second harvest conducted one to two months after the first (main) harvest. *Tahir* resins were produced by a hardened resin on the outside of the bark or around tapping holes.

The three lowest qualities are small pieces of resin. Grade IV is reddish yellow coarse flake resin that produced by the third harvest (intermediate harvest), while Grade V, light reddish-brown flake resin was produced by bark cleansing. This third harvest period was run one to two months later after the second harvest. The lowest one is ash quality (Grade VI), light reddish-brown fine flakes, or granules, resulting from cleansing barks or trunk. It contains a lot of impurities.

In local and regional trade, the first grade is highly desirable, but relatively more difficult to obtain, so it is valued much higher. Based on tapping of 20 sample trees with the application of 8-12 tapping holes per tree, the average of resin productivity based on the local grading was obtained in the following Table 2.

**Table 2**. The resin productivity based on the local grading from tapping of 20 sample trees

| Grading | Dried resin weight (grams) | | Proportion (%) | | Remarks |
|---|---|---|---|---|---|
| | Mean | Std. Dev | Mean | Std. Dev | |
| I (Coarse *Mata*) | 233.1 | 40.9 | 15.6 | 1.4 | First harvest |
| II (Fine *Mata*) | 280.0 | 33.6 | 18.8 | 1.2 | First harvest |
| III (*Tahir*) | 322.4 | 29.6 | 21.8 | 1.5 | Second harvest |
| IV (*Jurur*) | 249.7 | 31.0 | 16.8 | 1.3 | Third harvest |
| V (*Barbar*) | 219.6 | 29.2 | 14.8 | 1.9 | Bark cleansing |
| VI (*Ash - flakes*) | 180.3 | 29.9 | 12.1 | 1.6 | Cleaning remnants |
| Total | 1,484.9 | 142.7 | | | |

Overall, resin productivity is 1,484.9 grams per tree. The highest proportion (21.8%) is Grade III (Tahir) with productivity about 222.4 grams per tree. The highest quality (Grade I Coarse *Mata*) produces about 233.1 grams (15.56%), while the second quality (Fine *Mata*) at of 280.0 grams or 18.8%. If these two highest qualities were combined, it contributed 34.4% of total resin productivity.

**Table 3**. An overview of variance analysis of the parameters tested

| | | Sum of Squares | df | Mean Square | F | Sig. |
|---|---|---|---|---|---|---|
| Moisture content | Between Groups | 5.385 | 5 | 1.077 | 24.141 | 0.000 |
| | Within Groups | 2.409 | 54 | 0.045 | | |
| | Total | 7.794 | 59 | | | |
| Ash content | Between Groups | 18.008 | 5 | 3.602 | 726.407 | 0.000 |
| | Within Groups | .268 | 54 | 0.005 | | |
| | Total | 18.276 | 59 | | | |
| Impurities | Between Groups | 474.581 | 5 | 94.916 | 349.987 | 0.000 |
| | Within Groups | 14.645 | 54 | 0.271 | | |
| | Total | 489.226 | 59 | | | |
| Melting point | Between Groups | 3032.952 | 5 | 606.590 | 55.954 | 0.000 |
| | Within Groups | 585.408 | 54 | 10.841 | | |
| | Total | 3618.360 | 59 | | | |
| Balsamic acid | Between Groups | 999.824 | 5 | 199.965 | 51.029 | 0.000 |
| | Within Groups | 211.609 | 54 | 3.919 | | |
| | Total | 1211.433 | 59 | | | |
| Purity | Between Groups | 501.231 | 5 | 100.246 | 120.061 | 0.000 |
| | Within Groups | 45.088 | 54 | 0.835 | | |
| | Total | 546.319 | 59 | | | |

Furthermore, a complete randomized sampling was carried out to investigate the effect of quality differentiation on physico chemical properties. The parameters tested were water content, ash content, impurities, melting point, balsamic acid content, and purity. The design ran with 10 replications for each grading. An overview of the variance analysis of the parameters tested was described in Table 3.

The results of the analysis described that current quality grading affected the physico chemical properties of the resin. Based on the significant value of 0.000, the current classification of six resin qualities also influenced the moisture content, ash content, impurities, melting point, balsamic acid content, and purities. An overview of results of analysis of physico chemical properties of incense resin was listed in Table 4.

**Tabel 4**. The physico chemical properties of incense resin based on current quality grading

| Grading | N | Moisture content (%) | Ash content (%) | Impurities (%) | Melting point (°C) | Balsamic acid | |
|---|---|---|---|---|---|---|---|
| | | | | | | Content (%) | Purity (%) |
| I | 10 | 2.26±0.16 a | 0.07±0.02 a | 3.31±0.34 a | 57.45±1.21 a | 35.06±1.35 a | 96.53 a |
| II | 10 | 2.21±0.11 a | 0.08±0.01 a | 3.55±0.36 a | 57.86±1.12 a | 34.12±1.20 ab | 96.18 a |
| III | 10 | 2.29±0.13 a | 0.25±0.05 b | 3.98±0.22 a | 60.71±1.27 a | 30.94±0.81 b | 93.97 b |
| IV | 10 | 2.45±0.14 a | 0.50±0.03 c | 4.74±0.24 b | 67.27±1.78 b | 27.29±0.91 c | 93.28 bc |
| V | 10 | 2.47±0.10 a | 0.67±0.07 d | 6.08±0.22 c | 69.43±2.11 b | 26.47±1.58 d | 92.67 c |
| VI | 10 | 3.10±0.09 b | 1.66±0.03 e | 11.47±0.38 d | 77.33±3.15 c | 23.87±0.82 d | 87.77 d |

The mean difference is significant at the 0.05 level (Tukey HSD test)

Overall, first grade contains the best physico chemical properties, conversely the sixth grade is the lowest among others. This best quality (Grade I) resin has a balsamic acid content reaching 35.06%, the highest with purity 96.53%, resin impurities 3.31%, ash content 0.07% and 2.26% moisture content. The impurities and ash content were the lowest among others.

Furthermore, the Tukey HSD-test showed that the physico chemical properties of Grade II were similar to Grade I. The balsamic acid content reached 34.12% with purity 96.18%, resin impurities 3.55%, ash and moisture content were 0.08% and 2.21%, respectively. These similarities were a consequence of difference between these two highest grades only in shape and size and visual sorting that carried-out. Similar results were also reported by previous study [4] where the similarity in physics-chemical properties includes Grade I, II, and III.

In a single variable of balsamic acid content, the Tukey HSD-test showed that there were four levels of content quality. Grade I and II have the same balsamic acid content relatively, followed by Grade III at the second, then Grade VI at the third level. The two lowest grades (V and VI) have the similar balsamic acid content relatively even though both ash content and impurities were different. Levels of isolated cinnamic acid from resins Grade I to VI were above 11%, indicating that all grades tested meet the minimum requirements for these compounds in incense resin.

Isolated-cinnamic acid from various grades produces different purity levels. The highest purity (96.53%) was derived from the highest quality incense (Grade I) while the lowest was contained by Grade VI (87.77%). Based on the Tukey HSD-test, there were four different purities levels. The first grade purity was similar to the second, relatively, while Grade III has a purity of about 93.97%, statistically as similar as Grade IV (93.28%). These purities were similar regarding their same origin as outer bark resins. However, the purity of grade IV is also similar to Grade V (92.67%), due to higher impurities content. Grade VI contains the lowest purity resin (87.77%). The cinnamic acid purity tends to decrease at an increasingly lower quality also reported [4].

Ash content, impurities level, and the melting point of Grade I and II were lower than others. This is due to higher cleanliness of both grades. The resin comes from parts between the inner skin and stem, so these protected resins were not affected by an unfavorable environment. Conversely, resins in grades III, IV, and V derived from hardened resin outside the barks at the second and third harvesting, consequences both higher impurities and ash content. These characters were caused by the melt hardened resin which also traps various other organic and inorganic materials on the surface of the trunk.

Except for Grade VI, the others have similar moisture content. Even though it was granules or fine flakes, contained water in lowest grade was higher due to the high impurities content. These impurities include wood chips or other materials that contain higher moisture content. The highest ash content was also demonstrated by grade VI. Ash content shows the amount of minerals in an ingredient. Organic materials will ignite and evaporate when heating, while inorganic materials will be left as ashes. High impurities in this quality cause high ash content.

Referring to National Standards (Indonesian Industrial Standards (SII) No. 2044-87 (Processed Form), Grades I, II and III meet the First Quality (*Mutu* I) requirements (balsamic acid content > 30), while Grade IV, V and VI classified as Second Quality (*Mutu* II) (balsamic acid content > 25%). The processed resins also meet the requirements of water content, ash content, and soft spots for each quality.

### *3.2. Clustering*

Two-step cluster is an SPSS method that works with both numerical and categorical data. This method has an advantage in determining the number of proper clusters effectively. In this method, data clustering can form objects classification with similar characteristics.

The dataset that has been used for this study has been obtained from physico chemical analysis which was carried out in the previous stages of this study. The dataset has 420 records and the overview is presented in Table 5. This table contains information about the resin grade (visual classification), moisture content, ash content, impurities, melting point, balsamic acid content, and purity. Table 5 presents and overview of the data.

**Table 5**. Data and variable in this study

| | Variables | Category |
|---|---|---|
| *X1* | Grade | Categorical data |
| *X2* | Moisture content (%) | Continuous data |
| *X3* | Ash content (%) | Continuous data |
| *X4* | Impurities (%) | Continuous data |
| *X5* | Melting point (°C) | Continuous data |
| *X6* | Balsamic acid content (%) | Continuous data |
| *X7* | Purity (%) | Continuous data |

The database contains one categorical variable and six continuous variables. Continuous variables were standardized by default. Considering this study used a mixture of data, only *log-likelihood* option for distance measure. In the first running, *BIC* was chosen to determine the number of clusters, though it may override this and specify a fixed number.

According to the SPSS algorithm, the optimal number of clusters is four, because the largest ratio of distances is for four clusters. The cluster distribution is shown in Table 6.

**Table 6**. The cluster distribution

| | | N | % of combined | % of total |
|---|---|---|---|---|
| Cluster | 1 | 10 | 16.7 | 16.7 |
| | 2 | 10 | 16.7 | 16.7 |
| | 3 | 20 | 33.3 | 33.3 |
| | 4 | 20 | 33.3 | 33.3 |
| | Combined | 60 | 100.0 | 100.0 |
| Total | | 60 | | 100.0 |

The first and second clusters fill about 16.7% data, respectively. The important clusters are third and fourth clusters. Both are the largest cluster (33.3%, respectively). SPSS presents also the frequencies for each categorical variable. Table 7 shows the frequencies for the grade variable. The first and second clusters contain 100% Grade VI and III, respectively. Furthermore, the third cluster consists of 100% Grade IV and V. While the last cluster contains resins classified as Grade I and II.

**Table 7**. The frequencies for grade variable

| Cluster | Grade I | | Grade II | | Grade III | | Grade IV | | Grade V | | Grade VI | |
|---|---|---|---|---|---|---|---|---|---|---|---|---|
| | F | % | F | % | F | % | F | % | F | % | F | % |
| 1 | 0 | 0.0 | 0 | 0.0 | 0 | 0.0 | 0 | 0.0 | 0 | 0.0 | 10 | 100.0 |
| 2 | 0 | 0.0 | 0 | 0.0 | 10 | 100.0 | 0 | 0.0 | 0 | 0.0 | 0 | 0.0 |
| 3 | 0 | 0.0 | 0 | 0.0 | 0 | 0.0 | 10 | 100.0 | 10 | 100.0 | 0 | 0.0 |
| 4 | 10 | 100.0 | 10 | 100.0 | 0 | 0.0 | 0 | 0.0 | 0 | 0.0 | 0 | 0.0 |

The determinant variables for the formation of clusters were shown in Figure 1. Both ash content and impurities are the most important predictor for differentiating each segment. While the moisture content variable is the lowest predictor. This is similar to the results of the Tukey HSD-test analysis that applied previously where moisture content does not have much effect on grading differentiation.

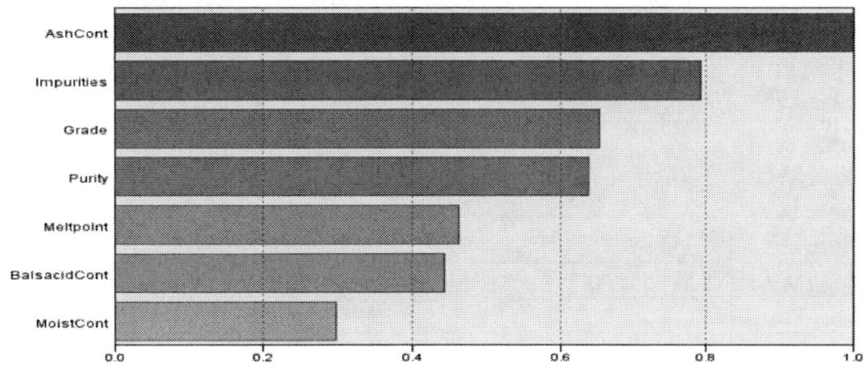


**Figure 1**. Predictor importance

A brief profile of each cluster with the characteristics differentiating each segment from the other groups as outlined below (Figure 2).

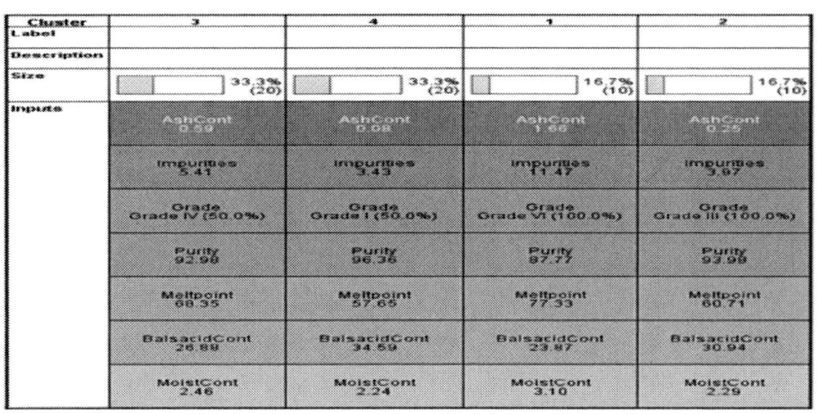


**Figure 2**. Clustering and physico chemical properties description

Briefly, a two-step cluster analysis with *BIC* identified four optimum clusters. The first cluster, which fills about 16.7% sampling data, contains resins with higher ash content (1.68%) and impurities (11.47%), but lower resin purity (87.77%). The melting point is about 77.33°C with lower balsamic acid content (23.97%) but higher moisture content (3.10%). The characteristics of these variables are represented by resins at the lowest current local Grade VI (Ash).

Furthermore, the second cluster covers about 6.7% observation data. This cluster contains resins with low to moderate of ash content (0.25%), low to moderate impurities level (3.97%), and moderate cinnamic acid purity (93.98%) with balsamic acid content (30.94%). The average melting point is about 60.71°C and moderate moisture content (2.29%). These properties are contained by the current local Grade III (*Tahir*).

The third cluster represents about 33.3% data which contains a moderate ash content resin (0.59%) and impurities (5.41%), respectively. This cluster also accommodates resins with low to moderate balsamic acid content (26.88%) with moderate purity (92.98%). The melting point is about 68.35°C with moisture content about 2.48%. The cluster is also represented by resins at the current local Grade IV (*Jurur*) and V (*Barbar*).

The most important cluster is the fourth. This is the largest cluster (33.3%) containing physico chemical properties resins with lowest ash content (0.08%) and impurities (3.43%), but highest balsamic acid content (34.59%) and purity (96.36%) respectively. The melting point is lower about 57.65°C as well as moisture content (2.24%). These properties are represented by both current local Grade I (Coarse *Mata*) and II (Fine *Mata*), resins with the highest economic value.

Furthermore, based on the four clusters identified above, the new classification of styrax resin grading was prepared as shown in Table 8. This new grading is simpler and more measurable compared to current local grading (6 classes) or SNI (5 grades).

**Table 8**. New classification of styrax resin grading

| Current grading | | Proposed new grading | | | | | | | | | |
|---|---|---|---|---|---|---|---|---|---|---|---|
| Local Grading | SNI* | Grade | Color | Sizes | Moisture content | Ash content | Impurities | Melting point | Balsamic content | Purities |
| I (*Coarse Mata*) | Grade I | Grade I | Pure white, yellowish white | Chunk, medium - small plates | 2.24 | 0.08 | 3.43 | 57.65 | 34.59 | 96.36 |
| II (*Fine Mata*) | Grade II | | | | | | | | | |
| III (*Tahir*) | Grade III | Grade II | Light yellow to reddish brown | Small plate, coarse flakes | 2.29 | 0.25 | 3.97 | 60.71 | 30.94 | 93.98 |
| IV (*Jurur*) | Grade IV | Grade III | Reddish yellow, light reddish brown | Coarse to fine flakes | 2.46 | 0.59 | 5.41 | 68.35 | 26.88 | 92.98 |
| V (*Barbar*) | | | | | | | | | | |
| VI (*Ash - flakes*) | Abu | Grade IV | Light reddish brown | Granules, flakes | 3.10 | 1.66 | 11.47 | 77.33 | 23.87 | 87.77 |

*Indonesian National Standard/Indonesian Industry Standards (SII) No. 2044-87

The proposed new grading consists of four grade resins. The new first grade represented by resins at the current local grade I (Coarse *Mata*) and II (Fine *Mata*). The highest grade contains high balsamic content and purities, but lower impurities and ash content (Table 8). The physical properties are indicated by a pure white to yellowish white color with a chunk, medium to small plate size.

The new second grade consists of resins at the current local Grade III (Tahir). This cluster contains moderate resin quality which indicated by low to moderate ash content and impurities, moderate balsamic acid content, and purity. The physical properties are indicated by a light yellow to reddish brown color with a small plate to coarse flake resin.

The third grade is a new cluster of resins which are classified in both current local Grade IV (*Jurur*) and V (*Barbar*). The physical properties are indicated by a coarse to fine flakes resin form with reddish yellow, light reddish color. The grade represents a low to moderate balsamic acid content and purity, but moderate ash content resin and impurities (Table 8).

The lowest quality, fourth grade, contains resin with higher ash content and impurities, but lower balsamic acid content and purity. The physical properties resins at this grade are granules or flakes form with a light reddish brown color. These characteristics are represented by resins at the lowest current local Grade VI (*Ash*), the lowest economic value [10].

Compared to current local grading with six classes, the new classification with four grades will encourage farmers easier to sort resins visually. Furthermore, even compared to Indonesian National Standard (SNI) with five grades, a simpler and measurable grade application will promote a more honest and consistent resin quality classification, so that pricing becomes fairer for all parties [10].

## 4. Conclusion

The results of the analysis showed that the current quality grading affects the physico chemical properties of styrax resin. The first grade contains the best properties, conversely the sixth is the lowest among others. The highest balsamic acid content and purity were derived from the highest grade (Grade I) while the lowest was contained by Grade VI.

Clustering methods can be applied in both numerical and categorical attributes. Clustering algorithms with a two-step method can be easily used and determines four optimal clusters. The most important profile (Grade I) contains higher balsamic content and purity, lower impurities, ash content, melting point, and moisture content. Ash content and impurity variables are the most important predictor for segment differentiating. These simpler and measurable grades application will encourage a more honest and consistent resin quality classification.

**References**

[1] Aswandi and Kholibrina CR 2017 *Pemulihan Ekosistem Danau Toba* (Medan: Bina Media Perintis)

[2] Dinas Perkebunan Sumatera Utara 2019 Statistik Perkebunan Dinas Perkebunan Sumatera Utara http://www.disbun.sumutprov.go.id/statistik_2019/web/index.php?r=site%2Flaporan-komoditas&komoditas=8&tahun=2015&kabupaten=20

[3] Aswandi and Kholibrina CR 2017 Growth and yield model for non-timber forest product of kemenyan (*Styrax sumatrana* J.J. Sm) in Tapanuli, North Sumatra *International Conference on Agriculture, Environment, and Food Security IOP Conf. Series: Earth and Environmental Science* **122** 012036

[4] Waluyo TK, Hastoeti P, and Prihatiningsih T 2006 Karakteristik dan sifat fisiko- kimia berbagai kualitas kemenyan di Sumatera Utara *Jurnal Penelitian Hasil Hutan* **24**(1) 47-61

[5] Şchiopu D 2010 Applying two-step cluster analysis for identifying bank customers profile *Buletinul* **62**(3) 66-75

[6] SPSS Inc. 2004 *Two-Step Cluster Analysis* (Chicago: Technical report) https://www.spss.ch/upload/1122644952_The%20SPSS%20TwoStep%20Cluster%20Component.pdf

[7] Mongi CE 2015 Penggunaan analisis two step clustering untuk data campuran *d'CARTESIAN* **4**(1) 9-19

[8] Waluyo TK and Setiawan E 2007 Isolasi asam sinamat dari berbagai kualitas kemenyan asal Sumatera Utara *Jurnal Penelitian Hasil Hutan* **25**(4) 319-326

[9] Kiswandono AA, Iswanto AH, Susilowati A, and Lumban TAF 2016 Analysis of the content of cinnamic acid and phytochemical screening sap incense type of Bulu (*Styrax benzoine var. hiliferum*) of North Tapanuli *Prosiding Seminar Nasional Kimia*

[10] Aswandi and Kholibrina CR 2020 Increasing added-value of styrax resin through post-harvesting techniques improvement and essential oil based product innovation (Paper presented in ICONART 2020)

**Acknowledgments**
The author thanks to SEAMEO Biotrop for research funding support and Dr. Supriyanto for his guidance and some ideas related to increasing the added value of styrax resin.

# Color and dimensional stability of fast growing teakwood by mild pyrolysis and combination process

**A Kartikawati[1], I Wahyudi[2*], G Pari[3], and L Karlinasari[2]**

[1]Forest Products Science and Technology Study Program, IPB University, Bogor 16680, Indonesia
[2]Forest Products Department, Faculty of Forestry, IPB University, Bogor 16680, Indonesia
[3]Forest Products Research and Development Center, Ministry of Environment and Forestry, Bogor 16680, Indonesia

*Email: imyudarw16@yahoo.com

**Abstract.** Fast-growing teakwood had been subjected to Mild Pyrolysis (MP) and combination (MP + steam) processes. Performa of treated wood was evaluated by surface color change, mass loss, and dimensional stability. The CIELab. system was used to evaluate the color change, while anti-swelling efficiency was used to determine dimensional stability. The two processes were conducted within pyrolysis reactor under atmospheric pressure in the presence of limited oxygen. MP process was applied by heating the wood for 2 hours at 170 and 190$^0$C without steam, while combination process was employed by flowing steam for 0.5 and 1 hour during the 2 hours heating period. The result showed total color change and dimensional stability increased along with the increase in temperature. Even though MP process could reduced water absorption and swelling, but combination process permits a better results. Treated wood by MP and steam has lower total color change and mass loss, but greater dimensional stability.

## 1. Introduction

Fast-growing teak (*Tectona grandis* L.f.) wood has different properties compared to conventional teakwood planted by Perum Perhutani. Pale color, low dimensional stability, and low durability are some significant differences [1-2]. These characteristics resulted in relatively less desirable and limiting wood utilization. Therefore, quality improvement of fast-growing teakwood needs to be done.

Thermal modification process was applied in this study for two reasons. Firstly, it is relatively simple and environmental friendly process, and secondly it is an alternative to coloring wood without adding paint or chemical. Thermal modification process could improve dimensional stability and durability of wood [3-5]. This process has several names such as le bois perdue, rectification, and Oil Heat Treatment (OHT). The main differences among them are related to material used, process condition applied, and the equipment [6]. The famous products from this process are thermowood and plato.

Pyrolysis, one of thermal modification process, is carried out at high temperature namely 402-702$^0$C without or with limited oxygen (inert atmosphere), and was originally intended to produce charcoal [7]. In order to keep the sample in the form of wood, the process must be carried out at a lower temperature, i.e. 160-260$^0$C. Such kind of process is known as MP. MP has been studied by several researchers and the result shows well improvement in dimensional stability, appearance, and durability of wood

[3,4,8,9]. Generally, nitrogen gas was used in the process in order to obtain inert atmosphere and prevent combustion of wood.

In previous study, MP process using a reactor without adding nitrogen gas was applied for fast-grown teakwood were in green condition [1]. The results showed that treated wood surface is darkening, while dimensional stability and its durability against termite are increase. The optimal process was MP at 200°C for 4 hours. Even though some improvements obtained, the existence of wood defect cannot be avoided. At 230°C for 6 hours treatment, Modulus of Rupture (MOR) reduced more than 50% and carbon layer existed on the wood surface. Some modifications, therefore, need to be done to minimize these negative effects.

According to Missio *et al.* [10], combined treatment could be utilized to reduce the negative impact of treated wood. Steam provides more effective heat transfer and acts as heating medium and shielding gas, and as the result drying defects was significantly reduced [11-13]. The main objective of this study was, therefore, to investigate effect of MP and combination (MP + steam) processes on color and dimensional stability of fast-growing teakwood.

## 2. Materials and Methods

### 2.1. Materials

Fast-growing three teak trees with 20-25 cm diameter at breast height was cut from plantation forest area in Bogor District, West Java Province, Indonesia. Four logs 3 m long from the basal area of the 8-year-old trees were used in this study. Logs were converted into wood sample and then air-dried for 2 months to obtain the moisture content of 18%. Wood samples of 5 (w) x 1 (t) x 15 (l) cm were used for color measurement, while sample of 2.5 (T) x 2.5 (R) x 10 (L) cm was used for mass loss and dimensional stability measurements. Number of replications for each parameter of each combination process was 5 times.

### 2.2. Methods

#### 2.2.1. MP process and combination processes

The process was carried out within a small pyrolysis reactor under atmospheric pressure in the presence of limited oxygen. MP process only was applied by heating the sample for 2 hours at different temperatures (170 and 190°C) without steam, while combination (MPS) process was employed by flowing steam into the reactor at different durations (0.5 and 1 hour) during the 2 hours heating period. After finishing the process, the reactor was turned off and allowed to cool naturally. Wood samples were then taken out at the next day and stored at room temperature before testing.

#### 2.2.2. Mass loss measurement

Mass loss was determined by comparing the mass of wood samples before ($M_o$) and after ($M_t$) process. The value was measured by formula by equation:

$$\text{Mass loss } (\%) = \frac{\text{Mo-Mt}}{\text{Mo}} \text{ x } 100 \quad (1)$$

#### 2.2.3. Color measurement

The color measurement of wood surface before and after process was carried out by the CIE Lab* system. The HP Deskjet Ink Advantage 1515 scanner was used to collect the images. The image was processed by Image-J Fiji software to determine L*a*b* for color change measurement. The L*, a*, and b* coordinates were represented degree of lightness, greenness and redness, and degree of blueness and yellowness, respectively. The measurement was carried out at ten different points (five points on each surface) of each sample. The total color difference before and after treatment (ΔE) was calculated based on equation 2. The ΔE value was then classified as small (1.0–3.0), moderate (3.0–6.0), and large (> 6.0) considered to Hunter Lab (1996).

$$\Delta E = \sqrt{\Delta L^{*2} + \Delta a^{*2} + \Delta b^{*2}} \quad (2)$$

### 2.2.4. Dimensional stability measurement

Anti-Swelling Efficiency (ASE) was used as indicator to evaluate dimensional stability [14]. The swelling was measured at nine different marked positions (@3 positions for T, R, and L direction) before and after soaking wood sample within water for 24 hours at room temperature. Water absorption (A), volumetric swelling (S), and ASE was determined using formula by equation 3, 4, and 5, respectively:

$$\% A = \left[\frac{Wt\text{-}Wo}{Wo}\right] x\ 100 \tag{3}$$

$$\% S = \left[\frac{Vt\text{-}Vo}{Vo}\right] x\ 100 \tag{4}$$

$$\% ASE = \left[\frac{Sc\text{-}St}{Sc}\right] x\ 100 \tag{5}$$

where: $W_t$ is sample weight after soaking, $W_o$ is sample weight before soaking, $V_t$ is sample volume after soaking, $V_o$ is sample volume before soaking, $S_c$ is swelling of control wood, and $S_t$ is swelling of heated wood.

## 3. Results and Discussion

Color change in wood surface due to processes was presented in Figure 1. It shows that temperature has significant effect on wood color, conversely for steam flow (Table 1). The surface treated teakwood tends to be darker and color uniformity is well increases compared to those of untreated wood. This result was similar to [5,12]. When steam is applied, the brightness will improve. Since total color change ($\Delta E$) of all treatments was greater than 6, so that they are classified as large color changes (Figure 2). Result also shows that the $\Delta E$ is greater in treated wood without steam (MP process). Compared to that of MP process, total color change of treated wood with steam flow (MPS process) decreased by 11.25 and 13.79% at temperature of 170 and 190$^o$C, respectively. This may be caused by the release of accumulated decomposition products of cell wall components on the wood surface. The dark of the board surface is caused by decomposition of the accumulated low-molecular carbohydrates [15].

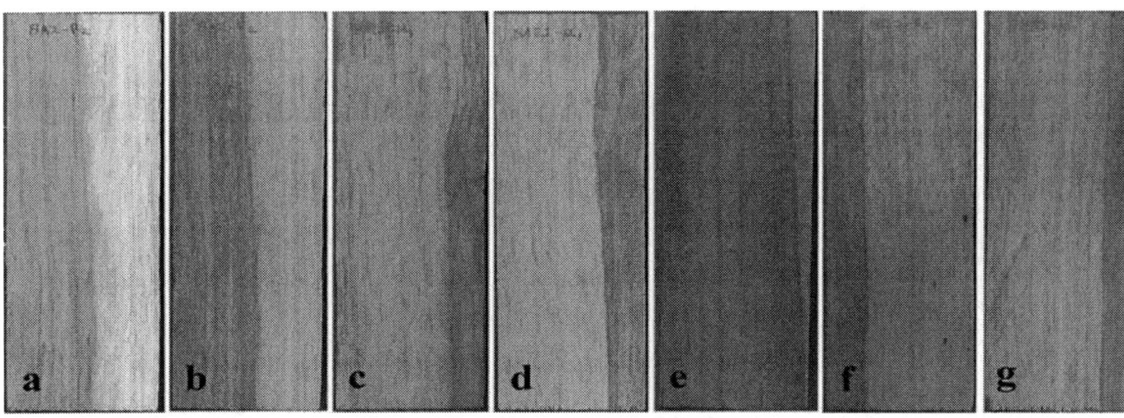


**Figure 1.** Wood color change of treated teakwood, (a) untreated, (b) 170$^o$C, (c) 170$^o$C + 0.5 hours steam, (d) 170$^o$C + 1 hours steam, (e) 190$^o$C, (d) 190 $^o$C + 0.5 hours steam, (g) 190 $^o$C + 0.5 hours steam

The color of wood surface is often used to estimate the severity of treatment applied. The darker of the treated wood indicates the higher intensity of treatment. The higher intensity treatment (higher temperature and longer duration) could have negative impact on wood properties [16,17].

Decomposition or/and modification of chemical component of wood plays important role in the change in color of wood surface [18].

**Table 1.** Two-way analysis of variance of effect of temperature and steam on mass loss, color change, and anti-swelling efficiency of treated teakwood at the 95% confidence level

| | Mass loss | | Color change (ΔE) | | Anti-swelling efficiency (ASE) | |
|---|---|---|---|---|---|---|
| | F | *P-value* | F | *P-value* | F | *P-value* |
| Temperature (T) | 199.849 | 0.000** | 263.706 | 0.000** | 36.464 | 0.000** |
| Duration of steam (S) | 9.901 | 0.001** | 4.384 | 0.24 ns | 3.291 | 0.55 ns |
| T x S | 8.684 | 0.001** | 0.877 | 0.429 ns | 2.244 | 0.128 ns |

*significant effect (P<0.05); **very significant effect (P<0.01); ns, not significant effect

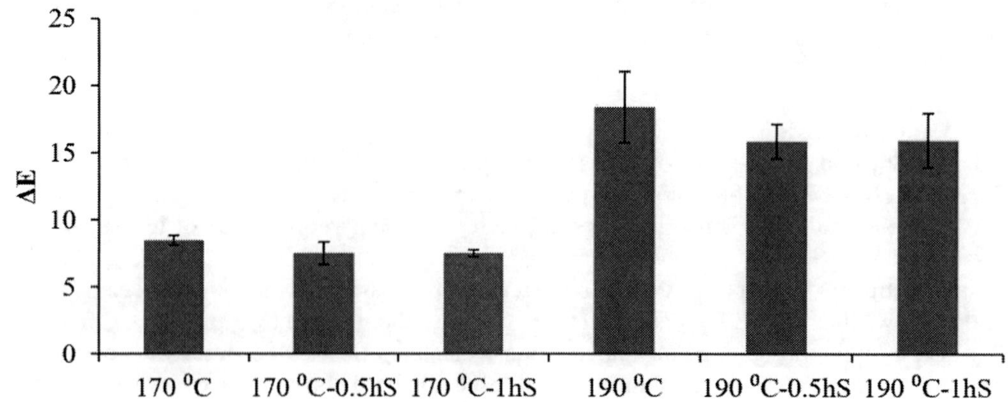


**Figure 2.** Total color change (ΔE) of MP and MPS treated teakwood

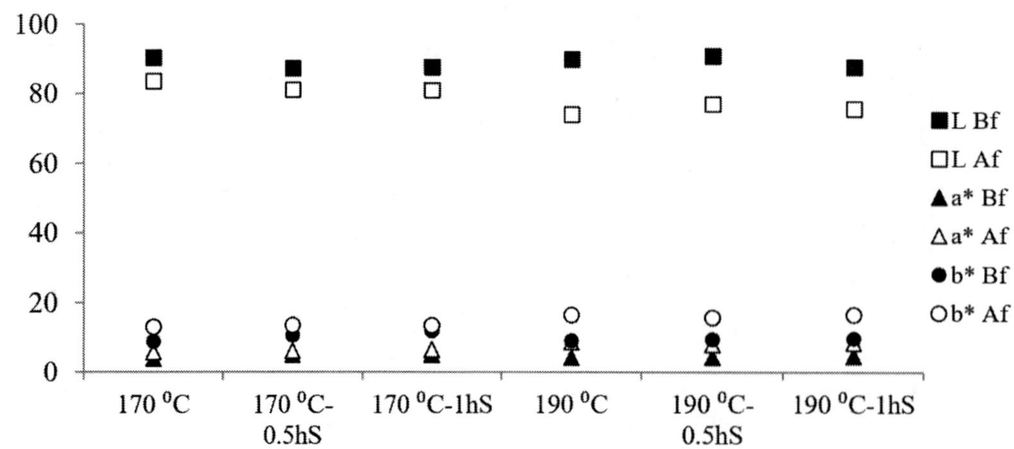


**Figure 3.** L*a*b* value of MP and MPS treated teakwood

Figure 3 shows the three color coordinates (L*, a*, and b*) before and after process. The lightness of wood surface decreased after treatment as shown by the decrease of L* value. This result indicates that wood color becomes darkening, and it is suitable to Figure 1. But, generally with steam inside the reactor, wood surface becomes more brightness. Result also shows that after treatment, wood surface becomes more redness and yellowness which indicated by a* and b* value increased, respectively. Similar results have been reported in previous study [5,18]. Thermal modification caused color change to yellow, brown, red or grey [18].

Usually, wood quality of treated wood is predicted through mass loss after the treatment. Mass percentage loss of temperature 170°C and 190°C was 14.99% and 18.46%, respectively (Figure 4). The higher the temperature, the greater the mass loss. Similar to the total color change, steam flow has positive effect to lowering the mass loss. Lower mass loss was obtained because of the removal of volatile degradation products from the reactor, especially acetic acid [19]. Therefore, further degradation of polysaccharides should be minimized.

Consequently, mechanical properties are expected slightly decreased after processing because mass loss is related to mechanical properties of wood. Result shows that modulus of elasticity and modulus of rupture reduced significantly when the mass loss above 16% [3,20]. Based on analysis of variance (Table 1), both temperature and steam have significant effect to mass loss.

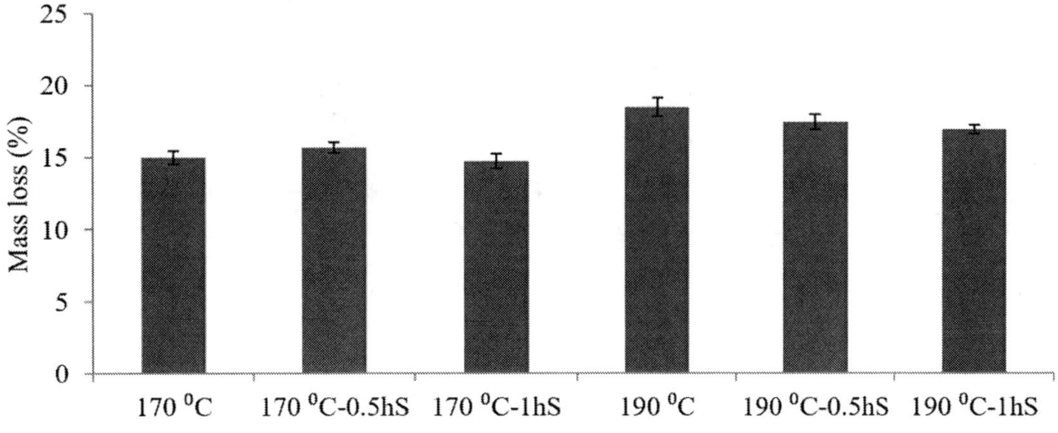


**Figure 4.** Mass loss of MP and MPS treated teakwood

In current study, dimensional stability was evaluated by water absorption, volumetric swelling, and ASE. Based on data obtained, these two process was able to improve dimensional stability because water absorption decreases (Figure 5) as well as volumetric swelling (Figure 6). Water absorption was reduced by 14% and 30% at temperature of 170°C and 190°C, respectively compared to untreated wood; while the swelling was only about 3-5%.

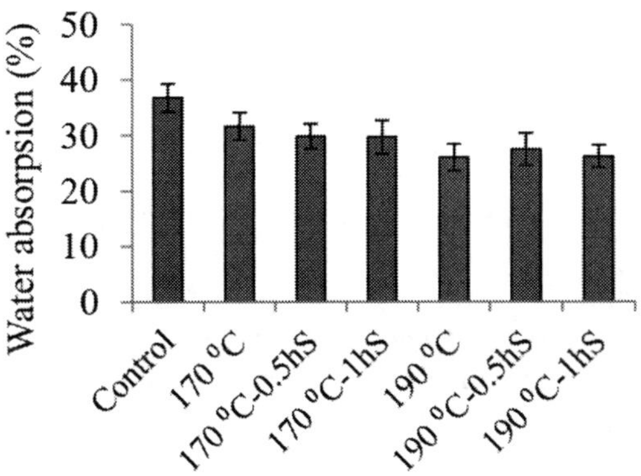


**Figure 5.** Water absorption of control and treated teakwood

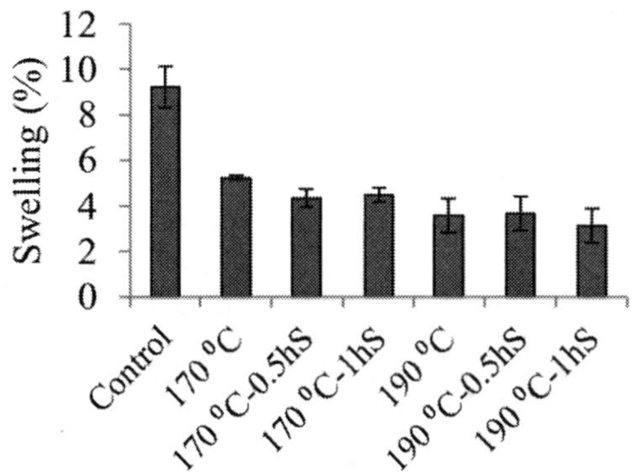


**Figure 6.** Swelling of control and treated teakwood

Result shows that ASE was increased along with increasing the temperature (Figure 7). This finding is different from [5]. He stated that ASE of heated young teakwood at temperature of 150°C and 180°C was 29.17% and 26.47%, respectively. Although combination temperature and steam has no significant effect on ASE (Table 1), the ASE of combinations process was slightly greater than that of MP process.

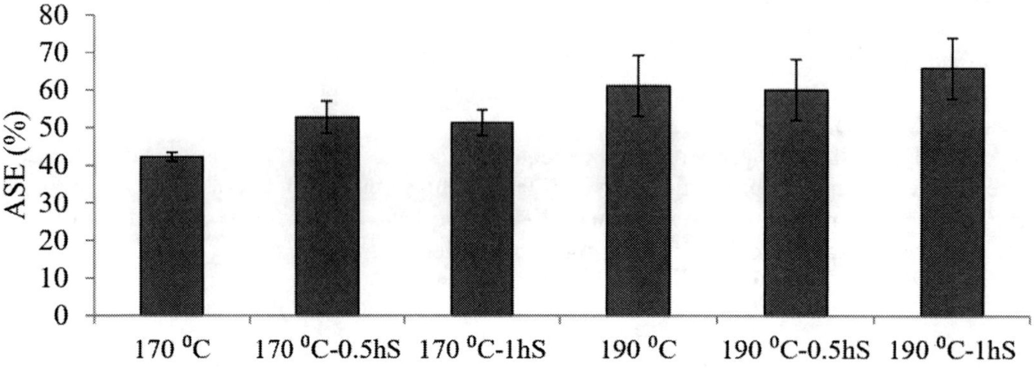


**Figure 7.** Water absorption, swelling, and ASE of MP and MPS treated teakwood

The increase in wood stability indicates a decrease in wood hygroscopicity. Previous study reported that the increased dimensional stability was a contribution from decomposition or transformation chemical component of cell wall especially hemicellulose [9], the increased in relative proportion of crystalline cellulose [21], as well as the decreased in the amount of hydroxyl groups of wood [22].

From the study, it was clear that fast-grown teakwood which is treated by MP or by combination process has better characteristics than control wood. The wood became more stable and its colour almost similar to old teakwood. From the two aspects, treated wood is a good quality raw material for various purposes. Therefore, both of these processes have great potential to be applied in wood industries. If doing so, it will strongly support community teak forest development activities.

## 4. Conclusion

Mild Pyrolysis process generated darkening of wood surface and increase dimensional stability. After processing, wood surface shifted towards red and yellow, while the brightness decreases. The highest dimensional stability was obtained at 190°C with combination of steam for 1 hour. At this process, mass loss reached 16.92%, while water absorption and swelling was low. Steam flow into the pyrolysis reactor during MP process could minimized the negative impact of thermal modification. This combination resulted in lower mass loss but higher dimensional stability of treated wood.

## References

[1] Kartikawati A 2019 Peningkatan mutu kayu jati cepat tumbuh dengan perlakuan mild pyrolysis (Bogor: IPB University)

[2] Wahyudi I, Priadi T, and Rahayu IS 2014 Karakteristik dan sifat-sifat dasar kayu jati unggul umur 4 dan 5 tahun asal Jawa Barat *JIPI* **19**(1) 50-6

[3] Pétrissans A, Hamada J, Chaouch M, Gérardin P, and Pétrissans M 2014 Wood thermodegradation: Experimental analysis and modeling of mass loss kinetics *Cienc. technol.* **16**(2) 133-48

[4] Méndez-Mejías LD and Moya R 2018 Effect of thermo-treatment on the physical and mechanical, color, fungal durability of wood of *Tectona grandis* and *gmelina arborea* from forest plantations *Mater. Sci.* **24**(1) 59-68

[5] Priadi T, Suharjo AAC, and Karlinasari L 2019 Dimensional stability and color change of heat-treated young teak wood *Int. Wood Prod. J.* **10**(3) 119-25

[6] Boonstra M 2008 *A Two-stage Thermal Modification of Wood* (Nancy: Ghent University and Université Henry Poincaré)
[7] Demirbas A 2010 *Biorefineries: For Biomass Upgrading Facilities* (London: Springer-Verlag)
[8] Rapp AO 2001 Review on heat treatment of wood. *Proceeding of Special Seminar* pp. 11-64
[9] Candelier K, Dumarçay S, Pétrissans A, Lyne D, Gérardin P, and Pétrissans M 2013 Comparison of chemical composition and decay durability of heat treated wood cured under different inert atmospheres: Nitrogen or vacuum *Polym. Degrad. Stabil.* **98** 677-81
[10] Missio AL, de Cademartori PHG, Mattos BD, Santini EJ, Haselein CR, and Gantto DA 2016 Physical and mechanical properties of fast-growing wood subjected to freeze-heat treatments *BioResources* **11**(4) 10378-90
[11] Yin Y, Berglund L and Salmén L 2012 Effect of steam treatment on the properties of wood cell walls *Biomacromolecules* **12** 194-202
[12] Cao Y, Jiang J, Lu J, Huang R, Jiang J, and Wu Y 2012 Color change of chinese fir through steam-heat treatment *BioResources* **7**(3) 2809-19
[13] Moya R, Berrocal A, Rodriguez-Solis M, and Muñoz F 2017 Effect of steam-drying treatment on moisture content, drying rate, color and drying defects in juvenile wood of *Tectona grandis* from fast-growth plantations *Drying Technology* **35**(15) 1832-42
[14] Basri E and Balfas J 2015 Stabilisasi dimensi kayu jati cepat tumbuh dan jabon dengan perlakuan pemadatan secara kimia *JPHH* **33**(4) 315-27
[15] Sehlstedt-Persson M 2005 Properties of Solid Wood: Responses to Drying and Heat Treatment [Thesis] Luleå University of Technology Swedia
[16] Esteves B and Pereira H 2009 Wood modification by heat treatment: a review *BioResources* **4**(1) 370-404
[17] Guller B 2012 Effects of heat treatment on density, dimensional stability and color of *Pinus nigra* wood *Afr. J. Biotechnol.* **11**(9) 2204-9
[18] Salca EA, Kobori H, Inagaki T, Kojima Y, and Suzuki S 2016 Effect of heat treatment on color changes of black alder and beech veneers *J. Wood Sci.* **62**(4) 297-304
[19] Candelier K, Dumarçay S, Pétrissans A, Gérardin P, and Pétrissans M 2014 Advantage of vacuum versus nitrogen to achieve inert atmosphere during softwood thermal modification *Proligno* **10** (4) 10-7
[20] Mburu F, Dumarcay S, Huber F, Pétrissans M, and Gérardin P 2007 Evaluation of thermally modified *Grevillea robusta* heartwood as an alternative to shortage of wood resource in Kenya: Characterisation of physicochemical properties and improvement of bio-resistance *Bioresource Technol* **98** 3478-86
[21] Bhuiyan TR, Hirai N, and Sobue N 2000 Changes of crystallinity in wood cellulose by heat treatment under dried and moist conditions *J. Wood Sci.* **46** 431-6
[22] Fabiyi JS and Ogunleye BM 2015 Mid-infrared spectroscopy and dynamic mechanical analysis of heat-treated obeche (*Triplochiton scleroxylon*) wood *Maderas-Cienc. Technol.* **17**(1) 5-16

## Acknowledgments

This research was supported by a grant from the Ministry of Research, Technology and Higher Education of Indonesian Republic through Master Degree Program leading to Doctorate Program for Excellence Students (PMDSU) Batch II. All authors have the same contribution.

IOP Conf. Series: Materials Science and Engineering **935** (2020) 012015 doi:10.1088/1757-899X/935/1/012015

# Potential uses of teak leaf litter for liquid smoke and of other utilizations: A review

**Gusmailina*, S Komarayati, and H S Wibisono**

Forest Products Research and Development Centre, Ministry of Environment and Forestry, Bogor, 16610, Indonesia

*Email: gsmlina@gmail.com

**Abstract.** Teak (*Tectona grandis*) is one of the high-quality wood-producing trees. Large trees, straight trunk, large leafy, with the height reach 30-40 m. The nature and characteristics of teak trees usually drop their leaves every dry season, and this will cause problems for the surrounding community because the dry leaves fall into waste and have not been utilized properly. Waste of dried teak leaves will accumulate under the stands of teak trees, and in the dry season can be a trigger for forest fires. the teak leaves contain 48.51% bounded carbon, 6.61% moisture content, 31.72% ash content, 13.17% volatile content, and density 0.65 g/cm. In addition, teak leaves also contain flavonoid compounds, protein, nine phenolic acid or tannin compounds as well as crude fiber content covered in lignin of 22.9. Based on the content in the dry teak leaf litter waste, this paper presents an effort to use it into a number of useful products, one of which is the utilization of dry teak leaf litter waste for liquid smoke, and several other uses carried out by communities around the teak forest such as for charcoal briquettes, compost, medicine, and animal feed. Analysis of liquid smoke was carried out at the Integrated Laboratory of the Forest Products Research and Development Center using the Pyrolysis GCMS method. Meanwhile, other results are a review of several studies conducted by the community.

## 1. Introduction

The teak forest is a commodity whose processed wood product have been widely used for various needs. Therefore, the sustainability of forest must be maintained because it has social, economic and environmental functions, therefore the main purpose of the forest management is to produce wood sustainably. The characteristic of teak trees that shed their leaves every dry season presents a new problem for the surrounding community which has an impact on environmental pollution and the risk of forest fires due to the large number of dry teak leaves that have not been utilized. Like the case of fire in Blora teak forest on September 2015, it was caused by natural fires due to the abundance of dry teak leaf litter and rubbing against each other.

Teak (*Tectona grandis*) is one of the high quality wood producing trees. Large trees, straight trunk, and can grow to reach 30-40 m high. Teak wood is one of the mainstays commodities as a source of carpentry wood. Broadleaf and large, and will shed leaves every dry season, and this will cause problems for the surrounding community because these fallen dead leaves become trash and have not been utilized optimally (Figure 1a, 1b). Waste of dried teak leaves accumulate under the stands of teak trees, and farmers often burn them for land clearing. It is very dangerous because it can burn other crops (Figure 1a). In addition, burning of teak leaves also destroy the surface layer of the soil (Figure 1c), so that it will reduce the soil moisture. Decreasing soil moisture will reduce the soil water content, and the soil becomes denser with lower soil porosity. The consequences, the lower soil porosity will increase the soil temperature and decrease soil moisture. The higher the soil density, the lower the porosity of the soil the lower the ability of the soil to bind and store and remove the water needed by plants, especially at the beginning of the rainy season. If burning organic litter which is still rather wet will not burn all of it, and will produce solid particles containing hydrocarbon compounds that will fly and are quite dangerous for humans if inhaled breathing [1,2]. Most of the people in Indonesia do not know the potential of these teak leaves and they prefer to burn teak leaves. According to Supriyono and Prehaten [3], chemical content of teak leaves contains 48.51% bounded carbon, 6.61% moisture content, 31.72% ash, 13.17% volatile, and density 0.65g / cm. In addition, teak leaves also contain flavonoid compounds, low protein content of 4.9%, and nine phenolic acid or tannin compounds and crude fiber content covered in lignin which is equal to 22.9% [3]. Therefore, based on the chemical content of the dry teak leaf litter waste, several efforts have been tried to utilize them into useful products, one of which is the utilization of dry teak leaf litter waste for liquid smoke and several other uses carried out by communities around the teak forest.

a b c

**Figure 1**. (a) Teak leaf litter waste, (b) under stands (c) which has the potential to trigger forest fires [1,2]

Liquid smoke or also known as wood vinegar is a by-product of combustion in the process of making charcoal with raw materials containing lignin and cellulose, it can be wood, coconut shell, small twigs or other lignocellulose wastes, such as dry leaf litter teak [4-9]. Smoke from process, if it cooling will turn into liquid, so it is called liquid smoke. The main component of liquid smoke is acetic acid, and other ingredients such as phenol, alcohol, benzoic acid and other chemical compounds that has function as natural pesticides instead of chemical pesticides. Although in relatively small amounts, liquid smoke also contains nutrients such as N, P, K, Ca, Fe, Na, Ca, Mg, and other elements that function as fertilizer. Actually, this liquid smoke has long been used by Indonesian ancestors to treat various diseases. Like toothache, various skin diseases caused by fungi, viruses, bacteria by rubbing on the affected part. It is also used to prevent acute water lice, phlegm, ringworm, ringworm, and herpes.

Liquid smoke is produced by pyrolysis of incomplete combustion which involves the decomposition of polymer constituents into organic molecules with low molecular weight due to the effects of heat which include oxidation, polymerization, and condensation reactions. The cooling media used in the condenser is water that is flowed through the inlet pipe that comes out of the incomplete combustion and then flowed through the condenser and condensed into a smoke distillate. Liquid smoke is known to contain 43.02% acetic acid and 36.22% phenol which can function to prevent plants attacked by pests, especially fungi, caterpillars, and root rot both on agricultural, forestry and plantation crops. In addition, teak leaf ash has a lot of nitrogen which can increase the availability of N elements in the soil so that it can be used by plants as macro nutrients in the growth process.

The application of teak leaf handling technology into liquid smoke is intended to provide more value to the function and market of a product. The use of N replacement sources with forestry sector biomass waste such as teak leaf litter processed into liquid smoke is an alternative solution that can be done. In this case, fertilizing with liquid smoke is one alternative to avoid chemical pollution and is considered suitable to be used as a bio pesticide. This bio pesticide is ready to be used in eradicating pests to increase the productivity of an agricultural land. Besides this liquid smoke can be promoted and sold to the market and even to the international market in the form of packaging so that it creates more value. The sale of liquid smoke in a product can increase employment opportunities so that the standard of living and welfare of the community can increase [10].

In the food industry sector, liquid smoke is used as a specific flavoring and aroma ingredient as well as a preservative due to its anti-microbial properties. Liquid smoke can also replace the traditional fish fogging process that was previously by directly fogging. Liquid smoke can also be used in food processing such as tofu, wet noodles, and meatballs. In the plantation industry, liquid smoke is used as a latex coagulant because of its anti-fungal, and anti-bacterial, as well as to improve the quality of rubber products produced [7].

In Japan, liquid smoke is now commonly used, for example for detoxification drugs. Liquid smoke is mixed with oak wood powder, then packed with filter paper bags, and then put on the soles of the feet before going to sleep. Usually the next day the bag which was originally white turns black to a sign that the poison in the body has been absorbed. Japanese residents have been using liquid smoke for hundreds of years. Some areas in Japan such as Shitara in Aichi Province and in the Togo Islands, are centers for producing liquid smoke. Not only for health, in Togo, liquid smoke is also used as fertilizer and pesticides. As fertilizer, liquid smoke is sprayed on the leaf surface with a concentration of 1:1000 (1 part of liquid smoke and 1000 parts of water) for young plants, for mature plants, liquid smoke solution is splashed around the plants every two weeks. A mixture of liquid smoke and water (1:300) can accelerate the decomposition of compost and prevent the formation of ammonia gas. Therefore, liquid smoke is also useful to eliminate unpleasant odors when sprayed on a pile of garbage, also able to kill certain pests in vegetable cultivation [11].

In the field of food, people in the European Union and the United States are accustomed to eating grilled meat that is first marinated in a liquid smoke solution. The liquid smoke used has been purified through multi-level distillation so that harmful oil and tar are lost. Meat that has been dipped in liquid smoke has a more tender meaty texture and smells delicious when grilled and durable. Liquid smoke functions as a preservative because of the acid, phenol, and carbonyl compounds. Phenol is considered the most important, although the ability of liquid smoke to preserve is the interaction of various compounds. Phenols are antibacterial and antifungal and can inhibit the fat oxidation. Other compounds are also the same as phenols. How to use sprayed on a place attacked by termites at a dose of 15-20 cc/L every three days. To accelerate vegetative plant growth, liquid smoke can be sprayed on leaves at a dose of 20 cc/L every 2 weeks. Sprayed through the roots at a dose of 20 cc/L every 2 weeks [10].

## 2. Method to Get Liquid Smoke

Material in the form of dry teak leaf litter/pieces of wood/twigs, coconut shell, or other lignocellulose material is roughly cut, measured, and weighed before being put into the furnace/kiln. Rough cutting is in order to facilitate the combustion process [11]. Adjusting the oxygen concentration is using an air regulating valve located at the top of the drum. The oxygen regulating valve is opened during the combustion process so that the combustion process takes place quickly, then closed if the material has been burned in the furnace/kiln (Figure 2).

**Figure 2**. Furnaces (kilns) for the production of charcoal and liquid smoke

Smoke from burning material is channeled through a condenser to convert smoke into liquid [11-13]. During the carbonization process, temperature monitoring is carried out using a temperature gauge mounted on the top of the furnace. The pyrolysis stage ends at a temperature of 400-450°C, where the smoke is thinner and has a clear blue color. The initial furnace section is then covered using bricks and sand to prevent complete combustion in the furnace. The pyrolysis charcoal and liquid smoke are then allowed to stand for 24 hours for the cooling and settling process. The final stage of the production of liquid smoke is filtering tar using filter paper or cloth.

## 3. Results and Discussion

### *3.1. Liquid smoke of teak leaf litter*

Teak leaf litter waste which is processed into liquid smoke has a yield that is very dependent on the moisture content of the material. The drier the material the less liquid smoke obtained. The average liquid smoke obtained in this study ranged from 15-20%. Based on the Py-GCMS analysis conducted at the Integrated Laboratory for Forest Products Research and Development Center, there are 15 chemical components contained in the Liquid Smoke of Teak Leaf Litter (ACSDJ) which can be seen in Figure 3. The dominance of the analysis shows that a total of 67.51% consists of acetic acid (acetic acid), 23.2% total phenol, and 3.63% Nitrogen oxide.

The component of alcohol, phenol and acetic acid in liquid smoke is an indication of a compound that has synergy function as a protein regulation and can hydrolyze lipids, so that it can damage the cell membrane of the fungus's body tissue and inactivate enzymes secreted by the fungus [7], Pelczar in [9]. Therefore, liquid smoke has the potential to be anti-germ or anti-microbial. While phenol is a compound that has a characteristic odor and antiseptic properties, the function of phenol among others is in regulating the activity of certain enzymes, is toxic to insects and to plant predators Wawoeruntu *et al.* in [14]. Phenols also affect the development of termites, anti-fungal and bacterial Cowan in [14]. Other compounds besides phenols, furfural compounds, function as pest control of the Nematode animal group. According to Wiyono in [15], liquid smoke contains alkaloids and secondary metabolites that can be used as pesticides. In addition, the high content of acetic acid in liquid smoke of teak leaf litter (67.51%), is more suitable as a stimulant and fertilizer for plant growth, compared to the phenol content and its derivatives of 23.2%, but if applied together, it will be more optimal both as a stimulant and plant fertilizer, as well as prevent the attack of plant diseases by microbes (fungi and bacteria).

IOP Conf. Series: Materials Science and Engineering **935** (2020) 012015 doi:10.1088/1757-899X/935/1/012015

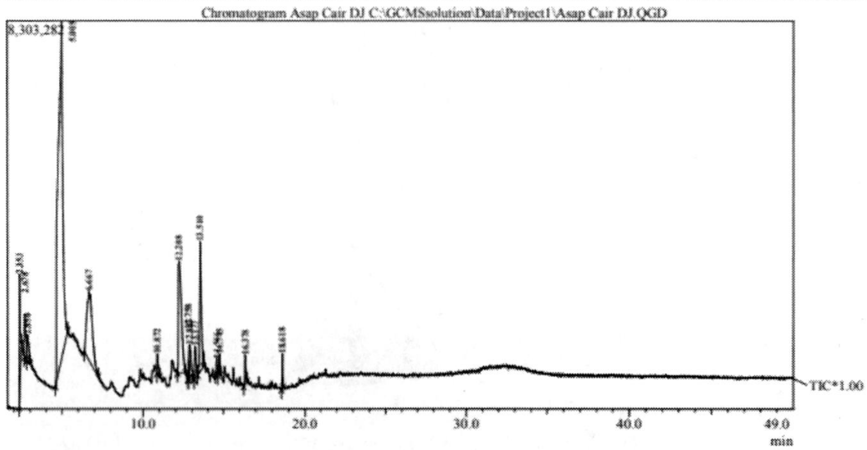


Peak Report TIC

| Peak# | R.Time | Area | Conc% | Name |
|---|---|---|---|---|
| 1 | 2.353 | 9867917 | 3.63 | Nitrogen oxide (N2O) (CAS) Nitrous oxide |
| 2 | 2.678 | 5116434 | 1.88 | Ethanol (CAS) Ethyl alcohol |
| 3 | 2.850 | 4186346 | 1.54 | Acetic acid, hydrazide (CAS) 374 |
| 4 | 5.005 | 144592996 | 53.16 | Acetic acid (CAS) Ethylic acid |
| 5 | 6.667 | 34836314 | 12.81 | Acetic acid (CAS) Ethylic acid |
| 6 | 10.872 | 1987463 | 0.73 | 2(3H)-Furanone, dihydro- (CAS) Butyrolactone |
| 7 | 12.208 | 31738433 | 11.67 | Phenol (CAS) Izal |
| 8 | 12.758 | 1330337 | 0.49 | 2-Cyclopenten-1-one, 2-hydroxy-3-methyl- (CAS) Corylon |
| 9 | 12.847 | 5662856 | 2.08 | 2,3-Dimethyl-2-cyclopenten-1-one |
| 10 | 13.177 | 4259552 | 1.57 | Phenol, 2-methyl- (CAS) o-Cresol |
| 11 | 13.510 | 20852701 | 7.67 | Phenol, 4-methoxy- (CAS) Hqmme |
| 12 | 14.566 | 1685094 | 0.62 | Phenol, 3-ethyl- (CAS) m-Ethylphenol |
| 13 | 14.745 | 2046080 | 0.75 | 2-Methoxy-4-methylphenol |
| 14 | 16.378 | 2513534 | 0.92 | Phenol, 2,6-dimethoxy- (CAS) 2,6-Dimethoxyphenol |
| 15 | 18.618 | 1315771 | 0.48 | Naphthalene, 1,4,6-trimethyl- (CAS) 1,4,6-Trimethylnaphthalene |
| | | 271991828 | 100.00 | |

**Figure 3.** Teak leaf liquid smoke chromatogram

### *3.2. Utilization of teak leaf litter by the community*

#### 3.2.1. Charcoal briquettes

The Partnership of the Institute of Research and Community Service (LPPM) of IPB, the Department of Physics-FMIPA IPB and the Blora Forest Service, carried out the use of teak leaf litter into charcoal briquettes (Figure 4), bio pellet, activated charcoal, and liquid smoke, which were processed in a simple process. Teak leaf charcoal is used as an enhancer of land fertility. Active teak charcoal is used to absorb pollutant waste, so that agricultural land is not easily damaged and also improve land quality. At present, more than 50 people who are members of farmer groups in Blora are carrying out this activity, but the products produced are still limited to meeting their own needs [17].

**Figure 4.** Charcoal briquettes of teak leaf [17]

In Purwodadi, Tepus, Gunung Kidul, there are around 1,219,059 teak trees. Teak trees that drop their leaves every dry season provide a new problem for the surrounding community. These dead leaves that fall into waste that has not been properly utilized and disrupt agriculture. Therefore, the manufacture of charcoal briquettes is an alternative use of this dry teak leaf litter [17]. Testing the calorific value of dried teak leaves is conductedin a laboratory Chemical processing and energy for non-timber forest products using a bomb-calorimeter. From the test results it was found that the calorific value of dried teak leaves ranged from 3902-4117 cal/gr. The process begins with making teak leaf charcoal waste, then mixed with adhesive and then printed. Dry teak charcoal briquette dried, and tested in accordance with national briquette standards. The parameters tested were the calorific value (cal/gr), vapor content (%), ash content (%), moisture content (%), vaporizer content (%), bounded carbon content (%), density ($g/cm^3$), and compressive firmness ($kg/cm^2$). The test results showed that almost all the quality parameters of dry teak leaf charcoal briquettes still did not meet the specified standards, except for the parameters of water content and briquette density, except after mixed with coconut shell charcoal which has high quality standards, the parameters of teak leaf charcoal briquettes meet national briquette standards. The calorific value has increased to more than 500 cal/gr to 5800 cal/gr, a relatively high value Laila and Arham on [17].

3.2.2. Teak leaf waste compost
CV Bahari Mitra Surya, has used teak leaf waste for compost made in Plembutan Village, Playen District. The company chose dry teak leaves because the company saw that Gunung Kidul's earth was mostly planted with teak trees. Dry teak leaves are easy to find, making it easy for farmers in Gunung Kidul to collect dried teak leaf waste. After the teak leaves are collected, farmers sell it to CV Bahari Mitra Surya at a price of IDR 325. Utilization of dried teak leaves has opened employment opportunities for local residents. Dry teak leaves are milled until smooth, then fermented for one week using water and sugar cane drops. After one week fermented, teak leaf waste is dried for two hours in the sun, then put into a sorting room. After the sorting process is complete, then packaged, then sent to Japan [29]. One week the company can produce 2560 compost fertilizer packages Wahyuni *et al.* in [2].

3.3.3. Benefits of Teak Leaves in the Health Sector
Some people who live around the teak forest, teak leaves are useful for treating health and reducing the symptoms of several diseases, although it has not been able to completely replace medical treatment. Some of the diseases that have benefited include the following:

a. Helps reduce asthma symptoms. Teak tree leaves have benefits to reduce and prevent asthma. Extracts from the leaves of teak plants had a significant effect as an anti-asthma
b. Help treat intestinal worms. Leaves from teak trees are believed to fight parasitic infections such as worms. Extracts from teak leaves can be used to treat intestinal worms. The study was conducted by determining the time of paralysis and death of the worms against standard piper zine citrate drugs. As a result, teak tree leaves have quite a powerful effect such as piper zine citrate against disease-causing worms
c. Skin care. Teak leaf extract can be used as an anti-inflammatory agent on the skin, which is taken by squeezing or grinding. Then the juice is used to treat various skin diseases due to inflammation, such as acne, and can help overcome itchy skin [3]
d. Diuretic agent. Teak tree leaves are believed to help diuretics in the body so that it becomes more frequent urination. According to Phalphale *et al.* [18] aqueous extract from teak tree leaves has the diuretic effect.
e. Rich in antioxidants. Teak leaves contain antioxidants that are good for the body against free radicals. Ramachandran *et al.* [19] found that phenolic components in teak leaves have excellent antioxidant agents. Free radicals themselves can be a cause of cancer cell growth and premature aging.

f. Accelerate wound healing. According to Majumder *et al.* [20], the front of the teak leaves can be used as a wound healer, especially for blisters or burns. This study evaluated hydrochloric extracts from teak leaves in rats. The results show that the leaves of teak tree can accelerate the repair of damaged cells and skin tissue so that the wound heals faster.
g. Stimulates hair growth. Ragasa *et al.* [21] found that oils from teak leaves can be used to accelerate hair growth. Then Jaybhaye [22] also found that seeds from teak plants can be used as hair tonics. So those who want to have long hair, and experience hair loss can benefit from teak leaves.
h. Antifungal. Astiti and Suprapta [23] suggested that the anti-fungal activity of *Arthrinium phaeospermum* the extracts of dried teak tree leaves showed that fungal growth could be prevented.
i. Laxative agent. Teak tree leaves can be used as a natural laxative or laxative. Teak leaves work as a stimulant and encourage the release of feces (dirt) from the intestine. Therefore, teak leaves can treat constipation.
j. Against disease-causing bacteria. Teak leaves have properties against *Listeria monocytogenes*, the bacteria that causes listeriosis, which is found in food. In addition, teak tree leaves can also inhibit the growth of *Staphylococcus aureus* bacteria and other bacteria causing infections.

3.3.4. Benefits of teak leaves for people with diabetes

Teak trees are not only the wood that is used to make household furniture, but teak leaves also have health benefits. The benefits of teak leaves in the health sector are widely used, one of which is to treat diabetes. The content contained in teak leaves are as follows:

- Saponin is a type of compound that functions as an antioxidant in the human body that is useful to fight free radical attack from outside. Antioxidants work to prevent premature aging of the skin. As is known antioxidants can prolong the aging process on the skin.
- Tannins in the health sector function as anti-bacterial agents against bacteria from the outside so that it can boost the immune system.
- Kuinon functions as an addition of vitamins to the body to meet the needs of vitamins for the body.
- Flavonoids are useful to fight allergies in the body so that the body is safe from the effects of allergies in the form of itching and swelling. Flavonoids are one of the most common groups of secondary metabolites found in plant tissue. Flavonoids are included in the group of phenolic compounds with the chemical structure of C6-C3-C6. The flavonoid framework consists of an aromatic ring A, an aromatic ring A, an aromatic ring B, and a middle ring in the form of oxygen-containing heterocyclic and oxidized form of the ring which is the basis for the division of flavonoids into the sub groups. The numbering system is used to distinguish the position of carbon around the molecule Cook in [24].
- Quercetin is used to prevent heart disease. Today's heart disease is one of the number one deadliest disease. So that awareness to care for the heart to stay healthy is needed early Cook in [24].
- The content contained in teak leaves is useful for treating diabetes. Diabetes is a complication of the body against the hormone insulin which cannot be used properly [25].

3.3.5. Benefits of teak leaves in other health areas

Not only to treat diabetes, but the decoction of teak leaves also has other health benefits such as treating diseases:

a. *Cholesterol*. The first benefit of teak leaves is to reduce cholesterol levels from the body. As it is known that high cholesterol levels in the body can cause complications for heart health. The heart is the most vital organ in human life. The heart's duty is to pump blood throughout the body so that all organs can work properly. Many factors can increase cholesterol levels in the body one of them is food such as fat, meat, cheese and fast food. Therefore, it is important to avoid foods that have adverse effects on health in the long run. Make it a habit to include fresh fruits and vegetables in your daily menu. In general, normal cholesterol levels between 160-200 mg. And not all cholesterol is bad for health. Cholesterol also has a good function in the form of protection for the body in the health sector. Cholesterol is good for the body as long as it is within the normal amount [26].

b. *Hypertension*. High blood pressure is a type of disease that is still associated with the heart. One of the benefits of drinking teak leaves is to normalize blood pressure in the body. Teak leaves contain saponin compounds that are useful for antioxidants that are sourced from outside the body [26].
c. *Anemia*. Anemia is caused by the reduced ability of hemoglobin to bind oxygen in the blood. If this ability is reduced, the body will become weak and tired easily. Anemia can be categorized as a non-communicable disease. This anemia if left unchecked can have fatal consequences such as sudden death. Anemia can be treated with natural ingredients such as stew from teak leaves. Teak leaf stew can be made at home and more practical than using other chemical drugs. Not to mention the natural benefits of this material do not cause side effects [26].

#### 3.3.6. Benefits of teak leaf waste for alternative animal feed

Farmers in Girisekar Village, Panggang District, Gunung Kidul Regency, DIY, succeeded in making alternative animal feed from teak leaf waste. Teak leaf waste is very easy to find in the breeders environment because in Gunungkidul there are indeed many teak plantations. This teak leaf waste is processed into animal feed by fermentation. Assisted by 28 KKN students (Kuliah Kerja Nyata) from Gadjah Mada University, the breeders in 3 hamlets in Girisekar Village namely Warak Hamlet, Sawah Hamlet, and Bali Hamlet succeeded in processing the environmental waste around them into valuable goods. Not only teak leaf waste can be processed into alternative feed, rice waste such as corncobs, tebon, yams, straw, and banana gedebog can also be utilized. From its geographical location, the Gunung Kidul area is categorized as hilly and barren land so that in the dry season it is difficult for farmers to find forage feed such as Kolonjono grass. The breakthrough to make alternative animal feed from environmental waste is certainly very helpful for farmers [27].

The process of making alternative animal feeds by fermentation is very easy. First of all, dry teak leaf waste and other waste is chopped using a chopper then mixed with bran. Next, the ingredients are moistened with a mixture of water, molasses, and liquid organic supplements (SOC). SOC is an organic supplement made from herbal ingredients that can stimulate the appetite and growth of farm animals. All ingredients that have been mixed are then put in a closed barrel for several days until they are fully fermented [27]. The process of making alternative animal feed from dried teak leaf waste is cost-effective because it only uses molasses and SOC which is more affordable than buying feed on the market. In addition, alternative feed is also environmentally friendly because it uses organic ingredients and reduces the amount of waste. Farmers in Girisekar Village, Panggang District, Gunung Kidul Regency, DIY, succeeded in making alternative animal feed from teak leaf waste. Teak leaf waste is very easy to find in the breeders' environment because in Gunung Kidul there are indeed many teak plantations. This teak leaf waste is processed into animal feed by fermentation [27].

#### 3.3.7. Benefits of teak leaf waste as soap material

Teak leaves are extracted by maceration method using two types of n-hexane and ethanol solvent. Liquid soap preparations from teak leaf extract with a concentration of 0.01%, 0.02%, 0.03% can inhibit the growth of *Escherichia coli* bacteria that is equal to 15 mm, 17 mm, 19 mm, whereas the inhibitory power of *Staphylococcus aureus* is 15 mm, 17 mm, and 19 mm. Increasing the concentration of teak leaf extract in liquid soap can increase the inhibition of bacterial growth. The results research of Agnes [28], showed that the quality of all teak extract liquid soap formulas proved to be in accordance with SNI standards consisting: organoleptic tests, pH measurements, homogeneity tests, and foam height measurements.

**Figure 5**. Process of making teak leaf waste for alternative animal feed [27]

## 4. Conclusion

The potential of community forests that have not been utilized at this time, such as teak leaves that can be processed into liquid smoke which is an alternative material for bio pesticides and stimulants and fertilizers in increasing the productivity of agricultural land, soil fertility, and opening new jobs. In utilizing forests, it does not have to be wood, but there are many potentials that have not been optimally utilized, such as teak leaf litter. Liquid smoke from teak leaf litter waste has a high content of acetic acid, which has the potential to be used for stimulants and plant fertilizers.

Teak leaf litter waste used for charcoal briquettes can be improved by mixing quality with coconut shell charcoal, as an alternative household energy.

Waste of teak leaf litter by the surrounding community is used to cure several kinds of diseases that are closely related to body health, used for compost which is exported to Japan, even teak leaf litter waste is also used for alternative animal feed.

## References

[1] Bapelluh Karang Anyar 2013 Daun jati kering mengapa harus dibakar (bapelluh_karanganyar.com)

[2] CV Bahari Mitra Surya 2015 Limbah daun jati bisa dimanfaatkan sebagai pupuk kompos (Playen Kh)

[3] Supriyono H and Prehaten D 2014 Kandungan unsur hara dalam daun jati yang baru jatuh pada tapak yang berbeda *Jurnal Ilmu Kehutanan* **8**(2)

[4] Nurhayati T, Roliadi H, and Bermawie N 2005 Production of mangium wood vinegar and its utilization *Journal of Forestry Research* **2**(1) 13-26

[5] Nurhayati T, Pasaribu RA, and Mulyadi D 2006 Produksi dan pemanfaatan cuka kayu dari serbuk gergaji kayu campuran *Jurnal Penelitian Hasil Hutan* **24**(5) 395-411

[6] Komarayati, S, Gusmailina, and Pari G 2011 Produksi cuka kayu hasil modifikasi tungku arang terpadu *Jurnal Penelitian Hasil Hutan* **29**(3) 234-47

[7] Wibowo S 2012 Karakteristik asap cair tempurung nyamplung *Jurnal Penelitian Hasil Hutan* **30**(3)

[8] Komarayati S and Wibowo S 2015 Karaketristik asap cair dari tiga jenis bambu *Jurnal Penelitian Hasil Hutan* **33**(2) 167-74

[9] Aisyah I, Juli N, and Pari G 2013 Pemanfaatan asap cair tempurung kelapa untuk mengendalikan cendawan penyebab penyakit antraknosa dan layu Fusarium pada ketimun *Jurnal Penelitian Hasil Hutan* **31**(2) 170-8

[10] Darmadji P 2002 Optimasi pemurnian asap cair dengan metode redistilasi *Jurnal Teknologi dan Industri Pangan* **8**(3) 267-71

[11] Ardila CM, Lopez MA, and gusman IC 2010 High resistance against clindamycin, metronidazole and amoxicillin in porphyromonas gingivalis and aggregatibacter actinomycetemcomitans isolates of periodontal disease *Med. Oral Patol. Oral Cir. Bucal.* **15**(6) 947-51
[12] Ariyani D, Mujiyanti DR, Umaningrum D, and Halianto YA 2015 Studi kajian: Kandungan senyawa pada asap cair dari sekam padi *Prosiding Seminar Nasional Kimia Universitas Negeri Surabaya*
[13] Karima R 2014 Karakterisasi sifat fisika dan kimia cuka kayu dari tandan kosong kelapa sawit *Jurnal Riset Industri Hasil Hutan* **6**(1)
[14] Darmawan UW 2014 Cuka kayu sebagai pengendali hama dan penyakit tanaman *FORPRO* **3**(1) 25-9
[15] Rahimah DS 2014 Asap Usir Elmaut *Trubus* **536** *XLV*
[16] Supriyanto and Cahyo 2016 *Grow Your Own Fruits: Monograf Panduan Praktir Menanam 28 Tanaman Buah Populer di Pekarangan* (Yogyakarta: Lily Publisher, Penerbit Andi)
[17] Laila, Nurul A, and Arham SD 2017 Pemanfaatan limbah daun jati kering menjadi briket di Desa Purwodadi, Tepus, Gunung Kidul https://enerbi.co.id/2017
[18] Phalphale SG, Ashish G, Biyani KR, Shete RV, Kore KJ, Chaudhari SR, and Samadhan M 2013 Evaluation of diuretic activity of *Tectona grandis* linn. in rats *World J. Pharm. Sci.* **2** 245-52
[19] Ramachandran A and Snehalata C 2009 *Diabetes melitus* (Jakarta: EGC) pp. 407-8
[20] Majumder R, Hasnat NE, Md Ashraf Md UZ, and Alam Md B 2014 In vivo evaluation of the pharmacological activities of *Syzygium samarangense* (Blume) Merr. *Advances in Biological Research* **8**(3) 107-15
[21] Ragasa, Consolacion Y, Tsai PW, and Shen CC 2009 Terpenoids and Steroids from the endemic and endangered Philippine trees *Ficus pseudopalmaand, Ficus ulmifolia Philipp. J. Sci.* **138**(2) 205-9
[22] Jaybhaye S, Sharon M, Ansaldo A, Ricci D, and Sing L 2009 Taguchi methodology to grow single-walled carbon nanotubes on silicon wafer. *Proceedings 2009 9th IEEE Conference on Nanotechnology (IEEE-NANO)*
[23] Astiti NPA and Suprapta DN 2012 Antifungal activity of teak (*Tectona grandis* L.f.) leaf extract against *Athrinium phaeospermum* (Corda) MB Ellis the cause of wood decay on *Albizia falcataria* (L.) Fosberg *Journal of ISSAAS* **18**(1) 62-9
[24] Redha A 2010 Flavonoid: Struktur, sifat antioksidatif dan peranannya dalam sistem biologis *Jurnal Belian* **9(**2) 196-20
[25] Firdaus, Yusra, and Arsanta MD 2014 Memanfaatkan 10 khasiat daun jati bagi kesehatan tubuh. https://beneficial.co.id/Benefit-daun-jati-for-diabetes
[26] Firdaus Y 2017 Memanfaatkan 10 khasiat daun jati bagi kesehatan tubuh https://hellosehat.com
[27] Loka Penelitian Sapi Potong Grati-Pasuruan 2014 Manfaatkan limbah daun jati untuk alternatif pakan ternak https://sekarmulyo.wordpress.com/2014/08/
[28] Agnes JC 2016 Uji aktivitas sabun cair ekstrak daun jati (*Tectona Grandis* L.F.) sebagai antibakteri terhadap *Staphylococcus aureus* dan *Escherichia coli* [Skripsi] Jurusan Kimia Fakultas Matematika dan Ilmu Pengetahuan Alam Universitas Negeri Semarang
[29] Suci TW and Korda W 2018 Gunung Kidul: pupuk organik daun jati diekspor ke Jepang *http://www.sapa.or.id/lp/118-diy/2813-penanggulangan-kemiskinan*

IOP Conf. Series: Materials Science and Engineering **935** (2020) 012016 doi:10.1088/1757-899X/935/1/012016

# The consumer preferences for new styrax based perfume products using a conjoint analysis approach

**C R Kholibrina* and A Aswandi**

Environment and Forestry Research Development Institute of Aek Nauli, Simalungun, 21174, Indonesia

*Email: rizlanicut@gmail.com

**Abstract**. Thousands of tons of raw styrax resin are exported, while various derivative products with millions of dollars in value are imported for centuries. *Tobarium Styrax Perfume* is the first domestic product that explores the styrax essential oil formula for signature perfumes. The objective of study is to determine consumer preferences for a new styrax based perfume product. A conjoint analysis approach was utilized to determine consumer preferences for these innovation products. Fifty consumers were asked to provide an assessment of the combination of perfume products. Data processing was assisted by computer with SPSS 14.0 program. The influence of five product profiles on consumer preference including scent variant, package, volume, grade of essential oil, and prices were examined. The results show that higher utility values indicate greater preference. The highest utility value is indicated by factor grade (essential oil concentration), this illustrates the respondents' greatest preference for the quality of signature perfume. The scent variant (importance value 53.958) has the most influence on overall preference. This means that there is a large difference in preference between product profiles containing the most and the least desired scents. This is because the preference for a particular scent is personal and unique which might have affected the perception of respondents. The highest consumer's preference is a perfume with scent *Rizla*, package spray, volume 25 mL, grade signature perfume and price IDR 120,000. The grade and price play a significant role but not as significant as scent variant. Two statistics, *Pearson's R* and *Kendall's tau*, provides high correlation between the observed and estimated preferences. This information is expected to encourage this domestic innovation product development.

## 1. Introduction

Styrax resin has been traded throughout the civilization in Nusantara. Although it contributes significantly to community's livelihood, the added value of endemic and historic non-timber forest products has not yet been explored [1]. Thousands of tons of raw resin are exported, while various styrax resin derivative products in the form of medicines, perfumes, beauty and food ingredients with millions of dollars in value are imported annually [2]. One of the added value schemes is product innovation of *Tobarium Styrax Perfume*. This perfume is the first domestic product that utilizes styrax essential oil as a base-note for signature perfumes [3].

The main consideration of styrax perfume innovation is multi-dimensional benefits of styrax resin as a source of community's livelihood. Increasing added value is implemented by processing the raw materials into of high economic perfume products [3-5]. Knowledge about aromatic compounds of styrax resin is very limited in domestic communities. The resin associated with the scent of burned incense and it is underutilized for high-value products for centuries. The consequences, their aromatic formulas are unexplored, and potential for higher prices does not met in the local market.

*Tobarium Styrax Perfume* is a research invention which explores the incense fragrance formulation that known as god's resin and used by almost all religions and beliefs in the world. The market prospect to be targeted is to meet a demand for imported perfume products which reaches USD 401 million per year [6,7]. The consumer awareness for organic products is also an opportunity to develop these quality products.

Since the last decade, consumers have become more selective for a personal care product that they will use. Although many similar products are available, high-quality products and competitive prices will fulfill customer's satisfaction and increase the loyalty to the product. To fulfill the customer satisfaction, Tobarium Styrax Perfume provides six main scent variants including *Rizla* (floral fresh), *Azwa* (woody), *Riedha* (fruity fresh), *Aphis* (blue oceanic), *Jeumpa* (cananga) and *Tiara* (oud oriental). These products are packaged in various sizes that can be adjusted to consumer's needs and desires. The market segments being targeted are middle up society peoples and young executives who like exotic and energetic aromas; tourists visiting Lake Toba; aromatherapy oil collector; organic product buyers; and hospitality industry [4,6,7].

The objective of study is to determine consumer preferences for a new styrax based perfume product innovation. The information about some important attributes considered by consumers and how the combination of perfume product attributes that meet consumer needs and desires are expected to encourage the development of this domestic innovation product. Impacts of product development include increasing the value added of styrax resin that driving the local community welfares. The innovation product will increase the product efficiency and national competitiveness and it will build the national pride on domestic products.

## 2. Materials and Methods

A conjoint analysis approach was utilized to determine consumer preferences for *Tobarium Styrax Perfume* products. This approach was also applied in several previous studies [8,9]. Conjoint analysis is a research tool for developing effective product design. Using this analysis, the researcher can answer questions such as: What product attributes are important or unimportant to consumer? What levels of product attributes are the most or least desirable in the consumer's mind? The main result of this analysis is a design of a product or service, or a particular object desired by most respondents [9].

The scope for research is *Tobarium Styrax perfume* product with an analysis unit of 50 customers of this new innovation product. Respondents were selected purposively and interviewed during October 2019 to March 2020. The consumers were asked to provide an assessment of the combination of perfume products including scent, packaging, volume, essential oil concentrations and prices of perfume products.

Data processing includes generate orthogonal design and running the conjoint analysis were assisted by computer with SPSS 14.0 program. Two statistics analyses, *Pearson's R* and *Kendall's* tau were measured for correlation between the observed and estimated preferences.

## 3. Results and Discussion

### *3.1. Generating an orthogonal design*

Generate orthogonal design generates a data file containing an orthogonal main-effects design that permits the statistical testing of several factors without testing every combination of factor levels. This design can be displayed with the Display Design procedure [9].

This study examined the influence of five factors on consumer preference-scent, package, volume, grade essential oil, and prices. There are six factor levels for scent, each one differing in scent variant (*Rizla, Azwa, Riedha, Aphis, Jeumpa* or *Tiara*); two package levels (roll on spray); two volume levels (15 mL and 25 mL); two grade levels (perfume or signature perfume) and three levels price (IDR 120,000; IDR 150,000; IDR 200,000) for last factor. Table 1 displays the variables used with their labels and values.

IOP Conf. Series: Materials Science and Engineering **935** (2020) 012016 doi:10.1088/1757-899X/935/1/012016

**Table 1**. Variables in *Tobarium styrax* perfume consumer's preferences

| Factor | Values | Label |
|---|---|---|
| Scent | 1, 2, 3, 4, 5, 6 | *Rizla, Azwa, Riedha, Aphis, Jeumpa, Tiara* |
| Package | 1, 2 | Roll on spray |
| Volume | 1, 2 | 15 mL, 25 mL |
| Grade | 1, 2 | perfume, signature perfume |
| Price | 1, 2, 3 | IDR 120,000; IDR 150,000; IDR 200,000 |

The first step is to create the combinations of factor levels that are presented as product profiles to the subjects. The generate-orthogonal design procedure creates an orthogonal array, also referred to as an orthogonal design. The combinations of factor levels are displayed in Table 2.

**Table 2**. Display of orthogonal design

| Card ID | Scent variant | The packaging | Volume | EO grade | The price |
|---|---|---|---|---|---|
| 1 | Rizla | spray | 25 mL | perfume | IDR 150,000 |
| 2 | Azwa | rollon | 15 mL | signature perfume | IDR 120,000 |
| 3 | Azwa | spray | 25 mL | perfume | IDR 200,000 |
| 4 | Rizla | spray | 15 mL | signature perfume | IDR 120,000 |
| 5 | Azwa | rollon | 15 mL | perfume | IDR 150,000 |
| 6 | Riedha | rollon | 15 mL | perfume | IDR 150,000 |
| 7 | Jeumpa | spray | 15 mL | perfume | IDR 120,000 |
| 8 | Tiara | spray | 15 mL | perfume | IDR 150,000 |
| 9 | Azwa | rollon | 15 mL | perfume | IDR 120,000 |
| 10 | Jeumpa | rollon | 25 mL | signature perfume | IDR 200,000 |
| 11 | Azwa | rollon | 25 mL | perfume | IDR 200,000 |
| 12 | Aphis | rollon | 25 mL | signature perfume | IDR 150,000 |
| 13 | Rizla | rollon | 15 mL | signature perfume | IDR 200,000 |
| 14 | Azwa | spray | 15 mL | signature perfume | IDR 150,000 |
| 15 | Riedha | rollon | 25 mL | perfume | IDR 120,000 |
| 16 | Rizla | rollon | 15 mL | perfume | IDR 200,000 |
| 17 | Riedha | rollon | 15 mL | signature perfume | IDR 150,000 |
| 18 | Rizla | rollon | 15 mL | perfume | IDR 120,000 |
| 19 | Aphis | rollon | 15 mL | perfume | IDR 120,000 |
| 20 | Aphis | spray | 15 mL | perfume | IDR 200,000 |
| 21 | Jeumpa | rollon | 15 mL | perfume | IDR 150,000 |
| 22 | Riedha | rollon | 15 mL | perfume | IDR 200,000 |
| 23 | Tiara | rollon | 15 mL | perfume | IDR 200,000 |
| 24 | Riedha | spray | 15 mL | signature perfume | IDR 200,000 |
| 25 | Riedha | spray | 25 mL | perfume | IDR 120,000 |
| 26 | Tiara | rollon | 25 mL | signature perfume | IDR 120,000 |
| 27 | Rizla | rollon | 25 mL | perfume | IDR 150,000 |
| 28 | Rizla | spray | 15 mL | signature perfume | IDR 200,000 |
| 29 | Aphis | rollon | 25 mL | perfume | IDR 120,000 |
| 30 | Jeumpa | rollon | 15 mL | signature perfume | IDR 200,000 |
| 31 | Riedha | rollon | 15 mL | signature perfume | IDR 120,000 |

Remarks:

```
*Generate Orthogonal Design.
SET SEED 2000000.
ORTHOPLAN
  /FACTORS=scent 'scent variant' (1 'Rizla' 2 'Azwa' 3 'Riedha' 4 'Aphis' 5 'Jeumpa' 6 'Tiara') package 'the packaging' (1 'rollon' 2 'spray') volume 'volume' (1 '15mL' 2 '25mL') grade 'EO grade' (1 'perfume' 2 'signature perfume') price 'the price' (1 'IDR120,000' 2 'IDR150,000' 3 'IDR200,000')
  /REPLACE
  /MINIMUM 18
  /HOLDOUT 4
  /MIXHOLD NO.
```

Once an orthogonal design was created, it necessary to create the product profiles to be rated by the subjects or respondents. A listing of the profiles in a single table can be obtained or display each profile in a separate table or plan card. The information for each product profile is displayed Figure 1.

**Profile Number 1**

| Card ID | scent variant | the packaging | volume | EO grade | the price |
|---|---|---|---|---|---|
| 1 | Rizla | spray | 25mL | perfume | IDR150,000 |

**Profile Number 2**

| Card ID | scent variant | the packaging | volume | EO grade | the price |
|---|---|---|---|---|---|
| 2 | Azwa | rollon | 15mL | signature perfume | IDR120,000 |

**Profile Number 3**

| Card ID | scent variant | the packaging | volume | EO grade | the price |
|---|---|---|---|---|---|
| 3 | Azwa | spray | 25mL | perfume | IDR200,000 |

**Figure 1**. Example of some plan cards for each product profile

### *3.2. Running the analysis*

After an orthogonal design was generated, consumer preferences for each product profile were obtained from 50 respondents as listed in Figure 2.

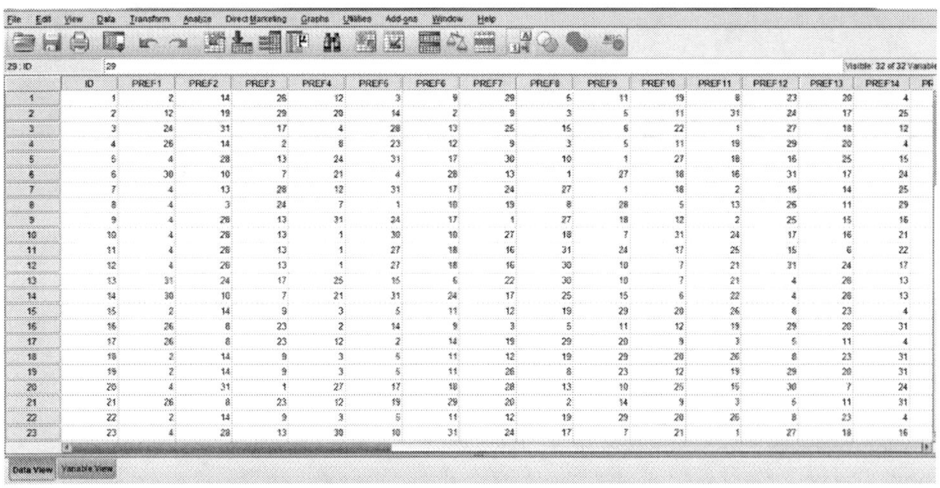

| | ID | PREF1 | PREF2 | PREF3 | PREF4 | PREF5 | PREF6 | PREF7 | PREF8 | PREF9 | PREF10 | PREF11 | PREF12 | PREF13 | PREF14 |
|---|---|---|---|---|---|---|---|---|---|---|---|---|---|---|---|
| 1 | 1 | 2 | 14 | 26 | 12 | 3 | 9 | 29 | 6 | 11 | 19 | 8 | 23 | 20 | 4 |
| 2 | 2 | 12 | 19 | 29 | 20 | 14 | 2 | 9 | 3 | 5 | 11 | 31 | 24 | 17 | 25 |
| 3 | 3 | 24 | 31 | 17 | 4 | 28 | 13 | 25 | 15 | 6 | 22 | 1 | 27 | 18 | 12 |
| 4 | 4 | 26 | 14 | 2 | 8 | 23 | 12 | 9 | 3 | 5 | 11 | 19 | 29 | 20 | 4 |
| 5 | 5 | 4 | 28 | 13 | 24 | 31 | 17 | 30 | 10 | 1 | 27 | 18 | 16 | 25 | 15 |
| 6 | 6 | 30 | 10 | 7 | 21 | 4 | 28 | 13 | 1 | 27 | 18 | 16 | 31 | 17 | 24 |
| 7 | 7 | 4 | 13 | 28 | 12 | 31 | 17 | 24 | 27 | 1 | 18 | 2 | 16 | 14 | 25 |
| 8 | 8 | 4 | 3 | 24 | 7 | 1 | 10 | 19 | 8 | 28 | 5 | 13 | 26 | 11 | 29 |
| 9 | 9 | 4 | 28 | 13 | 31 | 24 | 17 | 1 | 27 | 18 | 12 | 2 | 25 | 15 | 16 |
| 10 | 10 | 4 | 28 | 13 | 1 | 30 | 10 | 27 | 18 | 7 | 31 | 24 | 17 | 16 | 21 |
| 11 | 11 | 4 | 28 | 13 | 1 | 27 | 18 | 16 | 31 | 24 | 17 | 25 | 15 | 6 | 22 |
| 12 | 12 | 4 | 28 | 13 | 1 | 27 | 18 | 16 | 30 | 10 | 7 | 21 | 31 | 24 | 17 |
| 13 | 13 | 31 | 24 | 17 | 25 | 15 | 6 | 22 | 30 | 10 | 7 | 21 | 4 | 28 | 13 |
| 14 | 14 | 30 | 10 | 7 | 21 | 31 | 24 | 17 | 25 | 15 | 6 | 22 | 4 | 28 | 13 |
| 15 | 15 | 2 | 14 | 9 | 3 | 5 | 11 | 12 | 19 | 29 | 20 | 26 | 8 | 23 | 4 |
| 16 | 16 | 26 | 8 | 23 | 2 | 14 | 9 | 3 | 5 | 11 | 12 | 19 | 29 | 20 | 31 |
| 17 | 17 | 26 | 8 | 23 | 12 | 2 | 14 | 19 | 29 | 20 | 9 | 3 | 5 | 11 | 4 |
| 18 | 18 | 2 | 14 | 9 | 3 | 5 | 11 | 12 | 19 | 29 | 20 | 26 | 8 | 23 | 31 |
| 19 | 19 | 2 | 14 | 9 | 3 | 5 | 11 | 26 | 8 | 23 | 12 | 19 | 29 | 20 | 31 |
| 20 | 20 | 4 | 31 | 1 | 27 | 17 | 18 | 28 | 13 | 10 | 25 | 15 | 30 | 7 | 24 |
| 21 | 21 | 26 | 8 | 23 | 12 | 19 | 29 | 20 | 2 | 14 | 9 | 3 | 5 | 11 | 31 |
| 22 | 22 | 2 | 14 | 9 | 3 | 5 | 11 | 12 | 19 | 29 | 20 | 26 | 8 | 23 | 4 |
| 23 | 23 | 4 | 28 | 13 | 30 | 10 | 31 | 24 | 17 | 7 | 21 | 1 | 27 | 18 | 16 |

**Figure 2**. Preference data for parfume consumer

The preference data collected from survey of 50 customers who bought and will buy a Tobarium styrax perfume product. The data consist of responses from 50 subjects, each identified by a unique value of the variable ID. Subjects were asked to rank the 31 product profiles from the most to the least preferred. The variables PREF1 through PREF22 contain the IDs of associated product profiles. Analysis of the data is a task that requires the use of command syntax [10] as below:

```
CONJOINT PLAN='perfumeiwors'
/DATA='DataSet1'
/SEQUENCE=PREF1 TO PREF31
/SUBJECT=ID
/FACTORS=SCENT (DISCRETE)
PACKAGE (LINEAR MORE)
VOLUME (LINEAR MORE)
GRADE (LINEAR MORE)
PRICE (LINEAR LESS)
/PRINT=SUMMARYONLY.
```

IOP Conf. Series: Materials Science and Engineering **935** (2020) 012016 doi:10.1088/1757-899X/935/1/012016

### 3.3. *Utility scores*

Part-worth utilities (also known as attribute importance scores and level values or simply as conjoint analysis utilities) are numerical scores that measure how much each feature influences the customer's decision to make that choice [9]. Table 3 shows the utility (part-worth) scores and their standard errors for each factor level.

**Table 3**. Utility scores

| | | Utility Estimate | Std. Error |
|---|---|---|---|
| Scent | Rizla | 2.763 | .457 |
| | Azwa | .113 | .457 |
| | Riedha | .809 | .457 |
| | Aphis | -1.144 | .601 |
| | Jeumpa | -.291 | .601 |
| | Tiara | -2.251 | .601 |
| Package | rollon | .863 | .477 |
| | spray | 1.727 | .954 |
| Volume | 15 mL | .793 | .477 |
| | 25 mL | 1.587 | .954 |
| Grade | perfume | 6.437 | .477 |
| | signature perfume | 12.873 | .954 |
| Price | IDR 120,000 | -1.186 | .276 |
| | IDR 150,000 | -2.371 | .551 |
| | IDR 200,000 | -3.557 | .827 |
| (Constant) | | 5.171 | 1.255 |

Higher utility values indicate greater preference. The highest utility value is indicated by factor grade, this illustrates the respondents' greatest preference for the quality of signature perfume. Furthermore, the lowest utility value is obtained by the *Tiara* scent, this indicates the lowest preference for this variant. As expected, there is an inverse relationship between prices and utility, with higher prices corresponding to lower utility (larger negative values mean lower utility).

The presence of a grade or scent or volume corresponds to a higher utility, as anticipated. Since the utilities are all expressed in a common unit, they can be added together to give the total utility of any combination. For example, based on the statistics analysis, the total utility of a perfume with scent *Rizla*, package spray, volume 25 mL, grade signature perfume and price IDR 120,000 is:

utility(scent Rizla) + utility(package spray) + utility(volume 25 mL) + utility(grade signature perfume) + utility(price IDR120,000) + constant or 2.763 + 1.727+ 1.587 + 12.873+ (-1.186) + 5.171 = 22.935

If the perfume had package scent Azwa, package rollon, volume 15 mL, grade perfume and price IDR200,000, based on the statistics analysis, the total utility would be:

utility(scent Azwa) + utility(package rollon) + utility(volume 15 mL) + utility(grade perfume) + utility(price IDR200,000) + constant or 0.113 + 0.863 + 0.793+ 6.437+ (-3.557) + 5.171 = 9.82

### 3.4. *Coefficients*

Table 4 shows the linear regression coefficients for those factors specified as LINEAR. The utility for a particular factor level is determined by multiplying the level by the coefficient. For example, the predicted utility for a price of IDR 150,000 was listed as -2.371 in the utilities Table 3. This is simply the value of the price level, 2, multiplied by the price coefficient, –1.186.

**Table 4.** Coefficients

| Factor | Coefficient estimate |
|---|---|
| Package | 0.863 |
| Volume | 0.793 |
| Grade | 6.437 |
| Price | -1.186 |

### *3.5. Relative importance*

Running the analysis, it can determine both the relative importance of each attribute as well as which levels of each attribute are most preferred. The range of the utility values (highest to lowest) for each factor provides a measure of how important the factor was to overall preference [9]. Factors with greater utility ranges play a more significant role than those with smaller ranges. Table 5 shows the importance value for each factor as below:

**Table 5.** Importance values

| Factor | Value |
|---|---|
| Scent | 53.958 |
| Package | 3.490 |
| Volume | 3.876 |
| Grade | 27.938 |
| Price | 10.738 |

The results show that scent (importance value 53.958) has the most influence on overall preference. This means that there is a large difference in preference between product profiles containing the most desired scent and the least desired scents. This is because the consumer's preference for a particular scent is personal and unique [11].

The results also show that a package and volume (importance value 3.490 and 3.876, respectively) play the least important role in determining overall preference. The grade and price play a significant role but not as significant as scent. Perhaps there are because the range of grade and prices are narrow. In addition, consumers also consider that the price offered is still relatively affordable for a signature quality if compared to imported similar quality perfumes.

This result showed that it would be critical to consider how a product performs in a testing panel before launching it. Regardless of how effective a product is, if consumer does not like its scent, it may have a completely different perception of product, making the product fail miserably. Considering the fragrances used in this study are smoother, more refreshing, more striking or sweeter, which might have affected the perception of the respondents. Therefore, the scents are the subject of a complex discussion because its interpretation varies with context, and external factors can influence since the recognition of a particular scent as its personal interpretation, linked also to the sensory memory [11,12].

### *3.6. Correlations*

The Table 6 displays two statistics, Pearson's R and Kendall's tau, which provide measures of the correlation between the observed and estimated preferences. The table also displays Kendall's tau for just the holdout profiles. The conjoint procedure computed the correlations between the observed and predicted rank orders for these profiles as a check on the validity of the utilities. In this case, there are high significant correlations between the observed and estimated preferences.

**Table 6.** Correlation coefficients

| | Value | Sig. |
|---|---|---|
| Pearson's R | .968 | .000 |
| Kendall's tau | .829 | .000 |
| Kendall's tau for Holdouts | .667 | .087 |

a. Correlations between observed and estimated preferences

### *3.7. Reversals*

When specifying LINEAR models for package, volume, grade and price were chose, a linear direction (LESS or MORE) for the relationship between the value of variable and the preference for that value were expected. The conjoint procedure keeps track of the number of subjects whose preference showed the opposite of the expected relationship [9]. These cases are referred to as reversals.

**Table 7.** Number of reversals by factor and subject

| | | | |
|---|---|---|---|
| Factor | volume | | 3 |
| | price | | 1 |
| | grade | | 0 |
| | package | | 0 |
| | scent | | 0 |
| Subject | 1 | Subject 1 | 0 |
| | 2 | Subject 2 | 0 |
| | 3 | Subject 3 | 1 |
| | 4 | Subject 4 | 0 |
| | 5 | Subject 5 | 0 |
| | 6 | Subject 6 | 0 |
| | 7 | Subject 7 | 0 |
| | 8 | Subject 8 | 0 |
| | 9 | Subject 9 | 0 |
| | 10 | Subject 10 | 0 |
| | 11 | Subject 11 | 0 |
| | 12 | Subject 12 | 0 |
| | 13 | Subject 13 | 0 |
| | 14 | Subject 14 | 0 |
| | 15 | Subject 15 | 0 |
| | 16 | Subject 16 | 0 |
| | 17 | Subject 17 | 1 |
| | 18 | Subject 18 | 0 |
| | 19 | Subject 19 | 0 |
| | 20 | Subject 20 | 0 |
| | 21 | Subject 21 | 0 |
| | 22 | Subject 22 | 0 |
| | 23 | Subject 23 | 1 |
| | 24 | Subject 24 | 1 |
| | 25 | Subject 25 | 0 |
| | 26 | Subject 26 | 0 |
| | 27 | Subject 27 | 0 |
| | 28 | Subject 28 | 0 |
| | 29 | Subject 29 | 0 |
| | 30 | Subject 30 | 0 |
| | 31 | Subject 31 | 0 |
| | 32 | Subject 32 | 0 |

Table 7 displays the number of reversals for each factor and for each subject. There are three subjects showed a reversal for volume. That is, they preferred product profiles with lower volume. Furthermore, only one subject showed a reversal for price. This subject preferred product profiles with higher price. These reversals are probably due to complexity of consumer's preference for an aromatic product and the interpretation varies greatly since the recognition of a particular scent as its personal interpretation [11,12].

## 4. Conclusion

A conjoint analysis approach is reliable to determine consumer preferences for *Tobarium Styrax Perfume* products. The influence of five product profiles on consumer preference including scent variant, package, volume, grade essential oil, and prices were examined. The highest utility value is indicated by factor grade, this illustrates the respondents' greatest preference for the quality of signature perfume. As expected, there is an inverse relationship between price and utility, with higher prices corresponding to lower utility. The scent variant (importance value 53.958) has the most influence on overall preference. This means that there is a large difference in preference between product profiles containing the most and the least desired scents. This is because the preference for a particular scent is personal and unique which might have affected the perception of the respondents. The grade and price play a significant role but not as significant as scent variant. The highest consumer's preference is a perfume with scent *Rizla*, package spray, volume 25 mL, grade signature perfume and price IDR 120,000. Two statistics, Pearson's R and Kendall's tau, provides high correlation between the observed and estimated preferences.

## References

[1] Aswandi and Kholibrina CR 2017 *Pemulihan Ekosistem Danau Toba* (Medan: Bina Media Perintis)

[2] Dinas Perkebunan Sumatera Utara 2019 *Statistik Perkebunan* Dinas Perkebunan Sumatera Utara http://www.disbun.sumutprov.go.id/statistik_2019/web/index.php?r=site%2Flaporan-komoditas&komoditas=8&tahun=2015&kabupaten=20 (diakses 30 Oktober 2019)

[3] Aswandi and Kholibrina CR 2019 Obarium: parfum kemenyan Danau Toba *Forest Digest Magazine* Available from https://www.forestdigest.com/detail/392/tobarium-parfum-kemenyan-danau-toba

[4] Kholibrina CR and Aswandi 2019 Inovasi parfum kemenyan untuk peningkatan nilai tambah resin kemenyan Toba *Bunga Rampai Perakitan Teknologi Jenis Tanaman Hutan Unggulan Lokal Daerah Untuk Peningkatan Nilai Tambah* Editor Mindawati N and Yulianti (Bogor: IPB Press)

[5] Aswandi and Kholibrina CR 2017 Growth and yield model for non timber forest product of kemenyan (*Styrax sumatrana* J.J. Sm) in Tapanuli, North Sumatra *International Conference on Agriculture, Environment, and Food Security IOP Conf. Series: Earth and Environmental Science* **122** 012036

*[6]* Aswandi and Kholibrina CR 2020 Increasing added-value of styrax resin through post-harvesting techniques improvement and essential oil based product innovation *Paper presented in ICONART 2020*

[7] Kholibrina 2019 *Strategi Pemuliaan dan Inovasi Kemenyan Toba-Bunga Rampai Pengembangan Hasil Hutan Bukan Kayu Indonesia untuk Mendukung Sustainable Development Goals* Editor Tata HL (Bogor: IPB Press)

[8] SPSS Inc. 2005 *SPSS Conjoint™ 14.0.* (Chicago, IL: SPSS inc.)

[9] SPSS Inc. 2005 *SPSS 14.0 Command Syntax Reference* (Chicago, IL: SPSS inc.)

[10] Anindiaty RK and Putong I 2019 Analisis preferensi konsumen terhadap produk parfum the body shop dengan pendekatan analisis konjoin https://fdokumen.com/document/analisis-preferensi-konsumen-terhadap-produk-ringkasan001pdf-parfum-body-lotion.htmL

[11] Gonçalves GMS, Srebernich SM, Vercelino BG, and Zampieri BM 2013 Influence of the presence and type of fragrance on the sensory perception of cosmetic formulations *Brazilian Archives of Biology and Technology* **56**(2) 203-12

[12] Milotic D 2006 The impact of fragrance on consumer choice *J. Consumer Behaviour* **3**(2) 179-91

**Acknowledgments**
The author thanks to SEAMEO Biotrop for research funding support about increasing the added-value of styrax resin-based products in year 2020

# Consumer acceptance on product diversification of Bengkirai (*Shorea laevis*): Attitudes, expectations and perception

**R Maharani[*] and A Fernandes**

Dipterocarps Research Center, Ministry of Environment and Forestry, Indonesia

*Email: rizmaannisa@yahoo.com

**Abstract.** Bengkirai/Irai (*Shorea laevis*) is one of the potential species of Dipterocarps that is widely known as an antioxidant source that it can be utilized as a traditional herbal medicine for forest communities in East Kalimantan. However, no product diversification of Irai was introduced in the industrial scale yet. This study is therefore intended to introduced some products diversification of Irai as featured NTFPs products to give better understanding that herbal medicine is not a herbal product that it has been bitter tastes or old-fashioned. Besides, this study is also aimed to compare the difference in consumers' attitudes, expectations, and perceptions towards product diversification of Irai if they have or do not have experience with those products. In this study, some products diversification of Irai were introduced, and divide into beverages (Irai tea and Irai beer) and desserts (Irai Plain Pudding and Irai silky Pudding). Organoleptic or consumer sensory analysis of these products was tested on 50 adult panelists by using a Hedonic scale with 5 classifications (very dislike, dislike, neutral, like, and very like). Data analysis was performed descriptively by using Excel and presented in the graphic. If organoleptic assessment that significantly affected observational variables, then advanced test (T-test) followed by using SPSS 16. Sensory analysis for dessert products showed there was a statistically significant difference in taste (T-test = 0.005) and flavor (T-test = 0.01). While there was no significant difference in all sensory parameters in beverage products. A recent study was also indicated that these products provided 2-3 times the economic benefit. It means these products need to be developed into a new business by developing an entrepreneurial spirit for improved a local community's income.

## 1. Introduction

Bengkirai/Irai (*Shorea laevis*) is one of the potential species of Dipterocarps family that it mostly grows in tropical rain forests in Kalimantan [1]. Besides being known to produce good quality wood for buildings, furniture, etc. [2], Irai is also widely known as a traditional herbal medicine for forest communities in East Kalimantan. Forest communities utilize boiled Irai leaves to reduce high blood pressure, relieve neck aches, fatigue, and make body refresh. Bengkirai leaves contained antioxidants such as alkaloids and triterpenoids that can be inhibited free radicals entering the body, therefore the content of Bengkirai leaf antioxidant compounds has high potential as functional foods or beverages [3]. Meanwhile, there has been no other form of presentation for Irai products, the importance of a form of presentation that is applicable and diversified without affecting its efficacy.

Diversification of food products needs to be done to increase higher nutrients values and active compositions contained in these food or beverage products [4]. Food and beverage ingredients from Irai leaves are a good alternative source of antioxidants for the community and are expected to have added value both in terms of taste, economic value, and its benefits in human health. Product diversification strategy is also very important as an effort to increase non-timber forest products [5] and to reduce Irai wood harvesting. Such as product diversification will require social entrepreneurship so that it's not only benefits but also increases the number of people who want to work by gathering jobs that promise sustainability.

In order to support the diversification of the product from natural materials into industrial innovation, it is necessary to conduct a market and competitiveness test of these products. The industry of natural product needs to be developed and innovated in order to compete with ever tastier, healthier, safer, more convenient, more sustainable, cheaper, and more varied products on the market. All the developed product preparations were organoleptically evaluated using the Hedonic scale [6,7]. The organoleptic test used the scoring tests on taste, texture, color, and aroma/flavor while the overall acceptance was determined by the hedonic test [8,9].

Some alternative diversified products from natural ingredients which are predicted to be more quickly accepted and will be consumed regularly by ordinary people are drinks and snacks. Herbal drinks can be packaged in the form of herbal teas. Herbal tea is a form of tea, has a medicinal function, and can be taken daily [10]. As for snack products that are commonly consumed by all communities is pudding. Pudding is a dairy dessert, easy to make, with a variety of flavors [11]. Pudding is semi-solid food, contained starch, sugar, salt, gum such as carrageenan, and sometimes milk is added [12]. Milk also was one of the main components in desserts [13]. Adding milk to the pudding will increase nutritional value than traditional pudding [14].

This study is therefore intended to introduced some products diversification of Irai as featured NTFPs products to give better understanding that herbal medicine is not a herbal product that it has been bitter tastes or old-fashioned. Besides, this study is also aimed to compare the difference in consumers' attitudes, expectations, and perceptions towards product diversification of Irai if they have or do not have experience with those products.

## 2. Materials and Methods

### *2.1. Raw materials*

Some products diversification of Irai were produced fromIrai leaves which its origin from nursery, arboretum, and natural forest in East Kalimantan. In this study, Irai leaves was collected from natural forest in Labanan, Berau, East Kalimantan. Thus, Irai leaves were withered process that is air-dried at room temperature to reduce water content up to 70%. Air-dried samples were placed on metal trays and placed in a room temperature between 25-27$^0$C for 12 hours. After wilting and softening, samples have reduced the size using scissors and smoothed with a blender to obtained Bengkirai leaves wet powder. After that, the wet powder was oven-dried at 50$^0$C for 4 hours, then filtered to get a uniform size. In the next step, the filtered material is weighed and packed into a tea bag (± 2 g/bag). Furthermore, Irai leaves in a tea bag brewed with hot water as a usual tea, thus the "tea water" (± 200 cc/teabag) ready to be processed into the beverage (Irai tea and Irai beer) and dessert (Irai plain pudding and Irai silky pudding) products will be introduced in this study.

(a)

(b) (c)

**Figure 1**. Irai leaves origin from nursery (a), arboretum (b) and natural forest (c)

*2.2. Organoleptic or consumer sensory analysis*

Organoleptic or consumer sensory tests are highly relevant to assess the most important attributes of a product that are related to quality perception and acceptability [15]. Descriptive sensory analysis is traditionally conducted using panelists and has the purpose of developing the sensory profile of a product by evaluating the intensities of the main descriptors. This type of sensory analysis is usually used in new product development and to assess the quality of the same formulation in the different batches produced [15]. Based on observing, touching, and sniffing senses, the organoleptic analysis was carried out by a human panel [16]. Organic evaluation of herbal as drugs by means of the organ of sense (skin, eye, tongue, nose, and ear) or macroscopic evaluation of herbal drugs by color, odor, taste, size, shape, and special features, like touch-feeling, and texture [17]. In this study, organoleptic analysis of beverage products (Irai tea and Irai beer) and dessert products (Irai pudding and Irai silky pudding) were tested on 50 adult panelists by using Hedonic scale with 5 classifications (very dislike, dislike, neutral, like, and very like).

(a) (b)

**Figure 2**. Organoleptic tested some products diversification of Bengkirai: Irai tea (A) and Irai silky pudding (B)

*2.3. Data analysis*
Data analysis was performed descriptively by using Excel and presented in graphical form. Furthermore, if organoleptic assessment that significantly affected observational variables, then advanced test (T-test) followed by using SPSS 16.

## 3. Results and Discussion

*3.1. Irai beverage products*
The products diversification of Irai leaves were introduced as beverage products i.e. Irai tea and soft drink beverage namely Irai beer (Figure 3). These beverage products are made based on easy, inexpensive, and current considerations, without reducing their efficacies. These products are designed to suppress the perception of ordinary people that products with the aim of prevention or treatment will be like as usual herbal products that even certain people consume them. In addition, lay perception about the content of caffeine in tea beverage products and or caffeine-containing beverages, it actually has a negative effect on health. Coffee, tea, caffeinated soda, and energy drinks are important sources of caffeine in the diet but each present with other unique nutritional properties [18]. Irai Leaf has several uniqueness; besides its distinctive taste, Irai leaf does not contain caffeine so that the negative risk of caffeine can be avoided. Rigorous reviews of caffeine toxicity conclude that consumption of up to 400 mg caffeine/day in healthy adults is not associated with adverse effects [19,20,21].

In order to maintain quality, purity, potency, safety, and efficacy of herbal drugs (product of medicinal plants) need sensory/organoleptic analysis as consumer acceptance [22,23]. In this study, organoleptic analysis of Irai tea and Irai beer was tested on 50 adult panelists included taste, flavor, color, and an appearance by using Hedonic scale with 5 classifications (very dislike, dislike, neutral, like, and very like) will be shown in the following graphic, respectively.

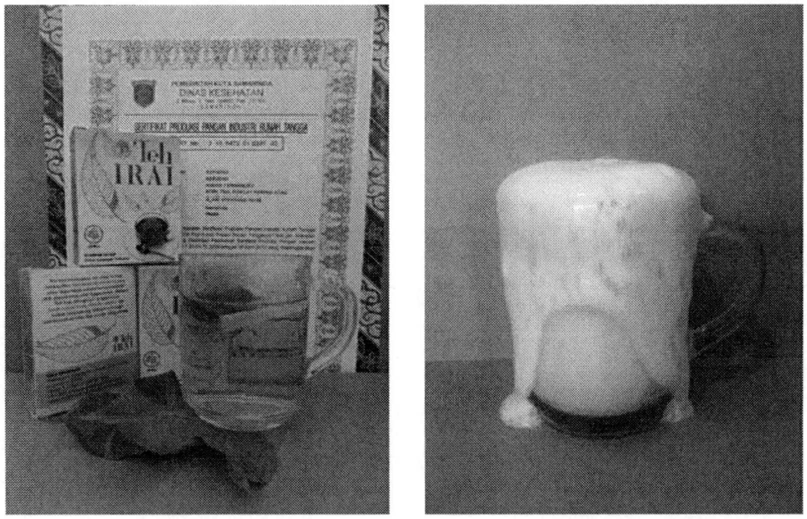


(a) (b)

**Figure 3**. Irai tea (a), and Irai beer (b)

3.1.1. Irai beverage products taste

As can be seen in Figure 4. there was no significant difference of taste between Irai tea and Irai beer (T-test = 0.33). Although not statistically significant, the difference percentage in taste preferences for Irai tea and Irai beer products showed different habits and perceptions of panelists. Panelists of Irai beer tend to have the same drinking habits as drinking of soft drinks, which is drunk directly when bubbling. Indeed, if it is drunk immediately it will have a fresh effect, but after the bubble ($CO_2$) runs out it will return to its original taste. When $CO_2$ is formed spontaneously, the taste of the drink increases, but after the $CO_2$ runs out, the taste returns to its origin [15].

| | very dislike | dislike | neutral | like | very like |
|---|---|---|---|---|---|
| Irai Beer | 0 | 0 | 0 | 55 | 45 |
| Irai Tea | 0 | 0 | 0 | 50 | 50 |

**Figure 4**. Organoleptic tested on Irai beverages taste

3.1.2. Irai Beverage products flavor

According to the organoleptic test, it can be shown that percentage of the flavor of Irai beer similar to Irai tea. The flavor of these products was not much different because even though it turns into bubbly, it does not change the original flavor of Irai beer which is similar to the flavor of Irai tea. Although some panelists have different expectations about the flavor of Irai beer, in this study, it is presented as original

Irai beer that caused response tends similar to irai tea. Herbal tea and sodas (carbonated drink) for consumption must avoid any flavoring agent [18]. Irai beer and Irai tea, both have brought distinctive and unique flavor. The tea flavor is caused by the presence of volatile aromatic compounds, the extraction process of chemical components of herbal teas such as carbohydrates, proteins, sugar reduction groups when brewing, as well as the oxidation of polyphenol compounds and their derivatives such as catechins to theaflavins and theaburigin which gives a distinctive flavor [24].

| | very dislike | dislike | neutral | like | very like |
|---|---|---|---|---|---|
| ■ Irai Beer | 0 | 0 | 0 | 55 | 45 |
| ■ Irai Tea | 0 | 0 | 0 | 55 | 45 |

**Figure 5**. Organoleptic tested on Irai beverages flavor

### 3.1.3. Irai beverage products color

There was no significant difference (T-test = 0.65) of color between Irai beer and Irai tea percentages (Figure 6.). The panelists showed almost similar response for two products, the difference in color was only Irai beer produced bubble. After the bubbling is finished (about 1-2 minutes), it will return to its original color as Irai tea. The habit of tea drinking was also predicted to be able to contribute an impression of the color of these two products. In this study, many panelists liked the color of tea which was not too thick or natural like these products. The level of color density of tea affected tannin levels. The higher concentration of a teabag caused lower in tannin content. This is due to several factors, one of which is if tannin compounds are exposed to light and air for longer, then tea colors will be changed to become more dark/concentrated [25]. The color of the drink was influenced by the basic ingredients in the drink [26]. Irai beer and Irai tea had the same basic ingredient, i.e. Irai leaves.

| | very dislike | dislike | neutral | like | very like |
|---|---|---|---|---|---|
| ■ Irai Beer | 0 | 0 | 0 | 55 | 45 |
| ■ Irai Tea | 0 | 0 | 0 | 65 | 35 |

**Figure 6**. Organoleptic tested on Irai beverages color.

3.1.4. Irai beverage products appearance

In general, these products showed similar responses in the appearance tested, thus no statistically analyzed for those as can be seen in Figure 7. Panelists liked the appearance of Irai beer at the beginning of bubbling even though after a few minutes it would return to normal, like as usual tea. bubbles provided a permanent effect when mixed with sugar in drinks [27]. Irai tea and Irai beer did not use sugar, so there was no difference in their appearance. The appearance of Irai tea, itself is considered normal by panelists, the appearance should be that Herbal can be well received, even if it will be marketed.

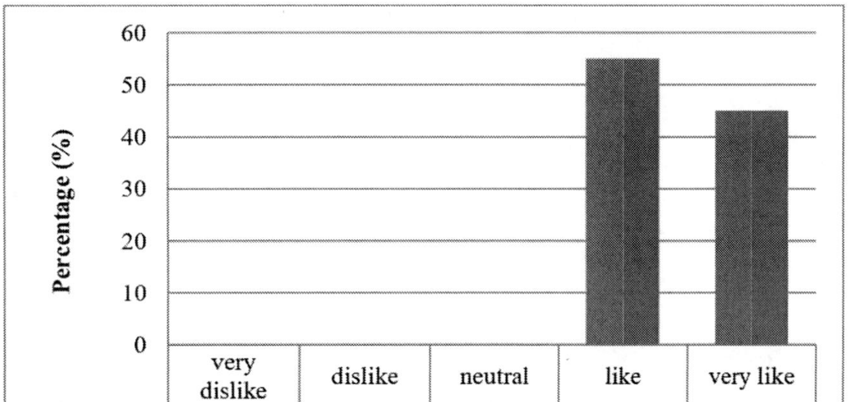


| | very dislike | dislike | neutral | like | very like |
|---|---|---|---|---|---|
| Irai Beer | 0 | 0 | 0 | 55 | 45 |
| Irai Tea | 0 | 0 | 0 | 55 | 45 |

**Figure 7.** Organoleptic tested on Irai beverages appearance

### *3.2. Irai dessert products*

According to the current government's strategy to confirm national food security, one of the priority programs is increasing diversification and community food security, especially from natural resources [28]. Therefore, this study introduced Bengkirai diversification products (Irai plain pudding and Irai silky pudding) which is origin from Dipterocarps forest ecosystem (Figure 8).

(a) (b)

**Figure 8.** Irai plain pudding (original) (a), and Irai silky pudding (b)

3.2.1. Irai dessert products taste

It can be concluded that there was a statistically significant difference (T-test = 0.005) of taste between Irai plain pudding and Irai silky pudding (Figure 9). The Panelists have very liked Irai silky Pudding significantly more than Irai plain pudding. Introducing of dessert products that are commonly consumed is an initial consideration taken to introduce Bengkirai product diversification as an alternative food ingredient. The taste-tested on these products showed that prevalence or habit of consuming pudding leads the panelists' perceptions of taste that they normally know and like.

| | very dislike | dislike | neutral | like | very like |
|---|---|---|---|---|---|
| Irai Plain Puding | 0 | 0 | 0 | 80 | 20 |
| Irai Silky Puding | 0 | 0 | 0 | 45 | 55 |

**Figure 9**. Organoleptic tested on Irai desserts taste

The presence of milk in the silky pudding is very popular among panelists, especially those who are younger. In adults, prefer low fat as milk taste than sweet or original taste of pudding [29]. Whereas plain/original flavors are mostly preferred by older, assuming that it contains more efficacies. The addition of flavor in this product is expected not to replace the efficacy that already exist, but rather to give impression that the efficacy of herbs is not necessarily as usual herbal taste.

3.2.2. Irai dessert products flavour

There was a statistically significant difference (T-test = 0.01) of flavour between Irai plain pudding and Irai silky pudding. Panelists were extremely like Irai silky pudding more compare to Irai plain pudding due to Irai silky pudding preferred milk flavor dominantly. The flavor of milk was preferred in pudding [30]. The tendency of younger people to favor silky pudding also influences assessment. While, the fresh flavor of herbs in plain pudding has more acceptable due to more health benefits for older people's perception.

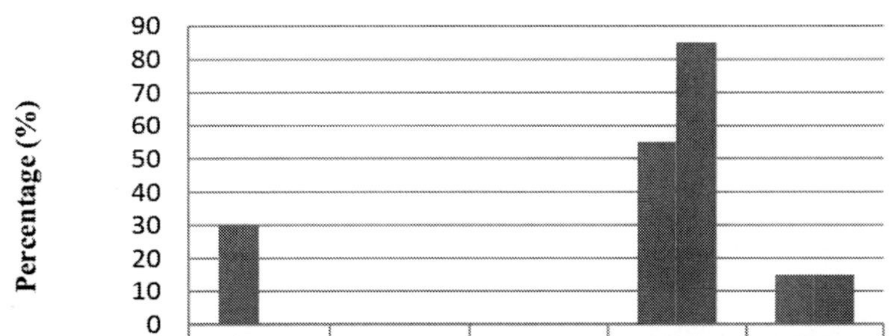


| | very dislike | dislike | neutral | like | very like |
|---|---|---|---|---|---|
| ■ Irai Plain Puding | 30 | 0 | 0 | 55 | 15 |
| ■ Irai Silky Puding | 0 | 0 | 0 | 85 | 15 |

**Figure 10**. Organoleptic tested on Irai desserts flavour

3.2.3. Irai Dessert products color

It can be seen in Figure 11, both of color tests showed similar results. Although the taste and flavor differences are very significant, the color matter tends to be ignored by panelists. It's probably due to the level of preference for a product is relative [31].

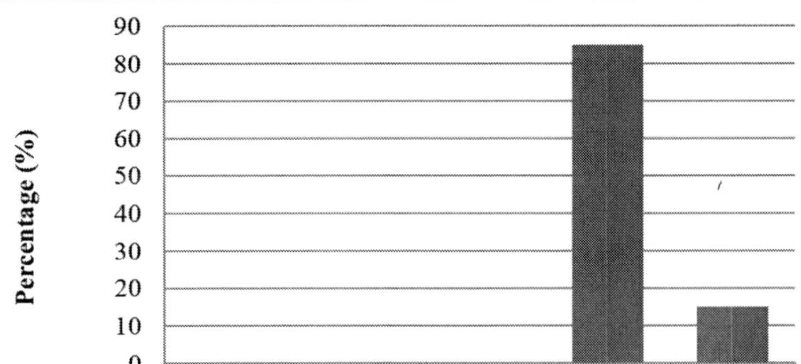


| | very dislike | dislike | neutral | like | very like |
|---|---|---|---|---|---|
| ■ Irai Plain Puding | 0 | 0 | 0 | 85 | 15 |
| ■ Irai Silky Puding | 0 | 0 | 0 | 85 | 15 |

**Figure 11**. Organoleptic tested on Irai desserts color

In general, the colors of the two products are similar even though there are additional ingredients. Adding milk to the herbal pudding cannot change the color of the pudding. As a comparison, in okra pudding even added by milk, okra color still dominant than milk [12].

3.2.4. Irai dessert products appearance

According to the appearance tested on these products, there was no significant difference (T-test = 0.163) between them. However, plain pudding and silky pudding have a different appearance. Irai pudding as conventional pudding was a solid gel that cannot be poured, while the silky pudding is more resilient like a gel due to milk additional. The addition of milk will change the texture of pudding [13] and produced high viscosity pudding but still can be poured or shipped out [14]. The test results showed that the majority of panelists can accept the different appearances, it is predicted due to mostly panelists are adults. The age level greatly influences the assessment that is directly related to their habits, expectations and perceptions of a product.

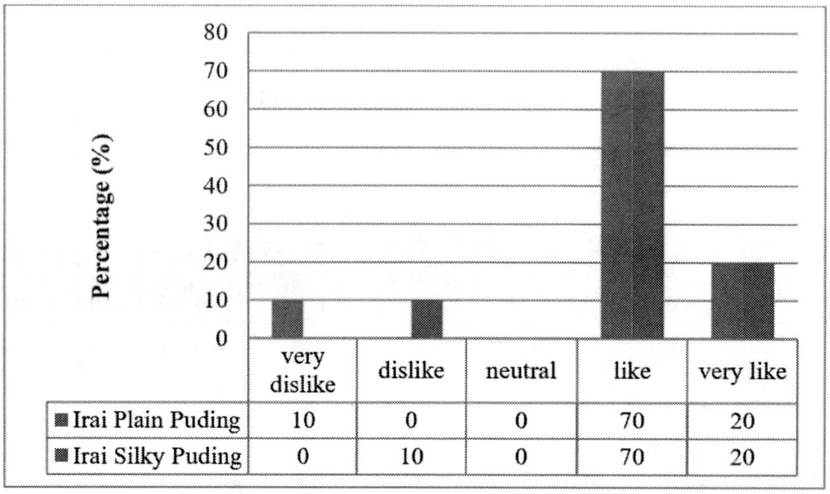


**Figure 12.** Organoleptic tested on Irai desserts appearance

### *3.3. Economic benefit of Irai derivate products*

As important as customer sensory analysis, economic considerations are needed to identified market opportunities for these products. The economic value was estimated according to calculation of the most economical price standard due to promotional prices yet (Table 1). Even though, it still benefits two to three times of production costs. This fact will provide great product development opportunities with promising continuity.

**Table 1.** Estimation of economic value for Irai derivate products.

| | Estimated Production Costs Per pcs (IDR) | Selling Price Per pcs (IDR) | Remarks |
|---|---|---|---|
| Teh Irai | 1,250* | 5,000 | * costs per pcs not including packaging cost yet |
| Bir Irai | 1,500* | 5,000 | |
| Irai Plain Pudding | 1,750* | 5,000 | * costs per pcs not including packaging cost yet |
| Irai Silky Pudding | 2,250* | 5,000 | |

Consumers buying decision is very complex. Usually purchasing intention is related to consumers behavior, perception, and attitude. Purchase behavior is an important key point for consumers during considering and evaluating of certain products [32].

## 4. Conclusion

The product diversification of Bengkirai was introduced as a beverage (Irai beer and Irai tea) and dessert (Irai plain pudding and Irai silky pudding) products. These products are made based on convenient, inexpensive, and contemporary considerations, without any reduction of its efficacies. Organoleptic or consumer sensory analysis of beverage products showed that younger consumer preferred Irai beer more than Irai tea. Meanwhile, consumer sensory analysis for dessert products described that elder panelists preferred Irai Plain Pudding more than Irai silky pudding. However, Irai silky pudding is a promising offer for younger, in addition to its function for health as well as for healthy snacks. According to their attitudes, expectations, and perceptions, it can be concluded that the level of sensitivity of panelists tends to be more sensitive, first in taste, flavor, color, and appearance, respectively. On the other hand, there was no statistically significant difference of taste, flavor, color, and appearance for those beverage products. While sensory analysis for dessert products showed that there was a statistically significant difference on taste and flavor.

A recent study was indicated that these products provided 2-3 times economic benefit which can be improved the local community's income. It means these products need to be developed into a new business by developing the entrepreneurial spirit of the farmer groups, especially women's involvement. The development of product diversification is expected to have competitiveness in marketing, and it will support a community's economy. These products diversification was required social entrepreneurship, so that it's not only benefits but also increases the number of people who want to work by gathering jobs that promise sustainability. This present study also provided an innovation of herbal medicine that it would be predicted in attracting consumers to the market to promote general positive image consumers for having more healthy food and beverages, and positive accepting about herbal medicine not bitter as usual.

**References**

[1] Budi SW, Siregar IZ, Siregar UJ, Sukendro A, Pamoengkas P, and Yunanto T 2012 Rarity status and habitat of *Shorea leavis* and *Shorea leprosula* in Muara Teweh, Central Kalimantan *JMHT* **XVIII**(2) 86-93

[2] Jemi R, Syafii W, Ferbrianto F, and Hanafi, M 2013 Aktivitas anti jamur 2,3-dihydroxyoctadecanoic acid dari kayu palepek baringin (*Shorea laevis* Ridl) *Prosiding SNKTI* 179-83

[3] Fernandes A and Maharani R 2018 *Kala Irai Ceria: Wajah Baru Aplikasi Herbal, Karamunting-Bengkirai-Kersen* (Bogor: Forda Press)

[4] Lourenco SC, Martins MM, and Alves VD 2019 Antioxidants of natural plant origins: from sources to food industry applications *Molecules* **24**

[5] Chilalo M and Wiersum KF 2011 The role of non-timber forest products for livelihood diversification in southwest ethiopia *Ee-JRIF* **3**(1) 44-59

[6] Meilgaard M, Civille GV, and Carr BT 1997 *Sensory Evaluation Techniques 3rd Edition* (Florida: CRC Press) pp. 247-48

[7] Hasani A, Kongoli R, and Beli D 2018 Organoleptic analysis of different composition of fruit juices containing wheatgrass *Food research* **2**(3) 294-98

[8] Soekarto ST and Hubeis M 1992 *Metodologi Penelitian Organoleptik* (Bogor: Pusat Antar Universitas Pangan dan Gizi Institut Pertanian Bogor)

[9] Hafid H, Patriani P Nuraini, Norma, Ananda SH, and Inderawati 2020 Organoleptic characteristcs of broiler chicken meat using juice of starfruits (*Averrhoa bilimbi* L.) *International Conference on Agriculture, Environment and Food Security (AEFS*)

[10] Park SJ, Bae YC, Choi NR, Ryu SY, Kwon YM, and Joo JC 2014 Clinical study on constitutional herbal tea for treating chronic fatigue *Journal of Pharmacopuncture* **17**(4) 55-60

[11] Chandan RC and Kilara A 2015 *Puddings and Dairy-Based Desserts Dairy Processing and Quality Assurance* (John Wiley & Sons Ltd) pp. 397-427

[12] Qasem AAA, Alamri MS, Mohamed AA, Hussain S, Ahmood K, and Ibraheem MA 2016 High soluble-fiber pudding: Formulation, processing, texture and sensory test *Journal of Food Processing and Preservation*

[13] Raymundo VG and Ruiz JFV 2018 Physicochemical and rheological properties of a dairy dessert, enriched with chickpea flour *Foods* **7**(25)

[14] Damian C 2012 Effect of different milk substitutes on rheological properties of pudding *Food and Environment Safety* **XI**(3) 62-7

[15] Viejo CG, Torrico DD, Dunshea RR, and Fuentes S 2019 Bubbles bubble formation, stability and consumer perception of carbonated drinks: a review of current, new and emerging technologies for rapis assessment and control *Food* **8**
[16] Xu DY, Lin R, Bauer HR, Chen RQ,Yang HQ, Zou, and Yang YH 2018 Organoleptic evaluation of amomi fructus and its further background verified via morphological measurement and GC coued with E-nose Evidence-based *Complementary and Alternative Medicine* 1-9
[17] Shulammithi R, Sharaya M, Tejaswini R, and Kiranmai M 2016 Standardization and quality evaluation of herbal drugs *IOSR Journal of Pharmacy and Biological Sciences* **11**(5-1) 89-100
[18] Reyes C and Cornelis MC 2018 Caffein in the diet: Country-level consumption and guidelines *Nutrients* **10**
[19] Wikoff D, Welsh BT, Henderson R, Brorby GP, Britt J, Myers E, Goldberger J, Lieberman HR, O'Brien C, and Peck J 2017 systematic review of the potential adverse effects of caffeine consumption in healthy adults, pregnant women, adolescents, and children *Food Chem. Toxicol.* **109** 585-648
[20] Nawrot P, Jordan S, Eastwood J, Rostein J, Hugenholtz A, and Feeley M 2003 Effects of caffeine on human health *Food Addit. Contam.* **20** 1-30
[21] EFSA 2015 Panel on dietetic products nutrition and allergies Scientific opinion on the safety of caffeine *EFSA J* **13** 4102
[22] Patil SG, Wagh AS, Pawara RC, and Ambore SM 2013 Standard tools for evaluation of herbal drugs: An overview *The Pharma Innovation Journal* **2**(9) 60-65
[23] Vanhonacker F, Kühne B, Gellynk X, Guerrero L, Hersleth M, and Verbeke W 2013 Innovations in traditional foods: impact on perceived traditional character and consumer acceptance *Food Research International* **54** 1828-35
[24] Hadi DK 2011 *Proses Pengolahan Teh* (Jakarta: Erlangga)
[25] Fadli A, Wahyuni S, and Rejeki S 2017 Analisis organoleptik produk teh celup tawaloho (*Spondias Pinata*) *J. Sains dan Teknologi Pangan* **2**(5)792-9
[26] Zahra BE, Shoaib S, and Iqbal RK 2019 An Overview of effects of carbonated drinks National *Journal of Health Sciences* **4** 80-4
[27] Bisclesanu CF, Manea S, Alexandrescu O, Nazem D, Pangica AM, Banita M, and Florescu A 2017 *Rev. Chim.* **68**(12) 2823-8
[28] Andoko E and Doretha AC 2019 Analysis of Indonesian government strategies to food security: Harnessing the potential of natural and human resources *FFTC agricultural policy platform: Food and fertilizer for the Asian and Pacific region* https://ap.fftc.org.tw/article/1588
[29] Mennella JA, Finkbeiner S, and Reed DR 2012 The proof is in the pudding: Children prefer lower fat but higher sugar than do mothers *Int. J. Obes.* **36**(10) 1285-91
[30] Kristanti D and Herminiati A 2019 Characteristics of physical, chemical and organoleptic properties of inulin-enriched pudding as a complementary food *2nd International Conference on Natural Products and Bioresource Science*
[31] Daroini O 2006 Kajian proses pembuatan teh herbal dari campuran teh hijau (*Camellia sinensis*), rimpang bangle (*Zingiber cassumunar* Roxb.) dan daun ceremai (*Phyllanthus acidus* (l.)Skeels.) [Skripsi] Fakultas Pertanian IPB Bogor
[32] Jaafar SN, Lalp PE, and Naba MM 2012 Consumers perceptions, attitudes and purchase intention towards private label food products in Malaysia *Asian Journal of Business and Management Sciences* **2**(8) 73-90

IOP Conf. Series: Materials Science and Engineering **935** (2020) 012018 doi:10.1088/1757-899X/935/1/012018

# The activated carbon produced from mayan bamboo (*Gigantochloa robusta* Kurz) and its application as dye removal

**L Efiyanti*, D A Indrawan, N Hastuti, and S Darmawan**

Forest Product Research and Development Center, Ministry of Environment and Forestry, Bogor, 16118, Indonesia

*Email: lisnaefiyanti@gmail.com

**Abstract.** Bamboo pyrolysis was carried out at a temperature of 400°C for 4 hours to obtain a carbon as a pyrolysis product. The carbon soaked using potassium hydroxide (KOH) 5.35 M for 24 hours. The carbon obtained heated in a reactor at a temperature of 800°C for 5 hours then washed and neutralized until activated carbon obtained. The characterization of activated carbon conducted based on the Indonesian National Standard (SNI 06-3730-1995). The methylene blue adsorption process using a concentration of 50, 100, 150, and 200 ppm for 120 minutes reaction with the contact time variation are 30, 60, 90, 120,150, and 180 minutes in optimum concentration. The result shows that activated carbon has good quality based on SNI 06-3730-1995. The optimum methylene blue removal by bamboo activated carbon reached 99.77% in 120 minutes reaction and the adsorption type follows Langmuir isotherm.

## 1. Introduction

Bamboo as lignocellulose biomass has wide benefits, including as a raw material for activated carbon. The application of bamboo activated carbon in this research as a dye removal in the methylene blue adsorption process. The methylene blue is a cationic dye that is water soluble with stable color [1, 2]. Methylene blue commonly used in textile, cosmetics, and paper industries. The content of methylene blue in the water based on KepMen LH No. 51/MENLH/10/1995 not recommended to exceed the threshold (5-10 mg/L) because it can give rise to the environmental pollution. If the excess dyes on water released into the environment such as water, soil, lakes, and oceans, it will affect serious environmental damage, not only resulting in undesirable color changes but also disrupting the health of aquatic and human organisms.

The adsorption method widely used and considered as one of the promising methods for processing dyes [3]. Several raw materials usually used as an adsorbent include zeolite [4,5], soil [6,7] clay [8], chitosan and activated carbon [9]. Graphene as an advanced material used as adsorbent [10], but the acquisition of this product complicated and requires a high cost therefore low-cost green technology and products are needed.

Activated carbon from biomass is relatively effective and low cost because it is abundant and not utilized properly. Activated carbon has a prospective industrial use due to superior characteristics, such as high surface area, good chemical stability, and has unique structural properties [11]. The raw materials for activated carbon obtained from forestry, agriculture, and waste biomass [12]. One used as raw material for activated carbon is bamboo. Bamboo is a tropical plant biomass, which is easily found in Indonesia [13] with the third largest quantity in the world [14]. Bamboo contains cellulose, hemicellulose, lignin, and silica [15] which can be converted into other products such as activated carbon or natural silica. Mayan bamboo (*Giganchlotoa robusta* Kurz) has the potential as a raw material for activated carbon because it has a low water content with high lignin content as well as with high iodine number.

The activated carbon from mayan bamboo as an adsorbent of HCl and NaOH resulting in adsorption capacity of 1.3 mmol/g and 6.1 mmol/g respectively, higher than the commercial activated carbon [16]. Activated carbon from andong and ater bamboo used as diazinon insecticide adsorbent with an adsorption capacity of 4.63 mg/g and 1.763 mg/g [17]. The production and application of activated carbon from mayan bamboo as methylene blue removal necessary to study to know the potential of bamboo activated carbon as an adsorbent for dye removal. Hopefully, it will be a low cost and green adsorbent in the future.

## 2. Materials and Methods

### *2.1 Materials*

The raw material used in this study is mayan bamboo (*Gigantochloa robusta* Kurz). The chemicals used are KOH pellets, 10% HCl, methylene blue, distilled water, and Whatman filter paper. The equipment used in this research is glass equipment (Pyrex, Iwaki), furnace, vacuum desiccator, oven, grinding machine, pyrolysis reactor with a capacity of 2 kg, electrical retort capacity of 0.5 kg, 100 mesh sieve, analytical balance (Libror EB330 Shimadzu), digital pH meter, 100 ml crucibles. The analysis instrument used was a UV-Vis spectrophotometer (Shimadzu-1700, Japan).

### *2.2 Methods*

#### *2.2.1* Activated carbon production

The bamboo is cut into a size of about 5 x 2.5 cm and pyrolyzed at a temperature of 400°C for 4 hours. The carbon obtained is then cooled and grounded. The carbon immersed in a 5.35 M KOH solution for 24 hours, drained and put into an electric retort to be heated at 800°C for 5 hours. Furthermore, the activated carbon neutralized using distilled water and dried so that the activated carbon is ready to use.

#### *2.2.2* Characterization and activated carbon application as methylene blue adsorption

The quality of bamboo activated carbon analyzed based on SNI 06-3730-1995 method including moisture content, ash content, volatile matter content, and fixed carbon. The process of methylene blue adsorption by activated carbon bamboo follows the research procedure of [18]. A standard solution of methylene blue dyes prepared by dissolving 0.1 g of methylene blue into 100 ml of distilled water to obtain a solution with a concentration of 1000 ppm. Then several variations concentration of methylene blue were made into 50 ppm, 100 ppm, 150 ppm, 150 ppm, and 200 ppm adsorbed by 0.1 g of activated carbon for 2 hours. After the adsorption process, the solution filtered and analyzed using a UV-Vis spectrophotometer at a wavelength of 665 nm. Furthermore, under optimum adsorption conditions, 25 ml of solution was taken and 0.1 g of activated carbon added with variations for 30, 60, 90, 120, 150, and 180 minutes. The resulting solution then analyzed following the same procedure.

The equation of adsorption process:

$$X/m = \frac{V\,(Co-Ca)}{m} \tag{1}$$

$$Q = \frac{Co - Ce}{Co} x\ 100\% \quad (2)$$

$$\frac{Ce}{Xe/m} = \frac{1}{ab} + \frac{1}{a} Ce \quad (3)$$

where:

X/m = adsorption capacity (mg/g)
V = solution volume (L)
Co = initial concentration of methylene blue (mg/L)
Ca = final concentration of methylene blue (mg/L)
m = adsorbent weight (g)
Q = methylene blue removal (%)
a = adsorption maximum capacity (mg/g)
b = equilibrium constant (l/mg)

## 3. Results and Discussion

*3.1 Characterization of activated carbon from mayan bamboo*

In general, the process of carbon production is through the pyrolysis method with limited oxygen and the activation process using chemical or physical activators. In this study, pyrolysis carried out at 400°C followed by chemical activation using KOH and reheated at 800°C so the activated carbon obtained.

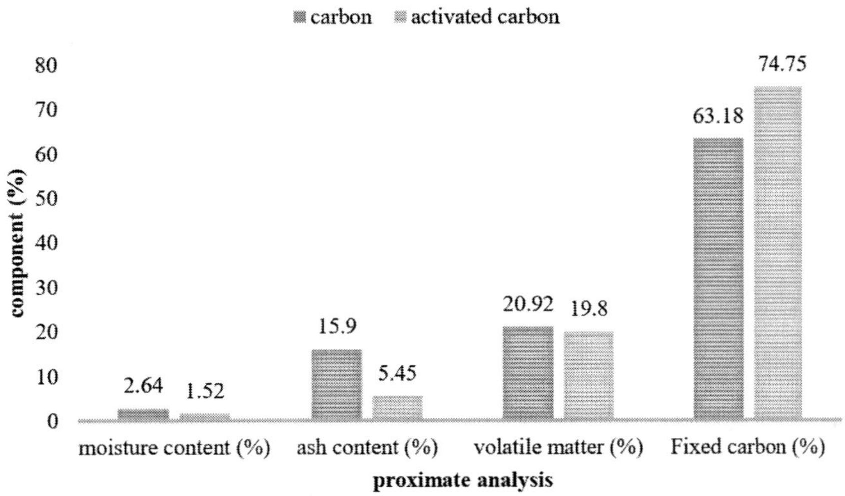


**Figure 1.** Activated carbon characterization based on SNI 06-3730-1995

The characterization of activated carbon fulfilled the requirements of SNI 06-3730-1995 (presented in figure 1). Indonesian National Standard of activated carbon requires the moisture content ≤15%, ash content ≤10%, volatile matter content ≤25%, and fixed carbon ≥65%. The water content, ash content, and volatile matter expected to be low so that the activated carbon is more able to adsorb the adsorbate. So, as it can block/close the surface and pore of the carbon. For the fixed carbon, this content indicates the purity of activated carbon, the higher fixed carbon expected that the quality will be better, and have more active sites for the adsorption process. The quality of activated carbon influenced by the carbon activation process. The activation process can remove impurities on the carbon, open pores, and enlarge the surface area so the activated carbon has better quality than without the activation process.

*3.2 Application of activated carbon on methylene blue adsorption*

The methylene blue adsorption process by bamboo activated carbon conducted to determine the potential of activated carbon as an adsorbent. Based on the calculation results, the final concentration of methylene blue, adsorption capacity, and the removal percentage of methylene blue can be determined as shown in Table 1.

**Table 1.** Methylene blue adsorption data at various initial concentrations

| Co (ppm) | Ce (ppm) | Ce/(Xe/m) | Q (%) | X/m (mg/g) | log Ce |
|---|---|---|---|---|---|
| 50 | 0.191 | 0.0139 | 99.65 | 13.7022 | -0.7189 |
| 100 | 0.230 | 0.0092 | 99.77 | 24.9425 | -0.6383 |
| 150 | 1.134 | 0.0305 | 99.24 | 37.2165 | 0.0546 |
| 200 | 1.690 | 0.0341 | 99.15 | 49.5775 | 0.2278 |

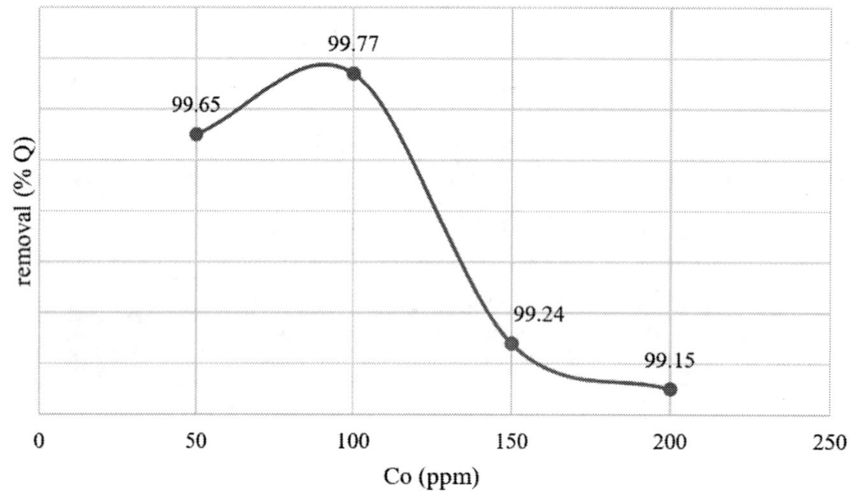


**Figure 2.** Effect of the initial concentration of methylene blue on the percentage of methylene blue removal

The percentage of methylene blue removal decreased after optimum state when the initial concentration of methylene blue increased (Figure 2). The highest percentage removal of methylene blue reached 99.77% at 100 ppm of initial concentration. The increasing of initial methylene blue until the optimum state will increase the adsorption capacity due to the activated carbon surface was occupied with methylene blue molecules [19,20]. The saturation occurs when the initial concentration of methylene blue increased above 100 ppm because, in this condition, the number of active sites is not enough to adsorb methylene blue with a higher concentration [21]. The higher dose of adsorbent required for higher adsorbate concentrations. The increasing initial concentration will decrease the methylene blue removal [22].

Based on the equation, we can determine the type of methylene blue adsorption. Through the plot of Ce/(Xe/m) and the final concentration of methylene blue, the Langmuir isotherm graph with an $R^2$ value of 0.932 while Freundlich isotherm with an $R^2$ value of 0.8521 calculated from Ce/(Xe/m) and Ce logarithm. These results can be shown that the Langmuir isotherm is more suitable for adsorption of methylene blue because of $R^2$ values close to 1. The adsorption isotherm used to analyze the interactions between the adsorbent and adsorbate. The Langmuir and Freundlich isotherms can describe the process of methylene blue adsorption on bamboo activated carbon. Langmuir isotherm shows the adsorption process of monolayer interaction, activated carbon has homogeneous surface and adsorbate evenly

IOP Conf. Series: Materials Science and Engineering **935** (2020) 012018 doi:10.1088/1757-899X/935/1/012018

adsorbed on the surface [11]. Based on the calculation of the $R_L$ value ($R_L$ = 0.0027-0.0108), it can be seen that the adsorption process is favorable, because of the value of $0 < R_L < 1$ ($R_L > 1$: unfavorable, $R_L = 0$: irreversible, $R_L = 1$: linear) [15]. K-value from the equation obtained 1.8295, is Langmuir's constant which states the relationship of methylene blue and adsorbent. The higher of K-value, the affinity of methylene blue to the adsorbent will be higher.

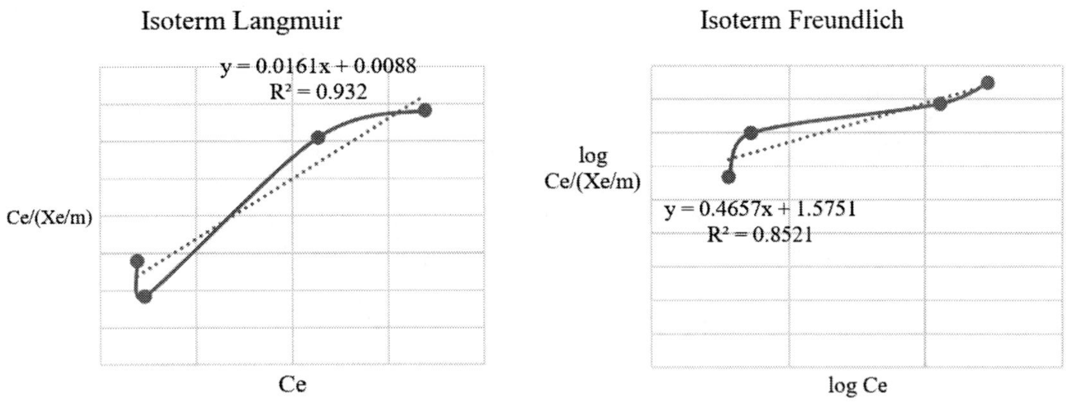


**Figure 3**. Methylene blue adsorption types by activated carbon

The research result is better than Latupeirissa *et al.* [18] who found that the adsorption capacity of methylene blue adsorption and removal percentage by activated carbon with the same condition are 23.18 mg/g and 92.72% respectively. Nuchar (commercial AC) used for methylene blue adsorption resulting in 21.5 mg/g adsorption capacity. Activated carbon from ashitaba waste and walnut peel on adsorption of methylene blue produce the adsorption capacities of 381.88 and 400.11 mg/g, respectively [24]. The conditions used were with a methylene blue solution of 200 ml, as much as 0.1 g adsorbent, agitation rate of 300rpm for 540 minutes at a concentration of 0-50 mg/L. Activated carbon/cellulose composite for methylene blue removal and the maximum adsorption capacity reached 103.66 mg/g for 24 hours adsorption process [25]. Several studies have a higher adsorption capacity compared with this research due to different conditions, such as higher initial concentration, adsorbent weight and longer contact time.

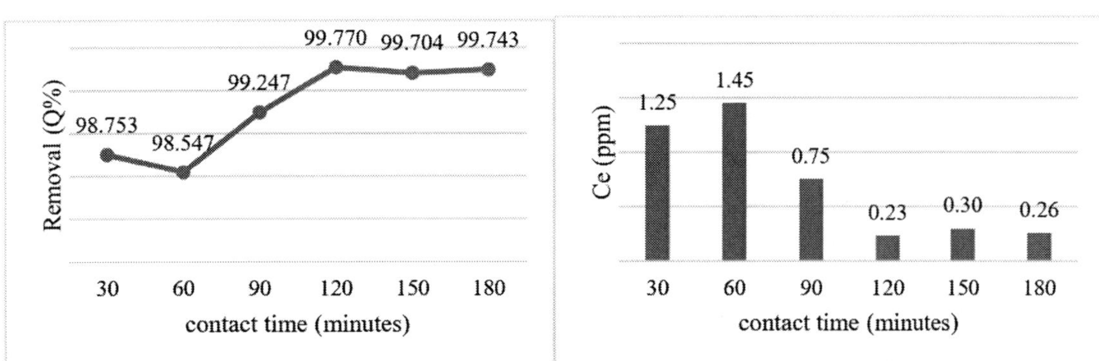


**Figure 4**. The effect of contact time to the removal percentage of methylene blue (4a) and the final concentration of methylene blue (4b).

As shown in Figures 4a and 4b, we can see the relationship between the use of contact time in the adsorption process with the final concentration and removal percentage of methylene blue. The increase of contact time in the adsorption process, the percentage removal of methylene blue increased to an optimum point of 99.77% with a final concentration of a solution of 0.23 ppm at 120 minutes. The longer contact time will make methylene blue adsorbed more on the adsorbent because it provides longer and intense interactions. However, when the contact time continuously increased, the percent of removal can decrease because of the surface and pore of the adsorbent filled with methylene blue [26]. This result is following Sahu *et al.* [23] which obtained an optimum contact time of 120 minutes with maximum removal of a percent was found 98.6%, and then the saturation occurred in the adsorption process.

Bamboo as a raw material for activated carbon by pyrolysis with the addition of nitrogen and $KHCO_3$ as activators for the methylene blue adsorption process with an adsorption capacity of 499.3 mg/g [27]. Adsorption of methylene blue is generally longer than phenol, and a decrease in the adsorption rate of methylene blue is due to the internal diffusion of large molecules besides filling the pores during the adsorption process. Methylene blue has a large molecular with an aromatic structure so that it has a positive interaction with the functional group and generally influenced by the adsorbent functional group so it can make $\pi$- $\pi$ bonds. The mechanism of methylene blue adsorption on activated carbon explained by 3 processes including $\pi$-$\pi$ stacking between aromatic of methylene blue and activated carbon, electrostatic interactions and hydrogen bonds between CO and OH groups of activated carbon with amine groups on methylene blue [10,28].

## 4. Conclusion

The activated carbon from mayan bamboo has fulfilled the requirements of the Indonesian National Standard. This activated carbon has a good potential as an adsorbent for dye removal (methylene blue), and the result shows that the optimum methylene blue removal by bamboo activated carbon reached 99.77% in 120 minutes reaction and the adsorption type follows Langmuir isotherm. To improve the quality and adsorption capacity of activated carbon, further research may use different activator or modify adsorption process with varying pH, temperature, and higher doses of adsorbent.

## References

[1] Li YH, Chang FM, Huang B, Song YP, Zhao HY, and Wang KJ 2019 Activated carbon preparation from pyrolysis char of sewage sludge and its adsorption performance for organic compounds in sewage *Fuel* **266** 117053

[2] Dhaouadi F, Sellaoui L, Dotto GL, Bonilla-Petriciolet A, Erto A, and Lamine AB 2020 Adsorption of methylene blue on comminuted raw avocado seeds: Interpretation of the effect of salts via physical monolayer model *J. Mol. Liq.* **305** 112815

[3] Sakr F, Alahiane S, Sennaoui A, Dinne M, Bakas I, and Assabbane A 2020 Removal of cationic dye (Methylene Blue) from aqueous solution by adsorption on two type of biomaterial of South Morocco *Mater. Today Proc.* **22** 93–96

[4] Handayani M, and Sulistiyono E 2009 Uji persamaan langmuir dan freundlich pada penyerapan limbah chrom (VI) oleh zeolit *Prosiding Semin. Nas. Sains dan Teknol. Nukl. PTNBR – BATAN* **VI** 130-6.

[5] Alver E, Metin AU, and Brouers F 2020 Methylene blue adsorption on magnetic alginate/rice husk bio-composite *Int. J. Biol. Macromol.* **154** 104-13

[6] Apriyanti H, Candra IN, and Elvinawati 2018 Karakteristik isoterm adsorpsi dari ion logam besi (Fe) pada tanah di Kota Bengkulu *ALOTROP Jurnal Pendidikan dan Ilmu Kimia.* **2**(1) 14-19

[7] Wijayanti A, Susatyo EB, Kurniawan C, and Sukarjo 2018 Adsorpsi logam Cr (VI) dan Cu (II) pada tanah dan pengaruh penambahan pupuk organik *J. Chem. Sci.* **7**(3) 242-8

[8] Sekewael SJ, Tehubijuluw H, and Reawaruw DR 2013 Kajian kinetika dan isoterm adsorpsi logam Pb pada lempung asal Desa Ouw teraktivasi garam ammonium nitrat *Ind. J. Chem. Res.* **1** 38-46

[9] Rahmi R, Fachruddin S, and Nurmalasari N 2018 Pemanfaatan limbah serat sagu (Metroxylon sago) sebagai adsorben iodin *Jurnal. Rekayasa Kimia Lingkungan* **13**(1) 70-7

[10] Rohaizad A, Shahabuddin S, Mehmood S, Rashid NM, Hir ZA, Ramly MM, Awang K, Siong CW, and Aspanut Z 2020 Green synthesis of silver nanoparticles from Catharanthus roseus dried bark extract deposited on graphene oxide for effective adsorption of methylene blue dye *J. Environ. Chem. Eng.* **8**(4) 103955

[11] Efimov MN, Vasilev AA, Muratov DG, Baranchikov AE, and Karpacheva GP 2019 IR radiation assisted preparation of KOH-activated polymer-derived carbon for methylene blue adsorption *J. Environ. Chem. Eng.* **7**(6) 103514

[12] Yao X, Ji L, Guo J, Ge S, Lu W, Cai L, Wang Y, Song W, and Zhang H 2020 Magnetic activated biochar nanocomposites derived from wakame and its application in methylene blue adsorption *Bioresource Technology* **302** 122842

[13] Himawanto DA 2012 Optimasi kondisi pirolisis sampah bambu guna mendapatkan energi terbesar *Buana sains* **12**(2) 35-38

[14] Abdul Khalil HPS, Bhat IUH, Jawaid M, Zaidon A, Hermawan D, and Hadi YS 2012 Bamboo fibre reinforced biocomposites: A review *Mater. Des.* **42** 353-68

[15] Sulastri S, Nuryono, Kartini ES, and Kunarti 2015 Kinetika dan keseimbangan adsorpsi ion kromium (III) dalam larutan pada senyawa silika dan modifikasi silika hasil sintesis dari abu sekam padi *Jurnal Penelitian Saintek* **19** 33-44

[16] Hastuti N, Pari G, Setiawan D, Daud D, and Godang M 2015 Tingkat keasaman dan kebasaan arang aktif bambu mayan (AABM) terhadap uap jenuh HCL dan NaOH acidity and alkalinity level of mayan bamboo activated charcoal (MBAC) on Saturated vapor of acid chloride and natrium hydroxide *Widyariset* **1** 41-50

[17] Murtihapsari, Mangallo B, and Handyani D 2012 Model isoterm freundelich dan langmur oleh adsorben arang aktif bambu andong (*G . verticillata* ( Wild ) Munro) dan bambu ater (*G. ater* (Hassk) Kurz Ex Munro) *Jurnal Sains Natural Universitas Nusa Bangsa* **2**(1) 17-23

[18] Latupeirissa J, Tanasale MFJDP, and Musa SH 2018 Kinetika adsorpsi zat warna metilen biru oleh karbon aktif dari kulit kemiri (*Aleurites moluccana* (L) Willd) *Indo. J. Chem. Res.* **6**(1) 12-21

[19] Ebadollahzadeh H, and Zabihi M 2020 Competitive adsorption of methylene blue and Pb (II) ions on the nano-magnetic activated carbon and alumina *Mater. Chem. Phys.* **248** 122893

[20] Jawad AH and Abdulhameed AS 2020 Mesoporous Iraqi red kaolin clay as an efficient adsorbent for methylene blue dye: Adsorption kinetic, isotherm and mechanism study *Surfaces and Interfaces* **18** 100422

[21] El Hajam M, Kandri NI, Harrach A, El khomsi A, and Zerouale A 2019 Adsorption of methylene blue on industrial softwood waste 'cedar' and hardwood waste 'mahogany': Comparative study *Mater. Today Proc.* **13** 812-21

[22] Üner O 2019 Hydrogen storage capacity and methylene blue adsorption performance of activated carbon produced from Arundo donax *Mater. Chem. Phys.* **237** 121858

[23] Sahu S, Pahi S, Tripathy S, Singh SK, Behera A, Sahu UK, and Patel RK 2020 Adsorption of methylene blue on chemically modified lychee seed biochar: Dynamic, equilibrium, and thermodynamic study *J. Mol. Liq.* **315** 113743

[24] Li Z, Hanafy H, Zhang L, Sellaoui L, Netto MS, Oliveira ML, Seliem MK, Dotto GL, Bonilla-Petriciolet A, and Li Q 2020 Adsorption of congo red and methylene blue dyes on an ashitaba waste and a walnut shell-based activated carbon from aqueous solutions: Experiments, characterization and physical interpretations *Chem. Eng. J.* **388** 124263

[25] Somsesta N, Sricharoenchaikul V, and Aht-Ong D 2020 Adsorption removal of methylene blue onto activated carbon/cellulose biocomposite films: Equilibrium and kinetic studies *Mater. Chem. Phys.* **240** 122221

[26] Mrunal VK, Vishnu AK, Momin N, and Manjanna J 2019 $Cu_2O$ nanoparticles for adsorption and photocatalytic degradation of methylene blue dye from aqueous medium *Environ. Nanotechnology, Monit. Manag.* **12** 100265

[27] Li Z, Xing B, Ding Y, Li Y, and Wang S 2020 A high-performance biochar produced from

bamboo pyrolysis with in-situ nitrogen doping and activation for adsorption of phenol and methylene blue *Chinese J. Chem. Eng.*

[28] Han Q, Wang J, Goodman BA, Xie J, and Liu Z 2020 High adsorption of methylene blue by activated carbon prepared from phosphoric acid treated eucalyptus residue *Powder Technol.* **366** 239-48

**Acknowledgements**

The authors highly acknowledge and pay gratitude to the Forest Products Research and Development Center, Ministry of Environment and Forestry, Indonesia for providing technical support and facilities, also Professor Gustan Pari for supervising this work.

IOP Conf. Series: Materials Science and Engineering **935** (2020) 012019 doi:10.1088/1757-899X/935/1/012019

# Some properties of laminated veneer lumber manufactured from oil palm trunk

**I M Sulastiningsih*, D R Trisatya, and J Balfas**

Forest Products Research and Development Centre, Ministry of Environment and Forestry, Bogor, 16610, Indonesia

*E-mail: ignasiasulastin@gmail.com

**Abstract**. Indonesia is the largest oil palm producer in the world and it was predicted that around 7 million metric tons was left in the field as a replanting waste, including palm frond and trunk. The utilization of this waste as a raw material of composite products may overcome the problem. The objective of this study was to determine the physical and mechanical properties of Laminated Veneer Lumber (LVL) made of oil palm trunk. Laboratory scale five-ply LVLs glued with urea formaldehyde adhesive were fabricated with five types of layer compositions i.e five-ply LVL with all layers made of oil palm trunk veneers (OP), five-ply LVL with face and back layers made of jabon veneers (J2), five-ply LVL with face, back and center core layers made of jabon veneers (J3), five-ply LVL with face and back layers made of mahoni veneers (M2), and five-ply LVL with face, back and center core layers made of mahoni veneers (M3). Results showed that the physical and mechanical properties of laminated veneer lumbers made from oil palm trunk were affected by layer compositions. Incorporating jabon or mahoni wood veneers in the structure of oil palm LVL improved the physical and mechanical properties of LVL.

## 1. Introduction

The increasing demand for timber has led to a shortage of timber supply in wood industries. To anticipate the shortage, a wide range of alternative substitute materials are being explored. Oil palm (*Elaeis guineensis* Jacq.) trunk appears to be the most potential substitute material. Indonesia is the largest oil palm producer in the world, with the total crude palm oil production of 36.59 million ton in 2018 [1]. It was revealed that in 2018 at least 12.76 million hectares of land are being planted with oil palm [1]. Further, Badan Pusat Statistik [1] reported that over the years oil palm plantation is significantly increasing.

The productive live span of oil palm tree is between 25-30 years. Thus, it would be replanting after this period. It was predicted that around 7 million metric ton was left in the field as a replanting waste [2], including palm frond and trunk. The utilization of this waste as a raw material of composite products, however, can give additional economic value of the useless by products. A wide range of research have been conducted to investigate the properties of composite products made of oil palm trunk [3–11]. It was suggested that oil palm LVL is suitable for panelling and non-structural purposes [8]. Furthermore, the strength of soft inner part of oil palm trunk could be improved by applying Close System Compression (CSC) method for resin impregnation and densification [9]. Low molecular weight phenol formaldehyde treatment with different resin solid content and initial moisture content of veneer improved the physical properties of oil palm veneer LVL [10]. The inclusion of rubberwood veneer significantly improved the bending and compression strength of the LVL [7]. This study examined the physical and mechanical properties of laminated veneer lumber made of oil palm trunk with combined layers of jabon (*Anthocephalus cadamba* Miq.) and mahoni (*Swietenia macrophylla* King.) wood.

## 2. Materials and Methods

### *2.1. Materials*

The materials used in this study were oil palm trunk (*Elaeis guineensis* Jacq.) collected from Malimping, Banten Province, jabon wood (*Anthocephalus cadamba* Miq.) and mahoni wood (*Swietenia macrophylla* King.) collected from Sukabumi, West Java Province, Indonesia. A commercial liquid urea formaldehyde was used as adhesive.

### *2.2. Methods*

#### 2.2.1. Preparation of oil palm trunk veneer

The oil palm trunk logs with average diameter of 36 cm and 140 cm in length were converted into veneer using rotary peeling machine with targeted veneer thickness of 5.5 mm. Rotary veneers were then cut into 40 cm long by 42 cm wide and then air drying at room temperature for one day and followed with sun drying to about 16% moisture content. Oil palm trunk veneers were then hot pressed at temperature of 120°C with specific pressure of 10 kg/cm$^2$ for 1 minute.

#### 2.2.2. Preparation of wood veneer

The logs of jabon wood and mahoni wood with average diameter of 28 cm and 125 cm in length were converted into wood veneer using rotary peeling machine with targeted veneer thickness of 2 mm. Rotary veneers were then cut into 40 cm long by 42 cm wide and then air drying at room temperature for one day and followed with sun drying to about 16% moisture content. Afterwards, it oven dried at temperature of 80°C for 48 hours to about 10% moisture content.

#### 2.2.3. Laminated Veneer Lumber (LVL) fabrication

Laboratory scale five-ply LVLs with the dimension of 40 cm x 40 cm x thickness were fabricated with five types of layer compositions i.e five-ply LVL with all layers made of oil palm trunk veneers (OP), five-ply LVL with face and back layers made of jabon veneers (J2), five-ply LVL with face, back and center core layers made of jabon veneers (J3), five-ply LVL with face and back layers made of mahoni veneers (M2), and five-ply LVL with face, back and center core layers made of mahoni veneers (M3). To manufacture five-ply LVL, the urea formaldehide adhesive (100 UF resin: 10 wheat flour: 10 water: 0.5 hardener) was applied to each face of the veneers, targeting a total spread rate of 200 g/m$^2$ per glue line. After the adhesive was applied, the assembly of five veneer sheets was cold pressed together parallel to each other for 15 minutes, with a nominal dimensions of 40 cm x 40 cm x thickness for each LVL board. The assembly then was further hot pressed at temperature of 120°C and specific pressure of 10 kg/cm$^2$ for 7 minutes. Five replications for each type of LVL were prepared. The LVLs produced were conditioned for two weeks before testing.

#### 2.2.4. Testing

The laminated veneer lumbers were cut into desired specimen dimensions and measured for density, moisture content, thickness swelling, water absorption, delamination, Modulus of Rupture (MOR), Modulus of Elasticity (MOE), compression strength, and hardness. The tests were performed using the American Standard ASTM D 1037-93 [12] and Japanese Standard for Laminated Veneer Lumber [13] for evaluating properties of LVL.

#### 2.2.5. Data analysis

A completely randomized design was used in the experiment with the layer compositions as the treatment factor. Five replications were prepared for each treatment. The effects of combined wood species on physical and mechanical properties of LVLs were computed by analysis of variance using Minitab Statistical and Qualitative Data Analysis Software.

## 3. Results and Discussion

### *3.1. Physical properties and bonding quality of laminated veneer lumbers*

The mean values of physical properties and bonding quality of laminated veneer lumbers (LVLs), the results of analysis of variance (ANOVA), and the results of HSD are presented in Table 1.

#### 3.1.1. Moisture content

Average value of moisture content of OP LVL is given in Table 1. The moisture content of laminated veneer lumbers made from oil palm trunk varied from 12.10% to 12.43% with an average of 12.28%. These values are still within the range of air-dried moisture content and meet the Japanese Standard requirement for laminated veneer lumber since the values are below 14%. The results of ANOVA in Table 1 showed that moisture content of LVL was not affected by layer compositions. The previous study Sulaiman *et al.* [4] reported that LVL made from oil palm trunk glued with urea formaldehyde had moisture content of 9.81% while those glued with phenol formaldehyde had moisture content of 9.90%.

#### 3.1.2. Density

As shown in Table 1, the density of laminated veneer lumbers produced in this study varied from 0.500 g/cm$^3$ to 0.591 g/cm$^3$ with an average of 0.535 g/cm$^3$. The results of ANOVA in Table 1 showed that density of LVL was significantly affected by layer compositions. It was observed that the values of LVL density increased with increasing number of wood veneer incorporating in the LVL structure. The lowest density of LVLs was produced from OP LVL while the highest density of LVLs was produced from M3 LVL. In Table 1 it can be seen that there is no different in density between OP LVL with J2 LVL and J3 LVL. Incorporating mahoni veneer in the structure of OP LVL, however, increased the density of OP LVL by 12% and 18.2% for M2 LVL and M3 LVL, respectively. This is in agreement with [7] who found the increased density as the incorporated rubberwood veneers in the LVL. The entirely oil palm trunk veneers LVL had the lowest density (0.545 g/cm$^3$), while the highest found in the LVL with the inclusion of three layers of rubberwood veneers (0.589 g/cm$^3$) [7]. It was argued that the incorporating of wood veneers in the layers could give a more uniform and denser LVL [7].

The previous study [4] reported that LVL made from oil palm trunk glued with urea formaldehyde had density of 0.57 g/cm$^3$ while those glued with phenol formaldehyde had density of 0.37 g/cm$^3$. Other previous study reported that the density of three-layer oil palm trunk LVL glued with either EPI or PVAc had density around 0.52-0.55 g/cm$^3$ [5]. The density of 0.597 g/cm$^3$ was found in the LVL from the bottom part and peripheral zone of oil palm trunk adhered with urea formaldehyde [8]. It was the highest density observed in the study compared to another portion and zone of the trunk. The density of seven-ply LVL made of poplar wood veneers glued with phenol formaldehyde was 0.419 g/cm$^3$ while the density of eucalyptus LVL was 0.616 g/cm$^3$ [14]. From this information it can be seen that the density of LVL is governed by the density of wood veneer raw materials. This finding is in agreement with that reported by Sulastiningsih *et al.* [15], that the density of wood composite product is governed by the density of its raw materials, the use of adhesive and the pressure applied during wood composite manufacture, which produced a denser product.

#### 3.1.3. Thickness swelling

Thickness swelling is one indicators of dimensional stability and it performs as an important property of composite products. The thickness swelling of laminated veneer lumbers made from oil palm trunk varied from 6.08% to 9.22% with an average of 8.38%. The data on thickness swelling of LVL were subjected to analysis of variance and the results showed that thickness swelling of LVL was significantly affected by layer compositions. It was observed that the lowest value of thickness swelling was resulted from five-ply oil palm trunk LVL which included three mahoni wood veneers in the LVL structure as face, back and center core layers (M3). The percentage of thickness swelling decreased with the addition of wood veneer layer in the OP LVL. This is in agreement with [8] who found the similar trend, where thickness swelling increased as the density decreased. Thickness swelling of seven-ply LVL made of poplar and eucalyptus glued with phenol formaldehyde were 2.09% and 2.59% respectively [14], while thickness swelling of five-ply LVL made of gmelina glued with phenol formaldehyde was 2.09% [16]. A recent study Masseat *et al.* [10] reported that the low molecular weight phenol formaldehyde treatment with 45% solid content at 15% initial moisture content resulted in 0.76% of oil palm veneer thickness swelling. Impregnation with phenol formaldehyde or non-resin (chemical) is the most commonly method to improve the oil palm trunk physical properties [11].

3.1.4. Water absorption

The results of ANOVA in Table 1 showed that water absorption values were significantly affected by layer compositions. The water absorption of laminated veneer lumbers varied from 64.13% to 85.88% with an average of 76.77%. It was observed that water absorption of laminated veneer lumbers decreased as the number of wood veneer increased in the LVL structure. Previous study also recorded the similar trend, where the water absorption increased as the density decreased [8]. The highest water absorption value, 94.33%, was found in the LVL made of top portion and inner zone, while the lowest, 63.03%, was found at bottom portion and peripheral zone [8]. It was also reported that the lowest and highest values of thickness swelling were found in the LVL made of bottom portion and peripheral zone (2.99%) and top portion and inner zone (5.62%) of the oil palm trunk, respectively [8]. Water absorption of seven-ply LVL made of poplar and eucalyptus glued with phenol formaldehyde were 59.56% and 23.23% respectively [14], while water absorption of five-ply LVL made of gmelina glued with phenol formaldehyde was 15.23% [16].

**Table 1.** Physical properties and bonding quality of LVLs, the results of ANOVA and HSD

| Properties | Layer compositions | | | | | ANOVA results |
|---|---|---|---|---|---|---|
| | OP | J2 | J3 | M2 | M3 | |
| Moisture content, % | 12.10 $^{a}$ (0.46) | 12.37 $^{a}$ (0.23) | 12.43 $^{a}$ (0.24) | 12.15 $^{a}$ (0.23) | 12.35 $^{a}$ (0.51) | 0.80$^{ns}$ |
| Density, g/cm$^3$ | 0.500 $^{c}$ (0.006) | 0.507 $^{c}$ (0.009) | 0.518 $^{c}$ (0.009) | 0.560 $^{b}$ (0.018) | 0.591 $^{a}$ (0.024) | 33.86** |
| Thickness swelling, % | 8.99 $^{a}$ (0.94) | 9.22 $^{a}$ (1.48) | 8.44 $^{a}$ (1.24) | 9.16 $^{a}$ (1.18) | 6.08 $^{b}$ (0.74) | 6.62** |
| Water absorption, % | 84.75 $^{a}$ (12.65) | 85.88 $^{a}$ (2.10) | 74.97 $^{ab}$ (4.15) | 74.14 $^{ab}$ (3.11) | 64.13 $^{b}$ (7.12) | 8.18** |
| Delamination,% | 0 | 0 | 0 | 0 | 0 | |

Each value was the average of five specimens
Values in parentheses are standard deviations
Values followed with the same letter within the same row are not significantly different
$^{ns}$ not significant
** highly significant

3.1.5. Delamination

The bonding quality of any composite product is very important. The simple way to determine the bonding quality of multilayers composite product in this study is delamination test. The results of the warm water delamination test showed that there was no delamination in all samples and, therefore, the bonding quality of the LVLs made from oil palm trunk and in combination with jabon veneer and mahoni veneer were considered acceptable and meet the Japanese Agricultural Standard requirement for laminated veneer lumber. Comparatively, the eight-ply oil palm trunk LVL made of bottom portion and peripheral zone and top portion and peripheral zone had superior bonding quality compared to the rubberwood LVL [8].

### *3.2 Mechanical properties of Laminated Veneer Lumbers*

The mean values of mechanical properties of laminated veneer lumbers (LVLs), the results of analysis of variance (ANOVA), and the results of HSD are presented in Table 2.

#### 3.2.1. Modulus of Rupture

The LVL produced with different layer compositions showed higher MOR values than the entirely oil palm veneer layers LVL. The MOR of laminated veneer lumbers (Table 2) varied from 227.2 kg/cm$^2$ (22.28 MPa) to 417.6 kg/cm$^2$ (41 MPa) with an average of 362.6 kg/cm$^2$ (35.6 MPa). The data on MOR of LVL were subjected to analysis of variance and the results showed that MOR of LVL was significantly affected by layer compositions. It was observed that the MOR values increased with the increase in number of wood veneers incorporated in the oil palm LVL structure. Incorporating wood veneers in the structure of OP LVL increased the MOR of OP LVL by 65.7% for J2 LVL, 83.8% for J3 LVL, 70.6% for M2 LVL and 78% for M3 LVL respectively. Additionally, a study by [7] reported that the inclusion of two and three layers of rubberwood veneers escalated the MOR of the oil palm trunk LVL to about 19.8% and 45.6%, respectively.

**Table 2.** Mechanical properties of laminated veneer lumbers, the results of ANOVA and HSD.

| Properties | Layer compositions | | | | | ANOVA |
|---|---|---|---|---|---|---|
| | OP | J2 | J3 | M2 | M3 | Results |
| MOR, kg/cm$^2$ | 227.2$^b$ (43.4) | 376.4$^a$ (73.3) | 417.6$^a$ (62.6) | 387.5$^a$ (68.2) | 404.4$^a$ (59.30 | 7.73** |
| MOE, kg/cm$^2$ | 36428$^{bc}$ (5574) | 33531$^c$ (4853) | 43192$^b$ (4352) | 53643$^a$ (1247) | 43241$^b$ (5006) | 15.05* |
| Compression strength, kg/cm$^2$ | 164.1$^d$ (9.12) | 222.6$^c$ (15.24) | 236.7$^c$ (5.79) | 274.7$^b$ (15.73) | 317.4$^a$ (15.1) | 100.24** |
| Hardness, kg/cm$^2$ | 178.7$^c$ (8.2) | 199.1$^{bc}$ (21.3) | 232.4$^b$ (14.0) | 234.8$^b$ (13.5) | 273.1$^a$ (30.3) | 18.16** |

Each value was the average of five specimens
Values in parentheses are standard deviations
Values followed with the same letter within the same row are not significantly different
$^{ns}$ not significant
$^*$ significant
$^{**}$ highly significant

The seven-ply LVL made of jabon veneers glued with phenol formaldehyde at glue spread rate of 200 g/m$^2$ with various adhesive compositions had MOR values varied from 38.33 MPa to 69.75 MPa [17]. Another study reported that laminated board made from oil palm trunk glued with isocyanate adhesive at glue spread rate of 300 g/m$^2$ had MOR values varied from 20.2 MPa to 56.98 MPa [18]. The MOR of eight-ply LVL made of three hybrid poplar clones glued with phenol formaldehyde at glue spread rate of 200 g/m$^2$ varied from 77.64 MPa to 95.82 MPa [19]. The MOR of five-ply LVL made of gmelina glued with phenol formaldehyde at glue spread rate of 220 g/m$^2$ was 61.39 MPa [16] while the MOR of seven-ply LVL made of surian glued with urea formaldehyde was 72.18 MPa [20]. The MOR of mahoni solid wood was 623 kg/cm$^2$ or 61.1 MPa [21] while the MOR of jabon solid wood was 691 kg/cm$^2$ or 67.77 MPa [22]. Another study recorded for static bending parallel to the grain the MOR of eight-ply MOR of oil palm trunk ranged from 11.05 to 19.29 MPa and 13.04 to 24.63 MPa for the edgewise position [8]. A markedly low MOR value was reported for static bending in perpendicular to the grain of oil palm trunk, ranged from 1.33 to 1.95 MPa [8].

3.2.2. Modulus of Elasticity

The assessed treatment was different with respect to the MOE. The MOE values of laminated veneer lumbers varied from 33531 kg/cm$^2$ or 3289 MPa to 53643 kg/cm$^2$ or 5261 MPa with an average of 42007 kg/cm$^2$ or 4120 MPa. The data on MOE of LVLs were subjected to analysis of variance (ANOVA) using completely randomized design and the result showed that the MOE of LVLs was affected by layer compositions. It was observed that the highest MOE was resulted from five-ply oil palm LVL which included two mahoni wood veneers in the LVL structure which used as face and back layers (M2). Similarly, an increase in the MOE value was found in the oil palm trunk LVL with the inclusion of two and three layers of rubberwood veneers to 61.8% and 80.4% compared to the entirely oil palm trunk layers, respectively [7].

The seven-ply LVL made of jabon veneers glued with phenol formaldehyde at glue spread rate of 200 g/m$^2$ with various adhesive compositions had MOE values varied from 5295.71 MPa to 8428.70 MPa [17]. Other study reported that laminated board made from oil palm trunk glued with isocyanate adhesive at glue spread rate of 300 g/m$^2$ had MOE values varied from 2.28 GPa to 10.01 GPa [18]. The MOE of eight-ply LVL made of three hybrid poplar clones glued with phenol formaldehyde at glue spread rate of 200 g/m$^2$ varied from 6773.78 MPa to 8363.78 MPa [19]. The MOE of five-ply LVL made of gmelina glued with phenol formaldehyde at glue spread rate of 220 g/m$^2$ was 12.86 GPa [16]. The MOE of seven-ply LVL made of surian glued with urea formaldehyde was 8569.86 MPa [20]. The MOE of mahoni solid wood was 92000 kg/cm$^2$ [21] while the MOE of jabon solid wood was 68000 kg/cm$^2$ [22]. These values were superior than the eight-ply oil palm trunk LVL in the previous study [8]. The MOE in parallel to the grain of oil palm trunk LVL ranged from 405.83 to 712.84 MPa and 816.47 to 1501.11 MPa, for flatwise and edgewise position, respectively [8]. For static bending in perpendicular to the grain, the MOE was around 106.68 to 117.39 MPa and 154.71 to 205.27 MPa for the edgewise and flatwise position, respectively [8].

3.2.3. Compression strength.

As shown in Table 2, the compression strength values of laminated veneer lumbers varied from 164.1 kg/cm$^2$ (16.1 MPa) to 317.4 kg/cm$^2$ (31.1 MPa with an average of 223.6 kg/cm$^2$ (21.93 MPa). The ANOVA showed that the compression strength of laminated veneer lumbers was affected by layer compositions. Incorporating wood veneers in the structure of OP LVL increased the compression strength of OP LVL by 35.6% for J2 LVL, 44.2% for J3 LVL, 67.4% for M2 LVL and 93.4% for M3 LVL respectively. Similar trend was also reported by [7], who found the escalated compression strength of oil palm trunk LVL with the inclusion of rubberwood veneers, compared to the nine-ply oil plam trunk LVL. The inclusion of two and three layers of rubberwood in the LVL gave compression strength values of 26.6 MPa and 31.6 MPa, respectively [7].

The highest compression strength value of oil palm LVL in this study was higher than the compression strength of five-ply LVL made of gmelina [14]. It was reported that the five-ply gmelina LVL glued with phenol formaldehyde at glue spread rate of 220 g/m$^2$ was 103.29 kg/cm$^2$ or 10.13 MPa [14]. Additionally, the compression strength of M3 LVL (317.4 kg/cm$^2$) was comparable to the compression strength of mahoni solid wood (360 kg/cm$^2$) [21] and jabon solid wood (374 kg/cm$^2$) [22]. It suggests that the inclusion of rubberwood veneers as face, core and back layers of the oil palm trunk LVL raised the compression strength.

3.2.4. Hardness

The results of ANOVA (Table 2) showed that the hardness of laminated veneer lumbers was affected by layer compositions. The hardness of laminated veneer lumbers varied from 178.7 kg/cm$^2$ to 273.1 kg/cm$^2$ with an average of 243.1 kg/cm$^2$. Based on the data in Table 2 it can be seen that OP LVL has the lowest hardness value. In addition, it was observed that the hardness of LVL increased with the increase in the number of wood veneers incorporating in LVL structure. The hardness of five-ply OP LVL increased by 11.4% for J2 LVL, 30.1% for J3 LVL, 31.4% for M2 LVL and 32.7% for M3 LVL, respectively. The side hardness of mahoni solid wood was 271 kg/cm$^2$ [21] while the side hardness of jabon solid wood was 268 kg/cm$^2$ [22].

## 4. Conclusion

The study concluded that the physical and mechanical properties of laminated veneer lumbers made from oil palm trunk were affected by layer compositions. Incorporating jabon or mahoni wood veneers in the structure of oil palm LVL improved the physical and mechanical properties of LVL. The results of the current study indicated that the produced LVL made from oil palm trunk in combination with jabon or mahoni veneer glued with urea formaldehyde adhesive meet the Japanese Agricultural Standard requirement for laminated veneer lumber.

## References

[1] [BPS] Badan Pusat Statistik 2019 *Statistik Kelapa Sawit Indonesia 2018* (Jakarta: Badan Pusat Statistik)

[2] Wahab R, Salam MA, Sulaiman O, Sudin M, Samsi HW, and Rasat MSM 2013 Properties of engineered oil palm composite boards from 32 year-old tree stems *J. Agric. Biol. Sci.* **8**(7) 541-45

[3] Sulaiman O, Hashim R, Wahab R, Wan H, and Mohamed AH 2008 Evaluation on some finishing properties of oil palm plywood *Holz als Roh-und werkstoff* **66**(1) 5-10

[4] Sulaiman O, Salim N, Hashim R, Yusof LHM, Razak W, Yunus NYM, Hashim WS, and Azmy MH 2009 Evaluation on the suitability of some adhesives for laminated veneer lumber from oil palm trunks *Mater. Des.* **30**(9) 3572-80

[5] Hashim R, Sarmin SN, Sulaiman O, and Yusof LHM 2011 Effects of cold setting adhesives on properties of laminated veneer lumber from oil palm trunks in comparison with rubberwood *Eur. J. Wood and Wood Prod.* **69** 53-61

[6] Suhaily SS, Gopakumar DNS, Samsul R, Paridah MT, and Khalil HA 2019 Evaluation of screw pulling and flexural strength of bamboo-based oil palm trunk veneer hybrid biocomposites intended for furniture applications *BioResources* **14**(4) 8376-90

[7] Nordin K, Samsi HW, Ahmad M, and Jamaludin MA 2004 Improved properties of oil palm trunk (OPT) laminated veneer lumber (LVL) through the inclusion of rubberwood veneers *Sci. Lett.* **1** 51-6

[8] Wahab R, Samsi HW, Mohamed A, and Sulaiman O 2008 Utilization potential of 30 year old oil palm trunks laminated veneer lumbers for non structural purposes *J. Sustain. Dev.* **1**(3) 109-13

[9] Hartono R, Hidayat W, Wahyudi W, Febrianto F, Dwianto W, Jang JH, and Kim NH 2016 Effect of phenol formaldehyde impregnation on the physical and mechanical properties of soft-inner part of oil palm trunk *J. Korean Wood Sci. Technol.* **44**(6) 842-51

[10] Masseat K, Bakar ES, Kamal I, Husain H, and Tahir PM 2018 The physical properties of treated oil palm veneer used as face layer for laminated veneer lumber *IOP Conference Series: Materials Science and Engineering* 368 012025

[11] Dungani R, Jawaid M, Khalil HA, Jasni, Aprilia S, Hakeem KR, Hartati S, and Islam MN 2013 A review on quality enhancement of oil palm trunk waste by resin impregnation: Future materials *BioResources* **8**(2) 3136-56

[12] [ASTM] American Society for Testing and Materials 1995 *Standard Test Methods for Evaluating Properties of Wood-Based Fiber and Particle Panel Materials ASTM D1037-93* (Philadelphia: American Society for Testing and Materials)

[13] [JAS] Japan Agricultural Standard 2008 *Japanese Agricultural Standard for Laminated Veneer Lumber* Notification No 271 (Ministry of Agriculture, Forestry and Fisheries)

[14] Bal BC 2016 Some technological properties of laminated veneer lumber produced with fast-growing poplar and eucalyptus *Maderas Cienc. y Tecnol.* **18**(3) 413-24

[15] Sulastiningsih IM, Santoso A, and Krisdianto 2016 Karakteristik balok bambu lamina susun tegak dari bilah bambu andong (*Gigantochloa pseudoarundinacea* (Steud.) Widjaja) *Jurnal Penelitian Hasil Hutan* **34**(3) 167-77
[16] Tenorio C, Moya R, and Munoz F 2011 Comparative study on physical and mechanical properties of laminated veneer lumber and plywood panels made of wood from fast-growing *Gmelina arborea* trees *J. Wood Sci.* **57** 134-9
[17] Eka S, Alamsyah M, Sulistyawati E, and Suheri A 2018 The potential use of wood waste ash nanofiller for improvement of laminated veneer lumber production made from jabon (*Anthocephalus cadamba*) *J. Indian Acad. Wood Sci.* **15**(2) 126-31
[18] Prabuningrum DS, Massijaya MY, Hadi YS, and Abdillah IB 2020 Physical-mechanical properties of laminated board made from oil palm trunk (*Elaeis guineensis* Jacq.) waste with various lamina compositions and densification *J. Korean Wood Sci. Technol.* **48**(2) 196-205
[19] Kurt R, Meric H, Aslan K, and Cil M 2012 Laminated veneer lumber (LVL) manufacturing using three hybrid poplar clones *Turk. J. Agric. Fork* **36** 237-45
[20] Alamsyah EM and Karliati T 2011 Kayu surian sebagai alternatif bahan baku produk perekatan kayu masa depan (II): Laminated Veneer Lumber (LVL) *Jurnal Ilmu dan Teknologi Kayu Tropis* **9**(2) 126-33
[21] Martawijaya A, Kartasujana I, Kadir K, and Prawira SA 2005 *Atlas Kayu Indonesia Jilid I* (Bogor: Pusat Penelitian dan Pengembangan Hasil Hutan, Badan Penelitian dan Pengembangan Kehutanan, Departemen Kehutanan)
[22] Martawijaya A, Kartasujana I, Mandang YI, Prawira SA, and Kadir K 2014 *Atlas Kayu Indonesia Jilid II* (Bogor: Pusat Penelitian dan Pengembangan Hasil Hutan, Badan Penelitian dan Pengembangan Kehutanan, Departemen Kehutanan)

**Acknowledgement**
Authors are wishing to acknowledge assistance from Kuswandi, Alit Prio Utomo and Nuryani on producing oil palm trunk LVL and data management. Authors are very thankful for research facilities from Forest Product Research and Development Center and financial support from the Palm Oil Plantation Fund Management Agency.

IOP Conf. Series: Materials Science and Engineering **935** (2020) 012020 doi:10.1088/1757-899X/935/1/012020

# Fiber dimension and anatomy of *Acacia mangium* wood from two mother trees

**Andianto[1]*, K Yuniarti[1], N A Saputra[1], I S Saputra[2]**
[1]Forest Products Research and Development Center, Ministry of Environment and Forestry, Bogor, Indonesia
[2]Caraka Nusantara Academy of Chemical Analysis, Depok, 16951, Indonesia

*E-mail: andiant068@yahoo.co.id

**Abstract.** Mangium *(Acacia mangium)* has been opted as one of potential species for the establishment of industrial timber estate in Indonesia, particularly to support the production of pulp and paper. In order to obtain mangium stands that produces high quality wood, one attempt can be carried out through the selection and development of potential mother trees from the selected provenances. This study aim to obtain information on fiber dimension and the anatomy of mangium wood from two different mother trees from Papua New Guinea provenance. The wood was collected from mangium estate in Wonogiri forest research which was established in 2012 and managed by the Center Forest Biotechnology and Tree Improvement. Maceration process followed *Forest Product Laboratory* guidance and the preparat preparation followed Sass method. The wood anatomical terminology referred to IAWA (International Association of Wood Anatomists) list of microscopic features for hardwood identification was used to scrutinize the samples. The results showed that the average fiber dimension of mangium from the mother tree group 1 was 1239 micron (length), 25.96 micron (fiber diameter), and 3.67 micron (cell wall thickness). Those from the mother tree group 2 posseses the dimension as follow: 1215 micron (length), 25.42 micron (fiber diameter), and 3.73 micron (cell wall thickness). The fiber quality of mangium fiber from both mother tree groups could be considered in the quality class II. The anatomical structure of mangium from both mother tree groups include vessels cells in radial multiples radial multiple of 2-3 or 4 cells; vestured intervessel pits; parenchyma vascicentric and aliform; all ray cells procumbent and prismatic crystals in chambered axial parenchyma cells.

## 1. Introduction

Mangium wood (*Acacia mangium* Willd.) is widely known for its use as a fiber-producing material as well as construction, furniture and energy materials. Currently, mangium is one of the species selected for the development of timber estate in Indonesia. The main purpose for the species development is to meet the raw materials demand from the pulp and paper industry.

The conventional development of mangium plantations still uses seeds which are not resulted from the breeding technology. Therefore, the timber production is still not optimal yet. Due to this, a number of tree breeding techniques has been investigated. Many efforts have been made to breed mangium species in order to attain superior quality seeds with good quality wood.

One of the breeding techniques developed is through the selection of mother trees from the same provenance. The mother tree selection technique could affect several wood qualities such as density, cellulose content, tangential and radial shrinkage ratio (T/R ratio), initial moisture content, etc. [1, 2]. Nevertheless, less attention is still given to understand that specific breeding technique on the anatomical structure of the wood. This paper thus aimed to investigate the fiber quality as well as the anatomical structure of mangium wood from 2 different mother trees from the same provenance (Papua New Guinea) which were originally planted in 2012 in Wonogiri research forest managed by the Center for Forest Biotechnology and Tree Improvement.

## 2. Materials and Methods

The main materials used were 4-year old mangium samples collected from the bottom and top parts of the original trunks (4 samples for each part). The diameter range of the wood was 400-500 mm. The original trunks were from two mother tree groups from Papua New Guinea provenance which were planted in Wonogiri research forest. Two trees were collected from each mother tree group.

The maceration process for determining fiber length, lumen diameter and wall thickness followed the method developed by Forest Product Laboratory in Rulliaty [3]. The derivatives factors values were counted following Nurrahman and Silitonga [4]. The determination of fiber quality class as a raw material for paper followed the guidance developed by Haroen [5]. The wood anatomical features were determined on transverse/cross, radial and tangential section which were prepared according to the Sass method [6]. IAWA list of microscopic features for hardwood identification was also used for the observation of wood anatomical features [7].

## 3. Results and Discussion

### *3.1 Mangium fiber dimension and quality*

The fiber dimension of wood is one of the factors affecting the quality of pulp and paper. Fiber length, fiber diameter, lumen diameter and cell wall thickness also affect their derivatives factors values such as Runkle ratio, felting power, Muhlsteph ratio, flexibility ratio and coefficient of rigidity. Table 1 and Table 2 show the fiber dimension and their derivative factors of mangium wood investigated in this study. According to the fiber quality classification grade for pulp and paper raw materials [5], mangium fiber dimension and its derivative factors values are categorized into quality class II.

**Table 1.** Fiber dimension of 4-year old mangium from 2 mother tree groups

| Mother tree | Fiber dimension, μm | | | |
|---|---|---|---|---|
| | Fiber length | Fiber diameter | Lumen diameter | Cell wall thicknees |
| I | 1239.5±53.92 | 25.96±0.19 | 18.12±0.14 | 3.67±0.17 |
| II | 1215.8±53.96 | 25.42±0.05 | 17.95±0.08 | 3.73±0.09 |

According to Haroen [5] and Kasmudjo [8], density and fiber dimension affect the pulp yield and quality resulted from the pulping process. Syafii and Siregar [9] also stated that the pulp properties of a wood can be predicted based on its fiber dimension and derivative factors values.

The fiber length of mangium wood studied ranges from 1215 μm to 1239 μm and cell wall thickness from 3.67 μm to 3.73 μm. The fiber length of *A. mangium* in present study was higher than average fibre dimension of 4-year-old and 8-year-old *A. mangium* grown in Malaysia (933.7 μm and 1017.5 μm) [10]. According to Lim et al. [11], the mean fibre length of trees (age: 20-year) was 1048 μm whereas for the 16-year the mean fibre length was 954 μm. In contrast, fibre wall thickness in this study was commonly smaller than which grown in Malaysia (3.3 μm, 4.3 μm, 8 μm and 9 μm). Based on IAWA classification [7], this mangium fiber length is in medium class (900-1600 μm). According to Kasmudjo [8], the common range of the fiber length of broad-leaf wood species is 800-1500 μm. The fiber length is a crucial parameter since it contributes to the quality of paper tearing strength [9].

**Table 2**. Quality grade of 4-year old mangium fibers for pulp and paper raw materials

| No. | Mother tree | Fiber length, μm | Derivative factors values | | | | | Total score | Fiber quality |
|---|---|---|---|---|---|---|---|---|---|
| | | | Runkle ratio | Felting power | Muhlsteph ratio, % | Flexibility ratio | Coefficient of rigidity | | |
| 1. | I | 1239.5± 53.92 | 0.40 | 47.74 | 51.24% | 0.69 | 0.14 | | |
| | Score | 50 | 75 | 50 | 75 | 75 | 75 | 400 | II |
| 2. | II | 1215.8± 53.96 | 0.41 | 47.81 | 49.17% | 0.70 | 0.14 | | |
| | Score | 50 | 75 | 50 | 75 | 75 | 75 | 400 | II |

According to the fiber quality grade shown in Table 3, the Runkle ratio of the studied mangium fiber is relatively low (0.4). A fiber with low Runkle rasio is a very good raw material for pulp production due to its thin wall, wide lumen diameter, easy to flatten, and possess high tensile and bursting strength [9]. Nevertheless, the felting power of the studied mangium is also low (47). A fiber with high felting power value is more flexible and has better tearing strength than that with low felting power [9]. Muhlsteph ratio of the studied mangium fiber is classified into medium class (49-51). The lower the Muhlsteph ratio is the higher the pulp sheet density and its strength are [9].

**Table 3.** Fiber quality grade [5]

| Unit | Quality grade | | | | | | | |
|---|---|---|---|---|---|---|---|---|
| | I | | II | | III | | IV | |
| | Limit | Value | Limit | Value | Limit | Value | Limit | Value |
| Fiber length (L), (micron) | 2200 | 100 | 1600-2200 | 75 | 900-1600 | 50 | 900 | 25 |
| Runkel ratio (2w/l) | 0.25 | 100 | 0.25-0.50 | 75 | 0.5-1.0 | 50 | 1,0 | 25 |
| Felting power (L/d) | 90 | 100 | 70-90 | 75 | 40-70 | 50 | 40 | 25 |
| Muhlsteph ratio (%) | 30 | 100 | 30-60 | 75 | 60-80 | 50 | 80 | 25 |
| Flexibility (l/d) | 0.80 | 100 | 0.60-0.80 | 75 | 0.40-0.60 | 50 | 0.40 | 25 |
| Rigidity coefficient (W/d) | 0.10 | 100 | 0.10-0.15 | 75 | 0.15-0.20 | 50 | 0.20 | 25 |
| Total | | 600 | | 450 | | 300 | | 150 |
| Ranges | 451 - 600 | | 301 - 450 | | 151 - 300 | | < 150 | |

The flexibility ratio of the studied mangium fiber is also in medium class (0.69-0.70). This parameter is the ratio between lumen diameter and fiber dimeter. Syafii and Siregar [9] stated that a fiber with high flexibility ratio has thin wall and a shape that is easy to change. The ability to change shape assists in improving the contact between the fiber surfaces. As a further result, better bonding between fibers is created and the pulp sheet produced will have good strength quality [9]. On the other hand, a fiber with low flexibility ratio has small lumen diameter. As a result, the pulp sheet and paper produced will be thick with uneven surface and low bursting strength [12].

The rigidity coefficient of the studied mangium fiber is also in medium class (0.14). The rigidity coefficient is the ratio between the fiber wall thickness and fiber diameter. The lower the coefficient value is, the higher the tensile strength of the paper [9]. Table 2 further shows that the fiber quality of the studied mangium is at II grade. It matches the research result of Nirsatmanto *et al.* [13] that Acacias species used were classified as level II. This result means that the quality of the studied fiber meets the requirement as raw material for pulp and paper production.

### *3.2. General features and anatomical structure of mangium wood*

The general features of the heartwood of mangium wood is indicated by its colour shades, from pale brown to dark brown. Its sapwood, on the other hand, has colour shades from pale yellow to yellow. Nevertheless, the heartwood has not developed yet in the studied mangium wood due to its young age. Other features are smooth to quite coarse texture; straight to intertwined fiber; quite shiny, plain, slippery and hard surface. The anatomical features are shown in Table 4 and Figure 1-3 below.

**Table 4.** The anatomical features of 4-year old mangium

| No. | Features | Information |
|---|---|---|
| 1 | Growth ring | indistinct |
| 2 | Vessels | diffuse, solitary and radial multiples of 2-3 and sometimes 4; the average length is 339.14 ± 17.91 μm, the average diameter is 201.12 ± 14.13 μm; simple perforation plates; intervessel pits alternate and vestured (Figure 1) |
| 3 | Parenchyma | vasicentric, aliform; four (3-4) cells per parenchyma strand length (Figure 1) |
| 4 | Ray cells | all ray cells procumbent; ray width 1 to 3 cells (Figure 2) |
| 5 | Fiber cells | Fibers with simple to minutely bordered pits; fibres thin-to thick-walled; the mean fibre lengths is 1239.5 ± 53.92 μm (for mother tree group 1) and 1215.8 ± 53.96 μm (for mother tree group 2); the mean wall thickness is 3.67 ± 0.17 μm (for mother tree group 1) and 3.73 ± 0.09 μm (for mother tree group 2) (Figure 3) |
| 6 | Mineral inclusion/deposit | Prismatic crystals present in chambered axial parenchyma cells |

Species of *A. mangium* belonging to the family Leguminosae. This species is native to Australia, Papua New Guinea, and Indonesia [11]. The sapwood and heartwood of *A. mangium* have distinct colours. The sapwood is often white or yellowish white and the heartwood is yellowish brown to golden-brown when fresh [10,11]. Growth rings of *A. Mangium* are absent, vessels is solitary and diffuse, in radial multiples of 2-3 (-4); parenchyma is mainly scanty paratracheal to very thinly vasicentric, occasionally aliform, strand length of 2 to 4 cells; perforation simple; intervessel pits alternate; tyloses absent or sparse; ray mostly 1-2 cells wide rarely 3 and prismatic crystals present in chambered axial parenchyma cells [10,11]. The observation carried out for the anatomical structure of the 4-year old mangium wood is in harmony with the above statements.

IOP Conf. Series: Materials Science and Engineering **935** (2020) 012020 doi:10.1088/1757-899X/935/1/012020

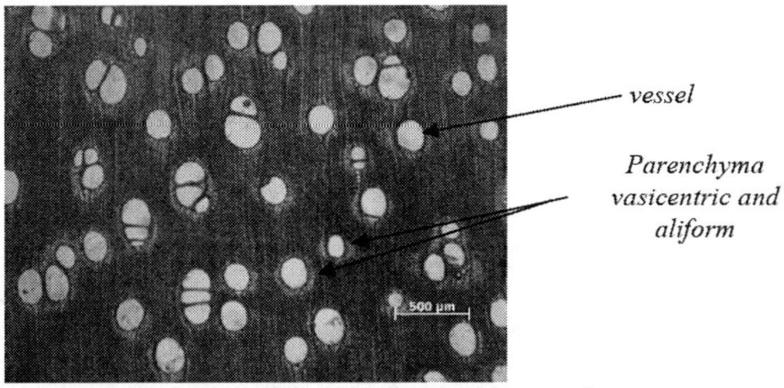


**Figure 1**. Transverse section (a) showing vessel and parenchyma

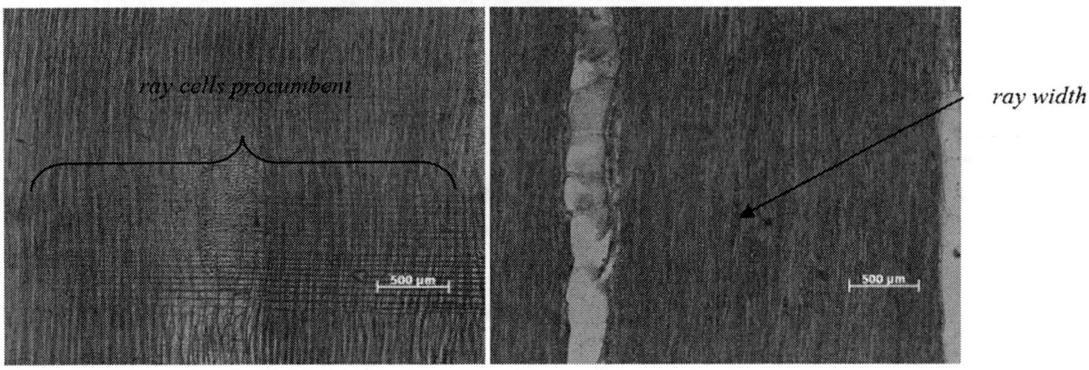


**Figure 2**. Radial (b) and tangential (c) section showing ray cells procumbent and ray width 1 to 3 cells

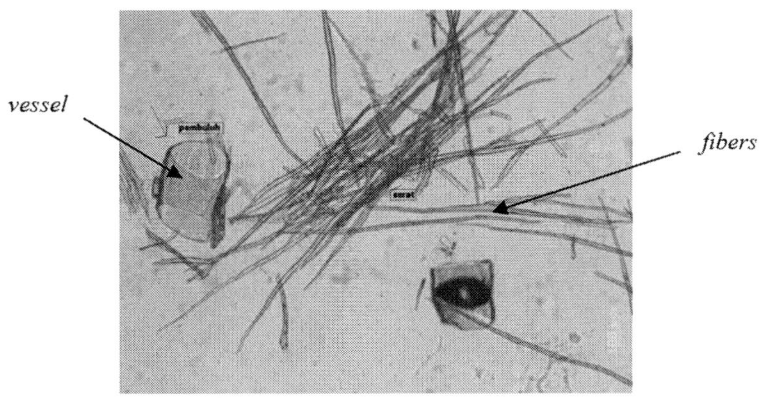


**Figure 3.** Vessels and fibers of *Acacia mangium* in maceration

## 4. Conclusion

The fiber of 4-year old mangium collected from Wonogiri research forest has II quality grade. Its quality meets the requirement as raw material for pulp and paper. The observation carried out for the anatomical structure of the 4-year old mangium wood supports previous record on the same species.

## References

[1] Quang, TH, Kien N., von Arnold, Jansson G, Thinh HH, and Clapham D 2010 Relationship of wood composition to growth traits of selected open-pollinated families of *Eucalyptus urophylla* from a progeny trial in Vietnam. *New Forests* **39**(3) 301-312

[2] Yuniarti K 2017 The influence of the mother tree and the position of the sample on the trunk on several physical properties of 4-year old mangium (*Acacia mangium* Willd). *Prosiding Seminar Nasional Masyarakat Peneliti Kayu Indonesia*, 25 September, Denpasar, Bali. Fakultas Kehutanan Universitas Gadjah Mada.

[3] Rulliaty S 1994 Wood quality indicators as estimators of juvenile wood in mahogany (Swietenia macrophylla King) from forest plantation in Sukabumi, West Java, Indonesia (Master thesis) University of the Phillipines, Los Banos College, Laguna, Phillipines.

[4] Nurrahman A, and Silitonga T 1973 Some fiber dimensions of several wood species from South Sumatra Forest Products Research Institute Report No. 13 (Bogor: Forestry general directorate Agriculture Department) p. 30

[5] Haroen WK 2016 *Fiber Technology of Paper Pulp Raw Material* (Jakarta: CV. Agung Ilmu)

[6] Sass J 1958 *Botanical Microtechnique* 3rd ed. (Ames, IOWA: The Iowa State University Press) pp. 3-77

[7] Wheeler EA, Baas P, and Gasson E 2008 *IAWA* list of microscopic features for hardwood identification. *IAWA Bulletin. n.s.* **10**(3*)* 219-332

[8] Kasmudjo 2010 *Technology of Forest Product* (Yogyakarta: Cakrawala Media)

[9] Syafii W and Siregar, IZ 2006 Chemical properties and dimensions of mangium wood fiber (*Acacia mangium* Willd.) *Jurnal Ilmu dan Teknologi Kayu Tropis* **4**(1) 28-32

[10] Sahri MH, Ibrahim FH, and Shukor NA 1993 Anatomy of *Acacia mangium* grown in Malaysia. *IAWA Journal* **14**(3) 245-51

[11] Lim SC, Gan KS and Tan YE 2011 *Properties of Acacia mangium planted in Peninsular Malaysia*. ITTO Project on improving utilization and value adding of plantation timbers from sustainable sources in Malaysia. Project No. PD 306/04(1) Perpustakaan Negara Malaysia pp 7-22

[12] Husein, N. (2004). Wood anatomy of palele (*Castanopsis javanica*) *Jurnal Ilmu dan Teknologi Kayu Tropis* **2**(2) 63-72

[13] Nirsatmanto A, Sunarti S, and Praptoyo H 2017 Wood anatomical structures of trofical Acacias and its implication to tree breeding. *International Journal of Forestry and Horticulture* **3**(3) 9-16

## Acknowledgments

This research was technical assisted by Mrs. Tuti and Mr. Romi, laborants in Forest Product R&D Centre. All authors are main contributor according to job description agreement.

IOP Conf. Series: Materials Science and Engineering **935** (2020) 012021 doi:10.1088/1757-899X/935/1/012021

# Polymerised merbau extractives as impregnating material for wood properties enhancement

**J Malik[1*], A Santoso[1] and B Ozarska[2]**

[1]Forest Products Research and Development Center, Ministry of Environment and Forestry, Bogor, Indonesia
[2]School of Environment and Forestry, The University of Melbourne, Burnley Campus, Australia

Email: jmalik@pustekolah.org

**Abstract.** This study aimed to investigate the polymerisation process of Merbau extractives (ME) and the characteristics of the phenolic resin polymers made from ME. These polymerised ME (PME) can potentially be utilised as an impregnating material to enhance the wood properties of young plantation timber. Selected PME were characterised via Fourier transform infrared (FTIR) spectroscopy, differential scanning calorimetry, thermogravimetric analysis, X-ray diffraction analysis, and ultraviolet-visible analysis. The prediction of the enhancement to the wood properties when treated with the obtained PME was discussed in relation to the physico-chemical and thermal characteristics of the polymeric materials. The results showed that the ME can be polymerised in its base condition with formaldehyde and resorcinol as the copolymer to produce the PME. The resin was classified as a resole and polymerisation can be done at room temperature. The physico-chemical tests and analyses, via boiling tests and FTIR spectra confirmed that the polymeric compound is a promising impregnating material that can enhance wood properties. This polymeric material is also eco-friendly as the low level of free formaldehyde.

## 1. Introduction

Merbau extractives (ME) have been shown to be a promising material for the enhancement of wood properties because of their high content of phenolic compounds, particularly resorcinol (R) [1]. A phenolic resin has been successfully made from ME in the past and used as an adhesive through polymerisation with the addition of a small quantity of R and formaldehyde (F) at a ME:R:F ratio of 100:5:10 (w/w). The adhesive quality fulfilled the exterior requirements [2]. Future research could determine if a similar formulation can be used to develop a wood impregnating material because of the initial hydrophilic nature of ME in an aqueous solution and it becoming water-insoluble or hydrophobic as a reaction proceeds, which occurred with the ME adhesive.

A phenolic resin is a compound that contains a wide variety of materials based on a broad range of phenols (P) and co-monomers [3]. Phenolic resins are obtained via a reaction of P and aldehydes. Both parent compounds, P and F, are by far the most important components in the production of commercial phenolic resins [4]. Phenolic resins are known to be a versatile resin system that feature either a stable, thermoplastic novolak composition that is cured with a latent source of F (hexa) or a heat reactive and perishable resole composition that is cured either thermally or under acidic or special basic conditions [5, 6]. Until the end of the 19th century, P was recovered primarily from coal tar. Phenolic resin was one of the first synthetic resins used commercially, both in plastics and surface coatings [7]. With the commercialisation of phenolic resins, the demand for P has grown remarkably. Currently, the cumene-to-P process, which is an industrial process for developing P and acetone from benzene and propylene, is the predominant synthetic route for the production of P[8].

There has been an increasing demand for natural products in industrial applications in recent years because of environmental issues, waste disposal, and depletion of non-renewable resources. These concerns encourage researchers to develop phenolic resins from renewable resources or bio-based phenolic resins [9]. According to Manjula et al. [10], the utilisation of renewable resources for polymer applications has been the source of much discussion since the early 1980s. This is partly because of the recognition that petroleum and coal supplies are limited and that the cost of petrochemicals is on the rise. Thus, biomass, which consists mostly of polymers and monomers, is a suitable renewable resource. The work of Manjula *et al.* [10] resulted in the successful creation of a novolak resin from 33.8 g (0.16 mol) of coconut shell tar that was mixed with 10.38 mL of a 37% aqueous solution (0.128 mol F and 1 g of oxalic acid) by stirring in a water bath at 100°C for 5 h. A vacuum was then applied to remove water via condensation. The yield of the phenolic resin was found to be 97%.

Lignocellulosic wastes (sawdust and cornstalks) have been converted to produce phenolic compounds that may be suitable for the production of green PF resins via direct liquefaction in hot-compressed water at temperatures of 250°C to 350°C in the presence of 2 MPa H2[11]. Phenolic resins are formed by either alkaline or acid catalysation, with the addition of F to P to give ortho- and para-substituted products. The properties of these products depend largely on the catalyst and P/F molar ratio. In resole formation, the excess F is reacted with P under basic conditions [5].

This paper aimed to investigate the polymerisation process of ME and the characteristics of phenolic resin polymers made from ME, further classified as polymerised ME (PME) that can be potentially utilised as an impregnating material to enhance wood properties

## 2. Materials and Methods

### *2.1. Material*

In the previous study [1], ME were obtained from a Merbau wood powder extraction using hot water. Similar procedures to produce the crude ME were used in this study for the polymerisation process to produce the impregnating material PME. In the polymerisation process, a technical grade R was used as an activating agent, 37% F was used as a cross-linker or curing agent, and 40% NaOH was used as the catalyst. The ME primarily contained R with a molecular weight of 110.11 g/mol, while the molecular weight of F is 30.03 g/mol. As the R present in ME is not free R, but bonded in a complex compound, pure R was added to activate the extractives.

### *2.2. Methods*

#### *2.2.1.* Polymerisation Process of ME

##### 2.2.1.1. Determining the basic and experimental proportions of ME:R:F

To determine the optimal formula, a typical resole resin with a F/P molar ratio in the range of 2:1 to 2.5:1 [3] was made with a 4% to 8% alkali content. According to this ratio, the proportions for the resole formula consisted of 27.65% P, 35.28% F, and 4.71% NaOH. Previous ME:R:F compositions with a lower portion of F (10%) and R (5%) have been successfully made for adhesive purposes [2]. Because of the different purpose of this study, a recalculation was required to determine a new composition. For the basic calculation, 4000 mL (density = 1.13 g/mL) of a liquid ME:R:F mixture was taken for composing proportions with consideration given to the densities and molecular weights of all three materials (Table 1).

**Table 1.** Molecular weight proportion to determine the ME:R:F mixture composition

| Description | ME | R | F |
|---|---|---|---|
| Density[a] (g/cm$^3$) | 1.02 | 1.29 | 1.09 |
| Molecular weight[b] (g/mol) | 110.11 | 110.11 | 30.3 |
| Reference proportion (%)[c] | 85 | 5 | 10 |
| Proportion of 4000 mL basis volume | 3400 | 200 | 400 |
| Weight (g) | 3468 | 258.4 | 436 |
| Molecular weight (g/mol) | 31.50 | 2.35 | 3.96 |
| Percent weight (wt.%) | 83.32 | 6.21 | 10.47 |
| Mole proportion | 1 | 0.1 | 0.2 |

[a] ME from [1]; R from [11]; F from [12]
[b] Molecular weight of ME was assumed as the same as R
[c] [12]

Based on the percentages of F (10.47%) and R (6.21%), a range of combinations for both substances were added to the liquid ME for the polymerisation process. Both values were assumed to be the median. For simplicity, the range for F was from 0% to 20% with an interval of 5% and the range for R was 0% to 10% with an interval of 2.5% (Table 2).

**Table 2.** Combination of R and F with ME

| Proportion | F (%) | 0 | 5 | 10 | 15 | 20 |
|---|---|---|---|---|---|---|
| R (%) | Code | 1 | 2 | 3 | 4 | 5 |
| 0 | 1 | PME11 | PME21 | PME31 | PME41 | PME51 |
| 2.5 | 2 | PME12 | PME22 | PME32 | PME42 | PME52 |
| 5 | 3 | PME13 | PME23 | PME33 | PME43 | PME53 |
| 7.5 | 4 | PME14 | PME24 | PME34 | PME44 | PME54 |
| 10 | 5 | PME15 | PME25 | PME35 | PME45 | PME55 |

*2.2.1.2. Polimerisation*

Based on the F (0%, 5%, 10%, 15%, and 20%) and R (0%, 2.5%, 5%, 7.5%, and 10%) that were added into 100 mL of aqueous ME, there were 25 combinations for the PME mixtures, which are shown in Table 2. The mixture was stirred until the R was dissolved. As a catalyst, 40% NaOH solution was also added to the mixture to increase the pH from 5 to 6 to a pH of 10 to 11. According to Detlefsen [3], the base solution results in the polymer becoming aqueous soluble when it is strong enough and causes a salt to form, which is the source of activation. After a homogeneous mixture was formed, 10 mL of 37% F was added to the mixture. It was stirred again until a homogeneous solution was obtained. The addition of F is known as methylation, and in this study the derivative compound that was expected to form was hydroxymethyl-ME. The above procedure was applied for all of the compositions, PME11 to PME55, at room temperature. The compounds were then set out at room temperature for 24 h to make sure that the reaction was completed, which was when condensation or chain extension occurred. A visual check of all of the compositions was conducted every hour.

2.2.2 Testings

*2.2.2.1. Polymer forming test*

The three components are hydrophilic compounds, and they can change to a hydrophobic material after the polymerisation process when in the proper proportion. A polymer is formed when the mixture changes from a liquid phase to a gel or solid phase. To accelerate the polymerisation process, 20 g of each mixture was placed into a petri dish and then oven-dried at 100°C ± 2°C to form a glassy solid on the petri dish, which is shown in Figure 1. All of the oven-dried mixtures with solid phases were allowed to sit for 24 h at room temperature (28°C). The polymer was formed when the solid phase did not absorb water vapour from the surroundings.

*2.2.2.2. Boiling test*

To make sure that the PME were permanently formed, a boiling test was performed. All of the PME that passed the polymer forming test were boiled in water at 100°C for 4 h. The PME were determined to have permanently formed if they did not dissolve in the boiling water.

**Figure 1.** PME mixtures: (a) polymer is not formed; (b) wet mixture after conditioning; oven-dried PME22 (c) and PME33 (d); (e) PME22 after the boiling test; and (f) PME33 after the boiling test

*2.2.2.3. Free formaldehyde test*

Formaldehyde is commonly used as a cross linker or curing agent [2, 13, 14]. Ideally, this compound should react perfectly with R and the extractive. However, it is normal for a small amount of reactant to not react. Lower amounts of free F in the PME are better for the polymers.

The test procedure to determine the free F content was as follows [2]. The amounts 0 mL, 5 mL, 10 mL, 20 mL, 50 mL, and 100 mL of a standard 3-mg/L F solution were pipetted into six bottles (100 mL each), and then they were diluted by adding water up to the 100-mL line. Each bottle contained F concentrations of 0 mg/L, 0.15 mg/L, 0.3 mg/L, 0.6 mg/L, 1.5 mg/L, and 3 mg/L, respectively. Samples of liquid PME were diluted in water up to 100 tiME, and then 10 mL of ammonium acetate (20%; 200 g in 1000 mL) were added into 10-mL PME samples, which was followed by 10 mL of acetyl acetone (0.4%; 4 mL in 1000 mL). All of the solutions were then put in a water bath at 40°C for 10 min and then cooled until room temperature was reached. The solution was measured with a photometer (UV-Vis 1700, Shimadzu, Kyoto, Japan) at a wavelength of 412 nm.

*2.2.2.4. Potlife test*
All of the PME mixture formulations were stored in covered cups and the time taken for the phase to change from liquid to gel was recorded. It is known that the potlife time is longer when it takes longer for the gel to form [2, 15].

*2.2.2.5. Solids content test*
Determining the solids content followed the procedure by Sluiter *et al.* [16] with some modifications regarding the sample. The weight of the empty dishes was measured. Then, 20 mL of ME and the ME:R:F polymer were weighed together with the dishes and put in an oven at 105 °C ± 3 °C for 1 h. The dishes with the polymer were conditioned at room temperature, stored in desiccators, and then weighed. The samples were then oven-dried and weighed to reach a constant oven-dried weight. The total solids content (*TSC*) was determined with the following equation:

$$TSC(\%) = \frac{(W_t - W_p)}{W_0} x100 \qquad (1)$$

where $W_p$ (g) = weight of dish; $W_o$ (g) = weight of polymer sample; and $W_t$(g) = $W_p$ + polymer residue.

*2.2.2.6. Selection of a proper PME composition*
The selection of the PME composition was based on the results of the polymer testing using the following criteria:

a. The polymer was formed when the mixture changed from a liquid phase to a gel or solid phase after conditioning.
b. The formed polymer was hydrophobic, and therefore it did not absorb water from the surroundings during conditioning at room temperature.
c. The polymer was not dissolved in boiling water (100 °C) after 4 h during the boiling test.
d. The amount of free F was less than 1%.
e. The potlife was greater than 18 h.

*2.2.2.7. Fourier Transform Infrared Spectroscopic Analysis*
Fifteen milligrams of PME from the selected samples (120 Mesh size) were directly taken for the FTIR spectroscopic analysis. The dried powder samples were taken from the PME samples, which represented formed and unformed polymers. They were then embedded in KBr pellets and analysed using a MB3000 spectrometer (ABB, City, Canada). They were scanned *via* the absorption mode in the range of 4000 $cm^{-1}$ to 500 $cm^{-1}$ with a resolution of 2 $cm^{-1}$.

## 3. Results and Discussion

*3.1. ME polymerisation*
The ME polymerisation is a process conducted on ME to form a polymer or resin that can be used for impregnating materials. When another component is added into the mixtures or an attempt is made to reduce the main component of the mixtures, this is referred to as modification of the phenolic resins [17]. In the current study, ME was used as the component addition, as well as for reducing the main component in the resin formulations, *i.e.*, R. Furthermore, reducing the R levels also means reducing the F levels. A RF-resin can be easily synthesised *via* the poly-condensation of R and F in an aqueous solution and the structure of the resulting resins with a base [1819].

In general, the reaction between R and F at basic catalysation conditions results in methylol R (Figure 2) [11]. Formaldehyde addition is known as methylation [3, 11]. In the current study, the derivative compound that was expected to form was hydroxymethyl-ME.

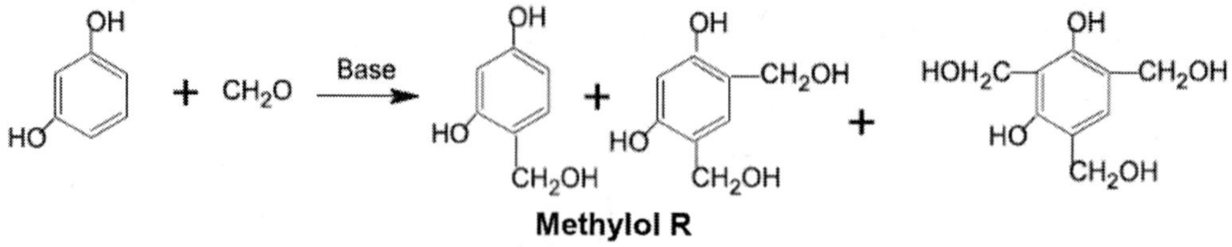


**Figure 2.** General reaction of the base-catalysed RF

The addition of a 40% NaOH solution as the catalyst forms a salt that is the source of activation. Referring to phenolic resins manufacturing, the salt has an enhanced electron density at the reactive ring positions. This is the source of activation [3]. In the current study, the reason R was treated with NaOH for ME polymerisation was to form the activated salt (Figure 3). The polymerisation process was base catalysation using NaOH, and so the pH increased (Table 3).

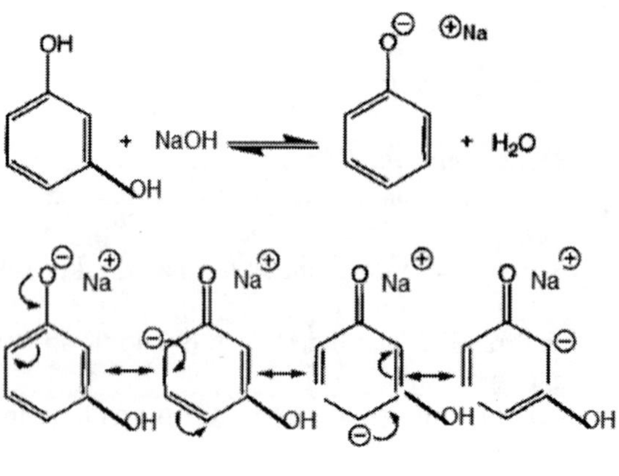


**Figure 3.** R activation by NaOH

The pH values of the ME and selected PME (*i.e.*, PME22 and PME33) are presented in Table 3. The pH of the PME doubled from 5.4 to approximately 10 to 11.

**Table 3.** pH of the crude ME and PME

| Sample | pH at Measuring Time (h) | | | | | | |
|---|---|---|---|---|---|---|---|
| | 0 | 1 | 3 | 5 | 8 | 12 | 24 |
| ME | 5.4 | 5.4 | 5.4 | 5.4 | 5.3 | 5.4 | 5.2 |
| PME22 | 10.3 | 10.6 | 10.5 | 10.4 | 10.8 | 10.5 | 10.2 |
| PME33 | 10.2 | 10.9 | 10.9 | 11.0 | 10.6 | 10.4 | 10.3 |

In addition to the provision of an aqueous solution and salt formation, manufacturing phenolic resins in a strong base with a very high pH (above 10) results in an increased gelation time [20]. Having a high pH could also prevent an extreme decrease in the resin pH when it comes in contact with wood during impregnation. The pH of the resin might decrease when the resin comes in contact with acids in wood [21].

The addition of F results in methylation [3,11]. The methylation and following reactions for ME polymerisation can be predicted by following the formation of resorcinolic resins from the RF reactions that generally take place in three stages, which are shown in Figure 4 [11]. The first step of the RF reaction is an addition reaction, where R combines with F to form hydroxymethyl (or methylol) R. In the second step, after methylation, the condensation of the methylol group occurs, which forms methylen and methylene ether (or benzyl ether) bridge structures. The last step is the disproportionation reaction in which the benzyl ether structures are decomposed to produce the methylene bridged structures and F. This is the reason why excess or free F emissions occur.

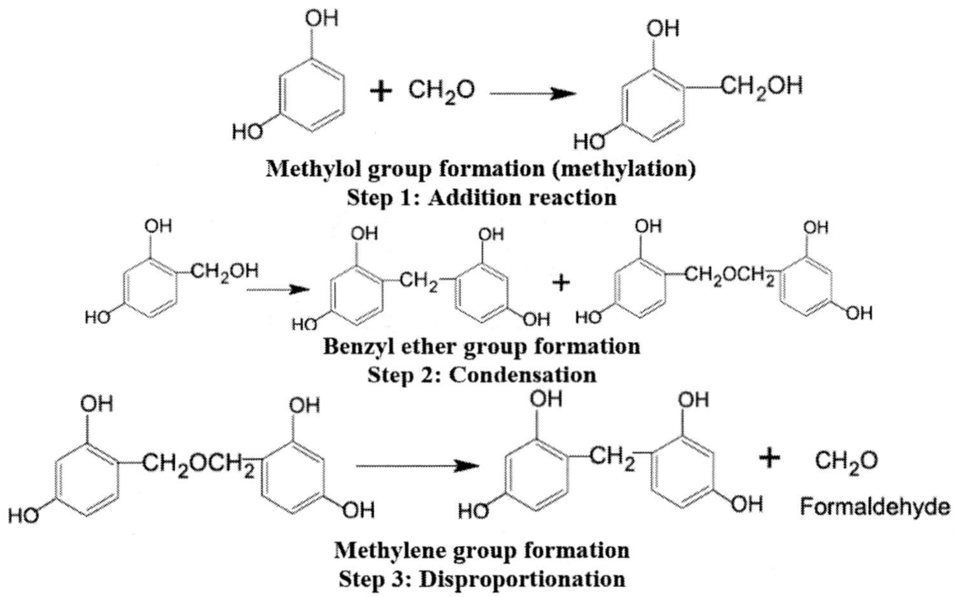


**Figure 4.** RF resin formation from the RF reaction

*3.2. PME formation*

Based on the results of the ME polymerisation process, which created 25 combinations of R and F polymers, and the data from the test results of all of the PME combinations, six different groups of PME formations were classified, which are as outlined below and summarised in Figure 5. The PME groups are as follows:

(a) Group 1: The polymers were not formed, as the mixture was still in the liquid phase after being conditioned at room temperature for 24 hours. These polymers had neither R nor F added to the mixtures. These polymers were PME11, PME21, PME31, PME41, PME51, PME12, PME13, PME14, and PME15.

(b) Group 2: The polymers were still wet after they were oven-dried and conditioned for 24 hours at room temperature because they absorbed water vapour from the surroundings. These polymers were PME23, PME24, and PME25.

(c) Group 3: The polymers were dissolved in hot water (80°C) after boiling for 4 hours. These polymers were PME34, PME35, and PME45.

(d) Group 4: The polymers were formed and were resistant to the boiling treatment. However, the level of free F produced was greater than or equal to 1%. These polymers were PME32, PME42, and PME52.

(e) Group 5: The polymers were formed and were resistant to the boiling test. The level of free F produced was less than 1%, but the storage time was less than or equal to 18 hours. These polymers were PME43, PME44, PME53, PME54, and PME55.

(f) Group 6: The PME fulfilled all of the criteria of Group 5, except they had a storage time greater than 18 hours. These polymers were PME22 and PME33.

The first step of ME polymerisation is the addition of R to the ME to activate the R contained in the complex compound of the extractives. Although the ME content is predominately R [1], the R is still bonded in the complex compound of the extractives. The addition of only R did not result in resin or polymer formation because methylation did not occur. As a result, polymers were not formed for PME12, PME13, PME14, or PME15 (Group 1). The addition of F is required for the second stage of polymerisation occur. The addition of only F into the aqueous ME also did not lead to resin formation, as was shown by PME21, PME31, PME41, and PME51 (Group 1). Based on the physical characteristics investigated during the polymerisation process, the Group 6 polymers (PME22 and PME33) could be recommended for use as an impregnating material for wood.

| | F (%) | 0 | 5 | 10 | 15 | 20 |
|---|---|---|---|---|---|---|
| R (%) | Code | 1 | 2 | 3 | 4 | 5 |
| 0 | 1 | PME11 | PME21 | PME31 | PME41 | PME51 |
| 2.5 | 2 | PME12 | PME22 | PME32 | PME42 | PME52 |
| 5 | 3 | PME13 | PME23 | PME33 | PME43 | PME53 |
| 7.5 | 4 | PME14 | PME24 | PME34 | PME44 | PME54 |
| 10 | 5 | PME15 | PME25 | PME35 | PME45 | PME55 |

= Group 1 (a)
= Group 2 (b)
= Group 3 (c)
= Group 4 (d)
= Group 5 (e)
= Group 6 (f)

**Figure 5.** Block graph of the PME compositions

*3.3. Effect of the F/R ratio*

The reaction of F with P can lead to either a heat reactive resole or a stable novolak and is dependent upon the mode of catalysis and the F/P molar ratio [4]. The mixtures of ME, R, and F found in Group 2 still showed similar properties to those of the original components of the mixtures, *i.e.*, hydrophilic or water soluble. This indicated that polymers were not formed. The ME is water soluble and can easily leach out from wood, which can lead to problems during wood processing. As a result, the extractive is easily extracted, even with only cold water. Resorcinol is also highly soluble in water [11]. To ensure the mixture remains hydrophobic, it should be polymerised with a proper R/F ratio, or the mixture will return to its initial properties, *i.e.*, hydrophilic.

Regarding the R/F ratio, Figure 5 shows that in Group 2, the proportions (%) of R/F were 5:5, 7.5:5, and 10:5 for PME23, PME24, and PME25, respectively. When these proportions were converted to a molar ratio, they were 1:0.59, 1:0.4, and 1:0.3, and the F/R ratios were 0.59, 0.40, and 0.30, respectively. This meant the R/F molar ratios were greater than 1. This was the reason why there was no polymer formation with these three mixtures. To make the mixtures polymerise at room temperature without heat treatment, the R/F molar ratio should be less than 1 or the F/R ratio should be greater than or equal to 1 [4,11]. These resins are classified as resoles [3,4,11,21]. Table 4 shows the molar ratios for all of the combinations of F and R added to the liquid ME in this study.

**Table 4.** F/R molar ratios to indicate polymer formation

| | | F (%) | | | | |
|---|---|---|---|---|---|---|
| | | 0 | 5 | 10 | 15 | 20 |
| R (%) | 0 | ~[a] | ~[a] | ~[a] | ~[a] | ~[a] |
| | 2.5 | 0[a] | **1.19**[f] | 2.37[d] | 3.56[d] | 4.74[d] |
| | 5 | 0[a] | 0.59[b] | **1.19**[f] | 1.78[e] | 2.37[e] |
| | 7.5 | 0[a] | 0.40[b] | 0.79[c] | 1.19[e] | 1.58[e] |
| | 10 | 0[a] | 0.30[b] | 0.59[c] | 0.89[c] | 1.19[e] |

Remarks: a, …., f = Group 1, …, Group 6 refers to Figure 5

Lower F/R ratios resulted in the PME failing the boiling test. The results indicated that PME34, PME35, and PME45 from Group 3 were dissolved in hot water. The F/R molar ratios for the three compositions were less than 1, *i.e.*, 0.79, 0.59, and 0.89, respectively. This meant that polymers were not formed for these three mixtures. Their solid states after the accelerated heating process were not permanent because of the high concentration of the mixture. The F concentration in the mixtures was not enough to function as the cross-linker. Possible polymerisation involving R and ME is indicated when no extractive is extracted after hot water treatment [2]. In contrast, the F/R ratio in Group 4 was much higher compared with the ratios of 2.37, 3.56, and 4.74 for PME32, PME42, and PME52, respectively. For these resins, the excess F was probably too high and resulted in free F levels of more than 1%. According to SNI 06-4567-1998 [22], free F levels are categorised as low if they are less than 1%.

The combination of R and F in the ME for PME43, PME44, PME53, PME54, and PME55 in Group 5 had a positive effect on polymer formation and the resistance to hot water, and resulted in low levels of free F. However, the storage time was shorter than 18 h, which was lower compared with the two formed resins PME22 (35 hours) and PME33 (28 hours). Additionally, the usage of R and F in Group 5 was higher than in the last two polymerised ME (Group 6).

The storage life is affected by the P/F ratio and pH [23]. Phenolic resins that are prepared *via* basic catalysts tend to have shorter storage lives than those prepared *via* an acidic process. Also, the lower portion of F in the resin resulted in a longer storage life. The storage lives for the resin with initial P/F ratios of 1:0.73 and 1:0.66 were 4 d to 5 d and 6 d to 7 d, respectively. In contrast, the portion of F must be higher than that of R (F/R greater than or equal to 1) so that polymerisation can occur at room temperature.

Table 4 shows the F/R ratios. It was observed that resins or polymers were formed with the ME mixtures when R and F were added to the aqueous ME with a F/R ratio greater than 1. The addition of F resulted in methylation [3]. The methylation and following reaction in the ME polymerisation process can be predicted by following the formation of resorcinolic resins from the RF reactions, which is shown in Figure 4.

### *3.4. Solids content*

The TSCs of the crude ME and two selected PME (PME22 and PME33) are presented in Figure 6. It was observed that PME33 had a higher TSC than PME22 because a higher proportion of the active compounds was added (5% R and 10% F), although the F/R ratios were the same (1.19), which is shown in Table 4.

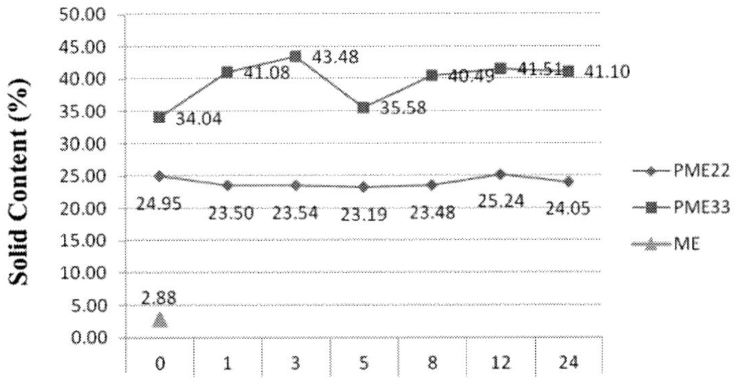


**Figure 6.** Total solids content of the crude ME and two selected PME

The solids content influences impregnated lignocellulosic materials, such as wood and rattan. Utilising the ability to alter the solids content, the weight gain and density of the impregnated material can be improved, as well as other properties of the impregnated material. A higher solids content improves the anti-shrink efficiency, specific gravity, hardness, and nail withdrawal of wood [24].

### *3.5. FTIR spectra*

Figure 7 and Table 5 show the FTIR spectra and band assignments of the peaks of the formed PME (PME22, PME33, PME44, and PME55), unformed PME (PME34, PME35, and PME45), and a formed PME with a free F greater than or equal to 1% (PME52). The spectra of PME22, PME33, PME44, and PME55 show relatively similar peak patterns. The FTIR spectra showed a broad absorption peak at approximately 3400 $cm^{-1}$, which could have been a stretching vibration of the hydroxyl linkage between the phenolic groups that appeared in all of the mixtures, as the major components of the ME are phenolic compounds [1]. However, it was obvious that the polymerisation of the ME caused the intensity of the OH absorption band (approximately 3400 $cm^{-1}$) to decrease by 22%, 31%, 44%, and 47% for PME33, PME22, PME44, and PME55, respectively, which were smaller compared with the 59%, 82%, and 86% decreases for PME35, PME34, and PME45, respectively. These decreases revealed that the hydroxyl group content was reduced after the polymerisation process. Furthermore, the reduction in the hydroxyl group content can explain why an improvement in the dimensional stability of the impregnated wood using the selected formed PME occurred. This reduction was caused by the occurrence of cross-linking reactions [25,26]. Therefore, it can be stated that the decrease of less than 20%, which occurred for PME52, was because of the higher F content, which can initiate cross-linkage.

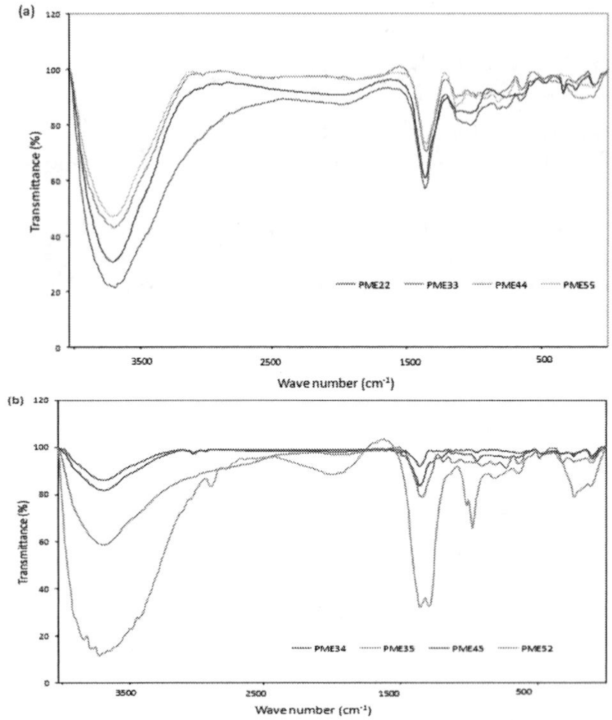


**Figure 7.** FTIR spectra of the PME: a) formed PME; and b) unformed and formed PME with a higher proportion of F/R

There were apparent differences between the two groups of PME (formed and unformed) with respect to the presence of the absorption bands of CH, $CH_2$, or $CH_3$ stretching in the methyl and methylene groups at 2924 $cm^{-1}$ in the formed PME. This could have been because of the polymerisation process that occurred in the ME after the methylation stage (Figure 4), in which condensation of the methylol group occurred to form methylen and methylene ether bridge structures [11,27]. A noticeable difference also occurred because of the existence of the alkyne (C≡C) peak in the formed PME group at 2100 $cm^{-1}$ to 2260 $cm^{-1}$. The peak associated with the C≡C structure is characteristic of the impact on the carbon-carbon bond stretching as an increasing function of the bond order for the series of single-, double-, and triple-bonded carbons [27]. This increase in the bond order produced a corresponding increase in the bond strength. As a result, it was concluded that the PME were bonded strongly and would not be leached out during the boiling test and the colour of impregnated wood would not be altered because of cross-linking between the PME and wood during curing. The existence of the functional groups that reacted with wood could improve the wood fixation or anti-shrinkage because of the hydrophobicity of the polymeric material. Cross-linking may be achieved *via* reactions with the functional groups [21].

The peak from 1600 $cm^{-1}$ to 1650 $cm^{-1}$ was identified as a prominent and strong C=C stretching absorption of an aromatic ring in the synthesised resins [28]. This typical peak indicated the presence of R because this compound is conjugated by two functional groups of OH and one aromatic ring of C=C. The aromatic aldehyde and CH deformation in the polymers appeared with small differences in the intensities of the absorption bands amongst both PME groups from 1300 $cm^{-1}$ to 1475 $cm^{-1}$. In the fingerprint area, the peaks represented ester CO groups at 1000 $cm^{-1}$ to 1300 $cm^{-1}$ and out-of-plane bending of alkene CH groups from 600 $cm^{-1}$ to 1000 $cm^{-1}$, which were in all the spectra. The presence of carboxylic acid esters, anhydrides, amides, and carbonates in the polymeric resin could be a special resole curing system that could accelerate the curing of resole resins [29].

**Table 5.** Peak assignments of the FTIR spectra of selected PME [26, 27, 30-32]

| Wavenumber ($cm^{-1}$) | | | | | | | | Band assignment |
|---|---|---|---|---|---|---|---|---|
| PME22 | PME33 | PME44 | PME55 | PME34 | PME35 | PME45 | PME52 | |
| 3456 | 3441 | 3418 | 3448 | 3441 | 3441 | 3448 | 3441 | Hydroxyl functional groups (OH) stretching |
| 2924 | 2924 | 2924 | 2924 | - | - | - | 2924 | C–H and $CH_3$ stretching in methyl and methylene groups |
| 2129 | 2214 | 2122 | 2114 | - | - | - | 2129 | Alkynes (C≡C) |
| 1713 | 1713 | - | - | - | - | 1744 | - | Carbonyls (CO) |
| 1643 | 1643 | 1636 | 1636 | 1643 | 1628 | 1643 | 1643 | Alkenes (C=C; Aromatic ring) |
| 1456 | 1381 | 1466 | 1466 | 1458 | 1458 | 1458 | 1458 | Aromatic aldehyde; C–H deformation in polymers |
| 1180 | 1219 | 1296 | 1296 | 1219 | 1296 | 1088 | 1219 | Esther groups (C-O-) |
| 849 | 849 | 764 | 725 | 957 | 972 | 949 | 957 | Alkene (C-H-) out-of-plane bending |

### *3.6. Free formaldehyde*

Because of global concerns about the health, safety, and environmental sustainability of phenolic resins, the effects of manufactured products that use these resins should be considered. Resole usually has excess aldehyde [3]. The manufacturing of PF resins, or any phenolic resin, could potentially jeopardize such an undertaking of developing an environmentally sustainable resin on both the laboratory and manufacturing scales with several levels of hazards. The hazards come from components or monomers that may be highly toxic, flammable, or corrosive to the human body. However, the resins themselves are generally rated as non-dangerous because of the excess F, as shown in the reaction above. Nevertheless, environmental factors should be considered.

**Table 6.** Concentrations of the standard F and absorbent

| | | | | | | | |
|---|---|---|---|---|---|---|---|
| Standard F (ppm) | 0 | 0.050 | 0.100 | 0.200 | 0.300 | 0.400 | 0.500 |
| Absorbent (ppm) | 0 | 0.082 | 0.171 | 0.380 | 0.596 | 0.818 | 1.034 |

Table 7 shows the measurement results of the free F based on the standard for F absorption at selected concentrations (Table 6) via a photometer at a wavelength of 412 nm. The relationship between the concentration ($X$) and absorbent ($Y$) was expressed by:

$$Y = 2.0927X - 0.0232 \quad (3)$$

The $R^2$ was 0.9987 (Figure 8). Therefore, this expression was reliable for calculating the F concentration in the samples.

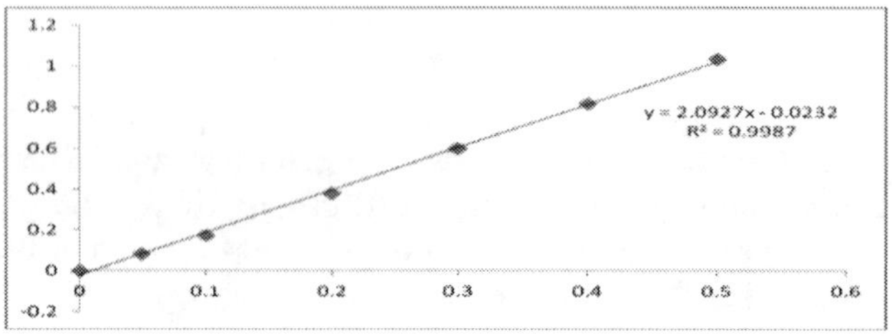


**Figure 8.** Concentration of the standard and absorbent

**Table 7.** Absorbent and F concentration in the tested samples

| Sample | Absorbent | | | Concentration (ppm) |
|---|---|---|---|---|
| | 1 | 2 | Average | |
| PME22 | -0.005 | -0.005 | -0.005 | 0.009 |
| PME23 | 0.003 | 0.003 | 0.003 | 0.013 |
| PME24 | 0.000 | 0.001 | 0.001 | 0.011 |
| PME25 | 0.001 | 0.001 | 0.001 | 0.012 |
| PME32 | 0.019 | 0.020 | 0.020 | 0.020 |
| PME33 | -0.001 | -0.001 | -0.001 | 0.011 |
| PME34 | 0.007 | 0.007 | 0.007 | 0.014 |
| PME35 | 0.000 | 0.000 | 0.000 | 0.011 |
| PME42 | 0.020 | 0.020 | 0.020 | 0.021 |
| PME43 | 0.010 | 0.010 | 0.010 | 0.016 |
| PME44 | -0.001 | -0.001 | -0.001 | 0.011 |
| PME45 | -0.001 | -0.001 | -0.001 | 0.011 |
| PME52 | 0.035 | 0.035 | 0.035 | 0.028 |
| PME53 | 0.016 | 0.016 | 0.016 | 0.019 |
| PME54 | 0.010 | 0.010 | 0.010 | 0.016 |
| PME55 | 0.005 | 0.005 | 0.005 | 0.013 |

In this study, the level of free F was considered to be one of the selection criteria for the PME formulation, with an extremely low level of maximum free F at 0.01 ppm. Table 9 shows that the lowest free F was 0.009 ppm (PME 22) and the second lowest was 0.011 ppm (PME24, PME33, PME35, PME44, and PME45). Based on the free F values, the polymerised ME that can be considered for use as an impregnating material were PME22 and PME33. PME33 also has other advantages that were explained previously. Free F and free P are the components that cause toxicity in phenolic or aldehydic synthetic resins. In general, PF resins with a low toxicity have poor bond strengths, while a great bonding strength is often accompanied by a high toxicity [33]. This study expected there to be no effect from lower free F levels on the bonding strength between the PME and impregnated wood.

## 4. Conclussion

Merbau extractives can be polymerised in basic conditions with F and R added as the copolymer to produce PME. The resin is classified as a resole because the R/F molar ratio should be less than 1 or the F/R ratio should be greater than or equal to 1, so that polymerisation can occur at room temperature. The results of the physico-chemical tests and analysis *via* FTIR spectroscopy confirmed that the polymeric compound is a promising impregnating material for the enhancement of wood properties. This material is also eco-friendly since extremely low level of free formaldehyde.

## References

[1] Malik J, Santoso A, Mulyana Y, and Ozarska B 2016 Characterization of merbau extractives as a potential wood-impregnating material *BioResources* **11**(3) 7737-53
[2] Santoso, A, Hadi YS, and Malik J 2014 Composite flooring quality of combined wood species using adhesive from merbau wood extract *Forest Prod. J.* **64**(5-6) 179-86
[3] Detlefsen WD 2002 Phenolic resins: Some chemistry, technology and history *in Adhesive Science and Engineering-2: Surfaces, Chemistry and Applications* Ed. Chadhury M and Pocius AV (Amsterdam, Netherlands: Elsevier) pp. 869-945
[4] Gardziella A, Pilato LA, and Knop A 2000 *Phenolic Resins: Chemistry, Applications, Standardization, Safety and Ecology* (Berlin, Germany: Springer-Verlag)
[5] Ebewele RO 2000 *Polymer Science and Technology* (Boca Raton, FL.: CRC Press)
[6] Pilato L 2010 *Introduction in Phenolic Resins: A Century of Progress* Ed. Pilato L (Berlin, Germany Springer-Verlag) pp. 1-8
[7] OCCA 1983 *Surface Coatings Vol I-Raw Materials and Their Usage* (Dordrecht, Netherlands: Springer)
[8] Weber M and Weber M 2010 *Phenols*, in *Phenolic Resins: A Century of Progress* Ed. Pilatio L (Berlin, Germany: Springer-Verlag) pp. 9-24
[9] Raquez JM, Deléglise M, Lacrampe MF, and Krawczak P 2010 Thermosetting (bio)materials derived from renewable resources: A critical review *Prog. Polym. Sci.* **35**(4) 487-509
[10] Manjula S, Sudha JD, Bera SC, and Pilw CKS 1985 Polymeric resin from renewable resources: Studies on polymerization of the phenolic component of coconut shell tar *J. Appl. Polym. Sci.* **30**(4) 1767-71
[11] Durairaj RB 2003 *Resorcinol: Chemistry, Technology and Application* (Berlin, Germany: Springer-Verlag)
[12] Sigma-Aldrich 2015 *Formaldehyde solution*
[13] Calve LR 1992 *Fast Cure And Pre-Cure Resistant Cross-Linked Phenol-Formaldehyde Adhesives and Methods of Making Same* (Canada)
[14] Sutherland BW, Toews J, and Kast J 2008 Utility of formaldehyde cross-linking and mass spectrometry in the study of protein–protein interactions *J. Mass Spectrom* **43**(6) 699-715
[15] Pizzi A and Stephanou A 1993 On the chemistry, behavior, and cure acceleration of phenol-formaldehyde resins under very alkaline conditions *J. App. Polym. Sci.* **49**(12) 2157-70
[16] Sluiter, Hames AB, Hyman D, Payne C, Ruiz R, Scarlata C, Sluiter J, Templeton D, and Wolfe J 2008 *Determination of Total Solids in Biomass and Total Dissolved Solids in Liquid Process Samples (NREL/TP-510-42621) Golden, Co.* (Colorado: National Renewable Energy Laboratory Golden)
[17] Dunky M, and Pizzi A 2002 Wood Adhesives *in Adhesive Science and Engineering-2: Surfaces, Chemistry and Applications* Ed. Chadhury AVPM (Amsterdam, Netherlands: Elsevier) pp. 1039-103
[18] Lu KT and Wu LY 2013 Substitution of phenol in Phenol-Formaldehyde (PF) resins by wood tar for plywood adhesives *Holzforschung* **67**(4) 413-19

[19] Zhang G, Ni C, Liu L, Zhao G, Fina F, and Irvine JTS 2015 Macro-mesoporous resorcinol–formaldehyde polymer resins as amorphous metal-free visible light photocatalyst. *J. Mater. Chem.* **3**(30) 15413-19
[20] Pizzi A and Stephanou A 1994 Phenol-formaldehyde wood adhesives under very alkaline conditions Part I: Behaviour and proposed mechanism *Holzforschung* **48**(1) 35-40
[21] Akay M 2012 *Introduction to Polymer Science and Technology* (London, UK: Bookboon)
[22] [BSN] Badan Standardisasi Nasional 1998 *Urea formaldehida cair untuk perekat kayu lapis* (Jakarta: Badan Standardisasi Nasional)
[23] Pizzi A 1994 *Advanced Wood Adhesives Technology* (New York: Marcel Dekker)
[24] Ariffin WTW, Koh MP, and Mustaf MT 1993 Improved rattan through phenolic resin impregnation: A preliminary study *J. Trop. For. Sci.* **5**(4) 485-91
[25] Tjeerdsma BF, Boonstra M, Pizzi A, Tekely P, and Militz H 1998 Characterisation of thermally modified wood: Molecular reasons for wood performance improvement *Holz Roh. Werkst* **56**(3) 149-53
[26] Tuong VM and Li J 2010 Effect of heat treatment on the change in color and dimensional stability of acacia hybrid wood *BioResources* **5**(2) 1257-67
[27] Coates J 2000 *Interpretation of infrared spectra, A practical approach*, in *Encyclopedia of Analytical Chemistry* Ed. Meyers RA (Hoboken, NJ: John Wiley & Sons Inc) pp. 10815-37
[28] Lee YK, Kim HJ, Rafailovich M, and Sokolov J 2002 Curing monitoring of phenolic resol resins via atomic force microscope and contact angle *International Journal of Adhesion and Adhesives* **22** 375–34
[29] Masson E, Pizzi A, and Merlin M 1997 Comparative kinetics of the induced radical autocondensation of polyflavonoid tannins II: Flavonoid units effect. *J. Appl. Polym. Sci.* **64**(2) 243-65
[30] Mattos BD, Lourençon TV, Serrano L, Labidi J and Gatto DA 2015 Chemical modification of fast-growing eucalyptus wood *Wood Sci. Technol.* **49**(2) 273-88
[31] Pandey KK 1999 A study of chemical structure of soft and hardwood and wood polymers by FTIR spectroscopy *J. Appl. Polym. Sci.* **71**(12) 1969-75
[32] Silverstein RM, Webster FX, and Kiemle DJ 2005 *Spectrometric Identification of Organic Compounds 7th ed.* (Hoboken, NJ: John Wiley & Sons)
[33] Lin R, Sun J, Yue C, Wang X, Tu D, and Gao Z 2014 Study on preparation and properties of phenol-formaldehyde-Chinese fir liquefaction copolymer resin *Maderas-Cienc. Tecnol.* **16**(2) 159-74

**Acknowledgments**
The authors gratefully acknowledge the support from the Australian Cooperation for Agricultural Research (ACIAR) and Forest Products Research and Development Centre for financial support.

IOP Conf. Series: Materials Science and Engineering **935** (2020) 012022 doi:10.1088/1757-899X/935/1/012022

# Potential use of fully wood based sandwich panel for house component in Indonesia for disaster mitigation and rapid built: A review

**J Malik***, **D R Trisatya and Abdurachman**

Forest Products Research and Development Center, Ministry of Environment and Forestry, Bogor, Indonesia

*Email: jmalik@pustekolah.org

**Abstract.** Indonesia is one of the most disaster-prone countries on the earth. The disasters, particularly earthquake and landslide, ussualy result in loss and damage of human shelters. In regards to the housing, mitigation or disaster risk reduction by selecting proper house component materials is important. This paper discuss on fully wood based sandwich panel that could replace other materials for wall part of the house. The study shows that wood in various forms of derivative products remain preferable building material for housing due to several advantages. Wood-based sandwich panels with veneer or plywood at the facing (face/back) and solid or laminated board for the core are promising products to be used to rapid-built houses for the mitigation and reconstruction of disaster areas.

## 1. Introduction

Natural disasters are catastrophic events coming from atmospheric, geological, and hydrological origins (e.g. droughts, earthquakes, floods, hurricanes, landslides) that can cause fatalities, property damage and social environmental disruption [1]. Simply, Indonesian National Board for Disaster Management (BNPB) cited Indonesian Law Number 24 year 2007 defining that natural disasters are disasters caused by events or a series of events caused by nature such as earthquakes, tsunamis, volcanic eruptions, floods, droughts, hurricanes, and landslides [2]. Trend disasters will increase in future with increasing global climate change and environmental degradation [3]. However, most disaster occurence cannot be predicted. The disasters have encouraged Indonesian people to accept the reality of coexistence with them. Earthquake disaster, for example, will likely occur in almost all regions in Indonesia.

Natural disasters often result in loss and damage. The biggest loss is - of course - the loss of life. Nevertheless, for those who survived, many victims also lost property and even their settlement. In the last three years (2018-2020), hundreds thousand houses loss and damage because of the disasters. In Lombok earthquake in 2018, the number of 83,392 units of houses were damage [4]. Throughout 2019 houses damaged by disaster as many as 72,992 housing units. It consisted of 15,747 units that were severely damaged (RB), 14,519 units were moderately damaged (RS), and 42,726 were slightly damaged (RR)[5]. Until mid-2020, the number of houses damaged by the disaster with the RB category amounted to 4051 units, RS by 2596 units and RR by 11,516 units [6]. The disasters lead to rise the need of houses in Indonesia highly besides due to new household growth. Currently, the need for houses in Indonesia is quite high at 7.6 million units [7] with an annual demand rate of 800,000 units house [8].

Government initiatives to build low-cost and fast-built public housing need to be supported by stakeholders including material and technical sectors providers. Regarding the material for housing, wood still retains many advantages over more industrial building materials like concrete or steel. Not only do trees absorb CO2 as they grow, they require much less energy-intensive methods to process into construction products [9]. As wood is the only significant building material that is grown, we have a natural inclination that building in wood is good for the environment [10]. Not only from environmental aspects, the construction of using wood as a building material is reliable, even for tall building. According to [10] timber is particularly structurally efficient material in structures, or parts of structures, in which a high proportion of the load to be resisted is the self-weight of the structure itself.

Although potentially available, wood has limited characteristics, especially timber from plantations and community forests which are dominated by fast growing species. They have some weaknesses i.e. small dimensions (diameter) and less in strength, durability as well as its stability. To overcome the weaknesses of these properties, it is necessary to input technology in the form of engineered wood and lignocellulosic materials, which include sandwich panels and composite beams that can withstand flexible loads.

There are still limited studies on wood-based sandwich panels (completely or partially consisting of wood materials) as transformable house constituent; however, several research projects have focused on design of facade panels and deck structures for traditional houses [11]. This paper reviews potential use of fully wood-based sandwich panel for house component for disaster mitigation with rapid rebuilt in Indonesia.

## 2. Disaster mitigation and housing

Due to its geographical and geological condition, Indonesia is one of the most disaster-prone countries on the earth [3,12]. Over the last 30 years, this tropical country has experienced an average of 290 significant natural disasters every year [13]. Almost all of type natural disasters occur in Indonesia, including volcanic eruptions, earthquakes, flooding, forest fires, and tsunami which have claimed lives and devastated communities in Indonesia. According to [14], the most frequent disasters in Indonesia are floods and earthquake.

The recent disasters that caused big loss and damage were summarized by the Center for Excellence in Disaster Management and Humanitarian Assistance (CFE-DM) [15] i.e. cyclone Cempaka on November 2017 in East Java and D.I. Yogyakarta provinces; volcanic eruption on August 2017 of Mt. Agung in Bali; earthquake on December 2016 in the Pidie Jaya District, Aceh Province; floods and landslides on May-July 2016 in the provinces of Kalimantan (South, West, and Central), Bengkulu and Gorontalo; floods and landslides on August-October 2016 in Ende, East Nusa Tenggara Province; floods and landslides on November 2015 in West Java, Central Java, West Sumatra and Jakarta; drought in 2015-2017 in eastern Indonesia; floods and landslides on November-December 2014 in multiple parts of Nanggroe Aceh Darussalam; Mt. Kelud Volcano Eruption on February 2014; Mt. Sinabung Volcano Eruptions on September 2013-November 2015; floods and landslides on November 2013-January 2014 and 2013 in Jakarta; and Lombok earthquake in 2018.

In general, earthquakes are nature's most unpredictable and one of the most devastating natural disasters. When high intensity earthquakes occurs they can cause thousands of deaths and billions of dollars in damaged property [16]. Since 1900 onward, [17] recorded that there are around 100 key earthquakes (or around 1% of damaging earthquakes) have caused around 93% of fatalities globally. In Indonesia, the total of 174,900 unit of houses were damaged with severe, moderate and slight damage categories [4-6]. While disasters are unavoidable and unpredictable, the elimination of all risk is also impossible. In spite of the increasing frequency and severity of the catastrophes that strike humankind, it would be better to increase possibility prevent and mitigate the effects of disasters [19].

Further, the literature on hazards and disasters is full of technical terms. Two mitigation and preparedness are commonly used to categorise the main methods of protecting communities against hazards and disasters. 'Mitigation' can be defined as any action to minimise the impact of a potential disaster; 'preparedness' refers to specific measures taken before a disaster strikes, usually to issue warnings, take precautions and facilitate a rapid response [19]. When a big earth quake attack a house, many possibilities can happen. If the house strong enough, then nothing happen and loss of human lives can be avoided. However, if it is in very bad condition as many houses on the earth, it can be broken or some of material waste could be felt down. It can be dangerous to the human under it. However, there is no building can be made 100% safe against earthquake forces. Instead buildings can be made earthquake resistant to certain extent depending upon serviceability requirements. Earthquake resistant design of buildings depends upon providing the building with strength, stiffness and inelastic deformation capacity which are great enough to withstand a given level of earthquake-generated force. This is generally accomplished through the selection of an appropriate structural configuration and the careful detailing of structural members, such as beams and columns, and the connections among them [20].

## 3. Wood based material for housing development

Forest serves a wide range of socio-economic and ecological benefits in our live [21,22]. It produces the most important renewable sources that can be utilized for building materials and other composite products [23]. Wood have been used for many years in Indonesia to build traditional houses [24-35]. Further, the environmental awareness has led modern community to utilize timber instead of the more energy intensive materials [23,36-38]. The recent study called the wood-based modern house as the environmentally and energetically efficient construction [37].

Previous study reported that the utilization of wood as a construction material in the New Zealand decreased about 20% in fossil fuel energy and 20% in atmospheric carbon emissions [36]. Nevertheless, the escalated wood utilization for construction material should be also coined with the sustainable timber stock [36]. This is inline with Ramage [10] who stated the urgency of the use of global certification schemes to ensure the sustainable use of legal wood.

The production trend of roundwood in Indonesia is increasing. It was recorded that the roundwood production from 2016 to 2018 was 23.12 millions $m^3$, 42.25 millions $m^3$ and 55.52 millions $m^3$, respectively [39-41]. It suggests that timber material is still demanding. However, the supply of large diameter timber is limited and therefore the utilization of wood-based composite products is inevitable. Wood-based composite product is defined as a multilayers materials that adhered together [42].

Varied technologies introduced in the wood composite manufacturing escalate its dimensional stability and durability, and give more homogenous mechanical properties [10]. Furthermore, [10] listed engineered timber products that can be produced from softwoods, such as glulam, Laminated Veneer Lumber (LVL), structural veneer lumber or beam, Cross Laminated Timber (CLT), I-joists, structural insulating panels and brettstapel. The application of these products are for interior and exterior usages: roof, wall, floor, beams, columns, cord, joist, long span and high loading of beam [10].

The wood-based modern house made of prefabricated sandwich panels was resulted in the greater life-cycle time assessment in contrast to a traditional masonry construction system [37]. It was argued that the use of wood-based composite product as a building material is such a promising way to an efficient and sustainable housing development.

## 4. Sandwich panel: a promising material for quick-built house

Sandwich panels is a layered construction formed by bonding two thin facings to a thick core [43] Sandwich panels have become more common in different engineering, including house buildings, and used either in structural or non-structural components in construction. Sandwich structures are usually made from two thin face sheets with high stiffness and strength, and a compliant and light-weight core that maintains the distance between the faces and sustains deformation, often with insulation properties.

By using variation of the material and thickness of core and face sheets, it is possible to obtain sandwich structures with different properties and performance [44]. Conventionally, sandwich panels are manufactured from steel, stainless steel and aluminium which are often used for face material. In many cases it is also suitable to choose fibre- or glass- reinforced plastics as face materials [45] The core can be made of a weak and low density material, separates and stabilizes the thin facings and provides most of the shear rigidity of the sandwich construction. Selecting properly of materials for facings and core, constructions with high ratios of stiffness to weight can be achieved. As a crude guide to the material proportions, an efficient sandwich is obtained when the weight of the core is roughly equal to the total weight of the facings [43].

Traditional housing construction industry usually relies on conventional well known building materials, like timber, concrete and steel due to attractive price and available design and manufacturing technologies. However in rapidly grooving market of transformable houses and shelters application of pre-fabricated wood based sandwich panels could provide significant savings on transportation and installing [46]. Regarding the economical consideration, Moody *et al.* [43] explained that sandwich construction is also economical because the relatively expensive facing materials are used in much smaller quantities than are the usually inexpensive core materials. The materials are positioned so that each is used to its best advantage.

Wood-based sandwich panels with low density fiberboard core were developed for structural insulated walls and floors, with different face material, panel thickness, and core density. The panel was found to be closest to the optimum design, which meant that its core and face thickness were optimum for stiffness with minimum density. The panel also provided enough internal bond strength and an excellent dimensional stability. The panel was the most feasible for structural insulation use with the weight-saving structure [18]. Thermal insulation and warmth-keeping properties of thick plywood-faced sandwich panels with low-density fiberboard Plywood Faced Sandwich (PSW), which were developed as wood-based structural insulation materials for walls and floors. The PSW provided thermal insulation properties superior to those of the boards and in terms of warmth keeping properties were greatly advantageous over the insulators. These advantages were due to the moderate densities of PSW panels. The PSW panel with sufficient thickness showed remarkably improved thermal resistance compared with those of the boards [47]. Table 1 summaries some advantages of using sandwich panel for house components.

**Table 1**. The benefits of using sandwich panel for house components

| No | Advantages | Sandwich panels (general) | Wood based sandwich panels | References |
|---|---|---|---|---|
| 1 | provide significant savings on transportation and installing | √ | √ | [11,43,46,52] |
| 2 | decorative effects | √ | √ | [43] |
| 3 | wide range of light material can be used for the core | √ | √ | [43] |
| 4 | wide range in density can be produced | √ | √ | [43] |
| 5 | higher specific stiffness compared to a solid plywood board | x | √ | [11] |
| 6 | better thermal insulation properties | x | √ | [18,47] |
| 7 | flexural and shear stiffness can be easily achieved by increasing the core thickness with little increase in weight | x | √ | [11,18,49] |
| 8 | renewable resources | x | √ | [48] |

Remarks: √ = applicable; x = not applicable

We can discuss further on each advantage of general sandwich panel. However, in this section the authors would like to elaborate on the most comparative advantage fully wood based sandwich panels compared to other materials of which can be applied for the wall parts of a house buildings. That is wood based sandwich panels that are made from veneer or plywood at the facings and solid or laminated board. In general, brick (*bata merah*) and concrete brick (*batako*) are most used traditionally for wall part of houses built in Indonesia. Nowadays, Aerated/Autoclaved Lightweight Concrete (ALC) (*bata ringan*) emerge to be used in house development in Indonesia (Table 2 and Figure 1).

**Table 2**. Common materials used for wall houses

| No | Wall material | Size (mm) |
|---|---|---|
| 1 | Brick | 25 x 12 x 5 |
| 2 | Concrete brick | 40 x 20 x 10 |
| 3 | Aerated Lightweight Concrete/ALC | 60 x 20 x 10 |

(a) (b) (c)

**Figure 1.** Common materials used for wall houses in Indonesia: (a) brick; (b) concrete brick; (c) lightweight brick/ALC

To build a type-36 house with two bedrooms, for instance, the number of bricks needed from those three types of them are obviously different because the difference of their sizes. The type-36 houses with two bedrooms generally have a total wall area of 154 $m^2$. Comparison of the amount of each brick needed, is presented in Table 2. In terms of the completion speed of the wall construction, the three types of material will definitely give a different speed. Building a wall using ALC will be faster than that using concrete brick, and using concrete brick will be faster than using brick. Now, imagine to build a house wall, we will replace the three types of bricks above with all-wood sandwich panel (Figure 2) with the size of 2440 x 1220 x 36 mm. Therefore, this means that using wood based sandwich panel composites can complete building the wall much faster than using all types of bricks. We made a simulation of the installation speed of each type of wall materials as presented in Table 3.

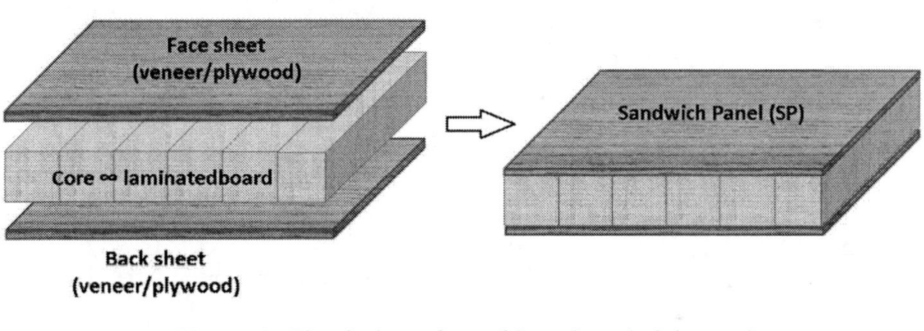


**Figure 2**. The design of wood based sandwich panel

**Table 3**. Simulation of installation speed comparison of four types of wall materials

| No | Wall material | Size (mm) | B (pcs/m$^2$) | L (m$^2$) | S (tu/m$^2$) |
|---|---|---|---|---|---|
| 1 | Brick | 250 x 120 x 50 | 70 | 154 | 1.000 |
| 2 | Concrete brick | 400 x 200 x 100 | 12 | 154 | 0.171 |
| 3 | Aerated Lightweight Concrete/ALC | 600 x 200 x 100 | 7 | 154 | 0.100 |
| 4 | Wood based sandwich panel | 2440 x 1220 x 36 | 0.34 | 154 | 0.005 |

Remarks: B = the number of materials needed (pcs) per square meter wall; L = total area of wall of type-36 house ; S = installation speed unit per square meter wall; pcs = pieces; tu = time unit.

Table 3 shows a simulation of the installation speed of four types of wall materials. The calculation based on a type-36 house with two bedrooms, one bathroom/toilet, kitchen, and living room, generally the total of wall area for this house type is 154 m$^2$ [50,51]. If using brick with 70 pcs bricks per square meter wall, we could complete the work for building the wall by 1.000 time unit, so when using concrete brick and ALC, the work would be finished faster, within 0.171 and 0.100 tu/m$^2$. Surprisingly, wood based sandwich panel could accomplish extremely fast. It's 0.005 tu/m$^2$ or only 5‰ compared to brick. In a house being developed, wood sandwich panels can be applied following the design as shown in Figure 3.

**Figure 3**. The design of wood-based sandwich applications: wall and beams structure (a); wall with fully-wood sandwich panel (b) interior views (c and d)

## 5. Conclusion

Wood in various forms of derivative products remain preferable building material for housing due to several advantages. Wood-based sandwich panels with veneer or plywood at the facing (face/back) and solid or laminated board for the core are promising products to be used to fast-built houses for the mitigation and reconstruction of disaster areas.

**References**

[1] Xu J, Wang Z, Shen F, Ouyang C and Tu Y 2006 Natural disasters and social conflict: A systematic literature review *International Journal of Disaster Risk Reduction* **17** 38-48
[2] BNPB(a) 2017 *Definition of Disaster* Available from: https://bnpb.go.id/definisi-bencana?
[3] Wibowo A, Surbakti I, and Yunus R 2013 *Indonesia Disaster Database Expert Group Meeting (EGM) on Improving Disaster Data to Build Resilience in Asia and the Pacific Tohoku University, Sendai, Japan*
[4] Septia K 2018 *BNPB: 32.129 Rumah Rusak akibat Gempa Lombok sudah Diverifikasi*, in *Kompas.com* PT. Kompas Cyber Media (Jakarta: Kompas Gramedia Digital Group)
[5] Rochimawati and Faris AF 2019 *72 Ribu Unit Rumah Rusak Akibat Bencana 2019* (Jakarta: Viva Network)
[6] BNPB(b) 2020 *Lebih dari 1300 Bencana Hingga Akhir Mei 2020* Available from: https://bnpb.go.id/berita/lebih-dari-1-300-bencana-hingga-akhir-mei-2020
[7] Puskim 2019 *Flagship Bangunan Tahan Gempa, Tahan Api, Cepat Bangun, dan Murah*
[8] Kurniawan A 2019 *Asumsi Kebutuhan Rumah di Indonesia, 800 Ribu Per Tahun* [cited 2019 12-12] Available from: https://suaraproperti.com/asumsi-kebutuhan-rumah-di-indonesia-800-ribu-per-tahun/
[9] Peckenham E 2016 *11 green building materials that are way better than concrete* [cited 2020 15-07]
[10] Ramage MH, Burridge H, Busche-Wicher M, Fereday G, Reynolds T, Shah DU, Wu G, Yu L, Fleming P, Densley-Tingley D, and Allwood J 2017 The wood from the trees: The use of timber in construction *Renewable and Sustainable Energy Reviews* **68** 333-59
[11] Labans E, Kalnins K, and Bisagni C 2019 Flexural behavior of sandwich panels with cellular wood, plywood stiffener/foam and thermoplastic composite core *Journal of Sandwich Structures & Materials* **21**(2) 784-805
[12] Djalante R, Garschagen M, Thomalla F, and Shaw R 2017 *Disaster Risk Reduction in Indonesia: Progress, Challenges, and Issues* (*Disaster Risk Reduction in Indonesia*) ed. Djalante R (Switzerland: Springer) pp 1-17
[13] Taylor M 2018 *With warning drums and river clean-ups Indonesian women head off disasters* 2018 [cited 2020 03 July] Available from: https://www.reuters.com/article/us-indonesia-disaster-women/with-warning-drums-and-river-clean-ups-indonesian-women-head-off-disasters-idUSKBN1I1036
[14] Djalante R, and Garschagen M 2017 *A Review of Disaster Trend and Disaster Risk Governance in Indonesia: 1900-2015* (*Disaster Risk Reduction in Indonesia*) ed. Djalante R (Switzerland: Springer) p. 638
[15] CFE-DM 2018 *Indonesia Disaster Management Reference Handbook* (Joint Base Pearl Harbor - Hickam, Hawai: Center for Excellence in Disaster Management & Humanitarian Assistance)
[16] Frontiers 2017 *The losses that come after the earthquake: Devastating and costly* [cited 2020 July 15]; Available from: www.sciencedaily.com/releases/2017/08/170825104002.htm
[17] Daniell JE, Schaefer AM, and Wenzel F 2017 Losses Associated with secondary effects in earthquakes *Frontiers in Built Environment* **3**(30)
[18] Kawasaki T, Zhang M, Wang Q, Komatsu K, and Kawai S 2006 Elastic moduli and stiffness optimization in four-point bending of wood-based sandwich panel for use as structural insulated walls and floors *Journal of Wood Science* **52**(4) 302-10
[19] Twigg J 2004 *Good Practice Review: Disaster risk reduction* (London: Humanitarian Practice Network)
[20] EAS-RR 2014 *Preventive and Mitigation Measures* (New Delhi: National Institute of Disaster Management)
[21] Führer, E 2000 Forest functions, ecosystem stability and management *Forest Ecology and Management* **132**(1) 29-38

[22] Fischer AP 2008 Forest landscapes as social ecological systems and implications for management *Landscape and Urban Planning* **177** 138-47
[23] Morrell JJ 2002 Wood-based building components: what have we learned? *International Biodeterioration & Biodegradation* **49**(4) 253-8
[24] Prihatmaji YP 2007 Perilaku rumah tradisional Jawa "Joglo" terhadap gempa *DIMENSI (Journal of Architecture and Built Environment* **35**(1) 1-2
[25] Aqli W 2011 Anatomi bubungan tinggi sebagai rumah tradisional utama dalam kelompok rumah Banjar *NALARs* **10**(1)
[26] Sardjono AB 2011 Respon rumah tradisional Kudus terhadap iklim tropis *Modul* **11**(1)
[27] Gosal PH 2012 Kearifan lokal masyarakat Minahasa membangun rumah tinggal yang hijau dan nyaman *Media Matrasain* **9**(3) 67-81
[28] Pranata YA, and William KM. 2013 Sistem base isolation pada rumah kayu tradisional Ammu Hawu, Pulau Sabu, Nusa Tenggara Timur *Sustainable Culture Architecture & Nature* 101-8
[29] Rumiati A, and Prasetyo YH 2013 Identifikasi tipologi arsitektur rumah tradisional Melayu di Kabupaten Langkat dan perubahannya *Jurnal Permukiman* **8**(2) 78-88
[30] Attaufiq M, and Waani JO 2014 Kenyamanan termal pada sebuah rumah adat tradisional Gorontalo *Media Matrasain* **11**(1) 55-65
[31] Hermawan I 2014 Bangunan tradisional Kampung Naga: Bentuk kearifan warisan leluhur masyarakat Sunda *Sosio-Didaktika: Social Science Education Journal* **1**(2) 141-50
[32] Rinaldi Z Purwantiasning AW, and Nur'aini RD 2015 Analisa konstruksi tahan gempa rumah tradisional Suku Besemah di Kota Pagaralam Sumatera Selatan *Prosiding Semnastek 2015*
[33] Gosal PH, and Makarau VH 2018 Optimalisasi Kenyamanan Thermal Pada Rumah-Rumah Kayu di Kampung Jawa Tondano *Frontiers: Jurnal Sains dan Teknologi* **1**(1)
[34] Siswanto A 2018 Studi pengembangan konstruksi rumah kayu sistem bongkar pasang berdasarkan konsep struktur kayu tradisional Sumatera Selatan *Jurnal Ilmu dan Teknologi Kayu Tropis* **2**(2) 103-9
[35] Baharta N, Dapas SO, and Pandaleke RE 2019 Evaluasi teknis kekuatan struktur rumah kayu tradisional produksi Desa Woloan terhadap gempa sesuai Standar Nasional Indonesia (SNI) *Jurnal Sipil Statik* **7**(5)
[36] Buchanan, Andrew H, and Levine SB 1999 Wood-based building materials and atmospheric carbon emissions *Environmental Science & Policy* **2**(6) 427-37
[37] Švajlenka J, and Kozlovská M 2018 Houses based on wood as an ecological and sustainable housing alternative-case study *Sustainability* **10**(5) 1502-22
[38] Tsai MT, and Wonodihardjo AS 2018 Achieving Sustainability of Traditional Wooden Houses in Indonesia by Utilization of Cost-Efficient Waste-Wood Composite *Sustainability* **10**(6) 1718
[39] BPS 2017 *Statistik Produksi Kehutanan* 2016 (Jakarta: Badan Pusat Statistik)
[40] BPS 2018 *Statistik Produksi Kehutanan* 2017 (Jakarta: Badan Pusat Statistik)
[41] BPS 2019 *Statistik Produksi Kehutanan* 2018 (Jakarta: Badan Pusat Statistik)
[42] Maloney 1996 The family of wood composite materials *Forest Products Journal* **46** (2) 19-26
[43] Moody RC, Hernandez R, and Liu JY 1999 *Glued structural members - Wood Handbook, Wood as an Engineering Material* (WI: USDA Forest Service, Forest Products Laboratory)
[44] Lakreb N, Bezzazi B, and Pereira H 2015 Mechanical behavior of multilayered sandwich panels of wood veneer and a core of cork agglomerates *Materials and Design* **65** 627–36
[45] DIAB 2005 *Sandwich Concept - DIAB Sandwich Handbook* (Sweden: LAHOLM)
[46] Edgars L, Kaspars Z, and Kaspars K 2017 Structural Performance of Wood Based Sandwich Panels in Four Point Bending *Procedia Engineering* **172** 628-33
[47] Kawasaki T, and Kawai S 2006 Thermal insulation properties of wood-based sandwich panel for use as structural insulated walls and floors *J. Wood Sci.* **52** 75–83

[48] Susainathan J, Eyma F, De Luycker E, Cantarel A, and Castanie B 2018 Experimental investigation of impact behavior of wood-based sandwich structures *Composites Part A: Applied Science and Manufacturing* **109** 10-9
[49] Srivaro S, Matan N, Chaowana P, and Kyokong B 2014 Investigation of physical and mechanical properties of oil palm wood core sandwich panels overlaid with a rubberwood veneer face *Eur. J. Wood Prod.* **72** 571-81
[50] Anonymous 2017 *Menghitung Biaya Membangun Rumah Tipe 36* https://depomaterialkebanggan.blogspot.com/p/menghitung-biaya-membangun-rumah tipe.html
[51] Berkah 2019 *Menghitung biaya bangun rumah sederhana sendiri type 36 2019* Available at https://kontraktorrenovasirumah.com/bangun-rumah-2/ Accessed 26 July 2020
[52] Teknopanel 2020 Sandwich Panel Features, Advantages and Application Areas Available at https://www.teknopanel.com/en-us/product-detail/sandwich-panels-about-sandwich-panels. Accessed 26 July 2020

**Acknowledgments**
The authors gratefully acknowledge the support from the LPDP for financial support.

IOP Conf. Series: Materials Science and Engineering **935** (2020) 012023 doi:10.1088/1757-899X/935/1/012023

# Utilization of biomass from Sago (*Metroxylon* sp) by local ethnic around the Sentani Lake Jayapura-Papua Province

**YC Ondikeleuw, Wahyudi* and M Arifudin**

Faculty of Forestry, University of Papua, Manokwari, 98314, Indonesia

*Email: w.sayutipono@unipa.ac.id

**Abstract.** Sago is multi-functional crops for local ethnics living at coastal and low land areas in Papua as main carbohydrate and raw-material of traditional houses. This research is designed to document the utilization and socio cultural values of sago biomass by local ethnic around the Sentani Lake, Jayapura. Field survey and structural interview were used to collect data at Yoboi and Sereh villages. The results indicate that carbohydrates were extracted from pith and other sago's biomasses were used for green material for various puIDRoses. Leaf are for floor mate, filter starch, carrying bags, housing roof and fishing trap. Sago fronds are for extraction chamber and sedimented starch. Rounded fronds are used for house wall, burning media for lime stone and planting media boxes. Bark of trunks are for flooring and decking their stilts, balcony and firewood. Residual waste from sago pulp are good media for mushroom, planting vegetables, imago inhibitor, and compost. Its fruits are for wall accessories and necklace. Sago lands is also supply animal protein by providing habitat for hunting animals. Socio cultural values mean that sago is ancestor heritage to be saved, utilized, managed proportionally for next generation. This crop is cultural identity contributed to socio cultural and economic values, but these wisdom values are pressuring recently.

## 1. Introduction

Sago (*Metroxylon sp*) is special palm and has special meaning to the local ethnics lived at the coastal and low land areas in Papua. Papua province has approximately 4,749,424 ha of sago area, where Jayapura district has 79,908 ha, mostly concentrated in Sentani [1]. It is communal sago forest belonging to the Sentani ethnics who live around the Sentani Lake. This palm resources has contributed to the multi values to the local ethnic from one to the next generation [2].

Sentani is the capital city of Jayapura district, where governmental offices, military bases, economy facilities and Sentani aiIDRort are located. These facilities are located at around the Sentani Lake. This is the biggest lake in Papua and West Papua province. Sago palm grows naturally around the Sentani Lake, and adjacent to the Sentani aiIDRort. Majority of local ethnics of Sentani living concentrate at lake area, and heavy depend on the resources around the lake. Therefore, existence of sago resources and its surrounding environment are vital for the local ethnic. This is applied not only for food security but also for their daily necessities of traditional material, sources of protein as well as economic income.

Many researches have been conducted to determine the diversity [3], productivity [4], growth development and characteristic of sago [5] in Sentani Lake area. Those researches include the potent utilization of residual sago processing for briquettes product [6], residual sago pulp for compost filler [7] and briquettes [8]. However, traditional utilization of biomass from sago in Sentani Lake have not been reported. Therefore, this research is to investigate the various traditionally utilization of biomass from sago by local ethnics live around the Sentani Lake, Jayapura. Socio culture and economical aspects of sago to the local ethnics are also investigated.

## 2. Materials and Methods

*2.1. Research site*
This research was conducted at two village around Sentani Lake, namely Yoboi and Sereh Village, and inhabited by local ethnic or Sentani tribe and completed for two months. This research site could be reached by car from Jayapura city for approximately 2 hours.

*2.2. Data collections*
Field work with survey was used to collect the research data, and structural interview were conducted to the selected respondents based on questioners prepared. PuIDRosive sampling was used to select respondents, where two different respondents, general and key respondents, were designed. General respondents are member of community living in the villages areas, actively involved and has right to extract and uses sago resources. 18 and 60 respondents were selected (representing 15% of total village population) for Yoboi and Sereh village, respectively. Key respondents are ethnic leaders, religious person, and village leaders.

*2.3. Research variables*
Research variables consist of variety of utilization of sago biomass from leaf, trunks, fronds, bark, and sago extraction tools used. Socio culture variables include traditional knowledge of sago, philosophical view of sago, and traditional conservation and transfer of knowledge to young generation. Volumetric starch collected from sago extraction and prices of sago starch marketed around Sentani Lake by the Sentani ethnic were also reported.

## 3. Results and Discussion

*3.1. Results*

*3.1.1. Sago's biomass utilizations*
Traditional utilization of biomass from sago by Sentani ethnic lived around the Sentani Lake are summarized in Table 1, and their corresponding pictures are presented in Figure 1, respectively.

**Table 1**. Traditional utilization of biomass from sago used by Sentani ethnic lived around the Sentani Lake Jayapura-Papua Province

| Sago Biomass *(local name)* | Local name *(Indonesian)* | Functions | Descriptions of utilization and pictured in Figure 1 |
|---|---|---|---|
| **Leaf** *(Fi yang)* | Khanggali *(sarangan)* | Sorter or filters | Stopping residual material from sago pulp extracting to enter the sediment chamber (1a) |
| | Olong *(wadah penampung)* | Handling of sago pulp | Mobile container to carry sago pulp from floor into extraction chamber (1b) |
| | Hougey *(tikar penjemur)* | Drying mattress | Multiple drying mattress used for drying agricultural commodity, such cocoa, coconut, etc (1c) |
| | Imea Yang *(Atap rumah)* | Roofing material | Roofing material of working house, huts of the local ethnics (1d) |
| | Phukhereng *(Perangkap ikan)* | Fish trap | Trapping fish for customary ceremony with tighten regulation (1e) |
| **Sago fronds** *(Fena)* | Fi ling *(wadah empulur)* | Chamber | This chamber is used to separate the starch from the sago pulp by hand squeezing (1f) |
| | Me'a *(tiang penyangga)* | Standing coloum | Supporting the separating chamber (1g) |
| | Meya *(tiang penahan saringan utama)* | Standing coloum for siever | Supporting for siever (1h) |

| | | | |
|---|---|---|---|
| | Hurlai *(penahan wadah ekstrak)* | Main coloum for starch *(sediment)* chamber | Supporting coloum for starch container (1i) |
| | Khetembi ha *(tali pengikat)* | Tinghten material *(rope)* | Outer part of the sago fronds used for wrapping or roping-materials (1j) |
| | Fena ha *(penahan saringan utama)* | Barrier material | Filtering sago sediment from outing with water (1k) |
| | Auma *(bedeng tanam)* | Planting boxes | Container for planting media, because all areal are standing on the top of the Sentani Lake (1l) |
| | Fena imea *(dinding rumah )* | House wall | Wall house material (1m) |
| | Au fena ye hanegokhate *(media bakar kulit bia/kerang)* | Calsium burning media | Calsium is need to consum pinang, and calsium is obtained by buring sea shells (1n) |
| **Sago bark** *(Rakhele)* | Wakhung *(lantai rumah)* | Flooring material | House flooring material (1o) |
| | Ye (kayu bakar) | Firewood | This wood energy (firewood) for local ethnic (1p) |
| **Sago pulp** *(Mele)* | Sago starch *(aci sagu)* | Sediment of sago pulp | Main source of carbohydrate |
| | Planting media *(media tanam)* | Soil supplement | fibrous material from residual sago pulp (1q) |
| | Compost *(kompos)* | Soil filler | Providing nutrition to the planting media (1r) |
| | Hem *(Ulat sagu)* | Imago | Sources of protein (1s) |
| | Mushroom | Compost mushroom | Source of protein |
| **Fruits** *(Fiole)* | Fiole neckless *(kalung adat)* | Customary accessories | Accessories for folk dance (1t) |

**Figure 1.** Various products made traditionally by Sentani ethnic from sago biomass around Sentani Lake

Table 1 highlights that biomass from sago have been used intensively by the local ethnic around the Sentani Lake for traditional material of daily usages. Three biomasses such as leaf, fronds, and sago pulp, could be simply processed into various green products. In total, there are 22 products derived from sago biomass (Table 1), which could be classified into three class, likes raw material of green products, sources of food, and green building material. One example in utilizing those material is sago fronds. The bottom of sago fronts, which is naturally similar to chamber (as illustrated by Figure 1f) is a perfect material naturally available for extraction chamber. It seems that this material is strong enough, elastic and easy to be used instantly.

3.1.2. Sago trunks characteristic and their starch production

Traditionally, Sentani ethnic lived around Sentani Lake has differentiated sago palms into two classes, sago with spiny and without spiny. Locally, sago spiny have five names, namely Bhara, Osokhulu, Rondo, Ruruma and Manno, while sago without spiny have five names, from Yebha, Follo, Wani, Phane, and Hobolo. These ten local names are the main sources of carbohydrate for the local ethnics. Processes of extraction to collect sago starch amongst those ten names are similar and the main differences are the colour and physical appearances (odour and taste) of the starch produced. Field survey for numbers of sago trunks felled per day, pieces of bucked stem per trunks, and volumes of fresh-sago starch collected and expressed in containers are summarized in Tabel 2.

**Table 2.** Local name, numbers of bucked trunks, and their volume of wet sago starch collected from Sentani ethnic live around Sentani Lake

| Sago | Local name | Numbers of sago processed by the local ethnic: Trunks/day | Pieces of bucked stems per trunks | Container*) per trunks | qualitative starch assessment |
|---|---|---|---|---|---|
| **Without spiny** | | | | | |
| | Yebha | 3 | 21-23 | 7-9 | Good |
| | Follo | 3 | 15-18 | 3-4 | Good |
| | Wani | 2 | 18-20 | 5-8 | Good |
| | Phane | 1 | 10 | 3-4 | Good |
| | Hobolo | 1 | 10 | 3-4 | Good |
| **Spiny** | | | | | |
| | Osokhulu | 3 | 15-18 | 5-6 | Good |
| | Bhara | 3 | 12-13 | 5-6 | Good |
| | Rondo | 2 | 8-10 | 3-4 | Good |
| | Ruruna | 1 | 8-10 | 3-4 | Good |
| | Manno | 1 | 15-17 | 3-4 | Bad |

*) an average weight for 35 kg of fresh sago starch

Table 2 highlights that both sago with spiny and without spiny are processed to collect sago starch for local ethnics. In average, 1-2 sago trunks are harvested per day in Yoboi and Sereh villages. After been felt using chain saw, sago trunks are bucked into 50 cm long and resulted 21-23 bucked stems of Yebha and 15-18 bucked stems of Osokhulu. From those bucked stems, 7-9 containers of wet sago starch were collected from Yoboi and 5-6 containers from Osokhulu. Both are reported with good qualitative starch sago assessment. Sago pulp were collected from sago pith using the grated sago engine [9].

3.1.3. Socio culture aspects of sago

According to their own tradition, Sentani ethnic called themselves as phuyakka, *phu* meaning water and *yakha* meaning clam and clean water. It has a meaning that the cleanest and calmness water is their ancestor symbol for their habitat, where the Sentani Lake is inhabited by 27 villages of the local ethnics. All Sentani ethnics had customary right (*mam*) and it is applied to all to those villages. Two basic customary right for Sentani ethnics are, *Erei Khamei* and *Holei Narei*, respectively. The first term has a meaning for full of respects and appreciation to all community members, whereas the last term has a meaning that all members have duty and responsibility to protect, maintain, save, and provide foods to their family members, customary leader, clan leader, widow and orphans.

A real man locally named as *Ro Khelu Fa* can be attributed to the man who has sago land, could manage his sago lands or orchards, extract sago's starch and to feed as many member community as he could. For Sentani's women *(Manggea Fa)* skills in preparing and serving Papeda (local food made from sago starch with fresh fish sauces) are customary measured for adultness. Sago forest or orchards (*Fiung akhla*) is acestor heritage and communal property that could not be offered or inherited directly into young generation, but it has to be managed, maintained and saved together and forever.

Sago starch extraction is identic to communal and women activity but all family members are involved in this activity. Collaboration in sago's trunks processing into starch collection is illustrated in Figure 2.

a b c

d e f

legends: a. peeling sago bark; b. ripping bucked stem; c. sago pulp production; d. standing coloum from sago's fronds; e. squeezing sago pulp; f. sago's starch or sediment chamber

**Figure 2**. Communal activity of sago's trunk processing into starch collection

Children, women and men are collaborated together to the sago orchards for extracting sago starch known well as *Pangkur sagu*. In general, the men (young and mature men) are responsibility for clearing the surrounding sago trunks area prior for cutting, felling sago trunks, bucking and ripping the bucked stem and mechanically processing sago pitch into sago pulp (Figure 2 a-c). Women then are constructed traditional supporting material and tools prepared for manual extraction (Figure 2c). Sago pulp are collected and squeezed intensively to obtain white browning sediment, namely sago starch (Figure 2 e-f).

3.1.4. Economic value of sago resources
When all water is running out from the sediment chambers, wet-muddy sago's starch was obtained (as illustrated in Figure 3a, after been steady for a couple hours, fresh-wet sago starch was collected into recycle rice bag (Figure 3b) and it is ready to be transported into home. For Sentani ethnic, the most common commodity from sago palm is fresh sago starch at approximately 20% MC. It is unique about sago starch that to prevent browning because of oxidation, it is must be soaked with full of clean water, and this water is frequently replaced with the fresh water for period of time such as fortnightly. Single container of wet sago starch, approximately 25 kg, are sold from IDR 300,000- 500,000 at local market or selling from home directly.

a b c

Remarks: a. wet-muddy starch; b.; filling fresh starch into container c. starch container ready for transporting or selling

**Figure 3**. Filling muddy starch into containers of recycle rice bags for transporting selling

Sago resources for local ethnic around Sentani Lake is invaluable resources. It means as cultural identity and philosophy. An example of this cultural identify is when proposing women, sago food (*natura*) has to be served to the visitor from the men family. Local cuisine Papeda is the main food for the ceremony.

Sago trunk growth and development are also applied to the philosophy of the Sentani ethnic. Sago palm has growing characteristics straight stem, strong, full of leaf and other epiphytes and all sago palm biomass could provide and supply multi beneficiary to the community. Each member of Sentani ethnic has to be strong as the sago palm, in line with the straightness of sago trunk, and provide beneficiary to all surrounding neighbourhoods.

When the family is extracting sago starch, the results has to be delivered to other members in their clan. It is because the sago forest and orchards belonging to the communal right. It means that this property right is not saleable, but it has to be utilised, managed, and shared together in their clan member. Togetherness, cooperation, kindness, and justice amongst all member of clan.

### *3.2. Discussion*

Sago is multi-functional crops for local ethnic around the Sentani Lake [10]. This palm could only grow in specific environmental condition, such as swam area closed to the river, estuarian sea, lake and mangrove area [11,12]. Area around the Sentani Lake is natural and suitable habitat for this carbohydrate producer palm that influence the socio culture of the local ethnics at around the Sentani Lake [13].

Sago palm around the Sentani Lake provide not only main source of carbohydrate for local ethnic but also supply protein from lake fish and animal hunting. It is clear that as multi-functional crops, sago, produces enormous quantity of biomass that could be utilized as raw material, which is totally sustainable and biodegradable. These various biomasses could be classified as green material that could be used as raw material for green construction.

Traditional utilizations of biomass from sago for daily usages from local ethnics around Sentani Lake are indigenous wisdom or traditional knowledge that has been in practice from acestor to next generation. This also are supported by the socio culture values of the local ethnics likes, *Erei Khamei* and *Holei Narei,* which are still being in practice until today. More importantly, the local philosophy for being member of community to have multifunctional contribution to the other is similar to sago as the multi-functional crops.

Recently Sentani is the capital city of Jayapura district, where intensive establishment of infrastructure from local and national government, transportation facilities, economic facilities, and building private houses are taken place. As a result, sago land is declining, and traditional utilization of sago biomass is being replaced with synthetic material such as plastics or their derivative products. These phenomena are serious pressures to the existence of traditional wisdoms in utilizing sago biomass at around the Sentani Lake.

## 4. Conclusion

It is concluded that sago palm offers various biomass, such as leaf, fronds, barks, and residual sago pulp, that could be simply processed into various traditional products for daily usages by local ethnic around the Sentani Lake. Even though, these utilizations are local scale, this practice has been existed for many generations and recently it is under pressure that should be taken into account. This reflects that sago is multi-functional crops for the local ethnics, not only for staple foods, but also for green material sources.

## References

[1] Bintoro HM 2020 Sagu sebagai alternatif pangan pasca pandemi covid-19 (lecture notes)

[2] Muyan F 2017 Hubungan beberapa variabel pertumbuhan terhadap kandungan tepung sagu di Sentani. Bachelor thesis University of Papua, Manokwari

[3] Yamamoto Y, Yanagidate I, Miyazaki A, Yoshida T, Irawan AF, Pasolon YB, Jong FS, Matanubun H, Arsy AA, and Limbongan J 2020 Growth characteristics and starch productivity of folk varieties of sago palm around lake Sentani near Jayapura, papua state, Indonesia *Trop. Agr. Develop.* **641** 2333

[4] Yamamoto Y, Katayama K, Yoshida T, Miyazaki A, Jong FS, Pasolon YB, Matanubun H, Rembon FS Nicholus, and Limbongan J 2020 Changes in leaf and trunk characteristics related to starch yield with age in two sago palm folk varieties grown near Jayapura, Papua, Indonesia *Trop. Agr. Develop.* **642** 6171

[5] Abbas B and Ehara H 2012 Assessment genetic variation and relationship of sago palm (*Metroxylon sagu* rottb.) in Indonesia based on specific expression gene (Wx genes) markers *Afr. J. Plant Sci.* **612** 314-20

[6] Agustinus 2013 Perlakuan uji kalor briket limbah sagu untuk pemanfaatan sebagai bahan bakar oleh masyarakat kampung Sentani *J. Sinergy* **112** 123-33

[7] Wahida and Limbongan AG 2015 Pemanfaatan ampas sagu sebagai bahan dasar kompos pada beberapa dosis pencampuran dengan kotoran sapi *Agricola* **51** 0108

[8] Wambrauw V and Agnesari L 2017 Pelatihan pembuatan bahan bakar briket ampas sagu sebagai alternatif energi dalam skala rumah tangga di distrik Sentani timur *J. Pengab. Papua* **12** 4752

[9] Reniana, Darma, and Kurniawan A 2017 Prototype of cylinder-type sago rasper machine powered by combustion engine *J. Tek. Pert. Lampung* **62** 8994

[10] Toyoda Y 2008 Anthropological studies of sago palm in Papua New Guinea *Osccasonal paper No.13* Centre for Asian area studies Rikkyo University Japan 1-10

[11]Dewi RK, Bintoro MH, and Sudradjat D 2016 Karakter morfologi dan potensi produksi beberapa aksesi sagu (*Metroxylon* spp.) di kabupaten Sorong Selatan, Papua Barat *J. Agron. Indones.* **441** 91-7

[12]Rostiwati T, Bogidarmanti R, Suripatty BA, and Bustomi A 2014 Potensi masak tebang lima tipe Sagu (*Metroxylon sagu* rottb.) di kawasan hutan sagu Sentani, Papua **821** 1014

[13]Widyastomo D, Faqih M, and Setijanti P 2015 The transformation of value and meaning of Sentani residential tribe in Papua Indonesia *Int. J. of Eng. and Sci.* **412** 2132

# DNA barcoding of *Pericopsis mooniana* from two different populations in Indonesia based on rDNA ITS (Internal Transcribed Spacer)

**I Prihatini[1*], A Y P B C Widyatmoko[1], I L G Nurtjahjaningsih[1], V Yuskianti[2], and S A Danarto[3]**

[1]Center for Forest Biotechnology and Tree Improvement, Sleman, Indonesia
[2]Forest Products Research and Development Center, Ministry of Environment and Forestry, Bogor, 16610, Indonesia
[3]Research Center for Plant Conservation and Botanic Garden Purwodadi, Pasuruan

*E-mail: istiana.prihatini@biotifor.or.id

**Abstract.** A DNA barcoding is a taxonomy method that provide fast and acurate species identification, thus it can be applied to anticipate when a species can not be identified morphologically. This study aimed to assess molecular characteristics of *Pericopsis mooniana* from two wild populations in Indonesia using a barcoding marker rDNA ITS (Internal Transcribed Spacer) as additional information for establishing conservation strategy of this species. Leaf samples of *P. mooniana* were collected from natural forests at South East Sulawesi and South Kalimantan, and a germplasm orchard at Purwodadi Botanic Garden. Molecular characteristic was evaluated based on a pair of ITS primers that produce between 634- 641 bp nucleotide. The sequence data were analyzed by Bioedit version 7.0.5.3 and Mega 7 software. Results showed that *P. mooniana* from South East Sulawesi and South Kalimantan were clustered together, while *P. mooniana* from Purwodadi Botanic Garden were separated with only one base difference from those two population. In conclusion, ITS barcoding in this study shows a diffences between *P. mooniana* collected from South East Sulawesi and South Kalimantan with *P. mooniana* that originally from Papua. A further study using more barcoding marker and more population could be approached to obtain more information on molecular characters of *P. mooniana* populations that will useful for establishing conservation and tree breeding program of this species.

## 1. Introduction

*Pericopsis mooniana* is one of Fabaceae species that categorized as a vulnerable economical tree species (Vulnerable Alcd) by the IUCN Red List of Threatened Species, since 1998 [1]. The genus Pericopsis consist of four species, while three species distributed in tropical Africa [2,3]. *Pericopsis mooniana* is the only species of Pericopsis that naturally distributed in Southeast Asia, Micronesia, and New Guinea. In Indonesia, it is distributed in Java, Sumatra, Kalimantan, Sulawesi and Papua. Recently, only small number of stands remain in South East (SE) Sulawesi (Lamedai Nature Reserve and Tanggetada) [4], in Pulau Laut South (S) Kalimantan [5] , Jambi [6] and Papua [7,8], where illegal logging and land conversion are the most common causes threatening natural species distribution [5]. This species tends to survive in the clusters and continuous forests [9]. The populations of *P. mooniana* had been fragmented by human activities such as overexploitation, and land conversion to agriculture and settlement. Low natural regeneration and lack of replanting also causing this species to decline thus encouraged the Indonesian Government to establish a conservation strategy to restore the natural habitat. The conservation strategy has been initiated by assigned Lamedai Nature Preserve as conservation areas since 1974 and many studies on several aspects to provide information for establishing conservation and plantation site to protect this species have been conducted. Studies on cultivation techniques and seed germination are developed to increase the survival rate in the field [4,10] and the prospect of planting this species in rehabilitating ex-mine program [11,12]. Wood properties study on *P. mooniana* plantation revealed that breeding program for wood quality is required to promote the establishing plantation [13].

Ministry of Environment and Forestry through Center for Forest Biotecnology and Tree Improvemet (CFBTI) has collected seed and seedling of *P. mooniana* from its remaining natural population in SE Sulawesi and S Kalimantan. These genetic materials were used to establih an ex-situ conservation plot in Java island and to provide materials for tree breeding program of this species in the near future. Conservation genetic should cover large genetic variation as possible and therefore studies on genetic diversity of these genetic materials prior to establish the ex-situ conservation plot were conducted. The observation on morphological characters such height and diameter growth on seedling of *P. mooniana* (collected from SE Sulawesi) in nursery shows high variability [14]. This finding was supported by an examination of genetic diversity using molecular marker (Random Amplification of Polymorphic DNA, RAPD) that has been applied on genetic material from SE Sulawesi [9]. Both studies show high genetic diversity of SE Sulawesi population, but currently no published data for genetic diversity of S. Kalimantan population. Despite of the decreasing of the number of population and the number of tree in each population, the previous studies revealed that material genetic collected from SE. Sulawesi population are important and optimum for establishing conservation plot of this species. The molecular genetic study also reported that genetic diversity in Lamedai Nature Preserve had the highest value among the tested populations. The genetic relationship analysis also indicated that Lamedai Nature Preserve represented a large continuous forest in the ancient period [9].

Establishing conservation planning required information on species taxonomy, species biology, species distribution, and current number in wild population [15]. Proper and accurate identification of species are important for recording the ecological data [16], especially for samples in mixed-species environment. The germplasm orchard of this species had been established to keep the genetic resources. Therefore, barcoding study was conducted to confirm the species identity and to observe the taxonomical position of *P. mooniana* individual collected from different population. Barcoding of plants have been conducted since a long time ago until recently, and no universal barcoding are available for plants [17], therefore it is still chalenging. Despite of its economical value and massive studies on this vurnerable species, the genetic information based on DNA barcoding of this species is currently very limited. Currently, only one sequence ribulose 1,5-bisphosphate carboxylase-oxygenase large subunit (rbcL) gene of *P. mooniana* was submitted in public database i.e. Genbank. Taxonomist classified plants based on morphological characters, such as leaves, bark, flowers and fruits. However, the characteristics might difficult when flowers and fruits are absent. DNA barcode are used to identify unknown species when morphological characters are incomplete [18]. The DNA barcoding is a useful tool for species identification, a dataset of DNA barcoding provide a foundation for bioforensic and provide information to support regulatory enforcement [17]. This study aimed to assess molecular characteristics of *Pericopsis mooniana* from two wild populations in Indonesia using a barcoding marker rDNA ITS (Internal Transcribed Spacer) as additional information for establishing conservation strategy of this species.

## 2. Materials and Methods

### *2.1. Materials*

2.1.1 Plant materials, morphological observation and DNA extraction
Leave samples were collected from 19 *P. mooniana* trees from natural populaton in South (S) Kalimantan (9 trees) and South East (SE) Sulawesi (10 trees). Two samples of *Pericopsis* from Purwodadi Botanic Garden (PBG) and two closely related species (*Pterocarpus* sp.) were included in this study (Table 1). Morphological characters of bark, leaf, infloresconse and pods were observed on *P.mooniana* collection in PBG (Purwodadi Botanic Garden). The DNA extraction, purification and DNA amplification were conducted in Molecular Genetic Laboratory CFBTI. The total genomic DNA was extracted from leave samples using a modified Cetyl Trimethyl Ammonium Bromide (CTAB) method [19]. The total DNAs were diluted 10 times and use as a template for PCR reaction to amplify ITS.

**Table 1**. List of individual samples used in *Pericopsis mooniana* molecular study

| *Sample Codes* | *Origins* | *Habitats* | *Species ID* |
|---|---|---|---|
| K1 | South Kalimantan | Inhutani mess | *P. mooniana* |
| K4 | S Kalimantan | Local community garden | *P. mooniana* |
| K5 | S Kalimantan | Seed orchard | *P. mooniana* |
| K37 | S Kalimantan | arboretum | *P. mooniana* |
| K38 | S Kalimantan | arboretum | *P. mooniana* |
| K39 | S Kalimantan | arboretum | *P. mooniana* |
| K40 | S Kalimantan | Stagen | *P. mooniana* |
| K41 | S Kalimantan | Stagen | *P. mooniana* |
| K42 | S Kalimantan | Stagen | *P. mooniana* |
| S10 | South East Sulawesi | Bali Jaya | *P. mooniana* |
| S11 | SE Sulawesi | Bali Jaya | *P. mooniana* |
| S12 | SE Sulawesi | Bali Jaya | *P. mooniana* |
| S106 | SE Sulawesi | Lamedai | *P. mooniana* |
| S107 | SE Sulawesi | Lamedai | *P. mooniana* |
| S110 | SE Sulawesi | Lamedai | *P. mooniana* |
| S131 | SE Sulawesi | Lamedai | *P. mooniana* |
| S132 | SE Sulawesi | Lamedai | *P. mooniana* |
| S50 | SE Sulawesi | Tangetada | *P. mooniana* |
| S51 | SE Sulawesi | Tangetada | *P. mooniana* |
| M18 | Unknown | Puwodadi Botanic Garden | *Pterocarpus echinatus* |
| M19 | Unknown | Puwodadi Botanic Garden | *P. echinatus* |
| M20 | Papua | Puwodadi Botanic Garden | *P. mooniana* |
| M21 | Papua | Puwodadi Botanic Garden | *P. mooniana* |

### *2.2. Methods*

2.2.1 PCR Amplification and sequencing of internal transcribed spacer DNA
Amplification of partial of internal transcribed spacer (ITS) region targeted sequence of ITS1, 5.8s and ITS2 were conducted using combination of ITS1 (5' TCC GTA GGT GAA CCT GCG G 3') and ITS4

primer (5' TCC TCC GCT TAT TGA TAT GC 3') [20]. The amplification reaction was performed in a total volume of 50 µl containing 10 µl of genomic DNA (final concentration was varying between 2.2 ng to 16 ng per µl), 0.5 µM of each primer, and 25 µl of 2x My Taq HS Red Mix (Bioline), 0.2µg of bovine serum albumin (BSA), and sterile water. DNA amplification was performed in a ProFlex PCR system (Applied Biosystem) with a program as follows: 94°C for 5 minutes, 30 cycles of 30 s at 94°C, 30 s at 55°C, and another 60 s at 72°C, followed by 7 minutes at 72°C. The PCR products were separated by electrophoresis in 1.5% agarose gel. Positive amplification products were sent to the 1st Base for sequencing following DNA purification and DNA extraction from gel.

DNA sequence chromatograms obtained from the 1$^{st}$ Base were viewed and edited to remove poor quality sequences at each end using Chromas software (http://technelysium.com.au/wp/chromas/). DNA sequence with good quality chromatograms was aligned using ClustalW [21] on Bioedit software version 7.0.5.3 [22]. DNA sequence with high similarity revealed by BLAST was retrieved from public DNA databases i.e. GenBank and were included in phylogeny analysis to verify the identity of species. One sequence from a more distantly related taxon was included as out group in the analysis. Pairwise distance to calculate the genetic distance within and between species of Pericopsis was created using Mega7 [23]. The phylogenetic trees were constructed using the Maximum Likelihood method based on the Tamura-Nei model [24] with 500 bootstraps in Mega7.

## 3. Results and Discussion

### *3.1. Results*

3.1.1. Morphological character of *Pericopis mooniana*

***Bark.*** Medium sizes to large trees with height up to 24 m (Figure 1A), grooved skin bark, bark soft and thin, bole straight or twisted, slightly grooved in the bark. ***Leaf.*** Leaf composite (imparipinnate), there are pelvinus at the base of the petiolus communis, petiolus communis is thin, number of leaflet is 6-8- 10, ovate to elliptical, alternate or rarely sub opposite, glabrous, the color in adaxial of leaf is dark green, rough surface like parchment, bright green abaxial, there is pelvinus at the base, acute to broadly accuminate at the tip, rounded at the base, wavy at the leaflet edge (margo folii), primary vein protude at the abaxial part, secondary vein alternate with each other, length: width ratio is 3:1, has stipules, there is a thick pelvinus at the base, rachis and petiol slightly grooved (Figure 1B).

**Figure** 1. A. *Pericopsis mooniana* that observed in PBG; B. Leaf composite of *P. mooniana* with ovate shape and dark green in colour.

***Pod.*** Pod-shaped fruit, long stalked, rounded at the base, flat linear to oblong, tapered at the tip, dark brown, hard, pod have 1-4 spaces, compressed ovate-orbicular, skin pod will break when ripe and seed will spread in surrounding of trees. *P. mooniana* has hard seed with dark brown colour. ***Inflorescense.*** We were observed of inflorescense on *P. mooniana* refer to publication by Bentham and Hooker [25] and Knaap-van Meeuwen [26]. Inflorescense racemose or paniculate, flower cream or white or purple, in axillary racemes or terminal panicles, bract and bracteoles minute, calyx large, campanulate, cleft to below the middle, lobes sub-equal or upper 2 shorter, sub connate, cadocous, standard broad to orbicular, wings and keel petals distinctly eared, wings falcate to obovate, keel incurved, blunt, petals free, stamens free, apically recurved, anthers versatile, ovary stalked, oblong, villous, encircled by conspicuous disk, few-ovuled, style long, filiform, awl-shaped, apically involute, stigma terminal, introrse.

3.1.2. DNA ITS sequences
Twenty three individual of *P. mooniana* were amplified by a set of primer (ITS1/ITS4) shown by clear strong bands around size 870bp (Figure 2). However, in the further analysis using ITS4 for DNA sequencing, only 17 individual were produced good quality product shown by the clean and clear chromatogram, while the chromatogram of the other four samples (S11, S13, S106, S110, and S50) were noisy and illegible. Only good quality sequences were used for further analysis (phylogenetic study). The ITS DNA sequences were started from 5'GATCATTG to CAGGCGGG 3' with total length from 634 to 650 bp. All of these good quality sequences were submitted to GenBank (Table 2).

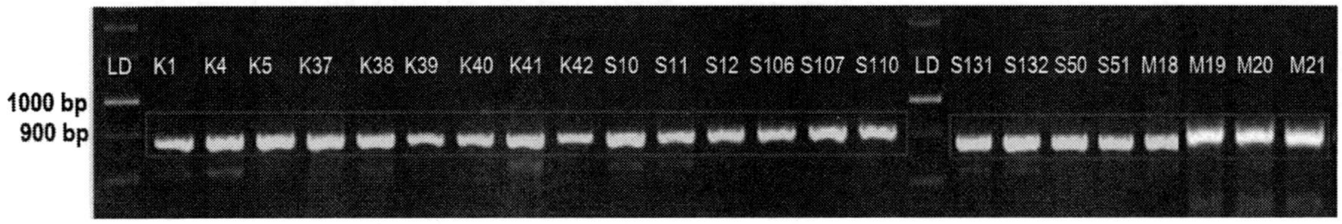


**Figure 2.** Purified amplicon of ITS DNA of 23 individual of *Pericopsis* that selected for DNA sequencing are marked with red squares. LD = DNA ladder

**Table 2.** List of individual samples with good quality of DNA amplicon that sequenced in *Pericopsis mooniana* study

| Individual | Band Length (bp) | Sequence chomatogram | Best match in BLAST | Genbank Accesion No |
|---|---|---|---|---|
| K1 | 634 | Clear | 573/651(88%) similar to *Pericopsis laxiflora* (KX057892) and 589/678(87%) to *P. angolensis* (KX584401) | MT093361 |
| K4 | 634 | Clear | 573/651(88%) similar to *P. laxiflora* (KX057892) and 589/678(87%) to *P. angolensis* (KX584401) | MT093362 |
| K5 | 634 | Clear | 573/651(88%) similar to *P. laxiflora* (KX057892) and 589/678(87%) to *P. angolensis* (KX584401) | MT093363 |
| K37 | 639 | Clear | 578/656(88%) similar to *P. laxiflora* (KX057892) and 594/683(87%) to *P. angolensis* (KX584401) | MT093364 |
| K38 | 639 | Clear | 578/656(88%) similar to *P. laxiflora* (KX057892) and 594/683(87%) to *P. angolensis* | MT093365 |

| Individual | Band Length (bp) | Sequence chomatogram | Best match in BLAST | Genbank Accesion No |
|---|---|---|---|---|
| | | | (KX584401) | |
| K39 | 638 | Clear | 577/656(88%) similar to *P. laxiflora* (KX057892) and 592/682(87%) to *P. angolensis* (KX584402) | MT093366 |
| K40 | 638 | Clear | 577/656(88%) similar to *P. laxiflora* (KX057892) and 592/682(87%) to *P. angolensis* (KX584402) | MT093367 |
| K41 | 639 | Clear | 578/656(88%) similar to *P. laxiflora* (KX057892) and 594/683(87%) to *P. angolensis* (KX584401) | MT093368 |
| K42 | 638 | Clear | 577/656(88%) similar to *P. laxiflora* (KX057892) and 592/682(87%) to *P. angolensis* (KX584402) | MT093369 |
| S10 | 639 | Clear | 578/656(88%) similar to *P.laxiflora* (KX057892) and 594/683(87%) to *P. angolensis* (KX584401) | MT093355 |
| S11 | unknown | Illegible | - | - |
| S12 | 639 | Clear | 578/656(88%) similar to *P. laxiflora* (KX057892) and 594/683(87%) to *P. angolensis* (KX584401) | MT093356 |
| S106 | unknown | Illegible | - | - |
| S107 | 641 | Clear | 578/658(88%) similar to *P. laxiflora* (KX057892) and 595/684(87%) to *P. angolensis* (KX584401) | MT093358 |
| S110 | unknown | Illegible | - | - |
| S131 | 640 | Clear | 578/657(88%) similar to *P. laxiflora* (KX057892) and 594/682(87%) to *P. angolensis* (KX584401) | MT093359 |
| S132 | 640 | Clear | 578/657(88%) similar to *P. laxiflora* (KX057892) and 594/682(87%) to *P. angolensis* (KX584401) | MT093360 |
| S50 | unknown | Illegible | - | - |
| S51 | 639 | Clear | 578/656(88%) similar to *P. laxiflora* (KX057892) and 594/683(87%) to *P. angolensis* (KX584401) | MT093357 |
| M18 | 578 | Clear | 578/578(100%) match to *Pterocarpus indicus* (MH487680) and 577/578(99%) to *P. macrocarpus* (MH487681) | MT127631 |
| M19 | 578 | Clear | 578/578(100%) match to *P. indicus* (MH487679) and 577/578(99%) to *P. macrocarpus* (MH487681) | MT093371 |

| Individual | Band Length (bp) | Sequence chomatogram | Best match in BLAST | Genbank Accesion No |
|---|---|---|---|---|
| M20 | 640 | Clear | 579/656(88%) similar to *P. laxiflora* (KX584403) and 594/682(87%) to *P. angolensis* (KX584402) | MT093372 |
| M21 | 640 | Clear | 579/656(88%) similar to *P. laxiflora* (KX584403) and 594/682(87%) to *P. angolensis* (KX584402) | MT093373 |

Sequencing of ITS region from 17 individuals produced between 634 bp to 650 bp of nucleotide and mostly 88% were similar to *Pericopsis* species *P. laxiflora* and *P. angolensis* (Table 2). There was no ITS DNA sequences of *P. mooniana* available in GenBank as a sequence reference to this study. This is the first study of *P. mooniana* barcoding based on ITS region and the first submission to GenBank. The individual M18 and M19 were 99% related to *Pterocarpus* species (*P. indicus* and *P. macrocarpus*). These two collection of PBG were identified as *P. echinatus* based on their morphological characters.

Based on their DNA ITS sequence, *P. mooniana* between SE Sulawesi and S. Kalimantan showed no significant differences (genetic distance= 0) calculated using Mega7, with only one nucleotide difference (an insertion of G base in 512 base position) for three individual from Lamedai S.E. Sulawesi. Population from Papua with only two individual shows small distance (0.003) to *P. mooniana* from both population, S.E. Sulawesi, and South Kalimantan with two insertion of G bases in 188 and 512 base position (data not shown).

3.1.3. Phylogeny of Pericopsis based on ITS DNA

Phylogeny analysis of *P. mooniana* in this current study (Figure 3) shows that DNA ITS sequences of S.E. Sulawesi population and S. Kalimantan population clustered together despite of a single nucleotide difference on individual S107, S131, and S132. Two individuals of *P. mooniana* from Papua population (M20 and M21) that planted in PBG were a separate cluster of SE Sulawesi population and S Kalimantan population. The reference sequences of other Pericopsis species (*P. laxifolia* and *P. angolensis*), were clustered together but separated from cluster *P. mooniana* of this study. The DNA ITS sequences of *P. elata* was not available from NCBI database (Genbank). Genetic distance of *P. mooniana* calculated based on pairwise distance matrices was 0.116 when paired with *P. laxifolia*, and 0.088 when paired with *P. angolensis,* while genetic distance between *P. laxifolia* and *P. angolensis* was 0.060.

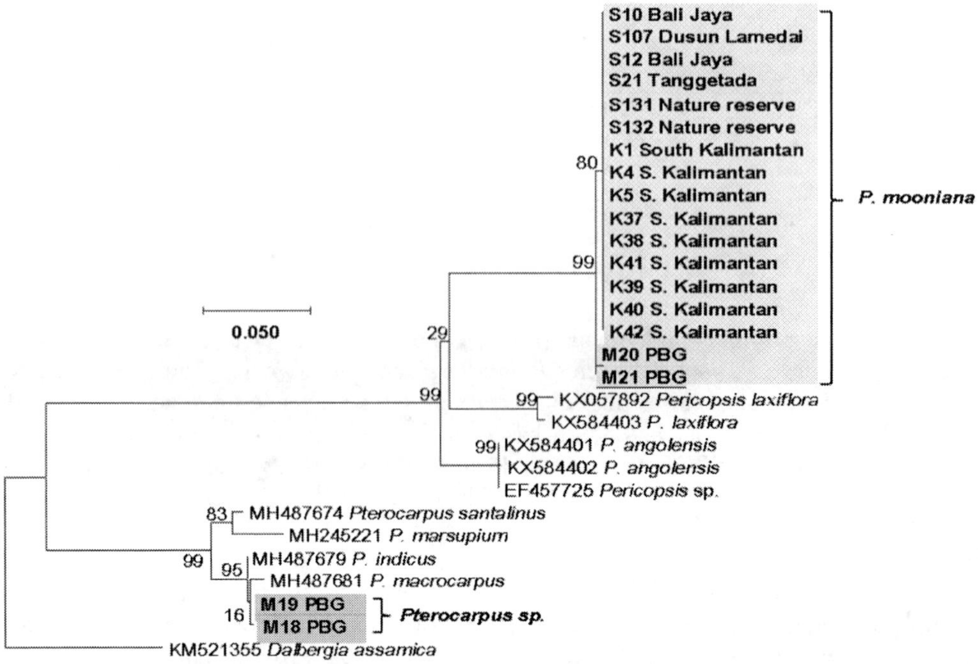


**Figure 3**. Dendrogram of *Pericopsis* based on ITS DNA sequence constructed using the Maximum Likelihood method in Mega7 with 1000 bootstrapping. The analysis involved 29 nucleotide sequences with a total of 693 positions in the final dataset. The tree with the highest likelihood is shown and drawn to scale, with branch lengths measured in the number of substitutions per site. The percentage of trees in which the associated taxa clustered together is shown next to the branches and *Dalbergia assamica* (KM521355) used as an outgroup.

Two individual of *Pterocarpus* from PBG collection also included in this study. Based on the morphological character, these two trees were identified as *P. echinatus* that belong to Fabaceae, the same family as Pericopsis. In the phylogeny dendogram these individual were clustered with two different species i.e. *Pterocarpus indicus* and *P. macrocarpus*. The cluster of Pterocarpus was separated with cluster of Pericopsis with genetic distance measured at 0.320.

### *3.2. Discussion*

Numerous barcoding studies have been conducting on Fabaceae species using various primer such as rbcL, matK, and trnH-psbA based on a single gene/DNA region [27] and multi gene combinations between different primers [28], but no ITS sequences of *P. mooniana* are reported from the previous studies. Application of a DNA barcoding marker (ITS) in this study was successfully amplified all of the samples studied (*P. mooniana* and *Pterocarpus* sp. ) and the sequencing proceess was successfully recovered from 74% of the total samples. A study conducted on Shorea and rattan also obtained high number of amplification and sequences product using ITS marker [29]. This result showed that ITS primer is an easy marker to be applied on *P. mooniana* and *Pterocarpus* sp. (used as a reference species).

The difference on nucleotide sequences of DNA ITS region between individual of *P. mooniana* found in this study is a good sign to prove that ITS marker is a powerfull tool for identification of *P. mooniana*. This study also found that barcoding based on DNA ITS able to discriminate *P. mooniana* from other species of *Pericopsis* (*P. laxifolia* and *P. angolensis*), although the differentiation with *P. elata* remain unknown as no ITS sequence of *P. elata* available. This study is the first to provide information on ITS sequences of *P. mooniana* on Genbank database. A barcoding study on flowering plant in Sumatra involving more than 2500 samples, was able to generate barcoding information of several individual of *P. mooniana* that was found in Jambi during the study [6], but unfortunatelly no ITS marker used in this study.

Contrary to Pericopsis, the discrimination power of ITS for Pterocarpus is low as this primer can not differentiate each species under genus of Pterocarpus. The limited number of each Pterocarpus species used in this study might be one of the factor that reducing the power of discrimination. Small number of Pterocarpus included in this analysis was used as a sequence reference from closely related genera under Faboidae subfamily. *Pericopsis mooniana* and *Pterocarpus indicus* have been selected by CFBTI as priority species for establishing ex situ conservation plot to provide material for tree breeding program in the near future. Therefore, barcoding study that focus on *Pterocarpus indicus* is required. This study should involve more individual collected from wild population along with the application of other barcoding markers such as chloroplast DNA marker, rbcL and matK, but combination of rbcL and matK has proven to improve the dicrimination rate [6].

Morphological observation of *P.mooniana* from SE Sulawesi and Papua planted in PBG revealed that SE Sulawesi has stiff leaflets like parchment with ratio of length : width is 3:1. Edge of leaflet is wavy rarely flat. Adaxial part of leaflets has dark green coloured. Pelvinus on leaflet is slightly thicker. While, *P. mooniana* from Papua has 5-10 leaflets, rotundate to slightly emarginate at base of leaflet, leaflet like membraneceous rarely stiff, ratio of length : width is 2:1, leaflets edge commonly flat rarely wavy, the colour of adaxial part of leaflets is light green rarely dark green. Unfortunatelly, the *P.mooniana* from SE Sulawesi planted in PBG was not included in DNA ITS characterization. Another genetic characterization study that involved the SE Sulawesi planted in PBG, collection from wild population in Papua along with Jambi population might be able reveal differentiation in their genetic characters as well.

The identification of plant species based on morphological characters is very challenging [30], as phenotipic character often resulted from a combination between genetic factors as well as environmental factors [31,32], therefore the ecological dan geographical variation can lead into variation of phenotypic character. This variation will causing the difficulty and uncertainty in identifying the species especially when collecting seedling from multispecies population [15] or collection was conducted by different person with different level of expertise [17], while misidentification of plant species will have serious implication for conservation effort and breeding program of those species [15]. The problem with species identification also occur in bioforensic when wood was only the material left for identification [33]. DNA barcoding has been proved to the misidentification applying barcoding marker for identification, as proven in previous studies that molecular identification will solve these problem and often found that plant species was inappropriately identified by morphological characters [6]. DNA barcoding also helps the rediscovery of extint species [34]. Despite of many studies conducted to generate DNA barcoding of different species, currently no universal barcoding marker is available for all different plant species [17,35,36]. Therefore barcoding study always challenging and requires proper selection of marker.

## 4. Conclusions

Barcoding of rDNA ITS of *Pericopsis mooniana* was generated in this study and produced between 634-641 bp nucleotides. Genetic distance between *P. mooniana* with *P. laxifolia* and with *P. angolensis* were 0.116 and 0.088 respectively, while genetic distance between *P. laxifolia* with *P. angolensis* was 0.060. Barcoding of rDNA ITS is a powerfull tool for dicrimination of *Pericopsis* spp. but not for *Pterocarpus* spp. The difference in morphological characters observed on *P. mooniana* different population planted in PBG, proved that morphological identification of *P. mooniana* need to be conducted properly, and identity confirmation can be conducted using rDNA ITS marker.

**References**

[1] Asian Regional Workshop 1998 Pericopsis mooniana The IUCN Red List of Threatened Species 1998: e.T31312A9623986 *Asian Regional Workshop Conservation and Sustainable Management of Trees*
[2] Roskov YR, Bisby FA, Zarucchi JL, Schrire BD, and White RJ 2006 *World Database of Legumes: Draft Checklist Version 10* (UK: ILDIS, Reading)
[3] National Academy of Sciences 1979 *Tropical Legumes: Resources for The Future* (Washington, D.C: National Academy Press)
[4] Suhartati and Alfaizin D 2020 Cultivation of *Pericopsis mooniana* the case study: KHDTK Malili, Luwu Timur Regency *IOP Conf. Ser. Earth Environ. Sci.* **486**
[5] Yuskianti V 2018 Genetic conservation of a threatened species, Kayu Kuku (*Pericopsis mooniana*) in Indonesia. *IUFRO-INAFOR Joint International Conference 2017 Promoting Sustainable Resources from Plantations for Economic Growth and Community Benefits"* ed Rimbawanto A, Krisdianto M, Turjaman HL, Tata H, Krisnawati T, Setyawati N, Sakuntaladewi and Muttaqim Z (Bogor: FORDIA Ministry of Environment and Forestry-Republic of Indonesia) pp. 403-10
[6] Amandita FY, Rembold K, Vornam B, Rahayu S, Siregar IZ, Kreft H, and Finkeldey R 2019 DNA barcoding of flowering plants in Sumatra, Indonesia *Ecol. Evol.* **9** 1858-68
[7] Anonymous 2018 *Ringkasan Rencana Pengelolaan PT Wijaya Sentosa 2015-2022* (Teluk Wondama Papua Barat)
[8] KPHP Sorong 2014 *Rencana Pengelolaan Hutan Jangka panjang, Kesatuan Pengelolaan Hutan Produksi Model Sorong di Kabupaten Sorong Provinsi Papua Barat*
[9] Nurtjahjaningsih ILG, Widyatmoko AYPBC, and Rimbawanto A 2019 Genetic diversity of kayu kuku (*Pericopsis mooniana* (Thw.)Thw.) population at Lamedai forest Southeast Sulawesi based on RAPD markers **13** 25-32
[10] Husna, Mahmud A, Tuheteru FD, Arif A, Nurdin WR, and Karepesiana S 2020 The effect of scarification and seed storage duration on the viability of Kuku Wood seeds (P*ericopsis mooniana* (Thw.) Thw.) *J. Kehutanan Indonesia Celeb.* **1** 31-40
[11] Husna 2015 Potential of Local Arbuscular Mycorrhizae Fungi (AMF) in Ex-Situ Conservation of Endangered Wood Species (*Pericopsis mooniana* (Thw.) Thw.) (Institut Pertanian Bogor)
[12] Suhartati and Alfaizin D 2018 Nursery technique of *Pericopsis mooniana* for clay post-mining land of reclamation *J. Faloak* **2** 103-14
[13] Ishiguri F, Wahyudi I, Takeuchi M, Takashima Y, Iizuka K, Yokota S, and Yoshizawa N 2011 Wood properties of Pericopsis mooniana grown in a plantation in Indonesia *J. Wood Sci.* **57** 241-6
[14] Yudohartono TP, Fiani A, and Yuliah 2019 Genetic diversity of kayu kuku (*Pericopsis mooniana* Thw.) from Southeast Sulawesi provenance observed at seedling stage *J. Pemuliaan Tanaman Hutan* **13** 11-7
[15] IUCN-SSC Species Conservation Planning Sub-Committee 2017 *Guidelines for Species Conservation Planning - version 1.0* (Gland, Switzerland)
[16] Austen GE, Bindemann M, Griffiths RA, and Roberts DL 2016 Species identification by experts and non-experts: comparing images from field guides *Sci. Rep.* **6** 33634
[17] Hollingsworth PM, Li DZ, Van Der Bank M, and Twyford AD 2016 Telling plant species apart with DNA: From barcodes to genomes *Philos. Trans. R. Soc. B Biol. Sci.* **371**
[18] Hartvig I, Czako M, Kjær ED, Nielsen LR, and Theilade I 2015 The use of DNA barcoding in identification and conservation of rosewood (*Dalbergia spp.*) *PLoS One* **10**
[19] Shiraishi S and Watanabe A 1995 Identifikasi of chloroplast genome between Pinus densiflora Sieb et Zucc and P. thumbergii Parl based on the polymorphism in rbct gene *J. Japanese For. Soc.* **77** 429-36

[20] White TJ, Bruns T, Lee S, and Taylor J 1990 *Amplification and Direct Sequencing of Fungal Ribosomal RNA Genes for Phyligenetics PCR Protocols: A Guide to Methods and Applications* ed. Innis MA, Gelfand DH, Sninsky JJ, and White TJ (San Diego, USA: Academic Press, Inc.) pp. 315-22
[21] Thompson JD, Higgins DG, and Gibson TJ 1994 CLUSTAL W: improving the sensitivity of progressive multiple sequence alignment through sequence weighting, position-specific gap penalties and weight matrix choice *Nuc.Acid. Res.* **22** 4673-80
[22] Hall TA 1999 Bio edit: a user-friendly biological sequence alignment editor and analysis program for Windows 95/98/NT *Nucl. Acids. Symp. Ser.* **41** 95-8
[23] Kumar S, Stecher G, and Tamura K 2016 MEGA7: Molecular evolutionary genetics analysis version 7.0 for bigger datasets. *Mol. Biol. Evol.* **33**(7) 1870-4
[24] Tamura K and Nei M 1993 Estimation of the number of nucleotide substitutions in the control region of mitochondrial DNA in humans and chimpanzees. *Mol. Biol. Evol.* **10** 512-26
[25] Bentham G and Hooker JD 1865 *Genera Plantarum* (London)
[26] Knaap-van Meeuwen MS 1962 Reduction of Afrormosia to Pericopsis (Papilionaceae) *Bull. du Jard. Bot. d'Etat* **32** 213-9
[27] de la Estrella M, Forest F, Wieringa JJ, Fougère-Danezan M, and Bruneau A 2017 Insights on the evolutionary origin of Detarioideae, a clade of ecologically dominant tropical African trees *New Phytol.* **214** 1722-35
[28] Bhagwat RM, Dholakia BB, Kadoo NY, Balasundaran M, and Gupta VS 2015 Two new potential barcodes to discriminate Dalbergia species *PLoS One* **10** 1-18
[29] Rangkuti AB 2016 *Aplikasi DNA Barcode untuk Identifikasi Jenis Meranti dan Rotan* (Institut Pertanian Bogor)
[30] Kirchoff BK, Delaney PF, Horton M, and Dellinger-Johnston R 2014 Optimizing learning of scientific category knowledge in the classroom: The case of plant identification *CBE Life Sci. Educ.* **13** 425-36
[31] Arnold PA, Kruuk LEB, and Nicotra AB 2019 How to analyse plant phenotypic plasticity in response to a changing climate *New Phytol.* **222** 1235-41
[32] Kusmec A, de Leon N, and Schnable PS 2018 Harnessing phenotypic plasticity to improve maize yields *Front. Plant Sci.* **9** 1-4
[33] Ng KKS, Lee SL, Tnah LH, Nurul-Farhanah Z, Ng CH, Lee CT, Tani N, Diway B, Lai PS, and Khoo E 2016 Forensic timber identification: A case study of a CITES listed species, *Gonystylus bancanus* (Thymelaeaceae) *Forensic Sci. Int. Genet.* **23** 197-209
[34] Lee SY and Mohamed R 2016 Rediscovery of *Aquilaria rostrata* (Thymelaeaceae), a species thought to be extinct, and notes on Aquilaria conservation in Peninsular Malaysia *Blumea J. Plant Taxon. Plant Geogr.* **61** 13-9
[35] Hollingsworth PM 2011 Refining the DNA barcode for land plants *Proc. Natl. Acad. Sci.* **108** 19451-2
[36] Hollingsworth PM, Graham SW, and Little DP 2011 Choosing and using a plant DNA barcode *PLoS One* **6**

**Acknowledgement**
This study was supported by the Ministry of Environment and Forestry of Indonesia through the Centre for Forest Biotechnology and Tree Improvement (CFBTI), Forest Research and Development Agency (FORDA). We thank Wahyunisari in CFBTI for her assistance in the lab work. All authors provided equal contribution to this study. IP conducted the experiment, analyzed the molecular data, and wrote the main manuscript, AW coordinating the umbrella project that provide fund to this study and provided advises on experimental design, IN and VY collected the main samples, and SD conducted morphological examination. All authors edited the manuscript.

IOP Conf. Series: Materials Science and Engineering **935** (2020) 012025 doi:10.1088/1757-899X/935/1/012025

# Adaptability of Ganitri (*Elaeocarpus ganitrus*) on degraded land of community forests in Tasikmalaya Distric, West Jawa Province

**A Sudomo* and B Dendang**

Research and Development Center of Agroforestry Technology, Ministry of Environment and Forestry, Ciamis, Indonesia

*Email: arisbpkc@yahoo.com.

**Abstract.** *Elaeocarpus ganitrus* is one of the prospective Non-Timber Forest Product (NTFP) species in the community forest. The wood can be used for building and music tool raw material. Specific products in the form of seeds are used as medicine, handicrafts, and traditional ceremonies. The aim of the research was to study the adaptability of *A. ganitrus* on degraded land of community forests in Tasikmalaya. Plot establishment was conducted in Gunungtandala Village, Kawalu Sub District and Tasikmalaya District, West Java Province. The spacing of this trial is 3 x 3 m with 2300 number of seedlings. The research showed that the survival of *A. ganitrus* in the total area reached 70.6%. The percentage of *A. ganitrus* life starting from the largest is 98% on flat land, 96% on slope land and 18% on steep land. Survival and growth prove that *A. ganitrus* are able to grow in this site. The slope factor greatly affects the vegetative and generative growth of *E. ganitrus.* There is a tendency that he stressed vegetative growth of *E. ganitrus* on steep land encourages generative growth with the appearance of flowers. Most of the *E. ganitrus* plants start flowering at the age of 3.5 years. The recommendation of this study is the use of *A. ganitrus* species in NTFP development on marginal lands in community forests. This species is also used in the restoration of the conservation area.

## 1. Introduction

*Alaeocarpus ganitrus* is a type of multipurpose tree species plant. In general, the use of *A. ganitrus* is a producer of seeds/fruit and some of that are used as carpentry wood [1]. Wood can be used as raw material for woodworking and materials for making musical instruments/guitars. *A. ganitrus* fruit has a hard petrified endocarp and attractive ornaments. The unique shape and size of *A. ganitrus* seeds can produce various jewelry products (bracelets, necklaces, beads) and seeds used for religious ceremonies and funeral ceremonies in India [2,3]. Hindus believe *A. ganitrus* seeds as Gods tears that drip into the earth then grow into trees. *A. ganitrus* seeds good for body health to overcome stress, control blood pressure and various mental illnesses, and affect the brain system when spreading bioelectronic stimulation. The marketing of *A. ganitrus* seed products has a specific route. It was exported to India and China [5].

*A. ganitrus* in Tasikmalaya community forest begins to bear fruit at the beginning of the dry season to the beginning of the rainy season. According Fitriani [6] the spread of *A. ganitrus* plants in Indonesia covers Central Java, Kalimantan, Bali, and Timor. The growth of wood is relatively straight with many branches and evergreens. *A. ganitrus* is relatively easy to grow on various types of regosol, andosol, podsolic, brown, latosol in the lowlands to highlands (0 to 1300 masl) and can be more than 1200 masl, with rainfall 3500-4500 mm/year [1,3,7].

The location of the research site is the degraded land of community forests. Land overgrown with reeds (*Imperata cylindrica* Raeusch ) with a slope starting flat, sloping, and steep. The soil conditions overgrown with reeds (*Imperata cylindrica* Raeusch ) are a reflection of degraded conditions. One of the government programs is to rehabilitate degraded land to return to its original function. Another benefit of the rehabilitation of degraded land is to restore the function of land, both economically and environmentally. Environmental functions are characterized by returning soil fertility, good land cover, and increasing land productivity. This can be carried out by vegetatively conserving land by planting species that can adapt to degraded land. *A. ganitrus* is a fast-growing type that is naturally easy to find and does not need a specific place of life [8] Trials of planting *A. ganitrus* at various levels of relative slope is rarely carried out. Moreover, the use of *A. ganitrus* for Forest Rehabilitation and Degraded Land has not been widely recommended. This is caused by a lack of information about the adaptation of *A. ganitrus* on degraded land. The purpose of this study was to test the adaptation ability of *A. ganitrus* plants to the level of the slope of the land in the Tasikmalaya community forest.

## 2. Materials and Methods

### *2.1. Location*

The research was carried out on the degraded land of *Imperata cylindrica* which administratively is in Gunungtandala Village, Kawalu Sub District, Tasikmalaya District of West Java Province. The study was conducted for 3 years 8 months from October 2015 to July 2019. Place height is 260 meters above sea level, with 18.8°C of minimum temperature and 28.2°C of maximum temperature. The humidity of air is between 19.7-26.4%. The average rainfall per month is around 302.67 mm [9].

### *2.2. Methods*

Planting is carried out on 3 different slope conditions, namely (1) flat land/0% of slope, (2) slope land/ 10-20% of slope, and (3) steep land/>45% of slope. The total plants planted are 2300 seedlings. Planting of *A. ganitrus* is carried out during the rainy season (October, 2015). The spacing of the research location is 3 x 3 m with 2300 plants. Planting holes size are 40 x 40 x 30 cm. Basic fertilizer at the time of planting is 2 kg cow manure/ planting hole. At the age of 8 months, weeding and fertilizing with 150 grams/planting holes of Urea + TSP + KCL (3:1:1) and 2.5 ounces/plant of furadan were carried out. At the age of 1.5 years, it's only weeding/cleaning of weeds that grow were applied.

### *2.3. Data and analysis*

Data collection was carried out at the age 3 years 8 months of *A. ganitrus*. The collected data covers life percentage, flowering/ fruit percentage, growth in height and diameter of the *A. ganitrus* plant. The data of the research were analyzed by analysis of variance and if it was significantly different, it would be continued by Duncan's further at 95% of test level. The SPSS 18 was used for data analysis.

## 3. Results and Discussion

### *3.1. Results*

Based on the analysis of variance, it was shown that the influence of land slope was significant different from height growth and the diameter of *A. ganitrus* at 44 months old as shown in Table 1. To find out the average difference among treatments the Duncan test was carried out.

**Table 1.** The variance analysis on land slope effect on growth of height and diameter of 44 months old *A. ganitrus*

| Parameter | FCal | signification |
|---|---|---|
| Height | 44.616 | 0.000* |
| Diameter | 16.355 | 0.000* |

The results showed that the average value of *A. ganitrus* growth in succession starting from the best was on flat land, the slope land (10-20%) and the steep land (>45%). The average growth rate of *A. ganitrus* diameter from the highest to the lowest is on the slope land (10-20%), flat and steep land (> 45%). The best percentage of flowering *A. ganitrus* obtained on the slope land (10-20%) is equal to the steep land (>45%) and higher than flat land. The highest percentage of the life of the *A. ganitrus* trees is obtained on flat land, the slope land (10-20%) and the steep land (>45%). The value of growth in height, diameter, percentage of life, and percentage of flowering *A. ganitrus* can be seen in Table 2.

**Table 2.** The average growth rate of height, diameter, life percentage and flowering percentage of 44 months old *A. ganitrus*

| Treatments (Slope of Land) | Height (cm) | Standard Deviation | Diameter (cm) | Standard Deviation | Life percentage (%) | Flowering percentage (%) |
|---|---|---|---|---|---|---|
| Flat (0% ) | 10.25 a | 2.63 | 9.73 a | 2.57 | 98 | 10 |
| Sloping (10-20% ) | 9.48 a | 2.06 | 10.79 a | 2.92 | 96 | 16 |
| Steep (>45% ) | 6.19 b | 2.12 | 7.69 b | 2.75 | 18 | 16 |

Remark: different letter shows significant different of 95% of level

Based on the Duncan test, the growth of *A. Ganitrus* had a significant difference. On the flat land the highest value was significantly different from the steep land (>45%). The growth of *A. Ganitrus* from the Duncan test results obtained the lowest value on the steep land (>45%) was significant different from the flat land and the slope land (10-20%). Based on Duncan Test, the lowest diameter growth was 6.19 cm on steep slope (>45%). It was significantly different to the flat land and the slope land (0% – 20%).

### *3.2. Discussion*

Growth of ganitri tested on land with various slopes shows a significant different in height, diameter, survival and flowering percentage. Plant growth is influenced by growth factors such as stand density (plant spacing), stand age characteristics, climate factors (temperature, precipitation, wind speed and air humidity) and soil factors (physical, chemical composition, and microbiological components soil) [10]. Growth of tree height can be influenced by differences in the speed of leaf formation which is very sensitive to the quality of site. The quality of site can be influenced by differences of land sloping. According Davis and Jhonson [10] there are three environmental factors and one genetic (internal) factor that are very significant in influencing height growth. These are nutrient content of soil minerals, soil moisture, sunlight, and the balance of genetic characteristics between the growth of tree height and diameter. Diameter growth is influenced by factors that influence photosynthesis (photosynthesis results for respiration, leaf replacement, root growth and high growth have been fulfilled).

The growth of ganitri on flat land is better than on sloping and steep land. This is caused by the slope factor of the land which is on the relatively flat land, it is more secure from the run-off. This can minimize nutrient leaching. The slope factor has a direct effect on soil erosion [11]. The higher the slope of the land from the horizontal field, the higher the occurrence of the erosion process (detachment, transportation, and sedimentation) [12]. On steep slopes, the potential for high surface erosion/ run off results in leaching which decreases soil fertility. The hight rainfall in the steep slope can cause damage to the soil [13]. At the same time sloping land with large erosion destroys the ganitri plant so that it is uprooted from its roots. 12-16% of slope caused nutrient availability for plants to decrease, decrease in productivity, decrease in crop quality, decrease infiltration rate and ability of soil to hold water down [14,15].

The effect of slope on the growth of ganitri diameter on the research plot showed significant results. On steep land, the smallest average diameter is obtained which is significant different from the diameter growth on flat and sloping land. Soekotjo [16] states that the growth of stem diameter depends on the moisture density, crown surface and root system, and the influence of climate and soil conditions. High air temperature will increase the transpiration rate. In a long time, it will disturb the water balance and it will disturb the plant growth. Growth of ganitri diameter on flat and sloping soil with 3 x 3 m of spacing is quite high and proportional while on steep land growth in height and diameter is smaller even though the light intensity is higher. The cause of the lack of growth of ganitri on steep land is high erosion. This is caused by the processing of land in the research plot so that the land is not covered by litter. It seems that having the ability to control erosion and surface flow instead of closing trees that are not balanced by the formation of litter will increase the rate of erosion because the kinetic energy of raindrops from trees more than meters taller is greater than raindrops which fall freely outside the forest [17]. The land cover factor has an exponential relationship with the contact surface and erosion produced that varies depend on the type of land use [18]. Erosion observation on sloping land with a slope of 50% to 60% during the rainy season in five types of land use obtained the lowest erosion rate on forest observations compared to observations made in coffee plantations with various plant ages [19].

The high erosion causes the land to be degraded. Degraded land is characterized by the presence of reeds which causes disturbed vegetative growth of ganitri. The success of planting on steep land is also relatively low. Agroforestry patterns are efforts for the success of planting on steep land. It can reduce run-off and leaching of nutrients. The presence of lower plants will cover the surface of the soil and reduce run-off. Construction of terraces will able to reduce soil damaged by surface run-off. Managing the minimum tillage land to reduce the disruption to the land can be applied as well. Another thing to applied is the combination of no-tillage and minimum tillage. In various cases, this requires careful processing of the soil to minimize soil damage and leaching of nutrients.

Generative growth on sloping land shows relatively higher results compared to flat areas. The best percentage of flowering starts on steep, sloping, and flat land. On steep land it causes low vegetative growth but encourages flowering. Although vegetative growth and the percentage of life on steep land are relatively lower, flowering works higher. With a low percentage of life, it gives open space. The abundance of sunlight that covers tree canopy due to lower canopy competition allows the flowering of ganitri plants. The abundance of light can reach the entire canopy surface that its gives a positive effect on flowering. As for ganitri, on flat land with a spacing of 3 x 3 m and a percentage of the life of 98%, the canopy is in contact so that the canopy is known to reduce light intensity. Canopy growth on flat land has intersected so that not all canopy surfaces receive sunlight intensity. On the steep land produce lowest vegetative growths but actually encourage generative growth with the appearance of flowers. Vegetative growth was not accompanied by generative growth (in contrast). Generative growth with the appearance of flowers is in fact opposed to vegetative growth. The physical appearance and survival of ganitri on sloping land is lower but produces more flowers. The process of stressing plants on steep land will inhibit vegetation growth but encourage generative growth. This happens because plants with high stressing levels for vegetative growth will encourage plants to grow generatively. Ganitri plants with low vegetative growth on steep land actually encourage generative growth with the appearance of flowers.

The research location at an altitude of 260 m asl, latosol soil, and with 18.8°C of minimum temperature and 28.2°C of maximum temperature and more than 2000 mm/year of rainfall. This is in accordance with the research [1] that the distribution of Ganitri can grow with latosol soil in the lowlands to the highlands (0-1300 masl). The success of ganitri planting which reached 70.6% showed that ganitri could grow on degraded land of Imperata with these biophysical conditions. Degraded land marked by reeds, latosol brown soil and topography varying from sloping, steep, and flat can be rehabilitated with *A. ganitrus*. The density of growth and the percentage of life on steep land can be anticipated by making frequent terraces and minimal tillage of agroforestry.

## 4. Conclusion

The slope factor greatly affects the vegetative and generative growth of *E. ganitrus*. Steep degraded land results in low vegetative growth but encourages generative growth with flowering. *A. ganitrus* can grow properly on flat and sloping slope levels (0-20%) *on ex Imperata cylindrica* fields. A slope of 10-20% can produce a higher balance of vegetative and generative growth than other slopes. *A. ganitrus* should be planted on flat or sloping land. If it is planted on steep land, a pattern for soil and water conservation should be applied. Planting *A. ganitrus* for the main products of the fruit should be planted with more than 3 x 3m of spacing.

## References

[1] Rohandi A and Gunawan 2014 Sebaran populasi dan potensi tanaman Ganitri (*Elaeocarpus Ganitrus* Roxb) Di Jawa Tengah *Jurnal Ilmu Kehutan* **8**(1)

[2] Heyne K 1987 *Tumbuhan Berguna Indonesia.* (Jakarta: Badan Penelitian dan Pengembangan Kehutanan, Departemen Kehutanan)

[3] Khan ML And Barst P 2003 Regeneration of rudraksh (*Elaeocarpus ganitrus* Roxb.) Athreatened tree species: Germination strategies *Int. J. Ecol. Environ. Sci.* **29** 255-60

[4] Wikidisastra K 1999 Ganitri Kai Panon Dewa In Baraya_*Sunda@yahoogroups.com* (Akses/4/11/2014)

[5] Trubus 2007 Mata Siwa Penyapu Polutan *Trubus Magazine* Edisi No. 456 November 2007/XXXVIII

[6] Fitriani 2011 *Jenis dan harga buah Ganitri* (Jakarta: Jenitri)

[7] Safitri 2011 Jenis komersial yang belum banyak dilirik usahawan *http://yuniarsafitri.blogspot.com/2011/06/1* Jakarta

[8] Rachman E, Rohandi A, and Hani A 2010 Evaluasi penerapan pola tanam jenis pohon potensial pada hutan rakyat. *Prosiding Seminar Peningkatan Produktivitas Hutan Rakyat Untuk Kesejahteraan Masyarakat*

[9] Anonim 2016 *Monografi Desa Gunung Tandala*

[10] Davis LS and Jhonson KN 1987 *Forest Management* (New York: MC Graw-Hill Book Company)

[11] Sitepu FSM and Harianto T 2017 Pengaruh intensitas curah hujan dan kemiringan lereng terhadap erosi yang berpotensi longsor. *J. JPE* **21** 23-27

[12] Wiradisastra 2010 *Geomorfologi dan Analisis Landskap* Laboratorium Penginderaan Jauh dan Kartografi Jurusan Ilmu Tanah Fakultas Pertanian Institut Pertanian Bogor

[13] Chen HLC and Law K 2004 Causative mechanismns of rainfall-induced fill slope failures. *Journal of Geotechnical and Geo-environmental Engineering* **130**

[14] Supirin 2001 *Pelestarian Sumber Daya Tanah dan Air* (Yogyakarta: Andi offset)

[15] Wati Y, Alibasyah MR, and Manfarizah 2017 Pengaruh lereng dan pupuk organik terhadap aliran permukaan, erosi dan hasil kentang di Kecamatan Atu Lintang Kabupaten Aceh Tengah. *Jurnal Manajemen Sumberdaya Lahan* **3** 496-99

[16] Soekotjo 2004 *Silvikultur Hutan Tanaman Prinsip-prinsip Dasar* Ed. Hardiyanto EB and Hardjono A (PT Musi hutan Persada)

[17] Agus F, Ginting N, and Noodwijk MV 2002 Pilihan teknologi agroforestri/konservasi tanah untuk areal pertanian berbasis kopi di Sumberjaya, Lampung Barat (Bogor: World Agroforestry Centre)

[18] Rose CW, Coughland KJ, Ciesiolka CAA, and Fentie B 1997 *The Role of Cover in Soil Conseervation In: A new Soil Conservation methodology and Application to Cropping Systems in Tropical Steeplands* (Canbera: Aciar Technical Reports)

[19] Widianto, Noveras H, Suprayogo D, Widodo RH, Purnomosidhi P, and Van Noordwijk M 2002 Konservasi hutan menjadi lahan pertanian: Apakah fungsi hidrologis hutan dapat digantikan agroforestry berbasis kopi *Seminar Himpunan Tani Indonesia*

# Assessing ultrastructure and density properties to predict wood hardness of young fast grown plantation teak

**R Damayanti*[1], L M Dewi[1], B Ozarska[2], J Illic[2], G Pari[1], Krisdianto[1]**

[1]Forest Products Research and Development Center, Research, Development and Innovation Agency, Ministry of Environment and Forestry, Bogor, Indonesia
[2]School of Ecosystem and Forest Sciences, Faculty of Science, The University of Melbourne, Burnley Campus, Richmond, Victoria, Australia

*E-mail: ratih_turmuzi@yahoo.com

**Abstract.** The use of plantation timber has been increasing to supply raw material for timber manufacturing. Super teak (*Tectona grandis* Linn. f), a fast grown teak planted in Indonesia was successfully harvested in very young age (5 years old). This frequently rises queries whether the young super teak could meet the minimum quality criteria for wood working. Investigation on wood properties should be undertaken to measure the required wood properties. However, physical and mechanical testings require a long time process and large size samples. Therefore, a non-destructive method to predict wood properties of standing trees need to be developed. This study aims to assess ultrastructure and density properties to predict wood hardness of young fast grown plantation teak. Samples were collected from super teak plantation in East and West Java. Crystallinity and its quantifiers were measured using X-Ray Diffraction. The hardness was determined using an Instron® universal strength testing machine. The result showed that wood density was significantly related to wood hardness for all structural directions; the power relationship explained 52-56% of the variation. Degree of crystallinity (DC) and crystallite width were found as parameters affecting wood hardness. DC was positively correlated to wood hardness for all structural directions ($r=0.2$; $r^2=0.04$), whereas crystallite width was negatively correlated to radial and end-grain hardness ($r=0.2$; $r^2=0.04$). The power of prediction increased only by 1-5% when crystallinity factors were included with density, thus the extra effort of using XRD is not warranted for non-destructive prediction of wood hardness.

## 1. Introduction

Young fast grown plantation teak has been widely developed to meet the high demand of timber for manufacturing in Indonesia. Super teak is one of fast grown teak (*Tectona grandis* Linn. f) derived from several clones. This has facilitated plantation rotations as short as 5 years. Harvesting of super teak started in 2012. Working with plantation timbers often raises questions if the trees have met minimum requirements for particular wood products. Such questions can be only answered if the information and data on wood properties are available. Generally, testing physical and mechanical properties necessitates time consuming destructive tests which are not suitable for predicting such properties from standing trees. Consequently, non-destructive techniques are desirable to enable low cost and rapid assessment of important wood properties.

The standard method for determining wood hardness requires destructive testing and the preparation of large standard size specimens 50 x 50 x 150 mm, using the standard Janka test method [1]. However, the standard method is time consuming and relatively costly and cannot be applied directly for predicting wood hardness of standing trees. Therefore, a non-destructive technique (NDT) to predict wood hardness in standing trees would provide significant benefits for evaluating optimal silvicultural treatments, and for selecting suitable trees for tree improvement programs.

Some non-destructive techniques have been described for measuring hardness, such as acoustic tomography [2] and nano indentation [3-5]. However, the results were inadequate because of many factors [6]: poor correlation between apparent acoustic velocity of the tomograms, the measured ultrasonic velocity, and end-hardness of small cubes was likely to have been associated with the wood anisotropy.

Another technique involving continuous nano indentation of single fibres for predicting wood hardness also experienced some problems because of the presence of extractives in the fibres [3]. With the hypothesis that any change in the smallest wood structure may reflect the wood properties at a higher level, Vincent *et al.* [5] applied nano indentation to the S2 layer and middle lamella at the secondary cell wall on early and late wood fibres to predict wood hardness obtained on the Brinell test [5]. The proposed model produced high adjusted $R^2$ (0.77) for predicting the average ring density. However, the authors concluded that the location of nano indentation measurement, moisture, and extractive content probably influenced the relationships. Finally they recommended that nano-measurement is too far removed from representing macro wood mechanical properties stemming from the complexity of wood structural hierarchy [5].

Other non-destructive testing approaches for wood hardness determination were further constrained by the complexity of the wood structure: the anisotropy behavior, the presence of moisture and the extractive content in the cell wall where nano indentation were noted [4-6]. It is well known that the physico mechanical properties of wood are controlled by three factors: the proportion of void volume (which can be determined from the wood density); the organization of the cell wall ultrastructure (in addition to the ultrastructure of the cell wall, variety and proportion of cell types), and the moisture content [7]. In addition, potentially wood extractives may influence wood density and need to be accounted for [7]. When density is used as a predictor of wood hardness, some of the aforementioned constraints may be overcome. Density, combined with the major cell wall structural component cellulose, was considered that it might provide additional information for predicting wood hardness.

As cellulose is the main component of wood and it acts as the skeletal polysaccharide of cell wall [8], it can be assumed from basic principles, that it should be related to wood hardness. The characteristics of cellulose can be quantified by crystallinity. Reportedly crystallinity affects significantly the physical, mechanical and chemical properties of cellulose fibres, including wood hardness [9,10]. This study considers the usefulness of crystallinity and its quantifiers (degree of crystallinity, dimension of crystallite and Micro Fibril Angle (MFA) as a non-destructive indicator of wood hardness.

Wood hardness is an important property for flooring, furniture products and structural utilization. It reflects a measure of resistance to indentation [11-13] and provides an indication of how well the wood performs in relation to wear and marking [14]. The standard test of hardness is empirical; it represents the force necessary to embed a hardened-steel hemisphere with diameter 0.444 inch [11,12]. Hardness is a practical mechanical property which is useful for identifying wood species that are suitable for flooring [15,16]. The selection of a hard timber can provide improved resistance to indentation and abrasion, but for domestic use, other aspect than hardness such as appearance sometimes determine the choice [17]. Therefore, this study aims to develop non-destructive test for predicting hardness based on ultrastructure and density.

## 2. Materials and Methods

### *2.1. Specimen preparation*

Eighteen trees ranging from small, medium and large diameter stems from super teak plantation forests in Bogor, West Java, Indonesia, and Magetan, East Java, Indonesia (**Error! Reference source not found.**), were used for hardness measurements. For developing non-destructive testing techniques, fifteen cm thick discs were taken from five fixed sampling heights (60, 80, 100, 120, and 130 cm above ground), at middle (50% of the tree height), and at the top of the tree, that is with minimum stem diameter of 7 cm. This sampling approach was similar to that prescribed by Raymond and Muneri [18] for the determination of optimal sampling heights.

The discs (seven discs from each tree) were used for density, crystallinity and hardness measurements. Hardness samples, measuring 25 mm (radial) x 25 mm (tangential) x 100 mm (longitudinal) were prepared from each disc aiming to preserve true radial and tangential directionality [19]. Specimen's radial position was also denoted as heartwood (HW), transition-wood (TZ) and sapwood (SW) accordingly. Six replications for each zone (depend on disc/stem diameter) were used.

**Table 1**. Tree selection used in the study

| No | Tree code | Diameter (cm) | Height* (m) | Clone** | Tree code | Diameter (cm) | Height* (m) | Clone** |
|---|---|---|---|---|---|---|---|---|
| | Site 1 (West Java) | | | | Site 2 (East Java) | | | |
| | Class Diameter 1: large ( > 26 cm) | | | | | | | |
| 1 | SA22 | 30.25 | 16.0 | A | 198 | 29.3 | 10.5 | B |
| 2 | AC18 | 31.40 | 18.2 | B | 12 | 29.3 | 12.0 | C |
| 3 | AC23 | 32.20 | 13.6 | A | 101 | 28.7 | 9.2 | C |
| | Class Diameter 2: medium (17-25 cm) | | | | | | | |
| 1 | S50 | 20.7 | 11.5 | - | 108 | 21.5 | 8.50 | - |
| 2 | S74 | 21.3 | 9.1 | - | 153 | 24.2 | 6.50 | - |
| 3 | S105 | 22.9 | 11.4 | - | 78 | 24.5 | 8.65 | - |
| | Class Diameter 3: small (10-16 cm) | | | | | | | |
| 1 | S89 | 14.3 | 9.65 | - | 27 | 14.3 | 6.4 | - |
| 2 | S53 | 11.9 | 8.90 | - | 2 | 16.6 | 6.7 | - |
| 3 | S120 | 14.6 | 10.10 | - | 10 | 16.2 | 5.8 | - |

*Economic height: height of stem at the top with minimum diameter of 7 cm
**At the time of sample collection, only three clones were available in the plantations. In this study, DNA testing to check the tree clone was conducted only on wood from the large diameter classes.

Specimens for crystallinity measurement were end-matched with the density and shrinkage specimens as shown in Figure 1. The samples were 50 mm (tangential direction) x 15 mm (longitudinal direction) x 30-40 μm (radial direction).

### *2.2. Testing and calculation*

Up to 42 samples (depending on the tree stem diameter), measuring 50 x 50 x 150 mm for Site 1 and 126 samples measuring 30 x 30 x 90 mm for Site 2, were prepared from each tree. There were difficulties in obtaining sufficiently large test specimens from the small diameter trees when preparing samples for hardness testing for Site 1 (wet site) where fewer specimens and sometimes no specimens especially at the top height were represented. Therefore, the size of hardness specimens at Site 2 (dry site) were decreased to 30 x 30 mm with length 90 mm (ratio 1:1:3). According to the Australian Standards for mechanically testing small clear specimens of timber [1], the standard dimension of hardness specimens are 50 x 50 x 150 mm, and the test may also be made on a specimen of any other size, as long as the thickness is at least 25 mm.

The samples in this study were identified as heartwood (HW), transition-wood (TZ) and sapwood (SW) on the basis of wood colour, and the hardness was determined using an Instron® type 3369 universal strength testing machine according to the Australian Standards for mechanically testing small clear specimens of timber [1]. Wood specimens were tested in the radial (R hardness), tangential (T hardness) and longitudinal (end hardness) grain directions. The reason for testing all wood faces was to ensure a more representative value of each specimen [20] and to enable radial and tangential hardness to be tabulated as side hardness for ready comparison with data for other species. Before and after testing, the weight and dimensions of each specimen were measured for determination of wood density and moisture content.

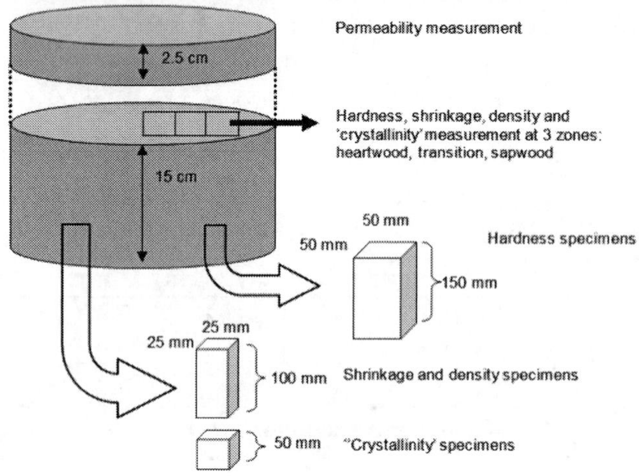


**Figure 1**. Cutting scheme for 15 cm discs at seven axial positions

Crystallinity and its quantifiers were measured using X-Ray Diffraction (XRD) (Shimadzu® QP-500). The X-ray beam was powered with a 40kV, 30mA source and scans made in the range from 0 - 40 degrees at a scan speed of 2 degrees per minute. The degree of crystallinity (DC) was calculated as the ratio between the crystalline and total non-crystalline and amorphous regions as shown in Figure 2. The dimension of crystallites (CW-width and CL-length) was determined using the Scherrer formula-Equation 1 [21] for diffraction intensities (002) and (040), usually in the angular range from 20 to 24 degrees and from 32 to 37 degrees, respectively [22]. Cellulose lattice spacing (d) was calculated using Bragg Equation (Equation 2) [22]. The number of crystallite planes for each of the diffraction intensities was calculated by dividing the crystallite dimension with the crystallite lattice spacing (d).
Equation 1 Scherrer formula for crystallite size determinations

$$B_{(2\theta)} = \frac{K\lambda}{\mathrm{L}\cos\theta'} \quad (1)$$

where, K = shape factor, 0.9 used to determine cellulose crystallite dimensions for the (200) reflection and 1.0 in the case of (004) reflection; λ = wave length of X-rays produced from a Cu target (0.154 nm); B ($2\theta$ ) = Peak width – Full Width at Half Maximum (FWHM), (radians); and θ' = ½ the diffraction angle, (radians); L = Crystallite size.

Equation 2 Bragg Equation calculation of lattice spacing

$$d = \frac{\lambda}{2 \sin \theta'} \quad (2)$$

where $\lambda$ = Wavelength of the X-ray beam (0.154 nm, CuK$\alpha$); $\theta$ = Bragg angle

Micro Fibril Angle (MFA) was determined from the (002) reflection plane, and calculated by using Cave's method (Equation 3). Abe and Yamamoto's equation (Equation 4) an alternative method was used for comparison [23]. The T parameter was calculated according to Stuart and Evans [23] by manually drawing tangents to the sides of plots of the diffraction arcs.

Equation 3 Cave's method for calculating MFA

$$\text{MFA} = 0.6\ \text{T} \quad (3)$$

Equation 4 Yamamoto's equation for calculating MFA

$$MFA = 1.575 \times 10^{-3}\, T^3 - 1.431 \times 10^{-1}\, T^2 + 4.693\, T - 36.19 \quad (4)$$

*2.3. Data analysis*

Saphiro-Wilk's W test (Statistica®) test was employed to determine if the density and wood hardness data were normally distributed. Variations between sites, tree growth rates and within trees were analyzed using T-test and ANOVA for normally distributed data, and the Mann-Whitney U test and Kruskal-Wallis ANOVA was applied for the above analyses when the data were not normally distributed.

MFA, Degree of Crystallinity (DC) and crystallite lattice spacing ($d_{002}$) were distributed normally according to Saphiro-Wilk's W test (Statistica®). However, crystallite width, crystallite plane ($N_{002}$) at diffraction intensity (002), crystallite length, crystallite lattice spacing ($d_{040}$) and crystallite planes ($N_{040}$) at the 040 diffraction intensity were not normally distributed according to the above test for normality. Statistical analysis was performed with T-test and ANOVA for normally distributed data and Mann-Whitney U-Test and Kruskal-Wallis ANOVA for non-normal data.

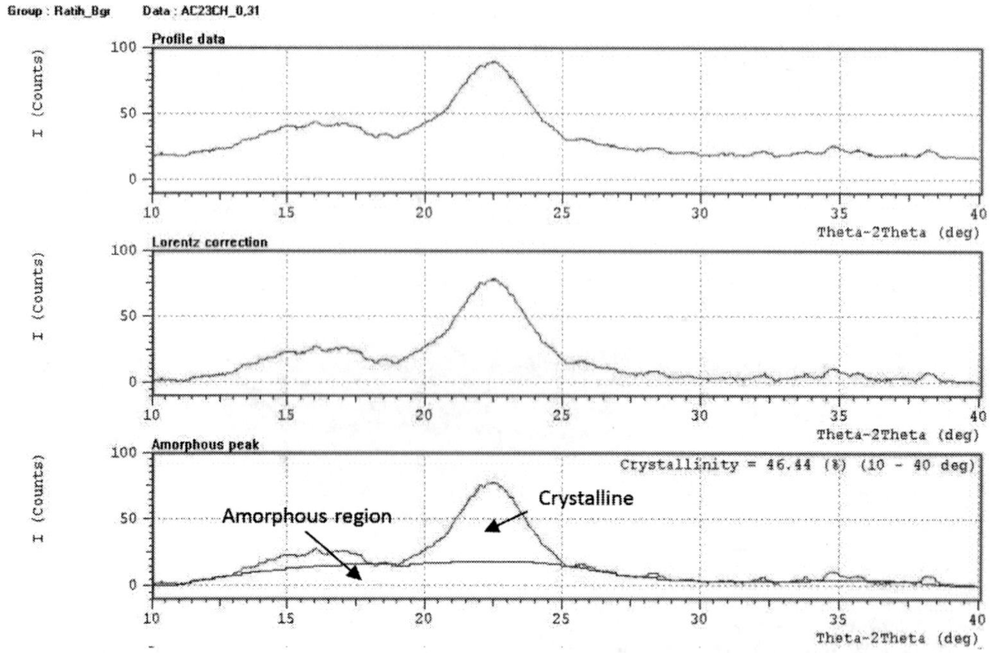


**Figure 2**. Typical data profiles for degree of crystallinity (DC) determinations

## 3. Results and Discussion

### *3.1. Wood hardness and wood density*

Wood hardness increases with density, but may not be linear [24]. Extensive tests of many wood species over a wide range of specific gravities showed that mathematical relationships were either parabolic or exponential of n degree [7,11,24]. However, mechanical properties are not all affected to the same degree by the changes in specific gravity [7], for example hardness of softwoods with n = 1.50 increases with specific gravity much more rapidly than MOE with n = 0.84. Equation 5 shows the general mathematical relationship between specific gravity and mechanical properties. Equation 5 General mathematical relationship between specific gravity and various strength properties [11].

$$S = aG^{n} \tag{5}$$

where, *S* is the strength property under consideration; *G* is specific gravity of the wood tested, and *n* is the slope of the curve.

Forest Product Laboratory in cooperation with the University of Wisconsin, Madison, developed formula for predicting hardness which has been presented in Equation 6 as follows:

$$H_{endgrain} = 21351\ G^{2.25};\ H_{radial} = 16574\ G^{2.25};\ H_{tangential} = 16992\ G^{2.25} \tag{6}$$

When the relationship in Equation 6 was applied to the super teak and compared with the actual wood hardness from standard testing, the result showed that 54% (46-62%) of the actual hardness varied between the limits of 87-111% of the computed hardness. The range 87-111% (specifically 88-111% for radial hardness, 90-111% for tangential hardness, and 87-111% for end-grain hardness) was determined based on the limits obtained from the study by Forest Products Laboratory [25] from a comparison of actual hardness and computed values. In that comparable study by Forest Products Laboratory [25], the actual values of approximately half of the specimens examined were observed to vary between the 87-111% of the computed values; the remainder, half (one-fourth of the total number) exceeded 111%, and the rest was below 87%. In this present study, by applying the **Error!**

**Reference source not found.**, the actual values were 20% (13-29%) below 87% and 23% (17-28%) exceeded 111% (Table 3) which is practically similar to the result obtained by Forest Products Laboratory [25].

The well accepted models (Equation 6) were developed based on empirical data fits for a large number of species. Because the numbers of species used to obtain the relationship by Forest Product Laboratory were large, they were collectively expected to exhibit a wider range in density than that for a single species; therefore the *K* and *n* coefficients are expected to be different for a species like super teak.

Table 2 presents Pearson correlations between wood hardness and specific gravity of super teak from both the wet and dry sites. The result showed that air dry specific gravity was significantly correlated with wood hardness positively, with r values 0.7 (0.72-0.75). Specific gravity in this study was adjusted based on the density at the test moisture content (approximately at 15% MC), using Equation 7. Equation 7 Method of computing specific gravity at any moisture content [11].

$$S = \frac{W_M}{V\,(1+\frac{M}{100})} \qquad (7)$$

where $S$ is specific gravity, $W_M$ is the weight of sample at $M$ moisture content, $V$ is the volume-equivalent of the sample, and $M$ is the moisture content expressed as the percentage of the oven dry weight.

**Table 2**. Simple correlation between wood hardness and air-dry specific gravity of super teak at both wet and dry sites showing the $r^2$ >0.5

| Wood Hardness | R |
|---|---|
| Radial | 0.73 |
| Tangential | 0.75 |
| End | 0.72 |

Correlations marked in red are significant at p <0 .05

To obtain new *K* and *n* coefficients for super teak, specific gravity (G) and measured hardness values were regressed and a trend-line was fitted (**Figure** 3-5). Comparable results of new parabolic equations for predicting hardness with the actual hardness values are presented in **Table 3**.

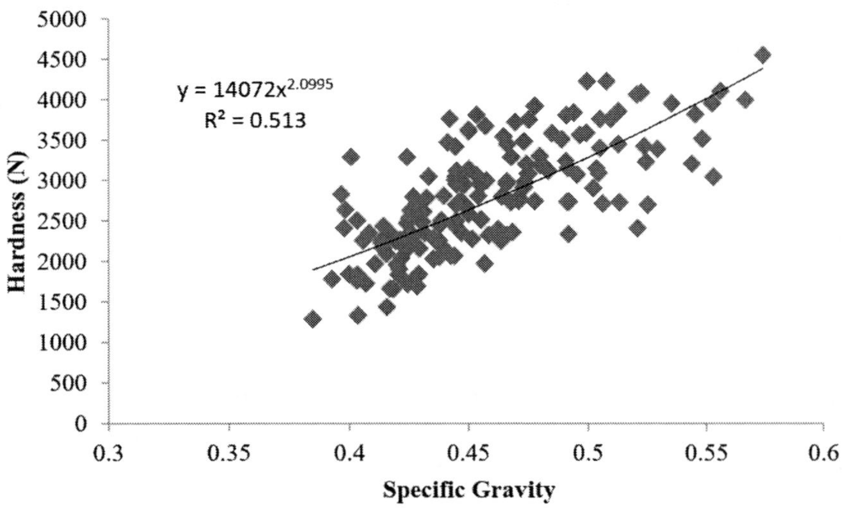


**Figure 3**. Relationship between radial hardness and specific gravity

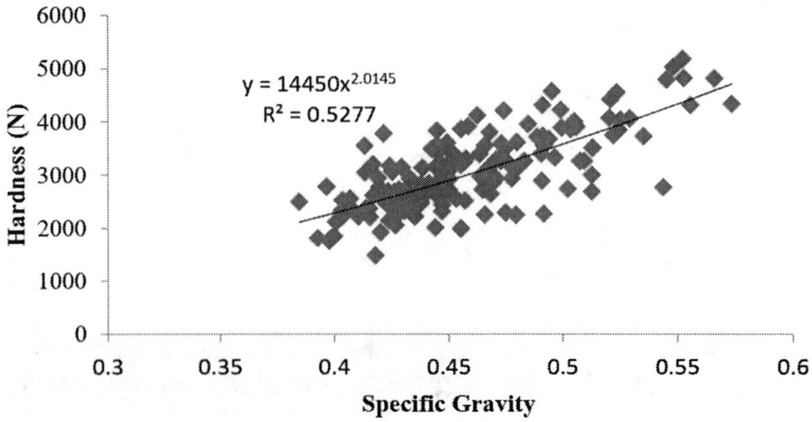


**Figure 4**. Relationship between tangential hardness and specific gravity

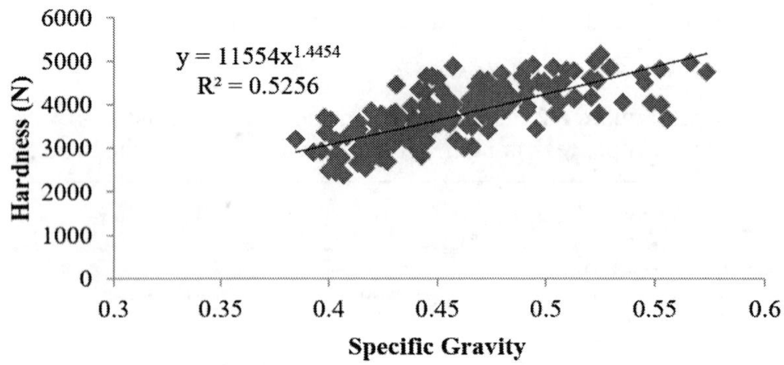


**Figure 5**. Relationship between end-grain hardness and specific gravity

**Table 3**. Comparable results of parabolic equations for predicting hardness of by using existing equations from Forest Products Laboratory and proposed equations for super teak with the actual hardness values

| Equations | Radial Hardness | | | Tangential Hardness | | | End-grain Hardness | | |
|---|---|---|---|---|---|---|---|---|---|
| | <88% | 88-111%* | >111% | < 90% | 90-111%* | >111% | <87% | 87-110%* | >110% |
| Existing equations<br>$H_r = 16574\ G^{2.25}$<br>$H_t = 16992\ G^{2.25}$<br>$H_e = 21351\ G^{2.25}$ | 29 | 46 | 17 | 19 | 53 | 28 | 13 | 62 | 25 |
| New equations<br>$H_r = 14072\ G^{2.10}$<br>$H_t = 14450\ G^{2.01}$<br>$H_e = 11554\ G^{1.45}$ | 22 | 51 | 27 | 20 | 56 | 23 | 13 | 65 | 22 |

*The range of ratio between actual and computed hardness values were determined based on the study by Newlin and Wilson [25]; Hr = Radial hardness; Ht= Tangential hardness; He = End-grain hardness

The results from **Table 3** show that the hardness at unit density (*K*) of super teak was lower than that for the previously published models developed by Forest Product Laboratory. Furthermore, the increases with specific gravity were smaller for super teak, possibly because the range of specific gravity of the 5 years super teak material is smaller; only 0.3 to 0.6. It is interesting to observe that in this case the relationship between wood hardness and specific gravity is almost linear (the n value is almost 1), for end-grain.

The new parabolic equations provided a better fit than the existing models (Equation 6), as shown by increases of the ratio between actual and computed hardness values for all hardness directions (**Table 3**). Consequently, the new *K* and *n* coefficients provided an improvement for predictions. The results obtained from this study are higher to those from Newlin and Wilson [25]. The result indicates that by using specific gravity as an approach, the maximum power of prediction for estimating wood hardness based on specific gravity is in the order of 65%.

The power of prediction of wood hardness was further tested by using the new coefficients by regression analysis from Statistica® (Table 4). The result confirmed that in general, the proposed models are significant but specific gravity on its own can only explain 52-56% of the variation in wood hardness. To improve predictability of wood hardness using additional properties, other wood properties are expected to also influence wood hardness, and consequently crystallinity was considered in that context as it has previously been reported that wood hardness increases with increasing crystallinity [9,10,23]. Table 4 Coefficients of determination ($R^2$) of proposed parabolic equations for predicting wood hardness based on specific gravity at both wet and dry sites.

**Table 4.** Simple correlation between wood hardness and air-dry specific gravity of super teak at both wet and dry sites

| Wood Hardness | r | $R^2$ | P Value |
|---|---|---|---|
| Radial | 0.72 | 0.52 | 0.00 |
| Tangential | 0.75 | 0.56 | 0.00 |
| End | 0.72 | 0.52 | 0.00 |

Correlations marked in red are significant at p <0 .05

### *3.2. Wood hardness and crystallinity*

Considering that the specific gravity only explained approximately half of the variation in wood hardness, and that according to Panshin and Zeeuw [26] the mechanical properties of wood can be influenced by other properties, crystallinity was considered as a potential characteristic based on the association between the amount and arrangement of cellulose crystallites on the physical, mechanical and chemical properties of cellulose fibres, including wood hardness [10,27].

Table 5 presents the Pearson correlation between wood hardness and crystallinity properties determined at the opposite-end matched specimens by X-ray Diffraction. The result showed that degree of crystallinity (DC) was significantly positively correlated to hardness for all hardness directions (r = 0.2). Furthermore, crystallite width was negatively correlated to all hardness direction, and significant only on radial and end-grain hardness (r = -0.2). All the correlations were significant but very low. The results from DC and crystallite width were consistent with the relationships published by Damayanti *et al.* [28] and Damayanti *et al.* [29].

Wood crystallinity (degree of crystallinity) is defined as the weight fraction of crystalline material to non-crystalline material in wood [10]. This crystalline region is considered to be in the state of relatively good three-dimensional order [9]. An increase in the degree of crystallinity is also expected to give rise to increases in wood hardness [8,9,15].

**Table 5**. Pearson correlation between wood hardness and crystallinity for super teak at both wet and dry sites

| Wood Hardness (N) | MFA (°) | DC (%) | Crystallite Width (nm) | d002 | N002* | Crystallite Length (nm) | d040 | N040* |
|---|---|---|---|---|---|---|---|---|
| Radial | -0.01 | 0.21 | -0.24 | 0.12 | -0.24 | 0.11 | -0.02 | 0.11 |
| Tangential | -0.10 | 0.16 | -0.07 | -0.05 | -0.06 | 0.05 | -0.02 | 0.05 |
| End | 0.06 | 0.17 | -0.19 | 0.03 | -0.18 | 0.00 | 0.03 | -0.01 |

Correlations marked in red are significant at p <0.05; * the variable is not used in the regression

The negative correlation obtained between crystallite width and hardness in this study meant that smaller crystallite size resulted in harder wood possibly because of reasons of homogeneity via the expectation that smaller crystallites are associated with greater homogeny and compact substance, thus potentially harder wood (Prof. Voichita Bucur, pers. comm.).

Considering the effect of MFA on wood hardness as another property of crystallinity it was found not to be usefully related to hardness. The result of this study is in good agreement with Gindl et al [30] who reported that hardness determined by nano-indentation was independent of MFA, and suggested that indentation hardness was governed by yield processes in the matrix of the middle lamella. While MFA is known to be related to the wood elastic modulus [31], it is not likely to be useful for predicting wood hardness as confirmed by Konnerth *et al.* [32].

When general regressions were developed, only radial and end-grain hardness were found to be significantly related to the DC and crystallite width (Table 6), with coefficient of determinations ($R^2$) 0.13 and 0.08, for radial and end-grain hardness respectively. This outcome implies that the crystallinity can only explain 8-13% of the variation in radial and end-grain hardness. With such small improvements it is hardly practically worthwhile using XRD. When crystallinity and specific gravity were combined, the power of prediction increased; from 52% to 57% for radial hardness, from 56% to 57% for tangential hardness, and from 52% to 54% for end-grain hardness.

**Table 6**. Coefficients of determination ($R^2$) from regressions for predicting wood hardness of super teak based on 'crystallinity' at both wet and dry sites

| Wood Hardness (N) | $R^2$ | Adjusted $R^2$ | P Value |
|---|---|---|---|
| Crystallinity (Degree of Crystallinity and crystallite width) | | | |
| Radial | 0.13 | 0.09 | 0.00 |
| Tangential | 0.05 | 0.01 | 0.26 |
| End | 0.08 | 0.05 | 0.03 |
| **Specific gravity** | | | |
| Radial | 0.52 | 0.52 | 0.00 |
| Tangential | 0.56 | 0.56 | 0.00 |
| End | 0.52 | 0.52 | 0.00 |
| 'Crystallinity' and specific gravity | | | |
| Radial | 0.57 | 0.55 | 0.00 |
| Tangential | 0.57 | 0.55 | 0.00 |
| End | 0.54 | 0.52 | 0.00 |

The effort of using crystallinity to predict wood hardness only improved the prediction of hardness by 1 - 5%. Therefore, it is suggested that, wood density is singly the most useful property for non-destructive prediction even though only approximately 60% of variation in wood hardness can be explained. It is possible that there might be other wood properties not yet proposed and considered in this study which may improve the predictability. Potentially they may include extractive content [26], cellulose and lignin content. According to Ates *et al.* [33], tangential, radial and cross section Brinell hardness of *Pinus brutia* Ten. exhibited a "high" Pearson's correlation with holocellulose and lignin content ($r > 0.8$), i.e. of a similar order to that obtained with density.

## 4. Conclusion

Wood density was significantly related to wood hardness for all structural directions; the power relationship explained 52-56% of the variation of the measured hardness. By using only crystallinity factors as independent variables, degree of crystallinity (DC) and crystallite width were found as parameters affecting wood hardness. DC was positively correlated to wood hardness for all structural directions (r = 0.2; $r^2$ = 0.04), whereas crystallite width was negatively correlated to radial and end-grain hardness (r = 0.2; $r^2$ = 0.04). The power of prediction increased only by 1-5% when 'crystallinity' factors were included with density, thus the extra effort of using XRD is not warranted for non-destructive prediction of wood hardness.

**References**

[1] Mack J 1979 *Australian Methods for Mechanically Testing Small Clear Specimens of Timber, Australia* (Commonwealth Scientific and Industrial Research Organization (CSIRO))
[2] Li L, Wang X, Wang L, and Allison RB 2012 Acoustic tomography in relation to 2D ultrasonic velocity and hardness mappings *Wood Sci. Technol.* **46** 551-61
[3] Tze W, Wang S, Rials T, Pharr G, and Kelley S 2017 Nanoindentation of wood cell walls: Continuous stiffness and hardness measurements *Composites* **38**
[4] Muñoz F, Valenzuela P, and Gacitúa W 2012 *Eucalyptus nitens*: Nanomechanical properties of bark and wood fibers *Appl. Phys. A* **108**
[5] Vincent M, Tong Q, Terziev N, Daniel G, Bustos C, Escobar W, and Duchesne I 2014 A comparison of nanoindentation cell wall hardness and Brinell wood hardness in jack pine (*Pinus banksiana* Lamb.) *Wood Sci Technol* **48**
[6] Li L, Wang X, Wang L, and Allison R 2012 Acoustic tomography in relation to 2D ultrasonic velocity and hardness mappings *Wood Sci. Technol.* **46**
[7] Bowyer J, Shmulsky R, and Haygreen J 2003 *Forest Products and Wood Science an Introduction,* (Iowa: Blackwell Publishing Company)
[8] Pereira H, Graca J, and Rodrigues J 2003 Wood chemistry in relation to quality *Wood Quality and Its Biological Basis* ed ed. Barnett J and Jeronimidis G (Oxford UK and Carlton Australia: Blackwell Publishing Ltd.)
[9] Lee C 1961 Crystallinity of wood cellulose fibers studies by X-ray methods *For. Prod. J.* **11** 108-12
[10] Jiang Z, Yang Z, So C, and Hse C 2007 Rapid prediction of wood crystallinity in *Pinus elliotii* plantation wood by near-infrared spectroscopy *J. Wood Sci.* **53** *449-53*
[11] Brown HP, Panshin AJ, and Forsaith CC 1952 Textbook of wood technology Volume II *The physical, mechanical, and chemical properties of the commercial woods of the United States*
[12] Ross RJ 2010 Wood handbook: wood as an engineering material *USDA For. Serv. For. Prod. Lab. Gen. Tech. Rep. FPL-GTR-190 509 p. 1 v.* **190**
[13] Winandy JE and Rowell RM 2005 11 Chemistry of Wood Strength *Handb. Wood Chem. Wood Compos.* **303**
[14] Thomas D, Henson M, Joe B, Boyton S, and Dickson R 2009 Review of growth and wood quality of plantation-grown *Eucalyptus dunnii* Maiden *Aust. For.* **72** 3-11
[15] National Wood Flooring Association 1994 *Wood species used in wood flooring, Technical Publication*
[16] Rautkari L, Laine K, Kutnar A, Medved S, and Hughes M 2013 Hardness and density profile of surface densified and thermally modified Scots pine in relation to degree of densification *J. Mater. Sci.* **48** 2370-5
[17] Australian Timber Flooring Association 2010 *Timber Flooring Hardness* (Queensland, Australia: Australian Timber Flooring Association Ltd).
[18] Raymond C and Muneri A 2001 Nondestructive sampling of *Eucalyptus globulus* and *E. nitens* for wood properties, I. Basic density *Wood Sci. Technol.* **35**
[19] Kingston R and Risdon C 1961 *Shrinkage and Density of Australian and other South-West Pacific Wood* (Australia: Division of Forest Product.)
[20] Green D, Begel M, and Nelson W 2006 *Janka hardness using non standard specimens.* (Wisconsin: Forest Product Laboratory)

[21] Peura M, Muller M, Vainio U, Saren M, Saranpaa P, and Serimaa R 2008 X-ray microdiffraction reveals the orientation of cellulose microfibrils and the size of cellulose crystallites in single Norway spruce tracheids *Trees* **22**
[22] Abe K and Yamamoto H 2005 Mechanical interaction between cellulose microfibril and matrix substance in wood cell wall determined by X-ray diffraction *J. Wood Sci.* **51**
[23] Stuart SA and Evans R 1994 *X-ray diffraction estimation of the microfibril angle variation in eucalypt increment cores* (CRC Publications Committee)
[24] Forest Products Laboratory 1955 *Wood Handbook: Basic Information on Wood as a Material of Construction with Data for Its Use in Design and Specification* (US Government Printing Office)
[25] Newlin JA and Wilson TRC 1919 *The relation of the shrinkage and strength properties of wood to its specific gravity* (US Department of Agriculture)
[26] Panshin AJ and Zeeuw C de 1964 Textbook of wood technology Vol. I *Structure, identification, uses, and properties of the commercial woods of the United States*
[27] Lee CL 1960 Studies on the crystallinity of wood cellulose fibres by X-ray methods
[28] Damayanti R, Ilic J, Ozarska B, Pari G, and Vinden P 2015 Crystallinity as a non-destructive indicator of wood hardness at standing trees *Applied Mechanics and Materials* **771** 232-41 (Trans Tech Publ)
[29] Damayanti R, Brodie G, Ilic J, Ozarska B, Pari G, and Vinden P Measures of crystallinity as a non-destructive indicator of wood hardness: Study on young fast grown plantation teak planted in dry and wet sites
[30] Gindl W, Gupta H S, Schöberl T, Lichtenegger HC, and Fratzl P 2004 Mechanical properties of spruce wood cell walls by nanoindentation *Appl. Phys. A* **79** 2069-73
[31] Bendtsen B A and Senft J 2007 Mechanical and anatomical properties in individual growth rings of plantation-grown eastern cottonwood and loblolly pine *Wood Fiber Sci.* **18** 23-38
[32] Konnerth J, Gierlinger N, Keckes J, and Gindl W 2009 Actual versus apparent within cell wall variability of nanoindentation results from wood cell walls related to cellulose microfibril angle *J. Mater. Sci.* **44** 4399-406
[33] Ates S, Akyildiz MH, and Ozdemir H 2009 Effects of heat treatment on calabrian pine (*Pinus brutia* Ten.) wood *BioResources* **4** 1032-43

**Acknowledgement**

Thanks to ACIAR (Australian Centre for International Agricultural Research) for providing funding of this study. Thank you are also presented to the technicians: Didik A. Sudika, Tutiana, Romi, and Ida who have helped in the data collection. Ir. Hariyono Soeroso, MS. (RIP) and Bpk. Muhammad Rafik (Unit Bagi Hasil KPWN Ministry of Forestry, Indonesia) are gratefully acknowledged for helping in providing the experimental materials – Jati Utama Nasional. All authors are equally contributed to the paper writing.

IOP Conf. Series: Materials Science and Engineering **935** (2020) 012027 doi:10.1088/1757-899X/935/1/012027

# Utilization of wood vinegar as plant based insecticide in mulberry (*Morus* sp)

**R Dewi[1*], N Hastuti[2], Y Nuraeni[1]**

[1]Forest Research and Development Center, Bogor, 16610, Indonesia
[2]Forest Product Research and Development Center, Bogor, 16610, Indonesia

[*]Email: rositadee@gmail.com

**Abstract**. Successful maintenance of silkworms (*Bombyx mori*) depends on the supply of mulberry leaf feed (*Morus* sp). In practice, mulberry plants are not safe from pests and diseases. In the early rainy season, mulberry plants are easily attacked by whitefly (*Trialeuroides vaporiorum* and *Bemissia tabaci*). Mulberry leaves that are attacked by whitefly can no longer be given to silkworms because the nutrients are reduced. The use of plant-based insecticides in the resolvent of whitefly pests has not been widely practiced. Wood vinegar is an organic material derived from wood waste that can be used as a plant-based insecticide. The objectives of this study are to determine the percentage of whitefly attacks on mulberry plants and the application of plant-based insecticides on mulberry plants towards whitefly attacks reduction. The study was employed by field observation for 12 weeks. The results showed that the percentage of whitefly attack was 33.34%. Application of wood vinegar as insecticide with a dose of 2.5% (v/v) increased the pest resistance and induced the growth of mulberry plants.

## 1. Introduction

Successful maintenance of silkworms (*Bombyx mori*) depends on the supply of mulberry leaf feed (*Morus* sp) [1]. Mulberry leaves are the main feed for silkworms [2]. Therefore, the availability of healthy mulberry leaves is an important factor in the activities of natural silk management. The quality and quantity of mulberry leaves affect both of the health of silkworms and the silk cocoon's quality [3] [4]. Mulberry plants were not free from various types of pests, including pests that attack the leaves [5]. The consequence of pest infestation are highly affect the leaves quality and health of silkworms. [6] Futhermore, these impacts are absolutely affecting both of the cocoon production and the silk threads.

One challenge in the cultivation of mulberry plants is the interference from plant-disturbing organisms such as whitefly. Initially, the whitefly only attack vegetables and crops, but now it is widespread in attacking mulberry plants [6]. The whitefly on mulberry plants occurs often evenly in almost all areas of natural silk management both in Java and Sumatra [5]. The whitefly occurs at the beginning of the rainy season until the middle of the dry season [6].

Whitefly is an insect from the family Aleyrodidae, sub order Sternorryhyncha, Hemiptera [7] reported that several species of whitefly infestation become plant pests in Indonesia, such: *Aleurolobus barodensis, Aleurolobus citripedus, Aleurolobus woglumi, Neomaskellia andropogonis, Aleurocanthus sp, Bemisia tabacci, Aleyrodicus dispersus, Trialeurodes vaporor sporum*. 4 species of whitefly found that attacked plants in Bogor, Cianjur and Sukabumi areas, namely: *Bemisia tabacci, Aleyrodicus dispersus, Trialeurodes vaporariorum* and *Dialeurodes* sp [8] The Whitefly that attacked mulberry in the Dramaga, Bogor is only one species *Trialeurodes vaporariorum* [5].

Losses caused by whitefly attack on mulberry plants is a failure in harvesting [5]. The whitefly attack reduced crop productivity to 80% [6,9]. Mulberry leaves that have been attacked by whitefly cannot be used as feed for silkworms due to the nutrients forfeit. Mulberry's rich of water, carbohydrate and protein [10]. Recently, handling of whitefly attack still using mechanical approach and chemical insecticides [9]. However, mechanical approach requires more time, energy, and cost compared to chemical insecticides. Meanwhile, the use of chemical insecticides are considered dangerous in terms of the environmental issue and the remaining chemicals left on the leaves endanger the health of silkworms [11]. Hence, the use of natural-based insecticides is needed to resolve those issues.

Wood vinegar as a product of wood pyrolysis. Wood vinegar is liquid smoke obtained from the charcoal production [12]. It is potential to be used as natural-based insecticide due to its chemical components. Wood vinegar chemical components such as acetic acid is effective in inducing growth and prevent pest attack and plant diseases. Methanol as an accelerator of growth, phenols and their derivatives as inhibitors or prevention of pests and plant diseases. The use of wood vinegar is expected to reduce the use of chemicals that are toxic and environmentally friendly. This study aims to determine the percentage of whitefly attack on mulberry plants and the response of infected mulberry plants by applying wood vinegar as insecticides.

## 2. Materials and Method

### 2.1. Materials

Materials used in this study were 1 month mulberry crop plot, wood vinegar solution with a concentration of 2.5% (v/v), equates, ropes, gauges, sprays, cameras, and stationery. This research was conducted at field research plot in Dramaga, Bogor, West Java from January to April 2016.

### 2.2. Method

Data collection consists of primary and secondary data. Primary data were obtained through field observations of the whitefly attack intensity on mulberry plants. Secondary data were obtained from literature studies related to the research. The observed parameters were: calculation of the whitefly attack intensity and physic observation of mulberry plants which were attacked by whitefly after 2.5% wood vinegar application (leaves counting and height measurement). Plant-based insecticide was applied in 2 times/week for 3 months of observation. The intensity of whitefly attack can be calculated based on the comparison of the number of infected mulberry plants with the total number of plants in the observation plot. The formulation of the Intensity of Attack (IS) as described in formula [13]:

$$IS = Nh/Nt \times 100\% \qquad (1)$$

where, IS: Intensity of attack (%), Nh: Number of infected plant, Nt: Total number of observed plants in the plot

## 3. Results and Discussion

### 3.1. Intensity of whitefly attack on mulberry plants

Whitefly develops through the stages of eggs, nymphs, pupae, and imago. Female insects generally lay eggs on the underside of the leaf by inserting their limbs into the leaf epidermal tissue so that the eggs are located perpendicular to the leaf surface. The whitefly eggs are elliptical with a size of 0.2-0.3 mm. the white lice eggs turn pale yellow and then turn dark brown. The whitefly has nymphs with three instar stages. The white nymphs are elliptical, green in color, and on the edges of the body of the nymph are fine hairs with a thin waxy coating. Instar nymph I lasted for 3.14 ± 0.24 days, Instar nymph II lasted for 3.21 ± 0.16 days and Instar nymph III lasted for 3.14 ± 0.16 days. Imago or adult insects are small between (1-1.5 mm) in white, and their wings are clear covered with a layer of starchy wax. Adult insects are usually clustered on the underside of the leaf, if the plant is touched will usually fly like fog (Figure 1).

**Figure 1**. Whitefly attached on mulberry leaves

Based on observations in the field, the symptoms observed on mulberry plants that attacked by whitefly are preceded by the number of white lice attached to the leaves and stems of mulberry plants. Mulberry leaves begin to blacken due to necrotic spots and sooty dew on the leaves. Mulberry leaves which are attacked by whitefly are generally cannot be given as silkworm feed. Whitefly infestation on mulberry plants will disturb plant growth if the attack is not handled properly.

The whitefly attack imposes direct damage to the host plant. Imago and nymphs attack plants by sucking the leaves and stems of liquid plants and cause necrotic spots on the leaves. These necrotic patches occur because of leaf tissue and damaged leaf cells. The whitefly excretion of honey liquid is a medium for soot dew formation. The whitefly attack caused photosynthesis is inhibited due to the black sooty dew covering the stomata (Figure 2).

**Figure 2**. Whitefly attack caused leaves damage and nutrients forfeit

Results indicated that the intensity of whitefly attack on mulberry plants aged 1 month after pruning was 33.34% (Table 1). The insenty of whitefly infestation on mulberry plants is in medium category where the intensity of pest attacks about 33.34 % of amount of mulberry plants [13,14]. Handling of whitefly attack is necessary to protect mulberry plants and prevent the risk of failure in harvesting mulberry leaves. The use of organic and environmentally friendly plant-based insecticides is expected to be a solution in handling whitefly attack on mulberry plants.

**Table 1.** Intensity of whitefly attack on mulberry plants

| Age of mulberry | Number of trees | | Intensity of attack (%) |
|---|---|---|---|
| | Good | Infected | |
| 1 month after prunning | 50 | 25 | 33.34 |

*3.2. Application of plant-based insecticides (Wood Vinegar)*

Plant-based insecticides have been widely used to replace chemical pesticides because they are considered more environmentally friendly. Wood vinegar is a plant-based insecticide that can be used to treat pests and plant diseases. Several benefits of wood vinegar as food preservatives, pest and plant diseases, organic liquid fertilizer, plant fertilizers, disinfectants, and microorganism inhibitors, and prevention of fungi and bacteria [12]. Wood vinegar used in this study was dissolved in water at a dose of 1:40 (2.5% v/v).

Solution of wood vinegar with a concentration of 2.5% was applied on all parts of the mulberry plants attacked by whitefly. Wood vinegar application was employed 2 times/week for 12 weeks of observations. Results showed that plant growth improved after wood vinegar application.

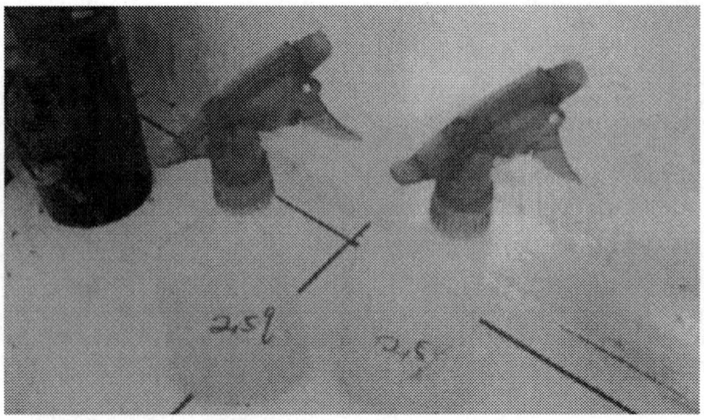

**Figure 3**. Wood vinegar with concentration 2.5% (v/v)

In the observation, after plant-based insecticide was applied on the attacked mulberry plants and physically the growth of plants gradually improved and became healthy again. After 12 weeks applied of wood vinegar with concentration 2.5 % (v/v) 25 mulberry plants that have been healthy have an average plant height of 61.24 cm and an average number of leaves of 40.92. (Figure 4 and 5).

Application of wood vinegar or pyroligneous acid made an impact on plant growth and health. Pyroligneous acid stimulate cell growth and acids as catalyst for growth microbes and enzymes activation. It is the key of biochemical and physiological processes in plants such as photosynthesis, nutrient absorption and cell growth [15].

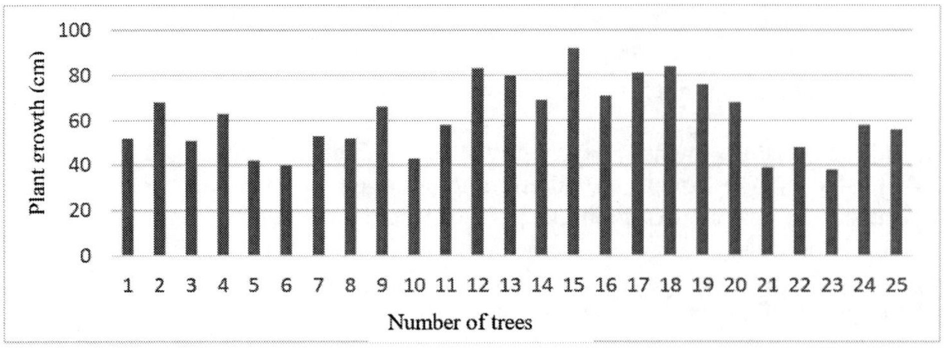


**Figure 4.** Height of mulberry after 12 weeks applied of wood vinegar with concentration 2.5 % (v/v)

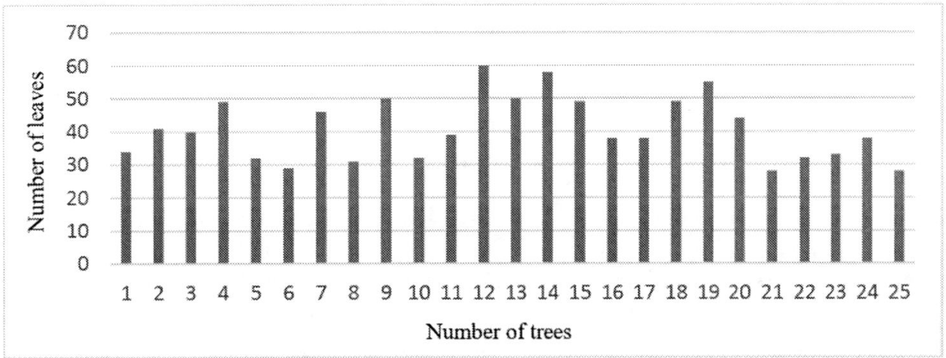


**Figure 5.** Number of mulberry leaves after 12 weeks applied of wood vinegar with concentration 2.5 % (v/v)

Whitefly is important pests in many crops worldwide [16]. Using synthetic chemical is the most common approach for handling whitefly infestation. Insecticides with active ingredients like profenofos, lamdasihalotrin, and tiamektosan are effectively suppress the whitefly population [14] but application of synthetic chemicals has negative impact on environment, ecological system and the health of all life form [15]. Wood vinegar from plants have been used in agriculture in an attempt to reduce the dependence on the use of chemicals [17]. The use of wood vinegar as a natural insecticide is very potential to be used for handling whitefly infestation. Mulberry leaf crop failure can be prevented, so mulberry leaves can be given to silkworms.

## 4. Conclusion

The application of wood vinegar on mulberry plants which infected by whitefly attack was effective at dose of 2.5%. Intensity of whitefly attack in the plot was 33.34% after applying of wood vinegar as natural-based insecticide. The results of this study induced the use of natural-based insecticide as a potential agent in whitefly attack handling.

## References

[1] Minarningsih, Dewi R, and Pudjiono S 2016 Pertumbuhan awal murbei hibrid baru pada jarak tanam yang berbeda *Prosiding Seminar Nasional Hasil Hasil PPM IPB* 40-53
[2] Andadari L, Minarningsih, and Dewi R 2017 Pengaruh jenis murbei terhadap produktivitas kokon dua hibrid ulat sutera *Bombyx mori* L *Widyariset* **3**(2) 119-30
[3] Andadari L, Dewi R, and Pudjiono S 2016 Uji adaptasi lima tanaman murbei hibrid baru untuk meningkatkan produktivitas persuteraan alam *Widyariset* **2**(2) 96-105
[4] Andadari L, Pudjiono S, Suwandi, and Rahmawati T 2013 *Budidaya Murbei dan Ulat Sutera* (Bogor: FORDA Press)
[5] Andadari L 2009 Identifikasi parasitoid dan predator kutu kebul pada tanaman murbei (*Morus* sp.) [Tesis] Sekolah Pascasarjana IPB Bogor
[6] Dewi R, Andadari L, and Maharani KE 2019 Tinjauan bioekologi dan pengendalian hama kutu kebul (*Bemissia tabaci* Genn.) *Prosiding Seminar Nasional Perhimpunan Entomologi Indonesia* pp. 40-5
[7] Kalshoven LGC 1981 *Pest of Crop in Indonesia* (Jakarta: PT Ichtiar Baru Van Houve)
[8] Yuliani 2002 Keragaman spesies kutu kebul (Hemiptera: Aleyrodidae) pada tanaman cabai, tomat dan kedelai di Bogor [Tesis] Sekolah Pasca Sarjana IPB Bogor
[9] Inayati A and Marwoto 2015 Kultur teknis sebagai dasar pengendalian hama kutu kebul *Bemissia tabacci* Genn pada tanaman kedelai *Buletin Palawija* **29**14-25

[10] Dewi R and Hastuti N 2017 Potensi senyawa kimia daun murbei (*Morus alba* var. kanva-2) sebagai anti bakteri dan anti oksidan *Prosiding Seminar Nasional Biodiversitas untuk Kesehatan dan Keberlanjutan Kualitas Ekosistem* pp. 205-14

[11] Nuraeni Y, Anggraeni I, and Dewi R 2017 Jenis-jenis mikroorganisme agensia hayati sebagai salah satu faktor pengendalian Organisme Pengganggu Tanaman (OPT) secara terpadu *Prosiding Seminar Nasional Biodiversitas untuk Kesehatan dan Keberlanjutan Kualitas Ekosistem* pp. 259-65

[12] Komarayati S and Santoso E 2011 Arang dan Cuka kayu: produk hhbk untuk stimulan pertumbuhan mengkudu *Jurnal Penelitian Hasil Hutan* **29**(2)

[13] Herdiana N 2010 Potensi serangan hama tanaman jati rakyat dan upaya pengendaliannya di Rumpin, Bogor *Jurnal Penelitian Hutan Tanaman* **17**(4) 177-85

[14] Inayati A and Marwoto 2012 Pengaruh kombinasi aplikasi insektisida dan varietas unggul terhadap intensitas serangan kutu kebul dan hasil kedelai *Jurnal Penelitian Pertanian Tanaman Pangan* **31**(1) 13-21

[15] Grewal A, Abbey L, and Gurupuru L R 2018 Production, prospects and potential application of pyroligneous acid in agriculture *Journal of Analytical and Applied Pyrolysis* (135) 152-59

[16] Wang XW, Li P, and Liu SS 2017 Whitefly interactions with plants current opinion *Insect Science* **19** 70-5

[17] Mungkunkamchao T, Kesmala T, Pimratch S, Toomsan B, and Jothtyangkoon D 2013 Wood vinegar and fermented bioextracts: natural products to enhance growth and yield of tomato (*Solanum lycopersicum* L.) *Scientia Horticulturae* (154) 66-72

IOP Conf. Series: Materials Science and Engineering **935** (2020) 012028 doi:10.1088/1757-899X/935/1/012028

# Chemical components content of seven Indonesian bamboo species

**M I Maulana[1], M Marwanto[1], D S Nawawi[1], S Nikmatin[2,3], F Febrianto[1*], N H Kim[4]**

[1]Department of Forest Products, Faculty of Forestry and Environment, IPB University (Bogor Agricultural University), Bogor 16680, Indonesia
[2]Department of Physics, Faculty of Mathematics and Natural Sciences, IPB University (Bogor Agricultural University), Bogor 16680, Indonesia
[3]Surfactant and Bioenergy Research Center, IPB University (Bogor Agricultural University), Bogor, Indonesia
[4]Department of Forest Biomaterials Engineering, College of Forest and Environmental Science, Kangwon National University, Chuncheon 200-701, Republic of Korea

[*]Email: febrianto76@yahoo.com

**Abstract.** Chemical components content of seven Indonesian bamboo species were investigated. Bamboo species used in this study including Andong bamboo (*Gigantochloa pseudoarundinacea*), Tali bamboo (*G.apus*), Hitam bamboo (*G. artoviolacea*), Betung bamboo (*Dendrocalamus asper*), Sembilang bamboo (*D. giganteus*), Ampel bamboo (*Bambusa vulgaris var. vulgaris*), and Kuning bamboo (*B. vulgaris var. striata*). Holocellulose, alphacellulose, Klason lignin, acid-soluble lignin, extractives content, and pH value were determined. The highest holocellulose and alphacellulose content were found in Ampel and Hitam bamboo, respectively. Sembilang bamboo has the highest lignin content. Generally, Ampel bamboo has the highest extractives content. The pH value of the seven bamboo species showed that all bamboo species tend to be acidic.

## 1. Introduction

Bamboo is a potential biomass resources as an alternative to wood. As a lignocellulosic plant, bamboo was grown with high species diversity, and its culm has similar characteristics to wood. In Indonesia, there are 160 bamboo species in which 122 species are endemic [1]. Bamboo has a short growing cycle and is easy to grow in various types of soil [2], so it has high productivity and can be classified as fast growing species. In 2018, Indonesia produced up to 20 million bamboo culms [3].

Indonesian bamboo has been widely used by the community. Bamboo can be used as a raw material for musical instruments, furniture, traditional handicrafts, and lightweight construction [4]. Some research also showed the use of bamboo as a raw material for biomaterials and composites such as ply-bamboo [5,6], bamboo oriented strand board [2,7,8], and bamboo zephyr boards [9,10]. Bamboo is also potentially used as raw material for pulp [11] and nanocellulose [12].

However, among the various species of bamboo, only a few species that have been used intensively. It can be caused by the lack of information about the characteristics of some species of bamboo. In order to optimize the use of bamboo in various applications and processing, the basic properties of bamboo need to be known. Chemical properties are one of the basic properties that can affect the material properties of woody materials and their processing. The cell wall chemical components content such as cellulose and lignin will affect the yield, efficiency, and properties of the product resulting from the pulping and extraction of nanofiber processes [11,13]. Meanwhile, the acidity and extractive content in woody material influence the adhesion process in processed wood and composites [7,14,15]. Therefore, in this study, the chemical components content and acidity of seven bamboo species in Indonesia were investigated.

## 2. Materials and Methods

### *2.1. Materials*

Seven bamboo species (4-5 years old) include Andong bamboo (*Gigantochloa pseudoarundinacea*), Tali bamboo (*G. apus*), Hitam bamboo (*G. atroviolacea*), Betung bamboo (*Dendrocalamus asper*), Sembilang bamboo (*D. giganteus*), Ampel bamboo (*Bambusa vulgaris Var vulgaris*), and Kuning bamboo (*B. vulgaris Var striata*) were harvested from bamboo arboretum at Center for Research and Development (P3) Biomaterials, Indonesian Institute of Sciences, Bogor, Indonesia. Bamboo culms were cut at the second node above the ground and the branched top parts were removed. One-third on the bottom part of the remaining bamboo culms were chopped, ground, and screened into 40-60 mesh size powders. All chemicals used in this study were analytical grade.

### *2.2. Methods*

Holocellulose and alphacellulose content determination were referred to the Browning methods [16]. Determination of lignin content referred to Dence [17]. Extractives soluble in water were determined according to TAPPI T-207 cm-99 [18]. Determination of Ethanol-benzene soluble extractive referred to ASTM D-1107-96 [19]. The solubility of bamboo in 1% NaOH solution were determined according to TAPPI T-212 om-02 [20]. The pH value was determined according to SNI 06-6989.11-2004 [21].

### *2.3. Data analysis*

Four replicates were carried out for each bamboo species. Analysis of Variance (ANOVA) was used to test the influence of bamboo species on the chemical component and pH value. If the results show significant influence, Duncan Multiple Ranged Test was carried out to determine the significantly different values.

## 3. Results and Discussion

### *3.1. Holocellulose*

Holocellulose is the total fraction of polysaccharides consisting of alphacellulose and hemicellulose. Alphacellulose refers to cellulose with a degree of polymerization above 200 and does not dissolve in a strong base solution. Holocellulose, alphacellulose, and hemicellulose content of seven bamboo species were 69.41-82.77%, 34.60-53.12%, and 24.30-48.17% (Figure 1). Andong bamboo has the lowest holocellulose content while Ampel bamboo has the highest holocellulose content. The highest and the lowest alphacellulose were found in Hitam bamboo and Andong bamboo, respectively. Analysis of variance showed that bamboo species significantly affected holocellulose, alphacellulose, and hemicellulose content.

All bamboo species in this study showed high holocellulose content. Meanwhile, the alphacellulose content of the seven bamboo species belongs to different classes. The alphacellulose content of Ampel bamboo and Andong bamboo were classified low (<40%). In comparison, Sembilang bamboo and Kuning bamboo have moderate alphacellulose content (40-45%), and others have high alphacellulose content (>45%) [22]. Li *et al.* [23] reported that five-year-old *Phyllostachys pubescens* planted in Louisiana, USA, had a holocellulose content of about 69.94% and alphacellulose of about 46.08%. High holocellulose and alphacellulose content were also found in *Gigantochloa scortechinii* from Kedah, Malaysia, with values of 81.4% and 55.2%, respectively [24]. High holocellulose content could be produced high pulping yield [25]. Ampel bamboo has high holocellulose content but low alphacellulose content. It shows that Ampel bamboo has very high hemicellulose content. Bamboo with high hemicellulose content is very susceptible to organism attacks. Febrianto *et al.* [26] reported that Ampel bamboo has low durability (class IV) against termites and powder post beetle.

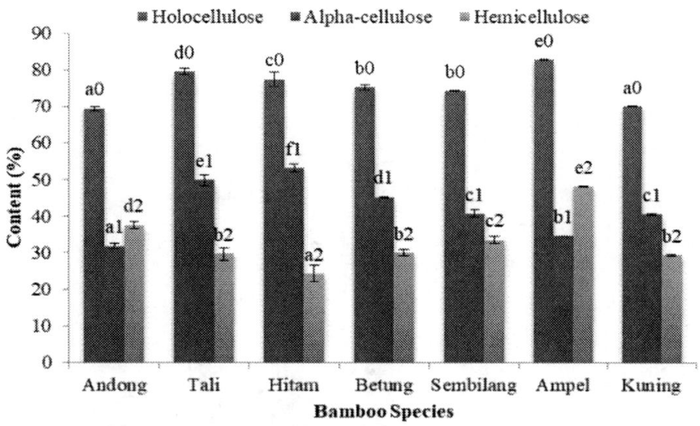


**Figure 1.** Holocellulose, alphacellulose, and hemicellulose content of seven Indonesian bamboo species

*3.2. Lignin*

In this study, lignin is expressed as Klason lignin and acid-soluble lignin. Klason lignin is a residue resulting from hydrolysis and condensation reaction using 72% and 3% sulfuric acid, while acid-soluble lignin is the fraction of soluble lignin during the reaction. Klason lignin and acid-soluble lignin of seven bamboo species were 21.86-30.43% and 0.94-2.37%, respectively (Figure 2). Sembilang bamboo has the highest lignin content, while Tali bamboo has the lowest lignin content. Analysis of variance showed that bamboo species has significant influence on lignin content.

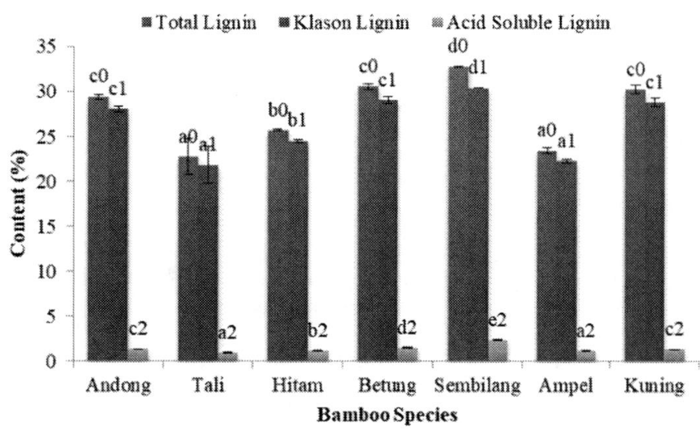


**Figure 2.** Lignin content of seven Indonesian bamboo species

The seven bamboo species have different lignin content values. Differences in the value of acid-soluble lignin can be caused by differences in the ratio of the lignin monomers. Bamboo Lignin is classified as syringyl-guaiacyl-*p* hidroxyphenyl lignin [27]. Nawawi *et al.* [28] reported that there is a positive correlation between the acid-soluble lignin values and the syringyl-guaiacyl ratio (S/G ratio) in the bamboo lignin. Bamboo with high lignin content has the potential to be used as bio energy raw materials. Lignin content has a positive correlation with calorific value [29]. However, in pulping and cellulose isolation processes, high lignin content can increase the energy and chemicals needed [30].

### *3.3. Extractives*

Extractives soluble in water consist of tannin, gum, low molecular weight carbohydrates, dyes, and starch. Meanwhile, ethanol-benzene solvent dissolves wax, fat, oil, and a little gum. The cold water and hot water-soluble extractives content of seven bamboo species were 3.30-17.34% and 5.33-23.34%, respectively (Figure 3). The ethanol-benzene soluble extractive of seven bamboo species were 4.45-10.15% (Figure 4). The highest water and ethanol-benzene soluble extractive were found in the Ampel bamboo and the lowest water and ethanol-benzene soluble extractive were found in Tali bamboo and Kuning bamboo, respectively. Analysis of variance showed that bamboo species significantly affected water and ethanol-benzene soluble extractive.

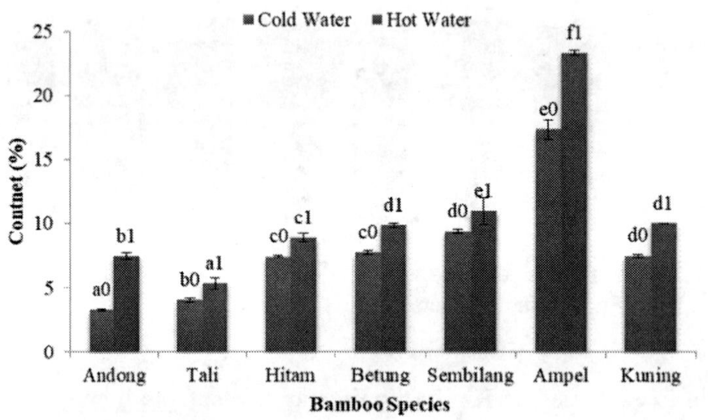


**Figure 3.** Water soluble extractives content of seven Indonesian bamboo species

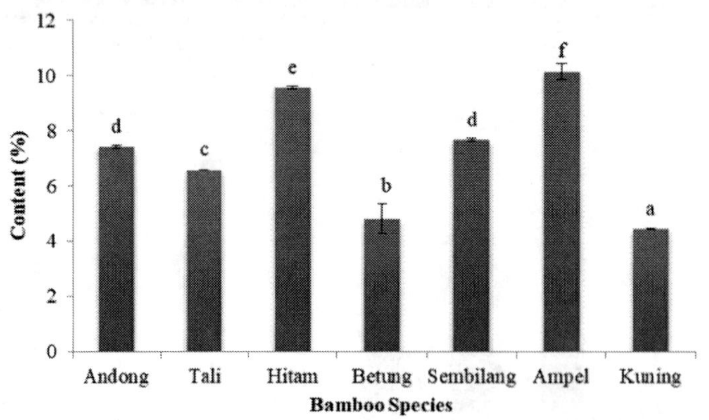


**Figure 4.** Ethanol-benzene soluble extractives content of seven Indonesian bamboo species

A solution of 1% NaOH dissolves extractive substances, some part of lignin, and low molecular weight hemicellulose. The degree of fungal decay or degradation of wood can be indicated by solubility in 1% NaOH solution [20]. Solubility in 1% NaOH solution of seven bamboo species were 21.64-30.99% (Figure 5). Tali bamboo has the lowest value and Sembilang bamboo has the highest value of solubility in 1% NaOH solution. Analysis of variance showed that bamboo species significantly influence the solubility in 1% NaOH.

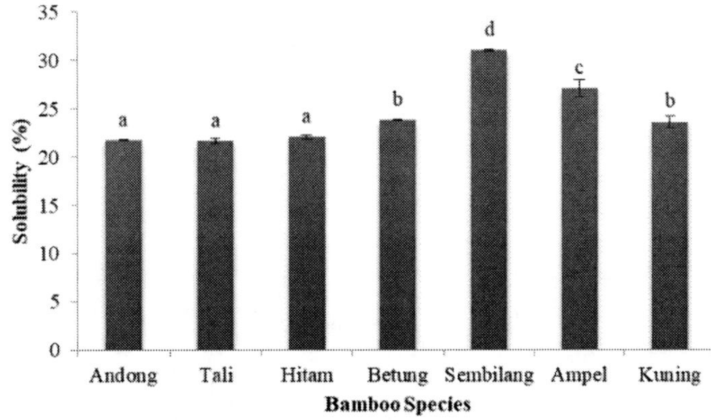


**Figure 5.** Solubility in 1% NaOH solution of seven Indonesian bamboo species

Bamboo extractive content varies between species. In the gluing process of composite products such as particle boards and plywood, high extractive content can inhibit adhesive penetration, resulting in low mechanical strength [31]. In the pulping process, the presence of extractives in the form of wax and fat will reduce the strength of bonds between fibers, increase consumption of alkali, and slow down delignification. The lower extractives content in the material, the easier chemicals to penetrate the materials, so that pulping is more optimum [11].

### *3.4. Acidity*

Bamboo acidity is determined based on the pH value of the bamboo. The pH value of seven bamboo species were 4.94-6.49 (Figure 6). Sembilang bamboo has the highest pH value while Tali bamboo has the lowest pH value. Analysis of variance showed that bamboo species affected the pH value. In general, bamboo has an acidic pH. The pH value of bamboo which tends to be acidic could be influenced by free acids and acid groups such as acetyl and other acids from hemicellulose [32]. The pH value of bamboo will affect the curing of the adhesive on certain adhesive types. The curing of urea formaldehyde adhesives will be optimum under acidic conditions [33]. Meanwhile, the curing of phenol formaldehyde adhesives is more optimum starting at pH 8 [34].

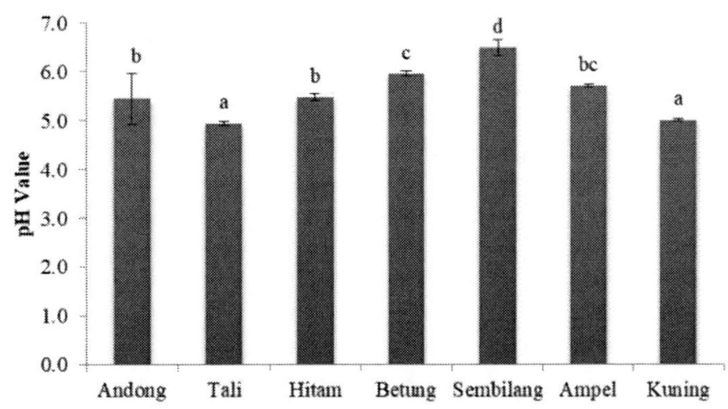


**Figure 6.** pH value of seven Indonesian bamboo species

## 4. Conclusion

The chemical components content of bamboo was different among species. The highest holocellulose and alphacellulose content were found in Ampel and Hitam bamboo, respectively. Ampel bamboo has the highest hemicellulose content. Sembilang bamboo has the highest lignin content. Generally, Ampel bamboo has the highest extractives content. The pH value of the seven types of bamboo shows that all bamboo species tend to be acidic.

## References

[1] Widjaya EA 2012 The utilization of bamboo: at present and for the future *Proc. of International Seminar Strategies and Challenges on Bamboo and Potential Non Timber Forest Products*
[2] Febrianto F, Sumardi I, Hidayat W, and Maulana S 2017 *Papan Untai Bambu Berarah Material Unggul Untuk Komponen Bangunan Struktur* (Bogor: IPB Press)
[3] [BPS] Badan Pusat Statistik 2018 *Statistik Produksi Kehutanan.* (Jakarta: Badan Pusat Statistik Indonesia)
[4] Sudarnadi H 1996 *Tumbuhan Monokotil* (Jakarta: Penebar Swadaya)
[5] Jasni J and Sulastiningsih IM 2018 The prevention of the powder post beetle dinoderus minutus farb. infestation on plybamboo and plywood *J Trop. Wood Sci. Technol.* **3**(2) 68-72.
[6] Suryana J, Massijaya MY, Hadi YS, and Hermawan D 2017 Fundamental properties of ply bamboo. *J Trop. Wood Sci. Technol.* **9**(2) 153-65
[7] Fatrawana A, Maulana S, Nawawi DS, Sari RK, Hidayat W, Park SH, Febrianto F, Lee SH, and Kim NH 2019 Changes in chemical components of steam-treated betung bamboo strands and their effects on the physical and mechanical properties of bamboo-oriented strand boards *Eur. J. Wood Prod* 1-9
[8] Maulana S, Gumelar Y, Fatrawana A, Maulana MI, Hidayat W, Sumardi I, Wistara NJ, Lee SH, Kim NH, and Febrianto F 2019 Destructive and non-destructive tests of bamboo oriented strand board under various shelling ratios and resin contents *J. Korean Wood Sci.Technol.* **47**(4) 519-32
[9] Gopar M and Sudiyani Y 2018 The change of physical and mechanical properties of zephyr bamboo panel after exposed to out-door weathering *J. Trop. Wood Sci. Technol.* **2**(2) 90-4
[10] Subyakto S, Gopar M, Ismadi I, Nugroho A, Sumarno A, Widodo E, and Sudarmanto S 2017 Pembuatan dan karakterisasi komposit zephyr bambu dengan perekat kempa dingin Journal of Lignocellulose Technology **1**(1) 32-6
[11] Fatriasari W and Hermiati E 2008 Analisis morfologi serat dan sifat fisis-kimia pada enam jenis bambu sebagai bahan baku pulp dan kertas *J. Forest Prod. Sci. Technol.* **1**(2) 67-72
[12] Jang JH 2015 Nanoscopic Morphology and Physicochemical Properties Of Lignocellulose Nanofibers and Their Utilization (Chuncheon: Kangwon National University)
[13] Seo PN, Han SY, Park CW, Lee SY, Kim NH, and Lee SH 2019 Effect of alkaline peroxide treatment on the chemical compositions and characteristics of lignocellulosic nanofibrils *BioResources* **14**(1) 193-206
[14] Maulana MI, Nawawi DS, Wistara NJ, Sari RK, Nikmatin S, Maulana S., Park SH, and Febrianto F 2018 Change of chemical components content in andong bamboo due to steam treatment. *J. Trop. Wood Sci. Technol.* **16**(1) 82-90
[15] Maulana MI, Murda RA, Purusatama BD, Sari RK, Nawawi DS, Nikmatin S, Hidayat W, Lee SH, Febrianto F, and Kim NH 2020 *J. Korean Wood Sci. Technol.*
[16] Browning BL 1967 *Methods of Wood Chemistry Vol. I* (New York: Wiley)
[17] Dence CW 1992 *The Determination of Lignin in Methods in Lignin Chemistry* Ed. Lin SY and Dence CW (Berlin: Springer-Verlag) pp. 33-61
[18] Technical Association of the Pulp and Paper Industry 1999 T-207 *Water Solubility of Wood and Pulp* (Atlanta: TAPPI)
[19] [ASTM] American Society for Testing and Materials *Standard Test Method for Ethanol-Toluene Solubility of Wood D1107-96 2013* (West Conshohocken: ASTM International)

[20] Technical Association of the Pulp and Paper Industry 2002 T-212 *One Percent Sodium Hydroxide Solibility of Wood and Pulp* (Atlanta: TAPPI)
[21] [BSN] Badan Standardisasi Nasional 2004 *pH meter SNI 06-6989.11-2004* (Jakarta: Badan Standardisasi Nasional)
[22] Departemen Pertanian 1976 Vademecum Kehutanan Indonesia (Jakarta: Departemen Pertanian)
[23] Li XB, Shupe TF, Peter GF, Hse CY, and Eberhardt TL 2007 Chemical changes with maturation of the bamboo species *Phyllostachys pubescens Journal of Tropical Forest Science* 6-12
[24] Salim R, Wahab R, and Ashaari Z 2008 Effect of oil heat treatment on chemical constituents of Semantan bamboo (*Gigantochloa scortechinii* Gamble) *Journal of Sustainable Development* **1**(2) 91-8
[25] Pasaribu G, Sipayung B, and Pari G 2007 Analisis komponen kimia empat jenis kayu asal Sumatera Utara *JPHH* **25**(4) 1–11
[26] Febrianto F, Gumilang A, Maulana S, Busyra I, and Purwantiningsih A 2014 Natural durability of five bamboo species against termite and powder post beetle *J. Tropical Wood Sci. Technol.* **12**(2) 146-56
[27] Sjostrom E 1991 *Wood Chemistry: Fundamentals and Applications* (San Diego: Academic Press)
[28] Nawawi DS, Sari RK, Wistara NJ, Fatrawana A, Astuti P, and Syafii W 2019 Lignin characteristic of four bamboo species *J. Tropical Wood Sci. Technol.* **17**(1) 1-7
[29] White RH 1987 Effect of lignin content and extractives on the higher heating value of wood *J. Wood Fiber Sci.* **19**(4) 446-52
[30] Mutia T, Risdianto H, Sugesty S, Hardiani H, and Kardiansyah T 2017 Serat dan pulp bambu Tali (*Gigantochloa apus*) untuk papan serat *Arena Tekstil* **31**(2)
[31] Maloney TM 1993 *Modern Particleboard and Dry Process Fibreboard Manufacturing* (Madison: Forest Products Society)
[32] Iswanto AH, Sucipto T, and Febrianto F 2011 Keasaman dan kapasitas penyangga beberapa jenis kayu tropis *J. Forest Prod. Sci. Technol.* **4**(1) 22-5
[33] Nawawi DS, Priadi T, and Murwentiano B 2017 The change of wood acidity during drying process *J. Tropical Wood Sci Technol.* **10**(2) 195-200
[34] Pizzi A 1983 *Wood Adhesives* (New York: Marcel Dekker INC)

**Acknowledgement**
We sincerely acknowledge the Deputy of Strengthening for Research and Development, Ministry of Research and Technology/National Research and Innovation Agency, Republic of Indonesia for the financial support through PMDSU (No. 4122/IT3.L1/PN/2020) and PDUPT (No. 4033/ IT3.L1/PN/2020) Grant scheme. This study was also supported by the Science and Technology Support Program through the National Research Foundation of Korea (NRF) funded by the Ministry of Science and ICT (MSIT) (NRF-2019K1A3A9A01000018) and the Basic Science Research Program through NRF funded by the Ministry of Education (No. 2018R1A6A1A03025582).

IOP Conf. Series: Materials Science and Engineering **935** (2020) 012029 doi:10.1088/1757-899X/935/1/012029

# Quality of rattan board with lamination process using tannin-based adhesives

**A Santoso*** **and R Pari**

Forest Products Research and Development Center, Ministry of Environment and Forestry, 16610 Bogor, Indonesia

*E-mail: profadisantoso@gmail.com

**Abstract.** The use of rattan with a diameter of more than 30 mm as rattan laminated board has prospective opportunities in the market. This paper presents research results on the quality of the *Calamus zollingeri* rattan board glued using tannin adhesive from *Acacia mangium* bark extract. Rattan boards were laminated with 3 glue spread (100, 150, and 200 g/m$^2$ surface) and 2 pressing temperature (60°C and 100°C). The quality of gluing was analyzed by testing the bonding strength and formaldehyde emission. Results showed that the best quality of rattan laminated board was observed in 150 g/m$^2$ surface glue spread and 60°C pressing temperature application. The formaldehyde emission of products was still within safe level comply with JAS 234-2003 standards.

## 1. Introduction

Rattan is one of the substitutions of furniture made from wood and is demanded by the market [1,2]. In addition, rattan is one of the trade commodities of non-timber forest products with high export value, and rattan furniture producing regions in Indonesia are Sumatra, West Java, East Java, Kalimantan, Sulawesi [3]. Rattan used by craftsmen in production is with a diameter of less than 30 mm, while rattan with a diameter of more than 30 mm is not utilized [1].

Development of gluing technology provides alternative for rattan-based products into composite products such as rattan boards [4]. Commonly used adhesives for composite wood products are synthetic non-renewable resources which produce formaldehyde emission [5]. The effort of using organic adhesives has been developed to reduce these impacts. One source of organic adhesive is bark waste that contains tannin and can be formulated as an adhesive [5,6]. The bark of the Acacia family such as acacia (*Acacia decurrens* Willd.) and mangium (*Acacia mangium* Willd.) contain many phenolic compounds which can be used as substitutes for resorcinol or phenol [6-8].

Tannin adhesives from mangium bark have been used in improving oil palm trunks [5] and laminated composite panels [9,10] quality as well as gluing rattan boards [4]. This research aims to obtain information on the quality of rattan laminated boards using mangium tannin from bark extract. The data can be determined by the appropriate treatment of pressing temperature and glue spread to produce rattan laminated board.

## 2. Materials and Methods

The materials used for this study were rattan *Calamus zollingeri* Bec. from Palu, Central Sulawesi, tannin-based adhesives, and tannins from bark extract of *Acacia mangium*. The equipment used were rattan split machine, pickling tub, measuring glass, scales, and universal testing machine.

The tannin adhesive is made from mangium bark extracted with a water : bark ratio of 1:3 (w/w) which is melted down macerated for 3 hours at 80°C as described in [5]. Dried rattan with 10–15% moisture content is used to make a five-layer rattan laminated board sized 1.5 x 7.5 x 90 cm which compressed for 3 hours with a specific pressure of 10 kg/cm$^2$, as done by Pari *et al.* [4].

The characteristics of tannin-based adhesives were analyzed by adhesion properties according to SNI 4567 [11] such as acidity, viscosity, specific gravity, solid content, free formaldehyde and gelatinous time. The quality of the rattan laminated board was examined through testing of the dry condition test sample with three replications. The test includes bonding strength and formaldehyde emission referring to the Japan Agricultural Standard (JAS) Glued Laminated Timber JAS 234-2003 [12].

The results of the test sample were analyzed using ANOVA with 2 factors which are pressing temperature and glued spread and their interaction on the bonding properties and formaldehyde emission of laminated rattan board.

## 3. Results and Discussion

### *3.1. Characterization of tannin-based adhesives*

Visually, the adhesive from mixing tannin extract from mangium bark with formaldehyde (tannin formaldehyde) produces a dark red liquid that is close to the color of phenol formaldehyde. The tannin-based adhesive characteristics of the formulations from mangium bark extract are presented in Table 1.

**Table 1.** The fisico-chemistry characterization of tannin-based adhesives.

| Properties | Tannin formaldehyde | Phenol formaldehyde* |
|---|---|---|
| Visual test | Dark red liquid | Red blackish liquid |
| Acidity (pH) | 10.50 | 10–13 |
| Viscosity (Poise) | 12.34 | 1.3–3.0 |
| Specific gravity | 1.02 | 1.165–1.200 |
| Solid content (%) | 8.63 | 40–45 |
| Free formaldehyde (%) | 0.054 | < 1 |
| Gelatinous time (minutes) | 300 | ≥ 300 |

Remarks: * source: SNI 4567 [11]

The results showed that the parameters of viscosity, specific gravity and solid content are different with phenol formaldehyde adhesive commercially (SNI 4567). Specific gravity and solid content lower than the standard indicate that tannin adhesive has a low solid content [5]. The adhesive with high viscosity has shorter pot life and will harden more quickly than low viscosity adhesive, therefore the quality of adhesion is relatively low [13]. However, the pot life of resin tannins as showed in the results of gelatinous times value is still safe because the values > 88 minutes [14]. Tannin-based adhesive with characteristics as mentioned above, has the ability to penetrate pores of lamina material and form a bond with good adhesion [15]. Besides, tannin-based adhesive produced relatively environment-friendly, indicated by the value of free formaldehyde that is lower than the required standard value [5].

### *3.2. The evaluation of rattan laminated boards*

The application of mangium tannin adhesives in rattan laminated boards manufacturing was examined through a bonding strength testing approach. Formaldehyde emission testing was conducted to determine the levels of formaldehyde from adhesives. Table 2 shows that the treatment of pressing temperature gave significant different results, while the treatment of glue spread and their interactions gave highly significant different results on bonding strength. In addition, both the treatment and the interaction have a highly significant effect on formaldehyde emission.

**Table 2.** Analysis of variances on bonding strength and formaldehyde emission.

| Source of variation | df | F-calculation | |
|---|---|---|---|
| | | Bonding strength | Formaldehyde emission |
| Pressing temperature, A | 1 | 5.62 $^{s}$ | 705.90 $^{hs}$ |
| Glue spread, B | 2 | 308.36 $^{hs}$ | 319.64 $^{hs}$ |
| A * B | 2 | 243.53 $^{hs}$ | 107.21 $^{hs}$ |

Remarks: hs = highly significant, s = significant, df = degree of freedom.

The quality of the rattan laminated boards from the test results is presented in Table 3. The highest bonding strength value (45.52 kg/cm$^2$) was obtained in the sample with a glue spread of 150 kg/cm$^2$ surface and pressing temperature of 60°C with a formaldehyde emission value of 1.04 mg/L. The bonding strength test value does not meet the requirements of JAS 234-2003 because the value is less than 54 kg/cm$^2$, however formaldehyde emission from the sample product meet the requirements of JAS 234-2003 with the category F*–F**. The bonding strength value of the product under study is relatively better compared to the results of the study of [4], which obtained the value of bonding strength from the rattan laminated board (40.36 kg/cm$^2$) using a mangium tannin adhesive of 200 g/cm$^2$ surface with pressing temperature 100°C, and formaldehyde emission reached 2.12 mg/L. Furthermore, the results of the study of [4] also showed that the same product using 100 g/cm$^2$ surface glue spread and pressing temperature at 60°C resulted in bonding strength and formaldehyde emission of 39.27 kg/cm$^2$ and 0.54 mg/L. Based on these results, the use of tannin adhesive glue spread with more adhesive per gluing surface unit (g/m$^2$ surface) and a higher pressing temperature, does not necessarily result in better adhesion, but has the opportunity to increase formaldehyde emission.

**Table 3.** The average value of bonding strength and formaldehyde emission.

| Pressing temperature, °C | Glue spread, g/m$^2$ surface | Test parameter | |
|---|---|---|---|
| | | Bonding strength, kg/cm$^2$ | Formaldehyde emission, mg/L |
| | | $\overline{X} \pm$ Sd | $\overline{X} \pm$ Sd |
| 60 | 100 | 39.39 ± 0.271 | 0.54 ± 0.023 |
| | 150 | 45.52 ± 0.542 | 1.04 ± 0.019 |
| | 200 | 34.71 ± 0.248 | 1.14 ± 0.143 |
| 100 | 100 | 38.53 ± 0.259 | 1.03 ± 0.003 |
| | 150 | 41.47 ± 0.115 | 2.05 ± 0.059 |
| | 200 | 41.02 ± 0.729 | 3.21 ± 0.172 |

Remarks: $\overline{X}$ = mean value; Sd = standard deviation.

This rattan laminated board is also better compared to laminated products made from types of low-density wood originating from plantations, such as sengon (*Falcataria moluccana*), jabon (*Anthocephalus cadamba*), mangium (*Acacia mangium*), sungkai (*Peronema canescens*), and rubberwood (*Hevea brasiliensis*) which use tannin as an adhesive with a glue spread of 140–170 g/cm$^2$ surface, with bonding strength of 22.19 kg/cm$^2$, 32.36 kg/cm$^2$, and 25.56 kg/cm$^2$ [9], 35.23 kg/cm$^2$, and 32.40 kg/cm$^2$ [16]. The bonding strength value of this rattan laminated board is also better than the laminated board made from bamboo andong, mayan, and betung using tannin adhesives from merbau wood powder extract (21.46–33.52 kg/cm$^2$) [17], and laminated products from a mixture of kecapi and sengon wood (18.99 kg/cm$^2$) using isocyanate adhesives (PI Bond) [18].

## 4. Conclusion

Rattan *Calamus zollingerii* with diameter more than 30 mm could be used as rattan laminated board using tannin of mangium bark extract as adhesive. The highest bonding strength values of rattan laminated board products in this research was observed in glue spread treatment of 150 g/m$^2$ surface and a pressing temperature of 60°C, with the formaldehyde emission at 1.04 mg/L this product is categorized as F** according to JAS 234-2003 standards.

## References

[1] Rachman O and Jasni 2013 *Rotan: Sumberdaya, Sifat dan Pengolahannya* (Bogor: Pusat Penelitian dan Pengembangan Keteknikan Kehutanan dan Pengolahan Hasil Hutan)
[2] Wahab R, Mokhtar N, Ghani RSM, and Sulaiman MS 2019 An overview of rattan industry status and its economic aspect in setting up rattan-based industry in Malaysia *Journal of Social Sciences and Humanities* **16**(3) 1-10
[3] Salim Z, Munadi E, Pujiati R, Budiarti FT, Laksani DD, Menanti S, and Sari NB 2017 *Bunga Rampai Info Komoditi Furnitur* ed. Salim Z and Munadi E (Jakarta: Badan Pengkajian dan Pengembangan Perdagangan)
[4] Pari R, Abdurachman, and Santoso A 2019 Keteguhan rekat dan emisi formaldehida papan lamina rotan menggunakan perekat tanin formaldehida *Jurnal Penelitian Hasil Hutan* **37**(1) 33-41
[5] Rachmawati O, Sugita P, and Santoso A 2018 Sintesis perekat tanin resorsinol formaldehida dari ekstrak kulit pohon mangium untuk peningkatan kualitas batang sawit *Jurnal Penelitian Hasil Hutan* **36**(1) 33-46
[6] Santoso A and Abdurachman 2016 Karakteristik ekstrak kulit kayu mahoni sebagai bahan perekat kayu *Jurnal Penelitian Hasil Hutan* **34**(4) 269-84
[7] Tahir PM, Halip JA, and Lee SH 2019 *Tannin-based bioresin as adhesives in Lignocellulose for Future Bioeconomy* (Elsevier) pp. 109-33
[8] Bharudin MA, Zakaria S, and Chia CH 2013 Condensed tannins from *Acacia mangium* bark: Characterization by spot tests and FTIR *AIP Conference Proceedings* **1571**(1)153-57
[9] Hendrik J, Hadi YS, Massijaya MY, Santoso A, and Pizzi A 2019 Properties of glued laminated timber made from fast-growing species with mangium tannin and phenol resorcinol formaldehyde adhesives *Journal Korean Wood Science and Technology* **47**(3) 253-64
[10] Hendrik J, Hadi YS, Massijaya MY, and Santoso A 2016 Properties of laminated composite panels made from fast-growing species glued with mangium tannin adhesive *BioResources* **11**(3) 5949-60
[11] [BSN] Badan Standardisasi Nasional 2000 *Fenol Formaldehida Cair untuk Perekat Kayu Lapis SNI 4567-2000* (Jakarta: Badan Standardisasi Nasional)
[12] [JAS] Japanese Agricultural Standard 2003 *Glued Laminated Timber JAS 234-2003* (Tokyo: Japanese Plywood Inspection Corporation)
[13] Santoso A 2003 Sintesis dan karakterisasi resin lignin resorsinol formaldehida untuk perekat kayu lamina [Dissertation] Institut Pertanian Bogor
[14] Pizzi A 1994 *Tannin Base Wood Adhesive and Preservative Chemistri and Tecnology* (New York: Marcel Dekker)
[15] Maloney T 1977 *Modern Particleboard and Dry-Process Fiberboard Manufacturing* (San Fransisco: Miller-freeman publication)
[16] Santoso A, Pari G, and Jasni 2015 Kualitas papan lamina dengan perekat resorsinol dari ekstrak limbah kayu merbau *Jurnal Penelitian Hasil Hutan* **33**(3) 253-60
[17] Santoso A, Sulastiningsih IM, Pari G, and Jasni 2016 Pemanfaatan ekstrak kayu merbau untuk perekat produk laminasi bambu *Jurnal Penelitian Hasil Hutan* **34**(2) 89-100

IOP Conf. Series: Materials Science and Engineering **935** (2020) 012029 doi:10.1088/1757-899X/935/1/012029

[18] Muthmainnah 2011 Pembuatan Cross Laminated Timber (CLT) dari kayu sengon dan kecapi [Thesis] Institut Pertanian Bogor

**Acknowledgments**
The author would like to express gratefully to Mr. Kuswandi, Mr. Aftoni, Mr. Supardiono, Mrs. Nuryani, Mr. Aris Sembiring, and Mr. Endang who has been assisting activities during the laboratory.

IOP Conf. Series: Materials Science and Engineering **935** (2020) 012030 doi:10.1088/1757-899X/935/1/012030

# Diversity and distribution of macrofungi in pine forest and mixed forest in Mount Merbabu National Park

**N Izati[1,2*], Sugiyarto[2], and T Purwoko[2]**

[1]Biodiversitas Study Club, Department of Biology, Faculty of Mathematics and Natural Sciences, Sebelas Maret University

[2]Department of Biology, Faculty of Mathematics and Natural Sciences, Sebelas Maret University

*Email: nailaizatii@gmail.com

**Abstract.** Pine forest and mixed forest in Mount Merbabu National Park (MMNP) have the different potential of macrofungi. However, the information of macrofungi in the pine and mixed forest is still lacking. The research aimed to determine the diversity and distribution of macrofungi in the pine forest and mixed forest MMNP. Retrieval of macrofungi diversity data used a plot method measuring 20 x 20 m which was divided into 16 subplots measuring 5 x 5 m. Subplots measuring 5 x 5 m was used to determine the distribution pattern of macrofungi. The results showed that there were 36 species of macrofungi found in the pine forest and 80 species in the mixed forest. The diversity of macrofungi in the pine forest belonged to moderate value with a diversity index of 2.84, whereas in the mixed forest belonged to high value with a diversity index of 3.64. There were 16 species macrofungi that can be found in the pine forest, 60 species that can be found in the mixed forest, and 20 species in both habitats. The most widespread macrofungi were the agaric type, which included Agaricales order, while the narrowest distributed macrofungi were the coral type, which include the Gomphales order. The distribution pattern of macrofungi species in the pine forest was obtained by 33 grouped species and three uniform species. The distribution pattern of macrofungi species in the mixed forest was obtained by 78 grouped species and two uniform species.

## 1. Introduction

Fungi have an important role in natural ecosystems through decomposition [1]. Fungi are actively involved in the soil biogeochemical cycles and nutrient cycling [2]. Fungi infect the roots of green plants by forming an association (mutualistic relationship) with green plants called mycorrhizal. Mycorrhizal get nutrients directly from plants, while plants depend upon mycorrhizal for healthy growth [3]. There are several species of fungi that are known as a part of parasitic associations such as *Ganoderma applanatum* (Pers.) Pat. which causes the decay of the heartwood [4].

Fungi are the second-most diverse group of organisms in the world after arthropods [5]. Fungi biodiversity worldwide has been estimated at 1,5 million [6]. However, only about 100,000 species of fungi have been identified [7]. Fungi in Indonesia has been estimated to be around 80,000 species consisting of 64,000 species of microfungi and 16,000 species of macrofungi. Macrofungi that are known in Indonesia are only around 1,200 species, consisting of 864 species of the Basidiomycota and 336 species of the Ascomycota [8].

The environmental factors such as biotic and abiotic factors influencing the macrofungi growth. The interaction from the host plant is one of the effects of the biotic factors on macrofungi growth. The suitable substrate that has nutrients is used by macrofungi to be a place to grow and develop. Substrates of macrofungi can be soil, litter, weathered wood, and host plant [9]. This makes the habitat type effects the growth of macrofungi in an ecosystem [10]. Abiotic factors such as temperature, humidity, light intensity, and pH which inside the ecosystem cause certain environmental conditions. In particular, macrofungi can grow nicely in humid environmental conditions such as mountain forests.

Mount Merbabu National Park (MMNP) is a mountain forest area with good conditions in Central Java Province. The humidity of MMNP is 90% [11]. Temperature conditions in this region around 15-25°C with 2000-3000 mm of rainfall each year. The intensity of the light entering the region is 10% for the shaded area and 15-25% for the unshaded area [12]. High rainfall and low light intensity cause the MMNP region to become moist and potentially found a variety of macrofungi species.

Pine forests and mixed forests are two types of habitats that can be found in MMNP. Both types of habitat have differences, as seen from the constituent species of vegetation and environmental conditions. Pine and mixed forest MMNP have different potential macrofungi diversity and distribution. However, until now there has been no research on the diversity and distribution of macrofungi in the pine forests and mixed forest MMNP. Research on the diversity of macrofungi was only carried out by Munandar [13] in the southern slope forest of MMNP, which found 23 species of wood-decay fungus. The lack of research on macrofungi has the potential of many macrofungi species that has not been found, especially in the pine forests and mixed forests. Therefore, research needs to be done on the diversity and distribution of macrofungi in the pine forests and mixed forest MMNP.

## 2. Materials and Methods

### *2.1 Location and time research*

The research was conducted in the pine forest and mixed forest MMNP on January 2020 which coincide with the rainy season. The area is located at the Pakis and Kopeng National Park Management Resort (NPMR) for pine forest areas and the Selo NPMR for mixed forest areas (Table 1). The map of the research location is shown in Figure 1.

**Table 1.** Location of the macrofungi sampling units

| Type of habitat | Location of the sampling units | | |
|---|---|---|---|
| | NPMR | Plot | Altitudes |
| Pine forest | Pakis | Goa Slandak (G) | 900 masl |
| | | Jambewangi (J) | 1200 masl |
| | Kopeng | Umbul Songo (U) | 1400 masl |
| Mixed forest | Selo | Hiking trails (A) | 1800 masl |
| | | Hiking trails (B) | 1900 masl |
| | | Hiking trails (C) | 2100 masl |

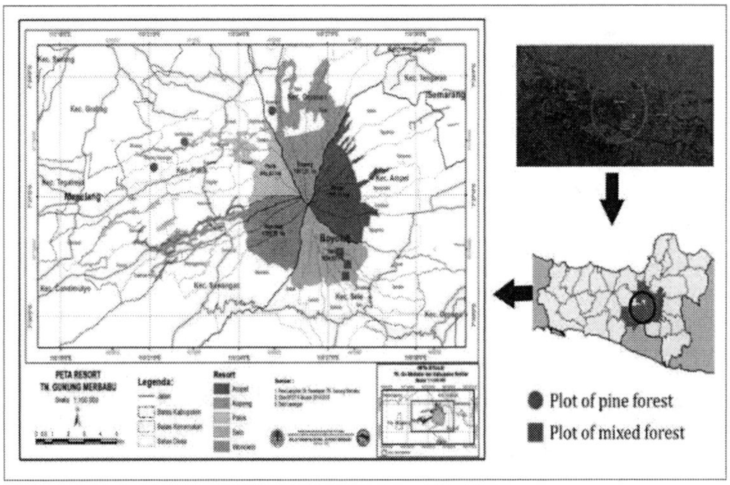


**Figure 1**. Map of research location in the pine forest and mixed forest in Mount Merbabu National Park

*2.2. Tools and research objects*
The tools used in this research include camera, cutter, mirrors, specimen container, hygrometer, multimeter, Global Positioning System (GPS), soil tester, and roll meter. The object of this research is a specimen of macrofungi, aluminum foil, label, flacon jar, alcohol 70%, a string of raffia, and a tally sheet.

*2.3. Procedures of research*
Retrieval of macrofungi diversity data used a plot method measuring 20 x 20 m which was divided into 16 subplots measuring 5 x 5 m. Subplots measuring 5 x 5 m were used to determine the distribution pattern of macrofungi such as random, uniform, and grouped (Figure 2).

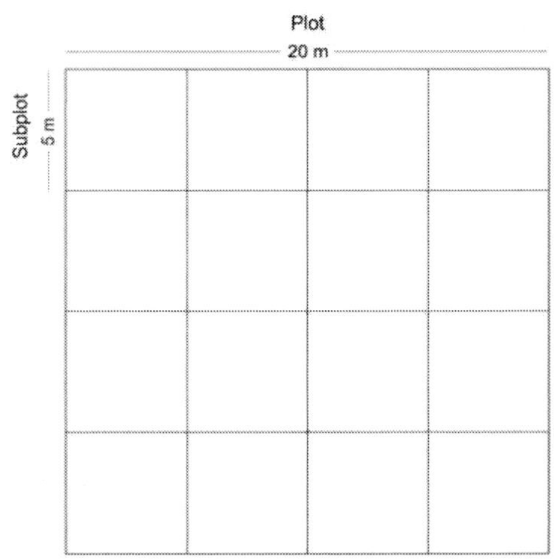


**Figure 2.** Design of research plot

Macrofungi that found at each location was collected, observed, and documented. Macroscopic observation conducted through the determination of shape and color of the fruit body, cap, stem, hymenophore, ring, and volva. The habitat of the macrofungi was also recorded. Macrofungi data that have been collected were analyzed by calculating species diversity index and distribution patterns. The formula used to calculate diversity index and Morisita's index, as follows:

1. Diversity Index (H')

Analysis of the diversity index (H') was calculated using the Shannon species diversity formula as follows:

$$H' = -\Sigma \, Pi \ln Pi \quad (1)$$

where:
H': Diversity index Shannon (Pi = ni/N)
ni: Density of each species
N: Total of density

According to Odum [14], the value of diversity index lower than 1 (H' < 1) means the diversity is low, while the value ranges from 1-3 (1 < H' ≤ 3) in diversity index is moderate, higher than 3 (H' > 3) means the diversity index is high.

2. Morisita's Index

Morisita's index was used to determine the distribution pattern of macrofungi species in the community which is divided into random, uniform, and grouped forms. The formula used to calculate Morisita's index as follows:

$$Id = n \frac{(\Sigma x^2 - \Sigma x)}{(\Sigma x)^2 - \Sigma x} \quad (2)$$

where:
Id: Degree of spread Morisita
n: Total of plot
$\Sigma x^2$: Total of the squares of the total individuals of a species in a community
$\Sigma x$: Total number of individuals of a species in a community

Analysis of the standardized Morisita's index (Ip) was calculated used the formula:

$$\begin{aligned}
Ip &= 0.5 + 0.5\left(\frac{Id - Mc}{n - Mc}\right) && \text{when } Id \geq Mc > 1 \quad (3)\\
Ip &= 0.5\left(\frac{Id - 1}{Mc - 1}\right) && \text{when } Mc > Id \geq 1\\
Ip &= -0.5\left(\frac{Id - 1}{Mu - 1}\right) && \text{when } 1 > Id > Mu\\
Ip &= -0.5 + 0.5\left(\frac{Id - Mu}{Mu}\right) && \text{when } 1 > Mu > Id
\end{aligned}$$

where:
Mu: Uniform Index
Mc: Clumped Index

According to Morisita [15], the value of standardized Morisita's index lower than 0 (Ip < 0) means the distribution pattern is uniform, while the value is 0 (Ip = 0) means the distribution pattern is random, higher than 0 (Ip > 0) means the distribution pattern is grouped.

## 3. Results and Discussion

### *3.1. Environmental conditions and characterization habitat of macrofungi*

There are several differences in environmental conditions in the pine forests and mixed forest in MMNP, especially on vegetation and its abiotic factors. Pine forests are classified as homogeneous forests that can be found only *Pinus merkusii* Jungh et de Vriese with distantly forest canopy. However, mixed forests are classified as covered and shadier areas because of diverse plant species with tightly forest canopy. The constituent dominant species are *Schima wallichii* (DC.) Korth., *Acacia decurrens* Willd., *Engelhardia serrata* Blume., and *Casuarina junghuhniana* Miq. The differences in the characteristics of vegetation in both habitats affect the humidity of each research location. The result of calculating abiotic factors is shown that pine forest drier than mixed forest (Table 2).

**Table 2.** Abiotic factors in the pine and mixed forest in MMNP

| | Pine forest | | | Mixed forest | | |
|---|---|---|---|---|---|---|
| | G | J | U | A | B | C |
| Air temperature (°C) | 24-25 | 23-25 | 21-24 | 19-20 | 18-20 | 17-18 |
| Air humidity (%) | 54-67 | 58-71 | 80-84 | 87-98 | 89-97 | 99 |
| Soil humidity | 1 | 1-2 | 2 | 2 | 2 | 3-4 |
| pH | 7 | 7 | 7 | 7 | 7 | 7 |
| Light intensity (Lux) | 1439-3343 | 1391-3333 | 1028-2720 | 1482-3203 | 2014-2100 | 1634-1803 |
| Altitudes (masl) | 923-938 | 1178-1229 | 1422-1433 | 1853-1878 | 1950-1961 | 2098-2189 |

Environmental conditions in the pine forests drier than mixed forests are also caused by the altitudes. Pine forests are at lower altitudes in the ranges of 923-1433 masl compared to mixed forests with a higher altitude in the ranges of 1853-2189 masl (Table 2).

The difference in vegetation types and abiotic factors in the pine and mixed forests will certainly affect the macrofungi microhabitat at each location and the availability of substrates for macrofungi growth. An example, the species of Polyporales are a group of cosmopolitan that are less affected by changes in environmental physical factors in their habitat [16]. However, for the species of Agaricales, changes in environmental physical factors will inhibit its growth [16].

### *3.2. Diversity of macrofungi*

The results showed that there were 36 species of macrofungi found in the pine forests from 20 families (Table 3). The twenty families were belonged to eight orders, i.e. Helotiales, Xylariales, Agaricales, Boletales, Dacrymycetales, Hymenochaetales, Polyporales, and Russulales. Different from the pine forest, the results showed that there were 80 species of macrofungi found in the mixed forest from 30 families (Table 3). The thirty families were belonged to twelve orders, i.e. Helotiales, Hypocreales, Xylariales, Agaricales, Auriculariales, Boletales, Dacrymycetales, Gomphales, Geastrales, Hymenochaetales, Polyporales, and Russulales.

Whole species were grouped in the division of Ascomycota and Basidiomycota. There are three in the pine forest and seven in the mixed forest of Ascomycota species and 33 species in the pine forest and 73 species of Basidiomycota in the mixed forest. The differences in the number of fungal are caused species of the Basidiomycota macrofungi is greater than Ascomycota [17]. The Ascomycota division has been a low frequency of encounters in nature. Families of Marasmiaceae, Psathyrellaceae, Agaricaceae, Mycenaceae, and Crepidotaceae are classified as Agaricales that dominant in the pine and mixed forest.

IOP Conf. Series: Materials Science and Engineering **935** (2020) 012030 doi:10.1088/1757-899X/935/1/012030

**Table 3.** Macrofungi diversity in the pine and mixed forest in MMNP

| Divisio | Order | Family | Species | Type of Habitat | |
|---|---|---|---|---|---|
| | | | | Pine | Mixed |
| Asco-mycota | Helotiales | Helotiaceae | *Ascotremella faginea* (Peck) Seaver | √ | |
| | | | *Bisporella citrina* (Batsch) Korf & S.E. Carp. | √ | √ |
| | | Dermateaceae | *Mollisia cinerea* (Batsch) P. Karst. | | √ |
| | | Sclerotiniaceae | *Ciboria batschiana* (Zopf) N. F. Buchw. | | √ |
| | Hypocreales | Nectriaceae | *Nectria cinnabarina* (Tode) Fr. | | √ |
| | Xylariales | Xylariaceae | *Xylaria hypoxylon* (L.) Grev. | √ | √ |
| | | | *Xylaria polymorpha* (Pers.) Grev. | | √ |
| | | | *Xylaria tentaculata* Berk. & Broome. | | √ |
| Basidio-mycota | Agaricales | Agaricaceae | *Agaricus semotus* Fr. | | √ |
| | | | *Lepiota atrodisca* Zeller | | √ |
| | | | *Lepiota castanea* Quél. | | √ |
| | | | *Lepiota rubrotinctoides* Murrill | √ | |
| | | | *Lepiota sequoiarum* Murrill | | √ |
| | | | *Leucocoprinus heinemannii* Migl. | √ | √ |
| | | | *Leucocoprinus fragilissimus* Murrill | √ | |
| | | Crepidotaceae | *Crepidotus applanatus* (Pers.) P. Kumm. | √ | √ |
| | | | *Crepidotus epibryus* (Fr.) Quél., | | √ |
| | | | *Crepidotus variabilis* (Pers.) P. Kumm. | √ | √ |
| | | Cortinariaceae | *Cortinarius anthracinus* (Fr.) Sacc. | | √ |
| | | Entolomataceae | *Entoloma bloxamii* (Berk. & Broome) Sacc. | | √ |
| | | | *Entoloma haastii* G.Stev. | | √ |
| | | Hydnangiaceae | *Laccaria laccata* (Scop.) Cooke | √ | √ |
| | | Hygrophoraceae | *Hygrocybe glutinipes* (J.E.Lange) R.Haller Aar. | | √ |
| | | | *Hygrophorus lucorum* Kalchbr. | | √ |
| | | Inocybaceae | *Inocybe rimosa* (Bull.: Fr.) P. Kumm. | √ | √ |
| | | Lyophyllaceae | *Asterophora* sp. | | √ |
| | | Marasmiaceae | *Campanella tristis* (G. Stev.) Segedin | √ | |
| | | | *Connopus acervatus* (Fries) R.H.Petersen | | √ |
| | | | *Gymnopus* sp. | | √ |
| | | | *Gymnopus dryophilus* (Bull.) Murrill | √ | √ |
| | | | *Gymnopus foetidus* (Sowerby) J.L. Mata & R.H. Petersen | | √ |
| | | | *Gymnopus fuscopurpureus* (Pers.) Antonín, Halling & Noordel. | | √ |
| | | | *Gymnopus ocior* (Pers.) Antonín & Noordel. | √ | √ |
| | | | *Gymnopus putillus* (Fr.) Antonín, Halling & Noordel. | | √ |
| | | | *Marasmiellus candidus* (Fr.) Singer | | √ |
| | | | *Marasmius androsaceus* (L.) Fr. | | √ |
| | | | *Marasmius calhouniae* Singer | | √ |
| | | | *Marasmius crinis-equi* F.Muell. ex Kalchbr | √ | √ |
| | | | *Marasmius siccus* Schwein. ex Fr. | √ | √ |
| | | | *Rhodocollybia maculate* (Alb. & Schwein.) Singer | | √ |
| | | Mycenaceae | *Mycena* sp. | | √ |
| | | | *Mycena adscendens* (Lasch) Maas Geest. | √ | √ |
| | | | *Mycena amicta* (Fr.) Quél. | | √ |
| | | | *Mycena flavoalba* (Fr.) Quél. | | √ |
| | | | *Mycena galopus* (Pers.) P.Kumm. | | √ |
| | | | *Mycena leaiana* (Berk.) Sacc. | | √ |
| | | | *Mycena leptocephala* (Pers.) Gillet | √ | √ |
| | | | *Mycena picta* (Fr.) Harmaja | | √ |
| | | | *Mycena ustalis* Aronsen & Maas Geest. | | √ |
| | | | *Panellus stipticus* (Bull.) P.Karst. | | √ |

| | | | | |
|---|---|---|---|---|
| | Physalacriaceae | *Armillaria mellea* (Vahl) P.Kumm. | | √ |
| | | *Oudemansiella melanotricha* (Dörfelt) M.M. Moser | √ | |
| | | *Oudemansiella mucida* (Schrad.) Hohn. | | √ |
| | Pluteaceae | *Pluteus cervinus* (Schäffer : Fr) P. Kumm. | | √ |
| | Psathyrellaceae | *Coprinellus disseminatus* (pers) j.e. lange | √ | √ |
| | | *Coprinus micaceus* (Bull.) Fr. | | √ |
| | | *Parasola plicatilis* (Curtis) Redhead | √ | √ |
| | | *Psathyrella candolleana* (Fr.) Maire | √ | √ |
| | | *Psathyrella conopilus* (Fr.) A.Pearson & Dennis | | √ |
| | Schizophyllaceae | *Schizophyllum commune* Fries | | √ |
| | Tricholomataceae | *Clitocybe cistophila* Bon & Contu | | √ |
| | | *Collybia eucalyptorum* Cleland | | √ |
| | | *Collybia tuberosa* (Bull.) P.Kumm. | √ | |
| | | *Omphalina pyxidata* (Bull.) Quél. | √ | |
| | Tubariaceae | *Tubaria furfuracea* (Pers.) Gillet | √ | √ |
| | Typhulaceae | *Macrotyphula juncea* (Alb. & Schwein.) Berthier | | √ |
| Auricularia-les | Auriculariaceae | *Exidia thuretiana* (Lév.) Fr. | | √ |
| Boletales | Rhizopogonaceae | *Rhizopogon* sp. | | √ |
| | Sclerodermataceae | *Scleroderma cepa* Pers. | √ | √ |
| | Strophariaceae | *Hypholoma fasciculare* (Lév.) Fr. | | √ |
| | Suillaceae | *Suillus bovinus* (L.) Roussel | | √ |
| | | *Suillus granulatus* (L.) Roussel | √ | |
| Dacrymyce-tales | Dacrymycetaceae | *Dacryopinax spathularia* Schwein. ex Fr. | √ | √ |
| Gomphales | Gomphaceae | *Ramaria stricta* (Pers.) Quél. | | √ |
| Geastrales | Geastraceae | *Geastrum pectinatum* Pers. | | √ |
| Hymenochae-tales | Hymenochaetaceae | *Coltricia cinnamomea* (Jacq.) Murrill | √ | √ |
| | | *Naucoria striatula* P. D. Orton | √ | |
| Polyporales | Ganodermataceae | *Ganoderma* sp. | √ | |
| | | *Ganoderma applanatum* (Pers.) Pat. | | √ |
| | Irpicaceae | *Byssomerulius corium* (Pers.) Parmasto | √ | |
| | Meruliaceae | *Phlebia rufa* (Pers.) M.P. Christ. | | √ |
| | | *Sarcodontia pachyodon* (Pers.) Spirin | | √ |
| | Polyporaceae | *Microporus xanthopus* (Fr.) Kuntze | √ | |
| | | *Pycnoporus cinnabarinus* (Jacq.) P.Karst | | √ |
| | | *Trametes ochracea* (Pers.) Gilb. & Ryvarden | | √ |
| | | *Trametes pubescens* (Schumach.) Pilát | | √ |
| | | *Trametes versicolor* (L.) Lloyd | √ | |
| | Steccherinaceae | *Irpex lacteus* (Fr.) Fr. | | √ |
| | | *Junghuhnia nitida* (Pers.) Ryvarden | | √ |
| | | *Steccherinum ochraceum* (Pers.) Gray | | √ |
| Russulales | Russulaceae | *Lactarius aurantiacus* (Pers.) Gray | | √ |
| | | *Lactarius obscuratus* (Lasch) Fr. | √ | |
| | | *Lactarius torminosus* (Schaeff.) Gray | √ | |
| | | *Lactarius spinosulus* Quél. & Le Bret. | | √ |
| | | *Russula emetica* (Schaeff.) Pers. | √ | |
| | Stereaceae | *Stereum ostrea* (Blume & T. Nees ex Fr. | | √ |

**Remarks:** √ = founded

The diversity of macrofungi in the pine forest belonged to moderate value with a diversity index (H') of 2.84, whereas in the mixed forest belonged to high value with a diversity index (H') of 3.64. The result corresponding to Odum [14] if the value of diversity index ranges from 1-3 ($1 < H' \leq 3$) in diversity index is moderate, and if the value of diversity index higher than 3 ($H' > 3$) means the diversity index is high. The difference of macrofungi diversity index values between pine and mixed forests is influenced by environmental factors, such as the availability of substrates, biotic factors, and abiotic factors.

The type of substrates in mixed forests is more abundant compared to pine forests as shown in Table 4. Litter substrate is not found in pine forests. It is suspected that *P. merkusii* leaf litter contains allelochemical compounds such as phenols and flavonoids which can inhibit macrofungi growth [18]. Phenols can inhibit the synthesis of chitin which is very important for the formation of macrofungi cell walls [19]. Flavonoid compounds have a mechanism by interfering with mitochondrial homeostasis and disrupting macrofungi cell membrane permeability [20]. Macrofungi are also not found growing on tree trunks because the stem of *P. merkusii* produces a sap that makes macrofungi difficult to grow. *P. merkusii* sap contains terpenoid compounds that can inhibit the growth of macrofungi, both through the cytoplasmic membrane and on the growth and development of spores [21].

**Table 4.** Substrate of macrofungi in the pine and mixed forest in MMNP

| No. | Substrate | Percentage of Species (%) | |
|---|---|---|---|
| | | Pine forest | Mixed Forest |
| 1. | Soil | 66.67 | 42.7 |
| 2. | Weathered wood | 33.33 | 19.1 |
| 3. | Litter | - | 29.2 |
| 4. | Host plant | - | 8.99 |
| | | 100 | 100 |

Macrofungi that recorded in mixed forests are thought to have an association with host plant substrate which is a biotic factor for the high diversity of macrofungi (Table 5).

**Table 5.** Host plant of macrofungi in the mixed forest MMNP

| No. | Species of host plant | Species of macrofungi |
|---|---|---|
| 1. | *Schima wallichii* (DC.) Korth) | *Psathyrella candolleana* (Fr.) Maire |
| | | *Mycena* sp. |
| | | *Crepidotus applanatus* (Pers.) P. Kumm. |
| 2. | *Casuarina junghuhniana* Miq. | *Trametes pubescens* (Schumach.) Pilát |
| | | *Sarcodontia pachyodon* (Pers.) Spirin |
| | | *Mycena ustalis* Aronsen & Maas Geest. |
| | | *Marasmiellus candidus* (Fr.) Singer. |
| 3. | *Engelhardia serrata* Blume | *Mycena picta* (Fr.) Harmaja |

The availability of substrate resources can determine the strength of the effects of the emergence of macrofungi species on the functioning of the ecological community. This can be seen from the number of macrofungi species grouped by ecological functions in each habitat (Figure 3). The number of macrofungi species as decomposers in pine and mixed forests has the highest order compared to symbionts and parasites. It is because of decomposers macrofungi have been abundant such as soil, litter, and weathered wood. The more substrate resources the higher the number of macrofungi species found.

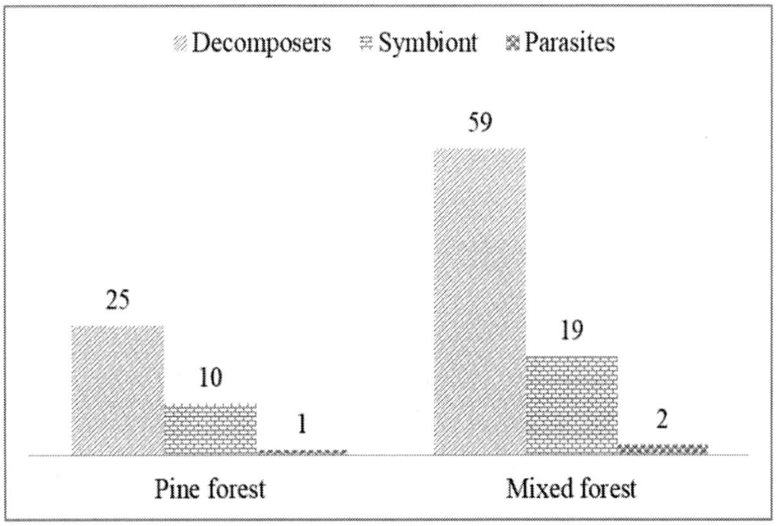


**Figure 3.** The number of macrofungi species based on ecological functions in the pine and mixed forest MMNP

Based on each plot, macrofungi have quite interesting diversity. The high and low diversity of macrofungi is directly proportional to the height of the plot in both pine forests and mixed forests. At an altitude of 1400 masl, pine forests have the highest diversity of macrofungi with a diversity index value of 2.65. The lowest macrofungi diversity was calculated at an altitude of 900 masl with a diversity index value of 1.76. The same results were found in mixed forests, the highest diversity of macrofungi was found at an altitude of 2100 masl with an index value of 3.17 including the high category. The lowest macrofungi diversity was calculated at an altitude of 1800 masl with an index value of 2.53, including the moderate category. It is because of abiotic factor that the higher the location, the more humid environmental conditions.

### *3.3. Distribution of macrofungi*

There were 16 species macrofungi that can be found in the pine forest, 60 species can be found in mixed forest, and 20 species in both habitats (Table 3). Macrofungi that can be found only in pine forests have an ectomycorrhizal role in *P. merkusii* such as *Russula emetica* (Schaeff.) Pers., *Suillus granulatus* (L.) Roussel and *Lactarius torminosus* (Schaeff.) Gray. [22]. Macrofungi that can be found only in mixed forests have a humid habitat characteristic such as *Armillaria mellea* (Vahl) P.Kumm. [23]. Macrofungi that can be found in pine and mixed forests include cosmopolitan fungi such as the Xylariaceae [24].

The distribution of macrofungi can also be seen by morphological type and order is shown in Table 6. The result relates to the spread spore of each macrofungi [25]. Environmental conditions cause the production of macrofungi fruit bodies to differ from one species to another [5]. The most widespread macrofungi were the agaric type, which included Agaricales order, while the narrowest distributed macrofungi were the coral type, include the Gomphales order.

**Table 6.** Distribution of macrofungi based on type of morphological and order

| No. | Type of morphological | Order | Frequency of species | | | | | | Total plot of Frequency species |
|---|---|---|---|---|---|---|---|---|---|
| | | | Pine forest | | | Mixed forest | | | |
| | | | G | J | U | A | B | C | |
| 1. | *Agaric* | Agaricales | 10 | 13 | 16 | 23 | 25 | 28 | 6 |
| 2. | *Boletes* | Boletales | 0 | 1 | 1 | 1 | 0 | 1 | 4 |
| 3. | *Leather* dan *Polypores* | Polyporales | 2 | 1 | 2 | 0 | 4 | 3 | 5 |
| 4. | *Puffballs* | Boletales, Geastrales | 1 | 0 | 1 | 1 | 1 | 2 | 5 |
| 5. | *Jelly* dan *flask* | Helotiales, Dacrymycetales, Auriculariales, Hypocreales, Xylariales | 1 | 0 | 2 | 1 | 2 | 4 | 5 |
| 6. | *Stinkhorn* | - | 0 | 0 | 0 | 0 | 0 | 0 | 0 |
| 7. | *Coral* | Gomphales | 0 | 0 | 0 | 0 | 1 | 2 | 2 |
| 8. | *Cup* | Helotiales | 1 | 0 | 1 | 1 | 1 | 3 | 5 |
| 9. | *Tooth* | Polyporaceae | 0 | 0 | 1 | 1 | 3 | 1 | 4 |
| 10. | *Truffle* | - | 0 | 0 | 0 | 0 | 0 | 0 | 0 |

The distribution pattern of macrofungi species in pine forest was obtained by 33 grouped species and three uniform species. The distribution pattern of macrofungi species in the mixed forest was obtained by 78 grouped species and two uniform species. The diagram of the distribution pattern of macrofungi species in pine and mixed forest in MMNP is shown in Figure 4.

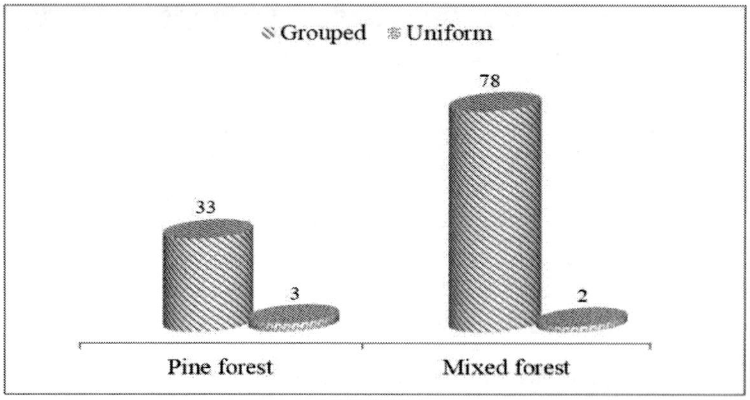


**Figure 4.** Distribution pattern of macrofungi in pine and mixed forest in MMNP

Macrofungi that can be found in pine and mixed forests that have group living behavior are *Bisporella citrine* (Batsch) Korf & S.E. Carp., *Crepidotus applanatus* (Pers.) P. Kumm., *Crepidotus epibryus* (Fr.) Quél., *Crepidotus variabilis* (Pers.) P. Kumm., *Scleroderma cepa* Pers., *Schizophyllum commune* Fries, and *Ramaria stricta* (Pers.) Quél. These species have grouped distribution patterns with Ip > 0. Three macrofungi species that have a uniform distribution pattern in pine forests are *Leucocoprinus heinemannii* Migl., *Leucocoprinus fragilissimus* (Berk. & MACurtis) Pat., and *Inocybe rimosa* (Bull.) P. Kumm. While the two macrofungi species that have a uniform distribution pattern in mixed forests are *Mycena amicta* (Fr.) Quél. and *Gymnopus fuscopurpureus* (Pers.) Antonín, Halling & Noordel. Distribution analysis through the Morisita approach that these species have Ip < 0 which means uniform. The presence of these fungi is suspected because it has a solitary growing behavior.

## 4. Conclusion

There were 36 species of macrofungi found in the pine forest and 80 species in the mixed forest. The diversity of macrofungi in the pine forest belonged to moderate value with a diversity index of 2.84, whereas in mixed forest belonged to high value with a diversity index of 3.64. There were 16 species macrofungi can be found in the pine forest, 60 species can be found in the mixed forest, and 20 species in both habitats. The most widespread macrofungi were the agaric type, which included Agaricales order, while the narrowest distributed macrofungi were the coral type, which include the Gomphales order. The distribution pattern of macrofungi species in the pine and mixed forest tends to be grouped.

## References

[1] Solle H, Klau F, and Nuhamara ST 2017 Keanekaragaman jamur di Cagar Alam Gunung Mutis Kabupaten Timor Tengah Utara, Nusa Tenggara Timur *Biota* **2**(3) 105-10
[2] Dighton J and White JF 2017 *The fungal community: its organization and role in the ecosystem. Fourth Edition* (Florida: CRC Press)
[3] Lindsay A, Robinson R, May T, and McMullan-Fisher SJM 2013 *Guide to Surveying Fungi in Australia* (Autralia: Fungimap)
[4] Garrett LG, Kimberley MO, Oliver GR, Parks M, Pearce SH, Beets PN, and Paul TSH 2019 Decay rates of above and below ground coarse woody debris of common tree species in New Zealand's natural forest *Forest Ecology and Management* **438** 96-102
[5] Moore S and O'Sullivan P 2014 *A guide to common fungi of the Hunter-Central Rivers region* (NSW: Hunter Local Land Services)
[6] Hawksworth DL 1991 The fungal dimension of biodiversity: Magnitude, significance, and conservation *Mycological Research* **95**(6) 641-55
[7] Campbell NA, Reece JB, Urry LA, Cain ML, Wasserman SA, Minorsky PV, and Jackson RB 2012 *Biologi Jilid 2 edisi ke delapan* Terjemahan Wulandari DT (Jakarta: Erlangga)
[8] Widjaja EA, Rahayuningsih Y, Rahajoe JS, Ubaidillah R, Maryanto I, Walujo EB, and Semiadi G 2014 *Kekinian Keanekaragaman Hayati Indonesia 2014* (Jakarta: LIPI Press)
[9] Priskila, Ekamawanti HA, and Herawatiningsih R 2018 Keanekaragaman jenis jamur makroskopis di kawasan hutan sekunder areal IUPHHK-HTI PT. Bhatara Alam Lestari Kabupaten Mempawah *Jurnal Hutan Lestari* **6**(3) 569-82
[10] Rohmaya, Mardji D, and Sukartiningsih 2011 Keanekaragaman jenis jamur ektomikoriza pada kondisi hutan dengan kelerengan yang berbeda di Hutan Wisata Bukit Bangkirai PT INHUTANI I Balikpapan *Jurnal Kehutanan Tropika Humida* **4**(2) 150-61
[11] Astuti FK, Murningsih, and Jumari 2017 Keanekaragaman jenis tumbuhan paku (Pteridophyta) di jalur pendakian Selo kawasan Taman Nasional Gunung Merbabu, Jawa Tengah *Jurnal Biologi* **6**(2) 1-6
[12] Mulyanto H, Cahyuningdari D, and Setyawan AD 2000 Kantung semar (*Nepenthes* sp.) di Lereng Gunung Merbabu *Biodiversitas* **1**(2) 54-8
[13] Munandar AA 2017 Inventarisasi Jamur Pelapuk Kayu dari Hutan Lereng Selatan Gunung Merbabu Jawa Tengah [Skripsi] Program Studi Pendidikan Biologi, Fakultas Keguruan dan Ilmu Pendidikan, Universitas Muhammadiyah Surakarta
[14] Odum EP 1993 *Dasar-dasar Ekologi* edisi Ketiga Terjemahan Samingan T (Yogyakarta: Gadjah Mada University Press)
[15] Morisita M 1959 Measuring of the dispersion of individuals and analysis of the distributional patterns *Memoirs Faculty of Science, Kyushu University, Seri E (Biology)* **40** 3-5
[16] Prasetyaningsih A and Rahardjo D 2015 Keanekaragaman dan potensi makrofungi Taman Nasional Gunung Merapi *The 2nd University Research Coloquium*
[17] Dwidjoseputro D 1978 *Pengantar Mikologi 2nd Edition* (Bandung: Penerbit Alumni)

[18] Christoper W, Natalia D, and Rahmayanti S 2017 Uji aktivitas antijamur ekstrak etanol umbi bawang Dayak (*Eleutherine americana* (Aubl.) Merr. Ex K. Heyne.) terhadap *Trichophyton mentagrophytes* secara in vitro *Jurnal Kesehatan Andalas* **6**(3) 685-9

[19] Wu X, Cheng A, Sun L, and Lou H 2008 Effect of Plagiochin E, an antifungal macrocyclic bis (Bibenzyl), on cell wall chitin synthesis in *Candida albicans Acta Pharmacol Sin.* **29**(12) 1478-85

[20] Kang K, Fonng W, and Tsang PW 2010 Antifungal activity of baicalein against *Candida krusei* does not involve apoptosis *Mycopathologia* **170** 391-6

[21] Lutfiyanti R, Ma'ruf WF, and Dewi EN 2012 Aktivitas antijamur senyawa bioaktif ekstrak *Gelidium latifolium* terhadap *Candidus albicans Jurnal Pengolahan dan Bioteknologi Hasil Perikanan* **1**(1) 1-8

[22] Darwo and Sugiarti 2008 Beberapa jenis cendawan ektomikoriza di kawasan hutan Sipirok, Tongkoh dan Aek Nauli, Sumatera Utara *Jurnal Penelitian Hutan dan Konservasi Alam* **5**(2) 157-73

[23] Islam MR and Aminuzzaman FM 2016 Macro fungi biodiversity at the central and northern biosphere reserved areas of Tropical Moist Deciduous Forest Region of Bangladesh *Journal of Agriculture and Ecology Research* **5**(4) 1-11

[24] Sir EB, Hladki AI, Parrado MF, and Romero AI 2012 Biodiversity of Xylariaceae (Ascomycota) and their hosts in protected areas from Tucuman (Argentina) *Tomo* **37**(2) 35-48

[25] Andrew EE, Kinge TR, Tabi EM, Thiobal N, and Mih AM 2013 Diversity and distribution of macrofungi (Mushrooms) in the Mount Cameroon Region *Journal of Ecology and the Natural Environment* **5**(10) 318-34

# Changes in the density, specific gravity and dimensional stability of candlenut wood (*Aleurites moluccanus* (L.) Willd.) from several variation temperatures with oil-heat treatment

**A Azis*, A D Yunianti, A Agussalim**

Departement for Forest Products, Forestry Faculty, Hasanuddin University, Makassar, 90245, Indonesia

*Email: frtsarjunazis04@gmail.com

**Abstract**. Quality of inferior woods can be improved with wood modification technology. There are some wood modification technologies available, one of them is oil-heat treatment. Candlenut wood (*Aleurites moluccanus* (L.) Willd.) is one of low-quality wood species with limited utilization. Efforts to improve the properties of wood are needed for more optimal utilization. This research was conducted to improve physical properties of candlenut wood namely density, specific gravity, and dimensional stability by applying oil-heating treatment at various temperatures (160, 180 and 200°C) for one hour. Physical properties of wood were tested according to ASTM D143-94. The result showed candlenut wood in oil-heat treatment as better than control. Density and specific gravity of wood increased by 25.13% and 26.97%, respectively, and the dimensional stability also increased as well, shown by the reduction of hygroscopic properties of wood at tangential and radial directions by 22%. The best treatment selected for candlenut wood with temperature of 160°C.

## 1. Introduction

The main use of candlenut plant is as a fruit-producing [1], and if it is no longer productive in producing fruit, the wood will be utilized. However, this species is known as low-quality wood both in terms of strength and durability. Candlenut wood is included in the strength class IV-V and durability class IV and V against wood rot and termite attack, respectively [2].

In order to optimize the candlenut wood utilization, the wood strength and durability should be increased. One way is through heat modification. Heat modification is a technology that is quite commercially developed compared to chemical modification and impregnation technology, especially in European countries [3]. A method in heat treatment is oil-heat treatment. This method is able to increase the durability of wood and is effective in improving the stability of wood dimensions [4].

Modification of wood with oil heat treatment has been studied recently and it is proven can improve the properties of wood, especially by giving heat temperatures higher than 140°C. Several studies have shown an increase and decrease in properties of wood after heating with oil. This shows that a wood species has different responses to the temperature and heating time, thus experiment is needed to obtain optimal results. Therefore, this study was conducted to find out the response of candlenut wood to oil heat treatment at several variations in temperature and time. Changes in physical properties of wood can be used as a reference response of treatment.

## 2. Materials and Methods

This research used candlenut wood that was collected from Limam Poccoe Village, Maros Regency. Bulk oil was obtained from traditional market in Maros. Small clear specimens were prepared based on modified ASTM D143-94 [5]: 2 x 2 x 2 cm for specific gravity, and 2 x 2 x 8 cm for shrinkage. The samples were then conditioned to reach at 65±5% relative humidity and temperature of 20±2°C.

There were three heat-oil treatments, namely 160°, 180°, 200°C, and control (untreated). Firstly, the weight and dimensions of all samples were measured. The treated-samples were gradually heated at 60° and 90°C in oven. The samples were then oil-heated for one hour, and conditioned for one month before measuring the weights and dimensions after treatment. The density, specific gravity, and shrinkage were then calculated. The data were analyzed using randomized completely design with four treatments and five replications. Tukey's Honestly Significant Difference (HSD) test at a significance level p>0.05 was used to compare significant differences in the means.

## 3. Results and Discussion

### *3.1. Density and specific gravity*

Density of candlenut wood after heating treatment with oil showed different results compared to control in Figure 1. The average density of controls was 0.38 g/cm³. The density of wood with oil-heat showed high increase compared to controls ranging from 18.85 to 25.13%. The highest density obtained at 160°C heating temperature of 0.48 g/cm³. Increased wood density generally occurs due to penetration of a substance in anatomical structure of wood and remains inside even drained and storage for long time. The density can increase because mass of oil-heated specimens increase that oil filled in the wood cells, while volume of wood just remains or only slightly changed.

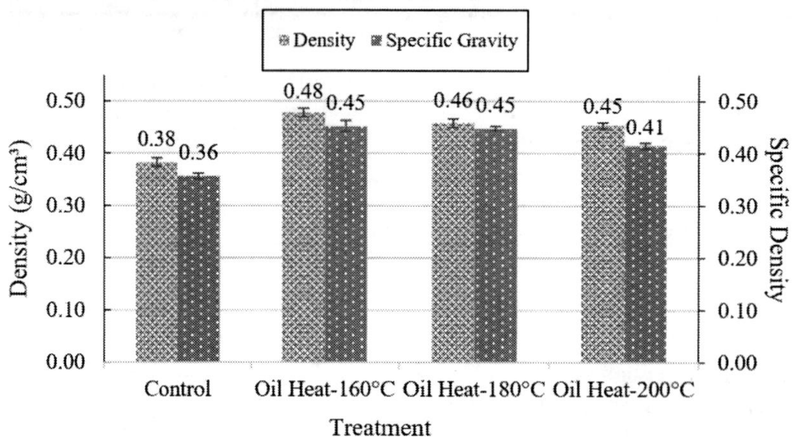


**Figure 1**. Density and specific gravity of candlenut wood after heat treatment

The increase of heat temperature affected to the decrease of wood density. High temperatures allow for an increase density due to extractive migration that occurs on surface of wood was not dissolved [6]. In addition, a longer heating time will cause more oil to fill in wood cells, thereby increasing the density of wood [7]. However, the decrease occurred in high temperatures can degrade chemical components of wood, especially hemicellulose [8]. Each characteristic is different for main components of wood. The degradation temperature in wood cellulose ranges from 275 to 350°C, hemicellulose by 150 to 300°C, and lignin by 250-500°C [9]. A decrease in hemicellulose content during heating with oil starting from 180°C [10].

The average specific gravity of controls was 0.36. The range was between 0.31 to 0.44, thus it fell to the strength class IV-V [2]. Specific gravity of wood with oil-heat treatment increased as much as 26.98%. Changes in value of specific gravity are closely related to the changes in wood density.

### *3.2. Dimensional stability*

The Dimension changes of wood in tangential direction were greater than at radial direction. This generally occurs even without treatment. Changes in tangential ranged from 1.5 to 3 times higher than radial direction. Oil-heat treatment at 160°C did not show significant differences in wood. This is because only secondary components of volatile and soluble wood were affected.

Figure 2 showed heat treatment with oil has a significant influence on decrease in dimensional stability of wood. Oil can be a good heating media for wood compared to water as oil can transfer the heat into the wood more quickly and evenly. Additionally, the oil can prevent oxygen penetrating the wood during treatment process. Oil absorption during treatment form a protective layer in surface of wood, thus it can maintain the wood stability [6].

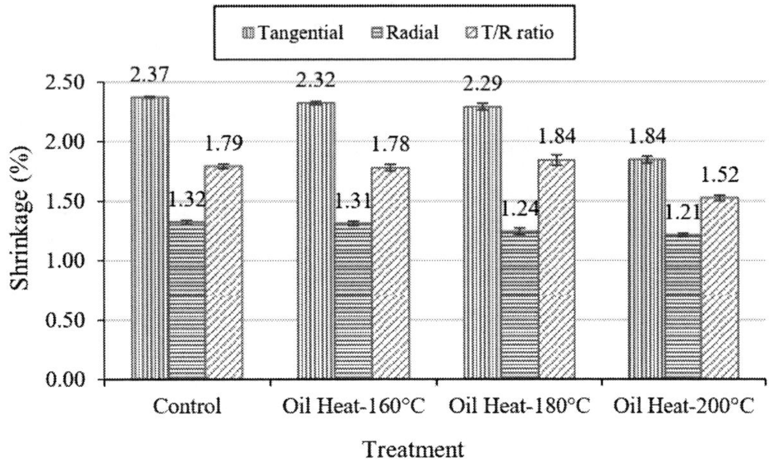


**Figure 2**. Shrinkage of candlenut wood after heat treatment from air-dry to oven-dry condition

The decrease in dimension by 0.96 to 22.19% occurred when the temperature increase by 200°C. Heat treatment in three species of Philippine bamboo (*Bambusa blumeana*, *B. vulgaris* and *Dendrocalamus asper*) decrease up to 80% compared to the control [11].

Wood have good stability in high temperature than lower. Heat treatments greater than 150°C can clearly decrease the shrinkage and change equilibrium moisture content (EMC) of wood [12]. The heating treatment can reduce water absorption in wood due to reduced hydroxyl groups (-OH). A large number of hydrophilic groups change to hydrophobic oxygen-acetyl (O-Acetyl) groups causing reduced -OH hands to bond with water molecules and wood hygroscopic decreases, resulting the increase in dimensional stability [8,13]. The heat treatment improves the quality of the timber for furniture industry, as well as improves the decorative property. The wood appearance is more attractive, apart from the resistance from fungi.

The ratio of tangential and radial direction (T/R ratio) of candlenut wood after heat treatment ranged below the value of 2. The T/R ratio more than 2 shows poor dimensional stability of wood because it can cause deformation in wood [14].

**Tabel 1**. Result of Tukey Honestly Significant Difference test for physical properties of candlenut wood

| Variables | Treatments | Mean | St. Dev. | R% |
|---|---|---|---|---|
| Density (g/cm³) | Control | 0.38a | 0.008 | - |
| | Oil Heat-160°C | 0.48c | 0.008 | +25.13 |
| | Oil Heat-180°C | 0.46b | 0.008 | +19.90 |
| | Oil Heat-200°C | 0.45b | 0.005 | +18.85 |
| Specific Gravity | Control | 0.36a | 0.005 | - |
| | Oil Heat-160°C | 0.45c | 0.011 | +26.97 |
| | Oil Heat-180°C | 0.45c | 0.004 | +25.84 |
| | Oil Heat-200°C | 0.41b | 0.005 | +16.29 |
| Shrinkage (Tangential) | Control | 2.37c | 0.008 | - |
| | Oil Heat-160°C | 2.32b | 0.012 | -2.11 |
| | Oil Heat-180°C | 2.29b | 0.029 | -3.44 |
| | Oil Heat-200°C | 1.84a | 0.030 | -22.19 |
| Shrinkage (Radial) | Control | 1.32b | 0.016 | - |
| | Oil Heat-160°C | 1.31b | 0.018 | -0.96 |
| | Oil Heat-180°C | 1.24a | 0.025 | -5.99 |
| | Oil Heat-200°C | 1.21a | 0.013 | -8.35 |
| T/R Ratio | Control | 1.79bc | 0.019 | - |
| | Oil Heat-160°C | 1.78b | 0.028 | -0.78 |
| | Oil Heat-180°C | 1.84c | 0.044 | +2.79 |
| | Oil Heat-200°C | 1.52a | 0.022 | -14.96 |

Notes: SD : Standard deviation
R% : Percentage change in variable after heat treatment from control
Positive mark (+) : Increase change variable from control
Negative mark (-) : Decrease change variable from control

## 4. Conclusion

In this study, oil heat-treatment improved the physical properties of candlenut wood. Heating with oil increased the wood density and specific gravity up to 25,13% and 26.97%, respectively. The increase of dimensional stability was indicated by the reduced hygroscopic properties of wood in tangential and radial directions by 0.96 to 22.19%, respectively. The use of 160°C was best treatment for improve the physical properties of candlenut wood, but better dimensional stability of wood was obtained in higher temperature (200°C).

## References

[1] Winarbowo S and Manoko IP 2006 Kemiri (*Aleurites moluccana* L Willd) tanaman industri potensial *Warta Penelitian dan Pengembangan Tanaman Industri* **12(2)** 13-15
[2] Martawijaya AI, Kartasujana YI, Mandang SA, Prawira K, and Kadir 1989 *Atlas Kayu Indonesia Jilid II* (Bogor: Pusat Penelitian dan Pengembangan Hasil Hutan)
[3] Evans P 2003 Emerging technologies in wood protection *Forest Products Journal* **53**
[4] Hidayat W and Febrianto F 2018 *Teknologi Modifikasi Kayu Ramah Lingkungan: Modifikasi Panas dan Pengaruhnya terhadap Sifat-sifat Kayu* (Bandar Lampung: Pusaka Media)

[5] [ASTM] American Society for Testing Material 2000 *Standard Test Methods for Small Clear Specimens of Timber: Annual Book of ASTM Standars ASTM D143-94 (reapproved 2000)* (West Conshohocken USA: American Society for Testing Material)

[6] Dubey MK, Pang S, dan Walker J 2011 Changes in chemistry color dimensional stability and fungal resistance of *Pinus radiata* D. don wood with oil heat treatment *Holzforschung* **66**(1) 49-57

[7] Daud M and Coto Z 2009 Peningkatan sifat fisis dan mekanis kayu durian (*durio* sp) dengan penggorengan *Simposium Forum Teknologi Hasil Hutan* 481-90

[8] Kocaefe D, Poncsak S, and Boluk Y 2008 Effect of thermal treatment on the chemical composition and mechanical properties of birch and aspen *BioResources* **3**(2) 517-37

[9] Fedyukhin A, Sultanguzin I, Gyul'Maliev A, and Sergeev V 2017 Biomass pyrolysis and gasification comprehensive modeling for effective power generation at combined cycle power plant *Eurasian Chemico-Technological Journal* **19**(3) 245-53

[10] Razak W, Izyan K, Hanim AR, Othman S, Aminuddin M, and Affendy H 2011 Effects of hot oil treatment on colour and chemical changes in 15 years old acacia hybrid *Journal of Tropical Forest Science* **23**(1) 42-50

[11] Manalo RD and Acda MN 2009 effects of hot oil treatment on physical and mechanical properties of three species of philippine bamboo *Journal of Tropical Forest Science* **21**(1) 19-24

[12] Syrjänen T and Oy K 2001 Production and Classification of Heat Treated Wood in Finland *Proceedings of the Trä Skydd- vä Rmebehandlat Trä- Egenskaper Och Anvä Ndningsomra Den* p. 1-8

[13] Lee SH, Ashaari Z, Lum WC, Halip JA, Ang AF, Tan LP, Chin LN, dan Tahir PM 2018 Thermal treatment of wood using vegetable oils: a review *Construction and Building Materials* **181**(1) 408-19

[14] Basri E 2000 *Bagan Pengeringan Dasar 12 Jenis Kayu dari Indonesia* (Bogor: Pusat Penelitian dan Pengembangan Hasil Hutan)

IOP Conf. Series: Materials Science and Engineering **935** (2020) 012032 doi:10.1088/1757-899X/935/1/012032

# Determination of optimum formulation of tusam (*Pinus merkusii*) tannin bark with resorcinol and formaldehyde

**S Hajriani[1] A D Yunianti[2*], and S Suhasman[2]**

[1]Magister Student of Forestry, Faculty of Forestry, Hasanuddin University, Makassar, Indonesia
[2]Laboratory of Forest Product Ulilization and Processing, Faculty of Forestry, Hasanuddin University, Makassar, Indonesia

*Email: dettiyunianti70@yahoo.com

**Abstract.**Adhesive is the one of main components that used for producing composite wood. The quality of the adhesive can affect the strength of the composite wood. However, adhesives which commonly used are synthetic adhesives that contain a lot of formaldehyde emissions and have negative impact on our health. To reduce its use, we need an alternative adhesive from natural raw materials. Tannin is a phenolic compound has been widely studied that can be used as a bio adhesive for wood composites. The main source of tannin for adhesive is derived from bark. One type of wood bark that contains tannins is tusam wood bark. Tannins have a high reactivity against formaldehyde. This study aims to determinate the optimum formula of tannin which is copolymerized with resorcinol and formaldehyde. Tannin is obtained by extracting wood bark with hot water. The optimum formula of tannins and resorcinol (TR) result is determined by the stiasny numbers, while the optimum formula of tannins and formaldehyde (TF) are determined by identification of solid content. The addition of resorcinol affects the percentage of stiasny number, the greater the formula of resorcinol added, the higher value of stiasny number, wherein the reactivity of tannin to formaldehyde increases. The addition of formaldehyde also affects the results of tannin formaldehyde solid content, the more addition of formaldehyde, the percentage of solid content decreases.

## 1. Introduction

Adhesive is the one of main components that required in composite wood industry. Adhesives that widely used are synthetic adhesives which are relatively expensive [1] because the raw material is derrivated from nonrenewable material, in addition to these adhesives also contain a lot of formaldehyde emissions which can negatively impact on health in the form of evaporated gas exposure at room temperature [2]. Adhesives are commonly commercialized by the most adhesive industries recently such as phenol formaldehyde and urea formaldehyde [3]. To reduce the use of synthetic adhesives, natural polymer compounds derived from plants and adapted for the same use as pure synthesis groups [4] are needed.

Tannin is one of the phenolic compounds that has great potential to produce bio adhesives because it has a high flavonoid content, especially in condensed tannins [5,6]. Tannins can be obtained in all parts of the plant, but the content of tannins in large quantities is obtained from wood bark extract which is studied as a bio adhesive. In recent years, condensed tannins have been widely studied as substitutes for phenol and resorcinol in their reactivity against formaldehyde [8,9]. Various types of pine bark have been studied and have tannin compounds that potential to be adhesives on composite wood [10-13].

Condensed tannins have a high affinity for resorcinol and formaldehyde so that they can copolymerize and used as adhesives [14]. Tannins have a high reactivity toward formaldehyde [15,16], but to reduce formaldehyde emission, the addition of resorcinol can be used to bind formaldehyde groups that do not react with tannins [17].

This study aims to react Tusam tannins with resorcinol and formaldehyde by determining optimum formulation of some modified ratios from several previous studies. Determination of the optimum tannin and resorcinol formulations based on stiasny number, and determination of tannin and formaldehyde formulations based on the percentage of solid content.

## 2. Materials and Methods

### *2.1. Bark extraction method*

Tusam bark is prepared to be chips and then dried at air temperature until it reaches the moisture content of about 12%, then it extracted with hot water at bark and water ratio of 1:4 (g/mL) in a waterbath for 4 hours. Then it filtered to get tannin filtrate. tannin liquid is then dried using a spray dryer to get tannin powder.

### *2.2. Determination of stiasny number of tusam tannin bark*

Determination of stiasny number aims to see the tannin reactivity toward formaldehyde which is shown based on the percentage of condensed tannins. Stiasny number is determinate by weighing the 2 g tannin powder sample then dissolved in 10 mL of water and reacted in a waterbath at 100° C with 37% formaldehyde (2 mL) that catalyzed by 10 N HCl 1 mL for 30 minutes. The resulting precipitate is accurately weighed and the Stiasny value is calculated as a percentage of the weight of dry extract [17].

### *2.3. Determination solid content of tusam tannin bark*

A total of 2 g of the extract sample was weighed (W1) and dried for 24 hours in an oven at 100° C then cooled in a desiccator and weighed (W2) [18]. Solid content is calculated by the formula:

$$\text{Solid content (\%)} = \frac{W2}{W1} x\ 100\% \qquad (1)$$

### *2.4. Determination of optimum tannin and resorcinol formulations*

The optimum formulation is carried out to see the effect of resorcinol addition to stiasny number. The purpose of adding resorcinol is to activate phenolic compounds contained in tusam bark [14] so that it is expected to increase the tannin's reactivity to resorcinol. Ratio of tannin and resorcinol is showed in Table 1.

**Table 1.** Ratio of tannin resorcinol

| Tannin | Resorcinol |
|---|---|
| 1 | 0.000 |
| 1 | 0.025 |
| 1 | 0.075 |
| 1 | 0.125 |

IOP Conf. Series: Materials Science and Engineering **935** (2020) 012032 doi:10.1088/1757-899X/935/1/012032

### *2.5. Determination of optimum tannin and formaldehyde formulations*
Addition of formaldehyde aims to react tannin with formaldehyde. The optimum formulation between tannins and formaldehydes is intended to see the effect of formaldehyde addition to the levels of solid content of tannins so as to form a new molecule comes from reaction of tannins with formaldehyde. Ratio of tannin and formaldehyde is determined from the results of modifications of previous studies [18-20] which are showed in Table 2.

**Table 2.** Ratio of tannin formaldehyde.

| Tanin | Formaldehida |
|---|---|
| 1 | 0.00 |
| 1 | 0.05 |
| 1 | 0.10 |
| 1 | 0.15 |

## 3. Results and Discussion

### *3.1. Optimum formulation of tannin resorcinol*
The optimum formulation produced from Tusam tannin bark and resorcinol is shown in Figure 1

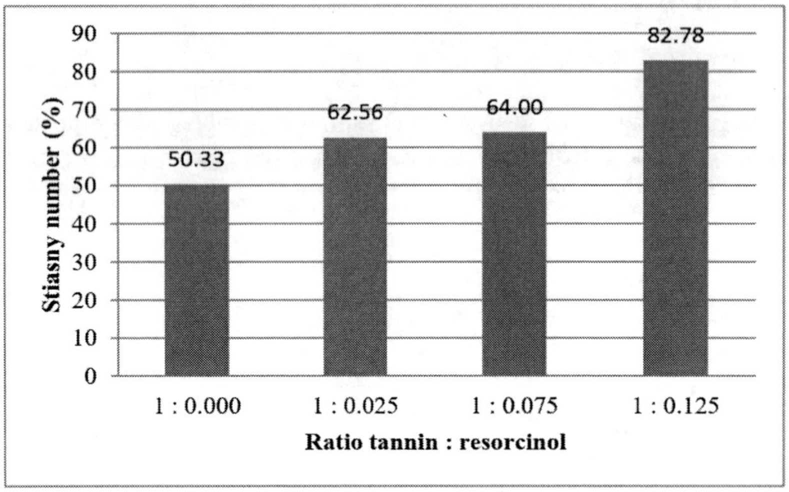


**Figure 1.** Results of solid content of tusam tannins bark and resorcinol

Figure 1 explains that the addition of resorcinol affects the percentage of the stiasny number where the stiasny number indicates the tannin reactivity of resorcinol. The higher levels of resorcinol are added, the greater the percentage stiasny number. The value of stiasny number of tusam tannin bark in figure 1 is 50.33%, but with the addition of resorcinol, the value is increase until the highest resorcinol addition reaches 82.78% whereas the resorcinol ratio which is the more reactive formulation of tannins. This result is almost same with previous research [1] on the percentage of the tannin stiasny number of 47.22% and after the addition of resorcinol, the percentage increased to 66.67%.

This also indicates that the addition of resorcinol to the adhesive can help increasing the complete copolymerization reaction, so that the molecules in resin are increasing [14]. In addition, resorcinol is also added with the aim to bind formaldehyde groups that do not react with tannins so as not to cause formaldehyde emissions [16].

*3.2. The optimum formulation of tannin formaldehyde*

The optimum formulation in tannin formaldehyde is shown in Figure 2. The results in Figure 2 show that the addition of formaldehyde affects percentage of formaldehyde tannin solid content. The higher addition of formaldehyde, the solid content decreases. The result of solid content in tusam tannin bark is 13.83%, but with the addition of formaldehyde by 5% of the amount of tannin, the value of solid content increased to 16.86%, but in the addition of formaldehyde 10% and 15%, the percentage of solid content decreased respectively also became 15.86 and 14.12%. This effect is not much different from tannin in research [18] whereas the level of solid content increases with the addition of formaldehyde by 5% and decreases with the increase in the addition of a number of formaldehydes. The same thing happened in mahogany tannins in the study [20] where an increase in solid content to 18.15% occurred in the addition of formaldehyde by 5% and decreased with the addition of formaldehyde by 20% to 17.77%.

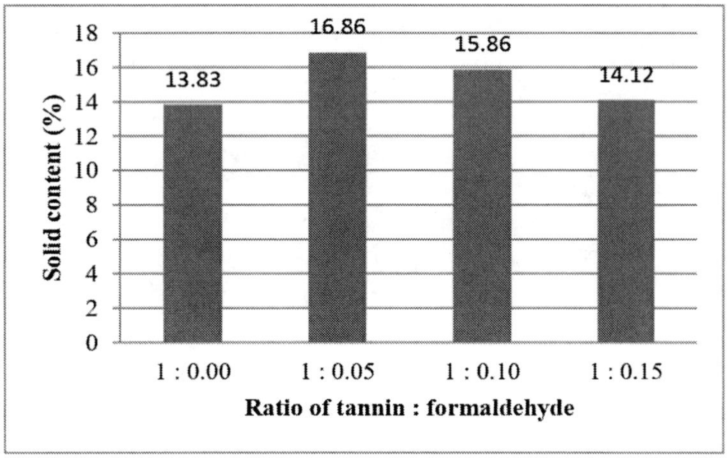


**Figure 2.** Results of optimum formulation of tannin and formaldehyde

The addition of formaldehyde to make polymerization bridges whose bonds tend to be more stable so that it increases the adhesion strength, but the addition of formaldehyde with high levels will cause the crosslinking process being weak and imperfect because there are a number of formaldehyde groups which do not react with tannin [18,20].

**References**

[1] Dradjat and Santoso A 2017 Identification of optimum formula on tannin adhesive by xrd and dta methods *Jurnal ITEKIMIA* **2**(1)1-14
[2] Raharyaningsih MA and Azizah R 2017 Formaldehyde emission levels and eye irritation in workers in the production area of wood adhesives industries in Surabaya *Enviromental Health Journal* **9**(2) 191-9
[3] Moubarik A, Allal A, Pizzi A, Charrier F, and Charrier B 2011 Characterization of a formaldehyde free cornstarch tannin wood adhesive for interior plywood *Eur. J. Wood and Wood Product* **68**(4) 427-33
[4] Santoso A 2005 Utilization of lignin and tannin as substitute alternative of glue raw material for wood composite *Proceedings of the national polymer symposium* 155-64
[5] Li T, Cao M, Liang J, Xie X, and Du G 2017 Mechanism of base-catalyzed resorcinol formal dehyde and phenol resorcinol formaldehyde condensation reactions: a theoretical Study *J. Polymers* **9** 426

[6] Yang I, Ahn S, Choi I, Kim HY, and Oh S 2009 Adhesives formulated with chemically modified okara and phenol resorcinol formaldehyde for bonding fancy veneer onto high-density fiberboard *Journal of Industrial and Engineering Chemistry* **15** 398-402
[7] Grisby W and Warnes J 2004 Potential of tannin extracts as resorcinol replacements in cold cure thermoset adhesives *J. Biomaterial Engineering* **62** 433-8
[8] Yang T, Dong M, Cui J, Gan L, and Han S 2019 Exploring the formaldehyde reactivity of tannins with different molecular weight distributions: bayberry tannins and larch tannins *Holzforschung*1-10
[9] Pizzi A 1982 Pine tannin adhesives for particle board *Holz als Roh-und Werkstoff* **40** 293-301
[10] Gaspar F, Cruz H, Gomes A, and Nunes L 2009 Production of glued laminated timber with copper azole treated maritime pine *Eur.J. Wood Prod.* **68** 207-18
[11] Saad H, Khoukh A, Ayed N, Charrier B, and Charrier F 2014 Characterization of Tunisian Aleppo pine tannins for a potentialuse in wood adhesive formulation *Industrial Crops and Product* **61** 517-25
[12] Derkyi NSA, Sekyere D, and Darkwa NA 2014 Effect of extraction solvent on tannin-formaldehyde adhesives for plywood production *JENRM* **1**(2) 120-6
[13] Santoso A, Pari G, and Jasni 2015 Quality of laminated boards glued with resorcinoladhesive from merbau wood extracts *J. Forest Product research* **33**(3) 253-60
[14] Santoso A and Abdurachman 2016 Characteristics of mahogany bark extract as wood adhesive *J. Forest Product research* **34**(4) 269-84
[15] Yusro F 2013 Active Tannin Concentration of jengkol bark (*PithecellobiumJiringa Jack*) extract and its reactivity to formaldehyde *J. Vokasi* **9**(1) 21-6
[16] Hernawati A and Sutoyo S 2018 Characteristics of adhesive firmness of tannin resorcinol formaldehyde from johar sawdust extract as adhesive on lamina wood *Proceedings of national seminar on chemistry* **171** 111-5
[17] Elbadawi M, Osman Z, Paridah T, Nasroun T, and Kantiner W 2015 Mechanical and physical properties of particleboards made from Ailanthus wood and UF resin fortified by Acacias tannins blend *J. Mater. Environ. Sci.* **6**(4) 1016-21
[18] Rachmawati O, Sugita P, and Santoso A 2018 Synthesis of tannin resorcinol formaldehyde adhesive from mangium bark extract for improving quality of oil palm trunks *J. Forest Product Research* **36**(1) 33-46
[19] Jessica 2018 Characterization and development of mangium (*acacia mangium* Willd.) extract tannin as an adhesive of laminated wood [Dissertation] Pascasarjana Institut Pertanian Bogor
[20] Lestari ASRD 2018 Mahogany tannin adhesive for glued laminated lumber [Dissertation] Pascasarjana Institut Pertanian Bogor

**Acknowledgement**
We acknowledge to the sponsorship of the Ministry of Research, Technology, and Higher Education for supporting this research well through the Master Thesis Research Grant Scheme 2020.

IOP Conf. Series: Materials Science and Engineering **935** (2020) 012033 doi:10.1088/1757-899X/935/1/012033

# Sound damping and sound absorption of *Acacia mangium* from various diameter

**A Rangga, Salmia, A D Yunianti*, and B Putranto**

Departement for Forest Product, Forestry Faculty of Hasanuddin University, Makassar, 90245, Indonesia

*Email: dettiyunianti70@yahoo.com

**Abstract.** Previous studies recommend acacia wood to be used as an acoustic component of the panel (sound dampening). This study wanted to obtain optimal diameter information in the use of *Acacia mangium* as raw material for acoustic panels. Given the diameter of the wood related with the development of cells that occur in trees that affect the other basic properties of the wood. The research tree of the acacia wood with diameter of 17 cm, 25 cm and 35 cm. The basic properties of the wood being tested are vessel diameter, density and MOE. While the acoustic properties are sound damping and sound absorption. The results showed the acoustic properties of acacia wood were not influenced by diameter, but related to the basic properties of wood such as vessel diameter, density and MOE.

## 1. Introduction

One coefficient of sound absorption shown by the proportion (percentage) of sound absorbed in the form of sound waves of the uses of wood is the main material of musical instruments because wood has acoustic properties [1]. Wood acoustics is a source of sound and has the ability to absorb sound well that measured by the coefficient of sound absorption shown by the proportion (percentage) of sound absorbed in the form of sound waves.

Imported wood trade for musical instruments has limited, so that the price of imported wood is increasingly expensive, musical instrument makers are looking for alternative types of wood in the country as substitute material. Research on the basic nature of acoustic wood originating from Indonesia (local) is limited by the quality of products of ready-made musical instruments, especially guitars made from local wood, without information that supports the acoustic [2-4]. Meanwhile, there are some types of local wood that have been known has a good acoustic properties and can be used as raw material for musical instruments such as mahogany wood (*Swietenia mahagoni*), pine (*Pinus merkusii*), sonokeling (*Dalbergia latifolia*),merah meranti (*Shorea pinanga*), sungkai (*Peronema canescens*), and acacia (*Acacia mangium*) [5,6]. Specially for acacia's wood, the result of the research showed that acacia's wood with a diameter of 30 cm had a high sound absorption value of 0.476 ns, sound speed of 0.672 ns, with a density of 0.58 kg/cm$^3$ than the other types of wood like mahogany, sonokeling and manii. So, the acacia's wood is recommended for acoustic panel components (dampening sound).

This study wanted to find out the acoustic's properties of wood which from small diameter acacia wood, with assuming differences of the tree's diameter can be expected has differ in age so that it can affect the acoustic properties of wood. The cells formation in a tree affects the basic properties of wood including the density and elasticity of wood, where these properties affect the acoustic properties of wood.

## 2. Materials and Methods

### *2.1. Materials*

The acacia trees are diameter with 17 cm, 25 cm and 35 cm. Each diameters of acacia tree are felled and cut short log 50 cm from a base. See the Picture 1 for more details.

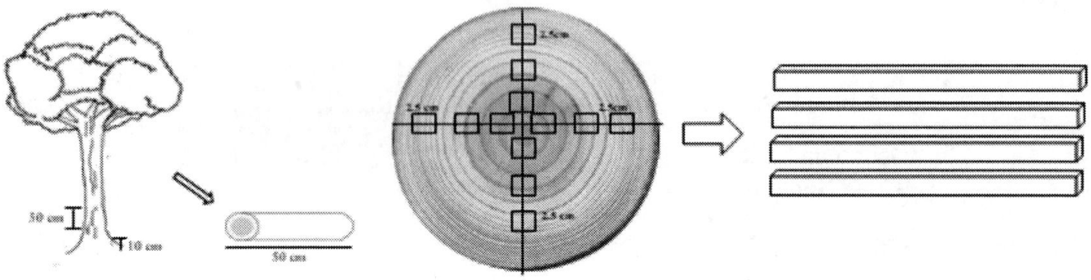

**Figure 1.** Cutting log tree's scheme and cutting sample's scheme

### *2.2. Methods*

2.2.1. Observation of anatomical structure

Wood anatomic structure observed sum and diameter of vessel at incision preparate from each sample test. For procedure of incision prepare make based on Purnawati *et al.* [7] and Sulistyobudi *et al.* [8] for vessel measurement results.

2.2.2. Test of physical and mechanical properties

Testing of mechanical properties are MOE based on ASTM D-143 [9] with a test sample measuring 2.5 x 2.5 x 41 cm. The loading speed used was 1.3 mm/second with a buffer distance (L) 36 cm. The physical properties of the wood calculated are moisture content, density and specific gravity of the wood. The sample of physical properties research comes from the mechanical properties test sample. The undamaged test sample section is used to measure moisture content, specific gravity and wood density based on [10]. The tools used for this test is Universal Testing Machine (UTM).

2.2.3. Test of acoustic properties

Testing of acoustic properties is carried out before other tests, namely mechanical properties, physical properties and anatomical structure of wood. Testing acoustic properties using sonic waves with a test sample measuring 2.5 x 2.5 x 41 cm refers to ASTM D-143 [9]. The test sample used was 32 units (3 wood diameters with different number of replications each diameter). The tools used for this test is digital oscilloscope Sonic wave testing is done by placing the test sample on a table. At the first side attached sound detector that has been connected with a digital oscilloscope. While the other side is tapped using a pendulum with the same distance that is as far as 12 cm from the test sample (Figure 2). Sonic waveform readings are seen on a digital oscilloscope screen connected to the sound detector using a variable frequency in the range of 1-8 kHz.

*2.2.3.1. Sound damping*

Sound damping (logarithmic decrement) can be calculated using the formula:

$$\tan\delta = \ln\frac{A1}{A2} \quad \text{with } \zeta = \frac{\tan\delta}{2\pi} \tag{1}$$

where, tan δ : logarithmic decrement, A1 : wave height on the first period (V), A2 : wave height on the second period (V), ζ : damping ratio/sound damping, π : 3.14.

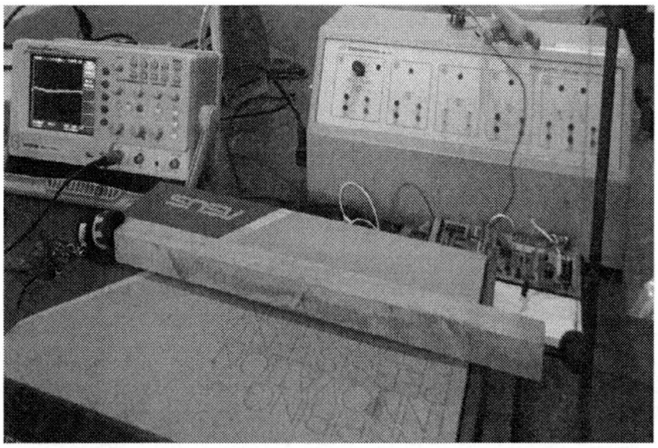

**Figure 2.** Testing of acoustic properties with oscilloscope screen

*2.2.3.2. Sound Absorption*

Sound absorption can be calculated using the formula:

$$SA = \frac{A_1 - A_2}{A_1} \times 100\% \quad (2)$$

where, SA : Sound absorption (%), A1 : wave height on the first period (V), A2 : wave height on the second period (V)

*2.3. Data Analysis*

Data analysis was performed with a simple descriptive analysis of the parameters of anatomical structure, physical properties and mechanical properties. Values displayed are average values, minimum values, maximum values and standard deviations. The results of wood acoustic properties were analyzed in oneway analysis using SPSS version 22 on all diameters.

## 3. Result and Discussion

*3.1. Characteristics of acacia wood*

The anatomical structure of wood observed was the number of vessel and vessel diameters. Based on Sulistyobudi *et al.* [8] the number of vessel of acacia wood in this study was classified as very rare to infrequently with a large vessel diameter. The number of vessel and vessel diameters of acacia wood at various diameters is presented in Table 1. The greater the vessel diameter, the fewer the number of vessel per $mm^2$.

Physical properties calculated in this study are moisture content, density and specific gravity of wood, while the mechanical properties are MOE. As the diameter increases, physical and mechanical properties also increase.

MOE values increase with increasing wood diameter. This is in line with research conducted by Tabet and Fauziah [11] and Susanti [12] which states that as the diameter of the tree increases, the MOE value increases.

MOE test results and specific gravity showed acacia wood in this study included in the strong class III [10]. Table 1 shows the number of vessel, vessel diameters, density and specific gravity of wood and acacia wood MOE at several diameters.

**Table 1.** Research data on the number of vessel, vessel diameter, moisture content, specific gravity, density, and MOE of acacia wood at several diameters

| Characteristics of acacia wood | d 17 | d 25 | d 35 |
|---|---|---|---|
| Anatomi Structre | | | |
| Number of Pores (per mm$^2$) | 5 | 7 | 4,6 |
| Min | 3 | 5 | 4 |
| Maks | 7 | 9 | 6 |
| Sd | 0.93 | 1.08 | 0.71 |
| Vessel Diameter (μm) | 139 | 127 | 150 |
| Min | 80 | 98 | 106 |
| Max | 213 | 206 | 263 |
| SD | 27 | 27 | 31 |
| Physical Properties | | | |
| Moisture Content (%) | 13.84 | 15.35 | 16.86 |
| Min | 13.27 | 12.27 | 9.94 |
| Max | 14.63 | 18.12 | 37.84 |
| SD | 0.45 | 1.52 | 7.48 |
| Specific Gravity | 0.56 | 0.65 | 0.65 |
| Min | 0.50 | 0.60 | 0.57 |
| Max | 0.63 | 0.68 | 0.73 |
| SD | 0.05 | 0.03 | 0.04 |
| Density (g/cm$^3$) | 0.63 | 0.75 | 0.78 |
| Min | 0.57 | 0.69 | 0.65 |
| Max | 0.72 | 0.78 | 0.90 |
| SD | 0.05 | 0.03 | 0.06 |
| Mechanical Properties | | | |
| MOE (kg/cm$^2$) | 48,533 | 62,186 | 79,330 |
| Min | 20,244 | 53,021 | 46,717 |
| Max | 64,219 | 70,875 | 102,778 |
| SD | 15,535 | 5564 | 12,556 |

### *3.2. Acoustic properties*

Acoustic materials can be divided into three basic categories, namely absorbing materials, barrier materials, damping materials [13]. In this study, the acoustic properties studied were sound damping and sound absorption with a frequency range of 1-8 kHz.

#### *3.2.1. Sound damping*

The results of the sound damping test on the diameters of 17, 25 and 35 cm of acacia wood are presented in Figure 3. Although there are differences in the value of damping sound from diameter 17-35 cm, but the value is relatively the same. The results of the analysis of variance showed that the difference in diameter of the wood did not have a real effect on the sound damping where the Sig. (0.281>0.05). Meanwhile for physical and mechanical properties do not have a significant correlation it is suspected that the difference in diameter, especially the density is not enough to give effect to the acoustic properties of sound damping. Trend of the number and diameter of vessel is not in line with the physical and mechanical properties of wood, this illustrates that the number and diameter of vessel does not greatly affect the density of wood the possibility of cell wall thickness being more influential but in this study were not analyzed. According to Baihaqi [14], the material used in the need for sound dampening is to have a low MOE value, while the value of density, sound damping, and sound absorption is high. In this study, a diameter of 17 cm is enough to be recommended as an acoustic raw material.

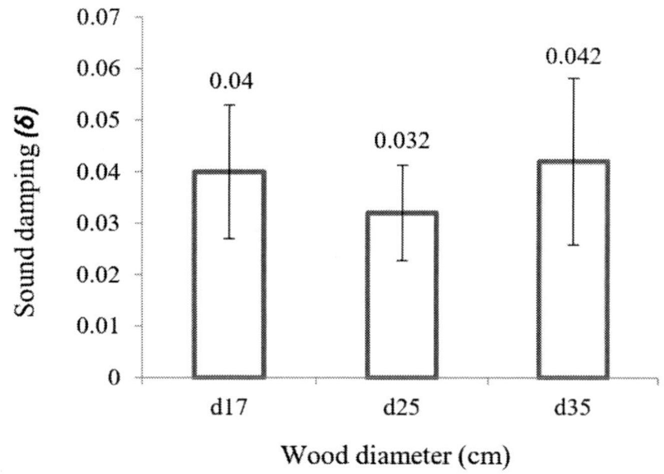


**Figure 3**. Sound damping seen from various diameters

*3.2.2 Sound absorption*

Part of the acoustic energy that affects wood partially or wholly can be absorbed, refracted and reflected. This acoustic energy will certainly cause molecular friction and cause energy changes from acoustic energy to thermal energy [15].

The ability of wood to absorb sound with the coefficient of sound absorption as indicated by the percentage of sound absorption. Sound absorption coefficient is influenced by density and other factors such as elasticity (MOE), moisture content, temperature, intensity, sound frequency and wood surface conditions. Wood with low density and elasticity at high temperature and moisture content absorbs more noise, greater absorption of sound with low frequency and lower on wood that is given a coat of varnish [1].

The results of sound absorption testing at diameters of 17, 25 and 35 cm are presented in Figure 4. The low sound absorption value at a diameter of 25 cm is suspected because at that diameter the pore diameter size is smaller than the other pore diameters, where one of the criteria for sound absorbing material is porous which functions as a cavity resonator. Through these pores, sound waves enter and vibrate the air molecules in the vessel [16]. The amount of sound absorption in the absorbent material is expressed by the absorption coefficient (α). Absorption coefficient (α) is expressed in numbers between 0 and 1. The absorption coefficient value 0 states that no sound energy is absorbed and the absorption coefficient value 1 expresses perfect absorption [17]. The results of the analysis of variance showed that the difference in diameter of the wood did not have a real effect on the sound damping where the Sig. ($0.281>0.05$) shows that differences in wood diameter do not have a significant effect on sound absorption.

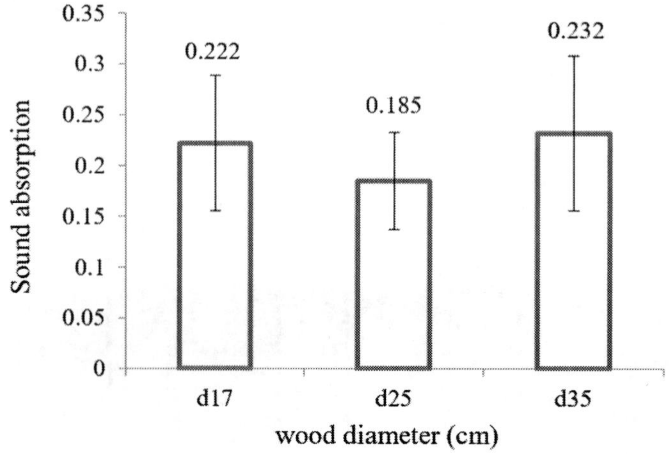


**Figure 4.** Sound absorption seen from various diameters

## 4. Conclusion

Based on the results of this study, the acoustic properties of acacia wood are not influenced by diameter, so the use of acacia wood between 17 cm to 35 cm in diameter can be recommended as a good acoustic raw material.

## References

[1] Tsoumis G 1991 *Science and Technology of Wood: Structure, Properties, Utilization* (New York, Van Nostrand Reinhold)

[2] Ardhianto N 2002 Study of making and quality assessment of acoustic guitar using mahogany wood (*Swietenia mahagoni* Jack.) and sonokeling (*Dalbergia latifolia* Roxb.) [Essay] Faculty of Forestry, IPB University

[3] Firmansyah R 2006 Study on the possible use of coconut wood (*cocos nucifera* l.) for guitar raw materials [Essay] Faculty of Forestry, IPB University

[4] Sejati KW 2008 Study of making acoustic guitar using agathis, pine, and sonokeling wood. [Essay] Faculty of Forestry, IPB University

[5] Widiyati M 2008 relation of anatomical and chemical characteristics of six types of wood to acoustic properties of wood [Essay] Faculty of Forestry, IPB University

[6] Karlinasari L, Nawawi DS, and Widyani M 2010 Study of the anatomical and chemical properties of wood in relation to the acoustic properties of wood *Bionatural Journal of Life and Physical Sciences* **12**(3) 110-6

[7] Purnawati R, Wahyudi I, and Priadi T 2012 Anatomical properties of wood *Flindersia pimenteliana* F. Muell from Teluk Wondama West Papua *Journal of Tropical Wood Science and Technology* **10**(2)

[8] Sulistyobudi A, Mandang YI, Damayanti R, and Rulliaty S 2008 *Microscopic Characteristics to Identify Wide Leaf Wood* (Bogor: Forest Products Research and Development Center)

[9] [ASTM] American Society Institute 2000 *Standard Test Methods for Small Clear Specimens of Timber ASTM D-143* (United State, Philadelphia: ASTM)

[10] [BSN] Badan Standardisasi Nasional 2002 *Standar Nasional Indonesia (SNI) SNI 03-6848-2002 Method of Testing the Density of Logs and Structural Timber* (Jakarta: BSN)

[11] Tabet TA and Fauziah AA 2013 Cellulose Microfibril Angle in Wood and Its Dynamic Mechanical Significance *Cellulose-fundamental aspects* **10** 51105

[12] Susanti J 2019 Relation between modulus of elasticity and nano structure of acacia wood (*Acacia mangium* Willd.) from various diameter [Essay] Hasanuddin University

IOP Conf. Series: Materials Science and Engineering **935** (2020) 012033 doi:10.1088/1757-899X/935/1/012033

[13] Lewis H and Douglas H 1993 *Industrial Noise Control Fundamentals and Application* (New York: Reyised)
[14] Baihaqi H 2009 Relation between acoustic properties with physical and mechanical properties of five types of wood [Essay] Faculty of Forestry, IPB University
[15] Zulkifli R, Nor MMJ, Tahir MMF, Ismail AR, and Nuawi MZ 2008 Acoustic properties of multi-layer coir fibres sound absorption panel *Journal of Applied Sciences* **8**(20) 3709-14
[16] Simatupang V 2007 Acoustic Test of Absorber Material with Variation of Core Configuration of Composite Material Based on Natural Fiber (Coconut Powder) [Thesis] Faculty of Forestry, IPB University
[17] Hayat W, Syakbaniah, and Yenni D 2013 Effect of density on absorption coefficient of pineapple leaf fiber particle board sound (*Ananas comosus L Merr*) *Pillar of Physic.* **1** 44-51

# Trypsin inhibitor activities as defense mechanism of sengon (*Falcataria moluccana*) against pest attacks

**U J Siregar*, I M Situmorang, F A Pasaribu, A Lestari, Y Istikorini, and N F Haneda**

Department of Silviculture, Faculty of Forestry, IPB University, Bogor, 16680, Indonesia

*Email: ulfahjs@apps.ipb.ac.id

**Abstract.** Sengon (*Falcataria moluccana*) plantation always suffers from a serious attack of stem borer, which is called Boktor (*Xystrocera festiva* Pascoe). This research aimed to investigate the presence of anti-pest substances, i.e. trypsin inhibitor on each part of a sengon tree, which might serve as a defense mechanism against the pest. Healthy and severely attacked sengon trees from Cianjur, Solomon, and Kediri provenances were sampled for their leaves, barks, and woods. Trypsin inhibitor activity from each sample was assayed using synthetic enzyme and substrate in a spectrophotometer. Data were analyzed under Complete Randomized Design, using SPSS 11.5. Different provenances, tree condition, and part of tree gave significant effects on trypsin inhibitor activity. Healthy trees had a higher Trypsin Inhibitor Activity (TIA) value compared to the pest-infected tree. The sengon tree tissue that had the highest TIA is the bark from healthy trees of Solomon provenance, with a value of 730.0403 TIU/mg. Obviously the bark is the first defense against the stem borer pest.

## 1. Introduction

One of tree species that are widely utilized in the development of plantation forest and community forest in Indonesia today is sengon (*Falcataria moluccana*). As legume tree [1] sengon (*F. moluccana*) could improve soil fertility, has an economical value that is quite high and prospective, also easily cultivated and adapted to marginal land. However, when planted in monoculture sengon always suffers from a serious attack of stem borer, which is called Boktor (*Xystrocera festiva* Pascoe). The increase of sengon populations as main food of the stem borer resulted in overbreed of the pest (*X. festiva*) and expanding pest population thereby leading to considerable economic loss. So far there is not any effective and efficient control methods for this pest. Therefore, a resistant line of sengon tree is needed to manage the pest attacks. Observation in the field showed that although most sengon trees are suffering to the pest, several individuals showed some resistance against the pest. These seemingly resistant individuals showed very few damages in the main stem and less suffering from the pest attacks. These resistant individuals are important for tree breeding program of sengon trees.

Similar to other insects, inside Boktor digestive tract has trypsin enzyme among others, which has a linear pattern of enzyme activity [2]. Trypsin enzyme plays a role in the digestion of proteins in the insect diet into simple molecules which is ready to be absorbed by the cells [3]. Boktor pest usually lays eggs at the wounded main stem of sengon. When the eggs hatch the larvae would start chewing the outer bark and move inside while eating the bark and eventually bore into the woody tissue. The larvae would eventually stay inside the wood until develop cocoon and hatch imago.

On the other hand, the seemingly resistant sengon trees are known to have compounds that can inhibit trypsin enzyme to work in boktor digestion process, known as trypsin inhibitor [4]. The inhibitor has different activities on each part of the sengon tree. Given the huge losses inflicted by Boktor's attack, it is interesting to investigate which part of the sengon tree, as well as which provenance has the largest trypsin inhibitor activity. Such information can be used to obtain superior sengon seeds that contain the largest trypsin inhibitor activity that can inhibit the work of trypsin enzyme inside boktor digestive tract.

This research aims to determine the trypsin inhibitor activity in sengon leaves, bark, and wood from Cianjur, as well as the differences in trypsin inhibitor activity between healthy sengon trees and pest-infected trees from Solomon and Kediri provenances.

## 2. Materials and Methods

### *2.1. Preparation of raw materials*

Observations on *trypsin* inhibitor are carried out using chemicals as substrate, i.e. synthetic benzoyl-DL-arginine-p-nitroanilid (BAPNA), buffer Tris-HCL, dimethyl sulfoxide (DMSO), NaOH, 30% acetic acid, trypsin enzyme from SIGMA 3.5 U/mg.

### *2.2. Determination of trypsin inhibitor activity*

This method is based on the inhibitory hydrolysis of a synthetic substrate (benzoyl-DL-arginine-p-nitroanilid) by a synthetic trypsin enzyme, in the presence of an anti-trypsin or trypsin inhibitor. The compounds that should be prepared including reagents and extracts, as followed.

A. Preparation of reagent solution:
   1) Tris-HCL buffer 0.05 M, pH 8.2 containing 0.02 M $CaCl_2$: 6.05 g Tris (methyl aminometen hydroxy) and 2.94 g $CaCl_2$. $H_2O$ dissolved in 900 mL distilled water. The pH was adjusted to 8.2 and the volumes of the solution were filled up to 1000 mL
   2) BAPNA solution (benzoyl-DL-arginine-p-nitroanilid): 40 mg BAPNA dissolved in 1 mL DMSO (dimethyl sulfoxide), then diluted to 100 mL by adding a Tris buffer that had a temperature of 37°C
   3) Trypsin Solution: SIGMA trypsin enzyme with the activity of 3.5 U/mg dissolved in 200 mL HCl 0.001 M
   4) 30% acetic acid solution (V/V)

B. Preparation of sample extracts:
   1) Leaf, bark, and wood samples were freeze-dried and then powdered
   2) As much as 0.2 g of each powdered sample was suspended in 50 mL of 0.1 N NaOH solution. After stirring for 3 hours at room temperature using magnetic stirrer then centrifuged at 2,000 x g or 5,000 rpm for 10 minutes at 5°C and the supernatant was separated
   3) The extract obtained was then diluted with distilled water as such that 1 mL of the extract will produce an inhibitory force about 40-60% of the trypsin enzyme activity used
   4) Procedures 1) and 2) were performed for leaves, bark, and wood powder from Cianjur, Solomon, and Kediri provenances

2.2.1. Procedure for analyzing trypsin inhibitor activity:

Series of extracts, i.e. 0; 0.2; 0.4; 0.6; 0.8; 1.0 mL were pipetted into the reaction tube. Distilled water was added to each reaction until the volume reached 2 mL. As much as 2.0 mL of trypsin solution was added to each reaction tube, then all tubes were incubated in a 37°C water bath and left for 5 minutes. As much as 5 mL of BAPNA solution 37°C was added into each tube, and the solution was mixed quickly using a vortex, and incubated in a 37°C water bath for 10 minutes. After 10 minutes, as much as 1 mL of 30% acetic acid is added to each tube and mixed quickly. When the solution was becoming clear, measurement on the spectrophotometer can be done immediately. Absorbance measurements of the extracts were done at a wavelength of 410 nm, using the above blank reagent containing 0 mL extract as a reference. If necessary, in the case of colored extracts, the blank samples can be made in the following ways: adding 5 mL of BAPNA solution into 2 mL of extracts, the mixture is incubated at 27°C for 10 minutes, then add 1 mL of 30% acetic acid and 2 mL of trypsin solution.

One Trypsin Unit (TU) is defined as an increase of 0.01 units of absorbance at 410 nm per 10 mL reaction mixture under the conditions used. Trypsin inhibitors activity is expressed as one Trypsin Unit Inhibited (TUI). The trypsin inhibitor activity is calculated as follows:

| | |
|---|---|
| TU | : absorbance/0.01 |
| TIU | : TU (S) – {TU (C) – TU (BC)} |
| TIU/mL | : TIU/2 |
| TIU/mg Example | : (TIU/mL x f.p)/sample weight |

where, C: example; S: Standard; BC: Example blanko; TU: trypsin unit; TIU: trypsin inhibitor unit; f.p: the dilution factor.

*2.3 Data analysis*
Data on trypsin inhibitor enzyme activity were analyzed using factorial complete randomized design. The equation form:

$$Y_{ijkl} = \mu + A_i + B_j + C_k + AB_{ij} + AC_{ik} + BC_{jk} + ABC_{ijk} + \varepsilon_{ijkl} \quad (1)$$

where:
i: 1 (Solomon), 2 (Kediri); k: 1 (leaves), 2 (bark), 3 (wood); j: 1 (healthy), 2 (attackted); l:1, 2, 3, 4, 5
$Y_{ijkl}$ : response provenance influence to-i, tree condition to-j, part of tree to-k, and repeat to-l
$\mu$ : average general value
$A_i$ : provenances to-i
$B_j$ : tree condition to-j
$C_k$ : parts of tree to-k
$AB_{ij}$ : interaction of factor A to-i and B to-j
$AC_{ik}$ : interaction of factor A to-i and C to-k
$BC_{jk}$ : interaction of factor B to-j and C to-k
$ABC_{ijk}$ : interaction of factor A to-i, B to-j, and C to-k
$\varepsilon_{ijkl}$ : trial error

## 3. Results and Discussion

*3.1 Study on trypsin inhibitor activity in every part of a sengon tree*
Trypsin inhibitor is compounds that have the ability to inhibit the proteolytic activity of trypsin enzymes. These compounds have been found in some crop plants, especially the types belonging to Leguminoseae family. The active compounds of this inhibitor are protein. Trypsin inhibitor has physiological functions as the plants' abilities to defend against natural pest attacks [5].

**Table 1.** Result of variance trypsin inhibitor activity using trypsin enzymes synthetic

| Factors | Various prints |
|---|---|
| Part of tree from Cianjur provenance | ns |
| Provenance | ** |
| Tree conditions | ** |
| Part of tree | ** |
| Provenance x Tree conditions | ns |
| Provenance x Part of tree | ns |
| Tree conditions x Part of tree | ns |
| Provenance x Tree Condition x Part of tree | ns |

Remarks: x: Interaction between factors; * *: highly significant; ns: not significant

Boktor pest attacks sengon trees when the trees have formed woody tissues, preferably after 3 – 4 years old. Imago of the pest usually lays eggs on wounded tree bark, and when larvae is hatched, the larvae will start feeding on the bark [6]. As the larvae are getting older they start to bore the woody tissue in order to establish the place for pupation. As sengon is a legume tree, it was assumed that sengon would have trypsin inhibitor in each part of the tree as defense mechanism against the pest.

Statistical tests at 99% confidence level (Table 1) showed that the provenance, tree condition, and part of the tree, with exception of those coming from Cianjur provenance, had a highly significant effect on trypsin inhibitors activity. Figure 1 showed that although part of tree from Cianjur provenance did not differ of their trypsin inhibitor activities, the bark still had the highest activity, while the leaves were lowest.

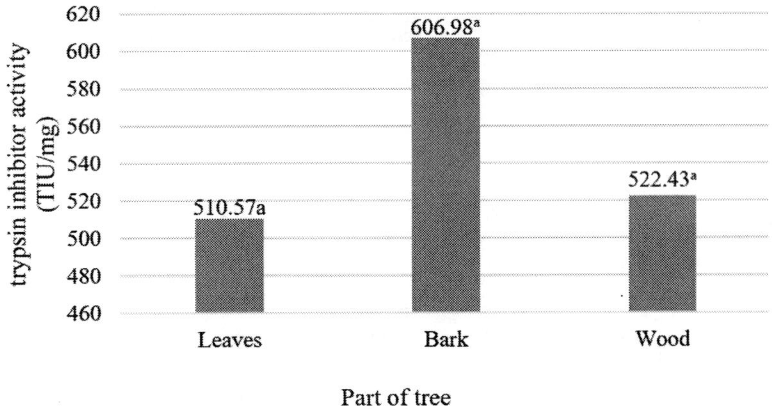


Remarks: ªThe same letter shows no significant difference at 95% confidence interval

**Figure 1.** Average trypsin inhibitor activity of each sengon tree part from Cianjur provenance

Figure 1 shows that the average value of TIA found in the bark is 606.98 TIU/mg, followed by the wood at a value of 522.43 TIU/mg, and the lowest is in leaves of 510.57 TIU/mg. This result corresponds to Winarni [4] which also examined the activity of trypsin inhibitor in plus trees from local community forest in West Java, of which activity and its diversity in the wood was lower than the bark. This indicates that the bark is the first main protection of sengon tree from the boktor pest attack. Higher trypsin inhibitor activity is present in the bark because the bark contains many living cells compared to wood. Living cells have organelles especially ribosomes, which is active sites for protein manufacturing, most probably including anti-protein trypsin enzymes that work on boktor's digestive system [7].

Figure 2 shows the results from experiments using different part ot the trees, coming from two different provenances, with two different tree conditions. Contrary to the results of different tree parts from Cianjur provenance (Figure 1), the different parts of trees coming from Solomon and Kediri provenances showed highly significant difference (Table 1 and Figure 2). The reason for this apparently discrepancy results were different sample sizes, of which on the later experiments more samples were examined, therefore the results were highly significant. The samples of the first experiment only consisted of one provenance without considering the trees condition, whether the tree was healthy or suffer from pest attacks. Albeit the statistical analysis gave different results, the trend however, was still similar. Trypsin inhibitor activity was found highest in the barks compared to other part of the trees, regardless of tree conditions and provenances. The only slight differences in the results of the first and second experiment were the order of amount of trypsin inhibitor activity in the woods and leaves. In the first experiment the wood had higher trypsin inhibitor activity compared to leaves (Figure 1), whereas in the second experiment the leaves had higher trypsin inhibitor activity than the wood (Figure 2). Nevertheless, both wood and leaves were much lower than barks for their trypsin inhibitor activities.

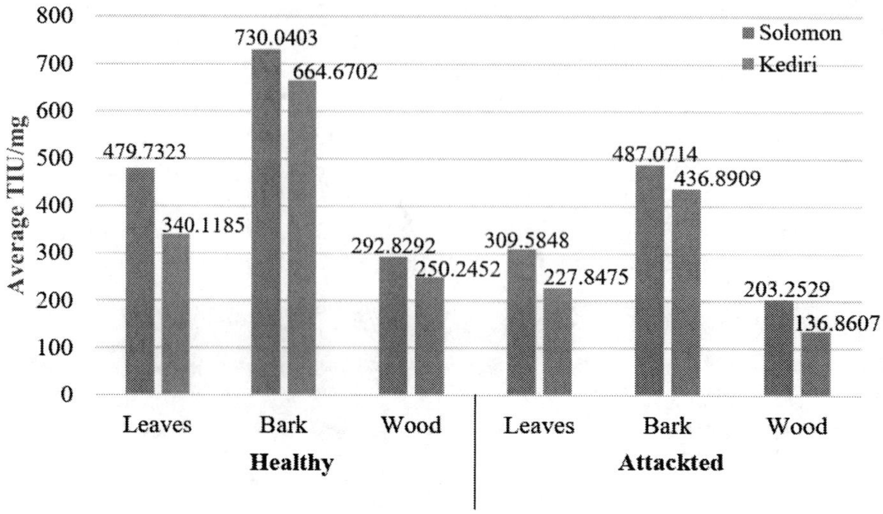


**Figure 2.** Histogram *trypsin* inhibitor activity on healthy trees and the pest-infected tree of Solomon and Kediri's provenance

Another important results as shown in Figure 2 are healthy trees, which were supposed to be resistant trees, always had higher value of trypsin inhibitor activities, regardless of the parts of tree and provenance origins, compared to attacked trees, which were assumed as susceptible trees. This experiment also showed that different provenance could have different resistance or susceptibility against pest attacks. As shown in Figure 2, Solomon provenance appeared to be more resistant to the pest attacks than Kediri provenance, as indicated by their higher trypsin inhibitor activities in every parts of the tree. The relative resistant of Solomon provenance compared to Kediri provenance still needs to be confirmed by comparison to some more other provenances.

The consistency of highest trypsin inhibitor activity in the barks of three provenances studied here, combined with higher activities in healthy, supposedly resistant trees compared to attacked, supposedly susceptible trees, has pin pointed to the role of trypsin inhibitor as defense mechanism in sengon tree against pest attacks. It would be interesting to find out later if this resistance is genetically based, by further identifying genes responsible for insect pest resistance, especially protease inhibitor genes.

## 4. Conclusion

Different provenances, tree conditions, and parts of the tree have highly significant effects on trypsin inhibitors activities. The highest Trypsin Inhibitor Activity (TIA) was consistently found in the barks from healthy, supposedly resistant trees. Trees from Solomon provenance apparently have higher TIA compared to trees Kediri provenance. Sengon tree that has the highest TIA is a healthy bark from tree of Solomon's provenance with the value of 730.0403 TIU/mg.

## References

[1] Heyne K 1987 *Tumbuhan Berguna Indonesia Jilid ke-II* (Jakarta: Badan Litbang Kehutanan Jakarta)

[2] Prasetya A 2007 Studi tentang enzim *trypsin* dan *α-amylase* pada hama boktor (*Xystrocera festiva* Pascoe) serta inhibitor *trypsin* pada pohon sengon (*Paraserianthes falcataria* (L) Nielsen) [Skripsi] Institut Pertanian Bogor

[3] Sunarti TC, Indah Y, and Nur R 2004 *Perubahan Komposisi pada Hidrolisis Enzimatis Pati Umbi-umbian Indonesia* (Bogor: IPB Press)

[4] Winarni I 2003 Studi keragaman *trypsin* inhibitor dan keragaman genetic isoenzim pohon plus sengon (*Paraserianthes falcataria* (L) Nielsen) pada hutan rakyat di Jawa Barat [Bachelor Thesis]. Program Pasca Sarjana Institut Pertanian Bogor
[5] Muchtadi 1993 *Metabolisme Zat Gizi* (Jakarta: Pustaka Sinar Harapan)
[6] Matsumoto K 1994 Studies on the ecological characteristics and methods of control of insect pests of trees in forested area in Indonesia Final Report submitted to Forestry Research and Development Agency, Ministry of Forestry, Jakarta, Indonesia
[7] Siregar UJ, Haneda NF, and Flowrensia L 2011 Correlation of enzyme inhibitor activity in sengon with boktor larval growth on artificial diet *J. Tropical Silviculture* **2**(2) 101-8

IOP Conf. Series: Materials Science and Engineering **935** (2020) 012035 doi:10.1088/1757-899X/935/1/012035

# Variation in density and shrinkage of six eucalyptus species and interspecific hybrid combinations at three sites in Northern New South Wales

**T Listyanto[1*], J D Nichols[2], K Glencross[2], and L Schoer[3]**

[1]Faculty of Forestry, Gadjah Mada University, Yogyakarta.
[2]Sustainable Forestry Program, School of Environmental Science and Management,
[3]Southern Cross University, Lismore, NSW Australia 2480, Lesley Schoer & Associates Pty Limited, PO Box 477 Coffs Harbour, NSW 2450 Australia

*Email: tomy.listyanto@gadjahmada.edu

**Abstract.** The study aimed to investigate the variation of density and shrinkage at radial and axial directions of nine eucalypt taxas at three different sites (Muller, Convery, and Ironbolt). The data used in this research was collected from six years old trials of nine *Eucalyptus* taxa at those three sites. Density and shrinkage were collected from 145 selected trees based on British Standard (BS 373 1957). A two-way analysis of variance and Tukey HSD were used to analysis density and shrinkage parameters. The effect of site, taxa, and interaction between sites and taxas on oven-dry density and shrinkage were statistically significant different ($p<0.05$). In terms of density, *Corymbia variegata* provenance Wodum had the highest oven-dry density at 781 kg/m$^3$ (CV=11.7%). There was a significant variation in oven-dry density at the radial and axial directions. The oven-dry density at Diameter at Breast Height (DBH) of all taxas, except *E. cloeziana* and the hybrid *E. grandis* x *E. camaldulensis*, tended to be lower than at the top point of measurement (TMS). In terms of variation in the radial direction, there were three types of trends of oven-dry density from pith outward. In terms of shrinkage, *C. variegata* provenance Richmond and Wodum, and *E. saligna* had the lowest radial shrinkage, tangential shrinkage and T/R ratio, while *E. dunnii* and *E. pilularis* showed the highest one. Another important finding is that the hybrid between *E. grandis* x *E. camaldulensis* and *E. urophylla* x *E. grandis* showed higher radial and tangential shrinkage than their parent species (*E. grandis*).

## 1. Introduction

Several species of Eucalyptus are in great demand for planting because of its fast growth, adaptability and potential purposes. There are several groups suitable for future genotype selection, including *E. dunniii*, *E. grandis*, *E. cloeziana*, *E. pilularis*, and *Corymbia variegata*, for breeding, improved seed production, and clone development [1]. Arnold [2] explains that *E. grandis* is an important parent for breeding inter-specific eucalypt hybrids (with *E. saligna*, *E. camaldulensis* or other species). On the other hand, *E. camaldulensis* is one of the hardwood species which has been shown to be well-adapted to low water availability, shows good stem-straightness, and is often without bifurcations [2]. Meanwhile *C. variegata* has desirable timber properties for appearance-grade products, good stem form, drought tolerance and adaptability [3]. Therefore, it is remained important to investigate the suitable euaclyptus species that have good growth performance and wood quality.

Species and environment are important factors in developing prescriptions for plantation forests. Different growth conditions significantly affect wood density and shrinkage of selected young poplar hybrid crosses [4]. Survival, growth, stem form and resistance to stem-boring insect larvae were significantly different between sites and between taxa [5]. The Convery site that had Red Ferrosol soil and the highest mean annual rainfall, 1400 mm showed the highest growth of all taxa [5]. It is necessary also to evaluate their wood quality for a range of species and hybrids in different soil types and annual rainfall.

Wood quality can be defined as the characteristics of wood that in turn affect the properties of the manufactured product [6]. Wood density and shrinkage are often used as general indicators of the suitability of wood for many purposes, including manufacture products [7, 8]. This is because the density of wood is a main determinant of strength; consequently, density and strength are directly related to each other [6,9,10]. Wood density also has an influence on wood processing, including on cutting, gluing, finishing, rate of drying and paper making [7], while shrinkage is important in terms of the stability of wood during and after drying [6,11,12].

In eucalyptus species (diffuse porous hardwood), wood density and shrinkage generally are controlled by a combination of environmental factors and silvicultural practice [7]. Understanding the wood density of a particular species allows the wood industry and the consumer to decide the appropriateness of wood uses. The variation of wood density between and within trees are considered to be caused by a combination of factors: growth rate, climate, silviculture and breeding [6,13]. Therefore, considering environmental factors and species is an important step in establishing plantations for commercial purposes with higher wood quality. Consideration of environmental conditions will help to produce a short list of candidate species [14]. Accordingly, different species or hybrids may give different responses for different sites. This research assesses wood quality of nine selected species of eucalypts in three different sites in northern New South Wales. This study examined whether there were differences in oven-dry density and shrinkage among nine selected species of Eucalyptus at three different sites namely Muller, Convery, and Ironbolt.

## 2. Materials and Methods

The data used in this research was collected from six-year-old trials of nine Eucalyptus taxa at three different sites (Muller, Convery, and Ironbolt). The nine taxa trials are seven Eucalypt species and two inter hybrid combinations. The six eucalyptus species were *Eucalyptus cloeziana* (CLO), *E. grandis* (GRA-5), *E. dunnii* (DUN-Y), *E. saligna* (SAL-56), *E. pilularis* (PIL-W), *Corymbia variegata* provenance Richmond (VAR-R), and *Corymbia variegata* provenance Wodum (VAR-W). Three inter hybrid combination are *E. grandis* x *E. camaldulensis* (GxC-11), *E. urophylla* x *E. grandis* (UxG-57), Codes indicate geographical origins of seed is presented in Table 1. Detail planting patterns are explained in Listyanto *et al.* [5].

**Table 1.** Seedlots, clones and taxa tested at three locations in northern New South Wales [5]

| Codes | Taxon | Seedlot/Provenance/ Clone no |
|---|---|---|
| Clo | *Eucalyptus cloeziana* | Natural provenance (identity was not known) |
| Dun-Y | *Eucalyptus dunni* | Safcol clonal seed orchard, South Africa |
| Gra5 | *Eucalyptus grandis* | Clone 5150 |
| SAL-56 | *Eucalyptus saligna* | Clone 5156 |
| PIL-W | *Eucalyptus piluaris* | Whian Whian elite, NSW |
| VAR-R | *Corymbia citriodora* subsp. variegata | Provenance Richmond Range, NSW |
| VAR-W | *Corymbia citriodora* subsp. variegata | Provenance Woondum, Queensland |
| GxC-11 | *E. grandis* x *E. camaldulensis* | Clone 11 |
| UxG-57 | *E. urophylla* x *E. grandis* | Clone 5157 |

### *2.1. Site locations and description.*

The detail of site location and description can be found in Listyanto *et al.* [5]. In general, the Muller site has annual rainfall of approximately 1200 mm and an elevation of 220 m above sea level. The soil is Brown Dermosol, which has good water holding capacity, organic matter levels and nutrient status. The Convery site has rainfall of approximately 1400 mm and an elevation of 450-500 m above sea level. The soils (Red Ferrosols) has a good water holding capacity and is well drained, with relatively high organic matter content and nutrient status. The Ironbolt site has annual rainfall of approximately 1100 mm and an elevation of 200 m. The soils in Ironbolt are also Brown Dermosols.

### *2.2. Data collection*

Oven-dry density and shrinkage were measured as a preliminary indication of wood quality. Samples of wood quality and shrinkage were taken by cutting down 145 trees or 5 trees each species. Straight stems were considered to minimize the probability of the presence of compression wood [15]. Cross sections, 14 cm thick, at DBH and at the top of the merchantable stem (TMS) or at diameter 10 cm of each log (approximately 70% of total height) were removed from the trees. Then the disks were cut into 20 x 20 x 20 $mm^3$ blocks for wood oven-dry density assessment [15,16]. A total of 120 samples for each species per site were collected. The method of determining oven-dry density and shrinkage were based on the British Standard [17]. Wood density and shrinkage of the wood sample were determined in the radial, tangential directions and T/R ratio was also recorded. Radial sections were measured along the rays or radius of the log, at right angles to the annual rings while tangential sections were measured perpendicular to the rays and tangential to the annual rings and face of the log.

### *2.3. Statistical analysis*

Two-way ANOVA was used to test whether there was a difference in the wood oven-dry density and shrinkage of nine taxa of eucalyptus between three different sites then continued by Tukey Honest Significant Different (HSD) as a post hoc test with using level of significance of 5%. These analyses were conducted using the Statistical Program for Social Science (SPSS) version 16.

## 3. Results and Discussion

### *3.1. Oven-dry density*

Variance analyses indicated that the mean values of oven-dry densities of different taxa were significantly different among the three sites. Details of oven-dry density of all taxa in three sites can be seen in Figure 1. The lower density was found significantly in the Convery compared to Ironbolt and Muller. The significance difference might be due to the distinction of the site quality among three sites was extensive [18]. One of important results is that there is a tendency for good sites to produce wood of lower density, among the taxa tested. It is possible that production within the tree of larger vessels with relatively thin fibers at a good site might produce lower oven-dry density of species [18-20]. The results also showed that the mean of oven-dry density decreased with increasing annual rainfall and the decreasing of elevation. For instance, the Convery site, which has annual rainfall higher than other sites (approximately 1400 mm) and has an elevation lower than other sites (220 masl), produced six-year old trees with the lowest mean oven-dry density. These results are also coincident with the summary of Blakmore [21] who worked with four low rainfall eucalypt species and concluded that low rainfall eucalypt species have much higher oven-dry density.

Interactions between species and site for oven-dry density were shown for nine species at three different sites. *Corymbia variegata* provenance Wodum had the highest oven-dry density at 808 kg/$m^3$ at Ironbolt, while the highest oven-dry densities at Convery and Muller were shown by *E. cloeziana* (665 kg/$m^3$) and *C. variegata* provenance Richmond (766 kg/$m^3$), respectively. In general, *C. variegata* provenances Wodum and Richmond, *E. cloeziana*, and *E. dunnii* behaved consistently as a higher group in oven-dry density. The significance of interaction between species and site was also in agreement with the summary of Lima [22] and Zobel and Buitjtenen [18]. Hybrids of *E. grandis* with *E. camaldulensis* or *E. urophylla* had higher densities than pure *E. grandis* as a parent species. The hybrid *E. grandis* x

*E. camaldulensis* showed oven-dry density at 679 kg/m$^3$ and *E. urophylla* x *E. grandis* showed oven-dry density at 613 kg/m$^3$, while *E. grandis* was only 554 kg/m$^3$. Subsequently, hybrids between *E. grandis with E. camaldulensis* or *E. urophylla* succeeded in increasing their oven-dry density. It is often found that important characteristics of eucalypt hybrids tends to be intermediate between their parent species [2].

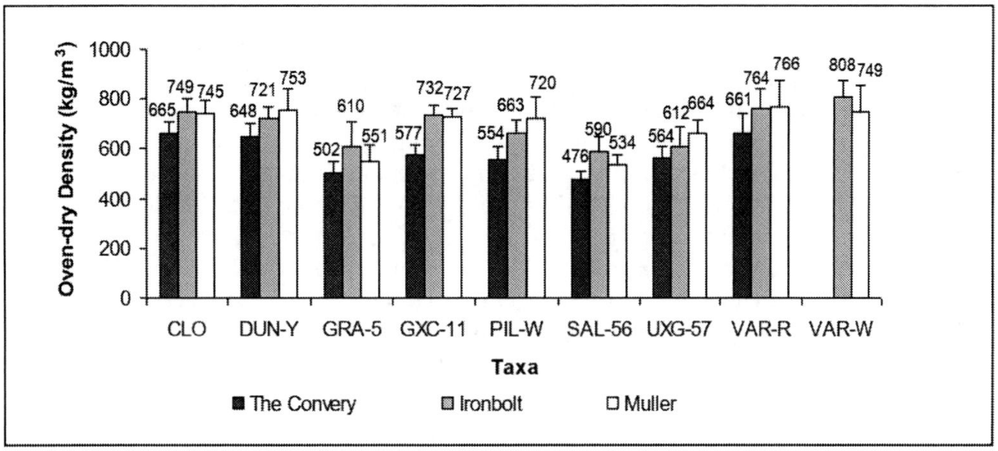


**Figure 1.** Wood oven-dry density of nine taxa at the Convery, Ironbolt and Muller sites, northern NSW.

Variations of oven-dry density in the radial and axial directions were also found. The oven-dry density at DBH of all taxa, except *E. cloeziana* and hybrid *E. grandis x E. camaldulensis*, tend to have a lower oven-dry density than at top of the stem (TMS). It appears that this type of variation of axial position may be due to the proportion of early wood at DBH, which is lower than in at the higher point (TMS) [19].

In terms of variation in the radial direction, there are three types of trends for oven-dry density from pith outward (the figure is not shown). In the first type, wood near the pith tends to have a lower oven-dry density than near bark. This occurs in *E. cloeziana*, *E. dunnii*, *E. grandis*, *E. saligna*, *E. urophylla* x *E. grandis*, and *C. variegata* provenance Richmond. In the second type, wood near the pith tends to have a higher oven-dry density than near bark and occurs in *E. grandis* x *E. camaldulensis*. In the third type, wood near the pith tends to have lower oven-dry density than near bark at DBH but have higher than near bark at top (TMS), which occur in *E. pilularis*. The opposite condition occurs in *C. variegata* provenance Wodum where wood near the pith tends to have higher oven-dry density than near bark at DBH but have lower than near bark at top (TMS). Hillis [7] also noted that the oven-dry density of eucalypts varies with the distance from pith outward. In the first type it appears that the lower oven-dry density of wood near the pith occurs because of the appearance of juvenile wood near the pith that tends to be have thinner cell walls than wood formed in later years [18]. In addition, the proportion of juvenile wood near the pith is slightly higher than near the bark creating the difference in oven-dry density between near pith and near bark in top (TMS) of *E. dunnii*, and *E. grandis*. *E. saligna*, *E. urophylla* x *E. grandis* tended to have lower density at DBH.

Variations in radial shrinkage, tangential shrinkage and T/R ratio were found among the three different sites, species and within individual trees. Details of radial shrinkage, tangential shrinkage and T/R ratio of all taxa in three sites can be seen in Figures 2-4. Interestingly, variance analysis indicated that radial shrinkage, tangential shrinkage and T/R ratio showed a trend in relation to the site quality. According to the results, radial and tangential shrinkage tend to have a higher percentage at good sites where oven-dry density showed a lower value. This result corresponds with the findings of Yang *et al.* [23] who worked with *E. globulus* Labil and concluded that shrinkage properties of *E. globulus* Labill were affected significantly by site. It has been noted that density may affect shrinkage behavior [6,19].

Variation in radial shrinkage, tangential shrinkage and T/R ratio are also shown between species (the figure is not shown). According to the results, *C. variegata* provenance Richmond and Wodum, and E. saligna had the lowest radial shrinkage, tangential shrinkage and T/R ratio, while *E. dunnii* and *E. pilularis* showed the highest. It has been noted that the shrinkage behavior can vary by individual tree [24]. Overall, mean values of radial and tangential shrinkage of *E. pilularis*, *E. urophylla* x *E. grandis*, and *E. dunnii* were higher than recorded by Spear and Walker [25] for medium density woods. These results are consistent with the report of McComb *et al.* [26] that tangential shrinkage of a young hybrid between *E. camaldulensis* x *E. globulus* was greater than in a mature *E. camaldulensis* x *E. globulus*. This might be due to the presence of juvenile wood which is usually found in early growth. A lower proportion of latewood and a thin cell wall in juvenile wood tend to increase the mean value of shrinkage of hardwoods [18]. Another important result was that hybrids between *E. grandis* x *E. camaldulensis* and *E. urophylla* x *E. grandis* showed higher radial and tangential shrinkage than did their parent species (*E. grandis*). These results were consistent with the report of McComb *et al.* [26] that young hybrid *E. camaldulensis* x *E. globulus* had higher levels of shrinkage than their parent species. Thus, in this case, hybrid species showed had higher value of shrinkage compared with one of their parent species.

Variations of radial and tangential shrinkage were also found within tree, especially in the radial and axial direction (the figure is not shown). In terms of radial direction, tangential shrinkage generally decreased from pith to bark, except for *E. grandis* and *E. pilularis*. This result was not consistent with the report of Kang and Lee [24] who worked on Korean larch and found that shrinkage generally increased from pith to bark. The change from juvenile to mature wood has been proposed to explain the variation of shrinkage from pith to the bark. The thinner and shorter fibers near the pith possibly explain the higher shrinkage at near pith than near bark [18]. In the axial direction, radial and tangential shrinkage varied at different heights in tree. For radial shrinkage, there were two patterns of variation. The first pattern was decreasing radial shrinkage from DBH to the top (TMS) such as in *E. dunnii*, *E. grandis* x *E. camaldulensis*, *E. pilularis*, *E. saligna*, and *E. urophylla* x *E. grandis*. The second pattern was that there was no significant difference between radial shrinkage at DBH and at top (TMS), such as in *E. cloeziana*, *E. grandis*, and *C. variegata* provenance Richmond and Wodum. For tangential shrinkage, there was only one pattern. The pattern is decreasing tangential shrinkage from DBH to the top point measured (TMS). A full understanding of wood shrinkage with height in tree can be achieved by studying the trend of juvenile and mature wood along the length of the stem [18].

Variation in radial shrinkage and tangential shrinkage of nine taxa eucalypts due to the anisotropic structure was also shown in this study. The ratio of tangential to radial shrinkage (T/R ratio) of all species was more than 1.6 (the figure is not shown). This result might be due to the hygro-expansion coefficient of tangential shrinkage which is usually greater than twice than coefficient of radial shrinkage [27,28]. This may be a function of fibril arrangement, especially the thickness of the S2 layer [29]. Less shrinkage in the radial direction shrinkage also has been proposed to be due to the greater microfibril angle in radial walls than in tangential and a high proportion of latewood, which forces weak earlywood to shrink more tangentially than independently of the latewood [25,28]. Furthermore, the T/R ratio of several species, such as *E. dunni* planted at the Convery and the Ironbolt sites, and *E. pilularis* planted at the Convery site, were more than 2.5. This is higher than the typical findings of Spear and Walker [25] who reported that the range of T/R ratio of normal shrinkage is 1.5-2.5:1. A high proportion of juvenile wood might be the possible reason for these results.

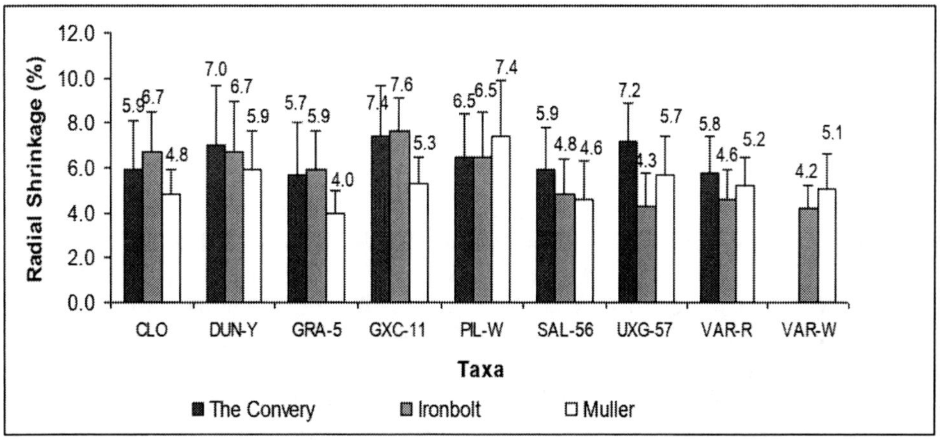


**Figure 2**. Radial shrinkage of nine taxa at the Convery, Ironbolt and Muller sites, northern NSW

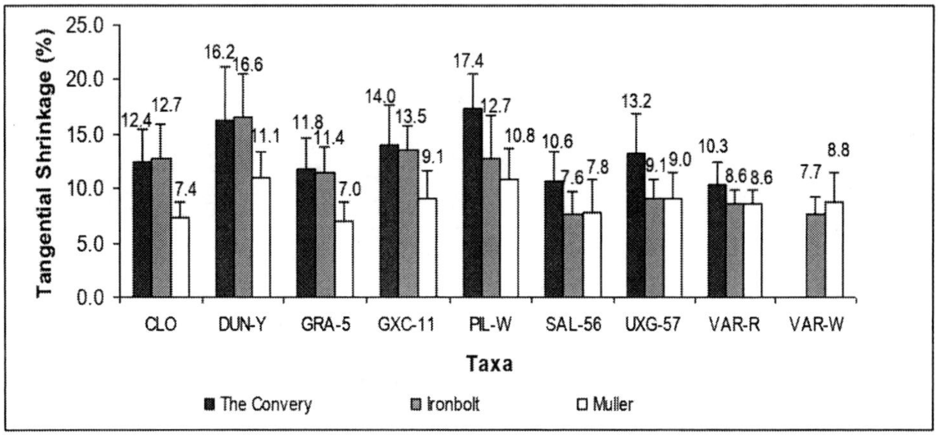


**Figure 3.** Tangential shrinkage of nine taxa at the Convery, Ironbolt and Muller sites, northern NSW

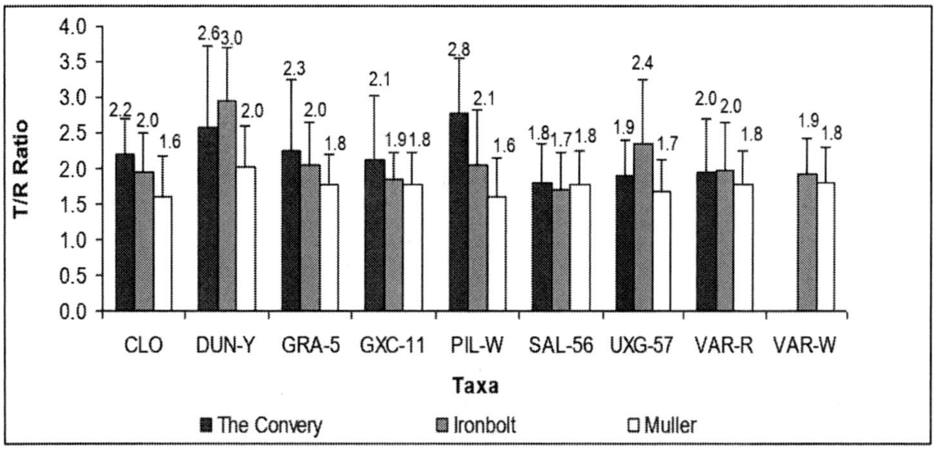


**Figure 4**. Tangential and radial ratio of nine taxa at the Convery, Ironbolt and Muller sites, northern NSW

IOP Conf. Series: Materials Science and Engineering **935** (2020) 012035 doi:10.1088/1757-899X/935/1/012035

## 4. Conclusions

Oven-dry density, tangential shrinkage, radial shrinkage and T/R ratio were significantly affected ($p<0.05$) by site, taxa and interaction between site and taxa on. In terms of sites, the plantation at Convery, where there is high soil quality, annual rainfall higher (approximately 1400 mm) than the other sites and an elevation lower than other sites (220 masl), showed the lowest oven-dry density and the highest radial shrinkage, tangential shrinkage and T/R ratio.

*C. variegata* provenance Wodum had the highest oven-dry density at 781 kg/m$^3$. Variations in oven-dry density in the radial and axial directions were also found. The oven-dry density at DBH of all taxa, except *E. cloeziana* and the hybrid *E. grandis* x *E. camaldulensis*, tended to be lower than at the top point of measurement (TMS). In terms of variation in the radial direction, there are three types of trends of oven-dry density from pith to outward.

In terms of shrinkage, *C. variegata* provenance Richmond and Wodum, and *E. saligna* had the lowest radial shrinkage, tangential shrinkage and T/R ratio, while *E. dunnii* and *E. pilularis* showed the highest. Another important finding is that hybrid between *E. grandis* x *E. camaldulensis* and *E. urophylla* x *E. grandis* showed higher radial and tangential shrinkage than their parent species (*E. grandis*). Another important result is that the oven-dry density of sampled species decreased in the better growth condition.

## References

[1] Eldridge K, Davidson J, Hardwood C, and van Wyk G 1993 *Eucalypt Domestication and Breeding* (Clarendon: Oxford)

[2] Arnold R, Bush D, and Stackpole D 2005 *Genetic variation and tree improvement* (New Forests: Wood production and environmental services) Ed. Nambiar S and Ferguson I (CSIRO: Victoria) pp. 25-50

[3] FAO 1979 *Eucalypts for planting* (Rome: FAO Forestry and Forest Product Studies)

[4] Pliura A, Yu Q, Zhang SY, MacKay J, Perinet P, and Bosquet J 2005 Variation in wood density and shrinkage and their relationship to growth of selected young poplar hybrid crosses *For. Sci.* **51** 472-82

[5] Listyanto T, Glencross K, Nichols J, Schoer L, and Harwood C 2010 Performance of eight eucalypt species and interspecific hybrid combinations at three sites in northern New South Wales, Australia *Aust. For.* **73** 47-52

[6] Bowyer JL, Shmulsky R, and Haygreen JG 2007 *Forest Products and Wood Science: An Introduction* (New Jersey: Wiley-Blackwell)

[7] Hillis WH 1978 *Wood Quality and Utilization* (*Eucalypts for Wood Production*) Ed. Hillis WH and Brown AG (Canberra: CSIRO)

[8] Listyanto T 2018 Wood quality of *paraserianthes falcataria* l. Nielsen syn wood from three years rotation of harvesting for construction application *Wood Res. Slovakia* **63**(3) 497-504

[9] Benddsten BA 1978 Properties of wood from improved and intensively managed forest *For. Prod. J.* **28** 61-72

[10] Bootle KR 1983 *Wood in Australia* (New South Wales: McGraw-Hill Book Company Australia)

[11] Listyanto T, Ando K, Yamauchi H, and Hattori N 2013 Microwave and steam injection drying of CO2 laser incised Sugi lumber *J. Wood Sci.* **59**(4) 282-9

[12] Listyanto T, Ando K, Yamauchi H, and Hattori N 2016 CO2 Laser-Incised Teak and Mahogany Lumber Dried by Microwave and Steam Injection *For. Prod. J.* **66**(7-8) 461-6

[13] Repola J 2006. Models for vertical wood density of Scots Pine, Norway Spruce and Birch Stems, and their application to determine average wood density *Silva Fenn.* **40** 673-85

[14] Booth T 2005 *Environment, species selection and productivity prediction (New Forests: wood production and environmental services)* Ed. Nambiar S and Ferguson XI (Victoria: CSIRO) pp. 5-24

[15] Olivia AG, Merino VB, Seco JIF-G, Garcia MC, and Prieto EH 2006 Effect of growth conditions on wood density of Spanish *Pinus nigra* *Wood Sci. Technol.* **40** 190-204

[16] Molteberg D 2004 Methods for the determintation of wood properties, kraft pulp yield and wood fibre dimensions on small wood sample *Wood Sci. and Technol.* **37** 395-410
[17] BS 373 1957 *Methods of Testing Small Clear Specimens of Timber* (London: British Standards Institution)
[18] Zobel BJ and Buitjtenen JPV 1989 *Wood Variation: Its Causes and Control* (Berlin: Springer-Verlag) pp. 418
[19] Panshin AJ and de Zeeuw C 1980 Textbook of Wood Technology (New York: McGraw-Hill Book Company)
[20] Raiskila S, Saranpää P, Fagerstedt K, Laakso T, Löija M, Mahlberg R, Paajanen L, and Ritschkoff AC 2006 Growth rate and wood properties of Norway spruce cutting clones on different sites *Silva Fenn.* **40** 247-56
[21] Blakemore P 2004 Density and shrinkage of four low-rainfall plantation-grown eucalypts *Aust. For.* **67** 152-5
[22] Lima JT, Breese MC, and Cahalan CM 2000 Genotype-environment interaction in wood basic density of eucalyptus clones *Wood Sci. and Technol.* **34** 197-206
[23] Yang JL, Fife D, Ilic J, and Blackwell P 2002 Between-site and between-provenance differences in shrinkage properties of 10 years old *Eucalyptus globulus* Labill *Aust. For.* **65** 220-226
[24] Kang W and Lee NH 2004 Relationship between radial variation in shrinkage and drying defects of tree disk *Journal of Wood Science* **50** 209-16
[25] Spear M and Walker J 2006 *Dimensional Instability in Timber* (*Primary Wood Processing: Principles and Practice*) Ed. Walker JCF (Netherlands: Springer)
[26] McComb JA, Meddings RA, Siemon G, and Davis S Wood density and shrinkage of five years old *Eucalyptus camaldulensis* x *E. globulus* hybrids: preliminary assessment *Australian Forestry* **67** 236-9
[27] Schulgasser K and Witztum A 2015 How the relationship between density and shrinkage of wood depends on its microstructure *Wood Sci. and Technol.* **49** 389-401
[28] Skaar C 1988 *Wood Water Relations* (Berlin, Heidelberg: Springer-Verlag) pp. 279
[29] Gu H, Zink-Sharp A, and Sell J 2001 Hypothesis on the role of cell wall structure in differential transverse shrinkage of wood *Holz Roh Werkst.* **59** 436-42

**Acknowledgements**
We thank to Forest Enterprises Australia for providing access to the trials and details of their establishment and management. John Grant provided information on soils at the three sites, and David Kleining provided taxa details.

IOP Conf. Series: Materials Science and Engineering **935** (2020) 012036 doi:10.1088/1757-899X/935/1/012036

# Comparing wood anatomy of root and stem of *Syzygium* sp from natural tropical forest of Papua

**Y Rahayu*, S B Husodo**

Forestry Faculty, Papua University, 98314, Papua

*Email: y.rahayu@unipa.ac.id

**Abstract.** Root and stem wood anatomy of *Syzygium*spfrom the natural tropical forest of Fakfak Districts, Papua Barat, Indonesia were studied. The study focused on vessel, fiber and parenchyma. Growth increments are not distinct, yet we observed that *Syzygium*sphas diffuse porosity and predominantly paratracheal with winged aliform of axial parenchyma in the wood of both root and stem. Many qualitative features were varying in the same way in both root and stem: simple perforation plates in vessel elements; large areas of axial parenchyma in both root and stem wood. Quantitative features differing between root and stem wood were evaluated using t-test, and vessel frequency, vessel element length, vessel diameter, percentage of tree wood's tissues, and vulnerability are differed. Root wood had vessels density/frequency, narrower and longer vessel elements. We assumed that root wood is less vulnerability of embolism compared to the stem wood as the narrower of the vessel diameter and the higher percentage of parenchyma area. Roots seem to be more susceptible to water stress and the stain than the stem as seen in the percentage and the dimension of the vessel and parenchyma.

## 1. Introduction

Indonesia has a massive potential of unknown wood/lesser known species for local timber purposes. There are more than 4000 wood species identified in Indonesia [1,2] and only 400 of them had specific name in local and regional trading [1]. The other species then only known as borneo consisting of mixed wood timber and the other species are neglected or limited used.

The lack of utilization of lesser known species and the lack of technical information about these species as well as the possibility of exploiting them resulted in only fixated on the use of certain commercial species. For example, In Papua, two most favorable commercial timber species extracted, traded and used extensively in domestic and international market which are Merbau (*Instia spp.*) and Matoa (*Pometia spp.*) respectively [3]. It is contrast to the facts, in Papua there are 28 commercial timbers have been identified and investigated their general utilization, comprising only a minor portion of 600 trees species known [4]. The rest of these timber species are neglected or even used in local and small scale or quantity. These timbers species are then known and grouped into less-used timber species or lesser-known species or even minor commercial species. In practices, however, these groups of timber have similar or even superior wood properties and natural durability with the existed-commercial timbers [5].

One of the endemics species of woody plant which is widely distributed in Indonesia is water gum (*Syzygium spp.*). This species is spread in Java Island which has nearly 52 species and in West Papua with nearly 132 species [6,7]. The availability of information about wood natural characteristics in

Papua's natural forests in particular and Indonesia in general has not yet been completed. Much information can be obtained to support the use of wood species from natural forests in Indonesia. One of the objectives is to collect the anatomical characteristics of stem wood and root wood. Both of these aspects affect tree growth and wood quality.

Generally discussed, that root wood anatomy received less attention compared to stem wood anatomy [8]. Most of the studies [9-14] of root wood anatomy, showed that the vessels in root wood is wider that in the stem wood of a tree physiologically and anatomically. The earlier studied also showed that the rood wood has higher vulnerability of the embolisms process, since it is directly connected to the effect of the drought [15]. The aim of this study was to compare root and stem wood anatomy of *Syzigium* sp. qualitatively and quantitatively.

## 2. Materials and Methods

### *2.1. Materials*
Samples were taken from natural tropical rain forest in Fakfak District. Fakfak lies in the Province of West Papua which has the highest rainfall occurring on January to May while the lowest rainfall is on September to December. The average air temperature recorded at the Meteorological, Climatology and Geophysics Station of Fakfak at normal temperatures is in the range of 20.70-32.50$^0$C. The lowest air temperatures occur in August and the highest air temperatures occur in March and December. The average Air Humidity is 84.8 percent, the lowest air humidity occurred in December which was 79.0 percent and the highest occurred in August which was 90.0 percent. The highest air pressure occurred in May of 996.1 mb and the lowest in February of 898.8 mb [16].

### *2.2. Methods*
Wood samples for anatomical observation of stem and main root were taken from adult tree. They were collected at 1.30 m (DBH) and 30 cm distal from root collar, respectively. *Non embedding* method was used to observe the cross-sectional feature of vessels, parenchyma and rays distribution [17]. The wood anatomical samples (root wood and stem wood) were cut and softened by immersing in a mixture of glycerol and alcohol in a ratio of 1:1 for 2 weeks [18,19].

After two weeks approximately, the wood sample were sliced using a sliding microtome in 15-25 μm thickness. The slice included the three incisions of wood which are transversally, radially and tangentially. Good sections were stained with safranin 1% for 24 hours. The next step is dehydration process in succession with alcohol levels of 70%, 80%, 90% and absolute alcohol respectively for 10 minutes. Furthermore, the sections were dried by soaking the samples in alcohol and xylol for about 10 minutes.Wood descriptions follow the microscopic features definitions of IAWA Committee [20]. The obtained data from presented in the form of visual microscopy and descriptive analysis.

## 3. Results and Discussion
Certain qualitative anatomical features of *Syzygium* sp. vary between stem and root wood (Figure 1), differences between the organs were found only in quantitative features (Table 1). The vessel diameter, vessel area, and parenchyma area (Table 4) are the features that different between roots and stem of *Syzygium* sp. We also found that the diameter of stem wood vessel is wider than in the root wood, it has negative correlation with the vessel. It is contrast to the finding of the earlier study which said that root wood vessel is wider than the stem wood [9-14].

If we correlate the hydraulic architecture and the tradeoff principle between safety and efficiency in terms of vessel diameter of the root and the stem within a tree, we found that root of *Syzygium*sp is not efficient in transporting water because of the narrower vessel diameter compared to the vessel of the stem. The positives things regarding this finding, it is assumed that the root is less vulnerable to the embolism occurrence. This less vulnerability of the embolism is also confirmed with larger area of the parenchyma of the root wood (37.5%) compared to the stem wood (31%). The larger parenchyma area indicates the higher reserved of metabolites in the parenchyma which will lead to increase the water transportation in vessels and the removal of embolisms [21]. This assumption would be better explained if we do further analysis on the pit membrane thickness as this data is not available at the moment. Since, the thickness of the pit membrane is known to be highly correlated to embolism resistance [22,23].

**Table 1.** Means and STDEV of quantitative anatomical features

| Sampel | | Varea (%) | Rayarea (%) | Parea (%) | Fibarea (%) | Vdens (n/mm2) | Vdiam (µm) | Flength (µm) | WT (µm) |
|---|---|---|---|---|---|---|---|---|---|
| STEM | Mean | 6.8 | 26.6 | 4.4 | 62.1 | 3.3 | 190.9 | 2935.5 | 9.0 |
| | STDEV | 7.7 | 12.6 | 0.6 | 14.9 | 1.0 | 23.2 | NA | NA |
| ROOT | Mean | 16.9 | 35.8 | 1.7 | 45.6 | 34.6 | 121.7 | 1790.2 | 8.3 |
| | STDEV | 2.9 | 9.6 | 1.0 | 10.1 | 10.3 | 13.4 | NA | NA |

Varea= Vessel area, Rayarea= Ray Area, Parea=Parenchyma area, Fibarea=Fiber area, Vdens=Vessel density, Vdiam=Vessel diameter, Flength= Fiber length and WT=fiber wall thickness

### *3.1. Growth rings*

Both in root wood and stem wood, they have indistinct growth rings; the pattern is diffuse-porous wood, with no clear earlywood-latewood pore arrangement, and no significant difference in pore size (Figure 1a-b). In addition, the axial parenchyma pattern varied through the growth ring, predominantly paratracheal with winged aliform (Figure 2 a-b).

### *3.2. Vessels*

Vessels exclusively solitary, almost 90% of the vessels are completely surrounded by other elements. It appears that every vessel is not to contact or connected one to another vessel, as viewed in Figure 1 and in multiples of 2-4 (7-9%) with the mean vessel diameter of stem wood is 190.9 mm and root wood 121.7, respectively; pit between vessel is alternately formed (Figure 2).

From earlier studies, root vessels are wider than stem vessels. Our study showed the opposite of the findings. It explains that the association of small vessel and low vessel density leads to the high mechanical stresses or strain and also affects the wood stiffness [24]. The existence of axial parenchyma (Figure 1 and 2) surrounding the narrow vessel in the root wood may influence the osmotic potential in the dry season and it will help the plants to refill the embolized vessel with water.

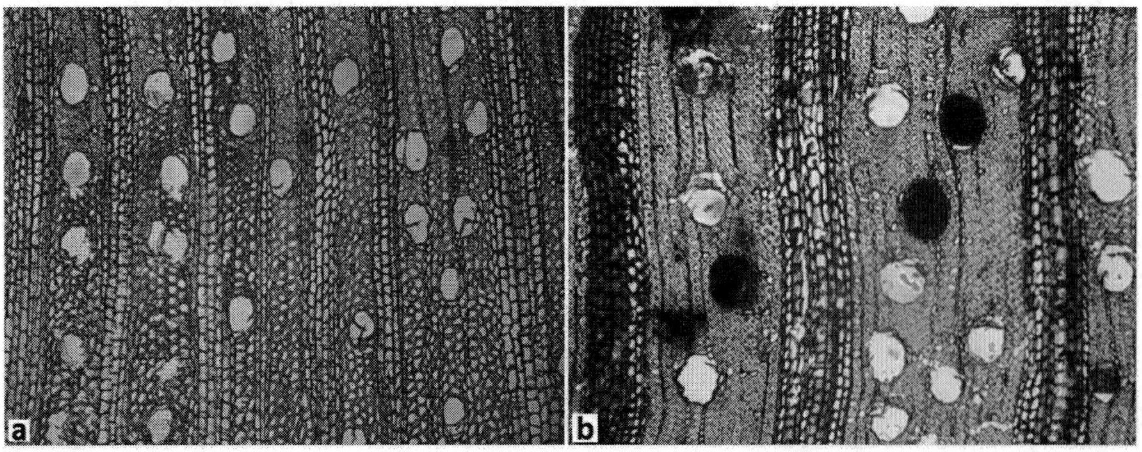


**Figure 1.** (a) Transversal section of root wood and (b) stem wood of *Syzygium* sp. in 140 magnification

### *3.3. Parenchyma*

The structure of axial parenchyma to the vessel both of root and stem is in predominantly paratracheal with winged aliform. Paratracheal parenchyma is parenchyma which is associated with pores while apotracheal is not associated with pores. Confluent paratracheal parenchyma consists of rows or diagonals. The axial parenchyma situated around the vessel. This will help plants on the water strain and stress [24].

### *3.4. Rays*

In root and stem wood, the rays are heterogeneous because it consists of procumbent cells and 2-4(6) upright cells and or quadratic cells. Uniseriate and multiseriate types are seen in the rays' formation with uniseriate portions. It has 1-3 series width, with an average width of 0.4482 mm. The perforation plates are in simple formation. It will help the water transported in the xylem.

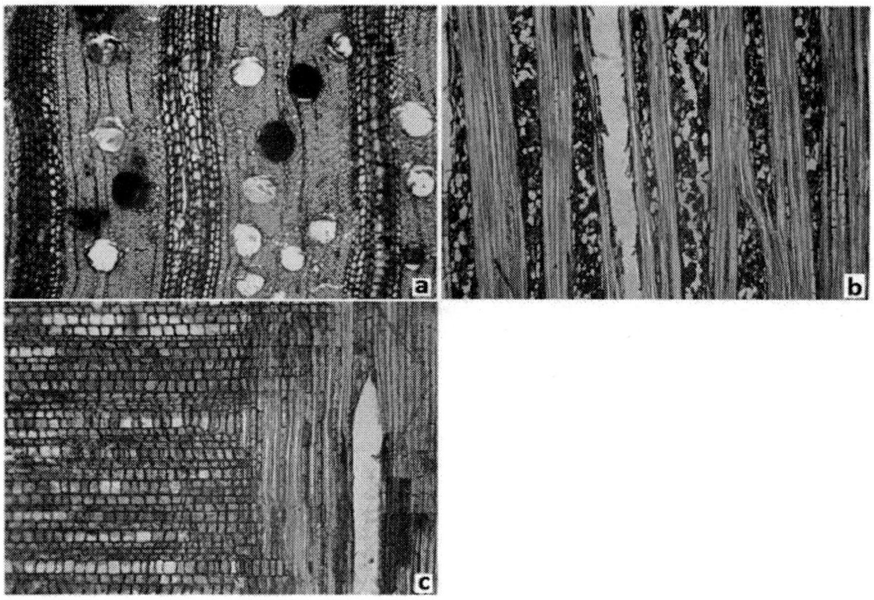


**Figure 2.** (a) Sections in Transversal, (b) Tangential and (c) Radial of Syxygium in 140X magnification.

### *3.5. Statistical analysis*

The correlation data by organ and the results of the t-test are summarized in Table 1 and 3. Vessel element parameters including length, diameter, and frequency, as well as ray height were significantly different between root and stem (Table 1) Root wood had higher vessel frequency and narrower vessel elements when compared to stem wood (Table 1).

The cross-sectional vessel area should be the product of vessel density and vessel diameter. Yet, in our observation (Table 3), the vessel area decreases with the vessel density and tends to increase with the vessel diameter, which is opposite to the expectations. This is probably because of the tradeoff between the vessel density and vessel diameter. The stem wood and root wood of *Syzygium* sp. shows us that there is positive correlative between Vessel area and Vessel diameter, cell wall thickness and fiber diameter, Lumen diameter and fiber diameter and between the area of parenchyma and Fiber area. This positive relation leads to the variation of wood density, water transportation and vulnerability to embolism.

The absence of growth rings in root wood in stem and root of *Syzygium* sp. is also found in most of the tropical forest tree species where the quantitative causes are not sufficiently studied. Yet, the rainfall seasonality is suspected has a high effect to the formation of the growth increment in the tropical forest [11,25-27]. The presence of the growth ring is also correlated with the distribution pattern of the vessel. It its most likely that for the porous wood will form the growth ring distinctively.

**Table 2.** t test for VDiameter = Vessel diameter in μm, Vdens = Vesel density/ frequency (vessels/mm2). CWT = ellwall thickness of fibres in μm., Varea = Vessel Area in %, Rarea = Ray Area in %, Parea = parenchyma area in % Farea = Fiber area in %.

| | Levene's Test for equality of variances | | T-test for equality of means | | | | | | |
|---|---|---|---|---|---|---|---|---|---|
| | F | Sig. | t | df | Sig. (2-tailed) | Mean difference | Std. error difference | 95% Confidence interval of the difference | |
| | | | | | | | | Lower | Upper |
| Vdiameter | 3.241 | .080 | 11.544 | 38 | .000 | 69.18700 | 5.99354 | 57.05371 | 81.32029 |
| | | | 11.544 | 30.445 | .000 | 69.18700 | 5.99354 | 56.95406 | 81.41994 |
| Vdens | 36.174 | .000 | -13.443 | 38 | .000 | -31.350000 | 2.332127 | -36.071144 | -26.628856 |
| | | | -13.443 | 19.413 | .000 | -31.350000 | 2.332127 | -36.224186 | -26.475814 |
| Varea | 14.408 | .001 | 5.481 | 38 | .000 | 10.09200 | 1.84135 | 6.36439 | 13.81961 |
| | | | 5.481 | 24.318 | .000 | 10.09200 | 1.84135 | 6.29428 | 13.88972 |
| Rarea | 1.419 | .241 | 2.574 | 38 | .014 | 9.13800 | 3.55010 | 1.95121 | 16.32479 |
| | | | 2.574 | 35.478 | .014 | 9.13800 | 3.55010 | 1.93439 | 16.34161 |
| Parea | 8.516 | .006 | -8.796 | 38 | .000 | -2.73250 | .31066 | -3.36141 | -2.10359 |
| | | | -8.796 | 28.064 | .000 | -2.73250 | .31066 | -3.36880 | -2.09620 |
| Farea | 2.101 | .155 | -4.084 | 38 | .000 | -16.49700 | 4.03893 | -24.67338 | -8.32062 |
| | | | -4.084 | 33.365 | .000 | -16.49700 | 4.03893 | -24.71085 | -8.28315 |
| CWT | .000 | 1.000 | .000 | 38 | 1.000 | .00000 | .64967 | -1.31519 | 1.31519 |
| | | | .000 | 38.000 | 1.000 | .00000 | .64967 | -1.31519 | 1.31519 |

In conclusion, *Syzygium* sp. root and stem wood were different in quantitative but not qualitative features. The root wood had higher frequency of vessels, with the higher percentage of axial parenchyma than the wood of the stem. It is assumed that with the narrower vessel diameter and the high percentage of axial parenchyma, the roots are suspected to be more susceptible to water stress than the stem.

Secondly, in this study we found that in the wood stem with high percentage of parenchyma tends to have a positives correlation with the wider vessel diameter and the parenchyma placed around the vessels. This leads us to the fact regarding the independent function of axial parenchyma.

**Table 3.** Pearson correlation of the organs between root and stem measurements. **. Correlation is significant at the 0.01*. Correlation is significant at the 0.05, VDiameter= Vessel diameter in μm, VHeight = Vessel Height in μm, Vdems= Vesel density/ frequency (vessels/mm2). CWT = cellwall thickness of fibres in μm., Rwidth = rays height in μm, Rdens = ray density (n/mm2), Varea = Vessel Area in %, Rarea= Ray Area in %, Parea=parenchyma area in % Farea = Fiber area in %, Flenght – Fiber length in μm, Fdiameter = Fiber diameter in μm, LumDiam = Lumen diameter in μm.

| | | Vdiameter | Vheight | Vdens | Rheight | Rwidth | Rdens | Varea | Rarea | Parea | Farea | Flength | Fdiameter | LumDiam | CWT |
|---|---|---|---|---|---|---|---|---|---|---|---|---|---|---|---|
| Vdiameter | Pearson Correlation | 1 | -.192 | -.827** | .141 | .095 | .105 | .638** | .281 | -.723** | -.467** | .363 | .246 | -.013 | .162 |
| Vheight | Pearson Correlation | -.192 | 1 | .221 | .104 | -.256 | .047 | -.030 | -.001 | -.190 | .025 | .033 | -.109 | -.108 | -.024 |
| Vdens | Pearson Correlation | -.827** | .221 | 1 | -.142 | -.396 | .319 | -.597** | -.340* | .708** | .495** | .022 | -.274 | -.262 | -.023 |
| Rheight | Pearson Correlation | .141 | .104 | -.142 | 1 | -.132 | .288 | .228 | .137 | -.056 | -.230 | .253 | .046 | .009 | .060 |
| Rwidth | Pearson Correlation | .095 | -.256 | -.396 | -.132 | 1 | -.590** | -.161 | -.157 | -.468* | .235 | .069 | -.093 | .112 | -.290 |
| Rdens | Pearson Correlation | .105 | .047 | .319 | .288 | -.590** | 1 | .107 | .140 | .310 | -.186 | .033 | .123 | .070 | .097 |
| Varea | Pearson Correlation | .638** | -.030 | -.597** | .228 | -.161 | .107 | 1 | .294 | -.494** | -.687** | -.227 | .369 | .355 | .116 |
| Rarea | Pearson Correlation | .281 | -.001 | -.340* | .137 | -.157 | .140 | .294 | 1 | -.469** | -.891** | .114 | -.178 | .003 | -.216 |
| Parea | Pearson Correlation | -.723** | -.190 | .708** | -.056 | -.468* | .310 | -.494** | -.469** | 1 | .512** | .043 | -.026 | -.195 | .021 |
| Farea | Pearson Correlation | -.467** | .025 | .495** | -.230 | .235 | -.186 | -.687** | -.891** | .512** | 1 | .019 | -.039 | -.177 | .110 |
| Flength | Pearson Correlation | .363 | .033 | .022 | .253 | .069 | .033 | -.227 | .114 | .043 | .019 | 1 | -.385 | -.381 | -.091 |
| Fdiameter | Pearson Correlation | .246 | -.109 | -.274 | .046 | -.093 | .123 | .369 | -.178 | -.026 | -.039 | -.385 | 1 | .764** | .533* |
| LumDiam | Pearson Correlation | -.013 | -.108 | -.262 | .009 | .112 | .070 | .355 | .003 | -.195 | -.177 | -.381 | .764** | 1 | -.138 |
| CWT | Pearson Correlation | .162 | -.024 | -.023 | .060 | -.290 | .097 | .116 | -.216 | .021 | .110 | -.091 | .533* | -.138 | 1 |

**. Correlation is significant at the 0.01 *. Correlation is significant at the 0.05

**Table 4.** Summary of Anova test for quatitative feature of VDiameter= Vessel diameter in μm, VHeight = CWT = cellwall thickness of fibres in μm, Varea = Vessel Area in %, Rarea = Ray Area in %, Parea = parenchyma area in % Farea = Fiber area in %.

| | Sum of Squares | df | Mean Square | F | Sig. |
|---|---|---|---|---|---|
| TreePart * Vdiameter | 9.333 | 16 | .583 | 20.125 | .000 |
| TreePart * Varea | 9.500 | 32 | .297 | 4.156 | .028 |
| TreePart * Rarea | 7.200 | 24 | .300 | 1.607 | .172 |
| TreePart * Parea | 9.333 | 19 | .491 | 14.737 | .000 |
| TreePart * Farea | 8.500 | 36 | .236 | .472 | .885 |
| TreePart * CWT | .000 | 6 | .000 | .000 | 1.000 |

**References**

[1] Mandang YL and Pandit IKN 2002 *Manual Series: Guidelines for Field Wood Species Identification* (Bogor: PROSEA Indonesia)
[2] Pandit IKN and Kurniawan D 2008 *The Characteristics Wood as A Raw Material and The Diagnostic Characteristics of Indonesian Timber Trade* (Bogor: Centium) p. 114
[3] Wahyudi M, Makrus, Antonius EB, and Susilo 2014 Wood machining properties of two less used species from West Papua *Jurnal Ilmu dan Teknologi Kayu Tropis* **12**(1) 74-81
[4] Directorate General of Forestry of Indonesia 1976 Knowing several timber wood species in Irian Jaya Indonesian (Jakarta: Ministry of Forestry)
[5] PROSEA 1998 Trees: lesser known timbers *Plant Res. of South East Asia Timber* **5**(3)
[6] Backer CA and Brink RC1963 Flora of Java (Spermatophytes only) **1**
[7] Hartley TG and Perry LM 1973 A provisional key and enumeration of species of Syzygium (Myrtaceae) from Papuasia *Journal of The Arnold Arboretum* **54** 160-227
[8] Gasson PE and Cutler DF 1990 Root anatomy of 17 genera growing in the British Isles *IAWA Bulletin* **11**3-46
[9] Alder NN, Sperry JS, and Pockman WT 1996 Root and stem xylem embolism, stomatal conductance and leaf turgor in *Acer grandidentatum* populations along a soil moisture gradient *Oecologia* **105** 293-301
[10] Ewers FW, Carlton MR, Fisher JB, Kolb KJ, and Tyree MP 1997 Vessel diameters in roots versus stems of tropical lianas and other growth forms *IAWA Journal* **18** 261-79
[11] Machado SR, Angyalossy-Alfonso V, and Morretes BL 1997 Comparativewood anatomy of root and stem in *Styrax camporum* (Styracaceae) *IAWA Journal* **18** 13-25
[12] Kavanagh KL, Bond BJ, Aitken SN, GartnerBL, and Knowe S 1999 Shoot and root vulnerability to xylem cavitation in four populations of Douglas-fir seedlings *Tree Physiology* **19** 31-37
[13] McElrone AJ, Pockman WT, Martínez-VilaltaJ, and Jackson RB 2004 Variation in xylem structure and function in stems and roots of trees to 20 m depth *New Phytologist* **163** 507-17
[14] Psaras GK and SofroniouI 2004 Wood anatomy of *Capparis spinosa* from an ecological perspective *IAWA Journal* **20** 419-29
[15] Kusumaningrum R 2017 The role of xylem and phloem in plant growth and development. *Prosiding Seminar Nasional Pendidikan Biologi dan Biologi Jurusan Pendidikan Biologi Fakultas MIPA Universitas Negeri Yogyakarta*
[16] Pemerintah Kabupaten Fak Fak Kondisi Geografis Wilayah (https://fakfakkab.go.id/?page_id=197) [Accessed 17 July 2020]
[17] Mandang YI and Wiyono B 2002 Wood anatomy of gaharu (*Aquilaria malaccensis* Lamk.) and several related species *Jurnal Penelitian Hasil Hutan* **20** 107-26
[18] Mandang YI and Aristien S 2003 Wood anatomy and wood fiber quality of utap-utap (*Aromadendron elegans* bi.) and seven other lesser known wood species *Jurnal Penlitian Hasil Hutan* **21** 111-27
[19] Pandit IKN 2004 *Wood Anatomy Structure of Kuku Wood* (*Pericopis mooniana* T.) (Bogor: IPB)
[20] IAWA Committee 1989 IAWA list of microscopic features for hardwood identification IAWA Bulletin n.s.**10** 219-332
[21] Eduardo L, Longui, Ana TD, Galão, Kishore S, Rajput, and Antônio CG de Melo 2018 Anatomical investigation of root, stem and branch wood in 10-year-old *Ingalaurina* in the context of anatomical adaptation to hydraulic and mechanical stresses *Anales de Biología* **40** 31-9
[22] Lens F, Tixier A, Cochard H, Sperry JS, Jansen S, and Herbette S 2013 Embolism resistance as a key mechanism to understand adaptive plant strategies *Current Opinions in Plant Biology* **16** 287–92

[23] Li S, Lens F, Espino S, Karimi Z, Klepsch M, Schenk HJ, and Jansen S 2016 Inter vessel pit membrane thickness as a key determinant of embolism resistance in angiosperm xylem *International Association of Wood Anatomists Journal* **37** 152-71
[24] Christensen D, Munos F, Ennos AR, and Barfod AS 2007 Changes in vessel anatomy in response to mechanical loading in six species of tropical trees *New Phytologist* **176** 610-22
[25] Nath CD, Munoz F, Pélissier R, Burslem DF, and Muthusankar G 2016 Growth rings in tropical trees: role of functional traits, environment, and phylogeny *Trees* **30**(6) 2153-75
[26] Metcalfe CR and Chalk L 1950 Anatomy of the Dicotyledons: leaves, stem, and wood, in relation to taxonomy, with notes on economic uses. pp. lxiv + 1500 pp.
[27] Goulart SL and Marcati CR 2008 Anatomia comparada do lenhoemraiz e caule de lippiasal viifolia cham. (verbenaceae) *Revista Brasil Bot.* **31**(2) 263-75

# Effect of breeding technology and trunk axial position on shrinkage and quality of 10 year old teak wood as a furniture's raw material

**Y Suranto**

Faculty of Forestry, Universitas Gadjah Mada, Yogyakarta, Indonesia

Email: suranto@ugm.ac.id

**Abstract**. Shortage on teak wood availability as material for furniture industry has been solved by developing tree breeding technology and decreasing on cut rotations. Three types of teak forest breeding technology being developed is called APB, KBK and JPP-PHT-1. Breeding technology was developed as an effort to get productive and hight quality of teak stands. Decreasing on cutting rotation encourages to increase young wood utilization. This study was aimed to determine the effect of breeding technology and axial stem position on teak wood shrinkages and wood quality of 10 years old. Nine trunk of 10 years old teak trees were harvested on forest areas of Nganjuk Forest District, East Java Province. These trees consisted of seedling grown by APB, KBK and JPP-PHT-1 breeding technology of which consisted of three trees on each kind of breeding technology. Three disks were taken from each trunk, namely from the butt, middle and upper parts. Wood shrinkages on tangential and radial and its ratio were measured based on British Standard method. Data were analyzed by using variance analysis arranged in blocked factorial and further testing were performed by using HSD Duncan. Wood quality was determined by using Suranto method. Results showed that interaction of twoo factors did not affect on tangential and radial shrikages and also tangential-radial ratio. Tree breeding technology very significantly affects on tangensial shrinkage and significanty effect on radial shrinkage. APB breeding technology produces the greatest tangential shrinkage (8.46), followed by KBK (6.62) and JPP-PHT-1 (6.59%). APB breeding technology produces the greatest radial shrinkage (5.51), followed by KBK (5.08) and JPP-PHT-1 (4.42%). The axial tree trunk position affects only on tangential and radial ratio. The stem base has a largest T/R ratio (1.84), followed by middle part (1.45) and upper part (1.26). In term of wood shrinkage, Breeding technology affects the wood quality. APB breeding technology produces the lowest timber quality (3.30), followed by JPP-PHT-1 (2.85), and KBK (2.78).

## 1. Introduction

Teak furniture industry faces many obstacles, among others are the limited availability of high quality wood [1]. This problem is solved by the use of young timber as a whole, either from the base, the middle or the end of the tree trunk. In addition, forestry companies of Perhutani has developed superior teak seedlings with rapid tree growth rates. The three types of seeds include seeds of APB (seeds from Seed Production Areas), seeds of KBK (seeds from the Clonal Seed Orchard) and seedlings of JPP PHT-1 (shoots from clonal seed orchard) [2]. These seeds have been planted together in the forest and there are

10 years old plantation. The tree growth rate has most likely affected the nature and quality of wood [3]. Therefore, the nature and quality of the wood produced from tree trunks derived from the seeds need to be studied [4].

The quality of the wood is a reflection of the level of compatibility between the properties of wood with the demands of the requirements needed for a particular use. The more suitable the properties of wood to the demands of specific requirements for the use of wood, the higher the wood quality [5]. The properties of wood covering the basic properties and the properties of wood processing. The wood basic properties are anatomical properties and physical properties of wood [6,7]. Three parameters measuring the wood physical properties are tangential shrinkage, radial shrinkage and radial-tangential shrinkage ratio.

The aims of these research were to achieve two things. First, to determine the effect of breeding technology for seed trees production and trunk axial position and their interaction to the wood shrinkage properties. Second, determine the quality of teak wood as a furniture material based on the parameters of wood shrinkages.

## 2. Material and Methods

### *2.1. Materials*

Research materials were 9 trunk of 10 years old teak trees harvested from research plots compartment located on forest areas of Nganjuk Forest District, East Java Province. These trees consisted of seedling grown by APB, KBK and JPP-PHT-1 breeding technology which consisted of three trees on each kind of breeding technology.

### *2.2. Methods*

The study was conducted by the following procedure, i.e. (1) cutting down nine sample trees. (2) separate the branch-free trunk into tree parts, namely butt, middle and upper part of the stem. (3) Take a disk in a thickness of 3 cm on every part of the trunk. (4) testing on tangential shrinkage and radial shrinkage based on British Standard 1957 method and calculating tangential- radial shrinkage ratio, (5) classifying wood quality of APB, KBK and JPP PHT-1 by using the standard of classification proposed by Suranto [8].

### *2.3. Data analysis*

The research involved two factors, namely the breeding technology and axial position of branch-free trunk. Factor of breeding technology in producing seed consist of three levels, namely APB, KBK and JPP PHT-1. The later breeding technology the newer one. Factor of axially wood position consisted of three levels, namely the butt, middle and upper of the trunk. Each treatment combination was replicated three times, so the research was obtained 27 units sample.

Results of research was analyzed by using variance analysis which arranged in completely randomized factorial design [9]. Further testing was done by using Honestly Significant Different (HSD). Classification method arranged by Suranto [8] was used to classify and to rank wood quality. Wood quality were classified into five classes, namely class 1 (very good), class 2 (good), class 3 (sufficient/moderate), class 4 (bad) and class 5 (very bad).

## 3. Results and Discussion

### *3.1. Results*

Research was carried out on three parameters, namely tangential shrinkage, radial shrinkage and the ratio of both shrinkages. The results are presented sequentially in Table 1, Table 2 and Table 3.

**Table 1**. Results of tangential shrinkage measurement (%)

| Breeding technology method | Axial position | Replication | | | Average |
|---|---|---|---|---|---|
| | | 1 | 2 | 3 | |
| APB | Butt | 10.70 | 8.87 | 7.24 | 8.94 |
| | Middle | 8.77 | 8.65 | 8.37 | 8.54 |
| | Top | 9.84 | 6.75 | 6.97 | 7.84 |
| KBK | Butt | 8.03 | 7.99 | 6.71 | 7.58 |
| | Middle | 7.13 | 5.82 | 6.29 | 6.41 |
| | Top | 5.23 | 5.20 | 7.20 | 5.88 |
| JPP PHT-1 | Butt | 5.59 | 6.47 | 8.02 | 6.69 |
| | Middle | 5.45 | 6.21 | 7.22 | 6.29 |
| | Top | 6.99 | 6.39 | 6.95 | 6.78 |

**Table 2**. Results of radial shrinkage measurement (%)

| Breeding technology method | Axial position | Replication | | | Average |
|---|---|---|---|---|---|
| | | 1 | 2 | 3 | |
| APB | Butt | 10.70 | 8.87 | 7.24 | 8.94 |
| | Middle | 8.77 | 8.65 | 8.37 | 8.54 |
| | Top | 9.84 | 6.75 | 6.97 | 7.84 |
| KBK | Butt | 8.03 | 7.99 | 6.71 | 7.58 |
| | Middle | 7.13 | 5.82 | 6.29 | 6.41 |
| | Top | 5.23 | 5.20 | 7.20 | 5.88 |
| JPP PHT-1 | Butt | 5.59 | 6.47 | 8.02 | 6.69 |
| | Middle | 5.45 | 6.21 | 7.22 | 6.29 |
| | Top | 6.99 | 6.39 | 6.95 | 6.78 |

**Table 3**. Results of tangential-radial ratio

| Breeding technology method | Position | Replication | | | Average |
|---|---|---|---|---|---|
| | | 1 | 2 | 3 | |
| APB | Butt | 1.84 | 1.31 | 1.49 | 1.55 |
| | Middle | 1.59 | 2.31 | 1.53 | 1.81 |
| | Top | 1.54 | 1.22 | 1.26 | 1.34 |
| KBK | Butt | 3.84 | 2.08 | 1.68 | 2.54 |
| | Middle | 1.28 | 2.26 | 1.15 | 1.23 |
| | Top | 1.09 | 1.09 | 1.55 | 1.24 |
| JPP PHT-1 | Butt | 1.55 | 1.27 | 1.46 | 1.43 |
| | Middle | 1.17 | 1.35 | 1.41 | 1.31 |
| | Top | 1.35 | 1.05 | 1.18 | 1.19 |

### *3.2. Results analysis*

3.2.1. Analysis of variance

Analysis of variance was ellaborated to determine the effect of breeding technology and trunk axial position as well as interaction of these two factors to three parameters. Analysis result on Tangential shrinkage, radial shrinkage and tangential-radial shrinkage ratio are presented consecutively in Table 4, 5 and 6.

**Table 4.** Analysis of variance on tangential shrinkage (%)

| Source | Sum of Squares | DF | Mean Square | F | Significantcy |
|---|---|---|---|---|---|
| Corrected Model | 27.475[a] | 8 | 3.434 | 2.877 | 0.030 |
| Intercept | 1409.056 | 1 | 1409.056 | 1180.443 | 0.000 |
| Method (M | 20.701 | 2 | 10.350 | 8.671** | 0.002 |
| Position (P) | 3.849 | 2 | 1.925 | 1.612 | 0.227 |
| M * P | 2.925 | 4 | 0.731 | 0.613 | 0.659 |
| Error | 21.486 | 18 | 1.194 | | |
| Total | 1458.016 | 27 | | | |
| Corrected Total | 48.961 | 26 | | | |

**Table 5.** Analysis of variance on radial shrinkage (%)

| Source | Sum of Squares | DF | Mean Square | F | Significantcy |
|---|---|---|---|---|---|
| Corrected Model | 14.913[a] | 8 | 1.864 | 3.438 | 0.014 |
| Intercept | 675.800 | 1 | 675.800 | 1246.259 | 0.000 |
| Method (M | 5.455 | 2 | 2.727 | 5.030* | 0.018 |
| Position (P) | 2.972 | 2 | 1.486 | 2.741 | 0.091 |
| M * P | 6.486 | 4 | 1.622 | 2.990 | 0.057 |
| Error | 9.761 | 18 | 0.542 | | |
| Total | 700.474 | 27 | | | |
| Corrected Total | 24.674 | 26 | | | |

**Table 6.** Analysis of variance on tangential-radial shrinkage ratio

| Source | Sum of Squares | DF | Mean Square | F | Significantcy |
|---|---|---|---|---|---|
| Corrected Model | 4.392[a] | 8 | 0.549 | 2.830 | 0.032 |
| Intercept | 61.956 | 1 | 61.956 | 319.348 | 0.000 |
| Method (M | 0.614 | 2 | 0.307 | 1.583 | 0.233 |
| Position (P) | 1.553 | 2 | 0.777 | 4.003* | 0.036 |
| M * P | 2.224 | 4 | 0.556 | 2.866 | 0.053 |
| Error | 3.492 | 18 | 0.194 | | |
| Total | 69.840 | 27 | | | |
| Corrected Total | 7.884 | 26 | | | |

Based on the analysis of variance in Table 4, it can be seen that breeding technology method influences very significantly to the mean value of tangential shrinkage, while the stem axial position and their interaction has no significant effect. Meanwhile, the analysis of variance in Table 5 shows that breeding technology method has a very significant effect on the mean of radial shrinkage value, while the axial position of the stem and their interaction have no significant effect. On the other hand, the analysis of variance in Table 6 shows that the stem axial position has a very significant effect on the mean value of the tangential-radial shrinkage ratio, while the breeding technology method and their interaction has no significant effect. Thus, further testing by using HSD Duncan method needs to be done on the factors that expressed a significantly affect.

3.2.2. HSD Duncan analysis

Duncan's HSD analysis was conducted to determine the difference in the level of breeding technology method to tangential shrinkage. The results are presented in Table 7 below.

**Table 7**. HSD Duncan test result for the effect of breeding technology on tangential shrinkage (%)

| Breeding Technology Method | N | Subset | |
|---|---|---|---|
| | | 1 | 2 |
| JPP PHT-1 | 9 | 6.59 | |
| KBK | 9 | 6.62 | |
| APB | 9 | | 8.46 |
| Significantcy | | 0.95 | 1.00 |

Based on the test results shows in Table 7 above, it is seen that tangential shrinkage on APB (8.46%) differs significantly to KBK (6.62%) and JPP PHT-1 (6.59%), but KBK and JPP IPM-1 breeding technology has no different value on tangential shrinkage.

Duncan's HSD analysis was conducted to determine the difference in the level of breeding technology method to radial shrinkage. The results wete presented in Table 8 below.

**Table 8.** HSD Duncan test result for breeding technology effect on radial shrinkage (%).

| Breeding Technology Method | N | Subset | |
|---|---|---|---|
| | | 1 | 2 |
| KBK | 9 | 4.42 | |
| JPP PHT-1 | 9 | 5.08 | 5,08 |
| APB | 9 | | 5.51 |
| Significantcy | | 0.07 | 0.23 |

The test in Table 8 above shows that radial shrinkage on JPP PHT-1 (5.08%) did not differ to KBK (4.42%) and APB (5.51%). Meanwhile, radial shrinkage on KBK is significantly different APB breeding technology.

Duncan's HSD analysis was performed to determine the difference in the level of stem axial position on tangential - radial shrinkage ratio. The results are presented in Table 9 below.

**Table 9.** HSD Duncan test result for influence of stem axial position on ratio of tangential - radial shrinkage

| Stem axial position | N | Subset | |
|---|---|---|---|
| | | 1 | 2 |
| Top | 9 | 1.26 | |
| Middle | 9 | 1.45 | 1.45 |
| Butt | 9 | | 1.84 |
| Significantcy | | 0.37 | 0.08 |

Based on the results of the test presented in Table 9, it is seen that the value of tangential/radial shrinkage ratio at the middle position of the stem (1.45) does not differ significantly to the both of stem butt position (1.84) and stem top position (1.26), but these latter two positions have a significant different on tangential / radial shrinkage ratios.

### *3.3. Discussion*

#### 3.3.1. Tangential shrinkage

The analysis of variance in Table 4 shows that the breeding technology method has a very significant effect, while stem axial position and the interaction between the two factors have no significant effect on tangential shrinkage. Thus, the total of tangential shrinkage of stem butt position (7.74%) and the middle stem position (7.10%) and the top of the stem (6.84%) were not different.

Based on the HSD Duncan test results presented in Table 7, it was seen that the tangential shrinkage in APB (8.46%) differs significantly to KBK (6.62%) and JPP PHT-1 (6.59%), but total tangential shrinkage on KBK and JPP PHT-1 were not different. Thus, the total tangential shrinkage of the wood grown from the both sources of clonal seed orchard did not difference. Thus, the generative (KBK) and vegetative planting material (JPP PHT-1) taken from clonal seed orchard have the same values in term of tangential shrinkage. Meanwhile, the total tangential shrinkage on wood grown from the clonal seed orchard was smaller than the wood grown from the seed garden area (APB).

The smaller tangential shrinkage in the APB (8.46%) compared to KBK (6.62%) and the JPP PHT-1 (6.59%) breeding technology may be due to the specific gravity factor. The value of specific gravity of wood grown by these three seeding methods also decreases respectively. Specific gravity of wood grown by APB was 0.50, KBK was 0.46 and JPP PHT-1 was 0.46 [10]. The linkage between specific gravity and tangential shrinkage was in line to the theoretical basis that the specific gravity affects tangential shrinkage [7].

#### 3.3.2. Radial shrinkage

The analysis of variance on Table 5 shows that breeding technology method significantly influences radial shrinkage, while the stem axial position and their interaction of the two factors were not significantly influenced. Thus, the radial shrinkage at stem butt (4.61%), the stem middle (4.97%) and the stem top position (5.42%) were not significantly different.

Testing result presented in Table 8 shows that wood radial shrinkage on JPP PHT-1 (5.08%) was not differ to KBK (4.42%) and APB (5.51%). Meanwhile, total radial shrinkage on KBK was different to APB. The total radial shrinkage of the wood grown from the clonal seed orchard, whether taken in vegetative form (JPP PHT-1) or generative form (KBK), has a lower value than that taken in generative form from seed production area (APB).

When the results of this research compared to the results of similar research conducted by Matumura and Kawasaki [11], both of the values were the nearly same. The results of Matumura and Kawasaki [11] research on young 16 years old teak growing in Gunungkidul Yogyakarta get radial shrinkage value of 3.06+0.49 and 51 years old teak growing in KPH Pemalang get radial shrinkage value of 2.75+0.44, while the results of this study get an average value of 3.27%.

3.3.3. Tangential-radial shrinkage ratio
The analysis of variance presented in Table 6 shows that the stem axial position has a very significant effect on the tangential/radial shrinkage ratio. Meanwhile, the factor of breeding technology method and its interaction to stem axial position has no significant effect. Thus, the ratio of tangential/radial shrinkage on wood grown by APB seedlings (1.57), KBK (1.67) and PHT-1 (1.31) were not significantly different.

The testing result presented in Table 9 shows that the value of tangential/radial shrinkage ratio at the stem middle position (1.45) is not significantly different both to the stem butt position (1.84) or to the stem top position (1.26). But the last two positions have different shrinkage ratios. The mean tangential/radial shrinkage ratio resulted in this study was 1.52.

If the results of this study were compared to the research results conducted by Matumura and Kawasaki [11], which were get the value of this ratio of 2.4 for 16 years old teak wood that grows in Gunungkidul Daerah Istimewa Yogyakarta and of 1.95 for 51 years old teak wood that grew in compartment of 46A, RPH Klapanunggal, BKPH Bantarsari, KPH Pemalang, it was seen that the results of this study get a relatively low value of this shrinkage ratio.

*3.4. Discriminant analysis of wood quality*
Discriminant analysis was conducted with the aim to know the effect of breeding technology method, and stem axial position and the interaction of these two factors to the wood quality class. Classification was determined by using parameters of tangential depreciation, radial shrinkage and tangential/radial arrangement ratio. The standard of wood quality determination was done based on Suranto method [8]. This classification standard is presented in Table 10.

**Table 10.** Quality Classification Standards for Teak Wood

| Wood quality class | Tangential shrinkage | Radial shrinkage | T/R shrinkage rasio |
|---|---|---|---|
| 1 | < 4.81 | < 0.83 | < 0.98 |
| 2 | 4.81-6.77 | 0.83-2,46 | 0.98-2.13 |
| 3 | 6.77-8.72 | 2.46-4,09 | 2.13-3.28 |
| 4 | 8.72-10.68 | 4.09-5,72 | 3.28-4.44 |
| 5 | >10.68 | > 5.72 | > 4.44 |

Based on the classification standard, the process of wood quality rating was done partially on the perspective of those three parameters. The result was then calculated to obtain average value. This average value describes the quality of wood from the shrinkage point of view. Resulted wood quality rating class were presented in Table 11, while the recapitulation was presented in Table 12.

Based on Table 12 presented above, it can be seen that the quality class of teak produced by APB was 3.30 and KBK was 2.78 and JPP PHT-1 seedling technology was 2.85. Meanwhile, the quality class of 10 years old teak wood taken from the stem butt position was 3.07 and the stem middle position was 3.85 and the top position of the stem was 3.

**Table 11.** Wood quality classification based on shrinkage parameter

| Sampel codifikation | Axial position | Total shrinkage (%) | | | Wood quality class on parameter of | | | Average quality class |
|---|---|---|---|---|---|---|---|---|
| | | T | R | T/R | T | R | T/R | |
| APB 1 | Butt | 10.70 | 5.82 | 1.84 | 5 | 5 | 2 | 4.00 |
| | Middle | 8.77 | 5.51 | 1.59 | 4 | 4 | 2 | 3.33 |
| | Top | 9.84 | 6.41 | 1.54 | 4 | 5 | 2 | 3.67 |
| APB 2 | Butt | 8.87 | 6.76 | 1.31 | 4 | 5 | 2 | 3.67 |
| | Middle | 8.65 | 3.75 | 2.31 | 3 | 3 | 2 | 2.67 |
| | Top | 6.75 | 5.51 | 1.22 | 2 | 4 | 2 | 2.67 |
| APB 3 | Butt | 7.24 | 4.85 | 1.49 | 3 | 4 | 2 | 3.67 |
| | Middle | 8.37 | 5.46 | 1.53 | 3 | 4 | 2 | 3.00 |
| | Top | 6.97 | 5.54 | 1.26 | 3 | 4 | 2 | 3.00 |
| KBK 1 | Butt | 8.03 | 2.09 | 3.84 | 3 | 3 | 4 | 3.33 |
| | Middle | 7.13 | 5.57 | 1.28 | 3 | 4 | 2 | 3.00 |
| | Top | 5.23 | 4.79 | 1.09 | 2 | 4 | 2 | 2.67 |
| KBK 2 | Butt | 7.99 | 3.83 | 2.08 | 3 | 3 | 2 | 2.67 |
| | Middle | 5.82 | 4.62 | 1.26 | 2 | 4 | 2 | 2.67 |
| | Top | 5.20 | 4.77 | 1.09 | 2 | 4 | 2 | 2.67 |
| KBK 3 | Butt | 6.71 | 3.98 | 1.68 | 2 | 3 | 2 | 2.33 |
| | Middle | 6.29 | 5.47 | 1.15 | 2 | 4 | 2 | 2.67 |
| | Top | 7.20 | 4.65 | 1.55 | 3 | 4 | 2 | 3.00 |
| JPP PHT 1 | Butt | 5.59 | 3.60 | 1.55 | 2 | 3 | 2 | 2.33 |
| | Middle | 5.45 | 4.64 | 1.17 | 2 | 4 | 2 | 2.67 |
| | Top | 6.99 | 5.16 | 1.35 | 3 | 4 | 2 | 3.00 |
| JPP PHT 2 | Butt | 6.47 | 5.08 | 1.27 | 2 | 4 | 2 | 2.67 |
| | Middle | 6.21 | 4.61 | 1.35 | 2 | 4 | 2 | 2.67 |
| | Top | 6.39 | 6.09 | 1.05 | 2 | 5 | 2 | 3.00 |
| JPP PHT 3 | Butt | 8.02 | 5.50 | 1.46 | 3 | 4 | 2 | 3.00 |
| | Middle | 7.22 | 5.13 | 1.41 | 3 | 4 | 2 | 3.00 |
| | Top | 6.95 | 5.89 | 1.18 | 3 | 5 | 2 | 3.33 |

**Table 12.** Recapitulation of wood quality class

| Breeding technology | Wood quality class on stem position | | | Average wood quality class |
|---|---|---|---|---|
| | Butt | Middle | Top | |
| APB | 3.78 | 3.00 | 3.11 | 3.30 |
| KBK | 2.78 | 2.78 | 2.78 | 2.78 |
| JPP-PHT 1 | 2.67 | 2.78 | 3.11 | 2.85 |
| Average | 3.07 | 3.85 | 3.00 | 3.30 |

## 4. Conclusion

Three points of conclusion was presented as follows:

a. The interaction of two factors, namely the breeding technology and stem axial position affect to radial shrinkage, but was not affect the tangential shrinkage and tangential/radial shrinkage ratio
b. Breeding technology method affects very significantly to tangential shrinkage and radial shrinkage. Seeds prepared by JPP PHT-1 resulted the smallest tangential shrinkage (6.59%), followed by KBK (6.62%) and APB (8.46%). Wood grown by KBK seeds resulted the smallest radial shrinkage (4.42%), followed by JPP PHT-1 (5.08%) and APB (5.51%). (b) The Stem axial position affected the tangential / radial shrinkage ratio. The stem butt position has the largest ratio (1.84), followed by the middle (1.45) and the top of the stem (1.26)
c. Breeding technology method resulted the highest wood class owned by KBK (2.78) followed by JPP PHT-1 (2.85) and APB (3.30). (b) Stem axial position produced a highest wood quality class owned by stem butt position (2.67), followed by the middle part (2.78) and top part of the stem (3.11).

One point to be recommended as follows, breeding technology of JPP PHT-1 method needs to be socialized as it produces the wood which classified on quality class of 2.85 at the age of 10 years and the highest wood volume compared to KBK and APB methods.

## References

[1] Seno A 2008 Pengembangan Industri Furnitur di Indonesia *Makalah Workshop Pengembangan Keunggulan Local Program Studi Teknologi Hasil Hutan Fakultas Kehutanan Universitas Gadjah Mada*
[2] Anonymous 2010 *Jati Plus Perhutani* http://perumperhutani.com/produk/benih-dan-bibit/jati-plus-perhutani/
[3] Zobel BJ and Buijtenen JP 1989 *Wood Variation. It's Causes and Control* (Berlin, Hiedelberg, New York: Springer-Verlag)
[4] Kollmann FEP dan Cote WA 1968 *Principle of Wood Science and Technology Volume I* (New York: Springer-Verlag)
[5] Prayitno 1995 *Pertumbuhan dan Kualitas Kayu* (Yogyakarta: Program Pasca Sarjana Fakultas Kehutanan Universitas Gadjah Mada)
[6] Panshin AJ and de Zeeuw C 1964 *Textbook of Wood Technology Vol. I: Structure, Identification, Uses and Properties of the Commercial Wood of United State* (New York, San Francisco Toronto, London: Mc. GrawHill Book Company)
[7] Shmulsky R and Jones D 2011 *Forest Product and Wood Science An Introduction 6 edition* (United Kingdom: Wiley-Blackwell Publication)
[8] Suranto Y 2015 Pengaruh umur pohon, bonita dan posisi aksial batang terhadap kualitas kayu sebagai bahan mebel: Studi kasus kesatuan pemangkuan hutan Kendal, Jawa Tengah [Disertasi] Program Doktor Fakultas Kehutanan Universitas Gadjah Mada Yogyakarta
[9] Snedecor GW dan Cochran WG 1964 *Statistical Methods: Applied to Experiments in Agriculture and Biology 5th Edition* (Iowa, US: The Iowa State University Press)
[10] Anonymous 2014 Pengaruh Metode Penyediaan bibit dan posisi aksial batang terhadap sifat dasar dan kualitas kayu jpp sebagai bahan furniture (Laporan Penelitian Kerjasama Fakultas Kehutanan Universitas Gadjah Mada dan Pusat Penelitian dan Pengembangan Kehutanan Perum Perhutani)
[11] Matumura J and Kawasaki Y 2012 Effect of rotation age on wood quality of Teak (*Tectona grandis* L.F.) planted in Indonesia. *Proceeding of Asia Sustainable Local Resouce Management Workshop Fokuoka* http://ffpsc.agr.kyushu-u.ac.jp/policy//scbrm/PP2012/Workshop_Proceedings.pdf

**Acknowledgement**
The author wants to express an appreciation to The Centre for Forest Research and Development at Cepu owned by "Perum Perhutani" Forest Company for wood materials and funds required to conduct this research.

IOP Conf. Series: Materials Science and Engineering **935** (2020) 012038 doi:10.1088/1757-899X/935/1/012038

# Effect of ethanol polarity on extraction yield, antioxidant, and sunscreen activities of phytochemicals from *Gyrinops versteegii* leaves

**R K Sari[1,4*], M Wahyuningrum[1], M Rafi[2,4] and I Wientarsih[3,4]**

[1]Department of Forest Products, IPB University, Bogor, 16680, Indonesia
[2]Department of Chemistry, IPB University, Bogor, 16680, Indonesia
[3]Department of Veterinary Clinic Reproduction and Pathology, IPB University, Bogor, 16680, Indonesia
[4]Biopharmaca Research Center, IPB University, Bogor, 16128, Indonesia

*Email: rita_kartikasari@apps.ipb.ac.id

**Abstract.** *Gyrinops versteegii* is one of the agarwood-producing species that is popularly cultivated in Indonesia. This research aimed to determine the yield, in vitro antioxidant and sunscreen activities of *G. versteegii* leaf phytochemicals (extracts) from the maceration with ethanol solvents at various polarities (E100%, E50%, E0%) and to analyze the total phenol content (TPC) and total flavonoid content (TFC) of the extracts. The result shows that the polarity of ethanol affects the yield, antioxidant and sunscreen activities, and TPC of the extract. The E50% dissolved extract is the highest yield (19.14%), antioxidant activity (DPPH and CUPRAC test value are 35.82 and 389.52 μmol trolox/g sample), sunscreen activity (sun protection value is 14.9) and TPC (89.35 mg/g AGE) with TFC 0.485%.

## 1. Introduction

Agarwood is one of the potential Non Timber Forest Products (NTFPs) in Indonesia. Agarwood is very expensive compared to other NTFPs commodities. The export value of agarwood in September 2017 to USD 11.92 million with a volume of 847,392 kg [1]. However, the increase in demand for agarwood cannot be fulfilled due to the decreased availability of agarwood in the natural forests. Therefore, CITES appendix II limits the export of agarwood [2]. For this reason, Indonesia is developing the cultivation of agarwood-producing trees. *Gyrinops versteegii* is one of the agarwood-producing species that is popularly cultivated in Indonesia. The *G. versteegii* trees that had been planted [3].

One of the prospective parts of *G. versteegii* is the utilization of leaf part as a natural antioxidant source. The antioxidant activity of the methanol extract of *G. versteegii* leaf is very strong ($IC_{50}$ 14.46 μg/mL). The extract also has a very high sunscreen effectiveness with sun protection factor (SPF) value of 16.28 [4] and according to Wilkinson and Moore [5] belong to the ultra-protection category. However, for commercial utilization, the extraction with ethanol is more recommended than methanol because methanol is flammable, toxic, and smelly compared to ethanol solvents [6]. Therefore, the research on the antioxidant activity and SPF of *G. versteegii* leaf extracts from the extraction using ethanol at various polarity needs to be done.

This research aimed to analyze the effect of solvent polarity on the yield, antioxidant and sunscreen activities, the Total Phenol Content (TPC), and Total Flavonoid Content (TFC) of *G. versteegii* leaf phytochemicals (extracts) from the maceration with ethanol solvents at various polarities. In order to achieve these objectives, the study was started with the determination of the extraction yield, antioxidant activity, SPF value TCF, and TFC.

## 2. Materials and methods

### *2.1. Raw material preparation*

The research sample is an old leaf from Depok, West Java. The dried leaves were then mashed by grinding and then sieved to be 40-60 mesh in size. The water content of samples was measured by taking a sample of ± 2 g and dried in an oven at a temperature of 103±2°C to achieve a constant kiln-dry weight.

### *2.2. Extraction*

*G. versteegii* leaf powder that has known water content was weighed 200 g x 3 replications. The sample extraction used the maceration method with ethanol at various polarities, such as the ethanol concentration of 100% (E100%), 50% (E50%), and 0% (E0%/only aquadest). Dilution of ethanol into an E50% solvent is done by mixing 50% ethanol with 50% aquadest. The extraction filtrate was evaporated with a rotary evaporator and dried at 40°C.

### *2.3. In vitro antioxidant and sun protection assay*

The antioxidant activity assay using the 2,2-diphenyl-1-pickrylhidrazyl (DPPH) method refers to Salazar *et al.* [7] and the Cupric ion reducing antioxidant (CUPRAC) method refer to Ozturk *et al.* [8]. The antioxidant capacity is expressed in μmol trolox/g dry powder. The higher value of μmol trolox/g values indicates the higher antioxidant activity. The determination of the SPF value refers to Kawira [9].

### *2.4. Phytochemical analysis*

The phytochemical analysis is carried out quantitatively. Analysis of total phenol content and total flavonoid content refers to Indrayani *et al.* [10].

### *2.5. Data analysis*

Data analysis using complete random design with treatment is ethanol solvent at different polarities (ethanol concentration), such as E100%, E50%, and E0%. Data analysis uses an analysis of variance (ANOVA). The responses analyzed were the yield, antioxidant activity, sunscreen activity, TPC, and TFC extracts which are being produced from the extraction process with ethanol solvents at various polarities. If the treatment has a significant effects then Duncan performs further tests using the SPSS application.

## 3. Results and Discussion

### *3.1. Extract yield*

The results of statistical analysis showed that the difference in polarity of ethanol had a significant effect on the extract yield. Duncan test results ($\alpha = 0.05$) showed that the highest yield is the E50% dissolved extract (E50E) and significantly different from the E100% dissolved extract (E100E) and E50% dissolved extract (E0E) (Fig 1). This shows that the difference in polarity affects the phytochemicals of the extracted leaves. This is confirmed by the results of research showing the difference in polarity of the solvent affects the yield of *Pluchea indicia* leaf extract [11]. Some chemical compounds can dissolve using a mixture of ethanol and water including amino acids, sugar, some phytochemical compounds such as alkaloids, flavonoids, flavonoid glycosides, and chlorophyll [12].

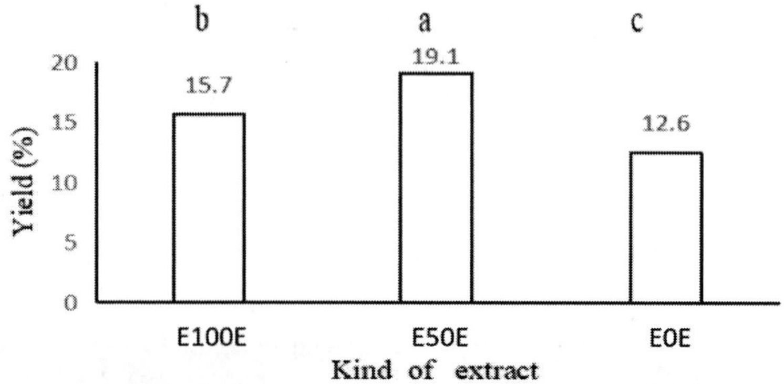


**Figure 1.** The yield of *G. versteegii* leaf extracts of E100E, E50E, and E0E from the maceration with the ethanol solvents at various polarities. The different letters in the lane show significantly different antioxidant activity values ($\alpha = 0.05$)

*3.2. Antioxidant activity*
The antioxidant activity of *G. versteegii* leaf extract using two antioxidant assays, namely the DPPH and CUPRAC methods. The antioxidant activity of testing with the CUPRAC method on all extracts has a higher value than the DPPH method (Table 1). This can be caused by the CUPRAC method which has more selective properties and more stable reagents than chromogenic reagents in measuring the antioxidant activity of DPPH. Also besides, the CUPRAC method can simultaneously measure antioxidant and lipophilic antioxidant compounds [13]. Besides that, according to Irawati [14], the DPPH method has a narrow range of absorbance and concentration linearity compared to the CUPRAC method which is able to react with large molecules so that it has a wide range of absorbance and concentration linearity.

**Table 1.** Antioxidant activity of *G. versteegii* leaf extracts from the maceration with the ethanol solvents at various polarities (concentration of ethanol)

| Kind of antioxidant assay | Kind of extract (µmol trolox/g sample)[1,2]) | | |
|---|---|---|---|
| | E100E | E50E | E0E |
| DPPH | 27.42±0.00$^b$ | 35.82±0.03$^a$ | 25.92±0.20$^c$ |
| CUPRAC | 342.95±0.00$^y$ | 389.52±0.05$^x$ | 286.06±0.06$^z$ |

Note: *) = 1): average of 3 replications, 2) the different letters in the lane show significantly different antioxidant activity values ($\alpha = 0.05$)

The results of statistical analysis showed that the type of solvent had a significant effect ($\alpha = 0.05$) on antioxidant activity based on tests using the DPPH and CUPRAC method. Duncan's further test results showed that the E50E had the highest antioxidant activity and was significantly different from other extracts and followed by the E100E. But, the antioxidant activity of the E100E was different from the E0E which had the lowest antioxidant activity (Table 1). If compared with the methanol extract of Mint leaves (*Mentha piperita*) which is known as an antioxidants source [15], the antioxidant activity of Mint leaves is almost the same (antioxidant value of 386 µmol trolox/g powder) with the E50% dissolved extract (Table 1). This shows the potential of *G. versteegii* leaves is developed as a source of antioxidants compounds.

*3.3. Sun protection factor value*
The results of the analysis of the variance showed that the differences in ethanol polarity influenced the sunscreen activity of the extract. Duncan's further test results showed that the sunscreen activity (SPF value) of the E50E is not significantly different from the SPF value of E100E. However, the SPF value of the E50E and E100E are higher and significantly different from the E0E (Figure 2). The difference

in the SPF value of these extracts is influenced by the content of secondary metabolites from phenolic groups such as flavonoids, isoflavonoids, and tannins which can absorb UV rays and potentially as sunscreens. Wolf *et al.* [16] stated that flavonoid compoundss have the potential to be a high sunscreen due to the presence of chromophore groups (conjugated double bonds) that are able to absorb UV so that it can reduce its intensity on the skin. Based on the sunscreen classification according to Wilkinson and Moore [5], all of the extracts are classified as maximum because the SPF value is in the range of 10-15.

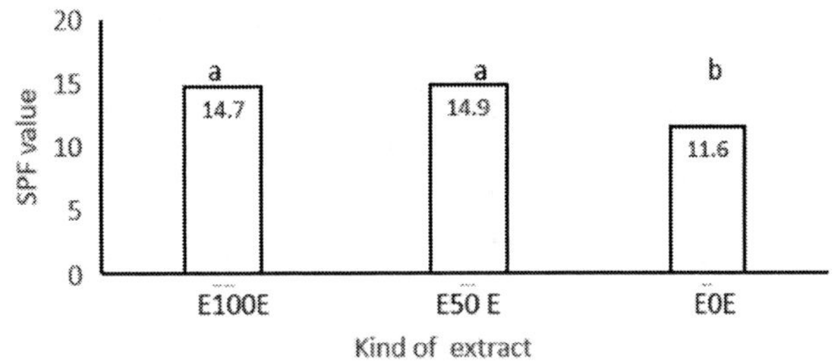


**Figure 2.** The SPF value of *G. versteegii* leaf extracts from the maceration with the type of solvent (E100, E50, and E0). The different letters in the lane show significantly different antioxidant activity values (α = 0.05).

*3.4. Total Phenol Content (TPC) and Total Flavonoid Content (TFC) of extract*

The results of the analysis of variance showed that the solvent type factor had a significant effect (α = 0.05) on the total phenol content of the extract. The TPC of the E50 extract is significantly higher than E100 extract and E0 extract (Table 2). But, the results of the analysis of variance showed that the solvent type factor had not a significant effect (α = 0.05) on the total flavonoid content of the extract. The difference in total phenol content can be due to the total concentration of phenol depending on the material and its solubility in the various polarities of the extraction solvent.

**Table 2.** TPC and TFC of *G. versteegii* leaf extracts from the maceration with the ethanol solvents at various polarities

| Phytochemical test | Kind of extract | | |
|---|---|---|---|
| | E100 | E50 | E0 |
| TPC (mg/g AGE)* | 79.840±0.004$^{b}$ | 89.350±0.001$^{a}$ | 34.070±0.006$^{c}$ |
| TFC (%b/b) | 0.496±0.001 | 0.485±0.004 | 0.490±0.003 |

Note: *) the different letters in the lane show significantly different antioxidant activity values (α = 0.05)

The E50 extract had a higher antioxidant activity than E100 (Table 1). This has the positive corelation with the total phenol content compared to E100% extract and E50% (Table 2). This can be caused by antioxidant activity was influenced by total phenol content. This is in line with Wolf *et al.* [16] research which reported that the strongest positive correlation between total phenolics and antioxidant activity of microalga.

## 4. Conclusion

The yield, antioxidant activity, SPF value, and TPC of the *G. versteegii* leaf extract were affected by the polarity of the solvent (ethanol concentration). The highest yield is the E50E (19.1%) and significantly different from the E100E (15.7%) and E0E (12.6%). The highest antioxidant activity is the E50E with DPPH and CUPRAC value of 35.8 and 389.5 µmol trolox/g sample and followed by E100E and E0E with DPPH values of 27.4 and 25.9 µmol trolox/g sample and CUPRAC value is 342.9 and 286.1 µmol trolox/g sample, respectively. The highest sunscreen activity is the E50E with SPF value of 14.9, and followed by E100E and E0E with SPF value of 14.7 and 11.6. The highest TPC is the E50E (89.4 mg/g AGE) and followed by E100E and E0E with TPC of 79.8 and 34.1 mg/g AGE. However, the TFC values of the three extracts are relatively the same (0.5%).

**References**

[1] Statistics Indonesia 2017 *Bulletin of Foreign Trade Export Statistics According to The September Harmonized System December 2017* (Jakarta: PT Citra Mawana Patamoro)
[2] Barden A, Anak NA, Mulliken T, and Song M 2000 *Heart of the Matter: Agarwood Use and Trade and Cites Implementation for* Aquilaria malaccensis (Cambridge: Traffic Network Report)
[3] Santoso E, Turjaman M, Irianto RSB, Agustini L, Efiyanti L, and Faulina SA. 2014. *Forestry Science and Technology Book Series V* (Bogor: Puslitbang Bogor)
[4] Wahyuningrum M, Sari RK, and Rafi M 2018 *J. Trop. Wood Sci.Technol.* **16**(2) 141-9
[5] Wilkinson JB and Moore RJ 1982 *Harry's Cosmeticology 7'th edition* (New York: Chemical Publishing Company)
[6] Romadanu, Rachmawati SH, and Lestari SD 2014 *Fishtech* **3**(1) 1-7
[7] Salazar AR, Perez LLA, Loppez AJ, Alanis GBA, and Torres NWA 2011 *Evid. Compl. Alterntat Med.* 536139 1-6
[8] Ozturk M, Ermin MD, Kivrak S, Mercan N, Turkoglu A, and Ozler MA 2011 *Food Chem. Toxicol* **49** 1353-60
[9] Kawira JA 2005 *Laboratory Procedures for Determining Sun Protection Factor* (Depok: University of Indonesia)
[10] Indrayani L, Soetjipto H, and Sihasale L 2006 *Hayati* **12** 57-56
[11] Sani RN, Nisa FC Andriani RD, and Maligan JM 2012 *J. Food Agro-Industry* **2**(2) 121-6
[12] Widyawati PS, Budianta TDW, Kusuma FA, and Wijaya EL 2014 *Internat. J. Pharm. Phytochem. Res.* **6**(4) 850-5
[13] Apak R, Guclu K, Demirata B, Ozyurek M, Celik SE, Bektasoglu B, Berker KI, and Ozyurt D 2007 *Molecules* **12** 1496-547
[14] Irawati I 2008 *Comparison of Methods for Determining the Antioxidant Activity of* Ginger Rhizome (Bogor: Bogor Agricultural University).
[15] Luminita P 2015 *J. Horticul. Forest Biotech* **19**(4) 9-12
[16] Wolf R, Wolf D, Morganti P, and Ruocco V 2001 *Clin. Derm.* **19** 252-59
[17] Shetty V and Sibi G 2015 *J. Food Res. Sci* **4**(1) 11-9

**Acknowledgment**
The author would like to thank the Directorate of Higher Education of the Ministry of Education and Culture of the Republic of Indonesia with the 'IPB Higher Education Research Grant' scheme 2019 (contract No. 3/EI/KP.PT.PTNBH/2019) for funding support for some parts of this research.

IOP Conf. Series: Materials Science and Engineering **935** (2020) 012039 doi:10.1088/1757-899X/935/1/012039

# Study on East Sumba-originated natural pigments for coloring woven fabrics

**G Pasaribu* and I Winarni**

Forest Products Research and Development Center, Ministry of Environment and Forestry, Bogor 16110, Indonesia

*E-mail: gun_pa1000@yahoo.com

**Abstract.** East Sumba Regency has long been renowned as the production site for tightly woven items. These woven items afford high economy values, since they are produced with natural-coloring stuffs (renewable-pigments). Pigment-producing plants serve as one of valuable natural sources, which afford beneficial potency for coloring textiles in Indonesia, especially in developing products that sound "back-to-nature". Relevantly, this research aimed to conduct technical assessment and processing of pigment-producing plants associated with natural coloring-agents for fabrics in East Sumba. The methods employed in-depth interviews with various parties, who played essential roles in processing of natural pigments from plant sources; pigment extraction; and chemical tests. Results revealed that utilization of natural pigments still kept going on nowadays by the community in East Sumba to produce tight-woven items. Extraction technique of natural renewable-pigments was still very simple by immersing pigment-producing plants in water. Likewise, so was the technique for coloring threads/yarns with natural pigments by dipping and immersing methods. There were three kinds of plant-derived natural pigments utilized most by the community, with their plant species origins comprising Wora/Nila (*Indigofera erecta* Thunb), Kambu (*Morinda cf.citrifolia* L.), and (*Symplocos cochinchinensis* (Lour.) S. Moore). Further, extraction technique using hot water brought out higher yields of extracted pigments and better fabric-coloring qualities. Activities of tight weaving were commonly used as additional income source. Besides taking considerable time for pigment production process, recipient markets were still not appropriately available. Moreover, currently raw material source for natural pigment became more difficult to obtain. Accordingly, a part of the community began mixing natural pigments with synthetic pigments. In future, pigment-producing plants should expectedly be cultivated more seriously to maintain natural-pigment stocks.

## 1. Introduction

Forests can provide various essential necessities for persons, beginning from food items to board products and clothes. One of the related supporting sources in producing clothes is coloring agents (pigments). Renewable natural pigments could be obtained from plants/trees; and unfortunately, most of the pigments have been produced synthetically. About natural pigments in accordance with the Regulation by Forestry Minister (No. 35/Menhut-II/2007) [1] regarding Non-Timber Forest Products, it was mentioned that there were 21 kinds of natural pigments originated from forests [2].

Natural pigments from plants could serve as renewable natural resources, which afford valuable potency in their uses as coloring agents for textiles in Indonesia particularly in developing products, associated with the motto "back to nature". Generally, products derived from renewable natural sources perform better than those produced synthetically, with respect to product qualities, yields, and environmentally friendly impacts [3].

In another side, the use of synthetic pigments in the coloring process for textiles/fabrics could bring about environment problems, because several synthetic pigments contain heavy metals which are harmful, dangerous, and severely polluted. Those metals are among others Cu, Ni, Cr, Hg, and Co [4]. Those pollutants finally might be just discarded to the public effluent stream, thereby contaminating the environments, particularly water-stream environments [5,6].

Natural pigments for textile items are usually obtained by extracting various parts of plants such as roots, woods, barks, leaves, and seeds as well as flowers, which further brings out the so-called extract results. Processing or removing natural pigments is usually conducted through two methods, which consist of extraction and fermentation [7].

One of pigment-producing plant species, from the natural pigments has been extracted and then utilized a lot by local community is the so-called Nila (*Indigofera tinctoria*). This Nila derived natural pigment is already popular and used as coloring agent for tight-woven fabrics in Sumba Island, as batik-fabric coloring in Jogyakarta, and as ulos coloring in North Sumatra. Despite those conveniences, unfortunately use of plant-derived natural pigments currently tends to be abandoned due to the presence of synthetic pigments as coloring agents. Actually, use of renewable natural pigments could bring several advantages, as they were more secure in their uses, more attractive to worldwide markets, and more environmentally friendly [8].

Amutha *et al.* [9] have identified two natural dyes extracted from the fruits of *Terminalia arjuna and Thespesia populnea* trees. Both the dyes gave bright yellow colour in silk and nylon fabrics and pale yellow shade in cotton. The dyes exhibited good fastness properties on silk and nylon fabrics even without mordant [9].

Among other kinds of plant-derived natural pigments are flavonoid, saponin, and tannin, which all belong to a group of wood extractive pigments. Flavonoid compounds can impart red, yellow, brown, or blue colors to particular wood species. Tannins are inherently complex organic compounds with their amorphously textured crystals, and soluble in water forming colored liquid. Such liquid colors and their color intensity (weak or strong) could be affected by acid, alkali or salt treatments [3].

Anthocyanins is one of the natural dyes which is a subclass within the class of flavonoids and, which are responsible for red till bluish colors in most flowers, higher plants, fruit and vegetables. Anthocyanins consist of a typical C6–C3–C6 aglycon, which is often glycosylated. Within plants, these sugar moieties are acylated. The most frequently occurring anthocyanidins are cyanidin, delphinidin, malvidin, pelargonidin, peonidin and petunidin. Anthocyanins are very prone to pH-fluctuations [10].

Relevantly, technical assessment and processing of pigment-producing plants/trees associated with the utilization of natural coloring agents for fabrics in East Sumba deserved carrying out to look into their related current situation as well as development potency. The practical detailed results are forthcoming [11].

## 2. Materials and Methods

### *2.1. Materials*

The main research materials were inherently natural pigments derived (extracted) from particular plant/tree species, which grow well and abundantly in East Sumba. The chemical stuffs which were used comprised among others methanol, ethanol, and distilled water. Meanwhile, the related devices and instruments consisted of beaker glasses, heating oven, analytic balance, GCMS (Gas Chromatography Mass Spectrometer) pyrolysis, pipettes, pyrex tubes, and other assisting devices/instruments.

*2.2. Methods*

2.2.1. Interview with the related parties
In-depth interview and discussion were performed with various parties who supposedly played essential roles in processing (extracting) natural pigments from plant sources. The targeted parties comprised farmers/crafters, enterprisers, self-assisting institutions, and regional government. From this discussion with those parties could then be obtained beneficial information especially about various kinds of renewable natural pigments commonly utilized for coloring fabrics and other woven items by the community.

2.2.2. Technical assessment on processing of plant-derived natural pigments
This assessment was conducted through the observation and recording (taking note) of every process which was carried out throughout the entire production of natural pigments.

2.2.3. Testing on the produced natural pigments
Natural pigments which were consecutively produced by the community; and also produced through the laboratory scale trial were then tested for comparing the properties of resulting-pigments between those produced by the community manner and through laboratory scale trial technique. The tested properties comprised natural-pigment yield and coloring qualities.

2.2.4. Pigment extraction and testing of phytochemical properties
Treatment that implemented several various extraction techniques could expectedly produce the pigments with satisfactory high yields and better qualities. In the pigment extraction process (in the laboratory scale trial), initially parts of the trees/plants indicatively potential to produce natural pigments were ground/milled to very small sized pieces using hammer mill or blender. Afterwards, the milled/ground tree parts (with their known oven dry weight equivalent) were sieved using staged screens such that the particles that passed through 40-60 mesh screen in sufficient amount could be obtained. The obtained screened particles were then extracted by refluxing in hot water and separately immersed (macerated) in methanol. The methanol immersed particles were screened; and then the obtained filtrate was concentrated using rotary vacuum evaporator. After all the methanol evaporated out, it left behind the concentrated extract (regarded/called as natural pigment), further put into the oven under mild heating (±50-60°C) for short duration, then and in this way its dry weight could be known. The yield (Y) of concentrated extract/natural pigment then could be calculated using the formula, as follows:

$$Y(\%) = \frac{\text{Dry weight of concentrated extract (natural pigment),grams}}{\text{Oven dry weight of milled/ground tree parts}}\ x\ 100 \qquad (1)$$

Meanwhile, a part of the concentrated extracts (natural pigments) was taken and then tested for phytochemical properties.

## 3. Results and Discussion

*3.1. Technical and economy assessment on processing of natural pigments*

3.1.1. Kinds of plant-derived natural pigments
Natural pigments of various kinds for coloring the woven fabrics were taken (extracted) by the community from their producing-plants, which grew well and abundantly in the forest vicinity. The forests which were explored a lot located around the National Park under the Region II in Wanggameti village. Kind of main natural pigments in the forest which was sought the most by the community was the one originated from Loba tree species, and the tree parts which were taken and extracted (for natural pigments) comprised leaves and barks.

Information obtained from the Industry and Trade Service at East Sumba regency mentioned that there were at least three main centers for production of tight-woven fabrics, among others in Raja (Rindi), Lambanapu, and Kaliuda villages. The data from the Office of Center for Statistics Bureau there listed that there existed 2548 units of tight-weaving enterprises with 4768 workers, which spread out at 12 subdistrict [12].

From the assessment results about natural-coloring technique for woven fabrics could be obtained beneficial information. The information mentioned that several pigment-producing tree (plant) species from which their pigments were utilized consisted of among others Wora (Nila) species as source for blue/violet color, Kambu species as pigment source for red color; and Loba species as source for strengthening the color, aided by the assisting chemical agents, such as lime and kemiri (*Aleurites moluccana)* seeds.

(a) (b) (c)

**Figure 1.** Plant/tree species from which the natural pigments were extracted and then used for coloring the fabrics; a) Nila species; b) Kambu species; c) Loba species

Results of identification on specimens of pigment-producing plant species which were performed by the Botany Division at the Biology Research Center, under the Indonesia's Science Institute (LIPI) are presented in Table 1.

**Table 1**. Species origins of pigment-producing plants/trees

| No | Regional/ Vernacular names | Scientific names | Family |
|---|---|---|---|
| 1 | Wora/ Nila | *Indigofera erecta* Thunb | Leguminosae |
| 2 | Kambu | *Morinda cf.citrifolia* L. | Rubiaceae |
| 3 | Loba | *Symplocos cochinchinensis* Lour.) S. Moore | Symplocaceae |

3.1.2. Coloring technique

The coloring-technique was performed by initially immersing in cold water, parts of particular pigment-producing plant species (Wora/Nila species). The immersion lasted for 24 hours to provide enough time for the natural pigment to elute out of the tree parts into the water. Afterwards, parts of the eluted plant/trees were removed from the water, leaving behind the colored aqueous solution (liquid) which appeared as black/blue/violet. The black/blue/violet-colored liquid (eluant) was let there for particular duration until it formed a precipitated mass; and the precipitated mass could further be used as natural pigments. The weaving thread/yarn was then dipped into the colored liquid (eluant) to impart it with blue/violet color. This process could overall take about 2 weeks. In the same way, to impart the weaving thread/yarn with red color, it should be dipped in the colored liquid, previously into which was immersed particular parts of Kambu plant/tree species. In coloring the thread/yarn with

Kambu natural pigment, it was usually performed by adding Loba-originated pigment to strengthen the red color at the thread/yarn.

This coloring technique of thread/yarn was performed after the tying (bundling) process finished. The tying/bundling of thread/yarn usually used raffia rope with the purpose that the tied part of thread/yarn would not be colored during the dipping process. The tying patterns could be varying, whereby it could shape figures like horses, chickens, flying-horses, etc. Usually, the first coloring process was conducted using Nira-originated pigments (affording blue color). The dipping of thread/yarn in Nila-originated pigment could last for one month, which should be performed repeatedly until the desired blue color was obtained. Afterwards, the tying of thread/yarn was opened (untied), and then colored with kemiri and Kambu-originated pigments as the sources for red color. All the tying ropes were completely removed. It then appeared that Nila originated pigment mixed Kambu-originated pigment would bring about black color. Meanwhile, Loba-originated pigment mixed with Kambu-originated pigments ended up with red color.

The preparation of natural pigments could be regarded as very time consuming, as it might take one month. For these reasons, the manufacture of tight woven fabrics became a very lengthy process. Moreover, for the weaving process itself, it could finish in the matter of one day until one week.

**Figure 2**. The dipping (coloring) process for yarn, fabrics, and other woven items

3.1.3. Economy assessment

In general, the community in East Sumba did not rely much their main occupation on fabric-weaving activities. In fact for their main occupation, the community there at Rindi Subdistrict which served as one of the assessment locations, performed other occupation activities such as farming, raising cattle, fishing, and trading [13].

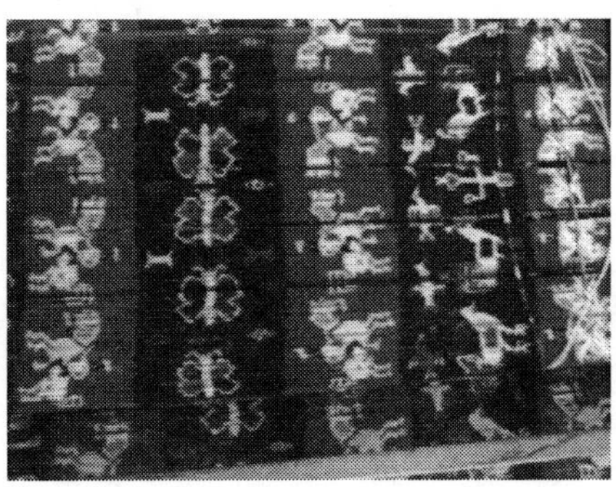

**Figure 3.** Fabric materials after being colored with various plant-derived natural pigments

In one week, duration, in average the fabric weavers could produce one sheet of woven fabrics that measured 20 cm by 200 cm, with the selling price as much as IDR 300,000–400,000 per sheet. Meanwhile for larger size of woven fabric sheets (50 cm by 200 cm), the weaving activities could take about 2 weeks.

### *3.2. Extraction of natural pigments and testing on their phytochemical properties*

#### 3.2.1. Pigment extraction

The extraction method as performed on particular parts of plant/tree species origins (to obtain the extracts/natural pigments) implemented the so-called maceration and refluxing techniques. Maceration technique was to separate active compounds from the plant/tree parts (and then to be included in the extracts), not only based not on the effectiveness, practical consideration, safety, and economy aspects in the technique use; but also for the purpose to avoid or minimize the possible destruction (degradation) to those compounds which could be vulnerable (not resistant) to particular heat-involving treatment. Meanwhile in refluxing technique, the selecting of methanol as the extracting solvent was because methanol could perform as multipurpose solvent that was very suitable to extract expectedly almost all the plant compounds.

In Table 2 was presented the yield of extracts (natural pigment) as obtained using two refluxing extraction techniques (in laboratory trial; using hot water and methanol solvents) and using the technique commonly performed by the community (cold water solvent). Meanwhile as described before, the maceration extraction technique was implemented to extract particular chemical compounds which were not resistant to heat.

In Table 2, it revealed that the extraction technique using hot water solvent brought out greater yield of the extract (natural pigment) than using cold water and methanol solvents. The extraction using cold water was commonly performed by the community (in East Sumba); and such lower pigment yield could expectedly be increased two times greater, if they used hot water solvent instead.

**Table 2.** Yield of extract (natural pigment)

| No | Species origin of pigment-producing plants | Yield, % / Extraction techniques | | |
|---|---|---|---|---|
| | | Cold water solvent **) | Refluxing extraction*) | |
| | | | Hot water solvent | Methanol solvent |
| 1 | Nila | 9.00 | 16.37 | 16.05 |
| 2 | Kambu | 4.04 | 17.23 | 14.87 |
| 3 | Loba | 5.42 | 9.69 | 7.96 |

Remarks: *) in laboratory-scale trial; **) commonly performed by the community in East Sumba

Efficiency of extraction of natural dyes largely depends on the type of sources and various processing variables such as pH, temperature, time, liquorratio, and particlesize. The methods of extraction of natural dyes can be classified based on the media used inextraction, such as aqueous, nonaqueous, organic, and mixture of aqueous-organic extraction [14].

#### 3.2.2. Phytochemical tests

Results of phytochemical tests on the extracts (natural pigments) were presented in Table 3 Phytochemical contents could detect or confirm the presence of particular pigment compounds (e.g. tannin, flavonoid, and saponin) [15] in the extracts from plant species origins.

**Table 3.** Results of phytochemical tests on plant derived pigments

| No | Species origins of pigment-producing plants | Kinds of natural pigments as detected (confirmed) | | |
|---|---|---|---|---|
| | | Tannin | Flavonoid | Saponin |
| 1 | Nila | +++ | + | + |
| 2 | Kambu | ++ | ++ | ++ |
| 3 | Loba | + | + | ++ |

Remarks: + = presence, ++ = strong presence; +++ = very strong presence

Tannin compounds could be used to impart specific colors to fabrics. The tannin used as coloring agents (natural pigments) acted as mordant, and commonly in combination with particular heavy metals. Prabhu and Teli have isolated tannin compounds from java tamarind (*Tamarindus indica*) and then used them as natural mordant, previously mixed with copper sulfate ($CuSO_4$). This natural mordant ($CuSO_4$-mixed tannin) could then be used as natural pigments for cotton, wool, and silk fabrics. In fact, this mordant was more resistant to leaching than the natural pigment alone (tannin without addition of mordanting agent, $CuSO_4$) [16].

In scrutinizing Table 3, it strongly indicated that Nila originated natural pigment exhibited greatest tannin contents, compared to the content in Kambu and Loba originated natural pigments. Accordingly, Nila originated pigment seemed better for use as mordant compared to the other pigments (Kambu and Loba). Also, with economy reasons, Nila originated pigments could be combined with alum ($Al_2[SO_4]_3$) and ash soda ($Na_2CO_3$) in order that the colored (pigmented) fabric fixed the colors (pigments) more strongly, thereby not easily leached out. Meanwhile, Loba originated pigments were already known to contain aluminum (Al) and iron (Fe) metals, which could also function as mordant. The presence of these metals could be found inside the leaf, root, and bark parts of the tree stems.

### *3.3. Tests on total phenol in plant-derived natural pigments*

Results of tests on total phenol in natural pigments originated from Nila, Kambu, and Loba plant species were presented in Table 4.

**Table 4.** Total phenol content in plant-derived natural pigments for coloring fabrics

| No | Species origins of pigment-producing plants | Total phenol % (w/w) |
|---|---|---|
| 1 | Nila | 228.20 |
| 2 | Kambu | 308.80 |
| 3 | Loba | 232.50 |

Total phenol content (Table 4) was strongly indicated to correlate linearly and positively with flavonoid presence in natural pigments (Table 3). It is necessary to know that flavonoid compounds typify as the greatest group that existed in nature. These flavonoid compounds signify as natural pigments that impart red, violet, and blue colors, and as pigments that brought out yellow colors in the particular parts of plants/vegetations.

### *3.4. Tests on coloring qualities of plant derived natural pigments*

The color qualities of natural pigments in this regard were expressed quantitatively as brightness values, as presented in Figure 4. From Figure 4, it mentioned that color qualities of natural pigments which were extracted using hot water solvent tended to be better (i.e. greater brightness values) than using other extraction techniques (e.g. using cold water and methanol solvents), as also confirmed through visual observation (Figures 5 and 6). This occurrence was quite interesting, because besides better coloring qualities (greater brightness) of natural pigments, their corresponding pigment yields tended also to be greater with refluxing-extraction techniques, particularly using hot water solvent (Table 2).

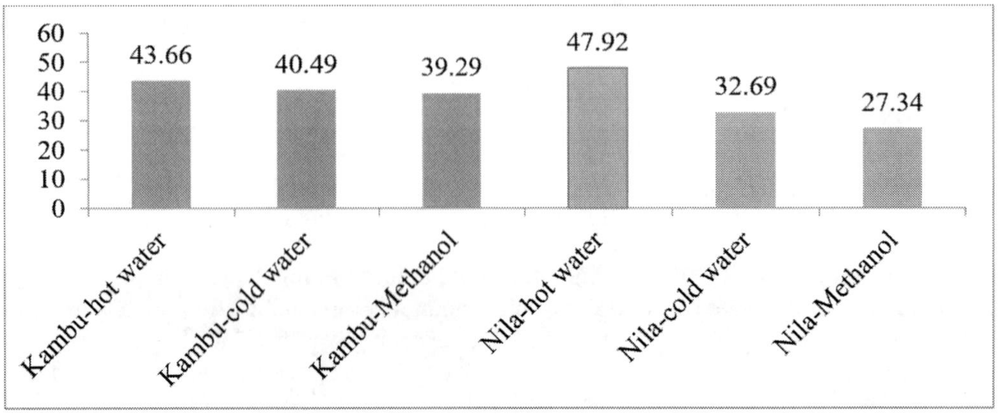


**Figure 4.** Color qualities of natural pigments extracted from particular plant species, expressed as brightness values (color qualities, Numbers 0 - 60 refer to the brightness values)

a b c

Remarks: a. Hot water; b. Cold water; c. Methanol

**Figure 5**. Visual appearance associated with color qualities of Nila originated natural pigments obtained using various extraction techniques (a, b, c)

a b c

Remarks: a. Hot water; b. Cold water; c. Methanol

**Figure 6.** Visual appearance associated with color qualities of Kambu originated natural pigments, obtained using various extraction techniques (a, b, c)

### *3.5. Analysis on chemical compounds in plant-derived natural pigments*

Results of chemical compound analysis were presented in Table 5. It tuned out that chemical compounds in Loba Nila and Kambu originated natural pigments were dominated by particular groups of fatty acids. In addition, dodecanoic acid dan hexadecanoic acid compounds were contained in all these investigated plant-derived natural pigments (Loba, Nila, and Kambu originated).

**Table 5**. Chemical compounds detected in plant derived natural pigments for coloring fabrics

| No | Species origins of pigment producing plants | Chemical compounds | Composition % |
|---|---|---|---|
| 1 | Loba | - Dodecanoic acid | 25.89 |
| | | - Hexadecanoic acid | 14.79 |
| | | - 9,12-Octadecadienoic acid | 11.15 |
| | | - 2-Propanone, 1-(4-hydroxy-3-methoxyphenyl)- | 7.79 |
| | | - 2-Furancarboxaldehyde, 5-(hydroxymethyl)- | 5.07 |
| 2 | Nila | - 2-Hexadecen-1-ol, 3,7,11,15-tetramethyl | 21.90 |
| | | - 8,11,14-Eicosatrienoic acid | 18.65 |
| | | - Hexadecanoic acid | 15.44 |
| | | - STIGMAST-5-EN-3-OL, OLEAT | 4.48 |
| | | - l-Limonene | 3.17 |
| 3 | Kambu | - Dodecanoic acid (CAS) | 20.57 |
| | | - 2-Furancarboxaldehyde, 5-(hydroxymethyl | 10.23 |
| | | - Acetic acid (CAS) Ethylic acid | 8.50 |
| | | - ISOSORBID | 5.37 |
| | | - 9-Hexadecenoic acid | 4.76 |

## 4. Conclusion

Utilization of plant-derived natural pigments has long been performed by the community in East Sumba as fabric coloring agents and to produce tightly woven fabrics; and apparently such undertaking would still keep going in the future. Unfortunately, the extraction technique implemented by the community was still very simple by just immersing the supposedly pigment containing parts of the plants/trees (e.g. roots, stems, leaves, and barks) in cold water. Likewise, so was the technique for coloring the fabric/thread/yearn with natural pigments, which was conducted by just dipping and immersing methods. There were three kinds of plant derived natural pigments used most by the community and those three corresponding plant species origins comprised Wora/Nila *(Indigofera erecta* Thunb), Kambu *(Morinda cf.citrifolia* L.), and Loba *(Symplocos cochinchinensis* (Lour.) S. Moore). Results of technical assessment on processing of pigment producing plants revealed that extraction technique using hot water solvent brought out the extracts (pigments) with higher yield and better coloring qualities (greater brightness values).

These weaving activities were generally performed by the community just as additional income sources. Moreover, the production process of woven fabrics was very time consuming and lengthy and also, the recipient markets so far were felt still inadequately and improperly available.

Nowadays, raw material sources for natural pigments tended to be more difficult to obtain. Accordingly, to cope with those difficulties, parts of the community began mixing natural pigments with synthetic pigments. In the future, expectedly those pigment producing plants should be cultivated seriously and thoroughly to maintain the stocks of natural pigments.

## References


[1] Kemeterian Kehutanan RI 2005 *Peraturan Menteri Kehutanan No. 35 Tentang Hasil Hutan Bukan Kayu* (Jakarta: Kementerian Kehutanan)

[2] Heyne 1987 *Tumbuhan berguna di Indonesia Jilid 1* (Jakarta: Badan Litbang Kehutanan)

[3] Sutara 2002 Jenis tumbuhan sebagai pewarna alam pada beberapa perusahaan tenun di Gianyar *J. Bumi Lestari* **9**(2) 217-23

[4] Sugiyana D 2003 *Pencemaran Logam Berat pada Limbah Industri Tekstil dan Alternatif Material Penyerap Ekonomis* (Bandung: Arena Tekstil 39 Balai Besar Tekstil)

[5] Wagner S 1993 Improvement in product and processing to diminish enviromental impact *COTTECH Conference Raleigh*

[6] Kismolo and Endro 1995 Teknik pemekatan rotavor R-151: Standar untuk limbah cair fase cair *Prosiding PP. Batan*

[7] Lemmens RH and Soetjipto NW 1999 *Sumber Daya Nabati Asia Tenggara, No 3: Tumbuhan Penghasil Pewarna dan Tanin* (Jakarta: Balai Pustaka)

[8] Andayani 2006 Citarasa tinggi batik alami *www.kabarejogya.com*

[9] Amutha K and Sudhapriya N 2020 Dyeing of textiles with natural dyes extracted from *Terminalia arjuna* and *Thespesia populnea* fruits *Ind. Crop. Prod.* **148** 1-8

[10] Phan K, Van Den Broeck E, Van Speybroeck V, De Clerck K, Raes K, and De Meester S 2020 The potential of anthocyanins from blueberries as a natural dye for cotton: A combined experimental and theoretical study *Dyes and Pigments* **176** 108180

[11] Triwi H 1998 *Pemanfaatan dan Pemberdayaan Zat Warna Alami (ZPA) untuk Industri Kecil dan Menengah* (Yogyakarta: Balai Besar Kerajinan dan Batik)

[12] [BPS] Badan Pusat Statistik Kabupaten Sumba Timur 2014 *Sumba Timur Dalam Angka 2014*

[13] [BPS] Badan Pusat Statistik Kabupaten Sumba Timur 2014 *Rindi Dalam Angka 2014*

[14] Zerin I, Farzana N, Muhammad Sayem A, Anang DM, and Haider 2019 Potentials of natural dyes for textile applications *Encyclopedia of Renewable and Sustainable Materials* 1-11

[15] Harborne JB 1987 *Metode Fitokimia, Penuntun Cara Modern Menganalisis Tumbuhan* (Bandung: ITB Press)

[16] Prabhu KH and Teli MD 2014 Eco dyeing using *Tamarindus indica* L. seed coat tannin as a natural mordant for textiles with antibacterial activity *Journal of Saudi Chemical Society* **18**(6) 864-72

IOP Conf. Series: Materials Science and Engineering **935** (2020) 012040 doi:10.1088/1757-899X/935/1/012040

# Anatomical structure and physical properties of the 13 years old solomon-clone teakwood planted in Bogor, Indonesia

**N V Riantin**[1], **I Wahyudi**[2*], **and T Priadi**[2]

[1]Forest Products Science and Technology Study Program, IPB University, Bogor 16680, Indonesia
[2]Forest Products Department, Faculty of Forestry, IPB University, Bogor 16680, Indonesia

*Email: imyudarw16@yahoo.com

**Abstract.** Anatomical structure, fiber dimension, and several physical properties of the 13 years old solomon-clone teak (*Tectona grandis*) wood planted in Bogor District, West Java Province, Indonesia have been studied. Macroscopic characteristics were observed directly by handy loupe, while microscopic characteristics were observed through microtome specimens. Fiber morphology was measured through maceration specimens, while physical properties were measured by modification of British Standard 373.1957. Results showed that anatomical structure and fiber length of this solomon-clone teakwood were not significantly different from those of common teakwood. The wood consists of 53% fiber, 29% vessel, 10% axial parenchyma, and 8% ray parenchyma; with 905.44-1199.36 μm fiber length and 2.50-6.72 μm cell wall thickness. Average of wood density and specific gravity was 1.04 g/cm$^3$ and 0.51, respectively. Compared to those of 29 years old conventional teakwood as well as 4 and 5 years old faster grown teakwood planted at the same area, specific gravity of this wood was higher. In general, this solomon-clone teakwood is suitable as raw material for wooden furniture manufacturing. This wood is classified as the Strength Class of III, has decorative figures, wood grain is straight, wood texture is fine and even, lower in volumetric shrinkage, and without crystal or silica.

## 1. Introduction

Characteristics of teak (*Tectona grandis*) wood especially from man-made forests need to be continuously investigated. This is not only to assure its quality as raw material for certain product, but also to evaluate wood characteristics variation among them. Since teakwood supply declines year by year, utilization of faster grown teakwood especially from plantation forest by wooden furniture industries increases every year. Among them is solomon-clone teak which has been planted in several sub districts in Central and West Java Provinces. According to Maskuro [1], solomon-clone teak tree is disease resistant, faster grown and slightly branched, while the stem is straight, round, and rarely broken. Previous study showed that heartwood portion of the 13 years old solomon-clone teakwood planted in Parung Sub-District, Bogor District was more than 70%. This indicates that this wood was durable enough and has potential to be used as raw material for timber industries.

Characteristics of solomon-clone teakwood until now have not been investigated yet. In fact, this information is needed specially to ensure the quality of furniture products produced, or at least to determine proper processing applied and end using purposed [2]. It is known that proper processing applied and appropriate end using purposed are determined by wood characteristics themselves, while wood characteristics of similar species are influenced by growing conditions, seed quality, and tree age.

Therefore, the objective of this study was to investigate anatomical structure, fiber dimension, and several physical properties of 13 years old solomon-clone teakwood planted in Bogor District in order to evaluate its suitability as raw material for wooden furniture manufacture. The results were then compared to similar characteristics found in another teakwood, either fast-grown or conventional one. Anatomical characteristics studied consisted of macro and microscopic characteristics, fiber dimensions (length, diameter, and thickness of the fiber wall), while physical properties consisted of wood density (ρ) and Specific Gravity (SG), as well as Volumetric Shrinkage (VS) from green to oven dried condition.

## 2. Materials and Methods

### *2.1. Materials*

The main material used was disc of solomon-clone teakwood 25.10 cm in diameter and 5 cm thick, 50% nitric acid ($HNO_3$), potassium chlorate ($KClO_3$), distilled water, safranin, glycerin, and filter paper. Wood disc was extracted from basal area of 13 years old tree at 130 cm above the ground from plantation forest area in Bogor, West Java. Three healthy trees were selected and cut.

Each disc was converted into 4 cm width wood strip including the pith, and then divided into 6 segments from pith to the bark representing heartwood (segment no.1 to 4), transition (segment no.5), and sapwood (segment no.6). Each segment was then cut: the upper part for anatomical structure observation and fiber dimension measurement, while the lower part for physical properties measurement. Sample dimension was 2 (l) × 4 (w) × 2 (t) cm per segment.

### *2.2. Anatomical structure observation*

Anatomical structures were observed macro and microscopically. Macroscopic characteristics were observed directly by using micro capture loupe, while microscopic features were observed through microtome specimens by using light microscope. Macroscopic characteristics observed consist of the existence of growth ring, wood color and figure, texture, wood grain, and luster [3], while microscopic features i.e. vessel elements, fibers, axial parenchyma, and ray parenchyma were observed IAWA List [4]. Microtome specimens of 25 μm thick were prepared following Forest Products Laboratory method [5] for cross-, radial-
, and tangential sections.

### *2.3. Fiber dimension measurement*

Fiber dimension was measured through maceration specimens prepared by Schultze's method with proper modifications. Fiber length, fiber diameter, and lumen diameter of 50 individual cell per segment were measured, while cell wall thickness was determined as a half of differences between fiber diameter and its lumen diameter. All measurements were carried out using the imageJ software.

### *2.4. Physical properties measurement*

Physical properties of wood i.e. ρ, SG, and VS were measured using the following the equations:

$$\rho\ (\text{g/cm}^3) = \frac{BA}{VA} \quad (1)$$

$$SG = \frac{BKT}{VA} / \rho_{water} \quad (2)$$

$$\%\ VS = \frac{VA-VB}{VA} \times 100 \quad (3)$$

where, BA = Sample weight in green condition (g), BKT = Sample weight in oven dried condition (g), VA = Sample volume in green condition ($cm^3$), and VB = Sample volume in oven-dried condition ($cm^3$).

## 3. Results and Discussion

Anatomical structures of solomon-clone teakwood are as follows: heartwood is reddish brown which is very easily distinguished from the yellowish white sapwood (Figure 1a). Growth ring is distinct. The wood is lustre and rather oily without specific odour. Texture fine and even, wood grain straight (Figure 1b), and have decorative pattern (Figure 1c). Vessels solitary and in radial multiples of 2-3 cells, the latter is dominant (Figure 2, 3, and 4); porosity ring porous in sapwood and transition zone, but diffuse porous in heartwood zone; pores somewhat circular; average tangential diameter 22.36 µm, 1-5 cells per $mm^2$; perforations simple; inter-vessel pits alternate; vessel-ray pits simple; deposits materials frequent in transition zone only; tyloses infrequent. Axial parenchyma paratracheal-zonate in earlywood and relatively sparsely vasicentric in latewood, 3-11 cells per strand (Figure 3, 4, and 5). Rays uni- to multiseriate (4-6 series) (Figure 3, 4, and 5), heterocellular, and without coloured content. Fibers exclusively septate; 993.35 µm long; 2.50-6.72 µm wide with tangential diameter of 8-14.53 µm. Crystals absent, but silica presents in some vessels. The wood consists of 53% of fiber, 29% of vessel, 10% of axial parenchyma, and 8% of ray parenchyma. In general, anatomical structure of solomon-clone teakwood is not different from that of common teakwood as [6,7].

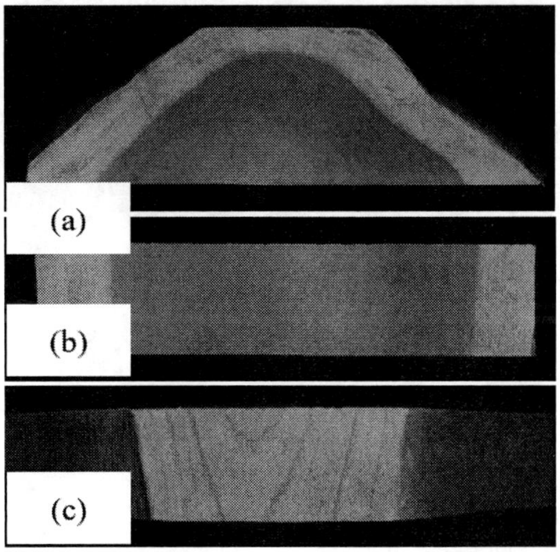


**Figure 1.** Macroscopic characteristic. (a) cross-, (b) radial-, and (c) tangential section

Result also showed sapwood and heartwood portions were 25.86 and 74.14%, respectively. Heartwood portion in this study was higher than that of 5-year-old fast growing teakwood as stated by [8,9] which are only 18 and 22%, respectively. However, it is lower than that of 76 years old conventional teakwood as stated by Wahyudi [10]. Because heartwood colour of this wood was darker than that of 4 and 5 years old fast growing teakwood planted at the same location [9] and not significantly different from 76 years old conventional teakwood planted at Cepu, Central Java Province [10], it can be deduced that solomon-clone teakwood is durable enough for destructive biological factors. The higher the proportion of heartwood, the higher the natural durability [11,12].

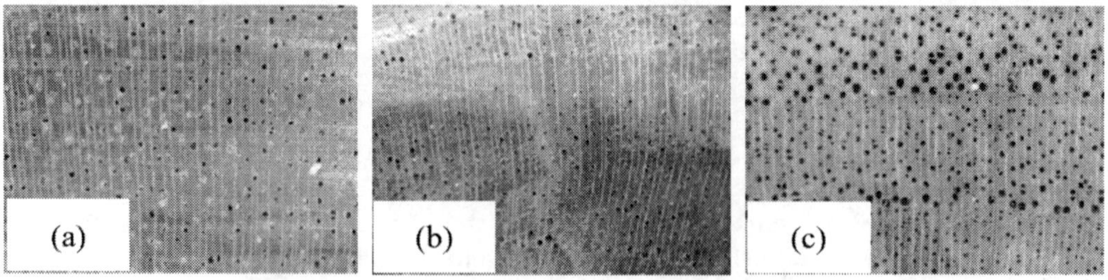


**Figure 2.** Cross-section of (a) heartwood, (b) transition, and (c), sapwood

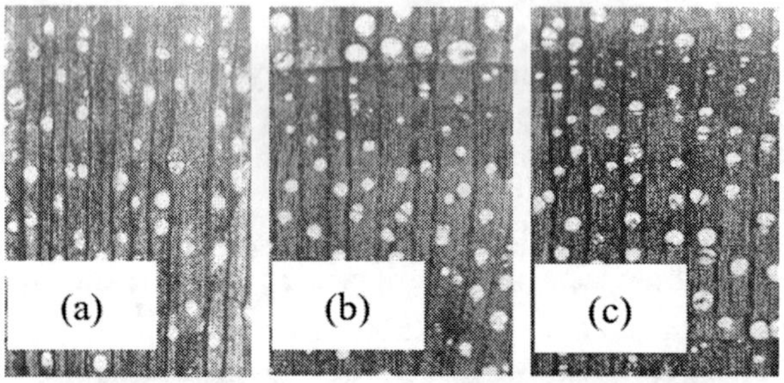


**Figure 3.** Cross-section in (a) heartwood, (b) transition, and (c), sapwood

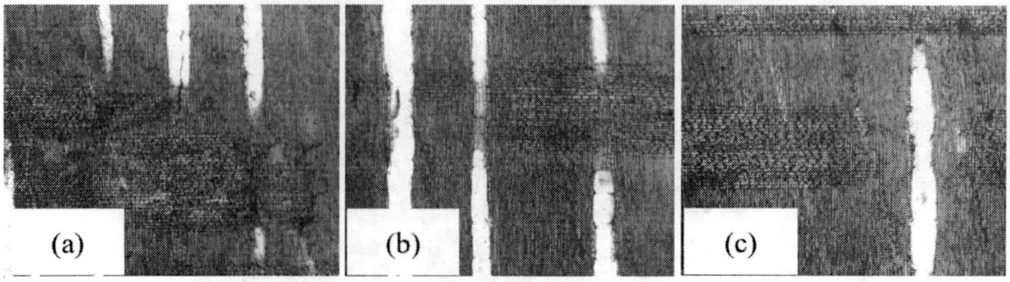


**Figure 4.** Radial section in (a) heartwood, (b) transition, and (c), sapwood

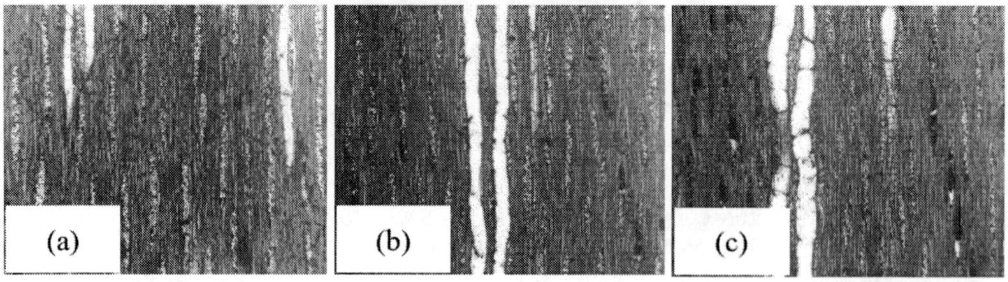


**Figure 5.** Tangential section in (a) heartwood, (b) transition, and (c), sapwood

Results also showed that fiber length tends to increase from pith (segment no.1) to bark (segment no.6) (Figure 6). From the ANOVA, it is known that fiber length is significantly different among segments. Average values of fiber length in heartwood, transition, and sapwood were 905.44, 1139.02, and 1199.36 μm, respectively. Since fiber length increases from pith to bark, it can be said that mature wood has not been produced yet within the trees. According to IAWA [13], fibers of this wood was classified as medium class (900-1600 μm).

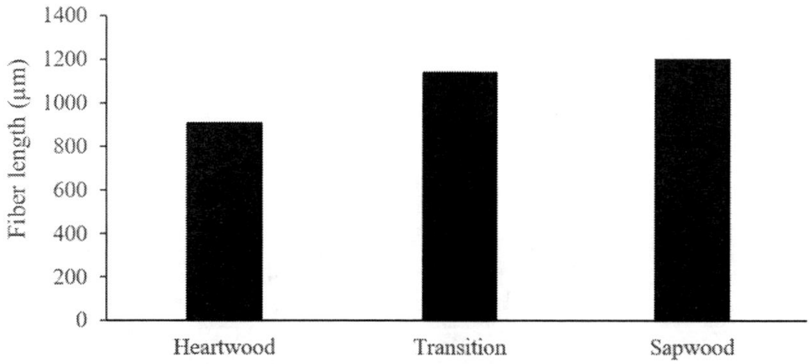


**Figure 6.** Radial variation in fiber length

Compared to 29 years old conventional teakwood planted in Bogor [14], our results are longer; but compared to 70 years old conventional teakwood [6,7], our results are shorter. These findings were similar to that of 14 years old conventional teakwood from Gunungkidul, Yogyakarta Province [15], and also to 4 and 5 years old fast growing teakwood planted in Bogor [8,16]. These differences indicate that fiber length is influenced by tree age and growing location.

Result showed that cell wall thickness was significantly influenced by sample location within stem (Figure 7). Fiber wall in transition zone was the thinnest (2.50 µm).

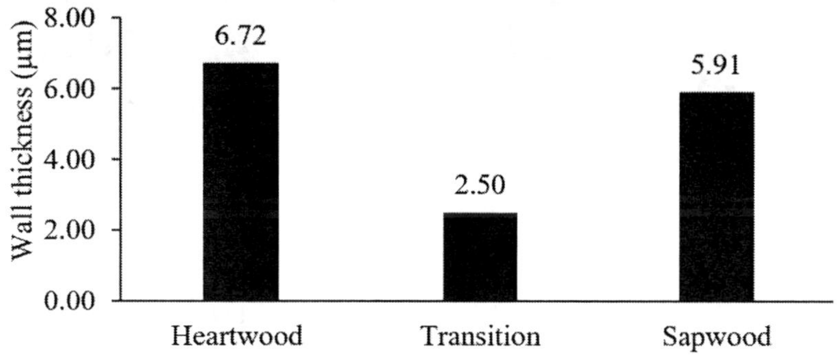


**Figure 7.** Radial variation in cell wall thickness

Wood density as well as SG in all parts (heartwood, transition, and sapwood zones) were almost similar (Figure 8 and 9). With an average of 0.53, therefore, solomon-clone teakwood studied was classified as the Strong Class of III [17]. Compared with [8,14,16], our values were higher; but compared to [6,15], our SG was lower.

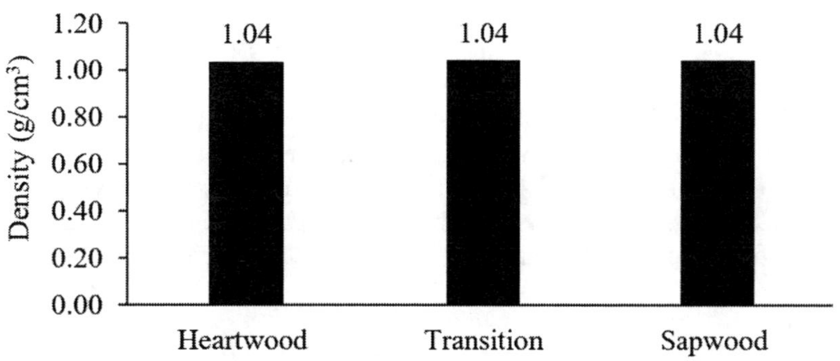


**Figure 8.** Radial variation in wood density

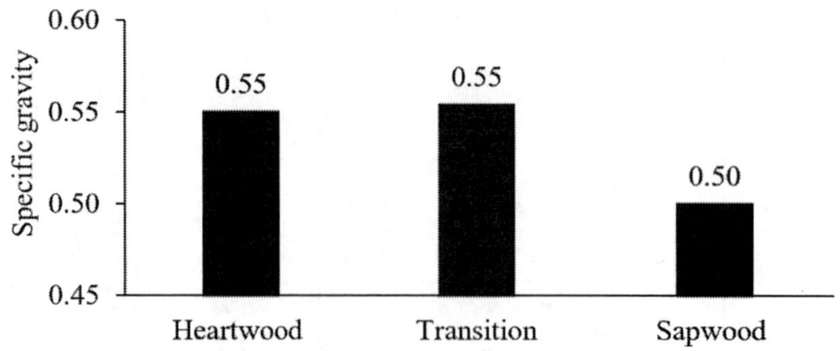


**Figure 9.** Radial variation in specific gravity

Average value of VS from green to oven-dried was 6.75% (Figure 10). The smallest was found in transition zone (3.47%), while VS in heartwood and the sapwood were 8.20 and 8.59%, respectively. The smallest shrinkage in transition zone was related to diameter vessel size and deposit materials within vessel elements. Compared to [6,16], our value was lower.

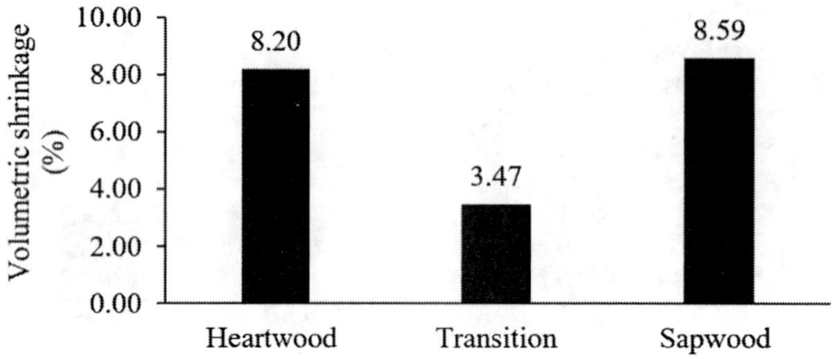


**Figure 10.** Radial variation in volumetric shrinkage

Suitability of wood as raw material for wooden furniture is depend on its decorative pattern, texture and wood grain, strength, dimensional stability, natural durability, and machining properties [2,18-20]. The glue-ability has to be good and the SG should be moderate (0.50-0.75). Considering to some indicators stated, therefore, the solomon-clone teakwood studied is suitable enough to be utilized as raw material. The wood has decorative figures, less deformation (VS from green to oven-dried condition < 10%), strong enough (strong class III), texture fine and even, straight grain, durable, and good in machining properties dues to medium in SG (0.50-0.55). Glue-ability will be studied in the next future.

## 4. Conclusion

Anatomical characteristics of 13 years old solomon-clone teakwood from Bogor are not different from those of teakwood in general. The wood consists of 53% fiber, 29% vessel, 10% axial parenchyma, and 8% ray parenchyma. Fiber length was 905.44-1199.36 μm, while cell wall thickness was 2.50-6.72 μm. Average of wood density and specific gravity was 1.04 g/cm$^3$ and 0.51, respectively. Specific gravity of this wood was higher than that of 29 years old conventional teakwood as well as 4 and 5 years old fast growing teakwood from similar location, but comparable to 14-year-old conventional teakwood from Gunungkidul, Yogyakarta Province. Generally, this solomon-clone teakwood studied is suitable as raw material for wooden furniture manufacturing.

## References

[1] Maskuro A 2012 *Deskripsi Tumbuhan Jati dan Peranannya dalam Kehidupan Sehari-hari* (Jember: Fakultas Keguruan dan Ilmu Pendidikan Universitas Muhamadiyah Jember)

[2] Wahyudi I 2013 *Hubungan Struktur Anatomi Kayu dengan Sifat Kayu, Kegunaan dan Pengolahannya* (Bogor: Fakultas Kehutanan Institut Pertanian Bogor)
[3] Pandit IKN and Ramdan H 2002 *Anatomi Kayu: Pengantar Sifat Kayu Sebagai Bahan Baku* (Bogor: Yayasan Penerbit Fakultas Kehutanan IPB)
[4] Wheleer EA, Baas P, and Gasson PC 1990 *IAWA List of Microscopic Features for Hardwood Identification* (IAWA, US: IAWA Bull.)
[5] Sass JE 1961 *Botanical Microtechnique* (Iowa: The Iowa State University Press.)
[6] Martawijaya A, Kartasujana I, Kadir K, and Prawira SA 2005 *Atlas Kayu Indonesia Jilid 1* (Bogor: Badan Penelitian dan Pengembangan Kehutanan)
[7] Ogata K, Fujii T, Abe H, and Baas P 2008 *Identification of the Timbers of Southeast Asia and the Western Pacific* (Japan: Kaiseisha Press)
[8] Muhran 2013 Kualitas pertumbuhan dan karakteristik kayu jati *(Tectona grandis L. f.)* hasil budidaya [Skripsi] Fakultas Kehutanan Institut Pertanian Bogor
[9] Wahyudi I, Priadi T, and Rahayu IS 2014 Karakteristik dan sifat-sifat dasar kayu jati unggul umur 4 dan 5 tahun asal Jawa Barat *JIPI* **19**(1) 50-6
[10] Wahyudi I 2000 Study on growth and wood qualities of tropical plantation species [Dissertation] Nagoya University Japan
[11] Butterfield BG 1993 *The Structure of Wood: An overview* Ed. Walker JCF *Primary Wood Processing, Principles and Practice* (Melbourne: Chapman and Hall)
[12] Darwis A, Hartono R, and Hidayat SS 2005 Presentase kayu teras dan kayu gubal pada Jati (*Tectona grandis* L.f.) *Jurnal Ilmu dan Teknologi Kayu Tropis* **3**(1) 6-8
[13] IAWA 2005 *Identifikasi Kayu: Ciri mikroskopis untuk identifikasi kayu daun lebar* (Bogor: Pustekolah)
[14] Zawawi Y 2014 Dimensi serat, sudut mikrofibril, dan beberapa sifat fisis kayu jati (*Tectona grandis*) umur 29 tahun [Skripsi] Fakultas Kehutanan Institut Pertanian Bogor
[15] Hidayati F, Fajrin IT, Ridho MR, Nugroho WD, Marsoem SN, and Na'iem M 2016 Sifat fisika dan mekanika kayu jati unggul "mega" dan kayu jati konvensional yang ditanam di hutan pendidikan Wanagama, Gunungkidul, Yogyakarta. *Jurnal Ilmu Kehutanan* **10**(2) 98-107
[16] Damayanti R 2016 Wood quality of young fast grow plantation teak and the relationship among ultrastructural and structural characteristics with selected wood properties [Dissertation] School of Ecosystem and Forest Sciences, Faculty of Science, the University of Melbourne
[17] PKKI-NI5 1961 *Peraturan Konstruksi Kayu Indonesia* (Jakarta: Yayasan Dana Normalisasi Indonesia)
[18] Bowyer JL, Shmulsky R, and Haygreen JG 2003 *Forest Products and Wood Science: An Introduction* (IOWA (US): IOWA State University Pr.)
[19] [BSN] Badan Standardisasi Nasional 2011 *Persyaratan Keamanan dan Pengujian Kayu dan Produk Kayu untuk Kursi Santai, Kursi Goyang, dan Kursi Teras* SNI 7555.16, 17, dan 18 (Jakarta: Badan Standarisasi Nasional)
[20] [BSN] Badan Standardisasi Nasional 2016 *Persyaratan Keamanan dan Pengujian Kayu dan Produk Kayu untuk Meja Tamu, Meja Makan, dan Meja Dapur* SNI 7555.1, 2, dan 3 (Jakarta: Badan Standarisasi Nasional)

IOP Conf. Series: Materials Science and Engineering **935** (2020) 012041 doi:10.1088/1757-899X/935/1/012041

# Characteristics of liquid smoke from several kinds of waste

**S Komarayati*, Gusmailina, D Hendra, and I Winarni**

Forest Products Research and Development Center, Ministry of Environment and Forestry, Bogor, 16610, Indonesia

*Email: skyati55@gmail.com

**Abstract.** Research into making liquid smoke has been carried out using a 3 in 1 model furnace. The material used is waste consisting of wide leaf litter, a mixture of small leaf litter and mahogany bark. The results showed that the pH ranged from 3.82 to 4.69; specific gravity of 0.99. The content of phenol and acetic acid, liquid smoke, mahogany peel 0.34%, 0.62% and 0.86%, leaf litter mixture width 0.01% and 3.74% and small leaf litter mixture 0.06% and 0.88 %. From the results of the GCMS pyrolysis test, it was known that the acid derivatives, phenols and other components. The highest acid derived compound was obtained from liquid smoke of mixed litter of 66.53% relative and the lowest from liquid smoke of mixed leaf litter which was 8.78% relative. The highest phenol derivative content is from liquid smoke of mahogany bark of 29.45% relative and the lowest of liquid smoke of mixed litter is 4.92% relative. Liquid smoke obtained can be used as a trigger for plant growth and prevention of disease pests inplants.

## 1. Introduction

Research into integrated charcoal production with liquid smoke has often been carried out using various kiln models, and raw materials from various materials, such as wood waste, coconut shell waste, nyamplung shell waste, bintaro shell waste, bamboo waste and others [1-6]. The materials and different kiln models will produce different characteristics and properties. In this research, a "three in one" model kiln is used, which is equipped with a cooling pipe.

Mahogany leaf litter is usually used as compost material with the addition of EM4. Several studies have shown that the results of macro nutrients, namely C-organic, N-total and K-total in the compost meet the compost quality standards in Indonesia as regulated in SNI 19-7030-2004 [7]. On the other hand, mahogony bark is also used as research material because the waste can be used as a source of energy/ fuel or made of liquid smoke. Where the liquid smoke produced is suitable for use as an organic material [8,9].

This paper presents the characteristics and content of the chemical components of liquid smoke from 3 types of waste consisting of wide leaf litter waste, small leaf litter waste and mahogany bark.

## 2. Materials and Methods

### *2.1. Materials*

The study was conducted at the Chemistry Laboratory and the Integrated Proximate Laboratory. Forest Products Research and Development Center, Bogor. The material used is 3 types of waste originating from West Java, namely a mixture of wide leaf litter, a mixture of small leaf litter, and mahogany bark. There was no replication in this study. The equipment which is used to make liquid smoke is a "three in one" kiln (Figure 1) made of brick inside and stainless on the outside is equipped with a chimney to drain into the open-air uncondensed gas formed during the cooking time.

**Figure 1**. "Three in one" kiln, for producing charcoal and liquid smoke

### *2.2. Methods*

The process of making liquid smoke took place at temperature of 400-450 C (Figure 1). The furnace is equipped with a smoke jet pipe, then the smoke is condensed. The resulting smoke condensation was called liquid smoke, which was then collected. Furthermore, the liquid smoke was testing / analyzed in laboratory.

Chemical components of liquid smoke such as acetic acid and phenol were analyzed using High Performance Liquid Chromatography (HPLC), while organic chemical components and their derivatives were analyzed by means of Gas Chromatography-Mass Spectrometer (GCMS) pyrolysis conducted at the Instrumentation and Proximate Integrated Laboratory, Bogor Forest Research Center.

## 3. Results and Discussion

The characteristic and content of the chemical components of liquid smoke can be seen in Table 1 and Table 2. The pH value of liquid smoke to the three types of waste ranges from 3.82 to 4.69. The range of pH values when compared with the quality of Japanese standard liquid smoke, all of them do not meet the requirements. The pH is too high, because the low pH of liquid smoke is the best, where the lower the pH of the liquid smoke, the higher the quality of the liquid smoke. Liquid smoke with low pH acts as an antibacterial and antioxidant [3].

The density of liquid smoke produced was around 0.99, lower than the quality of Japanese liquid smoke [10]. Japan's standard liquid smoke quality for specific gravity is above 1.05. The yield of liquid smoke from litter waste is between 26-30%, while the mahogany bark is only 17%. This difference is due to differences in the water content of the material. This is in accordance with the research conducted by Satriadi [11] which stated that if the water content of the raw material is too high, it will require more heat supply to evaporate water in the raw material before the pyrolysis process occurs. As a result, the quality of wood vinegar produced will be low, due to the mixing of the condensation of water vapor and the condensation of pyrolysis and the use of more raw materials [12].

The smell of the liquid smoke that was produced all had a very strong smell. The color varies from each liquid smoke, some are dark brown, brownish yellow, reddish brown and black. As for transparency, everything is murky. Means for the color and transparency of this liquid smoke does not matched with Japanese standards [10]. Meanwhile, till now the wood vinegar standard has not been made in in Indonesia.

**Table 1.** Liquid smoke properties

| Type of analysis | Sample Liquid/Smoke quality | | | Japanese standard*) |
|---|---|---|---|---|
| | 1 | 2 | 3 | |
| pH | 3.91 | 4.69 | 3.82 | 1.50-3.70 |
| Specific gravity | 0.99 | 0.99 | 0.99 | > 1.05 |
| Smell | strong | strong | strong | - |
| Colour | yellow brownish | reddish | black | yellow brownish |
| Transparency | murky | murky | murky | not murky |

Remarks: 1= Wide leaf litter mixture, 2= Litter mixture of small leaves 3= Mahogany bark [10]

The content of the chemical components of liquid smoke into three types of waste consisting of acetic acid ranged from 0.62 to 3.74% and phenol from 0.01 to 0.39% (Table 2). Liquid smoke of wide leaf litter mixture containing the highest acetic acid is 3.74%, while the highest phenol is produced from liquid smoke of mahogany peel which is 0.39%. The variation in the content of the chemical components of liquid smoke is influenced by the type of material and the chemical content characteristics of the materials used [4]. The content of acetic acid, liquid smoke, and litter mixture of large and small leaves is higher than liquid smoke of mahogany peel. This occurs in litter or leaves with little lignin content, whereas cellulose content is greater. According to Yatagai [10] in Komarayati and Wibowo [6], acetic acid from liquid smoke serves to accelerate the growth of phenol plants and their derivatives function to prevent pest and disease attacks.

Acetic acid contained in liquid smoke is an organic acid that is formed due to the pyrolysis process of chemical components such as lignin, cellulose, and hemicellulose [4]. In addition to acetic acid, phenol is also present in everyday life phenol and its derivatives can be used for disinfectants and inhibitors [2,6,10]. The content of phenol in the liquid smoke of the mahogany peel is higher than the liquid litter smoke, because the level of lignin in the bark is higher than in the leaves [13].

Liquid smoke contains a variety of chemical components including: alcohol, aldehydes, ketones, organic acids such as furfural, formaldehyde that function as preservatives, phenols, quinols and pyrogalols act as antioxidants, antiseptics and anti-bacteria. The acid group which plays an important role in liquid smoke is acetic acid which can stimulate plant growth, while propionic acid can prevent the growth of fungi and also as a fish preservative. Acid compounds contained in liquid smoke are organic acids formed by the pyrolysis process of chemical components of wood such as lignin, cellulose and hemicellulose.

Alcohols, phenols and acetic acid are indicated as compounds that have synergy functions as protein denominations and can hydrolyze lipids, so that they can damage the cell membrane of the fungus's body tissues and inactivate enzymes secreted by fungi [4,5]. Phenol is a compound that has a characteristic odor, has antiseptic properties, the function of phenol, among others, in regulating the activity of certain enzymes, is toxic to insects, is toxic to plant predators [14]. In addition, phenols affect termites, are antifungal and bacterial [14]. Other compounds besides phenols, furfural compounds, function as pest control of the Nematode animal group. According to Rahimah [15] stated that liquid smoke contains alkaloids and secondary metabolites that can be used as pesticides.

From the analysis using the GCMS pyrolysis tool, it can be seen the derivative of the chemical components of each type of liquid smoke from 3 types of waste as follows: In liquid smoke mixed with wide leaf litter, the highest acid content is acetic acid (CAS) ethylic acid of 8.78 %; The highest phenol is phenol, 2 methyl (CAS) o-creosol at 13.27% and the highest neutral component, 20.25%, is carbon oxide (CAS) dry ice. In the small leaf mixture liquid litter smoke, the highest acid component is acetic acid (CAS) ethylic acid by 66.53%; The highest phenol is phenol, 4 methoxy (CAS) hqmme of 4.92% and the highest neutral component is nitrogen oxide ($N_2O$) (CAS) nitrous oxide) which is equal to 1.68%. While in the liquid smoke of mahogany peel, the highest acid content is acetic acid (CAS) ethylic acid of 13.04%; The highest phenol is phenol (CAS) izal by 29.45% and the highest neutral component is carbon dioxide (CAS) dry ice by 10.89%. Of the three types of liquid smoke it appeared that containing the highest acid component was wide leaf little mixture (3,74%), the highest phenol was mahogany peel (29.45%) and the highest neutral component was wide leaf litter mix (20.25%).

**Table 2.** The main components of liquid smoke

| Chemical Component | Material | | |
|---|---|---|---|
| | 1 | 2 | 3 |
| Acetic acids, % | 3.74 | 0.88 | 0.62 |
| Phenol, % | 0.01 | 0.06 | 0.39 |

Remarks: 1= Wide leaf litter mixture, 2 = Litter mixture of small leaves, 3= Mahogany bark [10]

## 4. Conclusion

We can conclude that from the research results it turns out that liquid smoke from the two types of litter is more suitable for use as organic liquid fertilizer and liquid smoke of mahogany peel is suitable for controlling pests and plant diseases.

## References

[1] Nurhayati T, Roliadi H, and Bermawie N 2005 Production of mangium wood vinegar and its utilization *J. For. Res.* **2**(1) 13-26
[2] Nurhayati T, Pasaribu RA, and Mulyadi DP 2006 roduksi dan pemanfaatan cuka kayu dari serbuk gergaji kayu campuran *Jurnal Penelitian Hasil Hutan* **24**(5) 395-411
[3] Komarayati S, Gusmailina, and Pari G 2011 Produksi cuka kayu hasil modifikasi tungku arang terpadu *Jurnal Penelitian Hasil Hutan* **29**(3) 34-47
[4] Wibowo S 2012 Karakteristik asap cair tempurung nyamplung *Jurnal Penelitian Hasil Hutan* **30** (3)
[5] Aisyah I, Juli N, and Pari G 2013 Pemanfaatan asap cair tempurung kelapa untuk mengendalikan cendawan penyebab penyakit antraknosa dan layu Fusarium pada ketimun *Jurnal Penelitian Hasil Hutan* **31**(2) 170-8
[6] Komarayati S and Wibowo S 2015 Karaketristik asap cair dari tiga jenis bambu *Jurnal Penelitian Hasil Hutan* **3**(2) 167-74
[7] Hadiwidodo M, Sutrisno E, Handayani DS, and Febriani MP 2018 Studi pembuatan kompos padat dari sampah daun kering tpst undip dengan variasi bahan mikroorganisme lokal (mol) daun *J. Presipitasi Media Komunikasi dan Pengembangan Teknologi Lingkungan* **15**(2) 78
[8] Anggraeni I, Darwati W, and Intari SE 2006 *Hama Dan Penyakit Hutan Tanaman* (Pusat Penelitian dan Pengembangan Hutan Tanaman, Badan Penelitian dan Pengembangan Kehutanan)
[9] Utami S, Anggraeni, and Ismanto A 2009 Pemanfaatan cuka kayu (*wood vinegaar*) untuk pengendalian larva secara invitro
[10] Yatagai 2002 Utilization of charcoal and wood vinegar in Japan *Journal of Food Science Utilization of Charcoal and Wood Vinegar in Japan*
[11] Satriadi T 2016 Rendemen dan kualitas cuka kayu dari kulit tiga jenis meranti (*Shorea* Spp.). *EnviroScienteae* **8**(2) 102-7
[12] Yudoyono S, Pertiwi, and Munawwar 2007 Perbaikan proses produksi pada industri kecil cuka kayu di desa Sembawa kabupaten Banyuasin Sumatera Selatan *Prosiding seminar pembahasan hasil kegiatan Iptek dan Vuver LPM Unsri*
[13] Falah S, Suzuki T, and Katayama T 2008 Chemical constituens from *Swietenia macrophylla* bark and their antioxidant activity *Pakistan J. Biol. Sci.* **11**(6) 2007-12

IOP Conf. Series: Materials Science and Engineering **935** (2020) 012041 doi:10.1088/1757-899X/935/1/012041

[14] Darmawan UW 2014 Cuka kayu sebagai pengendali hama dan penyakit tanaman *Majalah FORPRO*
[15] Rahimah DS 2014 Asap usir elmaut *Majalah Trubus*

IOP Conf. Series: Materials Science and Engineering **935** (2020) 012042 doi:10.1088/1757-899X/935/1/012042

# Methanol soluble extractives of *Pinus merkusii* bark and its antioxidant activity

**Masendra, B A V Purba, and G Lukmandaru***

Department of Forest Products Technology, Faculty of Forestry, Gadjah Mada University, Yogyakarta, 55281, Indonesia

*Email: glukmandaru@ugm.ac.id

**Abstract.** The major function of bark in trees is to protect the living tissues. This is due to the chemical extractives in the bark such as polyphenols and terpenoids. In Indonesia, the bark of *Pinus merkusii* is of high economic value due to its use in the production of wood and oleoresins. Also, the presence of polyphenols in its bark has the potential of being used as antioxidant agents. Based on this, the aim of this study was to investigate antioxidant activity from the methanol extract of *P. merkusii bark.* This methanol extract of *P. merkusii* was subjected to gas chromatography mass spectrometric (GC-MS) method and antioxidant activity assay using 1,1-diphenyl-2-picrylhydrazyl (DPPH) method. The GC-MS analysis showed that the extract sample contained phenolics and sugars as minor compounds with fatty acids and alcohols as major components. The antioxidant activity ($IC_{50}$) of the sample was 1452.8 ± 208 μg/mL, which could be as a result of the presence of phenolic compounds, fatty acids and alcohols, as well as steroids and alkaloid in it.

## 1. Introduction

The major function of bark in plants is to protect the living tissues from extreme conditions such as low and high temperature, as well as from the attacks of microorganisms and herbivores. According to previous studies, the chemical compounds present in the bark are responsible for the protection ability [1-3].

In Indonesia, the plantation of *Pinus merkusii* occupies about 476,126 ha of forest area of Java. The needs for long fibers and oleoresins make this pine very important and it is the only species used from Pinaceae family. The production of oleoresins and wood from this species generally increase its demand. Moreover, the wood and bark of *P. merkusii* are very rich in both lipophilic and polyphenolic contents [4-7]. Hence, the use of this bark might be further developed for use in the pharmaceutical industries. A research also reported that the bark of pine could be used as antioxidant agents due to the presence of high polyphenolic compounds [8].

In addition, a study showed the difference between the ethanol extract of inner and outer bark of *P. merkusii* in terms of total phenolic, flavanol, and tannin-formaldehyde contents [7]. However, there is no study on the methanol (MeOH) soluble extract of *P. merkusii* and its antioxidant activity. Therefore, the purpose of this study was to investigate the constituents of MeOH fraction of *P. merkusii* bark and to evaluate its antioxidant activity.

## 2. Materials and Methods

### *2.1. Sample extraction*

The bark sample of *P. merkusii* was obtained from Magelang, Central Java, Indonesia. About 1 kg of the bark sample was grind to powder (40 mesh) and extracted with *n*-hexane, ethyl acetate (EtOAc), and MeOH successively for 2 weeks at room temperature. The extract solution was then evaporated and weighed; the percentage of dry bark sample calculated.

### *2.2. Antioxidant activity*

The antioxidant activity was measured with DPPH radical scavenging assay [9]. About 0.1 mL extract in methanol with different concentrations of 1000, 500, 250, and 125 µg/mL was reacted with 3 mL of 0.1 mM DPPH. This was incubated for 30 minutes at ambient temperature, after which the sample absorbance was read at 517 nm wavelength. The antioxidant activity was calculated through equation below:

$$\text{Antioxidant activity (\%)} = 100 \times (Ao - A_1)/\ Ao \quad (1)$$

where Ao is blank absorbance and $A_1$ is sample absorbance. In comparison, the standard of quercetin was assigned as the positive control. The value of antioxidant activity was shown in $IC_{50}$ (concentration of sample that inhibits 50% of DPPH radicals)

### *2.3. GC-MS analysis*

To detect phenolic compounds using GC-MS, the MeOH extract sample was silylated according to Wijayanto *et al.* [4]. Briefly, 1 mg of MeOH extract was dissolved into 100 µL of TMCS (trimethylchloroshilane) and BSA (N, O-bis(trimethylsilyl) acetamide) (8:2, v/v). The solution was incubated at room temperature for 1 h. The solvent was then evaporated and the dry extract was again dissolved with 1 mL of MeOH. For analysis, 1 µL of sample was injected to GC-MS. The GCMS-QP2010 (Shimadzu, Kyoto, Japan) was used to analyze MeOH extract constituents. The specification of GC-MS column: Rtx-5 capillary column (30 × 0.25 mm inner diameter and 0.25 µm; GL Sciences, Tokyo, Japan). The condition of GC-MS as follows: column temperature from 70°C (1 min) to 295°C at 5°C/min; injection temperature of 200°C; detection temperature of 295°C; acquisition mass range of 50 amu to 700 amu using helium as the carrier gas. The mass spectra of sample were compared to NIST library. The quantification of detected compounds was calculated with relative peak method.

### *2.4. Total phenol content (TPC)*

A modification method from Diouf et al. [10] was used for quantification of TPC. Briefly, A ten-fold dilution of Folin-Ciocalteu phenol reagent (2.5 mL) was mixed with 0.5 mL of sample (250 ppm). The mixture was stood for 2 min and to the solution, 2 mL of sodium carbonate ($Na_2CO_3$) was added, and the reaction mixture was again stood for 30 min at room temperature. Final solution was then read at 765 nm wavelength. The TPC results were expressed as gallic acid equivalents (mg GAE/g based on dry extract weight).

### *2.5. Total flavanol content (TVC)*

The TVC was conducted according to Richard *et al.* [11] with vanillin HCl assay. Briefly, half milliliter of ethanol solution sample (250 ppm), 3 mL of vanillin reagent with 4% concentration in methanol, and 37% of HCl (1.5 mL) were reacted and the mixture was incubated for 15 min at ambient temperature. At 500 nm wavelength, the final sample absorbance was read. The TVC results were calculated by performing regression standard of (+) catechin (mg CE/g based on dry weight).

### *2.6. Tannin-formaldehyde (TF)*

The TF was observed by literature of Humphreys [12]. The 2 g of bark sample was refluxed with distilled water in three times. The supernatant was filtered and the filtrate was cooled to 18°C. The solution was again diluted with distilled water until 500 ml. Further, 100 mL of the solution was used and reacted with 10 mL of formalin and 5 mL of HCl (37%). The mixture was then refluxed for 30 min. Final equipment was performed by filtering and weighing the precipitate of sample. The tannin-formaldehyde is calculated with equation 2 as follows:

$$\text{Tannin- formaldehyde (\%)} = (4.425 \text{ x A x } 10000)/(\text{W x } [100 - \text{sample moisture content}]) \quad (2)$$

where A is the weight of precipitate in gram and W is the weight of bark sample in gram.

### *2.7. Chemicals*

Quercetin (≥ 95%), (+)-catechin (≥ 95%), DPPH, TMCS and BSA were purchased from Sigma-Aldrich (Germany).

## 3. Results and Discussion

### *3.1. Extractive yield and antioxidant activity*

The extractive content of *n*-hexane has been reported in previous work [6], while that of EtOAc has not been reported. The MeOH extractive weighed 2.62 g or 0.26% of dried sample. Also, the antioxidant activity of MeOH soluble extract of *P. merkusii* bark is shown in Table 1. Based on the table, the antioxidant activity increased with increase in the sample concentration and the $IC_{50}$ value calculated was 1452 ± 208 μg/mL (means ± standard deviation). This activity was low compared to positive control of quercetin ($IC_{50}$ of 47.1 μg/mL).

The MeOH soluble extract yield from this study was lower than the ethanol soluble extractive of inner and outer bark of *P. merkusii* extracted with Soxhlet apparatus [7]. This could be as a result of difference in extraction methods. Similar result was also found in an earlier study on *Thymelaea hirsuta* [13]. Compared with the result of previous study on heartwood and knotwood of *P. merkusii*, the antioxidant activity of MeOH soluble extract from the bark in this study was higher than water soluble extracts but was lower than acetone soluble extracts [4]. Additionally, the MeOH extract of *P. merkusii* bark generally showed mild antioxidant activity in this study.

**Table 1.** DPPH scavenging activity of soluble MeOH extract from *P. merkusii*

| Sample concentration (μg/mL) | DPPH Scavenging activity (%) | $IC_{50}$ (μg/mL) |
|---|---|---|
| 1000 | 34.1 ± 4.7 | |
| 500 | 14.3 ± 3 | 1452 ± 208 |
| 250 | 4.7 ± 3.3 | |
| 125 | 1.6 ± 0.9 | |

### *3.2. GCMS analysis*

The GCMS analysis recorded the constituents of MeOH soluble extract of *P. merkusii,* as shown in Table 2. These constituents were categorized into phenolics, sugars, fatty acids and alcohols, steroids, and alkaloid. The fatty acids and alcohols group, composed of 53.68%, dominated the sample constituents while the predominant constituent was the steroids at 29.76%. Also, the concentration of lipophilic compounds extracted from the bark with *n*-hexane and ethyl acetate was high. The minor constituents detected by the GCMS method include alkaloid (10.68%), phenolics (2.79%) and sugars (1.91%). In these results, the phenolic constituent was detected in simple phenolic group, as shown in Table 2. In addition, the results showed that the phenolics and sugars detected were lower, possibly because some of these compounds were dissolved in EtOAc before MeOH extraction.

In the fatty acids and alcohols, methyl lignocerate and octacosoxy-trimethylsilane were the major compounds detected in the sample. However, the results were different from a previous study which involved analysing six different *Pinus* bark using the GCMS, without the detection of fatty acids [7]. In the case of steroids, the 25-hydroxycholesterol and stigmast-4-ene-3-one were the predominant constituents. These steroids were also found in bark of *P. caribeae*, *P. oocarpa*, and *P. elliotii* [5]. The alkaloid detected in higher amount from MeOH extract of *P. merkusii* was dodecahydro (1,2) oxazino

(2,3-b) isoquinoline. This is the first time this compound reported in the bark of *P. merkusii* as there were no detected previous study in relations to it.

The aromatic compound of 2-bromo-4-chloro-1-methoxybenzene was the dominant compound in the phenolics, while other monophenolics are from vanillin derivatives group. The presence of vanillin was also reported in *P. brutia* bark [14]. Lastly, the sugar compounds detected from the MeOH soluble extract include d-mannitol and beta-d mannofuranoside. This d-mannitol was also detected in bark of *P. elliotii*, *P. caribeae*, *P. montezumae*, and *P. insularis* [7].

**Table 2.** GC-MS derivative analysis of MeOH soluble extract from *P. merkusii*

| No | Ret. Time (min) | Constituents | Relative content (% of dried extract) | Similarity index (%) |
|---|---|---|---|---|
| | | **Phenolics** | **2.79** | |
| 1 | 21.2 | 2-Bromo-4-chloro-1-methoxybenzene | 0.83 | 65 |
| 2 | 21.3 | Vanilic acid hydrazide | 0.76 | 82 |
| 3 | 26.4 | Vanillin, acetate | 0.69 | 71 |
| 4 | 26.9 | Vanillyl methyl ketone | 0.51 | 74 |
| | | **Sugars** | **1.91** | |
| 5 | 21.4 | D-Mannitol | 0.78 | 89 |
| 6 | 22.8 | beta-d-Mannofuranoside, | 1.13 | 80 |
| | | **Fatty acids and alcohols** | **53.68** | |
| 7 | 26.6 | 9-Hexadecen-1-ol | 2.37 | 97 |
| 8 | 39.0 | Methyl lignocerate | 13.15 | 90 |
| 9 | 39.4 | trans-13-Octadecenoic acid | 0.65 | 89 |
| 10 | 39.5 | 3-Heptadecanol | 0.77 | 76 |
| 11 | 39.8 | Ethyl docosanoate | 2.92 | 88 |
| 12 | 40.0 | Heptadecyl acetate | 3.63 | 95 |
| 13 | 40.2 | Methyl pentacosanoate | 0.47 | 79 |
| 14 | 40.7 | 3-Hexadecanol | 1.93 | 81 |
| 15 | 41.1 | Ergost-5-en-3-yl acetate | 0.65 | 65 |
| 16 | 41.4 | Methyl hexacosanoate | 7.52 | 95 |
| 17 | 42.0 | Methyl heptadecyl ketone | 3.87 | 82 |
| 18 | 43.4 | Epoxy-linalooloxide | 1.71 | 78 |
| 19 | 43.6 | Methyl palmitate, (2,2-dimethyl-1,3-dioxolan-4-yl) | 3.66 | 70 |
| 20 | 47.4 | Octacosoxy-trimethylsilane | 10.38 | 70 |
| | | **Steroids** | **29.76** | |
| 21 | 42.2 | Dibromostigmasterol acetate | 3.64 | 80 |
| 22 | 43.7 | Stigmast-5-en-3-ol, oleate | 1.61 | 75 |
| 23 | 44.0 | Cholesta-4,6-dien-3-ol | 4.91 | 81 |
| 24 | 44.2 | beta.-Sitosterol | 0.72 | 65 |
| 25 | 44.3 | Stigmast-4-ene-3-one | 7.78 | 88 |
| 26 | 48.9 | gamma.-Sitosterol | 3.89 | 82 |
| 27 | 49.1 | 25-Hydroxycholesterol | 7.21 | 72 |
| | | **Alkaloid** | **10.68** | |
| 29 | 50.7 | Dodecahydro (1,2) oxazino (2,3-b) isoquinoline | 10.68 | 65 |

### *3.3. TPC, TVC, and TF*
The values of phenols contents of inner and outer bark of *P. merkusii* are shown in Table 3, as reported in a previous study by Masendra *et al.* [7]. Both the inner and outer bark contained higher TPC, while higher amount of TVC was observed in the inner bark. Then, the TF was found in outer bark sample. Based on a previous study, phenol contents were also found in the bark of *P. thunbergii*, *P. banksiana*, *P. radiata*, and *P.koraiensis* [8].

**Table 3.** Phenol contents of *P.merkusii* bark [7].

| Phenol contents | Inner bark | Outer bark |
|---|---|---|
| TPC | 421.4 | 486.5 |
| TVC | 367.4 | 55.2 |
| TF | - | 0.4 |

(-): not determined TPC: mg GAE/ g extract sample;
TVC: mg CE/ g extract sample; TF: % of dried bark sample

### *3.4. Correlation between antioxidants and related compounds*
The phenolics components of the MeOH soluble extract of this bark sample could be responsible for its antioxidant activity. However, these phenolic constituents were present in lower amount compared with other compounds. Hence, it was suggested that other major compounds such as fatty acids and alcohols played a major role in reducing the DPPH into DPPH-H. According to a previous literature, polyunsaturated fatty acids could act as antioxidant [15]. Additionally, other compounds capable of contributing to the antioxidant property are steroids and alkaloids. A previous paper showed that sterols from Turkish olive oils are potent for DPPH inhibitor [16]. Furthermore, the higher alkaloid such as quinoline derivatives could also act as antioxidant agents [17].

In comparison with phenols content, higher TPC was observed in the bark of *P.merkusii*. This suggested that TPC had a greater effect in inhibiting DPPH. However, the high concentration of TVC in the inner bark and TF in the outer bark indicated that flavan-3-ol structure (C4-C8 bonds) plays a major role in increasing the bark's antioxidant activity. These findings were in line with the results of the study conducted by Ku *et al.* [8] involving the bark of the following species; *P. densiflora*, *P. thunbergii*, *P. banksiana*, *P. contorta*, *P. rigida*, *P. taeda*, *P. serotina*, *P. radiata*, *P. koraiensis*, *P. parviflora*, and *P. rigida x taeda*.

## 4. Conclusion
Based on the findings, the antioxidant activity ($IC_{50}$) of the bark sample was 1452.8 ± 208 μg/mL, categorized under mild activity. Analysis of the bark sample through GCMS detected fatty acids and alcohols as the dominant compounds while phenolics and sugars were the minor compounds. The value of the antioxidant activity in the sample, as reported in this study, was attributed to the presence of phenolics and other compounds such as steroids, alkaloid, as well as fatty acids and alcohols.

## References
[1] Wittstock U and Gershenzon J 2002 Constitutive plant toxins and their role in defense against herbivores and pathogens *Curr. Opin. Plant. Biol.* **5**(4) 300-7
[2] Alfredsen G, Solheim H, and Slimestad R 2008 Antifungal effect of bark extracts from some European tree species *Eur. J. For. Res.* **127** 387-93
[3] Seki K, Orihashi K, Sato M, Kishino M, and Saito N. 2012. Accumulation of constitutive diterpenoids in the rhytidome and secondary phloem of the branch bark of *Larix gmelinii* Var. Japonica *J. Wood Sci.* **58** 437-45
[4] Wijayanto A, Dumarçay S, Gérardin-Charbonnier C, Sari RK, Syafii W, and Gérardin P 2015 Phenolic and lipophilic extractives in *Pinus merkusii* Jungh. et de Vries knots and stemwood *Ind. Crops. Prod.* **69** 466-71

[5] Masendra, Ashitani T, Takahashi K, and Lukmandaru G 2018 Lipophilic extractives of inner and outer barks from six different *Pinus* species grown in Indonesia *J. For. Res.* **29**(5) 1329-36
[6] Masendra, Ashitani T, Takahashi K, and Lukmandaru G 2018 Triterpenoids and steroids from the bark of *Pinus merkusii* (Pinaceae) *Bioresources* **13**(3) 6160-70
[7] Masendra, Ashitani T, Takahashi K, Susanto M, and Lukmandaru G 2019 Hydrophilic extracts of the bark from six *Pinus* species *J. Korean Wood Sci. Technol.* **47**(1) 80-9
[8] Ku CS, Jang JP, and Mun SP. 2007. Exploitation of polyphenol-rich pine barks for antioxidant activity *J. Wood Sci. 53* 524-8
[9] Baba SA and Malik SA 2015 Determination of total phenolic and flavonoid content, antimicrobial and antioxidant activity of a root extract of *Arisaema jacquemontii* Blume *J. Taibah. Univ. Sci.* **9**(4) 449-54
[10] Diouf PN, Stevanovic T, and Cloutier A 2009 Antioxidant properties and polyphenol contents of Trembling aspen bark extracts *Wood Sci. Technol.* **43**(4) 457-70
[11] Richard T, Broadhurst B, and Willian JT 1978 Analysis of condensed tannins using acidified vanillin *J. Sci. Food Agric.* **29** 788-94
[12] Humphreys FR 1956 Determination of tannin-formaldehyde in bark *Soc. Leather Trade Chemist J.* **40** 147-9
[13] Yahyaoui M, Ghazouani N, Sifaoui I, and Abderrabba M 2017 Comparison of the effect of various extraction methods on the phytochemical composition and antioxidant activity of *Thymelaea hirsuta* L. aerial parts in Tunisia.*Biosci. Biotechnol. Res. Asia* **14**(3) 997-1007
[14] Kivrak I, Kivrak S, Harmandar M, and Cetintas Y 2013 Phenolic compounds of *Pinus brutia* Ten.: chemical investigation and quantitative analysis using an ultra-performance liquid chromatography tandem mass spectrometry with electrospray ionization source *Rec. Nat. Prod.* **7**(4) 313-319
[15] Richard D, Kefi K, Barbe U, Bausero P, and Visioli F 2008 Polyunsaturated fatty acids as antioxidants *Pharmacol. Res.* 451-5
[16] Yorulmaz HO, and Konuskan, DB 2017 Antioxidant activity, sterol and fatty acid compositions of Turkish olive oils as an indicator of variety and ripening degree *J. Food Sci. Technol.* **54**(12) 4067-77
[17] Püsküllü MO, Tekiner B, and Suzen, S 2013 Recent studies of antioxidant quinoline derivatives *Mini Rev. Med. Chem.* **13** 365-72

**Acknowledgement**
We thank to BIDIK MISI Scholarship (Indonesia) for supporting this research, and Mr. Sukmono Edwi Susanto (Perum Perhutani) for providing the bark sample of *P. merkusii*.

IOP Conf. Series: Materials Science and Engineering **935** (2020) 012043 doi:10.1088/1757-899X/935/1/012043

# Characterization of ball-milled sago pith waste-based activated carbon treated with KOH and $KMnO_4$ as activating agent

**Qanytah[1*], K Syamsu[2], F Fahma[2], and G Pari[3]**

[1]Indonesian Center for Agricultural Postharvest Research and Development
[2]Department of Agroindustrial Technology, Faculty of Agricultural Engineering and Technology, IPB University, Bogor, Indonesia
[3]Products Research and Development Center, Ministry of Environment and Forestry

*Email: nita_mayall@yahoo.com

**Abstract.** The potential use of sago pith waste to produce activated carbon were studied. This study examined the effect of KOH or $KMnO_4$ as activating agent on producing ball milled sago pith waste–based activated carbon. Activation was done using KOH or $KMnO_4$ at 2.5% and 5% concentration, heated in a reactor to reach 800°C for 1 hour, and then continue steamed for 1 hour. Sample size was reduced to ≤500 nm using high-energy ball-milling at 500 rpm for 150 min. Analysis included the yield, water content, ash content, volatile matter, burnt-off weight percentage, morphology analysis, functional groups analysis (Fourier Transform Infrared Spectroscopy, FTIR), crystallinity analysis (X-ray diffraction, XRD), and surface area analysis (Brunauer, Emmelt, and Teller/BET). Ball-milling treatment produced sago pith waste activated carbon of 585.40 nm and a particle distribution index (PDI) score of 0.673. The activated carbon that met the criteria of SNI 06-3730-1995 for moisture content and volatile matter were sago pith waste with activating agent $KMnO_4$ 5%, and sago pith waste with activating agent $KMnO_4$ 5% & milling. The study showed the activated carbon that had the best Iodine number was sago pith waste with activating agent KOH 2.5% (619.40 mg/g). The activated carbon that met the criteria of SNI 06-3730-1995 for methylene blue adsorbtion capacity was sago pith waste with activating agent KOH 2.5% & milling and sago pith waste with activating agent KOH 5% & milling. The activated carbon with the highest surface area was sago pith waste with activating agent KOH 2.5% & milling. The best activated carbon based on all parameters was sago pith waste with activating agent KOH 2.5% & milling.

## 1. Introduction

Waste is one of environmental problems faced by the world. Agricultural and forest waste is considered as an important precursor because it is cheap, renewable, available in large amounts, rich in carbon, and low in ash (0.2-10%) [1-4]. One of the potential forest waste is sago pith. Sago pith waste contains of starch and lignocellulose [5]. The area of sago plantation in Indonesia covers 1.128 million hectares, and this contributes 51.3% to the world's sago [6]. The processing of sago produces waste in the form of sago pith waste amounting 75-83% [7]. Today sago pith waste is dumped into the nearest river along with the sago starch industry's water waste, these potentially causing environmental problems.

Converting such a high amount of sago pith waste into a suitable form for reuse is crucial both from an environmental and economic point of view. Converting these wastes into a form that can be used as an adsorbent will enable the production of cheap adsorbents. One of the potential adsorbed is activated carbon.

Activated carbon can be produced from various carbonaceous materials by physical or chemical activation, or by a combination of these two processes [8]. The physical activation method involves carbonization of a precursor at elevated temperatures (500-900°C) in an inert atmosphere, followed by activation of the resulting char at high temperatures (800-1000°C) in the presence of the $CO_2$ or steam atmosphere. In the chemical activation method, raw material is impregnated with an activating reagent and the impregnated material is heated under an inert atmosphere.

Activation process is generally conducted physically and chemically [9] using activating agent $CO_2$ [10], water vapor [11], KOH [12], $H_3PO_4$ [13], $ZnCl_2$ [14], and $CaCl_2$ [15]. The physical and chemical activation serves to break the hydrocarbon bond or to oxidize the surface molecule to enlarge the active carbon's pores. The chemical activation is usually preferred because it is simple, has a shorter activity time, renders a higher yield, needs a lower temperature, and develops a better porous structure. The chemical activation yields more porous carbon with a higher specific area and a greater pore volume [16]. Carbon's pores containing micropores are effective in increasing the capacity of specific adsorption [17], especially sub-nanometer-sized pores (0.7 nm) [18,19].

KOH constitutes one of activating agents which are the most frequently used. KOH will react with carbon to form new pores and to produce carbon dioxide which diffuses to the carbon's surface [20]. Some studies were conducted to find the optimal solution concentration of KOH as an activating agent in the making of active carbon from various materials [21-23]. Potassium permanganate has strong oxidizing property and could release nascent state oxygen when encountering organic matters. This was significantly improve surface oxygen-containing functional groups like carboxyl, carbonyl, phenolic hydroxyl group, and lactone group, etc., while improving the acidic and alkaline environments of activated carbon surface [24,25]. The researches on the use of $KMnO_4$ as an activating agent in activated carbon which have ever been carried out were those on granular activated carbon [26] and on the effect of the solution concentration of $KMnO_4$ [27,28].

The size of activated carbon's particles affects its adsorption performance. The ability of activated carbon as an adsorbent depends very much on various properties [29], namely, the size and the distribution of pore volume (pore structure), surface area, and the type and the quality of surface functional groups bond (surface chemistry). One of the methods to enlarge the contact surface area is by reducing the size of activated carbon's particles into nano size particles. The material of nanoparticles has a high ratio figure between the surface area and its volume so that it is more reactive in nature because only atoms on the surface will directly touch other materials. The aim of this study is to gain sago pith waste-based activated carbon with the best parameter using KOH and $KMnO_4$ as activating agents with ball-milling size reduction.

## 2. Materials and Methods

### *2.1. Materials*
The material used for this research was sago pith waste (*Metroxylon* sp.) and the chemicals used were $KMnO_4$, HCl provided from PT Merck Indonesia Tbk, Jakarta, Kalium hydroxide (KOH), and other chemicals for analysis purpose

### *2.2. Methods*
#### 2.2.1. Production of the activated carbon
Sago pith waste was rinsed 4 times and then dried. Furthermore, it was placed in furnace at 400°C for 4 hours and then crushed to 30 mesh. It was then activated using 2 activation methods, physical activation and physical-chemical activation using activating agent KOH and $KMnO_4$. Physical activation was conducted using heat in a reactor to reach 800°C and hot favor is streamed for steaming process for 1 hour. Chemical activation was conducted by soaking 100 g charcoal in activating agents KOH or KMnO4 solution of 5% and 2.5% concentration for 24 hours. After drying, the charcoal was put into reactor at 800°C and hot favor was streamed for steaming process for 1 hour. The resulted active carbon

was then cooled and soaked with HCl 1% solution for 1 hour then washed with distilled water to remove the chloride residue [30].

To acquire activated carbon with a 500 nm size, the sample size was reduced using high-energy ball-milling (Fritsch, pulverisette 7 premium line). Fifty grams of activated carbon was milled in the chamber at 500 rpm for 150 minutes which is a modification of Shan method [31]. Activated carbon particle size was measured based on particle distribution index (PDI), in which the score of less than 0.7 indicates homogenous sample size [32,33]. The activated carbon surface area and pore size was measured using BET (Brunauer, Emmelt dan Teller) surface analysis.

2.2.2. Measurements and observations of activated carbon parameters

The parameters of activated carbon examined were recovery, water content, ash content, volatile matter content, burn-off weight percentage, iodine number, methylene blue adsorption, and Braunanear, Emmelt and Teller (BET) surface analysis. Pyrolysis-gas chromatography/mass spectrometer (Py–GC/MS; Shimadzu GCMS-QP 2010, Kyoto, Japan) was used to analyze the functional group distribution of samples. The volatile fragments were analyzed by comparing the results with previous reports and by fitting with mass spectra in the NIST spectral libraries. Activated carbon morphology was observed using scanning electron microscopy (SEM; EVOIMA10, Zeiss, Cambridge, UK). Activated carbon was placed on the specimen holder and sputter-coated with gold at 20 mA sputter current for 60 seconds. The sample was observed with a voltage speed of 10 kV.

X-ray diffraction (XRD; D8 Advance A25, Bruker, Karlsruhe, Germany) was used to determine activated carbon crystallinity, inter-crystal distance, and crystal size. XRD was operated with copper anode X-ray tube at 40 kV and 30 mA. The monochromatic radiation wavelength was 0.154 nm. The sample holder dimension was 25 × 1 mm. Scanning area at diffraction angle (2θ) was set at 0° to 50°. Other operation conditions included: scanning speed 2°/min; interval 0.02°; and divergence and scatter slit, 1° [34]. Relative crystallinity percentage is defined as crystalline component peak area ratio and it is measured quantitatively. D-spacing and crystal size was measured using the Bragg equation (Equation 1) and peak width at half height of peak diffraction using the Scherrer equation (Equation 2) [35]:

$$\text{Bragg equation: } n\,\lambda = 2d \sin\theta \tag{1}$$

$$\text{Scherrer equation: } L = \frac{0.9\,\lambda}{H\cos\theta} \tag{2}$$

where, n is an integer, λ is wavelength of the X-ray used, d is distance between crystal, θ is diffraction angle, L is crystal size, and βis peak width at half maximum (FWHM). The activated carbon IR spectrum was measured by FTIR-UATR (Perkin Elmer Spectrum Two C106456, Waltham, MA, USA)

## 3. Results and Discussion

### 3.1. *Characteristic of sago pith waste-based charcoal and activated carbon*

Milling for 150 minutes resulted in activated carbon particles with an average size of 585.40 nm. The PDI score was 0.673. This PDI value indicates the particle size average distribution. If the PDI is lower than 0.7, the sample particle size is adequately homogenous [32,33].

Table 1 indicates that the water content of sago charcoal decreased rapidly compared to the sago pith waste raw material. Carbonation is the process to convert the organic material into charcoal. The process will release the flammable materials such as CO, $CH_4$, $H_2$, formaldehide, methene, formik and acetil acid and the unflammable materials such as $CO_2$, $H_2O$ and liquid tar. Sago pith waste-charcoal produced a yield of 28.36%. Activated carbon yield determined by time, activation temperature, and type of raw material. Increasing the temperature and activation time decreases the activated carbon yield. Burnt off weight percentage of sago pith waste-charcoal was 57.25%. Burnt-off as an indicator of activation level is one of the crucial factors that determines the quality of activated carbon [36].

**Table 1.** Characteristic of sago pith waste and sago pith waste charcoal

| Parameters | Sago pith waste |
|---|---|
| Water content of sago pith waste (%) | 19.46 ± 7.50 |
| Water content of sago pith waste charcoal (%) | 2.83 ± 0.15 |
| Yield of charcoal (%) | 28.36 ± 1.54 |
| Burnt off weight percentage (%) | 57.25 ± 18.78 |

Several essential parameters for activated carbon are moisture content, ash content, and volatile matter content, as displayed in Table 2. All activated carbon did not meet the criteria of the standard SNI 06-3730-1995 for ash content. Activated carbon that met the criteria of the standard SNI 06-3730-1995 for moisture content and volatile matter was sago pith waste which activated physically, sago pith waste with activating agent KOH 5% & milling, sago pith waste with activating agent $KMnO_4$ 2.5%, sago pith waste with activating agent $KMnO_4$ 2.5% & milling, sago pith waste with activating agent $KMnO_4$ 5%, and sago pith waste with activating agent $KMnO_4$ 5% & milling.

The moisture content is influenced by the activating agent used, concentration of activating agent, and activated carbon particle size. The result showed that activating agent $KMnO_4$ produced a lower water content than KOH. KOH possess a higher water absorbent capacity, so that the KOH in activated carbon can adsorb more water in the air.

The higher concentration of activating agent produces activated carbon with lower water content. Ash content is affected by the type of material, type of activating agent, and concentration of activating agent. The activated carbon prepared with KOH contained a lower ash content. Volatile matter content is determined by the type of activating agent. The activating agent $KMnO_4$ produced activated carbon with lower volatile matter content than that of activating agent KOH.

**Table 2.** Moisture content, ash content, and volatile matter of sago pith waste activated carbon

| Samples | Moisture content (%) | Ash content (%) | Volatile matter (%) |
|---|---|---|---|
| Sago pith waste charcoal | 2.832 | 20.429 | 32.398 |
| Sago pith waste & Physical activation | 5.557 | 18.965 | 18.538 |
| Sago pith waste+KOH2.5% | 9.553 | 25.220 | 34.534 |
| Sago pith waste+KOH2.5% & milling | 5.792 | 23.627 | 35.158 |
| Sago pith waste+KOH 5% | 7.229 | 27.217 | 27.859 |
| Sago pith waste+KOH 5% & milling | 6.780 | 26.259 | 20.806 |
| Sago pith waste+$KMnO_4$ 2.5% | 4.918 | 72.735 | 19.400 |
| Sago pith waste+$KMnO_4$ 2.5% & milling | 4.452 | 70.715 | 21.082 |
| Sago pith waste+$KMnO_4$ 5% | 4.771 | 32.951 | 15.359 |
| Sago pith waste +$KMnO_4$ 5% & milling | 5.010 | 33.166 | 17.713 |
| SNI 06-3730-1995 | Max 15 | Max 10 | Max 25 |

### *3.2. Iodine number and methylene blue adsorption*

Figure 1 presents the iodine number of charcoal and activated carbon. According to SNI 06-3730-1995, the minimum iodine number is 750 mg/g. Iodine number indicated huge number of micropores (the small size pores with size less than 2 nm). Activated carbon with micropores have adsorption capacity on smaller substance (ASTM 234 D 4607-94 1994). Haimour *et al.* [37] stated that the iodine number indicates the number of pores with size larger than 1.0 nm.

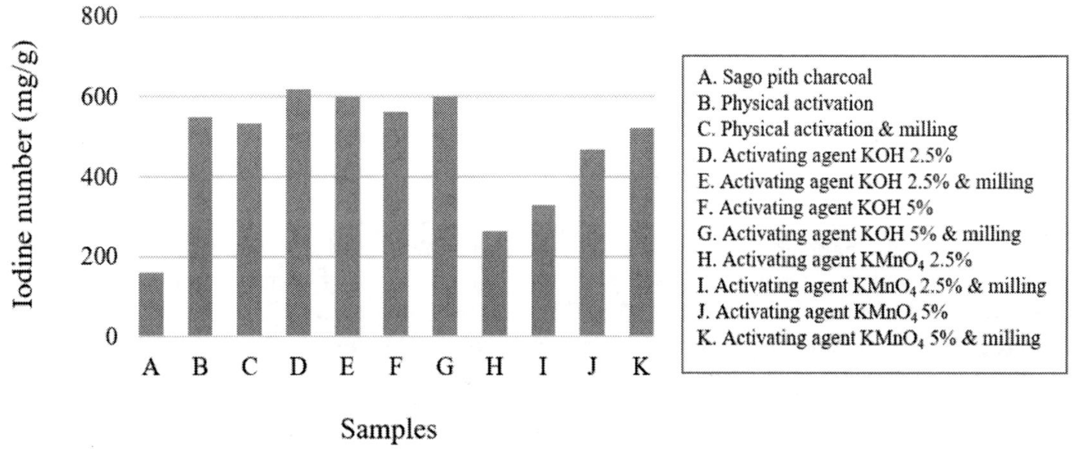


**Figure 1.** Iodine number of sago pith waste activated carbon

The results howed that all sago pith waste activated carbon that produced in this study did not met the criteria of the SNI for iodine number. The treatment that produced activated carbon with highest iodine number was sago pith waste with activating agent KOH 2.5% (619.4 mg/g).

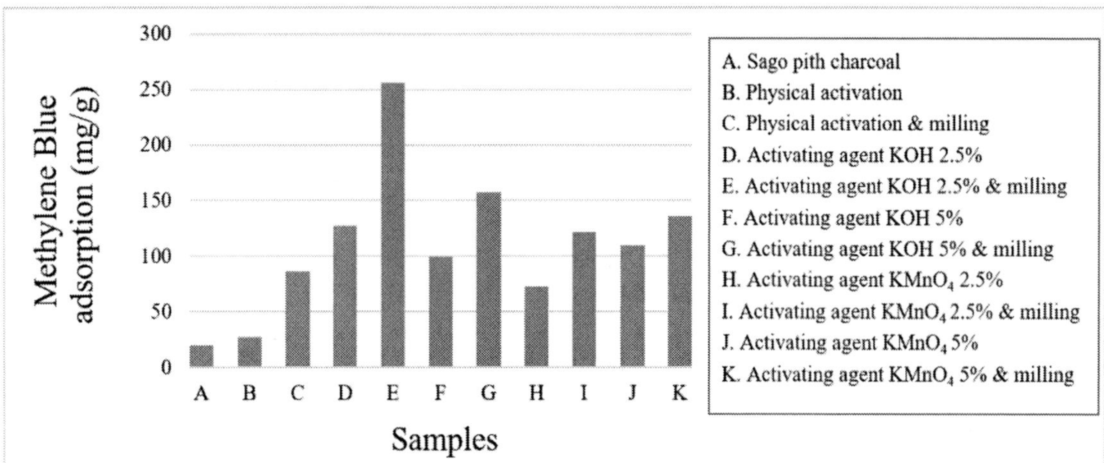


**Figure 2.** Methylene Blue adsorption of sago pith waste activated carbon

Figure 2 showed the Methylene Blue adsorption. The adsorption capacity towards methylene blue according to the Indonesia National Standards is minimum 120 mg/gL. The research result showed activated carbon that meet the criteria of the adsorption capacity towards methylene blue according to the Indonesia National Standards was that from sago pith waste with activating agent KOH 2.5%, sago pith waste with activating agent KOH 2.5% & milling, sago pith waste with activating agent KOH 5% & milling, sago pith waste with activating agent $KMnO_4$ 2.5% & milling, sago pith waste with activating agent $KMnO_4$ 5%, and sago pith waste with activating agent $KMnO_4$ 5% & milling. The result showed that ball milling effectively increased methylene blue adsorption capacity. This result was consistent with research conducted by Lyu *et al.* [38].

Based on the dimension of methylene blue's molecules, generally methylene blue adsorbed the substance on mesopores and macropores, though it did so on micropores but in very small amount. Compared with methylene blue's molecules, iodine's molecules had a smaller dimension, making it is

IOP Conf. Series: Materials Science and Engineering **935** (2020) 012043 doi:10.1088/1757-899X/935/1/012043

easier for them to penetrate micropores. It was found that the hydrothermal process at a low temperature produced a formation of fewer micropores than mesopores and macropores.

The methylene blue adsorption test was conducted to measure the material ability to adsorb colors and substance with large molecules connected to mesopores and macropores [39], even though it is also found in micropores in small amounts. Compared with methylene blue, iodine molecules have smaller dimensions which makes it is easier for them to penetrate micropores. The hydrothermal process at low temperature produces less micropores than mesopores and macropores.

*3.3. Surface* area analysis *(BET)*

Table 3 showed the result of the analysis of activated carbon's surface area (BET). This research produced nanopores activated carbon, the pore's radius being between 1.183-9.581 nm. The treatment producing the pore's radius greater than 9 nm was that of sago waste before activation. Activated carbon with surface area almost 500 $m^2/g$ was Sago pith waste+KOH 2.5% & milling. Activated carbon with the highest amount of nano-pore volume was Sago pith waste+$KMnO_4$ 5%.

**Table 3**. Surface area analysis (BET) of sago pith waste charcoal and sago pith waste activated carbon

| Samples | Surfce area ($m^2/g$) | Pore volume (cc/g) | Pore radius (nm) |
|---|---|---|---|
| Sago pith waste charcoal | 49.802 | 0.063 | 9.581 |
| Sago pith waste+Physical activation | 250.346 | 0.065 | 1.699 |
| Sago pith waste+KOH 2.5% | 170.101 | 0.040 | 1.879 |
| Sago pith waste+KOH 2.5% & milling | 496.723 | 0.180 | 2.565 |
| Sago pith waste+KOH 5% | 291.142 | 0.351 | 1.428 |
| Sago pith waste+KOH 5% & milling | 470.694 | 0.179 | 2.576 |
| Sago pith waste+$KMnO_4$ 2.5% | 255.895 | 0.267 | 2.083 |
| Sago pith waste+$KMnO_4$ 2.5% & milling | 342.765 | 0.328 | 2.876 |
| Sago pith waste+$KMnO_4$ 5% | 381.724 | 0.381 | 1.582 |
| Sago pith waste+ $KMnO_4$ 5% & milling | 441.272 | 0.133 | 2.576 |

Activated carbon's surface area after ball milling increased on average, but this did not affect the volume and the pore's radius. The pore's volume increased visibly as a result of activation treatment using KOH and KMnO4. The pore's radius after activation was far much smaller than that without activation. Surface area indicates the adsorption capacity of adsorbent or porous materials. Activated carbon covers a surface area of 300 to 4000 $m^2/g$ and it is the largest of all adsorbents [40].

The reduction of the size of activated carbon's particles was expected to increase the specific surface area. The treatment of ball milling could change the ratio of physisorbed gas into chemisorbed gas, and this conformed to the result of the previous research [41,42]. This condition showed that longer milling time caused the occurrence of agglomeration of smaller particles into bigger ones. The best size of active carbon's pores being between 2-50 nm showed the mesopore's structure.

Micropore (pore smaller than 2 nm) is the dominating adsorption mechanism, and the micropore size distribution affected the adsorption mechanism [43]. The type of pore resulted from the ball-milling process is mesopore. When smaller particles are produced, there are more pores formed, resulting in higher adsorption. Hydrothermal pyrolysis using KOH increased the surface area, and this is mainly caused by elimination of acid side chains, which block pore access [44]. KOH is a powerful reagent because it can interact with carbon atoms and serves as a catalyst for dehydrogenese reaction and oxidation, therefore increasing porosity [41].

Girgis and El-Hendawy [45] showed that the amount of KOH affects the activated carbon produced. Girgis and El-Hendawy [45] used KOH with the raw material to KOH ratio of 1:3. The higher ratio of KOH can cause the formation of an insulation layer and encapsulate particles, thus reducing the activation process and contact with the surrounding environment [45].

### 3.4. Morphology analysis (SEM)

SEM images of the charcoal and activated carbon (Figure 3) showed the size of the activated carbon pore. There were a number of nano-size pores (mesopore and micropore), along with some particles larger than 1 μm.

Sago pith waste activated carbon with KOH

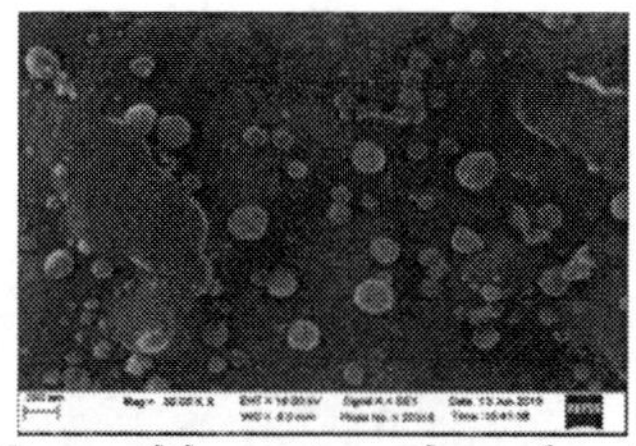

Sago pith waste activated carbon with $KMnO_4$

Sago pith waste activated carbon with KOH & milling

Sago pith waste activated carbon with $KMnO_4$ & milling

**Figure 3.** SEM images of sago pith waste activated carbon with KOH and $KMnO_4$

Activation treatment produced more pores compared with carbon with pyrolysis treatment without activation. The milling process using ball-milling reduced the size of particles. The process also produced smaller activated carbon with increased surface area. However, the reduced particle size can also cause some small particles to cover macropores and mesopores. According to Welham *et al.* [42] and Khalil *et al.* [41], ball-milling changes the ratio of physisorbed gas to be chemisorbed gas. The use of KOH activating agent produced active carbon's surface which was smoother, while $KMnO_4$ yielded carbon's surface which was rough. The covering of pores by smaller particles occurred more in active carbon with $KMnO_4$ activating agent.

The microstructure picture showed that the number of pores increased as a result of carbonation and activation, but the figure rendered qualitative data so that it was difficult to compare accurately [46]. The reduction of size using ball milling significantly reduced the size of active carbon's particles.

### 3.5. Functional group analysis (FTIR)

The FTIR test was conducted to identify the functional group of activated carbon. Increases and decreases in spectrum peaks indicate the impact of treatment to the material structure as presented in Fig. 4. The C=O bond and C-O stretching at wave number 1640 to 1750 $cm^{-1}$ is attributed to phenolic esther, carbocylic acid, and ketone. Peaks in wavenumbers 700 to 900 $cm^{-1}$ are related to aromatics [47]. The wave number 1375-1465 $cm^{-1}$ (C-H aliphatic bond) were more intense in sago pith waste activating with KOH.

The peak at wave number 1600 to 1400 $cm^{-1}$ showed the aromatic C=C bond from lignin. In this wave number range are benzene (1450 $cm^{-1}$), aromatic structures (1600 $cm^{-1}$), and methoxyle (1470 to 1430 $cm^{-1}$) [46]. The OC-O bond (carboxyl) at wave number 1250 to 1000 $cm^{-1}$ had a higher peak, which decreased after the hydrothermal process [47].

The adsorbency peak at the wave number of 500-800 $cm^{-1}$ was related to the aromatic substitution. Cellulose was shown by strong absorption on –OH and C–O, while hemicellulose was shown by the existence of the bond of C=O (aldehyde) [49]. Lignin was found at the wave number at 1830-730 $cm^{-1}$, and so were aromatic C=C ring, methoxyl –O–$CH_3$, and C–O–C stretching. The result of FTIR analysis showed that the lignin content was abundant enough and reduced only in small amount. Meanwhile, the cellulosic and hemicellulosic content was only a little. This result confirmed to that of the research by Sumtong *et al.* [50], which stated that only a little lignin content decomposed, while almost all cellulose and hemicellulose decomposed in the hydrothermal process.

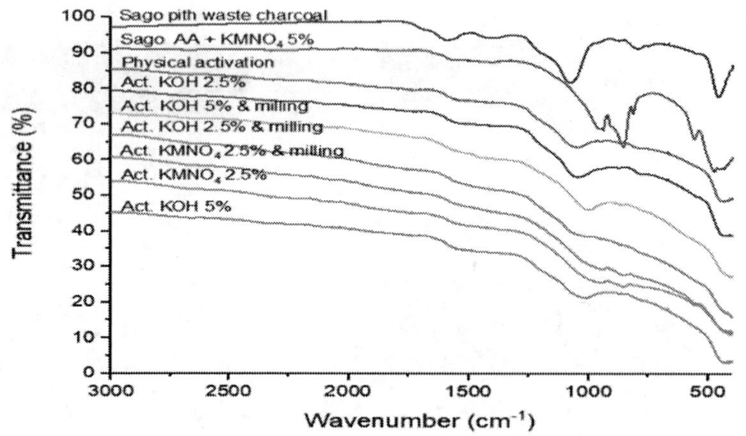


**Figure 4.** FTIR spectra of sago pith waste charcoal and activated carbon

Figure 4 illustrates the comparison of carbon FTIR spectrum before and after activation and size reduction. Carbon has C=C bond (wave number 1631 cm-$^{1}$), but it not emerges after activation. The new peak of the active carbon occurred at wave number 1573 $cm^{-1}$ which showed the stretching of C-C because the newly formed aromatic rings. This reinforced the fact that the activated carbon was formed because of the new pore formation [51,52]. The activation process also broke many chains either aliphatic or aromatic [53]. This effect was shown by the decrease in transmittance at the area of 1195 to 459 $cm^{1}$. The peak at 1195 $cm^{-1}$ indicated C-O stretching.

### *3.6. Analysis of crystallinity (XRD)*

Figure 5 showed the result of XRD of activated carbon sourced from sago pith waste. The peak of XRD spectrum of sago pith waste raw materials looked sharp, clear, and higher than that of activated carbon's spectrum. The diffractogram at 16 and 23° showed the A-type starch diffraction pattern of [54]. All solid had a peak which widened at the area of 20-30°, this constituting a peak characteristic of carbon. This was related to the asymmetric stretching (=C-O-C) vibration, which constituted a component of lignin. The peak of diffraction at 23.4° and 43.7° was characteristic of amorphous carbon. Some researchers stated that active carbon constituted an amorphous material [55-57].

The peak intensity at 23° decreased after activation process. It indicated that cellulose decomposed into carbon atom after the activation process. The broadening peak at 23.4° in carbon material indicated that atom bond gradually disappeared and established the hexagonal structures. Activation process causes inter aromatic layers not symmetrical so that the degree of crystallinity decreased. The activating process using KOH involved the carbon gasification catalyst and potassium metal intercalation, which would later disappear quickly [58]. The gasification process on the surface of carbon was the cause of the forming of macropores.

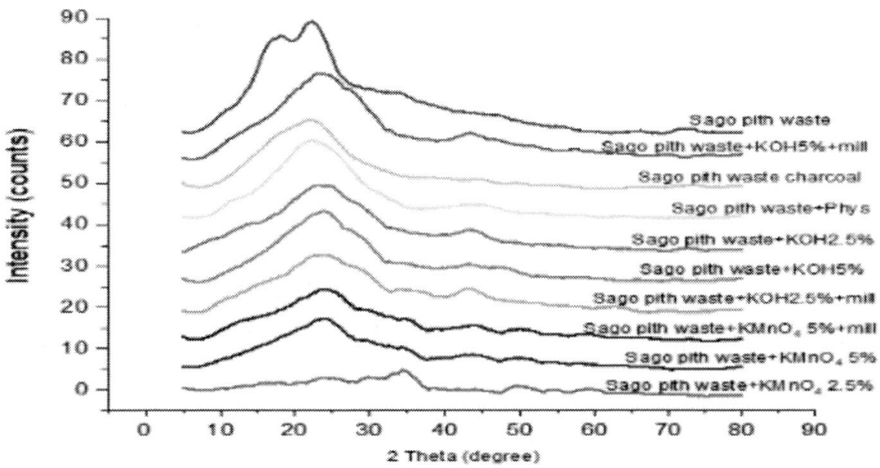


**Figure 5.** XRD analysis of sago pith waste charcoal and activated carbon

The treatment of ball milling produced activated carbon's particles which were smaller, and this was shown by the peak area which turned to widen, this related to the weakening of reflection intensity on the plane (002). It was believed that the carbon's original crystalline structure had changed because of the treatment of ball milling [59]. The peak widened at the area around 22-23°, this related to the reflection of the plane (002) from carbon [60,61]. This wide peak showed that active carbon was not crystalline in nature [60]. Meanwhile, the peak at the area around 43-44° became increasingly narrow with a low intensity. This peak was related to the plane reflection (100) [61].

Table 4 showed that activated carbon with the activating agent of KOH had lower crystallinity compared to that with the activating agent of $KMnO_4$. The treatment of ball milling also decreased the degree of active carbon's crystallinity, this meaning that active carbon was increasingly amorphous in nature. The structure of pores was formed during the process of carbonization and activation, namely, when the distance between basic crystals was cleaned from tar and other elements [63].

**Table 4.** Crystallite size and crystallinity of sago pith waste activated carbon

| Materials | d-spacing (nm) | | | crystallite size (nm) | | | Kristalinitas |
|---|---|---|---|---|---|---|---|
| | Peak 1 | Peak 2 | Peak 3 | Peak 1 | Peak 2 | Peak 3 | (%) |
| Sago pith waste charcoal | 0.38 | 0.38 | 0.53 | 147.608 | 95.469 | 123.632 | 23.9575 |
| Sago pith waste+Physical activation | 0.34 | 0.40 | 0.49 | 145.871 | 157.708 | 183.015 | 27.5028 |
| Sago pith waste+KOH 2.5% | 0.39 | 0.33 | 0.35 | 115.892 | 104.465 | 61.109 | 20.2982 |
| Sago pith waste+KOH 2.5% & milling | 0.46 | 0.38 | 0.38 | 138.301 | 121.692 | 152.319 | 36.3657 |
| Sago pith waste+KOH 5% | 0.34 | 0.30 | 0.25 | 101.962 | 136.976 | 87.855 | 20.4579 |
| Sago pith waste+KOH 5% & milling | 0.38 | 0.37 | 0.30 | 80.123 | 101.556 | 45.655 | 23.2472 |
| Sago pith waste+$KMnO_4$ 2.5% | 0.30 | 0.50 | 0.49 | 43.260 | 134.121 | 40.251 | 13.0651 |
| Sago pith waste+$KMnO_4$ 2.5% & milling | 0.18 | 0.26 | 0.29 | 102.054 | 83.357 | 79.326 | 69.9769 |
| Sago pith waste+$KMnO_4$ 5% | 0.36 | 0.36 | 0.29 | 135.579 | 147.939 | 77.314 | 24.1160 |
| Sago pith waste+ $KMnO_4$ 5% & milling | 0.33 | 0.39 | 0.31 | 97.828 | 125.213 | 64.195 | 29.6551 |

## 4. Conclusions

The study produced nano pore activated carbon sizing 585.40 nm and PDI score of 0.673 using ball-milling treatment for 150 min. The pore of activated carbon had pore radius ranging from 1.183 to 9.581 nm. The activated carbon which met the criteria of the SNI 06-3730-1995 for moisture content and volatile matter where the majority samples namely sago pith waste and Physical activation sago pith waste with activating agent KOH5% & milling, sago pith waste with activating agent $KMnO_4$ 2.5%, sago pith waste with activating agent $KMnO_4$ 2.5% & milling, sago pith waste with activating agent $KMnO_4$ 5%, and sago pith waste with activating agent $KMnO_4$ 5% & milling.

The activated carbon that had the best Iodine number was sago pith waste with activating agent KOH 2.5% (619.4 mg/g). The activated carbon that met the criteria of SNI 06-3730-1995 for methylene blue adsorbtion capacity was sago pith waste with activating agent KOH 2.5%, sago pith waste with activating agent KOH 2.5% & milling, sago pith waste with activating agent KOH 5% & milling, sago pith waste with activating agent $KMnO_4$ 2.5% & milling, sago pith waste with activating agent $KMnO_4$ 5%, and sago pith waste with activating agent $KMnO_4$ 5% & milling. The best activated carbon based on all parameters was sago pith waste with activating agent KOH 2.5% & milling.

## Refferences

[1] Dai J, Cui H, and Grace JR 2012 Biomass feeding for thermochemical reactors *Prog. Energy Combust. Sci.* **38** 716-36

[2] Mushtaq F, Mat R, and Ani FN 2014 A review on microwave assisted pyrolysis of coal and biomass for fuel production *Renew. Sustain. Energy Rev.* **39** 555-74

[3] Chayid MA and Ahmed MJ 2015 Amoxicillin adsorption on microwave prepared activated carbon from *Arundo donax* Linn: isotherms, kinetics, and thermodynamics studies *J. Environ. Chem. Eng.* **3** 1592-601

[4] Li Z, Zhai K, Wang G, Li Q, and Guo P 2016 Preparation and electrocapacitive properties of hierarchical porous carbons based on loofah sponge *Materials* **9**(11) 912-22

[5] Singhal RS, Kennedy JF, Gopalakrishnan SM, Kaczmarek A, Knill CJ, and Akmar PF 2007 Industrial production, processing, and utilization of sago palm-derived products *Carbohydrate Polymers* **72** 1-20

[6] Directorate General Forest Estate 2019 *Statistik Tanaman Perkebunan Indonesia 2017–2019: Sagu.* (Jakarta: Sekretariat Direktorat Jenderal Tanaman Perkebunan)

[7] Banu JR, Kaliappan S, and Beck D 2006 Treatment of sago wastewater using hybrid anaerobic reactor *Water Quality Research Journal* **41**(1) 56-62

[8] Kim MH, Jeong IT, Park SB, Kim JW, Kim MH, Jeong IT, Park SB, and Kim JW 2018 Analysis of environmental impact of activated carbon production from wood waste. *Environmental Engineering Research* **24**(1) 117-26

[9] Wang J, Nie P, Ding B, Dong S, Hao X, Dou H, and Zhang X 2017 Biomass derived carbon for energy storage devices *Journal of materials chemistry* **5**(6) 2411-28

[10] Song M., Jin BS, Xiao R, Yang L, Wu YM, Zhong ZP, and Huang YJ 2013 The comparison of two activation techniques to prepare activated carbon from corn cob *Biomass Bioenergy* **48** 250-6

[11] Deiana C, Granados D, Venturini R, Amaya A, Sergio M, and Tancredi N 2008 Activated carbons obtained from rice husk: influence of leaching on textural parameters *Ind. Eng. Chem. Res.* **47** (14) 4754-7

[12] Sun Y, Paul A, and Webley 2008 Preparation of activated carbons from corncob with large specific surface area by a variety of chemical activators and their application in gas storage *Chem. Eng. J.* **162** (3) 883-92

[13] Sun Y, Li H, Li G, Gao B, Yue Q, and Li X. 2016. Characterization and ciprofloxacin adsorption properties of activated carbons prepared from biomass wastes by $H_3PO_4$ activation. *Bioresources Technol.* **217** 239-44

[14] Boudrahem F, Soualah A, and Aissani-Benissad F 2017 Pb(II) and Cd (II) removal from mechanochemical treatment can reduce particle sizes and activate their surface chemical states, aqueous solutions using activated carbon developed from coffee residue activated with phosphoric acid and zinc chloride *J. Chem. Eng. Data* **56**(5) 1946-55

[15] Liu JJ, Deng YF, Li XH, and Wang LF 2016 Promising nitrogen-rich porous carbons derived from one-step calcium chloride activation of biomass-based waste for high performance supercapacitors *ACS Sustainable Chem. Eng.* **4**(1) 177-87

[16] Song M., Jin BS, Xiao R, Yang L, Wu YM, Zhong ZP, and Huang YJ 2013 The comparison of two activation techniques to prepare activated carbon from corn cob, *Biomass Bioenergy* **48** 250-6

[17] Raymundo-Piñero E, Kierzek K, Machnikowski J, and Béguin F 2006 Relationship between the nanoporous texture of activated carbons and their capacitance properties in different electrolytes *Carbon* **44**(12) 2498-507

[18] Chmiola J, Largeot C, Taberna PL, Simon P, and Gogotsi Y 2008 Desolvation of ions in subnanometer pores and its effect on capacitance and double-layer theory *Angew. Chem.* **120**(18) 3440-3

[19] Galhena DTL, Bayer BC, Hofmann S, and Amaratunga GAJ 2016 Understanding capacitance variation in sub-nanometer pores by in situ tuning of interlayer constrictions *ACS Nano* **10**(1) 747-54

[20] Pujiyanto. 2010 Pembuatan karbon aktif super dari batubara dan tempurung kelapa [skripsi] Departemen Teknik Kimia, Fakultas Teknik Universitas Indonesia

[21] Erlina E, Umiatin U, and Budi E 2015 Pengaruh konsentrasi larutan KOH pada karbon aktif tempurung kelapa untuk adsorpsi logam Cu *Prosiding Seminar Nasional Fisika (E-Journal)* **4** pp. SNF2015-VII

[22] Apriani R, Faryuni ID, and Wahyuni D 2013 Pengaruh konsentrasi aktivator kalium hidroksida (KOH) terhadap kualitas karbon aktif kulit durian sebagai adsorben logam Fe pada air gambut *Prisma Fisika* **1**(2) 82-6

[23] Melania MS 2012 Produksi karbon aktif dari bambu dengan aktivasi menggunakan Kalium Hidroksida [skripsi] Departemen Teknik Kimia, Fakultas Teknik, Universitas Indonesia

[24] Zhang X, Deng S, Liu Q, Zhang Y, and Cheng L 2011 Surface functional groups and redox property of modified activated carbons *Min. Sci. Technol.* **21** 181-4

[25] Watson VJ, Delgado CN, and Logan BE 2013 Improvement of activated carbons as oxygen reduction catalysts in neutral solutions by ammonia gas treatment and their performance in microbial fuel cells *J. Power Sources* **242** 756-61

[26] Rathnayake D, Korotta-Gamage SM, Kastal, and Sathasivan A 2017 Effect of KMnO4 treatment of granular activated carbon on the removal of natural organic matter *Desalin. Water Treat.* **71** 201-6

[27] Zhang G, Sun Y, Zhao P, Xu Y, Su AT, and Qu J 2017 Characteristics of activated carbon modified with alkaline $KMnO_4$ and its performance in catalytic reforming of greenhouse gases $CO_2/CH_4$ *Journal of $CO_2$ Utilization* **20**129-40

[28] Pang M, Liu B, Kano N, and Imaizumi H 2015 Adsorption of chromium (VI) onto activated carbon modified with $KMnO_4$ *J. Chem. Eng.* **9**(205) 280-7

[29] Ayg¨un A, Yenisoy-Karaka S, and Duman, I 2003 Production of granular activated carbon from fruit stones and nut shell sand evaluation of their physical, chemical and adsorption properties *Micropor. Mesopor. Mat.* **66** 189-95

[30] Garcia-Garcia A, Gregorio A, Boavida D, and Gulyurtlu I 2002 Production and characterization of activated carbon from pine waste gasified in a pilot reactor *National Institute of Engineering and Industrial Technology, Estrada do Paco do Lumiar.* **22** 1649-038

[31] Shan D, Deng S, Zhao T, Wang B, Wang Y, Huang J, Yu G, Winglee J, and Wiesner MR 2016 Preparation of ultrafine magnetic biochar and activated carbon for pharmaceutical adsorption and subsequent degradation by ball milling *J. Hazard Mater.* **305** 156-63

[32] Bera A 2015 Nanoporous silicon prepared by vapour phase strainetch and sacrificial techniquen *Proceedings of the international conference on microelectronic circuit and system (Micro)* pp. 42-5

[33] Sarmah S, Gogoi SB, Xianfeng F, and Baruah AA 2020 Characterization and identification of the most appropriate nonionicsurfactant for enhanced oil recovery *J. Pet. Explor. Prod. Technol.* **10** 115-23
[34] Cheetam NWH and Tao L 1998 Variation in crystalline type with amylose content in maize starch granules: An x-ray powder diffraction study *Carbohydr. Polym.* **36** 277-84
[35] Kim SH, Shon HK, and Ngo HH 2010 Adsorption characteristics of antibiotics trimethoprim on powdered and granular activated carbon *J. Ind. Eng. Chem.* **16** 344-9
[36] Lázaro MJ, Gálvez ME, Artal S, Palacios JM, and Moliner R 2007 Preparation of steam-activated carbons as catalyst supports *J. Anal. Appl. Pyrolysis*. **78**(2) 301-15
[37] Haimour NM and Emeish S 2006 Utilization of date stones for production of activated carbon using phosphoric acid *Waste Manag.* **26** 651-60
[38] Lyu H, Gao B, He F, Zimmerman AR, Ding C, Huang H, and Tang J 2018 Effects of ball milling on the physicochemical and sorptive properties of biochar: experimental observations and governing mechanisms *Environ. Pollut.* **233** 54-63
[39] Raposo F, Rubia MADL, and Borja R 2009 Methylene blue number as useful indicator to evaluate the adsorptive capacity of granular activated carbon in batch mode: influence of adsorbate/adsorbent mass ratio and particle size *J. Hazard Mater.* **165** 291-9
[40] Yang JB, Ling LC, Liu L, Kang FY, Huang ZH, and Wu H 2002 Preparation and properties of phenolic resin based activated carbon spheres with controlled pore size distribution *Carbon* **40** 911-6
[41] Khalil AHPS, Jawaid M, Firoozian P, Rashid U, Islam A, and Akil HM 2013 Activated carbon from various agricultural wastes by chemical activation with KOH: Preparation and characterization *J. Biobased Mater.* **7** 1-7
[42] Welham NJ, Berbenni V, and Chapman PG 2002 Increased chemisorption onto activated carbon after ball-milling *Carbon* **40**(13) 2307-15
[43] Hu Y, Zhang Y, Hu Y, Chu CY, Lin J, Gao S, Lin D, Lu J, Xiang P, and Ko TH 2019 Application of wasted oolong tea as a bio-sorbent for the adsorption of methylene blue *Journal of Chemistry* **2019** p. 10
[44] Lachos-Perez D, Brown AB, Mudhoo A, Martinez J, Timko MT, Rostagno MA, and Forster-Carneiro T 2017 Applications of subcritical and supercritical water conditions for extraction, hydrolysis, gasification, and carbonization of biomass: a critical review *Biofuel Res. J.* **14**(2017) 611-26
[45] Girgis BS, and El-Hendawy ANA 2002 Porosity development in activated carbons obtained from date pits under chemical activation with phosphoric acid *Micropor. Mesopor. Mat.* **52**(2) 105-17
[46] Negara DNKP, Nindhia TGT, Surata V, Sucipta M, and Hidajat F 2019 Activated carbon characteristics of tabah bamboo that physically activated under different activation *time IOP Conf. Series: Materials Science and Engineering* **539** 012011
[47] Cuhadaroglu and Uygun 2008 Production and characterization of activated carbon from bituminous coal by chemical activation *Afr. J. Biotechnol.* **7**(20) 3703-10
[48] Gao Y, Chen HP, Wang J, Shi T, Yang HP, and Wang XH 2011 Characterization of products from hydrothermal liquefaction and carbonation of biomass model compounds and real biomass *J. Fuel Chem. Technol.* **39**(12) 893-900
[49] Yang H, Yan R, Chen H, Lee DH, and Zheng C 2007 Characteristics of hemicellulose, cellulose and lignin pyrolysis *Fuel* **86** 1781-8
[50] Sumtong P, Chollacoop N, and Eiad-ua A 2017 Effect of temperature and times by hydrothermal carbonization process from sawdust and bagasse for carbon material supporter *J. Appl. Sci.* **16** 93-7
[51] Muniandy L, Adam F, Mohamed AR, and Ng EP 2014 The synthesis and characterization of high purity mixed microporous/mesoporous activated carbon from rice husk using chemical activation with NaOH and KOH *Microporous and Mesoporous Materials* **197** 316-23

[52] Thuan T. Thinh V, Quynh PV, Cong HT, Tam DT, and Bach LG 2016 Production of activated carbon from sugarcane bagasse by chemical activation with ZnCl2: Preparation and characterization study *Research Journal of Chemical Sciences* **6** 42-7
[53] Hesas R, Niya H, Daud A, and Sahu WJN 2013 Preparation and characterization of carbon from apple waste by microwave-assisted phosphoric acid activation: Application in methylene blue *BioResources* **8** 2950-66
[54] Jiang R, Zhu HY, Zeng GM, Xiao L, and Guan YJ 2010 Synergy of adsorption and visible light photocatalysis to decolor methyl orange by activated carbon/nanosized CdS/chitosan composite *J. Cent. South Univ. Technol.* **17** 1223-9
[55] Mohamed MA, Salleh WNW, Jaafar J, Rosmi MS, Mohd. Hir ZA, Mutalib MA, Ismail AF, and Tanemura M 2017 Carbon as amorphous shell and interstitial dopant in mesoporous rutile $TiO_2$: Bio-template assisted sol-gel synthesis and photocatalytic activity *Appl Surf.* **393**(30) 46-59
[56] Ip AWM, Barford JP, and Mc Kay G 2008 Production and comparison of high surface area bamboo derived active carbons *Bioresour. Technol.* **99**(18) 8909-16
[57] Jung MJ, Jeong E, Kim S, Lee SI, Yoo JS, and Lee YS 2011 Fluorination effect of activated carbon electrodes on the electrochemical performance of electric double layer capacitors *J. Fluor Chem.* **132**(12) 1127-33
[58] Marsh H, and Rodríguez-Reinoso F 2006 *Activated Carbon* (Amsterdam: Elsevier) pp. 350-8
[59] Zhang Q, Zeng Y, Xiao X, Deng P, He Q, and Zhang T 2019 Investigation on the preparation and adsorption performance of bamboo fiber based activated carbon *Fiber Polym.* **20**(2) 293-301
[60] Qu D 2002 Studies of the activated carbons used in double-layer supercapacitors *J. Power Sources 109* 403-11
[61] Bratek W, Świątkowski A, Pakuła M, Biniak S, Bystrzejewski, and Szmigielski R 2013 Characteristics of activated carbon prepared from waste PET by carbon dioxide activation *J. Anal. Appl. Pyrol.* **100** 192-8
[62] Welham NJ and Williams JS 1998 Extended milling of graphite and activated carbon *Carbon* **36** 1309-15
[63] Bansal RC and Goyal M 2005 *Activated Carbon Adsorption* (CRC Press)

IOP Conf. Series: Materials Science and Engineering **935** (2020) 012044 doi:10.1088/1757-899X/935/1/012044

# The potency of wood based electricity production from critical land in Indonesia

**B H Narendra*, C A Siregar, and A G Salim**

Forest Research and Development Center, Ministry of Environment and Forestry, Bogor, 16610, Indonesia

*Email: budihadin@yahoo.co.id

**Abstract**. Increasing demand for woody biomass as the fuel of renewable power plants must be anticipated by utilizing critical land through land rehabilitation activities using potential woody biomass species. However, there is a lack of information on the potency of such critical land. Therefore this study aims to explore the critical land area suitable for woody biomass cultivation and estimate the electricity production that can be generated from the utilization of the critical land. The study was conducted by determining the priority areas of integrated land to be rehabilitated, identifying the status and function of the critical land area, evaluating its suitability for the woody biomass cultivation, and calculating the potential output of electrical energy. The result shows that there are approximately 3.8 million hectares of critical land, which is suitable for producing woody biomass. This area consists of 2.3 million hectares of non state forest area and 1.5 million hectares of production forest area. The largest suitable area is located in the Sumatra region, while the smallest was in the Papua region. The suitable area potentially produces electricity as much as 102,953 GWh, should be able to supply electricity of 69.5 million households. Further, the biomass has the potential to be used as co-firing fuel in coal-based power plants to reduce the negative impacts of the use of coal.

## 1. Introduction

Energy resource is one of the national development capital, but the dominance of the use of energy sources derived from fossil fuels has increased the rate of $CO_2$ and other greenhouse gas emissions into the air as the leading cause of global warming and climate change [1,2]. In conjunction with the effort to reduce these global problems, the use of renewable energy sources should be initiated and put into priority. In 2018, the use of renewable energy globally outside the use of traditional biomass had reached 11% of the total final energy consumption (TFEC), and this magnitude has increased by 9.6% compared to the previous five years [3].

Indonesia, as one of the largest countries in the world, is also working to increase the use of renewable energy as an active role in addressing the global climate problem. Population and economic growth, as well as world oil price conditions, in fact, also affect energy consumption in Indonesia. In 2018, total energy use in Indonesia was dominated by the transportation sector (40%), industry (36%), households (16%), and 8% by others [4]. From the total energy use, only 9% originated from renewable energy. Comparing several uses of energy types, electricity demand has the most significant growth, with an average of 7% per year. It is projected that in 2050 the use of electrical power will reach nine times the current condition or become 2562 TWh. However, the current electricity generation from renewable energy is still around 12.4%, while the rest is fossil fuels, including 57% using coal [5].

The Government of Indonesia issued Regulation 79 of 2014, which aims to achieve 23% of energy production from new and renewable energy by 2025. This effort was taken by encouraging the production of electricity from renewable sources, among others from biomass, and reducing coal use to 32-41% by the year 2050. Like other countries, Indonesia, which has extensive land resources, is trying to catch up with the use of renewable energy from biomass. The Government, as stated in the General Plan of National Energy (*Rencana Umum Energi Nasional-RUEN*) is to encourage the improvement of non-food biomass cultivation as one of the principal renewable energy sources [6]. The use and cultivation of woody biomass-producing plants are also motivated by the ability of plants to absorb $CO_2$ so that it also plays an important role in reducing greenhouse gases [7]. Renewable energy sourced from woody biomass, in the future, is predicted to dominate renewable energy because it can be developed in every place and produced as desired [8].

For a greater use as a power plant fuel, the current obstacle often faced is the continuity of supply of large amounts of biomass and the low heating value of available biomass. To overcome this obstacle, the cultivation of woody biomass as a source of renewable energy began to develop replacing fossil fuels [9]. The demand for woody biomass as a raw material must be balanced with an increase in land area for cultivation purposes. On the other hand, the need for land as a source of food, animal feed, energy, and other economic aspects increases in line with population growth. Conversion of forest cover is frequently done to meet the needs of agricultural land, and if this step is not accompanied by sustainable land management practices, it has led to the occurrence of critical land [10,11]. The process of critical land formation is accelerated by climate change through the increase in the frequency of high-intensity rainfall, which causes a high level of erosion in a barren land and triggers hydrometeorological disasters [12,13]. Based on these considerations, land allocated for woody biomass plantation should not convert productive agricultural land or well-covered forest land. Instead, woody biomass plantation should be developed on the critical lands as rehabilitation efforts whenever possible.

In the global trend, many countries enlarge the effort to utilize the degraded land using fast-growing species to fuel wood-based power plants. Its operations are not only carried out by large corporations but also on a small scale by local communities to anticipate the economic fluctuations that commonly have an impact on changes in energy sources [14-18].

In Indonesia, data on critical land is released by the Ministry of Environment and Forestry, which covers the distribution and area of critical land existing within inside and outside of the state forest area. Not all critical land has the same potency to be developed as a producer of woody biomass. In this regard, scientific considerations must be put into an account for land use planning, namely through land suitability analysis [19]. Therefore, this study aims to explore the critical land area suitable for woody biomass cultivation and estimate the electricity production that can be generated from the utilization of the degraded land.

## 2. Materials and Methods

The framework of this study was designed by employing four main steps in sequences, namely determining the priority areas of degraded land to be rehabilitated; conducting spatial analysis to identify the status and function of forest area on the map of degraded land produced in the first step; evaluating the suitability of critical land for the wood biomass cultivation based on GIS analysis, and finally calculating the potential output of electrical energy from power plants that utilize woody biomass from critical land. All spatial analyses were performed using ArcGIS Desktop 10.

To determine the availability of critical land, a spatial analysis was carried out by delineation of the newest critical land map (2018) released by the Ministry of Environment and Forestry. In the first stage, delineation was carried out by eliminating the polygon in the map that have uncritical, potentially critical, or moderately critical status so that a map with only critical and severe critical levels will be obtained. In the second stage, delineation is carried out by overlaying the map of critical land that has been produced in the first stage, and the map of state forest functions. In Indonesia, the state forest areas are classified based on their function, namely protected forest areas, conservation forest areas, or production forest areas. The rest of the area is classified as non-state forest areas or commonly called as other use areas. The utilization of critical land as a development land of woody biomass is only possible in the production forest function areas and in the non-state forest area. Based on Government regulation, protected and conservation forest areas are prohibited from being used as locations for timber plantations. The second step of the procedure has produced critical land as the available for woody biomass plantation.

Based on this potential land, a suitability assessment was carried out based on the FAO [20] framework at the order level. Land suitability orders indicate global land suitability and provide an overview that indicates an area of the land is suitable or unsuitable for growing certain woody tree species. The analysis was carried out spatially by applying the maximum limitation method [21], which is the most stringent method in selecting the suitability status of land [19,22]. The woody tree species used in this study include *Leucaena leucocephala* (*lamtoro*), *Calliandra calothyrsus* (*kaliandra*), *Gliricidia sepium* (*gamal*), and *Acacia auriculiformis* (*akasia*). They are categorized as multipurpose legume tree species and included as the mainstay/superior species for woody biomass supply [23], as well as suitable for substituting the use of coal [24]. Suitability analysis in the maximum limitation method was then conducted by matching the land characteristics of the proposed critical land, compared to the land requirement of four tree species. The land characteristics consist of a set number of criteria in the form of land resource maps issued by the Indonesian Agency for Agricultural Research and Development, Ministry of Agriculture. These suitability criteria included factors of soil (soil type, solum thickness, texture, and slope) and climate (rainfall, wet month, dry month, air temperature). The land requirement of four species was formulated based on the range of criteria that meet the four species. A critical land area was categorized as suitable if all the plant requirements are in line with the characteristics of the land. On the other hand, if one of the criteria can not be met, then the area was categorized as unsuitable. This analysis produced a map of suitable areas along with the area information.

Woody-biomass yield and the potential result of electrical power can be calculated based on suitable critical land areas. The calculation is based on a number of assumptions from the references of the previous studies. In this study, there was a scenario for the area of suitable critical land that will be planted gradually in three years. The first woody-biomass harvesting will be carried out on the first stage of the plant at three years after planting. The next year's harvesting will be applied on the second stage plantation, which will also already reach three years old, and so on following the three-year cycle. After harvesting, the plants will sprout and grow become woody plants, and then the wood will be harvested again after the age of three years. For the four tree species, the average of stand growth was assumed at the rate of 39 $m^3$/ha/yr in 1x1 $m^2$ spacing [25-34], bulk density was 0.6 ton/$m^3$, and the calorific value was 4000 kCal/kg [23,32,35,36].

The potential electrical power that can be generated from wood biomass-fueled power plants is calculated annually based on woody biomass that can be produced annually at the age of three years after planting. The converting technology of woody biomass to electrical power can be done through various processes, including direct combustion, gasification, pyrolysis, and anaerobic/aerobic digestion. For solid biomass fuels, the process commonly uses direct combustion because it is the simplest, relatively cheap, and more reliable [37,38]. The calculation of energy conversion used the thermal efficiency assumption of a steam engine. Currently, the efficiency values of woody-biomass fueled power plants are in the range of 25%-40%. In this study, the efficiency rate was set at 25% [39-43]. Based on the amount of electrical output, it can be calculated the number of households that can be served. The calculation uses used the 2018 data of average electricity consumption for Indonesian households that amounting 1481 kWh/yr [44].

## 3. Results and Discussion

### *3.1. Critical land in Indonesia*

By definition, critical land indicates the condition of land that does not function properly as a medium for the production and growth of cultivated plants or other plants, as mentioned in Law No.37 of 2014 concerning soil and water conservation. A similar understanding is stated in the Decree of the Minister of Forestry No.52/Kpts-II.2001 concerning the management of watershed management, where the critical land is described as the physical condition of land that cannot function well for crop productivity or as a water system management. In Indonesia, the newest data and map of critical land has been released by the Ministry of Environment and Forestry through Decree No. 306/MENLHK/ PDASHL/DAS.0/7/2018 concerning on determination of critical land. The information was carried out using a geographic information system (GIS) involving land cover type, erosion level, and slope. Spatial analysis was done for the entire critical land in Indonesian, and the level has been classified into five critical levels, namely uncritical, potentially critical, moderately critical, critical, and severely critical. The results of these data analysis showed that from the total 190.5 million hectares land area of Indonesia, there are critical land areas with each category as displayed in Figure 1.

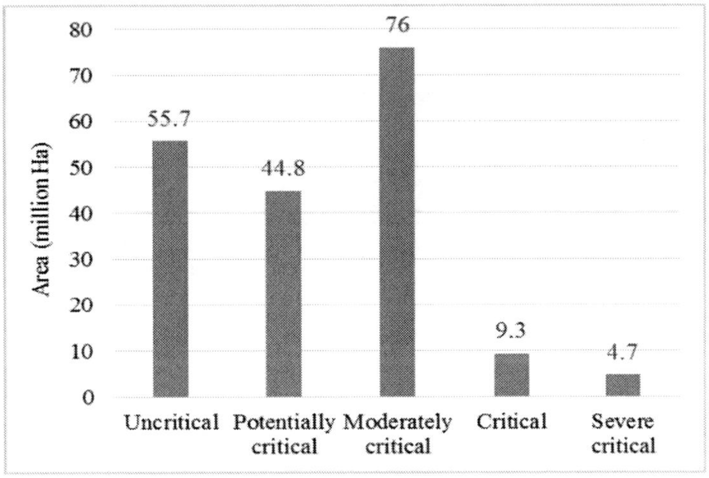


**Figure 1.** Classification area of critical land

Based on the criticality level, the land prioritized for rehabilitation is the area with a critical and severely critical level, and hence the total area is around 14 million hectares. However, not all of these areas are legally available for the activities of wood biomass production, because each part of the land has a functional status that will affect its management system. From the area of 14 million hectares, an analysis of the functional status of the area has been carried out based on forest area map issued by the Ministry of Environment and Forestry, and the results are presented in Table 1.

**Table 1.** Critical land area based on the function of forest area

| Level | Land Status Area (million Ha) | | | | Total |
|---|---|---|---|---|---|
| | Non-State Forest Area | Production Forest | Protected Forest | Conservation Forest | |
| Critical | 3.4 | 3.6 | 1.5 | 0.7 | 9.3 |
| Severe Critical | 2.1 | 1.5 | 0.8 | 0.3 | 4.7 |
| Total | 5.5 | 5.1 | 2.4 | 1.0 | 14.0 |

Table 1 shows that the critical lands that can be cultivated to produce woody biomass are in the status of the non-forest area covering 5.5 million ha and/or area with production forest function covering 5.1 million ha. The development of woody-biomass feedstock in the non-forest area in the context of critical land rehabilitation can be proposed using a community forest scheme. In the state forest area with the function of production forest, the cultivation of wood biomass can be done through an industrial plantation forest (HTI) scheme, community plantation forest (HTR), or using one of social forestry (PS) programs. Although critical land is not the ideal place for cultivation but the implementation of appropriate land management practices will be able to increase biomass production and thus land productivity. In addition to the selection of the appropriate species, optimization of critical land management is done by optimal fertilization, harvest time management, and water availability handling, especially in the early stages of tree growth [45].

### *3.2. Suitable area of the critical land*

Table 1 shows the area of degraded land in the non-state forest area and the production forest area is 10.5 million ha. Even though this area is quite large, in reality, not all of this area is suitable to be developed for the production of wood biomass. From the analysis of land suitability, several factors of soil and climate conditions are considered as limiting factors for land suitability analysis. The analysis shows that only about 3.8 million hectares of critical land are suitable for the production of woody biomass, which consists of the non-state forest area of 2.3 million Ha and in the production forest area of 1.5 million Ha with the distribution as shown in Figure 1.

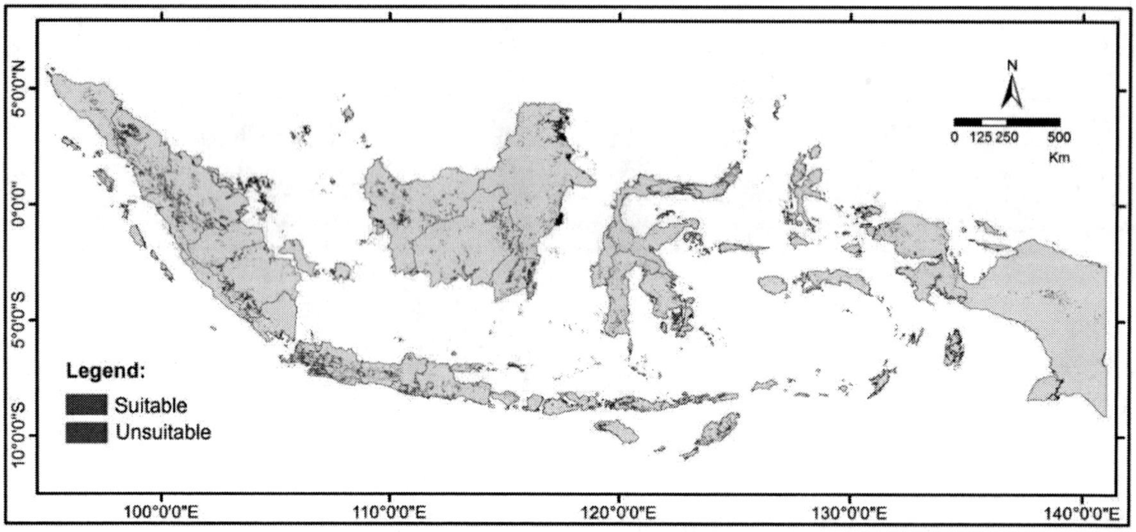


**Figure 2.** Distribution of critical land suitability for woody-biomass plantation

Figure 2 shows that the most widely distributed of degraded land has lain in the Sumatera region. The suitable land is mostly found in the Provinces of North Sumatera, Riau and South Sumatra with a total area of around 1.5 million hectares. North Sumatera has the largest suitable area, along with the degraded land in this province, which is the province with the largest critical land area, covering an area of about 1.3 million hectares. In addition to the Sumatra region, another region that has a quite large is the Java region with an area of around 1 million hectares. In this region, West Java Province has the largest area for suitable critical land in relation to biomass production. The region with the smallest suitable area is the Papua region with 169 thousand hectares, while the other regions have a suitable area at around 300 thousand hectares.

### *3.3. The potency of electricity production*

Wood biomass plantations are predicted to be the primary source of renewable energy on a global scale in the near future. Various studies have predicted that the use of woody biomass is a potential energy source with magnitude up to 115 EJ/yr in 2050 [46]. Among several wood biomass supply systems, short-rotation energy forests provide advantages over traditional wood-planting systems that require longer harvest time intervals. This study used fast-growing woody species, namely *Leucaena leucocephala* (*lamtoro*), *Calliandra calothyrsus* (*kaliandra*), *Gliricidia sepium* (*gamal*), and *Acacia auriculiformis* (*akasia*). In Indonesia, the wood of these species are commonly utilized as firewood, and this is true because they are bearing good properties in terms of wood properties (density and calorific value), sprouts quickly after pruning, dense branching, and good adaptability on marginal land [47].

The first woody-biomass harvesting was simulated and done in the third year after planting (could be in the second year for the better land characteristics). Based on annual yield, electricity power resulted from the biomass-based power plant is presented in Table 2. The last column indicates the number of households can be supplied based on recent national data of average electricity consumption.

**Table 2.** Potential woody-biomass and electricity resulted from suitable critical land

| Region | Suitable Area (Ha) [a] | Production ($m^3$/yr) [b] | Woody-Biomass (ton) [c] | Electricity (KWh) [d] | Household Supplied [e] |
|---|---|---|---|---|---|
| Sumatera | 1,554,067 | 60,608,594 | 36,365,156 | 42,292,676,983 | 28,556,838 |
| Jawa & Bali | 1,010,560 | 39,411,846 | 23,647,108 | 27,501,586,068 | 18,569,606 |
| Kalimantan | 399,382 | 15,575,902 | 9,345,541 | 10,868,864,321 | 7,338,869 |
| Sulawesi | 311,516 | 12,149,129 | 7,289,477 | 8,477,661,915 | 5,724,282 |
| Maluku & Nusa Tenggara | 338,223 | 13,190,715 | 7,914,429 | 9,204,480,693 | 6,215,044 |
| Papua | 169,303 | 6,602,823 | 3,961,694 | 4,607,449,954 | 3,111,040 |
| Total | 3,783,051 | 147,539,008 | 88,523,405 | 102,952,719,935 | 69,515,679 |

Remark: b=a*1/3*39m³/ha/yr*3yr; c=b*0.6 ton/m³; d=c*1,000kg*4,000kCal/kg*0.001163kWh; e=d/1481 kWh

Table 2 indicates that around 3.8 million hectares of suitable critical land are potentially available to support the electricity demand of 69.5 million households. The area of critical land, if fully optimized, can generate 102,953 GWh of electricity each year. This amount represents 44% of national electricity consumption for all sectors (234,618 GWh). By utilizing critical land to produce a renewable energy source, the electricity demand for all households totally in 2018 at a rate of 98,832 GWh can be fulfilled. However, the amount of electrical energy that can be produced from the critical land in Java and Bali regions is still below the region's total electricity demand, reaching as high as 61,594 GWh in 2018 [44]. For the Java and Bali regions, the use of renewable energy cannot rely solely on woody biomass due to limited land resource areas. Alternatively, the potential woody biomass will be used as an alternative fuel in coal-based power plants, knowing as co-firing or co-combustion.

The efforts to use biomass that are gradually replacing coal in power plants are starting to get a lot of attention from researchers due to several better advantages in terms of lower emission, ash content, sulfur, and classification of woody biomass as a carbon-neutral fuel [48]. The trial conducted by [49] showed that every MWh of electricity produced, there was a reduction in $CO_2$ emissions from 915 kg to 403 kg when co-firing with a ratio of 1:1 for woody biomass and coal. This research showed that substitution of woody biomass is significantly able to reduce the net $CO_2$ and $NO_x$ emission, without striking change in energy efficiency.

The Government of Indonesia is encouraging the development of biomass-based power plants through the issuance of the Ministry of Energy and Mineral Resource Regulation (Permen) Number 27 of 2014 and updated with Permen number 21 of 2016 which regulates the mechanism and tariffs for purchasing electricity from biomass power plants. By issuing this policy, the private sector is expected to be able to sell renewable electricity at more competitive prices. The latest policy of PT PLN (State Electricity Company) through its Board of Directors' Regulations has instructed to implement the co-firing of coal-fired steam power plants with biomass-based fuels. This effort is expected to accelerate the renewable energy mix without the need to build a new power plant [50]. This regulation should be supported by available wood biomass feedstock in sufficient quantities and competitive prices.

Increasing wood biomass production can be done by applying an intensive silviculture system and enhance with efforts in the form of using superior seeds and maintaining plants. In calliandra and acacia plants, the use of superior seeds can increase plant growth to 65 and 72 $m^3$/ha/year, respectively [29]. In addition, low soil fertility can be overcome by applying organic matter, effective mycorrhizae, and fertilization [45,51] In the case of Leucaena species, the use of superior cultivars (Taramba) accompanied by fertilization of NPK 250 kg/ha/yr can produce 64 $m^3$/ha/yr wood biomass [31].

## 4. Conclusion

From 14 million hectares of critical land area in Indonesia, there are 3.8 million hectares that are suitable for producing woody biomass. This area consists of 2.3 million hectares of non-state forest area and 1.5 million hectares of production forest area. The largest suitable area is located in the Sumatra region, while the smallest was in the Papua region.

The suitable area potentially produces electricity as high as 102,953 GWh and can supply 69.5 million households. The woody biomass resulted from critical land is also potential to be used as co-firing fuel in a coal-based power plant. Its application will have a significant impact on reducing $CO_2$ and $NO_x$ emissions without much-reducing energy efficiency.

## References

[1] Abhilash P C, Singh B, Srivastava P, Schaeffer A, and Singh N 2013 Remediation of lindane by *Jatropha curcas* L.: utilization of multipurpose species for rhizoremediation *Biomass Bioenergy* **5** 189-93
[2] Höök M and Tang X 2013 Depletion of fossil fuels and anthropogenic climate change : a review *Energy Policy* **52** 797-809
[3] REN21 2020 *Renewables 2020 Global Status Report* (Paris: REN21 Secretariat) p.103
[4] [MEMR] Ministry of Energy and Mineral Resources 2019 *Handbook of Energy and Economic Statistics of Indonesia 2018* (Jakarta: MEMR) p.129
[5] [MEMR] Ministry of Energy and Mineral Resources 2020 *Indonesia Energy Outlook 2019* (Jakarta: MEMR) p.75
[6] Pirard R, Bär S, Cahyat A, and Dermawan A 2017 *Prospects for wood-based electricity for the Indonesian National Energy Policy* (Bogor: Center for International Forestry Research) p. 29
[7] Edrisi S A and Abhilash P C 2016 Exploring marginal land degraded lands for biomass and bioenergy production: an Indian scenario *Renew. Sust. Energ. Rev.* **54** 1537-51
[8] Clair L and Fordham P 2009 *Biomass: an Emerging Fuel for Power Generation* (Florida: Power Engineering) pp 1-8
[9] Asfar JA, AlShwawra A, Shaban NA, Alrbai M, Qawasmeh BR, Sakhrieh A, Hamdan MA, and Odeh O 2019 Thermodynamic analysis of a biomass-fired lab-scale power plant *Energy* **194** 116843
[10] Gibbs HK and Salmon JM 2015 Mapping the world's degraded lands *Appl. Geogr.* **57** 12-21
[11] Wijitkosum S 2016 The impact of land use and spatial changes on desertification risk in degraded areas in Thailand *Sustain. Environ. Res.* **26** 84-92
[12] [WMO] World Meteorological Organization 2005 *Climate and Land Degradation* (Geneva: WMO)
[13] Lal R 2012 Climate change and soil degradation mitigation by sustainable management of soils and other natural resources *Agric. Res.* **1** 199-212
[14] Saha M and Eckelman MJ 2015 Geospatial assessment of potential bioenergy crop production on urban marginal land *Appl. Energy* **159** 540-47
[15] Xue S, Lewandowski I, Wang XY, and Yi ZL 2016 Assessment of the production potentials of Miscanthus on marginal land in China *Renew Sustain. Energy. Rev.* **54** 932-43

[16] Mehmood MA, Ibrahim M, Rashid U, Nawaz M, Ali S, Hussain A, and Gull M 2017 Biomass production for bioenergy using marginal lands *Sustain. Prod. Consum.* **9** 3-21
[17] Saha M and Eckelman MJ 2018 Geospatial assessment of regional scale bioenergy production potential on marginal and degraded land *Resour. Conserv. Recy.* **128** 90-7
[18] Shortall OK, Anker HT, Sandøe P, and Gamborg C 2019 Room at the margins for energy-crops? A qualitative analysis of stakeholder views on the use of marginal land for biomass production in Denmark *Biomass and Bioenergy* **123** 51-8
[19] Sharififar A 2012 Assessment of different methods of soil suitability classification for wheat cultivation *J Agrobiol* **29** 47-54
[20] [FAO] Food and Agriculture Organization 1976 *A framework for land evaluation* (Rome: Soil Resources Management and Conservation Service Land and Water Development Division FAO)
[21] Sys C, Van Ranst E, and Debaveye J 1991 *Land Evaluation* Part 2 methods in land evaluation (Brussels: Agricultural Publications)
[22] Biox LR and Zinck JA 2008 Land-use planning in the Chaco Plain (Burruyacu, Argentina) Part 1: Evaluating land-use options to support cop diversification in an agricultural frontier area using physical land evaluation. *Environ. Manage.* 42 1043-63
[23] Mulyasari T 2013 Characteristics of Some Wood Species as Biomass Energy Raw Materials (Bogor: Faculty of Forestry IPB)
[24] Cahyono TD, Coto Z, and Febrianto F 2008 Heat value analysis and economic feasibility of wood utilization as coal substitution in cement factory *Forum Pascasarjana* **31** 105-76
[25] Narendra BH 2019 Model of critical land rehabilitation for the sustainable raw material supply of biomass energy in East Lombok Regency (Bogor: IPB)
[26] Singh B, Dhillon GPS, Gill RIS, and Kaur J 2019 Biomass production of high density *Leucaena leucocephala* plantation under different levels of nutrients *Indian For.* **145** 34-7
[27] Hendrati RL and Hidayati N 2014 *Budidaya Kaliandra (Calliandra calothyrsus) untuk Bahan Baku Sumber Energi* (Bogor: IPB Press) p. 28
[28] Hendrati RL, Nurrohmah SH, Susilawati S, and Budi S 2014 *Budidaya Acacia auriculiformis untuk Kayu Energi* (Bogor: IPB Press) p. 33
[29] Hendrati RL 2019 *Pengembangan tanaman unggul untuk bahan baku kayu energi* (cited 2019 Juli 9) http://www.biotifor.or.id.
[30] Hudaedi D, Hariyadi and Anwar S 2018 Potensi gamal (*Gliricidia sepium*) sebagai bahan baku pembangkit listrik tenaga biomassa studi kasus: Kabupaten Manggarai Timur (NTT) *J. Env. Eng. & Waste Manage.* **3** 13-20
[31] Chotchutima S, Kangvansaichol K, Tudsri S, and Sripichitt P 2013 Effect of spacing on growth, biomass yield and quality of Leucaena (*Leucaena leucocephala* (Lam.) de Wit.) for renewable energy in Thailand *J. Sustain. Bioenergy Syst.* **3** 48-56
[32] Ty HX, Hernawan E, Liyanage M, Sila M, Ramdan H, Gintings AN, Hidayat Y, Setijoprodjo A, Roothaert R, Arias R, and Macqueen D 2001 *Produksi dan pemanfaatan kaliandra (Calliandra calothyrsus)* ed. Stewart J, Mulawarman, Roshetko JM and Powell MH (Bogor and Arkansas: International Centre for Research in Agroforestry and Winrock International)
[33] Karim AB and Savill PS 1991 Effect of spacing on growth and biomass production of *Gliricidia sepium* (Jacq.) Walp. in an alley cropping system in Sierra Leone *Agrofor. Syst.* **16** 213-22
[34] Ramos-Trejo OS, Canul-Solís JR, Alvarado-Canché AR, Castillo-Sánchez LE, Sandoval-Gío JJ, Campos-Navarrete MJ, Piñeiro-Vázquez AT, Chay-Canul AJ, and Casanova-Lugo F 2020 Growth, forage yield and quality of *Morus alba* L. and *Gliricidia sepium* (Jacq.) Walp. in mixed and pure fodder bank systems in Yucatan, México *Agrofor. Syst.* **94** 151-7
[35] [ICRAF] World Agroforestry Centre 2019 *Tree functional and ecological databases* (cited 2019 Juli 9) http://worldagroforestry. org/output?field_type_ tid=63

[36] Rengsirikul K, Kanjanakuha A, Ishii Y, Kangvansaichol K, Sripichitt P, Punsuvon V, Vaithanomsat P, Nakamanee G, and Tudsri S 2011 Potential forage and biomass production of newly introduced varieties of leucaena (*Leucaena leucocephala* (Lam.) de Wit.) in Thailand *Grassl. Sci.* **57** 94-100

[37] Gebreegziabher T, Oyedunb AO, Lukb HT, Lamb TYG, Zhangb Y, and Hui CW 2014 Design and optimization of biomass power plant *Chem. Eng. Res. Des.* **92** 1412-27

[38] Datta A, Ganguly R, and Sarkar L 2009 Energy and exergy analyses of an Externally Fired Gas Turbine (EFGT) cycle integrated with biomass gasifier for distributed power generation *Energy* **35** 341-50

[39] El-Sattar HA, Kamel S, Vera D, Jurado F 2020 Tri-generation biomass system based on externally fired gas turbine, organic rankine cycle and absorption chiller *J. Clean. Prod.* **260** 121068

[40] Gielen D, Saygin D, and Rigter J 2017 *Renewable Energy Prospects: Indonesia, a REmap analysis* (Abu Dhabi: International Renewable Energy Agency) p. 97

[41] Kacprzak A, Kobylecki R, Wlodarczyk R, and Bis Z 2016 Efficiency of non-optimized direct carbon fuel cell with molten alkaline electrolyte fueled by carbonized biomass *J. Power Sources* **321** 233-40.

[42] Elleuch A, Boussetta A, Yu JS, Halouani K, and Li YD 2013 Experimental investigation of direct carbon fuel cell fueled by almond shell biochar *Int. J. Hydrogen Energ.* **38** 16590-604

[43] Fuchs EF and Masoum MAS 2011 *Power conversion of renewable energy systems* (New York: Springer Science+Business Media)

[44] [MEMR] Ministry of Energy and Mineral Resources 2019 *Statistik Ketenagalistrikan Tahun 2018* (Jakarta: MEMR) p. 107

[45] Stolarski MJ, Krzyżaniak M, Szczukowski S, Tworkowski J, and Bieniek A 2014 Short rotation woody crops grown on marginal soil for biomass energy *J. Environ. Stud.* **23** 1727-39

[46] Hämäläinen S, Näyhä A, and Pesonen HL 2011 Forest biorefineries: A business opportunity for the Finnish forest cluster *J. Clean. Prod.* **19** 1884-91

[47] Jaung W, Wiraguna E, Okarda B, Artati Y, Goh CS, Syahru R, Leksono B, Prasetyo LB, Lee SM, and Baral H 2018 Spatial assessment of degraded land for biofuel production in Indonesia *Sustain.* **10** 1-17

[48] Ndou NR, Badaa SO, Falcona RMS, and Weiersbye IM 2020 Co-combustion of *Searsia lancea* and *Tamarix usneoides* with high ash coal *Fuel* **267** 1172-82

[49] Li J, Brzdekiewicz A, Yang W, and Blasiak W 2012 Co-firing based on biomass torrefaction in a pulverized coal boiler with aim of 100% fuel switching *Appl. Energy* **99** 344-54

[50] [PLN] Perusahaan Listrik Negara 2020 *Peraturan Direksi PT PLN nomor 001.P/DIR/2020 tentang Pedoman pelaksanaan cofiring pembangkit listrik tenaga uap berbahan bakar batubara dengan bahan bakar biomassa* (Jakarta: PT PLN)

[51] Lee KL, Ong KH, King PJH, Chubo JK and Su DSA 2015 Stand productivity, carbon content, and soil nutrients in different stand ages of Acacia mangium in Sarawak, Malaysia *Turk. J. Agric. For.* **39** 154-61

IOP Conf. Series: Materials Science and Engineering **935** (2020) 012045 doi:10.1088/1757-899X/935/1/012045

# Preparation and characterization of *Gigantochloa robusta* activated carbon to reduce COD levels of pharmaceutical waste

**N A Saputra[1,2*], I S Saputra[3], K Yuniarti[1] and Andianto[1]**

[1]Forest Product Research and Development Center, Ministry of Environment and Forestry, Bogor, 16610, Indonesia
[2] Doctoral Candidate, Departement of Forest Product Science and Technology, Faculty of Forestry, IPB University, Bogor 16680, Indonesia
[3] Study Programme Cosmetic Technology, Science-ITERA, South Lampung, 35365, Indonesia

*Email: nadisaputra@gmail.com

**Abstract.** Activated carbon prepared from *Gigantochloa robusta* was studied to reduce Chemical Oxygen Demand (COD) levels in pharmaceutical-liquid waste. Physical activator was used to produce Activated Carbon (AC) in temperature regulation of 750°C retort for 60 minutes. Dosage of AC powder were 0.25%, 0.5%, and 0.75% of 250 mL of pharmaceutical waste, respectively. Retention time variations of 30, 60, and 90 minutes were used in laboratory scale experiments. Proximate analysis including volatile substances, water content, ash content, fixed carbon and $I_2$ adsorption were observed to characterize AC. This research shows that the produced AC successfully meets the SNI 06-3730-1995 standards. The study results show variations in COD removal between 9-23%. The best result was performed by the AC dosage of 0.75% over a 60 minutes retention time.

## 1. Introduction

The presence of effluents in water causes a negative influence on aquatic ecosystems. Various methods have been offered to overcome the pollution of aquatic ecosystems, one of which is activated carbon, both in powder or granules. Even before being activated, charcoal form, which has been used to prevent environmental degradation [1]. Charcoal or activated carbon displayed as an alternative by considering its inherent character. The pore volume of activated carbon is believed to be responsible for improving adsorption properties. In addition to water treatment, activated carbon can be used in handling air pollution. While the requirements for the nature of pore volume against liquid phase adsorption applications are not as strict as for the gas-phase purification. The pore volume radius for gas adsorption is estimated from micro (< 20 Å) to macropores (> 500 Å), while liquid adsorption applications can be greater than that interval [2]. Another fact is the precursor of activated carbon from biomass or biomass waste becomes an advantage in reducing the cost of restoration of aquatic ecosystems. Activated carbon can be synthesized from green precursors such as coconut shells [3], oil palm shells [4], candlenut shells [5], corn cobs and bamboo betung [6].

Activated carbon has been noted for reducing level of COD. Effectiveness of COD removal varies based on the nature and dosage of activated carbon against effluent-liquid waste. The percentage of removal in COD levels was recorded between 47% and 68%, namely: 47.31% (200 g/L), 55.22% (300 g/L), 62.33% (400 g/L) and a value of 68.37% at of 500 g/L [7]. Another phenomenon, even with better effectiveness, was found in textile effluent treatment. The application of activated carbon from teak sawdust was declared successful in reducing COD of the liquid waste. The result demonstrated a decrease down to 84% when 1.15 grams of activated carbon adsorbent was used [8]. Adjacent study noted the consistency of activated carbon adsorption properties. It was stated that COD and BOD reduction levels were 78.96% and 33.51%, respectively, by activated carbon of coffee grounds [9]. A more recent study was carried out by including the retention time of activated carbon with effluents. Corn cob precursor was involved in the production of activated carbon. It successfully reduced the COD and BOD levels in domestic wastewater [10]. The emphasis is then directed towards pharmaceutical industry, a sector that has abundant chemical effluent. Advantage of activated carbon in the handling of pharmaceutical effluent has been pronounced through various technologies and application of activated carbon [11]. The general objective of this work is to investigate the properties of activated carbon from green precursor of bamboo mayan (*Gigantochloa robusta* Kurz.) and its effectiveness against COD removal of pharmaceutical effluent.

## 2. Materials and Methods

### *2.1. Materials*

Mayan bamboo (*Gigantochloa robusta* Kurz.) was obtained from Sukabumi, while pharmaceutical effluents were obtained from crude-pharmaceutical waste. $Na_2S_2O_3$ (Merck type, CID 24477), $HgSO_4$ (CID 24544), $H_2SO_4$ $K_2Cr_2O_7$ 0.25 N, 0.25 N ferrous ammonium sulphate solution, iodine ($I_2$), indicator (starch, ferroin) pH paper, distilled water and filter paper was also used for testing material. Apparatus used were analytical balance, oven, reactor, desiccator, furnace (VULCAN A-550), laboratory sieve (BBS), beaker, chalice, Erlenmeyer, funnel and burette.

### *2.2. Methods*

#### *2.2.1.* Activated carbon preparation

The initial stage of activated carbon production was the carbonization procedure. Each 100 mm long of *Gigantochloa robusta* Kurz stack were fed into a stainless-steel reactor. The carbonization duration was 5 hours with the applied temperature of 500°C. The activated carbons were prepared by one-step pyrolysis of *Gigantochloa robusta* Kurz, in the presence of steam. The treatment temperature was set at 750°C, and held for 60 minutes at the final temperature. After the treatment, the samples were left to cool down. The activation product was then ground with mortar and pastel, manually. The powder form was then retained by a 60 mesh filter system. The sample was ready to be tested for pharmaceutical COD removal.

#### *2.2.2.* Proximate analysis

The proximate analysis is based on the principles and procedures of SNI 06-3730-1995 concerning Technical Active Carbon which is not used as medicines raw material. The technical quality requirements for activated charcoal referred to this study consist of: volatile substances, water content, ash content, adsorption index $I_2$ and fixed carbon.

#### *2.2.3.* Effectiveness of pharmaceutical COD effluent removal

A total of 250 mL of pharmaceutical effluent was added to the Erlenmeyer. The activated carbon was added and stirred for 10 minutes, then filtered to get the filtrate. Three treatments were applied for activated carbon concentration (*% w/v*) and contact time (minutes), respectively. The active carbon concentration variations were 0.25%, 0.5%, and 0.75% while the retention time variations were 30 minutes, 60 minutes and 90 minutes.

As much as 0.4 gram of $HgSO_4$ crystal was added into 250 mL of filtrate, followed by the addition of 5 mL of concentrated $H_2SO_4$ to dissolve $HgSO_4$. The mixture was then cooled. Approximately 25 mL of $K_2Cr_2O_7$ 0.25 N solution and 30 mL $H_2SO_4$ were added subsequently. The mixture was refluxed for 2 hours. Afterward, the mixture was cooled. Further, approximately 50 mL of distilled water and 3 drops of the indicator ferroin were added into the mixture. The 0.1 N ferrous ammonium sulphate solution was for the titration process until the colour changes from green-blue to exactly red-brown.

## 3. Results and Discussion

### *3.1. the proximate analysis*

*Gigantochloa robusta* Kurz. was cut 100 mm in size, serves to increase effectiveness of the carbonization process. Previously, pieces of material were cleaned from dirt and dried in room for a week until moisture content reached a value of less than 40%. Steam activation is responsible for the formation of activated carbon pores. The endothermic reaction was involved in the stim activation process carried out at 750°C [12].

$$CO + H_2O \rightarrow H_2 + CO_2 \quad (1)$$
$$C + 2H_2 \rightarrow CH_4 \quad (2)$$
$$C + H_2O \rightarrow H_2 + CO \quad (3)$$

Granular activated carbon was then made into powder with a size of 60 mesh. The activated carbon powder was analysed by proximate-analysis before being applied to pharmaceutical effluent. Table 2 shows that activation process has successfully met the quality requirement for parameters: volatile, moisture content, ash content, adsorption capacity of $I_2$ and fixed carbon. The present study is consistent with Lam *et al.* [13] who observed dramatic increase of carbon content and activated carbon increase dramatically, as well as the significant decrease of the volatility properties.

**Table 1.** The proximate analysis of activated carbon compared to standard [14]

| No | Parameter | Unit | *Gigantoclhoa robusta* Kurz[a] | Present study | Standard |
|---|---|---|---|---|---|
| 1 | Volatile | % | | 10.30 | Max 25 |
| 2 | Moisture content | % | 9.68 | 7.40 | Max 15 |
| 3 | Ash content | % | 2.67 | 8.80 | Minimal 10 |
| 4 | $I_2$ Adsorption | Mg/g | | 796.98 | Minimal 750 |
| 5 | Fixed carbon | % | | 80.80 | Minimal 65 |

#### *3.1.1.* Volatile matter

Volatile matters were the dominant substances found in precursor, around 55-84% and decreased dramatically to 4-15% [15], 73% to 5-6% [16], and 84% to 16-22% [12] after carbonization. Volatile content indicates the amount of substance that evaporates at the treatment temperature. Volatile content, thus, can be an indicator of the success degree of treatment temperature. The higher level of activated carbon volatile substances is, the adsorption properties will be lower. High or low levels of volatile substances affect the porosity of charcoal or activated charcoal produced. The release of volatile biomass in carbonization stage enriched carbon composition and form porosity [17], even the activation process increases this phenomenon. Furthermore, activated carbon with volatile values and a low O/C ratio makes material more resistant to the oxidation processes. The atomic ratios of O/C and H/C were affected by the decarboxylation reaction, the loss of carboxyl groups and the release of $CO_2$, the H/C ratio notified the double bonding of hydrocarbon groups [12]. Multiple bonding (double/triple/etc) causes strong-stable bonds that can only be broken at high temperatures. Volatile matters generally burn at 530°C [18]. The mass left after volatile evaporation are pure carbon and ash. Volatile content in this study was 10.3% and fulfilled SNI 06-3730-1995 standard. This fact shows that organic matter evaporates during the temperature manipulation, while non-volatile ash remains in the product.

#### *3.1.2.* Moisture content

Moisture content is related to the hygroscopic properties of surface and activated carbon pores. Superior quality activated carbon was characterized by having a low moisture content, thus the activated carbon is very suitable for liquid effluent treatment [13,17], given that the surface was not clogged and an activated carbon pore has been formed. The evaporation of water begins immediately after reaching its boiling point, marked by the significant loss of mass. The evaporation of water began after the carbonization temperature reaches 120°C, followed by rapid volatile evaporation at temperatures of 400°C [16]. The result shows the moisture content of the sample of this study was 7.4%, which met the SNI 06-3730-1995 standard.

*3.1.3.* Ash content

Ash is composed of minerals contained in activated carbon precursors. Ash content is one of the parameters that determine the profile characteristics and adsorption performance. Mineral content may cover activated carbon pore thereby reducing activated carbon adsorption properties. Mineral properties trigger catalytic processes caused by inorganic content and reduce the adsorption of surface activated charcoal [17]. Temperature treatment does not burn precursor minerals and remains in the product. Increasing of ash content actually occurs after the carbonization stage is implemented. Ash content increased from 2% to 5% after carbonization [16], 10-25% to 28-51% [15] and relatively 4% to 3-4% [12]. Treatment of strong acids before activation is one of the efforts to reduce ash content. Typically, solutions of up to 10% $H_2SO_4$ were applied in the immersion of precursor (charcoal), in anticipation of acid washing by water. Ash content in this study was 8.8% and fulfilled SNI 06-3730-1995 standard.

*3.1.4.* Fixed carbon

The nature profiles of the precursors are low carbon content and high volatility. Research by Gonsalvesh *et al.* [15] showed an increasing fixed carbon after the carbonization stage, from 5-20% to 43-67%. Other study notified changes in the carbon content from 24% to 87-89% [16] and 34% to 73-77% [12]. Fixed carbon content is a component of the carbon fraction (C) contained in activated carbon material. Fixed carbon content is influenced by the value of ash content and volatile levels of activated carbon. The lower the ash content and volatile matter is, the higher the fixed carbon content us, in addition, it is also influenced by the content of cellulose and lignin. The decrease in elements (H,N,O) and volatile material caused by the degradation of cellulose, hemicellulose and lignin into organic molecules such as $CO_2$ and $CH_4$ [13]. The typical peaks of organic chemical degradation were 220, 300, and 350°C [16]. The general temperature to pure the fixed carbon was above 650°C [18]. The fixed carbon in this study was 80.8%, above the SNI 06-3730-1995 standard. The facts show that the carbonization and activation cause an increase in C content and decrease other content.

*3.1.5.* Iodine ($I_2$) number

The determination of $I_2$ index was conducted to observe the adsorption ability of activated carbon. The adsorption value of activated carbon indicates the amount of pore volume of activated carbon. The determination of iod adsorption used iodometric titration method. The activated carbon powder of *Gigantoclhoa robusta* Kurz. reacted with iodine solution. At first Na-thiosulfate reacted with iodide ions to produce sodium iodide and sodium tetrathionate as below.

$$2Na_2S_2O_3 + I_2 \rightarrow 2NaI + Na_2S_4O_6 \tag{4}$$

The reaction between Na-thiosulfate with iodide ion produced a yellow colour. After the addition of the starch indicator, the titration with Na-thiosulfate produces a clear colour. Figure 1 shows the colour changes before and after the titration. The iodine adsorption by the activated carbon in this study amounted to 796.981 mg/g.

*Gigantoclhoa robusta* Kurz. has solubility characteristics in different solvent as follow: in benzene alcohol 3.24%; 9.63% hot water; cold water 6.68%; and NaOH (1%) 23.95%; cellulose 57.55%; holocellulose 63.32%; lignin 31.66%; pentosane 18.60%; starch 9.42%; water content 9.68%; ash 2.67%, and silica 1.48% [14].

IOP Conf. Series: Materials Science and Engineering **935** (2020) 012045 doi:10.1088/1757-899X/935/1/012045

**(a)** **(b)**

**Figure 1.** (a) Colour alteration before titration and (b) after titration

### *3.2. Chemical Oxygen Demand Analysis*

3.3. The COD analysis was carried out by using 3 dosages of activated charcoal, which were 0.25%, 0.5% and 0.75% of the 250 mL effluent of pharmaceutical waste. The determination of COD index involves HgSO4 compounds to eliminate chloride disturbance in pharmaceutical wastewater and potassium bichromate as an oxidizing agent (source of oxygen) titration with ammonium sulphate. The colour alteration from orange to brighter colour indicates a decrease in COD index (Figure 2). The highest COD removal was 23.3% which was provided by 0.75% activated carbon for 60 minutes retention time (

3.4.
Figure *3*). The trend of COD removal by activated carbon and retention time in this study occurred gradually. Increasing the dosage and retention time caused the decrease in COD.

### *3.5. Effectiveness of activated carbon against COD removal*

The COD analysis is often used to determine the amount of oxygen required to oxidize organic particles and dissolved compounds in aquatic. The COD analysis is very important related to the determination of the effect of disposal of industrial effluents into aquatic ecosystem. The COD index indicates the level of oxidized organic matter in water. The higher value of COD is, the higher the amount of oxidation of inorganic chemicals into nitrate/ nitrite is at high level and potential to reduce dissolved oxygen (DO). Decreasing DO is responsible for triggering anaerobic conditions that alter the aquatic ecosystem.

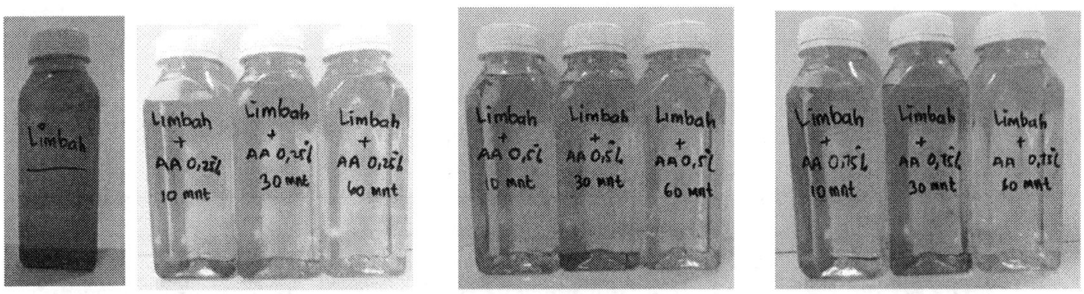


(a) (b) (c) (d)

**Figure 2.** The colour alteration of pharmaceutical effluent notifies COD decrease by activated carbon treatment in 250mL of each (a) Pharmaceutical effluent (b) activated carbon 0.25% (c) activated carbon 0.5% (d) activated carbon 0.75%

The principle of COD analysis is a strong oxidizing agent under the acidic condition to oxidize organic compounds into carbon dioxide and $H_2O$. The oxidizer used in this study was potassium dichromate ($K_2Cr_2O_7$) combined with boiling sulfuric acid ($H_2SO_4$) and mercury sulfate ($HgSO_4$). $Cr_2O_7^{2-}$ to oxidize organic compounds into carbon dioxide and water. the intervention of oxidation by chloride ions is limited by $HgSO_4$ assisted by concentrated $H_2SO_4$ solvent.

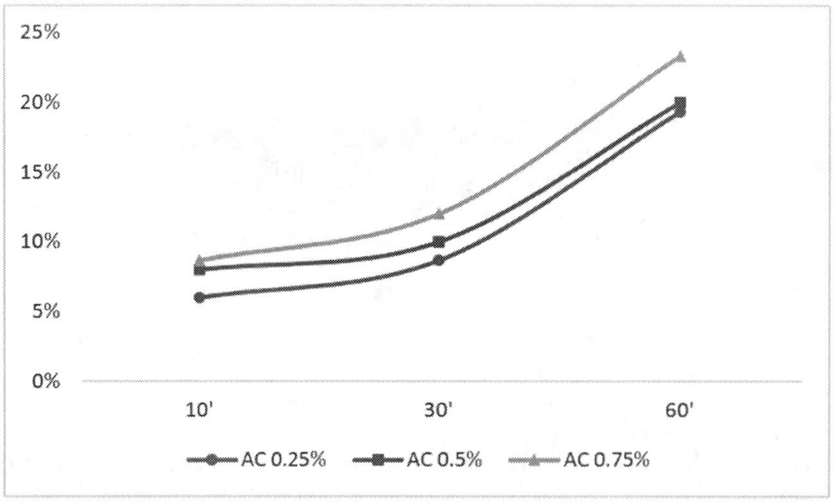


**Figure 3.** Percent of COD removal by activated carbon dosage and retention time (minutes)

3.6. This study uses activated carbon with $I_2$ value of 796.98 mg/g, and measurement of COD alteration (ppm) was observed on pharmaceutical effluent. The result was then compared to the pharmaceutical effluent water quality standard based on the Minister of Environment Regulation No. 5 of 2014. This study noted a decreasing trend in COD index by activated carbon from *Gigantoclhoa robusta* Kurz. (Table 2). Increasing in activated carbon concentration (*% w/v*) and retention time-intensity (minutes) positively correlated to COD index. An increasing amount of activated carbon and contact time will reduce the COD index [10]. The highest COD removal by 0.23% was attained in the present work (

3.7.

Figure *3*). AC 0.75 leading this achievement by 60 minutes retention, equal to COD removal as much as 280ppm compared to crude pharmaceutical effluent. Although the present work does not comply with the standard, this effort demonstrates activated carbon adsorption property to purify water waste.

The optimum retention time for a decrease in the pharmaceutical effluent COD index in this study was consistent with Karunya *et al.* [19] and Sivakumar and Muthukumar [20], which was 60 minutes. An initial conclusion from these studies that, the first optimal time to remove pharmaceutical effluent COD occurs after 60 minutes. This phenomenon was predicted related to effluents which cannot be further degraded or the presence of $OH^*$ and $H_2O_2$ oxidant [20]. Another result shows more distant retention with a better COD removal, between intervals of 180 and 210, which COD removal along 78-80% [21]. The present study was in line with previous studies, where the optimum contact time was achieved after 60 minutes. Activated carbon behaviour against waste water notifies an adsorbent function.

Effluent intermolecular adsorbed into active carbon pore reach a saturation point after 60 minutes. This process is possible through the electrostatic effect between surface particles of adsorbent and adsorbate in pore space of activated carbon: micro, meso and macro pores. The surface area is an excellent absorbent, so that adsorbate rapidly, then into pore volume inside. The adsorption performance is dependent on the carbon profile of activation. However, the time limit and effectiveness of activated carbon may be modified using multilevel technology by additives such as hydrogen peroxide-H2O2, carbon tetrachloride-$CCl_4$, potassium dichromate-$K_2Cr_2O_7$[20], continuous reactors, or membrane coating technology.

The effect of activation carbon of 0.625 g, 1.25 g and 1.87 g against 250 mL pharmaceutical liquid waste were 6-23%. A distinct dosage of 2 g/L (0.2%) successfully reduced the pharmaceutical effluent COD by 60% [20] for 120 minutes of retention. Higher concentration was applied to industrial effluents by 50 g/L, where the decline in COD index was around 56, 68 and 74% [21].

**Table 2.** the effects of activated carbon to COD removal by precursor, effluent, dosage and retention time

| Study/material | Effluent | Dosage | Retention time | COD removal (%) |
|---|---|---|---|---|
| The present study<br>*Gigantochloa robusta* Kurz | Pharmaceutical | 0.25% | 10 min, 30 min, 60 min | 6; 8.67; 19.33 |
| | | 0.5% | | 8; 10; 20 |
| | | 0.75% | | 8.67; 12; 23.33 |
| Bansode *et al.* [22]<br><br>Pecan shell | Municipal | 0.01-1.5 g/100 mL | 2 hours | above 80 |
| Delgado *et al.* [23]<br>Vegetable origin | Carbamazepine, sildenafil citrate | 100 mg/L | 72 hours | Promising agent by 3 adsorption isotherm model |
| Dalahmeh *et al.* [24]<br>Commercial biochar | carbamazepine, metoprolol, ranitidine and caffeine | Filter | Over 22 weeks | 7-99 |
| Li *et al.* [25]<br>Granular activated carbon | DEET, paracetamol, caffeine and tri- closan | Sandwich filter | 3 weeks | 65.8 |
| Karunya *et al.* [19]<br>Date seed | Pharmaceutical industry wastewater | 0.2 g | 60, 90, 120, 150 and 210 min | Above 80% |
| Sivakumar and Muthukumar [20]<br>Commercial activated carbon | Pharmaceutical Wastewater | 1, 1.5, 2, 2.5, 3 g/L | Interval 10 min | % significant of COD reduction until 60 minutes |
| Badmus and Audu [21]<br>Periwinkle shell | Industrial wastewater | 20, 30, 40, 50, 60 and 70 g | 30-300 min | % significant of COD reduction until 50 minutes |
| Saleem [26]<br>Commercial activated carbon | Pharmaceutical Wastewater | 1 g/L | filtration method | Granular activated carbon was effectively used in the final stages of filtration |

A decrease in the COD index of 23% by 1.875 gr/ 250 mL was obtained in this study. The ability of activated carbon adsorption in this study has successfully demonstrated a decrease in dissolved oxygen as a reducing factor for the COD index, in the context of the adsorption process. When a solution containing effluent materials contact with surface area, it triggers adsorption stage. Effluent organic molecules attach to activated carbon adsorbent and transform to adsorbates the surface area of product. This phenomenon alters activated carbon into a catalyst for oxidation of organic materials until it attains a saturation point. The efforts to increase the performance of COD removal by activated carbon adsorbents can be increased through the interpolation of 79% of $H_2O_2$, 83% of $CCl_4$ and 68% of $K_2Cr_2O_7$ respectively, and to COD removal by 60% [20]. Furthermore, Sivakumar & Muthukumar [20] stated that the decomposition of $H_2O_2$ into hydroxyl groups triggering the oxidation of adsorbates, while $K_2Cr_2O_7$ additive was responsible for pH improvement. The $CCl_4$ was a hydrogen atom binder and balances reaction of hydrogen atoms and hydroxyl radicals, following the formula [20]:

$$H_2O \rightarrow \bullet H + \bullet OH \quad (5)$$
$$CCl_4 + \bullet H \rightarrow HCl + \bullet CCl_3 \quad (6)$$

The activated carbon ability was tested to improve chemical handling of pharmaceutical effluent. Activated carbon was applied in the final stage of filtration and successfully reduced the COD index to 71.1%, using continuous reactors [26]. The best achievement of this study succeeded in reducing COD by 23% but it was still above the regulatory requirements regarding the wastewater quality standard. Thus, the chemical content of the pharmaceutical effluent was suppressed by the adsorption properties of activated carbon in a single product application, without the aid of other catalyst additives or in filtration system. Saleem [26] filtration technique may be applied in the future studies.

## 4. Conclusion

This study has succeeded in producing activated carbon which satisfactorily meet the requirements according to SNI 06-3730-1995. The proximate analysis (MC, ash content, volatile, pure carbon) were within the standard. On the other hand, the adsorption capacity of $I_2$ was above the required standard of 796.98 mg/g. The activated carbon profile applies as an adsorbent function.

This work demonstrated the highest COD reduction of 23% with the dosage of activated carbon was 0.75%, although was still above the minimal requirement standard for pharmaceutical wastewater. The retention optimum in this study was 60 minutes, indicating activated carbon performance in short period. An effort to increase activated carbon profile and filtration technique including a sequence study is required in the future.

## References

[1] Saputra NA and Pari G 2019 *Innovatioan of environmental friendly agricultural technology in supporting sustainable food self-suffeciency*
[2] Perrich JR 2018 *Activated Carbon Adsorption for Wastewater Treatment*
[3] Pambayun GS, Yulianto RE, Rachimoellah M, and Putri EMM 2013 Pembuatan karbon aktif dari arang tempurung kelapa dengan aktivator $ZnCl_2$ dan $Na_2CO_3$ sebagai adsorben untuk mengurangi kadar fenol dalam air limbah *J. Tek. Pomits.* **2**(1) 116-20
[4] Meisrilestari Y, Khomaini R, and Wijayanti H 2015 Pembuatan arang aktif dari cangkang kelapa sawit dengan aktivasi secara fisika, kimia dan fisika dan kimia *Konversi* **2** 45-50
[5] Agustina TE and Mardi T 2011 Pembuatan karbon aktif dari tempurung kemiri *Tek. Kim.* **17**(5) 66–74
[6] Rijali A, Malik U, and Zulkarnain Z 2015 Pembuatan dan karakterisasi karbon aktif dari bambu betung dengan aktivasi menggunakan activating agent $H_2O$ *Jom. FMIPA* **2**(1) 102-7
[7] Wicheisa, Hanani Y, and Astorina N 2018 Penurunan kadar Chemical Oxygen Demand (COD) pada limbah cair laundry orens tembalang dengan berbagai variasi dosis karbon aktif tempurung kelapa *J. Chem. Inf. Model* **6**(6)
[8] Akhsanti RY 2009 Pemanfaatan karbon aktif serbuk gergaji kayu jati untuk menurunkan Chemical Oxygen Demand (COD) Limbah Cair Industri Tekstil *Jurnal Kimia Sains dan Aplikasi* **13**(2) 66-70
[9] Irmanto I and Suyata S 2010 Optimasi penurunan nilai BOD, COD dan TSS limbah cair industri tapioka menggunakan arang aktif dari ampas kopi *Molekul* **5**(1) 22
[10] Wirosoedarmo R, Tunggul A, Haji S, and Hidayati EA 2018 pengaruh konsentrasi dan waktu kontak pada pengolahan limbah domestik menggunakan karbon aktif tongkol jagung untuk menurunkan BOD dan COD *J. Sumberdaya Alam dan Lingkungan* **3**(2) 30-7
[11] Gadipelly C, Pérez-González A, Yadav GD, Ortiz I, Ibáñez R, Rathod VK, and Marathe KV 2014 Pharmaceutical industry wastewater: review of the technologies for water treatment and reuse *Ind. Eng. Chem. Res* **53**(6) 11571-92
[12] Lam SS, Yek PN, Ok YS, Chong CC, Liew RK, Tsang DC, Park YK, Liu Z, Wong CS, and

Peng W 2020 Engineering pyrolysis biochar via single-step microwave steam activation for hazardous landfill leachate treatment *J. Hazard. Mater.* **390** 121649

[13] Lam SS, Liew RK, Wong YM, Yek PNY, Ma NL, Lee CL, and Chase HA 2017 Microwave-assisted pyrolysis with chemical activation, an innovative method to convert orange peel into activated carbon with improved properties as dye adsorbent *J. Clean. Prod.* **162** 1376-87

[14] Sutardi SR 2015 *Informasi Sifat Dasar dan Kemungkinan Penggunaan 10 Jenis Bambu* [Bogor: IPB Press Bogor]

[15] Gonsalvesh L, Yperman J, Carleer R, Mench M, Herzig R, and Vangronsveld J 2016 Valorisation of heavy metals enriched tobacco biomass through slow pyrolysis and steam activation *J. Chem. Technol. Biotechnol.* **91**(6) 1585-95

[16] Akila E, Pugalendhi S, Subramanian P, Duraisamy MR, and Surendrakumar A 2019 Production and characterization of activated carbon from date stones by single step steam activation *Environ. Ecol.* **37** 194-7

[17] Yek PN, Liew RK, Osman MS, Lee CL, Chuah JH, Park YK, and Lam SS 2019 Microwave steam activation, an innovative pyrolysis approach to convert waste palm shell into highly microporous activated carbon *J. Environ. Manage.* **236** 245-53

[18] Bardestani R and Kaliaguine S 2018 Steam activation and mild air oxidation of vacuum pyrolysis biochar *Biomass and Bioenergy* **108** 101-12

[19] Karunya S, Feroz S, Al Harassi S, and Sakhile K 2015 Treatment of oman pharmaceutical industry wastewater using low cost adsorbents *J. Multidiscip. Eng. Sci. Technol.* **2** 339-41

[20] Sivakumar R and Muthukumar K 2011 Sonochemical degradation of pharmaceutical wastewater *Clean - Soil, Air, Water* **39**(2) 136-41

[21] Badmus MA and Audu TO 2009 Periwinkle shell: Based granular activated demand (cod) in industrial wastewater *The Canadian Journal of Chemical Engineering* **8** 69-77

[22] Bansode RR, Losso JN, Marshall WE, Rao RM, and Portier RJ 2004 Pecan shell-based granular activated carbon for treatment of chemical oxygen demand (COD) in municipal wastewater Bioresource Technology **94**(2) 129-35

[23] Delgado N, Capparelli A, Navarro A, and Marino D 2019 Pharmaceutical emerging pollutants removal from water using powdered activated carbon: Study of kinetics and adsorption equilibrium *J. Environ. Manage* **236** 301-8

[24] Dalahmeh S, Ahrens L, Gros M, Wiberg K, and Pell M 2018 Potential of biochar filters for onsite sewage treatment: Adsorption and biological degradation of pharmaceuticals in laboratory filters with active, inactive and no biofilm *Science of the Total Environment* **612** 192-201

[25] Li J, Zhou Q, and Campos LC The application of GAC sandwich slow sand filtration to remove pharmaceutical and personal care products *Science of the Total Environment* **635** 1182-90

[26] Saleem M 2007 Pharmaceutical wastewater treatment: A physicochemical study *Journal of Research (Science)* **18**(2) 125-34

**Acknowledgments**
This work was technical assisted by Miss. Erni Monita, student of Caraka Nusantara Academy, and technicians in Forest Product Research and Development Center. All authors are main contributors according to job description agreement.

IOP Conf. Series: Materials Science and Engineering **935** (2020) 012046 doi:10.1088/1757-899X/935/1/012046

# Microwave assisted dilute organic acid pre-treatment of oil palm empty fruit bunch to improve enzyme accessibility

**N N Solihat*, E Hermiati, and D T N Adi**

[1]Research Center for Biomaterials, Indonesian Institute of Sciences, Cibinong, 16911, Indonesia

*Email: nissa.nurfajrin.solihat@lipi.go.id

**Abstract.** The depletion of fossil fuel is intriguing for researchers to find sustainable energy. Oil palm empty fruit bunch (OPEFB) is abundant biomass in Indonesia which is potential to be converted into bioethanol because of its high cellulose content. Suitable pre-treatment could increase the effectiveness of cellulose conversion into sugars. The aim of this study is to evaluate the effectiveness of Microwave-assisted Oxalic Acid pre-treatment (MOxA) and Enzymatic Hydrolysis (EH) to improve enzyme accessibility through variation of heating temperature, irradiation time, and EH incubation time. The higher the microwave heating temperature, longer irradiation time, and longer EH incubation time, the higher the glucose yield per pulp (GY) and reducing sugar yield (RSY) were observed. After MOxA at 200°C-15 min, as much as 43.22 % GY per initial biomass was obtained by EH for 10 days. SEM images showed more ruptures on the surface area of OPEFB fibre after MOxA that supported in increasing of enzyme digestibility. The XRD spectra demonstrated that the crystallinity index (CI) increased after MOxA because pre-treatment reduced amorphous components in the biomass. In addition, the FTIR spectra showed that peak intensity at 1714 $cm^{-1}$ corresponded to hemicellulose was gradually reduced after MOxA while peaks corresponded to lignin was more discernible.

## 1. Introduction

Oil Palm Empty Fruit Bunch (OPEFB) is a biomass waste formed after palm oil production. In 2018, Indonesia produced 47 million tons of palm oil [1] and approximately it produced 10.81 tons of OPEFB as waste [2]. OPEFB contains 4.25% of ash, 1.95% of extractives, 17.66% of acid insoluble lignin, and 67.98% of holocellulose, which consists of 43.18% of α-cellulose and 24.80% of hemicellulose [3]. The high α-cellulose content in the OPEFB makes it a potential feedstock for bioethanol production. However, a major drawback in the conversion proses is the recalcitrant of biomass because it impedes cellulose conversion into bioethanol. Therefore, the pre-treatment of OPEFB is needed as a prior step. Physical, chemical/physical-chemical, and biological treatments or a combination of them are common methods to breakdown a strong cell wall of biomass [4]. Microwave is a physical-chemical method that was reported as an effective method in degrading lignin and hemicellulose, eventually improved accessibility of cellulose to enzyme attack [5]. Besides, the microwave has short reaction time to degrade hemicellulose and lignin, hence it was reported as green technology pre-treatment [4].

Microwave irradiation has been combined with chemical solutions such as organic or inorganic acid to increase pre-treatment effectiveness. Fatriasari *et al.* [6] reported that microwave-sulphuric acid pre-treatment released 49.2% reducing sugars from jabon kraft pulp [6] while glucose 156 mg/g was obtained from wheat and rye stillages after microwave-sulphuric acid pre-treatment [7]. Sulphuric acid is known as an effective acid to overcome the lignocellulose structure. However, there are some disadvantages of using sulphuric acid in the pre-treatment, for example, it generates toxic compounds that can inhibit the fermentation process [8]. To overcome this limitation, an organic acid such as oxalic acid has been evaluated as a promising alternative to sulphuric acid pre-treatment due to its selectivity on hemicellulose hydrolysis and less corrosive behavior [9]. In simultaneous fermentation with pre-treated corncob pellets, ethanol concentration could reach 16.9 g/L when oxalic acid was used as a pre-treatment catalyst [10]. As much as 92% of glucose was converted from cellulose in bagasse after microwave-oxalic acid pre-treatment. Proper pre-treatment and enzymatic hydrolysis (EH) conditions play important roles in converting cellulose to sugars. Generally, the enzymatic reaction was conducted for 24 to 72 hours [11]. Longer incubation of enzymatic reaction resulted in higher reducing sugar yield (RSY), however, after reach optimum condition the RSY decreased due to sugar dehydration to side product [12,13]. Recently, Anita *et al.* [14] reported 34.6% of reducing sugar yield was obtained from microwave assisted oxalic acid pre-treatment (MOxA) of OPEFB which was then continued with EH for 72 hours [14]. Further research is needed to evaluate longer EH incubation time on releasing sugars. Besides observing sugar yield after pre-treatment, evaluating the change of structural properties is also important to assess the effectiveness of pre-treatment [15].

Therefore, in this study pre-treatment of OPEFB using various microwave temperatures, irradiation times, and EH incubation period were conducted to evaluate the effectiveness of MOxA pre-treatment determined by Glucose Yield (GY) and Reducing Sugar Yield (RSY) after saccharification. Alteration of ultrastructure, functional groups, and crystallinity in OPEFB fibre before and after pre-treatment was also studied.

## 2. Materials and Methods

### *2.1. Materials*

The OPEFB used in this study was the same as that in the previous report. The OPEFB was obtained from Sukabumi, West Java with chemical components (in percentage): moisture 8.44±0.04, ash 2.91±0.03, extractive 2.69±0.10, acid insoluble lignin (AIL) 23.62±0.08, cellulose 41.23±0.77, and hemicellulose 26.55±0.57 [16]. Oxalic acid from Merck (Darmstadt, Germany) was used as organic acid for pre-treatment.

### *2.2. Methods*

#### 2.2.1. Microwave-assisted oxalic acid pre-treatment (MOxA) of OPEFB

The pre-treatment procedure was following the steps in the previous report [3]. Briefly, three grams of dry weight OPEFB was added with 1% oxalic acid in the microwave Teflon tube with solid to liquid ratio of 1:10, and subsequently degassed at -20 bars for 5 minutes. The closed tube was put in the microwave oven (START D Microwave Digestion System, Milestone, Sorisole, Bergamo, Italy). The parameters of microwave pre-treatment conditions were set as follows: stirring at 195 rpm, temperatures at 160, 180, and 200°C, irradiation times for 5, 10, 15 minutes excluded preheating for 12 min and cooling for 10 minutes. After the process finished, the tube was put in an ice bath for approximately 30 minutes. Then, the solid fraction was separated from the liquid fraction through filter paper by vacuum filtration (GAST DOA-P504-BN, USA). The solid fraction or pulp was washed with distilled water until neutral condition. The pulp was weighed and was analysed for moisture content. Some pulp from each treatment was dried for its characterization. The rest of the pulp was stored in a freezer below 0°C before used for further processes or analyses. The pulp recovery (PR) was calculated using Equation 1 [14]. The experiment was carried out in duplicate.

$$PR\,(\%) = \frac{\left(\frac{100-PMC}{100} \times P\right)}{ODI} \times 100 \qquad (1)$$

where, PMC is moisture content of the pulp after pre-treatment, P is weight of pulp after pre-treatment, ODI is oven dry weight of initial biomass.

2.2.2. Enzymatic hydrolysis and sugar analysis
The MOxA residues or pulp were subjected to Enzymatic Hydrolysis (EH) by Meicellase from *Trichoderma viride* (Meiji Seika, enzyme activity 200 FPU/g). The reaction was conducted in glass bottles. Each glass bottle was filled with 1 g (oven dry weight) of each pulp, 10 mL of 0.05 M citrate buffer pH 4.8, 1 mL of 2% sodium azide, and 60 mL of 1% Meicellase in DI water. Finally, 0.1 M citrate buffer pH 4.8 was added into the glass bottle to get 100 g of total weight. The bottles were tightly closed and incubated at 50°C for 5 and 10 days in an incubator shaker (WIS-30R Wise Cube, witeg Labortechnik GmbH, Wertheim, Germany) at 150 rpm [15]. After it is completed, the samples were heated in boiling water for 5 minutes to inactivate the enzyme. After cooling down to room temperature the hydrolysate was separated from the residue by filtration.

Glucose concentration after enzymatic saccharification was analysed using the Glucose CII test kit (Wako Junyaku, Co., Osaka, Japan), and was quantified by a Paradigm P2 Spectramax microplate reader (Molecular Devices LLC, San Jose, California, USA) at a wavelength of 505 nm. Glucose yield based on the weight of pulp and based on the weight of initial biomass was calculated using Equation 2 and 3, respectively, while cellulose conversion was calculated using Equation 4 [15].

$$Glucose\ yield\ \ (\%\ pulp) = \frac{(GC \times 0.1)}{ODP} \times 100 \tag{2}$$

where, GC stands for glucose concentration in the hydrolysate (g/L), 0.1 is volume of the hydrolysate (L), and ODP is oven dry weight of pulp sample (g).

$$GY\ \ (\%\ initial\ biomass) = \frac{(GC \times 0.1)}{ODP} \times \frac{PR}{100} \tag{3}$$

where, GC is reducing sugar concentration (g/L), 0.1 is a total final volume of EH (L), ODP is oven dry weight of pulp sample (g), and PR is pulp recovery (%).

$$Cellulose\ conversion\ (\%) = \frac{(GC \times 0.1 \times 0.9)}{\left(\frac{CT}{100}\right) \times ODI} \times 100 \tag{4}$$

where, GC is glucose concentration in the hydrolysate (g/L), 0.1 is volume of the hydrolysate (L), 0.9 is conversion factor of glucose to cellulose, CT is cellulose content of the initial biomass (%), and ODP is oven dry weight of pulp sample (g).

The reducing sugars was analyzed by DNS method [17] and the absorbance was measured by Hitachi U-2001 spectrophotometer (Hitachi Corp., Tokyo, Japan) at a wavelength of 540 nm. The reducing sugars yield (RSY) was calculated based on the weight of pulp and the weight of initial biomass using the calculation method for GY as in Equation 2 and 3, by only changing the glucose concentration by reducing sugar concentration.

2.2.3. Morphological analysis of OPEFB
The dried untreated and pre-treated OPEFB were put on a 10 mm diameter specimen stub. Prior to load the stub in the scanning electron microscope (SEM), the samples were coated with 400-500 Å thick of gold using coating unit Ion Coater iB2. Then, the stub was loaded into a Hitachi SU-3500 SEM (Hitachi Corp., Tokyo, Japan). The images were captured at 1000 magnification and 20 KV accelerating voltage, while the work distance of SEM was maintained at 5 mm.

2.2.4. Functional groups analysis of OPEFB
The changes of functional groups in OPEFB before and after pre-treatment were observed by Fourier Transform Infra-Red (FTIR) Universal Attenuated Total Reflectance (UATR) Spectrometer (Spectrum

two Perkin Elmer, Shelton, USA). About 0.1 mg of dried sample was placed on a diamond plate and force gauge (FG) was pressed by hand until the obtained spectra appear (approximately FG not more than 80). The spectra were recorded at a wavelength of 4000-500 nm, resolution 16 $cm^{-1}$, and 32 scans.

2.2.5. Crystallinity analysis of OPEFB
Before crystallinity measurement, the samples were dried in a freeze dryer (Alpha 1-2 LDplus, Martin Christ Gefriertrocknungsanlagen GmbH, Germany) for about 9 hours. Then, the samples were placed on sample plate X-Ray Diffractometer (XRD) (Rigaku Smartlab Instrument, Japan) with Cu (λ=1.54 A°) as a radiant source. The analysis was conducted at a scanning angle of 10-40° with a scanning speed of 2.0°/min step size of 0.02°, the energy of 40.0 kV, and an electric current of 30.0 mA. The crystallinity index (CI) was automatically calculated by using the PANalytical X-pert Pro Diffractometer software.

## 3. Results and Discussion

### *3.1. Effect of MOxA on PR of OPEFB*
The PR of OPEFB pretreated by MOxA pre-treatment was in the range of 59-70%, where the highest PR was obtained after MOxA pre-treatment at 160°C for 5 minutes (70.57%) and the lowest was achieved from pre-treatment at 200°C for 15 minutes (59.68%). The effect of temperature and irradiation time on PR can be seen in Figure 1. The temperature has a higher effect on pulp recovery compared to irradiation time. There was approximately up to 9% of PR decrease due to 20°C increase in temperature, while it was only about 3% of PR decrease due to 5 min increase of irradiation time. It might be due to the significant effects of thermal pre-treatment on degrading large molecules in OPEFB, such as polysaccharide and lignin, into small molecules [14]. Besides, organic acid pre-treatment was reported effective to hydrolyse and solubilize hemicellulose in biomass, so that the solid fraction was reduced after pre-treatment [18]. The highest PR of OPEFB after MOxA in this study was slightly lower compared to PR of OPEFB after microwave assisted 1% maleic acid pre-treatment at 160°C for 2.5 minutes (80.32%) [19].

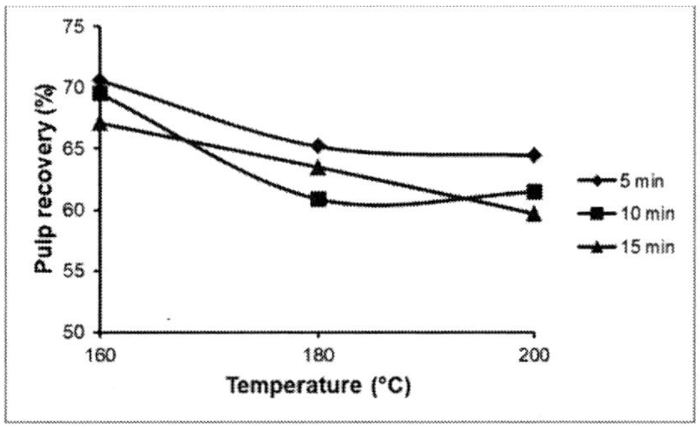


**Figure 1**. Pulp recovery of OPEFB after microwave assisted oxalic acid at 160, 180, and 200°C for 5, 10, and 15 minutes

### *3.2. Glucose yield obtained after MOxA pretreatment*
The solid fraction of OPEFB after MOxA was subjected to 5 and 10 days EH and the glucose yields (GY) based on the weight of pulp are presented in Figure 2a and GY per initial biomass in Figure 2b. It can be seen that MOxA had a positive effect on both GY per pulp and per initial biomass. For instance, only 9.99% GY was obtained from OPEFB without pre-treatment after 5 days EH. The GY was rapidly increased to 44.10% (per pulp) and 31.17% (per initial biomass) after pre-treatment at 160°C for 5 minutes. Generally, the GY per pulp and per initial biomass continuously increased as increasing in temperature. However, at heating temperature 160 and 180°C GY was decreased after 10 minutes duration of heating. The GY per pulp was increased by 10.15 % after pre-treatment at 160°C

for 10 minutes and it was slightly decreased to 53.48% after 15 minutes duration of heating. The effects of extending the duration of heating at temperature of 180ºC on GY per pulp have the same trend as those of heating at 160°C. After extending 5 minutes of heating from 5 to 10 minutes, increasing GY from 48.10 to 60.2% was observed, but extending another 5 minutes of heating was decreased the GY to 57.4%. On the contrary, GY obtained from pre-treatment at temperature 200°C was steadily elevated as the increase of heating time. A long heating time at high temperature was typically more effective in reducing biomass recalcitrant [20]. The same trend was also observed in GY per initial biomass (Figure 2b) which shows the increase of GY from 31.17 up to 43.22%. The cellular matrix of biomass that has been disrupted by MOxA pre-treatment was sent to EH for making cellulose accessible to the enzyme. The highest GY, 61.83% or 618.3 mg/g pre-treated OPEFB, was released after MOxA at 200°C for 15 minutes. This result proved that MOxA effectively increased enzyme susceptibility of OPEFB and disintegrated the lignocellulosic structure [21]. Ultimately, it facilitated the conversion of cellulose to glucose. Previous study indicated that the steam explosion and microwave irradiation (SE–MI) for 5 minutes at 180ºC were favourable to obtain 72.1 g per 100 g glucose and xylose in feedstock [22]. Another study reported that 75% of reducing sugar yield was achieved from Norway spruce after sulphuric acid-microwave irradiation at 200ºC for 5 minutes [23]. The GY obtained in this study was higher than the previous study by Mikulski and Kłosowski who achieved glucose concentration 156 mg/g dry weight of wheat after dilute sulphuric acid-microwave pre-treatment for 15 minutes [7].

The effect of long enzymatic reaction (10 days) on GY was also evaluated in this study because one of the important factors in bioethanol production from lignocellulosic biomass is selecting the optimal duration of EH. It was obviously seen in Figure 2a and 2b that extending duration of EH to 10 days increased GY as much 5-11% compared to GY after 5 days of EH. The highest GY (72.35% based on the weight of pulp and 43.22% based on the weight of initial biomass) was obtained after MOxA 200ºC for 15 minutes. EH is a heterogeneous reaction in binding enzyme with substrate to convert cellulose into glucose. In this study, GY obtained corresponded with longer hydrolysis time where the longer EH incubation time, the higher glucose yield released. The trend indicated that the enzyme was more susceptible to longer reaction times. This finding was similar to the EH of OPEFB after organosolv pre-treatment [24]. Nevertheless, the increase of GY and RSY from 5 to 10 day EH was lower than the GY and RSY obtained from 5 days EH. Therefore, prolonging the EH until 10 days did not give too much benefit in the glucose or sugars recovery from OPEFB.

The cellulose conversion in this study was in the range of 68.04 - 94.33%. The highest cellulose conversion in Figure 2c (94.33%) was obtained at 200ºC – 15 minutes after 10 days EH, which was the same conditions for obtaining the highest GY. Moreover, the tendency of cellulose conversion due to increase of pre-treatment temperature and heating time as well as the duration of EH (Figure 2c) was correlated with that of GY.

Reducing sugar yield (RSY) is another common parameter to evaluate the effectiveness of pre-treatment and EH. After 5 days EH untreated OPEFB only resulted in 16.07% RSY, while the MOxA pre-treated OPEFB resulted in 51.84-59.89% RSY per pulp (Figure 3a) or 31.83-39.60% RSY per initial biomass (Figure 3b). The RSY was steadily increased by 5-10 % after extending another 5 days of EH (Figure 3a and 3b). These increases were much lower than the RSY obtained from the first 5 days EH. The highest RSY (69.84% per pulp) was obtained from enzymatic saccharification of OPEFB pre-treated by MOxA at 200ºC for 15 minutes after 10 days EH but the highest RSY per biomass (46.13 %) was obtained from that pretreated at 160ºC for 10 minutes. However, the trend of RSY due to variations of irradiation times and heating temperatures was similar with that of GY, which showed that the increase of temperature and irradiation time tended to increase the GY as well as RSY. However, the irradiation time of longer than 10 minutes at 160 and 180ºC decreased the RSY. It was probably due to the degradation of sugars into lower molecular weight compounds, such as furan aldehydes [19].

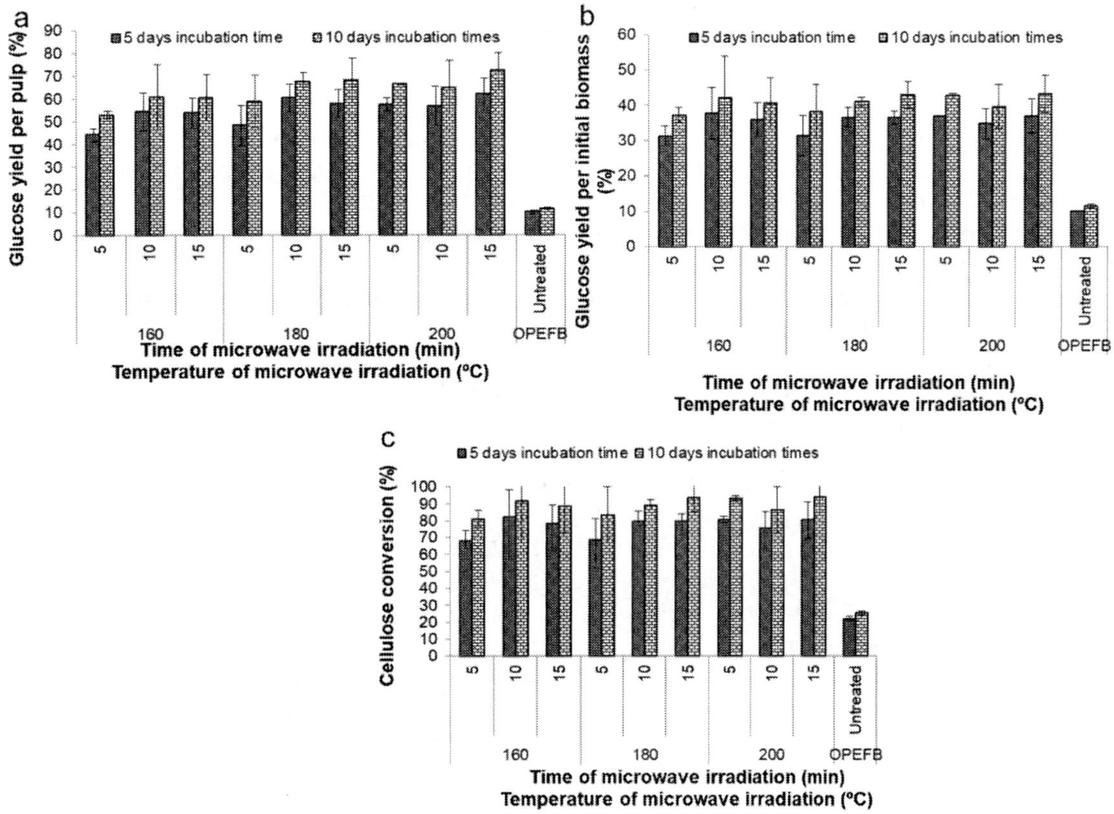


**Figure 2**. (a) Glucose yield per pulp, (b) glucose yield per initial biomass, and (c) cellulose conversion obtained from enzymatic saccharification of untreated and microwave assisted oxalic acid (MOxA) pre-treated OPEFB at different temperatures and heating times.

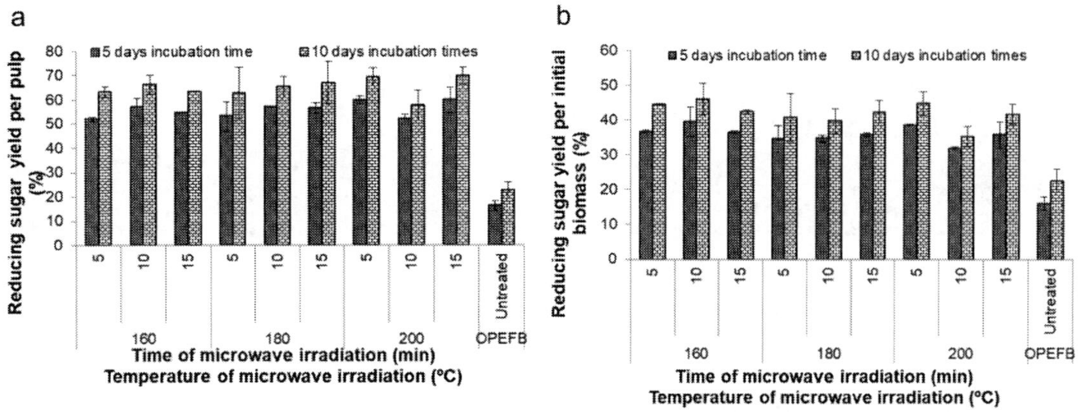


**Figure 3.** (a) Reducing sugar yield (RSY) per pulp and (b) RSY per initial biomass obtained from enzymatic saccharification of untreated and microwave assisted oxalic acid (MOxA) pretreated OPEFB at different temperatures and heating times.

### *3.3. Morphological changes on EFB fiber after MOxA pretreatment*

SEM is a common tool to investigate the effect of pre-treatment on surface, morphology, and microstructure analysis of biomass [20]. However, the samples for SEM are needed to be conductive to avoid sample damage by the electron beam. The gold coating was used in this study to make the surface conductive. Another drawback before SEM capture image is an unreasonable conclusion of SEM picture due to the high surface tension of samples, thus drying of the samples before loading them on the SEM plate is needed. Oven drying was not recommended, because it could be resulted in significant surface deformation [25]. Therefore, freeze drying was used in this study before SEM analysis.

The morphological changes after MOxA were observed in the OPEFB surface. Micrographs of the surface of untreated EFB in Figure 4a show that some silica bodies, the white dot, still embedded, but some of them were removed resulted in perforated craters on the surface. This was probably due to the effect of reducing the sizing process of raw OPEFB using a hammer mill. The removal of silica bodies and disruption of surface are more obvious in samples exposed to higher heating temperatures and longer irradiation time. The silica bodies were gradually disappeared at high temperature (Figure 4b-f) and almost completely vanished after heating at temperature and duration of higher than 180°C, 15 minutes (Figure 4g-j). Besides that, cracks of fibre were more pronounced at samples exposed to longer heating time and higher heating temperatures. When the silica bodies were removed and the circular craters were cracked, the enzyme could easily penetrate into the OPEFB fibre to breakdown the cellulose into cello-oligomers, and finally into glucose. In addition, at the highest temperature and longest duration of heating, more disrupted surface area was typically seen. The SEM images were in agreement with GY obtained (Figure 1a): the higher the temperature, the higher the GY was obtained because the enzyme was more accessible to attack the lignocellulose structures. Therefore, it can be concluded that high temperature and long heating time during MOxA pre-treatment has a significant effect on the disruption of morphological structures of OPEFB, so that it enhances enzyme susceptibility and eventually higher amount of glucose is released. This complies with previous studies that observed rougher biomass surface after microwave-acid pre-treatment [26,27].

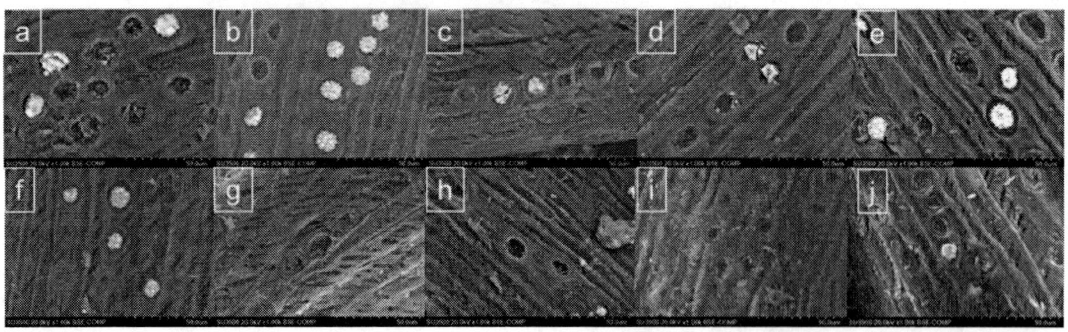


**Figure 4.** Fibre surface images of OPEFB before pre-treatment (a) and after treatment of MOxA at 160°C for 5 min (b); 160°C for 10 min (c); 160°C for 15 min (d); 180°C for 5 min (e); 180°C for 10 min (f); 180°C for 15 min (g); 200°C for 5 min (h); 200°C for 10 min (i); 200°C for 15 min (j).

### *3.4. Crystallinity Index (CI) measurement by XRD*

Crystallinity Index (CI) is a parameter to indicate the relative amount of crystalline and amorphous region of a cellulosic material [25]. CI is one of factors that can influence enzyme accessibility to the biomass during EH. Higher CI means that the biomass contained more cellulose or cellulose of crystalline structure due to the loss or solubilization of some amorphous parts of biomass, such as lignin and hemicellulose. The increase of biomass susceptibility to enzyme was due to the removal of those amorphous parts from the biomass rather than because of the increase of CI. The XRD analysis was performed for every sample. However, only representatives of the diffraction spectra of OPEFB were shown in Figure 5. There are two major peaks at 2θ of 14-18° and 20-25° which correspond to amorphous portion and crystalline portion, respectively. The major peak at 14-18° was slightly shifted in samples pretreated at higher temperature due to removal of hemicellulose and lignin, which are the major amorphous components in the biomass [28]. The shifting of peaks of crystalline region was also observed in the spectra that was believed to increase biomass digestibility [29].

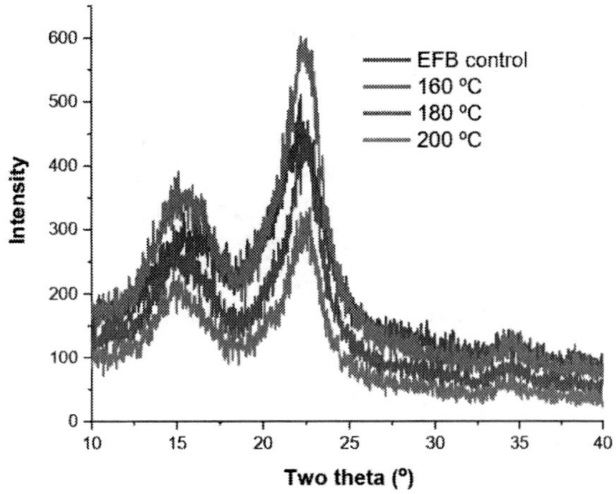


**Figure 5.** XRD pattern of OPEFB before and after treated at 160°C, 180°C, and 200°C for 10 min

Table 1 shows that CI gradually increased as increase of heating temperature and irradiation time which indicates degradation of amorphous substance. Untreated OPEFB has CI of 13.96% and there was 5% increased after microwave irradiation at 160°C-5 minutes. The highest CI (21.33 %) was obtained after pretreatment of OPEFB at 180°C-10 minutes, while CI of 20.08% was acquired from pretreatment at 200°C-15 minutes, the condition that the highest GY was achieved. In general, the result of XRD in this study was in correlation with data of PR, GY, and SEM image when higher temperature and longer duration of heating during microwave-acid pretreatment induced the deconstruction of larger molecules in the OPEFB fiber to make the fiber more vulnerable for cellulose degradation. This finding was in agreement with previous research such as dilute acid pretreatment of *Phalaris aquatica L.* [30], sulphuric acid pretreatment of sugarcane bagasse [31], and acetic acid pretreatment of *Eucalyptus* chips [18].

**Table 1.** Crystallinity index (CI) of untreated and microwave-oxalic acid pre-treated OPEFB at various temperatures and heating times

| Pre-treatment conditions | | Crystallinity Index (%) |
|---|---|---|
| Untreated | | 13.96 |
| 160°C | 5 min | 18.87 |
| | 10 min | 18.45 |
| | 15 min | 20.54 |
| 180°C | 5 min | 19.34 |
| | 10 min | 21.27 |
| | 15 min | 19.60 |
| 200°C | 5 min | 20.29 |
| | 10 min | 21.33 |
| | 15 min | 20.08 |

### *3.5. Functional groups of OPEFB*

Although the main effect of pre-treatment is removal of hemicellulose and lignin, the changes of other functional groups were also observed by FTIR. Structural changes of OPEFB before and after all MOxA conditions were observed. The representative FTIR spectra are presented in Figure 6, while the summary of FTIR bands from all samples is available in Table 2. The peak intensity corresponds to C=O ester linkage lignin-xylan and CH deformation in hemicellulose, peaks at bands 1714 and 1370 $cm^{-1}$ (no 3 and 9) were gradually decreased after MOxA. Moreover, the decrease of peak intensities was also obviously seen at bands 1646 $cm^{-1}$ (4) and 1241 $cm^{-1}$ (12) that were associated with lignin structure. It can be inferred that organic acid pre-treatment is effective in degrading lignin and hemicellulose compounds. However, some peaks associated with lignin (1600 (5), 1507 (6), 1456 (7), 1270 (11) $cm^{-1}$) were more noticeable after pre-treatment that may suggest releasing and co-precipitating of lignin on the pulp. This result was similar with pulp of *Eucalyptus* after acetic acid pretreatment [18]. More pronounced bands at 1425 (8), 1323 (10), 1105 (14), 1054 (15), and 897 (17) $cm^{-1}$ which represent stretching vibration of $–CH_2$ shear, C-H, C-O of secondary alcohol, C-C, and β-glyosidic linkage in cellulose were clearly discernible, implying an increase of the crystallinity of cellulose. The FTIR result is confirmed by XRD data that higher temperatures reduce amorphous substance in OPEFB, and eventually increases the CI.

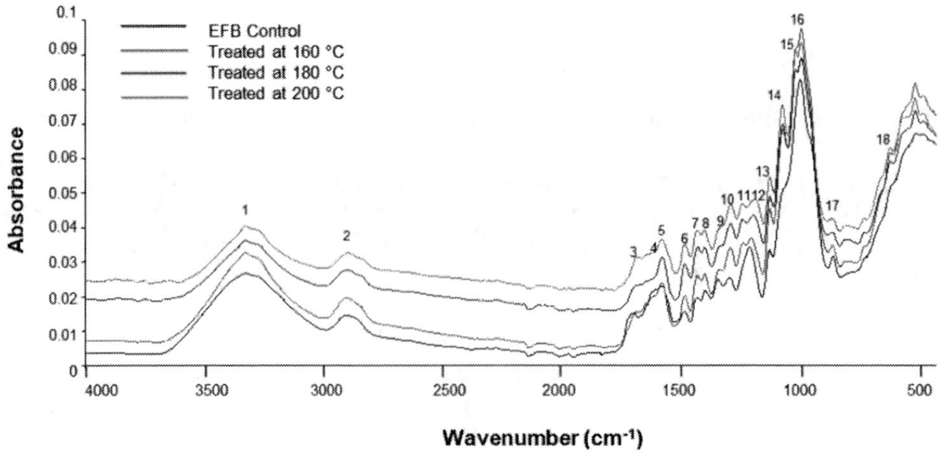


**Figure 6**. FTIR spectra of OPEFB before and after pre-treatment at 160, 180, and 200°C for 10 min

**Table 2.** FTIR bands functional group assignment of EFB before and after MOxA pre-treatment

| No | Absorption bands wavenumber ($cm^{-1}$) | | | | | | | | | | Assignment |
|---|---|---|---|---|---|---|---|---|---|---|---|
| | Untreated | 160°C | | | 180°C | | | 200°C | | | |
| | | 5 min | 10 min | 15 min | 5 min | 10 min | 15 min | 5 min | 10 min | 15 min | |
| 1 | 3335 | 3336 | 3335 | 3336 | 3334 | 3335 | 3335 | 3336 | 3335 | 3336 | O-H stretching of cellulose and hemicellulose [3] |
| 2 | 2909 | 2906 | 2909 | 2906 | 2902 | 2905 | 2908 | 2910 | 2902 | 2910 | C-H symmetric stretching of cellulose and hemicellulose [32] |
| 3 | 1714 | 1706 | 1708 | 1704 | 1704 | 1704 | 1704 | 1704 | 1704 | 1704 | C=O unconjugated stretching of xylan [33] |
| 4 | 1646 | - | - | - | - | - | - | - | - | - | -CHO, C=O, -COO stretching of lignin [34] |

| | | | | | | | | | | | |
|---|---|---|---|---|---|---|---|---|---|---|---|
| 5 | 1600 | 1597 | 1599 | 1598 | 1597 | 1598 | 1599 | 1601 | 1601 | 1600 | C=O (ketonic) stretching of lignin [35] |
| 6 | 1507 | 1505 | 1507 | 1507 | 1507 | 1507 | 1507 | 1508 | 1508 | 1509 | C=C and C=O stretching of aromatic lignin [35] |
| 7 | 1456 | 1456 | 1456 | 1456 | 1456 | 1454 | 1453 | 1455 | 1452 | 1456 | C-H deformations and aromatic ring vibrations of lignin [36] |
| 8 | 1424 | 1424 | 1425 | 1424 | 1424 | 1422 | 1422 | 1424 | 1426 | 1422 | $CH_2$ assymetric stretching of cellulose [14] |
| 9 | 1370 | 1362 | 1362 | - | - | - | - | - | - | - | CH deformation of cellulose and hemicellulose [37] |
| 10 | 1323 | 1321 | 1318 | 1319 | 1320 | 1318 | 1318 | 1318 | 1318 | 1318 | C-H vibration in cellulose and C1-O vibration in syringyl derivatives [38] |
| 11 | - | 1270 | 1270 | 1268 | 1270 | 1270 | 1270 | 1272 | 1266 | 1270 | C-O stretching of acetyl In lignin [3] |
| 12 | 1241 | 1232 | 1234 | 1230 | 1230 | 1227 | 1226 | 1221 | 1220 | 1218 | Guacyl ring and C-O strech of lignin [11] |
| 13 | 1158 | 1158 | 1158 | 1158 | 1157 | 1157 | 1157 | 1157 | 1157 | 1157 | C-O-C asymmetric stretching of cellulose and hemicellulose [39] |
| 14 | - | 1105 | 1105 | 1106 | 1105 | 1106 | 1106 | 1106 | 1107 | 1107 | C-O stretching vibration of secondary alcohol (cellulose) [26] |
| 15 | - | - | - | - | - | 1054 | 1054 | 1056 | 1058 | 1056 | C-C Stretching vibration [26] |
| 16 | 1033 | 1034 | 1029 | 1029 | 1028 | 1029 | 1029 | 1030 | 1030 | 1030 | deformation of primary alcohol in lignin [15] |
| 17 | 897 | 902 | 894 | 894 | 896 | 902 | 898 | 896 | 894 | 902 | β- glucoside linkage of cellulose [18] |

Comparison of functional groups changes before and after EH on OPEFB fibre after MOxA at 200°C-15 minutes was also observed (Figure 7). This condition was selected for comparison due to the highest GY was released at that condition, indicating most of the cellulose was converted into glucose. In Figure 7, although there was no functional group transformation after EH, distinction of peaks intensity was obviously observed and was marked by red asterisks. Depicting symmetric C-H bond and C-O-C stretching in cellulose were visibly appear at 2927 $cm^{-1}$ and 1114 $cm^{-1}$ after EH, while hemicellulose peak at 1714 $cm^{-1}$ was diminished. Peaks corresponding with lignin at 1600, 1507, 1456, and 1220, and 834 $cm^{-1}$ were sharper. This finding might be attributed to expose of linkage of lignin after degrading amorphous cellulose substance (band 1425 $cm^{-1}$ and 897 $cm^{-1}$) and hence lignin peaks were more visible after EH [40]. Previous research stated that the presence of lignin impeded enzymatic digestion of cellulose, thereby limiting conversion of cellulose to sugars [41]. Based on the results of this study, it can be concluded that hemicellulose was easier to be degraded than was lignin. However, the determination of syringyl/guaiacyl (S/G) ratio of lignin in OPEFB pulp before and after EH in the future is needed to evaluate the effectiveness of MOxA in degrading lignin.

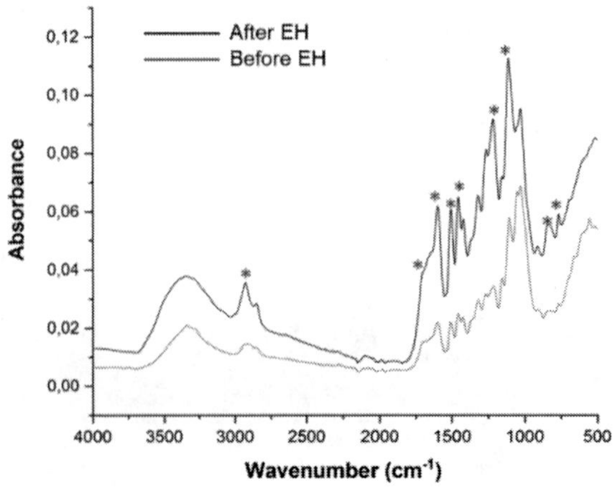


**Figure 7.** FTIR spectra of OPEFB after pretreatment at 200°C for 15 min and after enzymatic hydrolysis (EH) of the pretreated OPEFB.

## 4. Conclusions

This study revealed that microwave assisted oxalic acid (MOxA) pre-treatment can be used as an alternative method to increase enzyme accessibility in enzymatic hydrolysis (EH) of OPEFB. Glucose yield (GY) obtained after saccharification of OPEFB was increased about 4 times when the biomass was pre-treated by microwave heating at 160°C for 5 minutes, while the highest GY per initial biomass (43.22%) were obtained from that pre-treated at 200°C for 15 minutes. The higher microwave heating temperature and the longer the irradiation period, the higher the GY and CI were observed. In addition, the ultrastructure of OPEFB fibre had more ruptures at high temperatures and longer irradiation time which could better facilitate in increasing enzyme penetration into the OPEFB fibre and eventually improved hydrolysis of cellulose to glucose. MOxA pre-treatment reduced amorphous components in the OPEFB, such as hemicellulose, lignin, and amorphous cellulose, which was indicated by increasing of CI from 13.96% to 21.33% after MOxA pre-treatment at 200°C for 10 minutes. Alteration of the functional groups shown by FTIR spectroscopy also confirmed and supported the results of GY, morphological analysis, and CI determination.

## References

[1] Nikmatin S 2020 Dosen IPB University berikan solusi pengolahan limbah tandan kosong kelapa sawit (IPB University: IPB University)

[2] Kamal N 2014 Karakterisasi dan potensi pemanfaatan limbah sawit (Bandung: ITENAS)

[3] Solihat NN, Sari FP, Risanto L, Anita SH, Fitria F, Fatriasari W, and Hermiati E 2017 Disruption of oil palm empty fruit bunches by microwave-assisted oxalic acid pretreatment *J. Math, Fund. Sci.* **49 (3)** 244-257

[4] Maurya DP, Singla A, and Negi S 2015 An overview of key pretreatment processes for biological conversion of lignocellulosic biomass to bioethanol *3 Biotech.* **5** 597-609

[5] Diaz AB, Moretti MM, Bezerra-Bussoli C, Carreira Nunes CdaC, Blandino A, da Silva R, and Gomes E 2015 Evaluation of microwave-assisted pretreatment of lignocellulosic biomass immersed in alkaline glycerol for fermentable sugars production *Bioresource Technol.* **185** 316-23

[6] Fatriasari W, Fajriutami T, Laksana RPB, and Wistara NJ 2018 Microwave assisted-acid hydrolysis of jabon kraft pulp *Waste and Biomass Valori.* **10** 1503-17

[7] Mikulski D and Kłosowski G 2020 Microwave-assisted dilute acid pretreatment in bioethanol production from wheat and rye stillages *Biomass Bioenerg.* **136** 105528

[8] Tomás-Pejó E, Alvira P, Ballesteros M, and Negro MJ 2011 *Biofuels,* ed. Pandey *et al.* (Amsterdam: Academic Press) pp. 149-76
[9] Yan Y, Zhang C, Lin Q, Wang X, Cheng B, Li H, and Ren J 2018 Microwave-assisted oxalic acid pretreatment for the enhancing of enzyme hydrolysis in the production of xylose and arabinose from bagasse *Molecules* **23** 862
[10] Lee JW and Jeffries TW 2011 Efficiencies of acid catalysts in the hydrolysis of lignocellulosic biomass over a range of combined severity factors *Bioresour. Techl.* **102** 5884-90
[11] Qi B, Chen X, Shen F, Su Y, and Wan Y 2009 Optimization of enzymatic hydrolysis of wheat straw pretreated by alkaline peroxide using response surface methodology *Ind. Eng. Chem. Res.* **48** 7346-53
[12] María Teresa García-Cubero, Marcos M, Bolado S, and Coc M 2010 Optimization of operating conditions in enzymatic hydrolysis of pretreated lignocellulosic materials *Chem. Eng. Trans.* **21** 1285-90
[13] Daorattanachai P, Khemthong P, Viriya-empikul N, Laosiripojana N, and Faungnawakij K 2011 The effect of catalyst types and starting materials on furan production in hot compressed water *Energy Procedia* **9** 515-21
[14] Anita SH, Fitria, Solihat NN, Sari FP, Risanto L, Fatriasari W, and Hermiati E 2020 Optimization of microwave-assisted oxalic acid pretreatment of oil palm empty fruit bunch for production of fermentable sugars *Waste Biomass Valori.* **11** 2673-87
[15] Hermiati E, Solihat NN, Adi DTN, and Ghozali M 2018 The ultrastructure and enzyme susceptibility of oil palm empty fruit bunch pretreated by oxalic acid under microwave irradiation *5th International Symposium on Innovative Bio-Production Indonesia* pp. 92-9
[16] Risanto L, Fitria, Fajriutami T, and Hermiati E 2018 Enzymatic saccharification of liquid hot water and dilute sulfuric acid pretreated oil palm empty fruit bunch and sugarcane bagasse *Earth Envi. Sci.: IOP Conf. Series* **141** 1-6
[17] Adney B and Baker J 2008 Measurement of cellulase activities laboratory analytical procedure (LAP) *Technical Report NREL/TP-510-42628,* (Colorado National Renewable Energy Laboratory)
[18] Chen Y, Yan Z, Liang L, Ran M, Wu T, Wang B, Zou X, Zhao M, Fang G, and Shen K 2020 Comparative evaluation of organic acid pretreatment of eucalyptus for kraft dissolving pulp roduction *Materials (Basel)* **13** 361
[19] Fatriasari W, Raniya R, Oktaviani M, and Hermiati E 2018 The Improvement of sugar and bioethanol production of oil palm empty fruit bunches (*Elaeis guineensis* Jacq.) through microwave-assisted maleic acid pretreatment *bioResour.* **13** 4378-403
[20] Zhu Z, Macquarrie DJ, Simister R, Gomez LD, and McQueen-Mason SJ 2015 Microwave assisted chemical pretreatment of Miscanthus under different temperature regimes *Sustainable Chemical Processes* **3 (15)** 1-13
[21] Kashaninejad M and Tabil LG 2011 Effect of microwave–chemical pre-treatment on compression characteristics of biomass grinds *Biosystems Engineering* **108** 36-45
[22] Pang F, Xue S, Yu S, Zhang C, Li B, and Kang Y 2012 Effects of microwave power and microwave irradiation time on pretreatment efficiency and characteristics of corn stover using combination of steam explosion and microwave irradiation (SE-MI) pretreatment *Bioresource Technol.* **118** 111-9
[23] Kuittinen S, Rodriguez YP, Yang M, Keinänen M, Pastinen O, Siika-aho M, Turunen O, and Pappinen A 2015 Effect of microwave-assisted pretreatment conditions on hemicellulose conversion and enzymatic hydrolysis of norway spruce *BioEnergy Res.* **9** 344-54
[24] Pasma SA, Daik R, Maskat MY, and Hassan O 2013 Application of box-behnken design in optimization of glucose production from oil palm empty fruit bunch Cellulose *Int. J. Polym. Sci.* **2013** 1-8
[25] Karimi K and Taherzadeh MJ 2016 A critical review of analytical methods in pretreatment of lignocelluloses: Composition, imaging, and crystallinity *Bioresour. Technol.* **200** 1008-18

[26] Zhu Z, Simister R, Bird SJ, McQueen-Mason SD, Gomez L, and Macquarrie JD 2015 Microwave assisted acid and alkali pretreatment of Miscanthus biomass for biorefineries *AIMS Bioeng.* **2** 449-68

[27] Sorn V, Chang KL, Phitsuwan P, Ratanakhanokchai K, and Dong CD 2019 Effect of microwave-assisted ionic liquid/acidic ionic liquid pretreatment on the morphology, structure, and enhanced delignification of rice straw *Bioresour. Technol.* **293** 121929

[28] Cao S 2012 Effect of surfactants on the pretreatment of sugarcane bagasse with dilute ammonia. In: *The Department of Food Science,* (Louisiana: Louisiana State University) p. 92

[29] de Assis T, Huang S, Driemeier CE, Donohoe BS, Kim C, Kim SH, Gonzalez R, Jameel H, and Park S 2018 Toward an understanding of the increase in enzymatic hydrolysis by mechanical refining *Biotechnol. Biofuels* **11** 289

[30] Karapatsia A, Pappas I, Penloglou G, Kotrotsiou O, and Kiparissides C 2016 Optimization of dilute acid pretreatment and enzymatic hydrolysis of *phalaris aquatica* L. lignocellulosic biomass in batch and fed-batch rocesses *BioEnergy Res.* **10** 225-36

[31] Flórez Pardo LM, Salcedo Mendoza JG, and López Galán J E 2019 Influence of pretreatments on crystallinity and enzymatic hydrolysis in sugar cane residues *Braz. J. Chem. Eng.* **36** 131-41

[32] Shrestha B, le Brech Y, Ghislain T, Leclerc S, Carré V, Aubriet F, Hoppe S, Marchal P, Pontvianne S, Brosse N, and Dufour A 2017 A Multitechnique characterization of lignin softening and pyrolysis *ACS Sustain. Chem. Eng.* **5** 6940-9

[33] Gan L, Zhou M, Yang D, and Qiu X 2013 Preparation and evaluation of carboxymethylated lignin as dispersant for aqueous graphite suspension using turbiscan lab analyzer *J. Disper. Sci. and Tech.* **34** 644-50

[34] Kucharska K, Słupek E, Cieśliński H, and Kamiński M 2019 Advantageous conditions of saccharification of lignocellulosic biomass for biofuels generation via fermentation processes *Chem. Pap.* **74** 1199-209

[35] Uraki Y, Ishikawa N, Nishida M, and Sano Y 2001 Preparation of amphipilic lignin derivative as cellulose stabilizer *J. Wood Sci.* **47** 301–7

[36] Wang K, Xu F, and Sun R 2010 Molecular characteristics of Kraft-AQ pulping lignin fractionated by sequential organic solvent extraction *Int. J. Mol. Sci.* **11** 2988-3001

[37] Wang R, Yue J, Jiang J, Li J, Zhao J, Xia H, Wang K, and Xu J 2020 Hydrothermal CO2-assisted Pretreatment of Wheat Straw for Hemicellulose Degradation Followed with Enzymatic Hydrolysis for Glucose Production *Waste and Biomass Valori*

[38] Lima MA, Lavorente GB, da Silva HK, Bragatto J, Rezende CA, Bernardinelli OD, de Azevedo ER, Gomez LD, McQueen-Mason SJ, Labate CA, and Polikarpov I 2013 Effects of pretreatment on morphology, chemical composition and enzymatic digestibility of eucalyptus bark: a potentially valuable source of fermentable sugars for biofuel production: part 1 *Biotechnol. Biofuels* **6** 1-17

[39] Ben Taher I, Fickers P, Chniti S, and Hassouna M 2017 Optimization of enzymatic hydrolysis and fermentation conditions for improved bioethanol production from potato peel residues *Biotechnol. Prog.* **33** 397-406

[40] Wang C, Yang J, Wen J, Bian J, Li M, Peng F, and Sun R 2019 Structure and distribution changes of Eucalyptus hemicelluloses during hydrothermal and alkaline pretreatments *Int. J. Biol. Macromol.* **133** 514-21

[41] Li X, Li M, Pu Y, Ragauskas A J, Klett A S, Thies M, and Zheng Y 2018 Inhibitory effects of lignin on enzymatic hydrolysis: The role of lignin chemistry and molecular weight *Renew. Energ.* **123** 664-74

**Acknowledgment**
This research was supported by JST-JICA Satreps Project Innovative Bio-production in Indonesia (iBioI): Integrated Bio-refinery Strategy to Promote Biomass Utilization using Super Microbes for

IOP Conf. Series: Materials Science and Engineering **935** (2020) 012046 doi:10.1088/1757-899X/935/1/012046

Fuels and Chemicals Production (2013-2018). The authors would like to thank Research Center for Biotechnology, Indonesian Institute of Sciences, for providing a microplate reader for glucose analysis, Research Center for Physics, Indonesian Institute of Sciences, for providing XRD and SEM facilities, and Dr. Ratih Damayanti for assisting in crystallinity index calculation.

IOP Conf. Series: Materials Science and Engineering **935** (2020) 012047 doi:10.1088/1757-899X/935/1/012047

# Biopellet properties of agathis wood fortified with its peeled-off bark

**N J Wistara[1*], S Bahri[1], GPari[2]**

[1]Department of Forest Products, Faculty of Forestry, IPB University, Bogor, 16680, Indonesia
[2]Forest Research and Development Center, Ministry of Environment and Forestry, Bogor 16610, Indonesia

*Email: nwistara@gmail.com

**Abstract.** The agathis bark is expected to increase the energy content and durability of the resulting biopellet. The present research was aimed to analyze chemical components of agathis and to determine the quality of agathis wood biopellet fortified with it loosen bark. The klason lignin, extractives, holocellulose and α-cellulose content of wood and bark of the agathis were determined following the standard procedures of TAPPI T 222 0m-88, TAPPI T 204 0m-88, and Browning (1967), respectively. Biopellets of 15 mm diameter were prepared with pelletizing pressure of 1500 psi. The moisture content, ash content, density, durability, and calorific value of the resulting biopellet were compared to those required by DIN EN 14961-2 and 51731 standards. The α-cellulose and holocellulose content of the agathis bark were found lower than these of its wood. However, its klason lignin, extractive, and ash content were higher than these of its wood. The resulting biopellet retained the density of 0.61-0.71 g/cm$^3$, moisture content of 0.42-4.52%, ash content of 0.79-2.73%, durability of 21.26-44.59%, and calorific value of 4,524-4,628 kcal/kg. Except for that of the biopellet density and durability, all parameters satisfied the requirements of DIN EN standards. Thermal analysis of biopellet showed that water loss occurred at 36-100°C and significant mass decomposition between 100 up to 475°C with the weight loss of 77.34-80.51%. No further decomposition was found at above 480°C.

## 1. Introduction

The reserve of fossil-based energy such as petroleum, natural gas, and coal were predicted will last only for 40, 60, and 200 years to come, respectively [1]. Therefore, the search for fossil fuel alternatives is paramount. Biomass could be the most appropriate source for a renewable and environmentally friendly energy of the future. Biomass based energy is considered free of green house gases and $CO_2$ emission [2,3]. Its renewable nature will ascertain the sustainability of energy reserve.

In the 2012, consumption of biomass-based fuels in Indonesia was about 56.12 million tones, which was higher than that of coal with about 28.97 million tones [4]. The source of biomass energy can be from wood, kenaf, bamboo, and the waste of sawmill, forest industries, agriculture, paper products, sugar cane, and other lignocellulosic materials [5].

Biomass can be converted into gas, liquid and solid bio-fuels. The most common form of solid bio-fuel are briquette, biopellet, and cubes. Biopellet, which is prepared by pelletizing wood meals, is gaining popularity mainly in the European countries [6]. Biopellet demand in Sweden for instance, increased to 240% from 1997 to 2006 with a 45% increase in its price [7].

Wood is the most abundant terrestrial biomass. Agathis wood, for example, is a versatile biomass widely cultivated in Indonesia. It is used for raw material of woodworking, fuel-wood and its resin is tapped for various derivative non-wood forest products. Production of agathis wood in 2010-2011 just by Perum Perhutani, a state own enterprise, was of 305.728 $m^3$ [8]. In general, trees consist of about 10-20% bark [9], thus upon harvesting, Perum Perhutani wastes approximately of 30.573-61.146 $m^3$ wood bark. Furthermore, agathis wood produces a naturally peeled off bark that thicken when it's growing to maturity. Bark is rich in lignin and extractives that relatively higher in energy content compared to that of polysaccharide. Lignin and extractives content of agathis wood were reported to the amount of 24.7 % and 2.0%, respectively, with calorific value of 4,762 kcal/kg [10].

It is a common knowledge that the energy density of wood is relatively low compared to coal or charcoal. Considering that the relatively high content of lignin and extractives in the bark, agathis bark could be also used to fortify its wood in the production of biopellet. The use of bark could also result in a durable biopellet. Bark fortified Scott pine biopellet was reported to reach a durability of 95-97% [11].

Even though the quality requirement of biopellet varied according to different standards, however, they all are dependent on its moisture content, ash content, durability, and calorific value. Biopellet durability is important during its transportation. Densification pre-heating, moisture content, raw material, particle size, peletizing pressure, peletizing temperature, peletizing time, and biopellet diameter [12,13] influence the durability of biopellet. Particle size determines biopellet density, combustion period, and durability [13,14]. Smaller particle tends to increase density and combustion time [15]. Biopellet produced from particle size of less than 1.4 mm resulted in a higher durability compared to that produced from particle size of 1.4-3.2 mm [13]. In the present works, smaller particle size of agathis bark was used with an expectation of increasing biopellet density and durability. Despite the fact that its high lignin content is beneficial for pellet durability and calorific value, the use of bark in biopellet production tends to increase its ash content [11]. The high ash content of biopellet will bring about the formation of crust in combustion chamber (furnace). Currently, in order to cultivate the beneficial effect (calorific value and durability) of using bark, the percentage of bark used was varied to the maximum of 30%.

The present research was intended to determine the quality of agathis wood biopellet fortified with its peeled-off barks densified at various temperature. Thermal behavior of agathis wood peeled-off bark and its biopellet was also analized.

## 2. Materials and Methods

### *2.1. Materials*

A 50 years old agathis wood was chipped and air-dried. The moisture content of the air-dried chips was then determined. The moistre content of biopellet raw material should be approximately at 8-12%. The chips were milled and sieved to obtain wood meal of 20-40 mesh (0.84-0.42 mm) for biopellet production and 40-60 mesh (0.42-0.25 mm) for chemical and proximate analysis. Figure 1 indicates general outline of the present research.

Biopellet was prepared by thouroughly mixing the wood meal (W) and bark meal (B) based on their oven dry weight. The compositions (% B/W) of biopellet were varied at 0/100, 10/90, 20/80, and 30/70 referring to Filbakk *et al.* [11] which used pinewood.

IOP Conf. Series: Materials Science and Engineering **935** (2020) 012047 doi:10.1088/1757-899X/935/1/012047

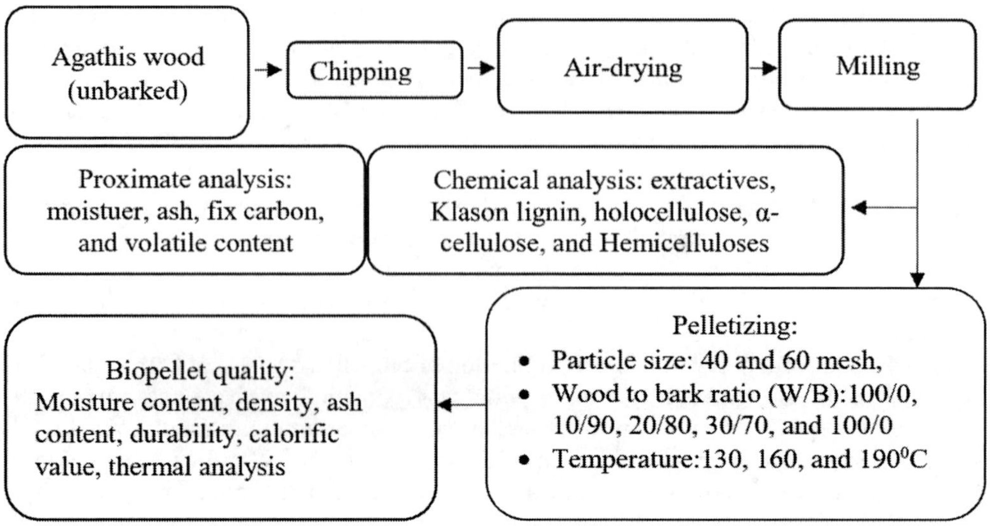


**Figure 1**. Generela outline of the present research

*2.2. Methods*

2.2.1. Chemicals and proximate analysis

The chemical contents of wood and bark were determined in accordance with TAPPI [16] T 204 om-88 (EtOH-Benzene extractives), TAPPI T 222 om-88 (Klason lignin), and Browning (1967) for α-cellulose and hemicellulose content. Proximate analysis was done following the ASTM E-871 [17] for the determination of moisture content, and ASTM D-1102 [18] for the determination of ash content, volatile content, and fixed carbon.

2.2.2. The preparation of biopellet

Biopellets with a diameter of 15 mm were prepared in a pelletizer using pelletizing pressure of 10.34 MPa. Pelletizing temperature varied at 130, 160, and 190°C following those applied previously by Lee *et al.* [13] Pelletizing process was carried out for 6 min consisting of 3 min pre-heating and 3 min of heating at the predetermined pelleting temperatures. Upon pelletizing, the biopellets were conditioned for one day before the determination of the biopellets moisture content, density, ash content, durability, and the calorific value. TG-DTA of the raw materials (wood and bark) and selected biopellet (biopellet with bark content of 30% was also conducted to evaluate their thermal behavior.

2.2.3. The measurement of biopellet quality

*2.2.3.1. Moisture content*

Moisture content was calculated as a ratio between moisture weight in a sample and oven dry weight of the biopellet. Oven dry weight of biopellet was procured by oven drying at 103±2°C to a constant weight. DIN EN 14961-2 [19] requires the biopellets moisture content of 10%. The moisture content was calculated by the following formulae:

$$\text{Moisture Content (\%)} = \frac{\text{Green weight (g)} - \text{Oven dried weight (g)}}{\text{Oven dried weight (g)}} \times 100 \quad (1)$$

*2.2.3.2. Density*

The density of biopellet was determined in accordance with the standard of DIN EN 51731 [20], which requires the biopellets density of 1.00-1.40 g/cm$^3$. The density of biopellet was calculated by the following formulae:

$$\text{Density (g/cm}^3\text{)} = \frac{\text{The weight of biopellet (g)}}{\text{The volume of biopellet (cm}^3\text{)}} \tag{2}$$

*2.2.3.3. Ash content*

Ash content (in percent) is the residue of sample combustion at 600°C for 4 hours expressed as percentage. DIN EN 14961-2 [19] requires the ash content of less than or equal to 3%. The ash content was calculated by the following formulae:

$$\text{Ash content (\%)} = \frac{\text{The weight of ash (g)}}{\text{Oven dry weight of sample (g)}} \text{ x } 100 \tag{3}$$

*2.2.3.4. Durability*

A durability tester was used to determine the durability of the biopellet. In which, 100 g of biopellet were shaken at 50 rpm for 10 min. Upon shaking, the biopellets were sieved in a 20 mesh screen. The biopellets remained on the 20 mesh screen were then weighed. DIN EN 14961-2 [19] requires the minimum durability of 96.5%. The durability of the biopellet was calculated following the formulae of:

$$\text{Durability (\%)} = \frac{\text{The weight of biopellet retained on 20 mesh sieve (g)}}{\text{100 (g)}} \text{ x } 100 \tag{4}$$

*2.2.3.5. The calorific value*

Bomb Calorimeter Parr 6400 was used to determine the calorific value of the biopellet. The analysis was involving the use of oxygen (450 psi), nitrogen (80 psi) and benzoic acid. The calorific value of the biopellets was calculated following the formulae of:

$$\text{Calorific value (kcal/kg)} = \frac{W.t - e1 - e2 - e3}{m} \tag{5}$$

where:
- W = equivalent energy value of benzoic acid
- t = temperature increase differential (°C)
- $e_1$ = chrom nickel wire correction
- $e_2$ = sulphur correction
- $e_3$ = nitric acid correction
- m = sampel mass (gram)

2.2.4. Thermal analysis

Thermographimetric analysis (TGA) and differential thermal analysis (DTA) using Thermal Gravimetric/ Differential Thermal Analyzer (TG/DTA) SII EXTAR 7300 (Hitachi High-Tech Science Corporation, Tokyo) were carried out to analyze the thermal behavior of the biopellet and its raw materials. Approximately 3-5 mg of sample was put in the sample container made of Alumina/Pt, and the analysis was done with the air flow rate of 50 $cm^3$/min and the heating rate of 10°C/min until the temperature of 1200°C.

*2.3. Data analysis*

Data were analized using a factorial completely randomized design with 3 factors, i.e. particle size (A) with 2 levels (20-40 mesh and 40-60 mesh), percentage of peeled off bark (B) with 5 levels (0, 10, 20, 30, and 100% of bark content), and pelletizing temperature (C) with 3 levels (130, 160, and 190°C). All experimentation was done in three replicates. The general odel of the design used to analyze the quality of biopellet such as density, moisture content, ash content, and durability was:

$$\mathbf{Y_{ijk} = \mu + A_i + B_j + C_k + (AB)_{ij} + (AC)_{ik} + (BC)_{jk} + (ABC)_{ijk} + \varepsilon_{ijk}} \tag{6}$$

where:

| | |
|---|---|
| Yijk | : the score of the ith factor A (particle size), the jth factor B (percentage of peeled-off bark), and the kth factor C (peletizing temperature) |
| $\mu$ | : the overall population mean |
| $A_i$ | : effect of ith level of A |
| $B_j$ | : effect of jth level of B |
| $C_k$ | : effect of kth level of C |
| $(AB)_{ij}$ | : joint effect of the ith level of A and the jth level of B |
| $(AC)_{ik}$ | : joint effect of the ith level of A and the kth level of C |
| $(BC)_{jk}$ | : jont effect of the jth level of B and the kth level of C |
| $\varepsilon ijk$ | : the error effect associated with $Y_{ijk}$ |

Calorific value was analyzed with completely randomized factorial design with one factor A (the percentage of peeled-off bark in biopellet) at four levels (0, 10, 20, 30, and 100 %). The general model used was:

$$\mathbf{Y_i = \mu + B_i + \varepsilon_i} \quad (7)$$

where:

| | |
|---|---|
| Yi | : the score of the ith level factor A (the percentage of peeled-off bark) |
| $\mu$ | : mean value |
| $B_i$ | : effect of the ith level factor A |
| $\varepsilon i$ | : the error associated with Yi |

P-value was used as the test criteria. P-value $\leq$ 0.05 indicates that the effect of treatment is significant at the level of confidence of 95% and P-value > 0.05 indicates that the effect of treatments is insignificant at the level of confidence of 95%. Duncan's multiple range test was carried out to evaluate the differences among treatments.

## 3. Results and Discussion

### *3.1. Chemical components of agathis wood*

The content and types of the chemical components of wood influence its energy characteristics. Table 1 indicates the chemical compositions and proximate analysis of agathis wood and its bark. It can be seen (from Table 1) that Klason lignin, extractives, and fixed carbon of the bark are higher than those of wood are. The Klason lignin and extractives content of the peeled-off bark of agathis is higher than the lignin and extractives content found in the bark of *Pinus sylvestris* and *Picea abies* by Miranda *et al.* [21] that was 26.8-32.9 % and 5.0-5.3%, respectively.

Carbon content of biomass determines its calorific value, in which increasing carbon content tends to increase the calorific value. Lignin is a phenolic compound consist of phenylpropane units with a relatively high content of carbon. Extractives such as wax, protein, sugar, pectin, resin, terpene, starches, and saponin are also carbaneous compound [22]. Ma *et al.* [23] reported that the carbon content of lignin, cellulose, and non-structural cellulose was 63-66%, 44%, and 44%, respectively. Therefore, calorific value of lignin and extractives is commonly higher than that of cellulose and hemicellulose. The calorific value of cellulose, lignin and extractives is of 4150-4350 kkal/kg, 6100 kcal/kg dan 8500 kcal/kg, respectively [24]. Based on the content of its lignin, extractives, and fixed carbon, the peeled-off bark of agathis should be a good fortifier for increasing the calorific value of its wood biopellet. However, the benefit of higher lignin content of the bark to the calorific value of biopellet could be counterweighed by its higher ash and moisture content. Ash and moisture content adversely influence the calorific value of biomass.

**Table 1**. Chemical components and proximate analysis of the peeled-off bark and wood of agathis

| No. | Components* | Unit | Bark (B) | Wood (W) |
|---|---|---|---|---|
| Chemical components | | | | |
| 1 | Extractives | % | 12.47 | 6.65 |
| 2 | Klason Lignin | % | 42.64 | 30.94 |
| 3 | Holocellulose | % | 54.53 | 71.06 |
| 4 | α-cellulose | % | 32.54 | 49.80 |
| 5 | Hemicellulose | % | 21.99 | 21.26 |
| Proximate analysis | | | | |
| 6 | Ash content | % | 2.63 | 1.08 |
| 7 | Moisture content | % | 13.29 | 11.73 |
| 8 | Volatile Content | % | 68.13 | 78.79 |
| 9 | Fixed Carbon | % | 30.79 | 18.58 |

note: (*) the values are the average of three replicates

### 3.2. *The quality of biopellet*

#### 3.2.1. Moisture content

Moisture content is an important determining quality of biopellet due to its direct relation with calorific value. Table 2 indicates that biopellet with the highest moisture content was found on the biopellet produced from bark with 40 mesh particle size and peletized at 130°C, while the lowest was from biopellet prepared from wood with particle size of 60 mesh peletized at 190°C. All biopellet fulfiled the DIN EN 14961-2 [19] standard that require the moisture content of equal or less than 10%.

Analysis of variance (Table 3) indicates that particle size, bark content, pelletizing temperature and all interaction of factors significantly influenced the biopellet moisture content. Further evaluation using mean range test of Duncan indicates that the moisture content of biopellet produced from raw material with particle size of 60 mesh, 20% bark content and pelletized at 130°C was significantly lower than that produce with the same particle size and bark content, but pelletized at 160°C and 190°C. The moisture content of biopellet with 160°C and 190°C pelletizing temperature was insignificantly different. Therefore, biopellet produced from particle size of 60 mesh, 20% bark content and pelletized at 130°C is the most acceptable in term of moisture content, despite all pelletizing temperature applied fulfilled the DIN EN 14961-2 [19] standard requirement. It seemed that producing biopellet with smaller particle size and pelletizing at higher temperature tended to lower the moisture content. The moisture content of biopellet prepared from teak, acacia and sengon wood also decreased with increasing pelletizing temperature [14].

#### 3.2.2. Density

Biomass is bulky and relatively low in energy content base on its volume. These properties bring about the biomass transportation and storage is inefficient. Densification or pelletizing will increase the density and the energy content of biomass. The density of biomass influences the density of the resulting biopellet. The density of agathis wood is in the range of 0.47-0.48 $g/cm^3$ [10]. In the present works, the density of the produced biopellet was of 0.61-0.71 $g/cm^3$. It is far below the DIN EN 51731 standard [20] requirement of 1.00-1.40 $g/cm^3$.

The influence of particle size, bark content, pelletizing temperature and interaction of factors can be seen in Table 3. The effect of bark content on the density of biopellet reported by several researchers is opposing, somehow. The present research indicates that bark tended to increase biopellet density; however, Filbakk *et al.* [11] found that bark insignificantly influenced the density of pine wood biopellet. The highest density of biopellet was resulted from biopellet made of 100% bark with particle size of 60 mesh pelletized at 160°C, and the lowest density was resulted from biopellet made of 100% bark with particle size of 40 mesh pelletized at 130°C. Higher lignin content of bark and higher surface area of smaller particle size could be the origin of this occurrence. Duncan MRT indicated that the biopellet density using 20% and 30% bark content with particle size of 60 mesh pelletized at 130°C was insignificantly different. Therefore, considering the relatively limited amount of bark, using 20% bark as fortifier is more efficient than that using 30% bark. Furthermore, using lower pelletizing temperature will be less costly in term of energy input during biopellet production processes. Refining process and pelletizing at high temperature are the most energy consuming in the production of biopellet. Data in Table 3 also indicates that temperature in interaction with bark content and particle size tends to be more influential on the density of biopellet at higher bark content.

**Table 2**. Quality parameters of the currently produced biopellet

| Treatments | | | Moisture content (%)* | Density (g/cm³)* | Ash content (%)* | Durability (%)* |
|---|---|---|---|---|---|---|
| Particle size (Mesh) | Bark content (%) | Pelletizing temperature (°C) | | | | |
| 20-40 | B0 | 130 | 3.59±0.31 | 0.64±0.03 | 0.82±0.10 | 26.10±1.07 |
| | | 160 | 2.19±0.05 | 0.65±0.01 | 0.79±0.07 | 21.26±0.83 |
| | | 190 | 1.34±0.19 | 0.61±0.02 | 0.82±0.02 | 24.65±0.28 |
| | B10 | 130 | 3.40±0.02 | 0.64±0.02 | 0.79±0.09 | 30.22±0.62 |
| | | 160 | 1.93±0.21 | 0.67±0.01 | 1.19±0.14 | 22.17±1.22 |
| | | 190 | 1.16±0.16 | 0.68±0.02 | 0.99±0.16 | 25.57±1.71 |
| | B20 | 130 | 3.00±0.28 | 0.61±0.02 | 1.11±0.19 | 34.20±0.92 |
| | | 160 | 2.65±0.19 | 0.61±0.01 | 1.03±0.07 | 26.37±1.06 |
| | | 190 | 2.11±0.09 | 0.63±0.01 | 1.36±0.15 | 30.01±1.06 |
| | B30 | 130 | 3.96±0.34 | 0.69±0.04 | 1.48±0.09 | 35.87±1.63 |
| | | 160 | 3.30±0.24 | 0.62±0.01 | 1.50±0.07 | 26.26±0.79 |
| | | 190 | 1.32±0.27 | 0.65±0.02 | 1.42±0.09 | 31.60±1.10 |
| | B100 | 130 | 4.52±0.20 | 0.61±0.00 | 2.48±0.05 | 30.54±0.74 |
| | | 160 | 2.57±0.29 | 0.63±0.03 | 2.58±0.04 | 25.65±0.64 |
| | | 190 | 1.57±0.17 | 0.65±0.02 | 2.73±0.06 | 29.65±0.56 |
| 40-60 | B0 | 130 | 1.43±0.31 | 0.67±0.03 | 1.21±0.09 | 36.31±0.92 |
| | | 160 | 1.13±0.24 | 0.63±0.00 | 1.11±0.12 | 26.28±0.75 |
| | | 190 | 0.42±0.14 | 0.68±0.02 | 1.03±0.14 | 33.62±1.07 |
| | B10 | 130 | 1.45±0.13 | 0.64±0.02 | 1.22±0.04 | 40.39±0.85 |
| | | 160 | 0.77±0.16 | 0.65±0.02 | 1.24±0.07 | 27.80±0.58 |
| | | 190 | 0.42±0.21 | 0.68±0.02 | 1.45±0.17 | 36.44±1.07 |
| | B20 | 130 | 1.44±0.30 | 0.64±0.03 | 1.43±0.08 | 44.12±1.39 |
| | | 160 | 0.63±0.07 | 0.62±0.01 | 1.44±0.14 | 31.25±1.47 |
| | | 190 | 0.55±2.31 | 0.65±0.04 | 1.46±0.06 | 41.38±0.83 |
| | B30 | 130 | 2.31±0.16 | 0.63±0.03 | 1.79±0.06 | 44.59±0.54 |
| | | 160 | 1.02±0.12 | 0.66±0.02 | 1.65±0.01 | 38.91±1.01 |
| | | 190 | 0.56±0.03 | 0.62±0.01 | 1.64±0.17 | 42.54±0.78 |
| | B100 | 130 | 3.25±0.24 | 0.66±0.01 | 2.25±0.05 | 34.33±0.74 |
| | | 160 | 1.39±0.26 | 0.71±0.02 | 2.48±0.08 | 40.84±0.86 |
| | | 190 | 0.90±0.29 | 0.70±0.03 | 2.50±0.06 | 39.63±0.53 |

note: (*) the average of tree replicates

Table 3 shows that the R-squared for biopellet density is 69%. This implies that in addition to particle size, bark content, and pelletizing temperature, other factors (at the level of 31%) also influenced the density of the biopellet. Pelletizing pressure is an important influencing factor that determines biopellet density. In the current works, pelletizing pressure used was considered low, i.e. at 10.3 MPa as compared to optimum pressure of 100-150 MPa according to Kaliyan and Morey [12].

**Table 3**. Analysis of variance of the biopellet quality parameters

| Components | Moisture content | Density | Ash content | Durability |
|---|---|---|---|---|
| *Dependent Variables (Models)* | | | | |
| R-*squared* | 0.97 | 0.69 | 0.98 | 0.99 |
| RMSE | 0.21 | 0.02 | 0.10 | 0.98 |
| *Effect Test (P-Values)* | | | | |
| Particle size (A) | < 0.0001 | 0.0006 | < 0.0001 | < 0.0001 |
| Bark content (B) | < 0.0001 | 0.0003 | < 0.0001 | < 0.0001 |
| Pelletizing temperature (C) | < 0.0001 | 0.1915 | 0.1327 | < 0.0001 |
| A * B | 0.0001 | < 0.0001 | < 0.0001 | 0.0006 |
| A * C | < 0.0001 | 0.6028 | 0.0089 | < 0.0001 |
| B * C | < 0.0001 | 0.0024 | 0.0006 | < 0.0001 |
| A * B * C | < 0.0001 | 0.0008 | 0.0071 | < 0.0001 |

Note: *P-Value* ≤ 0.05 significantly different and *P-Value* > 0.05 insignificantly different at confidence level of 95%.

### 3.2.3. Ash content

Ash reduces the quality of biopellet. The ash content of the presently produced biopellet was in the range of 0.79- 2.73 % (Table 2), well below the requirement of DIN EN 14961-2 standard [19] that requires an ash content of less than 3%. The ash content was mainly influenced by the bark content used in the preparation of the biopellet. Increasing bark content increased the ash content of the resulted biopellet. Using a 100% bark to produce biopellet increased ash content nearly to the same level of the ash content of bark (2.7%). The relatively low ash content of the presently produced biopellet could be a new source of renewable and environmentally friendly solid fuel. It is a better solid fuel compared to coal-based fuel [2], where coal combustion emits 18.1-19.2 % ash of its weight [25].

Analysis of variance (Table 3) idicates that pelletizing temperature insignificantly influenced the ash content of biopellet. Biopellet with 30% bark content made of 60 mesh particle size at pelletizing temperature of 130°C was higher than that with 20% bark content prepared under similar particle sice and pelletizing temperature. Therefore, using 20 bark content will be a better choice in prpeparing biopellet from agathis wood. Refining the raw material to a higher degree also significantly increased the ash content. Biopellet made from particle size of 60 resulted in a higher biopellet ash content compared to that of biopellet made from particle size of 40. Similar results were reported by Miranda *et al.* [21] in the production of biopellet with *Pinus sylvestris* and *Picea abies* woods. These authors concluded that ash is biopellet commonly related to the presence of nitrogen, kalsium and potasium in the raw materials.

### 3.2.4. Durability

Durability is indicative of the biopellet resistance to an external impact during transportation process. Biopellet durability is very important in storage and for long distance transportation. The durability of the presently produced biopellet was in the range of 21.26-44.59% (Table 2), which was far below the requirement of DIN EN 14961-2 standard [19] that required the durability of 96.5% or higher. Biopellet durability is related to pelletizing pressure and density. A 10.34 MPa pressure used in the present works brought about a low density and consequently low biopellet durability. The pelletizing pressure used was assumed inadequate to form a good interparticle bonding in the biopellet.

Table 3 shows that particle size, bark content, pelletizing temperature, and interaction of factors significantly influenced the durability of biopellet. The durability of biopellet prepared with 60 mesh particle size was higher than that prepared from particle size of 40 mesh, which is in accordance with

IOP Conf. Series: Materials Science and Engineering **935** (2020) 012047 doi:10.1088/1757-899X/935/1/012047

the previous finding of Lee *et al.* [13]. Higher durability of biopellet produced with 60 mesh particle size is possibly caused by the higher surface area of the finer particle that enable the particle to form a better interparticle bonding. Biopellet with 20% bark content pelletized at 130°C is the most preferable in term of its durability. The glass transition (Tg) temperature of lignin is slightly varied among lignocellulosic materials. The Tg of eucalypt and pinewoods for instant was found in the range of 120-140°C [26]. Pelletizing at 130°C might be the right temperature to melt lignin without degrading its polymer structure, thus it forms a good adhesive for interparticle bonding. Possible degradation of lignin structure at higher pelletizing temperature could be the origin of the reduced biopellet durability.

**Table 4.** Mean range test (MRT) of Duncan for the biopellet quality parameters

| Treatments | | | Moisture content | Density | Ash content | Durability |
|---|---|---|---|---|---|---|
| Particle size | Bark content | Pelletizing temperature | | | | |
| 20-40 | B0 | 130 | C | CDEFGH | KL | LM |
| | | 160 | GH | CDEFGH | L | O |
| | | 190 | IJK | H | KL | M |
| | B10 | 130 | C | DEFGH | L | JK |
| | | 160 | H | ABCDE | GHI | N |
| | | 190 | JKL | ABC | JK | M |
| | B20 | 130 | DE | GH | HIJ | G |
| | | 160 | EF | GH | IJ | LM |
| | | 190 | GH | EFGH | FG | JK |
| | B30 | 130 | B | AB | DEF | GH |
| | | 160 | CD | GH | DEF | LM |
| | | 190 | IJK | CDEFGH | F | J |
| | B100 | 130 | A | H | B | JK |
| | | 160 | F | EFGH | AB | M |
| | | 190 | I | CDEFGH | A | K |
| 40-60 | B0 | 130 | IJ | ABCDE | GH | G |
| | | 160 | JKLM | DEFGH | HIJ | LM |
| | | 190 | O | ABCD | IJ | I |
| | B10 | 130 | IJ | CDEFGH | GH | DEF |
| | | 160 | MNO | CDEFGH | GH | L |
| | | 190 | O | ABC | F | G |
| | B20 | 130 | IJ | CDEFGH | F | AB |
| | | 160 | NO | FGH | F | JK |
| | | 190 | NO | CDEFGH | EF | CD |
| | B30 | 130 | FG | EFGH | C | A |
| | | 160 | KLM | BCDEFG | CD | F |
| | | 190 | NO | GH | CDE | BC |
| | B100 | 130 | CD | ABCDEF | B | HI |
| | | 160 | IJK | A | B | DE |
| | | 190 | LMN | A | B | EF |

Note: the same letter indicates an insignificantly different; different letter indicates a significantly different

### 3.2.5. The calorific value

Calorific value is the main factor to be considered for the use of biomass as the source of fuel. In order to evaluate the highest possible calorific value of the present experiments, the measurement of the calorific value was carried out for the biopellet prepared with 60 mesh particle size pelletized at 190°C. At this pelletizing temperature, the moisture content of biopellet was the lowest. The calorific value of the tested samples was in the range of 4,524-4,628 kkal/kg (Table 5) that satisfied the requirement of DIN EN 14961-2 standard [19] of 3,821-4,538 kkal/kg. Table 5 shows that agathis wood biopellet with an acceptable calorific value can be produced by the addition of 10-30 % bark, particle size of 60 mesh, and pelletizing temperature of 190°C. Even though a very high calorific value can be acquired by preparing biopellet purely with bark, the availability of the bark would prevent the sustainability of the biopellet production.

**Table 5.** Calorific value of biopellet prepared from 60 mesh particle size and pelletizing temperature at 190°C.

| No | Bark content | Calorific value (kkal/kg) * | Duncan MRT** |
|---|---|---|---|
| 1 | B0 | 4524±54 | C |
| 2 | B10 | 4566±17 | BC |
| 3 | B20 | 4578±07 | AB |
| 4 | B30 | 4616±19 | AB |
| 5 | B100 | 4628±07 | A |

Note: (*) average of tree replicates; (**) the same letter indicates insignificantly different and different letter indicates a significantly different

It can be assumed from the Duncan MRT in Table 5 that biopellet with 20% bark content is preferred for biopellet production, despite its calorific value is lower than that of biopellet prepared merely of bark (100% bark content). As stated previously, the relatively limited quantity of bark produce by the agathis stands hinders the production of biopellet merely made of the barks. The increase of calorific value by increasing bark content in the biopellet has been well known. Bark contains a higher amount of lignin and extractives compared to that of the wood. Lignin and extractives are carbonous substances with relatively high calorific value. Therefore, calorific value of biopellet is positively correlated to the lignin Klason and extractives content [27].

### 3.3. *Thermal analysis*

Thermal analysis was done for wood biopellet (0% bark content/B0), biopellet with 30% bark content (B30) and biopellet of bark (100% bark content/B100). The thermogravimetric of the B0, B30, and B100 is depicted in Figure 2, 3, and 4, respectively. Degradation pattern of biopellet follows the commonly occurred for biomass that consists of three steps [28-30]. Evaporation of water (moisture loss) occurred at the first step, and followed by the decomposition of cellulose, hemicellulose and lignin in the second step. The third step is the residual combustion (burnout residue) characterized by a constant mass loss. Decomposition in the second step consists of decomposition I (volatilization) and decomposition II (burning) [28].

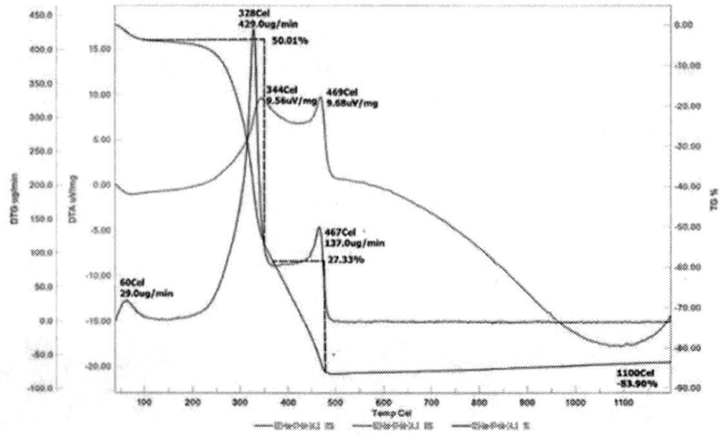


**Figure 2**. Thermografimetric of the agathis wood biopellet (B0)

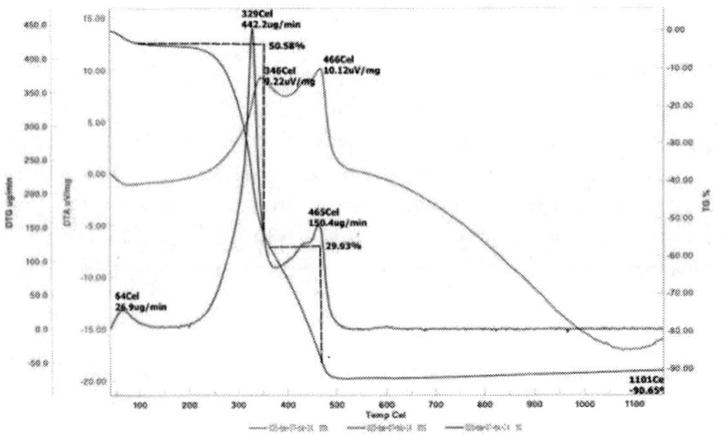


**Figure 3.** Thermografimetric of the biopellet with 30% bark content (B30)

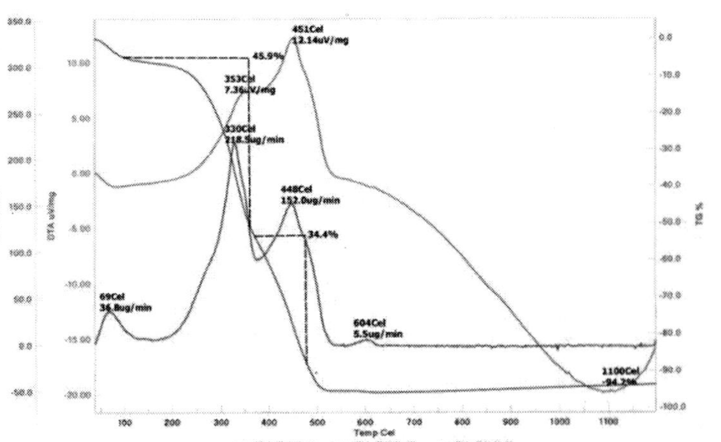


**Figure 4.** Thermografimetric of the biopellet with 100% bark content (B100)

**Table 6**. Thermal analysis of biopellet

| Note | Sample | | |
|---|---|---|---|
| | B0 | B30 | B100 |
| Water loss temperature (°C) | 36-100 | 36-100 | 37-100 |
| Peak temperature (°C) | 60 | 64 | 69 |
| Mass loss (%) | | | |
| Decomposition I | 50.01 | 50.58 | 45.90 |
| Decomposition II | 27.33 | 29.93 | 34.40 |
| Mass loss temperature (°C) | | | |
| Decomposition I | 100-350 | 100-352 | 100-359 |
| Decomposition II | 350-476 | 352-468 | 359-474 |
| Peak temperature (°C) | | | |
| Decomposition I | 328 | 329 | 330 |
| Decomposition II | 467 | 465 | 448 |
| Constant mass loss temperature (°C) | >476 | >468 | >474 |

Thermogravimetric in Figure 2, 3, and 4 and Table 6 indicate that water loss occurred at 36-100°C. That was a little lower than 125°C as reported by several previous works [28-31]. The peak temperature of B100 was higher than these of B0 and B30, which could be brought about by the higher moisture content of biopellet made of 100% bark. Decomposition of biopellet occurred at 100-476°C with total mass loss of 77.34-80.15%. Table 6 shows that mass loss at the first decomposition was of 45.90-50.58% and at the second decomposition the mass loss was of 27.33-34.40%. The mass loss temperature at the first decomposition occurred at 100-359°C, where all of carbohydrate decomposed. Hemicellulose decomposes at the temperature of below 275°C and cellulose at 200-340°C and lignin at 250-360°C [30]. Mass loss of B100 at the first was the lowest due to its lowest carbohydrate content. At the second mass loss that occurred at 350-474°C, lignin is much decomposed. Because of the lignin content of B100 is higher than these of B0 and B30, the highest mass loss was experienced by B100. Mass loss pattern in the present results is similar than that found by Gil *et al.* [29] for pinewood, eucalypt, chestnut, and cellulose residue which was of 60-70% at the first decomposition and of 25-30% at the second decomposition. Mass loss of biomass generally occurred up to 500°C [31]. In the present results, a constant mass loss occurred at 468-476°C. Considering the temperature of mass loss, pelletizing temperature should be carried out at the temperature of below 200°C in order to increase energy density and avoid significant mass loss.

## 4. Conclusion

The peeled-off bark of agathis wood can be used to fortify agathis wood biopellet with an acceptable quality and calorific value at 20% bark content. Except for the biopellet density and durability, all biopellet quality parameters satisfied the requirement of DIN EN 14961-2 and 51731 standard. Considering the bark resources and eficiency of pelletizing temperature, the most acceptable biopellet quality could be prepared from particle size of 40-60 mesh with 20% of bark content and pelletizing teperature of 130°C. Based on thermal properties of the bark, wood and biopellet of the agathis wood, pelletizing temperature should be approximatly at 140°C.

## References

[1] Quan VA 2006 Degradation of the solar cell dye sensitizer N719 Preliminary building of dye-sensitized solar cell [tesis] Roskilde University Denmark
[2] Demirbas A 2004 Combustion characteristics of different biomass fuels *Prog. Energy Combust. Sci.* **30** 219-30
[3] Lehmann B, Schroder HW, Wollenberg R, and Repke JU 2012 Effect of miscanthus addition and different grinding processes on the quality of wood pellets *Biomass Bioenergy* **44** 150-9

[4] [ESDM] Energi dan Sumber Daya Mineral 2012 *Handbook of Energy and Economic Statistics of Indonesia* (Jakarta: Kementerian Energi dan Sumber Daya Mineral)
[5] Demirbas A 2001 Biomass resources facilities and biomass convertion processing for fuels and chemicals *Energ. Convers. Manage* **42** 1375-8
[6] Toscano G, Riva G, Pedretti EF, Corinaldesi F, Mengarelli C, and Duca D 2013 Investigation on wood pellet quality and relationship between ash content and the most important chemical elements *Biomass Bioenergy* **56** 317-22
[7] Nilsson D, Bernesson S, Hansson PA 2011 Pellet production from agricultural raw materials – a systems study *Biomass Bioenergy* **35** 679-89
[8] [DEPHUT] Departemen Kehutanan 2012 *Data dan Informasi Kehutanan Perum Perhutani* (Jakarta: Pusat Inventarisasi dan Statistik Kehutanan Badan Planologi Kehutanan Departemen Kehutanan)
[9] Fengel D, and Wegener G 1984 *Wood: Chemistry, Ultrastructure, Reactions* (Berlin: Walter de Gruyter)
[10] Martawijaya A, Katasujana I, Prawira SA, and Kadir K 1981 *Atlas Kayu Indonesia Jilid I* (Bogor: Badan Penelitian dan Pengembangan Kehutanan, Departemen Kehutanan)
[11] Filbakk T, Jirjis R, Nurmi J, and Hoibo O 2011 The effect of bark content on quality parameters of Scots pine (*Pinus sylvestris* L.) pellets *Biomass Bioenergy* **35** 3342-9
[12] Kaliyan N, and Morey V 2009 Factor affecting strength and durability of densified biomass products *Biomass Bioenergy* **33** 337-59
[13] Lee SM, Ahn BJ, Choi DH, Han GS, Jeong HS, Ahn SH, and Yang I 2013 Effect of densification variables on the durability of wood pellets fabricated with *Larix kaempferi* C. and *Liriodendron tulipifera* L. sawdust *Biomass Bioenergy* **48**1-9
[14] Hendra D 2012 Rekayasa pembuatan mesin pellet kayu dan pengujian hasilnya *Jurnal Penelitian Hasil Hutan* **30**(2) 144-54
[15] Saptoadi H 2008 The best briquette dimension and its particle size *Asian J. Energy Environ.* **9** 161-75
[16] [TAPPI] Technical Association of Pulp and Paper Industry 1996 *TAPPI Test Methods* (Atlanta: TAPPI Press)
[17] [ASTM] American Society for Testing Material 2013 *Test Method for Moisture in the Analysis of Particulate Wood Fuels* ASTM E-871 (USA: American Society for Testing Material)
[18] [ASTM] American Society for Testing Material 2013 *Test Method for Ash in Wood* ASTM D-1102 (USA: American Society for Testing Material)
[19] [DIN] Deutsches Institut für Normung 2010 *Certification Scheme Industry Pellets* DIN EN 14961-2 (Berlin: DIN CERTCO)
[20] [DIN] Deutsches Institut für Normung 1996 *Testing of Solid Fuels, Compressed Untreated Wood. Requirements and Testing* DIN 51731 (Berlin: DIN CERTCO)
[21] Miranda I, Gominho J, Mirra I, and Pereira H 2012 Chemical characterization of bark from *Picea abies* and *Pinus sylvestris* after fractioning into different particle sizes *Ind. Crops Prod.* **36** 395-400
[22] Sjostrom E 1991 *Wood Chemistry: Fundamentals and Applications* (New York: Academic Pr.)
[23] Ma S, He F, Tian D, Zou D, Yan Z, Yang Y, Zhou T, Huang K, Shen H, and Fang J 2018 Variations and determinants of carbon content in plants: a global synthesis *Biogeosciences* **15** 693-702
[24] Haygreen JG, Bowyer JL, and Schmulsky R 2003 *Forest Products and Wood Sciences an Introduction* (Ames: IOWA State University Press)
[25] Cahyono TD, Coto Z, and Febrianto F 2008 Analisis nilai kalor dan kelayakan ekonomis kayu sebagai bahan bakar substitusi batu bara di pabrik semen *Forum pascasarjana* **31**(2) 105-16
[26] Lisperguer J, Perez P, and Urizar S 2009 Structure and thermal properties of lignins: characterization by infrared spectroscopy and differential scanning calorimetry *J. Chil. Chem. Soc.* **54**(4) 460-3

[27] Telmo C, and Lousada J 2011 The explained variation by lignin and extractive contents on higher heating value of wood *Biomass Bioenergy* **35** 1663-7
[28] Wang C, Wang F, Yang Q, and Liang R 2009 Thermogravimetric studies of the behavior of wheat straw with added coal during combustion *Biomass Bioenergy* **33** 50-6
[29] Gil MV, Oulego P, Casal MD, Pevida C, Pis JJ, and Rubiera F 2010. Mechanical durability and combustion characteristic of pellets from biomass blends *Bioresour. Technol.* **101** 8859-67
[30] Pasangulapati V, Ramachandriya KD, Kumar A, Wilkins MR, Jones CL, and Huhnke RL 2012 Effect of cellulose, hemicellulose and lignin on thermochemical conversion characteristics of the selected biomass *Bioresour. Technol.* **114** 663-9
[31] Gaur S, and Reed TB 1995 *An Atlas of Thermal Data for Biomass and Other Fuels* (Colorado: National Renewable Energy Laboratory, Departement of Energy)
[32] Jeguirim M, Dorge S, and Trouve G 2010 Thermogravimetric analysis and emission characteristics of two energy crops in air atmosphere: *Arundo donax* and *Miscanthus giganthus* *Bioresour. Technol.* **101** 788-93

IOP Conf. Series: Materials Science and Engineering **935** (2020) 012048 doi:10.1088/1757-899X/935/1/012048

# Fuel substitution by wood gasification for diesel electricity generator

**U J Siregar[1*], A Lestari[1], L Rusniarsyah[1], and C A Siregar[2]**

[1]Department of Silviculture, Faculty of Forestry, IPB University, Bogor, 16680, Indonesia
[2]Forest Research and Development Center, Ministry of Environment and Forestry, Bogor, 16610, Indonesia

*Email: ulfahjs@apps.ipb.ac.id

**Abstract.** Gasification is a technology for converting biomass into energy in the form of gas. The gas produced can be used as fuel in a diesel generator with dual fuel system. The purposes of this study were to identify the gasification characteristics of sengon, pine, and caliandra woods, as well as wood pellet to produce burnable gas and to analyze the role of burnable gas produced in reducing diesel input in a dual system generator engine. In this study, a locally designed downdraft gasifier and a commercially available diesel engine generator with 12 kW maximum power output were used. The study investigated wood consumption rate, color of fire, and percentage of diesel savings from each wood type. Gasification process was running well, of which burned woods and wood pellet produced an orange-stable flame. The consumption rate of wood was 18 kg/hour for sengon, 15 kg/hour for calliandra, 8 kg/hour for Pine, while wood pellet was 6 kg/hour. Diesel savings with gas co-firing was 33%, 50% and 95% for sengon, calliandra and pine respectively. The low efficiency of diesel savings may due to heat radiation in the reactor and heat loss in the gas purification process or a decrease in efficiency of the diesel generator engine.

## 1. Introduction

Human needs for energy annually increases with increasing population, national development, and technological developments. Petroleum and coal are fossil-based energy sources that still serve as the main energy source in the country. According to the British Petroleum [1], in 2017 Indonesia energy consumption reached 175.2 million tons of fossil fuels. The use of petroleum continues to increase while its reserves are becoming limited, causing its management should be done judiciously. Owing to this fact, it is about right time to encourage the utilization of new renewable energy as a solution to reduce dependence on fossil energy [1].

Various renewable energy sources developed include solar, biodiesel, geothermal, wind, and biomass. Biomass energy has high potential for development in Indonesia with the amount of biomass reserves reaching as high as 261.99 million tons or an equivalent of 49.81 GW of electricity. The abundance and distribution of existing biomass throughout the country are important factors that made biomass very suitable to substitute fossil-based energy sources to produce electrical power for use in the entire territory of Indonesia. In addition to its supplementary merit with considerable amounts, also less likely to cause a negative impact on the environment, the use of biomass as a fuel has another positive influence i.e. recycling the $CO_2$ so that the net emissions of $CO_2$ into the atmosphere is zero [2].

As forest resources in Indonesia is the third largest resources in the world, it is very attractive to utilize wood biomass as renewable energy source. Sengon (*Paraserianthes falcataria*), pine (*Pinus merkusii*), and caliandra (*Calliandra calothyrsus*) are among recommended fast growing species that have promising potency as wood energy. Meanwhile the biomass gasification is considered as the proper technology and indispensable innovation of renewable energy for local community, especially in a remote island in the near future.

Gasification is a conversion technology from biomass to energy in the form of gases for further utilization, such as electric power generation. The gasification process starts from the imperfect combustion of wood in the reactor to produce a burnable gas, then the gas is cooled and purified, and enters into the diesel engine to be further converted into electrical energy [3]. This gas coming from gasification process streamed into a dual fuel engine system, technically designed to decrease the diesel consumption. Although the gasification gases are able to generate electric power, it is important to consider to what extent the energy coming from wood biomass can affect the amount of electrical power produced. Therefore, it is necessary to learn the burnable gas usage on the performance of diesel engines by looking at the amount of diesel consumption spent to produce certain amount of electricity.

The purpose of this study was to identify the gasification characteristics of sengon, pine, and caliandra woods, as well as wood pellet to produce burnable gas and to analyze the role of burnable gas produced in reducing diesel input in the dual system generator engine.

## 2. Materials and Methods

### *2.1 Materials*

Biomass sources were 20 kg each of sengon wood chips size 4 x 2 x 1 cm; Pine wood chips size 3 x 1.5 x 1.5 cm; while Caliandra wood chips had a diameter of ± 2-5 cm and 3 cm length (Figure 1).

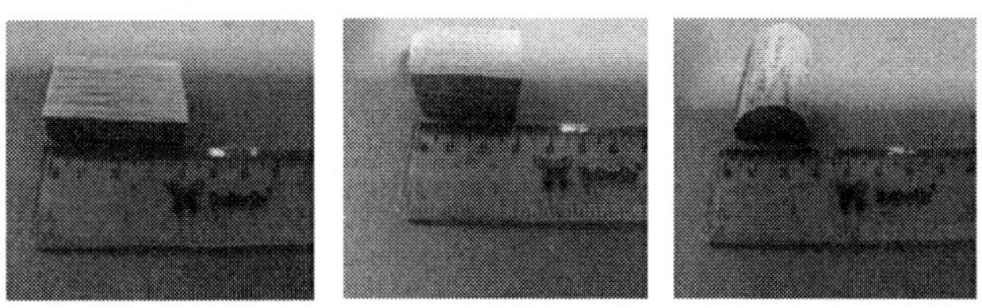

a b c

**Figure 1**. The raw materials of wood gasification (a) sengon; (b) pine; and (c) caliandra wood

Wood chips of sengon, pine, and caliandra then were dried up for five consecutive days under sunlight for 7 hours/day. This drying was done to reduce the moisture content of biomass reaching <14% [4]. According to Hardiansyah *et al.* [2] drying wood chips with an average wood length of 2.8-4 cm in cylindrical shape can be done using an oven with a temperature 70°C for 2 days.

### *2.2. Methods*

2.2.1. Testing thermal gasification process

Testing thermal gasification process was is done with the following stages:

a. Preparing gasifier unit with reactors measuring 30 x 30 x 60 cm, filling in water on the top water seal, preparing the blower, Ash grating is positioned at the ready to fill position and the Ash box was closed at the bottom
b. Sengon wood chips as much as 4 kg were inserted into the reactor, and then burned using a little amount of kerosene followed by turning on the blower
c. When the flame is evenly distributed at least 2/3 extents of the reactor area, the blower was turned off and the reactor top is closed. In this process the time is required for initial startup was measured.

d. Blower was returned on again, then waited until Syngas come out of the thermal nozzle, of which upon Syngas exit from thermal nozzle then the syngas was burned.
e. Observing and documenting the resulting flame color, then measuring the duration of the gasification of sengon wood to obtain the rate of biomass consumption during the gasification process
f. Once the raw material is depleted, the charcoal is ejected out through the bottom box, usually still hot and smoldering then quenched with water
g. The same procedure is done for raw materials of pine, caliandra wood, and wood pellet

2.2.2. Dual fuel engine system operation
Gasifier unit with a reactor measuring 32 x 33 x 116 cm and a 24 HP diesel engine are prepared. The steps needed to connect the gasifier unit to the diesel engine including (1) removing the nozzle and closing with the blind flange, (2) turning on water pump to flow water in a flow water scrubber with water discharge between 25-36 L/minute, (3) preparing gas clean up by filling the four compartment of gas washing tubes with water-water-diesel-diesel; The amount of diesel was 8 L for running the machines for 500 hours, (4) resulted syngas then will flow into the connecting flow tube towards diesel engine, (5) rwitching diesel engine on and setting the power voltage to 220 volts, (6) follow from the starting step as a test of the thermal gasification process up to closure of reactor, (7) slowly closing the syngas valve on the diesel engine until the diesel engine sucks syngas that has been cleaned through the gas clean up.

The amount of each type of wood used was 10 kg. After the combustion process then syngas was flowing into the diesel engine through the connecting tube. Measurements included how much diesel required when combined with a certain syngas from each type of wood, compared to diesel consumption on single fuel operations at the same length of operating time.

### *2.3. Data Analysis*

2.3.1. Wood consumption rate

$$V = \frac{\text{Amount of wood usage (kg)}}{\text{Duration of burning (hours)}} \qquad (1)$$

The percentage of diesel saving on diesel engine can be calculated using the following equation [2]:

$$\%\text{ Saving} = \frac{\frac{V1}{\text{electric power output}} - \frac{V2}{\text{electric power output}}}{\frac{V1}{\text{electric power output}}} \times 100\% \qquad (2)$$

where, $V_1$ = diesel consumption on single fuel engine, $V_2$ = diesel consumption on dual fuel engine

2.3.2. *Spesific gasification Ratio* (SGR)
SGR is the rate of biomass gasification per reactor unit area [5]. SGR can be calculated using the following equation:

$$SGR = \frac{m_{bb}}{A} \qquad (3)$$

where, SGR= *Specific Gasification Ratio* (kg/m$^2$ hour), $m_{bb}$= Biomass consumption rate (kg/hour), A= Cross-sectional area of the reactor (m$^2$)

2.3.4. Electrical Energy Potential
Every 1 kg of dry wood has a different calorific value, and one Kcal contains 4,200 Joule of energy. The higher the heat value, the greater the energy content. Estimation of potential electrical energy can also be done using the biomass conversion approach to wood pellet (WP), as shown in Table 1.

**Table 1** Conversion of biomass to electrical energy

| Energy (J) | Energy (MJ) | electical (kWh) |
|---|---|---|
| 1 kg dry wood = wood calorific value x 4200 | 1 kg WP = 19.8 MJ | 3.6 MJ = 1 kWh |

In producing 1 ton of WP it takes as much as 1.5 tons of biomass [6]. According to Payne [7], every kilogram of WP contained energy amounted to 19.8 MJ. The energy in MJ units was converted into kWh units (1 kWh = 3.6 MJ) with the efficiency of the machine used at 75%.

## 3. Results and Discussion

### *3.1 Gasification*

Utilization of biomass as a fuel needs pre-processing in advance to facilitate its utilization, commonly known as biomass conversion [3]. According to Susanto [8], biomass can be converted into energy through three processes, i.e. thermochemistry, biochemistry and oil-containing seed extraction. One example of thermal or thermochemical conversion techniques is gasification.

The gasification process to produce burnable gas was conducted through four phases i.e. drying, pyrolysis, oxidation, and reduction (Figure 2). Drying phase aims to remove water contained within the biomass material and turn it into moisture. Subsequently biomass materials would undergo pyrolysis process until it decomposes, which will produce tar, $CH_4$, and charcoal. At the combustion phase, the carbon will be burned to carbondioxide, and hydrogen will be burned into moisture. At the reduction phase some reactions occurs that produce CO gas and $H_2$ as the main ingredient of burnable gas.

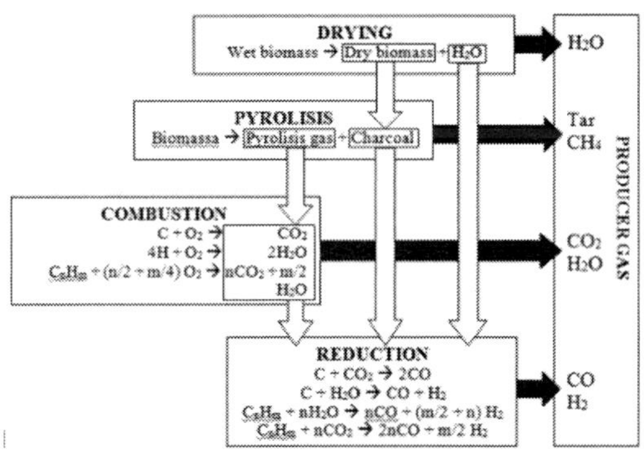


**Figure 2.** Phases of gasification process [9]

The burnable gas produced from this process still contains moisture, tar, nitrogen gas, ash particles, and at high temperatures. Efforts that can be done to remove the unwanted contents is passing the burnable gasses through gas clean up.

Figure 3 shows the biomass gasification unit used in this experiment equipped with a series of gas clean up consisting of cyclone, water condenser, water scrubber, and gas washing. The burnable gas from the reactor will flow through the cyclone to reduce its solid particles (charcoal and ash). Furthermore, gas is flowed into water condenser to be cooled down, reduce its tar content, and condensed its water content into liquid smoke. Liquid smoke is a byproduct of the gasification process that is useful for fish smoking, pesticides, as well as liquid fuels [10]. After passing the water condenser, burnable gas flowed into water scrubber to be cooled down. Burnable gas further was flown into the gas washing that is the water-water-diesel-diesel compartment to be cleaned. The water in the compartment serves to capture the particles and the remnants of tar, while the diesel serves to capture the remnants of the particles. Eventually the cooled and cleaned burnable gas can be supplied to a diesel engine to be converted into electrical energy.

The gasification process will produce heat and flammable syngas. The gas can be used as fuel in diesel engine dual fuel system through the connection pipeline. According to Hardiansyah *et al.* [2] dual fuel system is a technology that aims to minimize diesel consumption although the use of a little diesel is still needed as an ignition tool. The application of dual fuel system will save diesel use up to 70% or more.

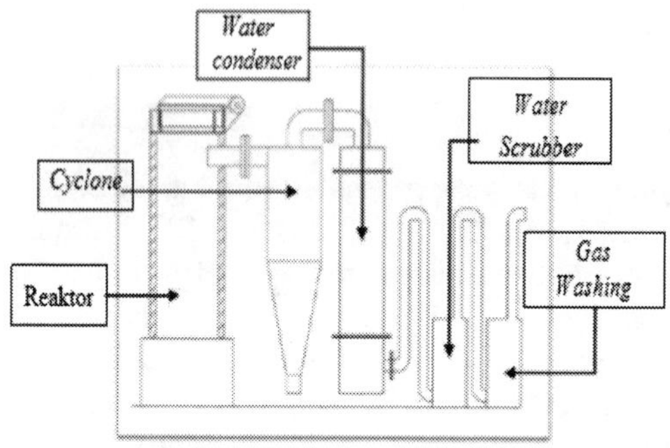


**Figure 3**. Locally designed biomass gasifier

*3.2 Thermal gasification characteristics*

Biomass conversion process of sengon, pine, and caliandra wood as well as wood pellet was carried out in a reactor. The amount of biomass consumption materials used is measured during a certain operating time, thus obtained the rate of biomass consumption as presented in Table 2. The initial starting time for each of the wooden biomass is 7 minutes long, while for wood pellet was 10 minutes. According to Jamilatun [11], the speed of ignition is depended on level of biomass water content, also the length of initial ignition can be affected by the shape of materials. When material was tightly compact, hard and heavy, then it would be slowly ignited and even though highly burned, it is not quickly exhausted that it could save the amount of biomass required.

**Table 2.** Evaluation results of gasification process

| Type of biomass | Rate of wood consumption (kg/hours) | SGR ($kg/m^2hours$) | Energy (MJ) | Electical (kWh) |
|---|---|---|---|---|
| Sengon | 18 | 10.53 | 321.15 | 66.9 |
| Caliandra | 15 | 8.77 | 289.80 | 60.4 |
| Pine | 8 | 4.68 | 156.58 | 32.6 |
| *Wood pellet* | 6 | 3.51 | 118.80 | 24.8 |

Consumption rate of sengon, caliandra, pine wood, and wood pellets, respectively were 18 kg/h; 15 kg/h; 8 kg/h; and 6 kg/h. The differences in consumption of wood can be influenced by the heat/calorific value contained in the biomass material. Judging from the calorific value, pine wood has a caloric value of 4660 kcal [12], is the largest among the three types of wood used. In addition, pine wood is also known as resin containing tree. The presence of extractive substances such as resin and tannins can increase its calorific value [13]. The higher calorific value of the material causes the combustion to be more efficient so as to save the needs of wood used.

From these data and then calculation of the SGR (Specific Gasification Ratio) value which is the rate of biomass gasification per reactor unit area [5] it was obtained a value of 10.53 $kg/m^2$ hours for sengon wood and caliandra wood amounting to 8.77 $kg/m^2$ hours while the SGR for pine wood was 4.68 $kg/m^2$ hours. The difference in the value of SGR for each biomass material used could be influenced by the weight of biomass, the weight of charcoal formed, and the duration of the tool operation. The higher the SGR value the faster the gasification rate so that the consumption rate of biomass and the gas flow rate of burnable gas will be greater.

Meanwhile, wood biomass has the potential to increase its efficiency if the material is in uniform processed forms such as wood pellet. Based on the results in Table 2, among all the biomass materials used, wood pellet is the most efficient material. Although 6 kg of wood pellet produced only 24.8 kWh electricity but gasification using wood pellet resulted in a tool efficiency of 48.39%. Compared with other wood materials, i.e sengon wood required as much as 18 kg, the flow efficiency was only 17.93%. Similarly, 15 kg of caliandra wood and 8 kg of pine wood resulted in respectively tool efficiency of 19.87% and 36.81%. This was because wood pellet has uniform shape, dry, solid, and has a higher calorific value than wood, that the amount needed would be less as the combustion becoming more efficient, resulting that the energy contained in the wood pellet was not much wasted. According to Adrian *et al.* [14], wood pellet as a result of uniforming an existing biomass, has drier properties and smaller forms, easing combustion, storage, and distribution, and has greater heat energy when compared to woods before processing. Yokoyama [15] stated that wood pellet has a higher superiority compared to fossil fuels. Some of its advantages include: handling, ignition, and easy combustion; uniform fuel shape and properties; high energy content; and fewer toxic gas emissions.

Electricity needs per day at Gunung Walat Education Forest (GWEF) IPB University, Sukabumi, Indonesia amounted to 29.3 kWh [16]. Based on the results of this research, the needs of electricity in GWEF could be fulfilled with the installation of similar gasification tool with a capacity of 12 kWh. As shown in Table 2, as much as 18 kg of sengon wood contains an energy of 321.51 MJ or equal to 66.9 kWh when converted into units used in electricity with engine efficiency by 75%. The GWEF electricity could be provided by 15 kg caliandra wood, which would produce energy amounting to 289.80 MJ equivalent 60.4 kWh, or even with only 8 kg of pine wood, which gives 32.6 kWh of electricity in. According to Prayitno *et al.* [17] one factor that need to be considered in wood utilization as an energy material is the growth rate with the nature of dense branches, e.g. in the case of caliandra (*C. calothyrsus).* Based on caliandra biomass estimation at GWEF [16] of 116.37 tons/ha, the amount of biomass could potentially supply electricity at GWEF through wood gasification equipment for 2 years without replanting.

a b c

**Figure 4.** The flame of gasification from (a) pinus wood, (b) caliandra wood, and (c) wood pellet

Qualitative analysis was also done based on flame color and flame stability as a standard reference for the syngas produced before flowing into the generator set. The flame color resulting from the burning of pine and caliandra wood tends to be orange, while the combustion of wood Pellet produces a bluish orange flame color. According to British Petroleum [1] an orange flame can be caused by a hydrocarbon content that is not condensed.

The results of this study were able to produce a stable flame, as shown in Figure 8. This phenomenon suggested that burnable gas produced by each biomass material have a fairly good quality and are worthy as fuel in the dual fuel engine system. In accordance with the statement by Pranolo *et al.* [18], biomass material that is burned and produces a stable flame, would produces gas from combustion that is suitable for use as fuel.

### *3.3 Operation of diesel engine dual fuel system*

Prior to the experiments using syngas from woods on the dual fuel engine system, reference data was collected using the single fuel engine system using diesel as fuel. The electricity generator power provided a maximum of 12 kWh with a voltage setting at 220 volts to produce an energy of 43.2 MJ. Result of single fuel system operating machine is displayed at second column of Table 3. This data is needed for comparison with data from dual fuel system operation at the third column for calculation of diesel consumption savings.

**Table 3** Diesel consumption for 1 hour usage of a diesel engine

| Wood type | Diesel fuel without gasification (L) | Diesel fuel with gasification (L) | Diesel fuel saving (%) |
|---|---|---|---|
| na | 0.60 | na | na |
| Sengon | | 0.40 | 33 |
| Pine | | 0.30 | 50 |
| Caliandra | | 0.03 | 95 |

Table 3 shows a comparison of diesel consumption for single fuel systems and when dual fuel systems were operated. In general diesel consumption decreases with the addition of burnable gas. This means that burnable gas that enters the diesel engine combustion chamber can replace any amount of diesel to obtain the required power. The decrease in diesel consumption is possible due to a decrease in the flow rate of combustion air, characterized by a decrease in opening of syngas valve that causes the flow rate of burnable gas into diesel engine combustion chamber is getting bigger. According to Pranolo *et al.* [18] the increase in flow rate of burnable gas can increase fuel substitution and decrease diesel required.

Decreased diesel consumption can occur due to the addition of electrical loads. The higher electrical load is given, the less diesel consumption while the syngas that enters the diesel engine is increasingly larger. This is in line with results of Pranolo *et al.* [18] that the rate of diesel consumption is increasingly small with the increase in electrical power. Increased electric load causes increased room temperature on diesel engines, and the high temperature in combustion chamber can increase the fuel evaporation while the air condition inside become better. This phenomenon causes the combustion process to become more effective so that the rate of diesel consumption is getting smaller.

The greater the burnable gas flow rate into the engine caused the lower percentage of diesel injected into combustion chamber to keep the engine speed constant, meaning diesel replacement will be even greater. Please note that in this case, the burnable gas role is still secondary to diesel, meaning it could not replace diesel 100%. However, syngas has the advantage of achieving homogeneity of air-fuel mixture, that expected combustion delay period in combustion chamber will be getting shorter [19].

Percentage of diesel savings with burnable gas from gasification of sengon, pine, and caliandra wood were respectively 33%, 50%, and 95% (Table 3). According to Sudarmanta and Dirgantara [19], efficiency of the diesel-saving dual fuel system is generally more than 40%. These results indicate that pine and caliandra wood are suitable for use as biomass material in gasification process to be converted into electrical energy because it produces a diesel saving >40%.

The low value of diesel saving efficiency with gas gasification of sengon wood can be caused by heat radiation in the reactor and heat loss in the gas purification process, or the decrease in efficiency of the diesel engine. A diesel engine power can decrease [19] because of wasted energy from fuel. This happens because the air-fuel mixture has become very rich that combustion in combustion chamber is no longer perfect. Considering its benefit, utilization of syngas alone as fuel to generate electricity has additional benefit, because gas does not pose an environmental problem. Nevertheless, when efficiency in the utilization of syngas for power generation is increased, problems of electrification in remote areas of Indonesia could somehow be overcome.

## 4. Conclusions

Gasification process of sengon, pine, and caliandra woods in a locally designed gasifier unit to produce syngas has been running well. The higher calorific value of certain wood type causes the combustion to be more efficient, that lesser amount of the wood is needed. Percentage of diesel savings with syngas from gasification of caliandra wood amounted to 95%, is the highest while diesel savings with syngas of Sengon wood was only 33%. The low value of diesel savings efficiency with syngas can be caused by heat radiation in the reactor and heat loss in the gas purification process, or the decrease in efficiency of the diesel engine after certain period of operation.

## References

[1] [BP] British Petroleum 2017 *BP statistical review of world energy 2017* [Internet] Available on http://www.bp.com/content/dam/bp/pdf/Energyeconomics/statisticalreview2017/bp-statistical-review-of-world-energy-2017-full-report.pdf

[2] Hardiansyah MF, Firdha J, Navitri AM, Alfianto D, Wibowo WA, and Pranolo SH 2016 Substitusi bahan bakar genset 5 kW dengan gas hasil gasifikasi gamal dan kaliandra *Seminar Nasional Industri Kimia dan Sumberdaya Alam Universitas Lambung Mangkurat* pp. 50-58

[3] Arhamsyah 2010 Pemanfaatan biomassa kayu sebagai sumber energi terbarukan *Jurnal Riset Industri Hasil Hutan* **2**(1) 42-8

[4] Pakiti D, Pangkerego F, Tooy D, and Ludong D 2013 Analisis konsumsi bahan bakar pembangkit listrik tenaga diesel dengan substitusi gasifikasi sabut kelapa [Skripsi] Universitas Sam Ratulangi Manado

[5] Pranolo SH, Ratmana IF, and Pratama NS 2015 Kinerja pembangkit listrik dual fuel kapasitas 5 kW berbasis gasifikasi sekam padi berunggun tetap *Ekuilibrium* **14**(2) 51-6

[6] Sjoding D, Kanoa E, and Jensen P 2013 *Developing A Wood Pellet/Densified Biomass Industry in Washington State: Opportunities and Chalenges* (Washington: Washington University)

[7] Payne GA 1980 *The Energy Managers Handbook* (Surrey UK: Westbury House)

[8] Susanto H 2018 Pengembangan teknologi gasifikasi untuk mendukung kemandirian energi dan industri kimia *Makalah. dalam Sidang Terbuka Forum Guru Besar ITB*

[9] Suyitno T 2007 *Teknologi Gasifikasi Biomasa untuk Penyediaan Listrik dan Panas Skala Kecil Menengah* (Surakarta: UNS Press)

[10] Bridgwater T 2006 Review biomass for energy *Journal of the Science of Food and Agriculture* **86** 1755-68

[11] Jamilatun S 2008 Sifat-sifat dan penyalaan briket, briket biomassa, briket batubara, dan arang kayu *Jurnal Rekayasa Proses* **2**(2) 37-40

[12] Martawijaya A, Kartasujana I, Kadir K, and Prawira SA 1989 *Atlas Kayu* (Bogor: Pusat Penelitian dan Pengembangan Hasil Hutan)

[13] Haygreen JG, Bowyer JL, and Schmulsky R 2003 *Forest Product and Wood Sciences: an Introduction* (Ames US: Iowa State University Press)

[14] Adrian A, Sulaeman R, and Oktorini Y 2015 Karakteristik wood pellet dari limbah kayu karet (*Hevea brasilliensis* Muell. Arg) sebagai alternatif sumber energi terbarukan *Jurnal Jom Faperta* **2**(2)

[15] Yokoyama S 2008 *The Asian Biomass Handbook: A Guide for Biomass Production and Utilization* (Tokyo: The Japan Institute of Energy)

[16] Siregar UJ, Damayanti A, and Dwiyanti FG 2020 Potency estimation of forest stands biomass in Gunung Walat Educational Forest, Sukabumi, West Java as fuel for electricity generation *IOP Conf. Series: Earth and Environmental Science* **528** 012062

[17] Prayitno TA 2007 *Pertumbuhan Pohon dan Kualitas Kayu KTT 667* (Yogyakarta: Fakultas Kehutanan UGM)

[18] Pranolo SH, Ratmana IF, and Pratama NS 2015 Kinerja pembangkit listrik dual fuel kapasitas 5 kW berbasis gasifikasi sekam padi berunggun tetap *Ekuilibrium* **14**(2) 51-56

[19] Sudarmanta B and Dirgantara EW 2011 Aplikasi sistem dual fuel bensin dan syngas hasil gasifikasi biomassa pada motor bensin stasioner untuk pembangkit listrik *Seminar Nasional Teknik Mesin 6* Institut Teknologi Surabaya

[20] Hartoyo and Syachri TN 1986 Gasifikasi arang kayu untuk pembangkitan tenaga listrik *Jurnal Penelitian Hasil Hutan* **3**(2) 33-8

**Acknowledgement**

This research was funded by USAID through SHERA Program, given to Center for Development of Sustainable Region (CDSR) Project led by UGM and IPB University as affiliate.

# A comparison of tree biometric performance of four tropical energy wood species

**M Yulianti*, W C Adinugroho, and H Krisnawati**

Forest Research and Development Center, Ministry of Environment and Forestry, Bogor, 16610, Indonesia

Email: mira_solvay@yahoo.com

**Abstract.** Four energy wood species, i.e. Weru (*Albizia procera* (Roxb.) Benth.), Akor (*Acacia auriculiformis* Benth), Pilang (*Acacia leucophloea* (Roxb.) Willd), and Kaliandra, (*Calliandra calothyrsus* Meisn.) that are suitable for substituting fossil energy in Indonesia were tested in experimental plots in Majalengka, West Java, Indonesia. In this paper, we assessed the performance of the planted species of up to 3.5 years old by measuring and analyzing the significant biometric parameters (height, diameter, volume, and biomass) of the trees growing on the plots. In each measurement, tree diameter and height were recorded. Volume and biomass were calculated by using allometric equations. The results showed that *A. auriculiformis* performed better than the other three species in terms of height, diameter, volume, and biomass. The performance was then followed by *C. calothyrsus*, *A. procera,* and *A. leucophloea*. The mean annual increment of volume and biomass of *A. auriculiformis* was 12.90 $m^3$/ha/yr and 7.32 tonnes/ha/yr, respectively.

## 1. Introduction

Fossil fuel is one of the energy sources that is still widely used by global community [1]. However, the availability of fossil fuels in the world is running low because fossil fuels used are still non-renewable natural resources [2]. If the situation continues without any replacement of alternative energy, it will lead to energy crisis [3]. The establishment of energy wood-based forests is currently implemented in Indonesia by developing energy-producing timber plantations as an effort to use energy resources sustainably, which, on the other hand, will have implications for the time expiration of fossil energy sources [4]. New and renewable energy should be the main concern, not only as an effort to reduce the use of fossil energy but also to realize clean or environmentally friendly energy [5]. The benefits of new renewable energy can also reduce emissions, mitigate climate change and provide a clean environment and clean energy for all and future generations [6].

In forestry sector, renewable energy can be in the form of energy wood biomass [7]. Many potential wood species have been tested for energy woods. In Indonesia, there are several species potentials for energy woods that are suitable for use as renewable fossil energy substitutes. The species are usually fast-growing species and have high heat value. One of the factors considered for establishing plantation forest for energy is species selection, i.e. the species which has characteristics of energy wood that are ideal for planting. The characteristics include fast growth with dense branching, high specific gravity, high increment, able to live in a variety of growing conditions, and high ability to sprout after pruning. High heat is also a criterion for selecting the species that is expected to gain a high amount of energy in a short crop rotation [8].

This study assessed the biometric performance of four energy wood species to compare the productivity of the species that are good for energy wood production. The four species being studied included Weru (*Albizia procera* (Roxb.) Benth.), Akor (*Acacia auriculiformis* A. Cunn ex Benth.), Kaliandra (*Calliandra calothyrsus* Meisn.), and Pilang (*Acacia leucophloea* (Roxb.) Willd). Weru is widely used as firewood, building wood, fodder, live fences, ropes of wood bark, and chosen as one of the energy-producing species [9]. The advantages of Weru as an energy wood are partly because the species has the character of fast-growing, dry resistance, and can be trimmed. Besides, the species is easily cultivated because it has a useful life process and fast initial growth and is widely cultivated in agroforestry systems [10]. For energy wood, this species has an increment of 25 $m^3$/ha/yr, a specific gravity of 0.67 with an energy production of 301.5 GJ/ha/yr, and a calorific value of 7382 kcal/kg [9].

Akor is one of woody species that can be used as a source of biomass energy for fuel wood, charcoal and pellet making [11]. The shape of the stem is generally not good, bent with low branching and a lot, so planting this species is mainly directed as energy wood [9]. This species produces high quality firewood and charcoal with heat value for dry wood kiln of 4711 kcal/kg and for wood charcoal of 7639 kcal/kg increment of 17 $m^3$/ha/yr, specific gravity of 0.77 with energy production of 235.6 GJ/ha/yr and a calorific value of 7322 kcal/kg [10].

Kaliandra is a plant species that has a potential as a biomass-based energy source (wood pellet). This species has a calorific value of 4600 kcal/kg and is usually used for charcoal. The heat generated at 7200 kcal/kg can be used as wood production [11] and has the best performance in producing leaves (as animal feed and green fertilizer). The species is a forage tree legume that can be used as an alternative to animal feed as a food source of protein as well as for wood production (for firewood) [12].

Pilang has a high calorific value of 5218 kcal/kg with energy productivity as large as 258.3 GJ/ha/yr. This is a fast-growing species, which can grow on dry land, savanna and marginal land (critical land) with a low rainfall between 400 and 500 mm per year and dry season of 9-10 months. Its natural distribution covers the arid regions of India, Sri Lanka, Bangladesh, Burma, Thailand, and in Indonesia, it grows naturally in Java, Bali and Timor [13].

## 2. Material and Methods

### *2.1. Materials*

This study was carried out in four experimental plots with the plot size of 25 x 50 m each. The planting sites are in RPH Sukajaya, BPKH Cibenda, KPH Majalengka. Administratively, they are located in Kertajati Sub-District, Majalengka District, West Java Province. The average temperature condition ranges from 26.3°C to 29.8°C and the rainfall is around 1974 mm/yr with 109 rainy days. The sites are in a lowland area with an altitude of approximately 50 meters above sea level and soil conditions contain clay, limestone, and marl. The plants were measured at ages of 20, 25, 31 and 38 months for diameter at breast height (DBH) and height. The description of data from the measurement plots are shown in Table 1.

### *2.2. Methods*

#### 2.2.1. Diameter at Breast Height (DBH) and height

Diameter and height were recorded at measurement times of 20, 25, 31, 38 months since planting. Tree diameter was measured at 130 cm above ground level. In the case of multi-stemmed trees (several trunks at ground level), each stem circumference at 130 cm was measured. Tree height was measured for total height from the ground to the top of tree.

**Table 1.** Description of data from measurement plots

| Species | Age at measurement time (yr) | N* | Diameter (Min-Max ) (cm)* | Height (Min – Max ) (m)* |
|---|---|---|---|---|
| *Albizia procera* | 1.7 yr; 2.1 yr; 2.6 yr; 3.2 yr | 80 | 1.3- 6.4 | 1.5-7.5 |
| *Acacia auriculiformis* | 1.7yr; 2.1 yr; 2.6 yr; 3.2 yr | 122 | 2.5-15 | 1-10 |
| *Acacia leucophloea* | 1.7 yr; 2.1 yr; 2.6 yr; 3.2 yr | 84 | 0.8-8.5 | 0.4-5.5 |
| *Calliandra calothyrsus* | 1.7 yr; 2.1 yr; 2.6 yr; 3.2 yr | 126 | 1.5-8.5 | 2-7 |

*trees measured at 3.2 yr

### 2.2.2. *Volume estimation*
Estimation of total tree volume was carried out using a general formula:

$$V = \frac{1}{4} \pi \times D^2 \times H \quad (1)$$

where, V = tree volume ($m^3$), D = DBH (cm), H = total tree height (m), $\pi = 3.14$

### 2.2.3. *Biomass estimation*
Biomass potential was estimated using allometric equations. In this study, the biomass measured was above-ground biomass. Schematic protocol based on woody physiognomy types was recommended for applying general aboveground biomass (AGB) models for woody individuals [11]. Multi-stemmed trees or shrub species have different biomass characteristic with single principal stem trees. Therefore, two types of allometric equations were used in this study.

Allometric equation: $AGB_{est} = 0.0673\ (\rho \times DBH^2 \times H)^{0.976}$ (2) was used to calculate aboveground biomass of Weru, Akor, and Pilang. While allometric equation $AGB_{est} = e^{(2.474\ \ln D) - 2.757} \times 1.0787$ (3) was used to calculate aboveground biomass of Kaliandra, where, $AGB_{est}$ = estimated aboveground biomass (kg), $\rho$ = tree density ($kg/cm^3$), DBH = diameter at breast height (cm), H = Height (m)

## 3. Results and Discussion
Tree growth performance can be measured quantitatively based on tree biometric parameters by identifying the tree from its physical characteristics, such as height, diameter, volume, and biomass. Tree growth will be determined by the ability of the tree species to obtain sunlight, water, nutrition, and air in the range of variability in specific climate and soil conditions. The growth performance of the four energy wood species in these environmental conditions can be explained as follows:

### 3.1. *Height performance*
Height is one of the growth performance parameters that can be used to observe the ability of tree growth at a location, especially during periods of initial growth. We found that the fastest-growing stands in the experimental plot were Akor stand. It has a height of up to 10 m at 3.2 years old. Pilang stand was the slowest growing tree that has a height of only 3 m. The height growth of this species looks stagnant since the age of 2.1 years old (Figure 1). In multi-stem vegetation or shrub species like Kaliandra, these characteristics with low and wide canopies are a more efficient way of growth.

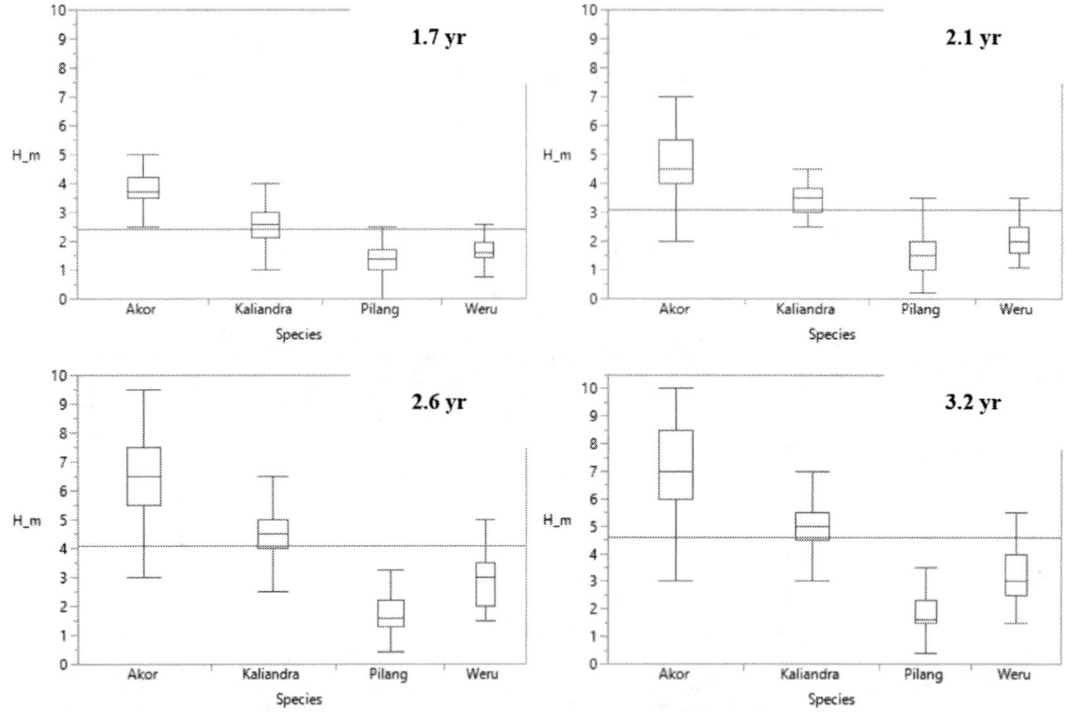


**Figure 1.** Height of four energy wood species: a) 1.7 yr, b) 2.1 yr, c) 2.6 yr, and d) 3.2 yr.

The average height of the four species in the experimental plots increased with age. Akor stand was the highest stand with an average height value of 7.02 m at 3.2 years old, followed by Kaliandra stand of 4.89 m, Weru stand of 3.23 m, and Pilang stand of 1.87 m (Table 2). The difference between four species on height indicates that Kaliandra stands with the smallest CV shows the height less dispersed than the other stands with the larger CV (Akor, Weru, and Pilang).

**Table 2.** Statistics of height growth of four energy wood species

| Species | Age = 1.7 yr | | | Age = 2.1 yr | | | Age = 2.6 yr | | | Age = 3.2 yr | | |
|---|---|---|---|---|---|---|---|---|---|---|---|---|
| | Mean | Std Dev | CV | Mean | Std Dev | CV | Mean | Std Dev | CV | Mean | Std Dev | CV |
| Akor | 3.76 | 0.83 | 22.05 | 4.66 | 1.18 | 25.26 | 6.49 | 1.48 | 22.85 | 7.02 | 1.53 | 21.87 |
| Pilang | 1.32 | 0.62 | 47.23 | 1.57 | 0.74 | 47.23 | 1.78 | 0.89 | 49.98 | 1.87 | 0.85 | 45.44 |
| Weru | 1.65 | 0.42 | 25.45 | 2.12 | 0.59 | 27.92 | 2.79 | 0.79 | 28.10 | 3.23 | 0.97 | 29.95 |
| Kaliandra | 2.54 | 0.59 | 23.40 | 3.31 | 0.77 | 23.34 | 4.41 | 1.06 | 23.96 | 4.89 | 0.91 | 18.54 |

### *3.2. Diameter performance*

Diameter is one of the important factors for determining stand volume. This parameter is an indicator that can be used to describe tree size because the diameter is the most easily used for measurement [15]. In the experimental plots, the Akor stand has the largest diameter compared to the others. The Akor stand reached a diameter of 13 cm at 3.2 years old, followed by Kaliandra stand of 9 cm and Weru stand of 6 cm. The smallest diameter was the Pilang stand of 4 cm (Figure 2).

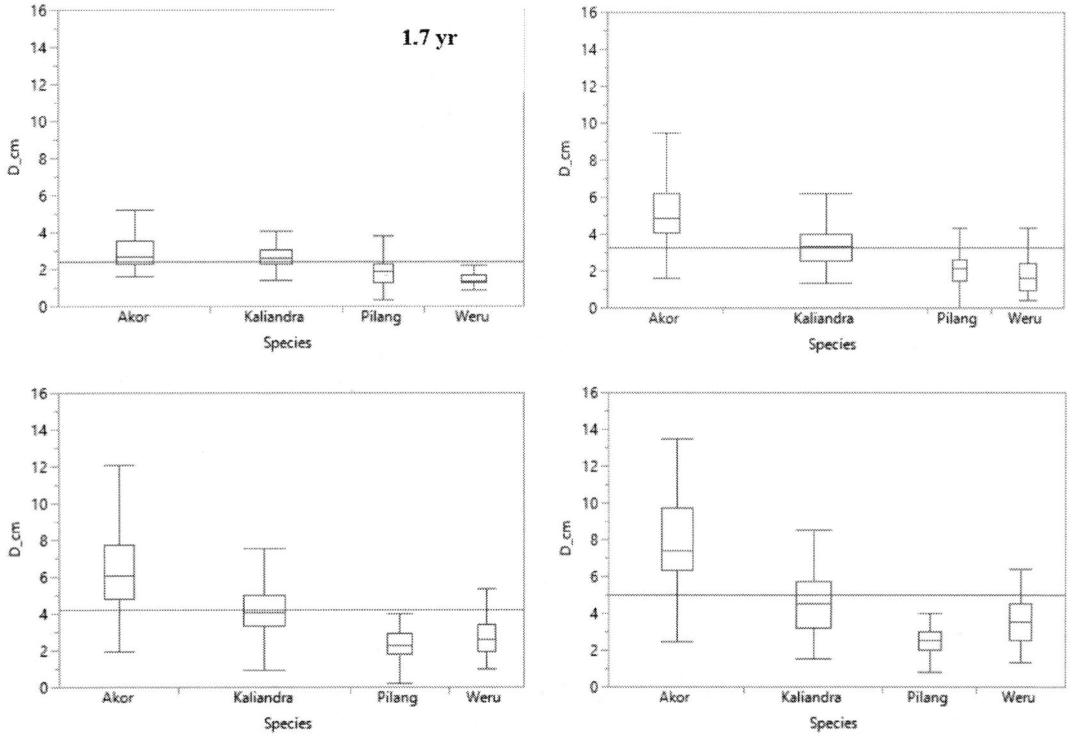


**Figure 2.** The diameter of four energy wood species: a) 1.7 yr, b) 2.1 yr, c) 2.6 yr, and d) 3.2 yr

Akor stand was the largest with an average diameter of 7.95 cm at 3.2 years old, followed by Kaliandra (4.52 cm), Weru (3.52 cm), and Pilang (2.76 cm). The average diameter of the four species in the experimental plots increased linearly over time, indicating continued growth. Akor stands with the smallest CV shows the diameter of the tree less dispersed than the other stands with the larger CV, Kaliandra, Weru, and Pilang (Table 3).

**Table 3.** Statistics of the diameter of four energy wood species

| Species | Age = 1.7 yr | | | Age = 2.1 yr | | | Age = 2.6 yr | | | Age = 3.2 yr | | |
|---|---|---|---|---|---|---|---|---|---|---|---|---|
| | Mean | Std Dev | CV | Mean | Std Dev | CV | Mean | Std Dev | CV | Mean | Std Dev | CV |
| Akor | 3.05 | 1.12 | 36.72 | 4.98 | 1.73 | 34.78 | 6.37 | 2.12 | 33.30 | 7.95 | 2.47 | 31.13 |
| Pilang | 1.87 | 0.75 | 40.24 | 2.24 | 1.04 | 46.40 | 2.62 | 1.86 | 71.02 | 2.76 | 1.29 | 46.90 |
| Weru | 1.47 | 0.36 | 24.71 | 1.74 | 1.02 | 58.33 | 2.73 | 1.03 | 37.84 | 3.52 | 1.36 | 38.69 |
| Kaliandra | 2.70 | 0.56 | 20.58 | 3.16 | 1.36 | 42.88 | 4.17 | 1.36 | 32.63 | 4.52 | 1.57 | 34.74 |

### *3.3. Volume potential*

At the last observation, we found that the Akor stand has the highest survival rate. The total number of individuals in the Akor stand was 113/plot, Kaliandra (89/plot), Pilang (84/plot), and Weru (81/plot). The potency of the total volume of these stands was: Akor > Kaliandra > Weru > Pilang. Akor was the stand with the largest volume compared to the others. It has a volume of 40.88 $m^3$/ha at 3.2 years old, followed by Kaliandra stand of 9.25 $m^3$/ha, Weru stand of 2.69 $m^3$/ha, and Pilang stand of 1.28 $m^3$/ha. The potency of the volume reflects how large the potential wood biomass for green energy feedstock in which the volume parameters correlate with the biomass.

In general, the growth of volume for four energy wood species has linear trends up to < 3.5 years old. Akor stands to show the highest slope compared to the others. It is shown that the Akor stand is the energy wood species that has the fastest volume growth, among others (Figure 3).

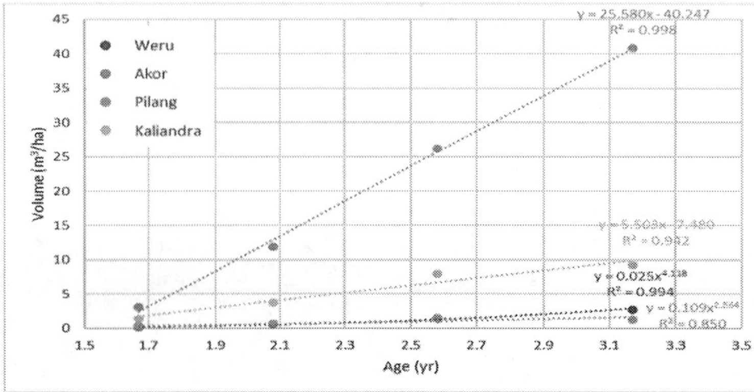


**Figure 3.** The trend of volume for four energy wood species at < 3.5 yr

### *3.4. Biomass production*

Biomass is the total amount of organic matter that lives above the ground in trees, including main stems, leaves, branches, twigs, and bark. Calculation of forest biomass can be used to measure the number of resources available in the forest, such as timber, and to assess the changes in forest structure. Biomass is an important variable of forest ecosystems in various environmental conditions in terms of their structure and function. Besides, the calculation of biomass also illustrates the role of forests in the carbon cycle [16].

Akor was the stand that produced the largest biomass in the experimental plot. It produced 1.86 tonnes/ha at 1.7 years old, followed by Kaliandra stand of 0.74 tonnes/ha, Pilang stand of 0.18 tonnes/ha, and Weru stand of 0.11 tonnes/ha. At 2.1 years, Akor stands produced biomass of 6.94 tonnes/ha, Kaliandra stand of 2.81 tonnes/ha, Pilang stand of 0.35 tonnes/ha, and Weru stand of 0.29 tonnes/ha. At 2.6 years old, Akor stands produced biomass of 15.01 tonnes/ha, Kaliandra stand of 4.19 tonnes/ha, Pilang stand of 0.82 tonnes/ha, and Weru stand of 0.74 tonnes/ha. At the end period of the observation, Akor was the species that was able to produce the highest biomass. Akor stand produced biomass of 23.20 tonnes/ha at 3.2 years old, followed by Kaliandra stand of 3.52 tonnes/ha, Weru stand of 1.41 tonnes/ha, and Pilang stand of 0.71 tonnes/ha. The biomass trend of four species was linear at < 3.5 yr, as shown in Figure 4.

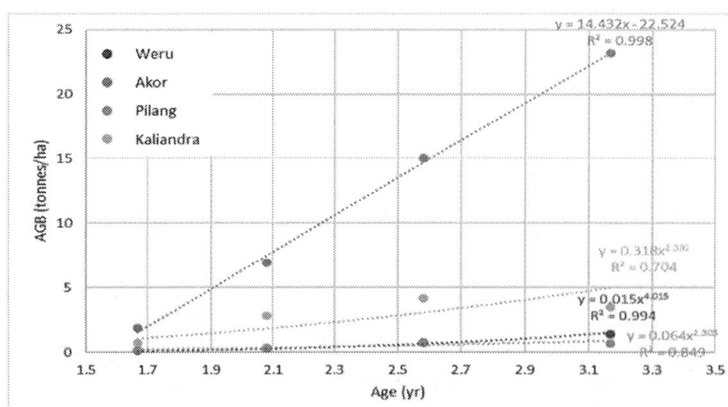


**Figure 4.** The trend of biomass for four energy wood species at < 3.5 yr

Comparison of diameter, height, volume, and accumulation of biomass in stands that grow on soil types and relatively similar environmental conditions shows that each energy wood species has different abilities to extract and use nutrients available for growth [17]. The amount of biomass accumulation is not only influenced by the physical characteristics of the tree (diameter, height). The magnitude of wood density also influences it. The density of wood shown a significant negative correlation, although weak, trees with a specific density of stems (including bark) of 0.2-0.49 show that the growth of stem diameter is relatively faster than trees with a specific density of more than 0.5 [18].

### *3.5. Volume and biomass increment*

The biomass increment of Akor stand on this site (7.32 tonnes/ha/yr) is not significantly different compared with the biomass increment found in Hutan Rakyat Desa Nglanggeran, Gunungkidul, Yogyakarta, which is 7.04 tonnes/ha/yr [19]. The Kaliandra stand has a volume increment of 2.92 $m^3$/ha/yr and a biomass increment of 1.11 tonnes/ha/yr. These are comparable with those recorded in Punak Ngegas-KPHK Unit Batulanteh, Sumbawa, which found the potential of Kaliandra biomass 2.7 tonnes/ha/yr [20]. The volume increment of the Weru stand was 0.85 $m^3$/ha/yr, and for biomass, the increment was 0.44 tonnes/ha/yr. The Pilang stand has volume and biomass increments of 0.41 $m^3$/ha/yr and 0.22 tonnes/ha, respectively. Akor was the stand with the highest increment on volume and biomass. It has been because the tree has the largest diameter and the highest height compared to the other species. Forest biomass differs significantly from each stand depending on species structure and composition, contributing to carbon accumulation and carbon density. Accumulation of carbon stored in standing biomass and organic soil in the experimental plots and competition of individual trees from medium and large diameter classes contribute dominantly to biomass accumulation [21]. The biomass of the tree species is also affected by the growth of the species [22].

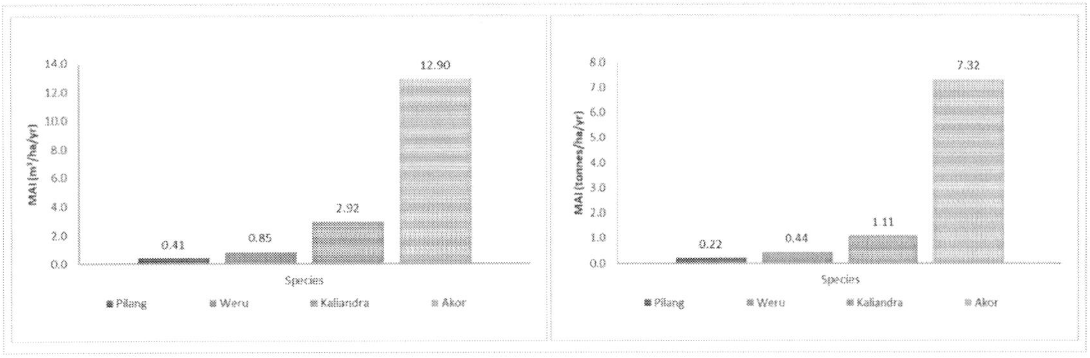


**Figure 5.** Volume and biomass stand increment of four energy wood species

## 4. **Conclusion**

Biomass-based energy wood as new renewable energy from the forestry sector is needed. There is not much growth information, and inventory data are available for energy wood species in Indonesia. It was considering the importance of the growth information necessary to develop experimental plots and conduct periodic measurements to get the information required on the stand growth of the energy wood species. The experimental plots of four energy wood species show that Akor is the fastest growing species and has the largest volume and biomass. However, this is an initial growth, and further research is needed to develop energy wood cultivation.

**Reference**

[1] Kholiq I 2015 Pemanfaatan energi alternatif sebagai energi terbarukan untuk mendukung subtitusi BBM *Jurnal IPTEK* **19** 75-91
[2] Noviyanti AR, Yuliyati YB, Solihudin, Eddy DR, and Tjokronegoro R 2016 Silika sekam padi *Jurnal Material dan Energi Indonesia* **6**(2) 1-6
[3] Manieniyan V, Thambidurai M, and Selvakumar R 2009 Study on energy crisis and the future of fossil. *Proceedings of SHEE* **10** 2234-3689
[4] Sudrajat R 1983 Pengaruh bahan baku, jenis perekat, dan tekanan kempa terhadap kualitas briket arang (*Laporan LPHH*) **165**
[5] Yudha SW 2017 Pemerintah perlu mengoptimalkan pemanfaatan energi baru terbarukan. (Yogyakarta: Humas UGM) *Https://Ugm.Ac.Id/Id/News/13754/*
[6] Owusu PA and Asumadu-Sarkodie S 2001 A review of renewable energy sources, sustainability issues and climate change mitigation *Cogent Engineering* **3**(1) 1167990
[7] Yulianti M, Nurrochmat DR, and Kuncahyo B 2018 Simulasi model pengembangan gasifikasi listrik berbasis biomasa hutan tanaman energi *Risalah Kebijakan Pertanian dan Lingkungan* **5**(1) 56-77
[8] Koeslulat E, Prayitno TA, Sutapa JPG, and Irawati D 2016 Karakteristik energi tiga jenis pohon cepat tumbuh pada tiga akelas diameter *Jurnal Agroforestri* **11**(6) 23-31
[9] Bustomi S 2010 *Pengelolaan Hutan Tanaman Kayu Energi* (Bogor)
[10] Orwa C, Mutua A, Kindt R, Jamnadass R, and Anthony S 2009 Agroforestry Database: a tree reference and selection guide version 4 (*World Agroforestry Centre*)
[11] Permatasari P, Susanto D, and Kusuma R 2018 Pertumbuhan stek pucuk kaliandra (*Calliandra Calothyrsus* Meissn.) dengan beberapa komposisi media tanam dan konsentrasi root up. *Bioprospek* **13**(2)
[12] Nurjannah S, Ayuningsih B, and Hernaman IS 2016 Penggunaan kaliandra (*Calliandra Calothyrsus*), *Indigofera* Sp. dan campurannya dalam ransum sebagai pengganti konsentrat terhadap produktivitas domba garut jantan *Jurnal Ilmiah Peternakan Terpadu* **11** 23-31
[13] Suita E and Bustomi S 2014 Berkecambah benih pilang techniques for increasing seed viability of *Acacia leucophloea* (Roxb.) Wild. *Jurnal Penelitian Hutan Tanaman* **11**(1) 45-52
[14] Conti G, Gorné LD, Zeballos SR, Lipoma ML, Gatica G, Kowaljow E, and Fernandes P 2019 Developing allometric models to predict the individual aboveground biomass of shrubs worldwide *Global Ecol Biogeogr*. **28**(7) 961-75
[15] Li YF, Yang HP, Wang HX, Ye SM, and Liu WZ 2019 Assessing the influence of the minimum measured diameter on forest spatial patterns and nearest neighbourhood relationships *Journal of Mountain Science* **16**(10) 2308-19
[16] Brown S 1997 Estimates biomass and biomass change of tropical forest (*FAO Forestry Paper*) **134**
[17] Purwanto RH, Rohman, Maryudi A, Teguh Yuwono, Permadi DB, and Sanjaya M 2015 Potensi biomasa dan simpanan karbon jenis-jenis tanaman berkayu di hutan rakyat desa Nglanggeran, Gunungkidul, Daerah Istimewa Yogyakarta *Jurnal Ilmu Kehutanan* **6**(2) 128-41
[18] Siarudin M and Indrajaya Y 2020 Adaptation and productivity of kaliandra for biomass energy source *IOP Conference Series: Earth and Environmental Science* **415**
[19] Hu Y, Su Z, Li W, Li J, and Ke X 2015 Influence of tree species composition and community structure on carbon density in a subtropical forest *PLoS ONE* **10**(8) 1-9
[20] Daniel TW, Helms JA, and Barker FS 1980 *Prinples of Silvikultur* (New York: Mc. Graw Inc.)
[21] Suzuki E 1999 Diversity in specific gravity and water content of wood among Bornean tropical rainforest trees *Ecological Research* **14** 211-24

[22] Wang D, Bormann FH, Lugo AE, and Bowden RD 1991 Comparison of nutrient-use efficiency and biomass production in five tropical tree taxa *Forest Ecology and Management* **46** 1-21

**Acknowledgment**
Data used in this paper was part of the research funded through the Research and Development Center for Forest Productivity Improvement. The authors thank to the Integrative Research Coordinator on Plantation Forest Management for Energy Wood for the period 2010-2014, Ir. Sofwan Bustomi, M.Si. and his team for initiating and developing activities.

IOP Conf. Series: Materials Science and Engineering **935** (2020) 012050 doi:10.1088/1757-899X/935/1/012050

# Examination of fecal pellet physical characteristics of an invasive drywood termite, *Cryptotermes dudleyi* (Isoptera: Kalotermitidae): A potential approach for species marker and non-destructive monitoring method

**S L D Zega [1], A Fajar [2]*, S K Himmi [2], D S Adi [2], D Tarmadi [2], D Nandika[1], S Yusuf [2]**

[1]Department of Forest Products, Faculty of Forestry & Environment, IPB University, Bogor 16680, Indonesia
[2]Research Center for Biomaterials, Indonesian Institute of Sciences (LIPI), Cibinong 16911, Indonesia

*Email: anugerah.fajar@lipi.go.id

**Abstract.** Cryptotermes, one of the major drywood termite genera in Indonesia, is a cryptic species that isolates their presence inside a wood piece. Due to its cryptic lifecycle hidden away inside wood, monitoring its presence and also identifying the corresponding species has been a difficult process. One of the Cryptotermes species, native to Java island, Indonesia, is *Cryptotermes dudleyi*. In this preliminary study, we used *C. dudleyi* as a species model to find out whether fecal pellet physical characteristics can be used as a stable species marker to assist in non-destructive monitoring surveillance. The characteristics used were maximum diameter, diagonal width of maximum diameter, and 2D surface area and area perimeter. The study used fecal pellets from orphaned and mature colonies and three different dietaries (grass, hardwood, and softwood) to check whether the characteristic value of fecal pellets is narrow over various influences. The results showed that each characteristic tends to have unique mean and also its unique value range which depends on their collection site condition. Due to unique means and value range, the species marker become not so accurate and not robust enough as consequences in accommodating these unique means and value range. On the other hand, employing completely new clustering based on like-like axiom on individual fecal pellets may results in robust species marker as long as adequate data comparison from other species fecal pellets to validate the species marker is available.

## 1. Introduction

Drywood termite has been characterized for its habit to dwell and feed in a single dry wood (a one-piece nester) which put the colony in a protected environment [1,2]. Most of the time, this protected environment is a part of human living space which makes drywood termites a pest since they can attack various objects made of wood in human space such as furniture, flooring, up to the body of house itself [3,4]. This condition makes monitoring the drywood infestation to be important especially for the prevalent environment like wooden building or various materials made of cellulose [4]. Several approaches have been developed in the monitoring of drywood termite infestation. These methods ranging from finding fecal pellet next to the wood, finding termite's wings near the wood, conducting focused light over the wood to wood scan using x-ray tomography [4-9]. The various development in monitoring method makes the monitoring of drywood termite has been easier than before. On the other hand, the main issue of identifying the species of drywood termites remain a difficult process.

The concern for identifying the species of drywood termite has become more relevant nowadays due to the drywood termite strict infestation that favored wood low moisture level and the surrounding environment warm temperature [10-12]. While normally such conditions limit drywood infestation and its potential to spread to other region, current era of globalized trade with modern housing equipped with controlled temperature has turned our house into the most viable place for drywood termite to breed. This situation makes drywood termites that were distributed fragmentally based on regions seasonal weather and general climate to be unrestricted and can cross over to other regions easily [4,13,14]. To combat this issue, the ability to identify and monitor the species of drywood termites has become important especially in the context of goods transfer across the world and also the biosecurity of the corresponding area.

The easiest way to identify termite down to species level is through the morphological features. Regarding the drywood termite, the morphospecies method might not be the easiest to conduct, not because of the difficulty in distinguishing the morphological features between species, but due to the difficulty in procuring the specimen in the first place. The only feasible method to get the specimen is through wood dissection [4,15]. While wood dissection manages to get the specimen out, our past experience results in a high number of destroyed specimens due to dissection shear force effected to termite inside the gallery in the wood. This situation makes the viable number of specimens to be used become smaller. Adding the fact that soldier was found in low abundance in drywood termite makes the wood dissection sometimes yielded no viable specimen to be identified. To counter this issue, a method to identify drywood termite without the specimen need to be developed.

One of the promising ideas is through fecal pellet as a mean to identify the species that may be useful for non-destructive approach for drywood termite monitoring and inspection. Earlier studies had showed that fecal pellet hydrocarbon composition can track the termite species without the specimen [16]. Another past study also showed, albeit ambiguously, that the fecal pellet of *Cryptotermes dudleyi* and *Cryptotermes cynocephalus* can be distinguished based on length and width dimension [17]. While this result might indicate the viability of fecal pellet physical characteristics as species differentiator, we are interested on whether these physical characteristics can be relatively unchanging under various condition and feed type. On this basis, we try to verify and conduct in-depth study regarding the suitability of physical characteristics of fecal pellets of drywood termite to be a species marker. To do so, we used a native drywood termite of Java Island, *C. dudleyi*, a well-known dry-wood termite wood pest as the species model in our study.

## 2. Materials and Methods

### *2.1 Termite source and fecal pellets collections*

The colony of drywood termite *C. dudleyi* that was used for short-term laboratory scale experiment with various dietaries subjection, was collected from Bogor, Indonesia. The mature colony of drywood termite that was used for long term field scale was collected from Termite Rearing Laboratory, Research Center for Biomaterials LIPI, Cibinong, Bogor, Indonesia.

### *2.2 Fecal pellets observation*

#### 2.2.1 Orphaned colony-laboratory scale experiment.

The experiment was conducted using three different feed sources (bamboo, rubber tree, and pine tree) representing grass, hardwood, and softwood. All of the feed sources were cut into blocks of 2 x 1 x 1 cm. For each feed source, 100 workers and 5 soldiers of orphaned colonies of *C. dudleyi* were introduced into 10 blocks of feed source and kept in a dark room. After 6 months, accumulated pellets were collected to be measured. An orphaned colony of at least two years old that was fed with hardwood was also added to the test as a comparison data, later referred as L.A1 in the study.

IOP Conf. Series: Materials Science and Engineering **935** (2020) 012050 doi:10.1088/1757-899X/935/1/012050

2.2.2 Mature colony-field scale experiment.
The collection of *C. dudleyi* fecal pellets were from four different types of hardwood timbers that has been infested for years, estimated more than 10 years. The termites were collected through through chipping the wood. The fecal pellets were collected from inside the infested wood and then put into sealed zipper bags for further measurement.

### *2.3 Fecal pellets measurement*
The collected fecal pellets from the laboratory scale and field collection was counted and labeled as indicated in Table 1.

**Table 1**. Fecal pellets identity

| ID label | Experiment set-up | Termite feed | Fecal pellets (pieces) |
|---|---|---|---|
| F.C1 | Field scale | Hardwood | 434 |
| F.C2 | Field scale | Hardwood | 470 |
| F.C3 | Field scale | Hardwood | 392 |
| L.A1 | Laboratory scale | Hardwood | 546 |
| L.B | Laboratory scale | Grass | 284 |
| L.R | Laboratory scale | Hardwood | 378 |
| L.P | Laboratory scale | Softwood | 168 |

These fecal pellets were observed using Keyence VHX 6000 (Keyence Corp., Osaka, Japan) using 150x magnifications over 225 stitched collated pictures for each type of feed. The measurement was conducted using *auto grain* function from Keyence Communications Tools (Keyence Corp., Osaka, Japan) with measured characteristics of area, perimeter, maximum diameter, minimum diameter (widest width), and diagonal width of maximum diameter.

### *2.4 Data analysis*
The measured characteristics were then turned into monotonic variables as described below,

$$\text{Shape Ratio(SR)} = \frac{\text{diagonal width}}{\text{maximum length}} \tag{1}$$

$$\text{Enclosure Ratio (ER)} = \frac{\text{perimeter}}{2(\text{max diameter} + \text{diagonal width})} \tag{2}$$

The Shape Ratio (SR) score indicates the general outline of fecal pellets. The closer the number to 1, the pellet's shape will be like a ball or cube while closer to 0 will indicate extreme ellipse or rectangle like shape. The Enclosure Ratio (ER) score indicates further classification of the shape ratio score in which closer to one might indicates a cube or rectangle while closer to 0 will indicates a ball or ellipse. The ER score of more than 1 indicates perimeter that was too wavy and meandering that might not well represent the perimeter.

At first, these monotonic variables, area and perimeter were tested according to several factors such as the generalized effect of colony condition between orphaned-laboratory scale, and mature field scale, the generalized effect of feeding materials (grass, softwood and hardwood) and specially for the condition of the orphaned colonies, the effect of different hardwood, albeit not yet identified in the field case, and the effect of consuming hardwood in different colony condition. We were interested in the effect of these factors in shaping the physical characteristics of the produced fecal pellets. These comparison tests were conducted using *Kruskal Wallis* test. To complete the comparison at collection site level, we also conducted site-to-site comparison using *Wilcox* test.

The data calculation was conducted in R (version 3.6.2, R Core Team, https://www.R-project.org/). The *Kruskal-Wallis* and *Wilcox test* were conducted using *rstatix* R package (version 0.6.0, A.

Kassambara, https://CRAN.R-project.org/package=rstatix). The test was grouped based on factors to consider as mentioned previously and direct site to site comparison. The test also compared each collection site to the base mean of the pooled data for each characteristic. Data presentation was using *ggpubr* (version 0.4.0, A. Kassambara, https://CRAN.R-project.org/package=ggpubr) and *ggplot2* r packages [18].

## 3. Results

The study managed to collects 1,296 fecal pellets from field scale collection (F.C1 = 434, F.C2 = 470, F.C3 = 392) and 1,376 fecal pellets from laboratory scale experiment (L.A1 = 546, L.R = 378, L.B = 284, L.P = 168). All of these fecal pellets were measured and then presented in boxplot together with the result of Wilcox test against the base mean of the corresponding characteristics are shown in Figure 1.

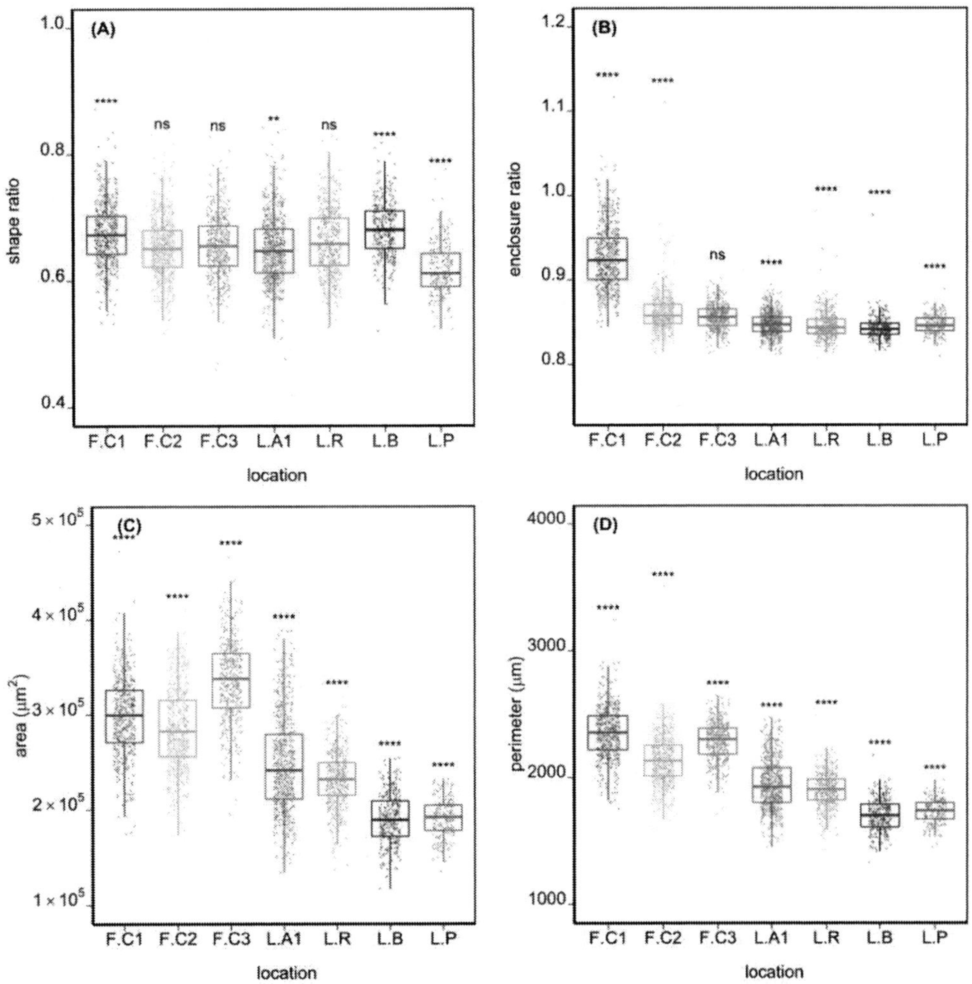


**Figure 1.** The boxplot of characteristics value of observed characteristic (A) Shape Ratio (SR), (B) Enclosure Ratio (ER), (C) Area, and (D) Perimeter from each collection site. The post-hoc marker was the *p value* adjusted significance generated using unpaired *Wilcox* test against the base mean of the corresponding characteristics, ns: not significant

IOP Conf. Series: Materials Science and Engineering **935** (2020) 012050 doi:10.1088/1757-899X/935/1/012050

Based on Figure 1, the spread of the characteristics value was not equally distributed with a lot of outlier, especially in enclosure ratio (ER) data that showed anomaly. The anomaly was indicated by some ER value from C1 and C2 that exceed 1 which indicated the perimeter has extreme waviness. This condition made generalized interpretation of fecal pellets shape in these collection site to be prone to mistake once closing to score 1 (Figure 1B).

Another observation also showed that there was a clear difference ($p = < 2 \times 10^{-16}$) in fecal pellets 2D surface area between the laboratory experiment (L.A1, L.R, L.B, and L.P) and the field experiment (F.C1, F.C2, and F.C3), see Figure 1C. A similar observation was also observed in perimeter category ($p = < 2 \times 10^{-16}$). The only parameter that was not significantly different between the two experiment was the shape ratio ($p = 0.34$). An observation from Figure 1C also showed increasing size of fecal pellet from orphaned colony (L.R and L.A1) to mature colony (F.C1-3).

Due to the limitation of the graphical marker, only the *Wilcox* test against the base mean of all of the corresponding value was used as the reference in deriving the significant difference in Figure 1. The rest of the comparison between factors was tested and the results are presented in Table 2 below.

**Table 2**. The result of comparison test using *Kruskal Wallis*

| Factors | Character | n | Statistic | df | *p* | *p*.adj | *p*.adj. signif |
|---|---|---|---|---|---|---|---|
| Global effect from different colony condition (orphaned - laboratory and mature-field colony) | Shape Ratio (SR) | 2672 | 6.900685223 | 1 | 0.00862 | 0.00862 | ** |
| | Enclosure Ratio (ER) | 2672 | 798.9699715 | 1 | 9.04E-176 | 9.04E-176 | **** |
| | Area | 2672 | 1207.134831 | 1 | 1.72E-264 | 1.72E-264 | **** |
| | Perimeter | 2672 | 1404.987676 | 1 | 1.73E-307 | 1.73E-307 | **** |
| Global effect from different feeding material regime (hardwood, softwood and grass) | Shape Ratio (SR) | 2672 | 185.7544109 | 2 | 4.61E-41 | 4.61E-41 | **** |
| | Enclosure Ratio (ER) | 2672 | 226.3501713 | 2 | 7.06E-50 | 7.06E-50 | **** |
| | Area | 2672 | 849.1043295 | 2 | 4.16E-185 | 4.16E-185 | **** |
| | Perimeter | 2672 | 849.6756433 | 2 | 3.13E-185 | 3.13E-185 | **** |
| Feeding effect of different hardwood in mature-field colony | Shape Ratio (SR) | 1296 | 54.38302957 | 2 | 1.55E-12 | 1.55E-12 | **** |
| | Enclosure Ratio (ER) | 1296 | 757.6002249 | 2 | 3.08E-165 | 6.16E-165 | **** |
| | Area | 1296 | 260.4235504 | 2 | 2.82E-57 | 2.82E-57 | **** |
| | Perimeter | 1296 | 267.7196189 | 2 | 7.34E-59 | 7.34E-59 | **** |
| Feeding effect between hardwood, softwood, and grass in orphaned-laboratory colony | Shape Ratio (SR) | 1376 | 186.8519923 | 3 | 2.92E-40 | 5.84E-40 | **** |
| | Enclosure Ratio (ER) | 1376 | 37.21191773 | 3 | 4.15E-08 | 4.15E-08 | **** |
| | Area | 1376 | 474.9808627 | 3 | 1.26E-102 | 2.52E-102 | **** |
| | Perimeter | 1376 | 463.8393194 | 3 | 3.27E-100 | 6.54E-100 | **** |
| Feeding hardwood effect in different colony state (orphaned - laboratory and mature-field colony) | Shape Ratio (SR) | 2220 | 9.648243082 | 1 | 0.0019 | 0.0019 | ** |
| | Enclosure Ratio (ER) | 2220 | 598.4695654 | 1 | 3.60E-132 | 3.60E-132 | **** |
| | Area | 2220 | 746.6913517 | 1 | 2.10E-164 | 2.10E-164 | **** |
| | Perimeter | 2220 | 953.9219429 | 1 | 1.86E-209 | 1.86E-209 | **** |

The result in Table 2, clearly demonstrated that for each factor accounted in this study, all of them showed a stark contrast in characteristic value which unwittingly expands the range for physical characteristics value that can be attributed to *C. dudleyi*. To better understand the scope of range existed for *C. dudleyi* according to our measured physical characteristics, multiple random, direct site to site comparison under *Wilcox* test was used to test the limit of physical characteristics that belong to *C. dudleyi*. The result was presented in Table 3.

IOP Conf. Series: Materials Science and Engineering **935** (2020) 012050 doi:10.1088/1757-899X/935/1/012050

**Table 3.** *Wilcox* test results for all of the possible combination in two measured characteristics and two monotonic characteristics

| Combination | | Shape Ratio (SR) | | | Enclosure Ratio (ER) | | | Area (μm$^2$) | | | Perimeter (μm) | | |
|---|---|---|---|---|---|---|---|---|---|---|---|---|---|
| gr.1 | gr.2 | *P* | *p*.adj | *p*.adj. signif | *p* | *p*.adj | *p*.adj. signif | *p* | *p*.adj | *p*.adj. signif | *p* | *p*.adj | *p*.adj. signif |
| F.C1 | F.C2 | 3.47E-12 | 3.82E-11 | **** | 1.01E-123 | 1.82E-122 | **** | 2.54E-06 | 7.62E-06 | **** | 7.04E-51 | 5.63E-50 | **** |
| F.C1 | F.C3 | 1.86E-08 | 1.86E-07 | **** | 1.13E-125 | 2.15E-124 | **** | 1.28E-33 | 5.12E-33 | **** | 2.71E-07 | 8.13E-07 | **** |
| F.C1 | L.A1 | 1.41E-14 | 1.97E-13 | **** | 1.58E-154 | 3.32E-153 | **** | 2.30E-50 | 1.84E-49 | **** | 1.76E-111 | 3.17E-110 | **** |
| F.C1 | L.R | 0.000228 | 0.001824 | ** | 5.62E-127 | 1.12E-125 | **** | 1.54E-87 | 2.46E-86 | **** | 4.15E-117 | 8.72E-116 | **** |
| F.C1 | L.B | 0.013 | 0.078 | ns | 3.72E-111 | 6.32E-110 | **** | 1.11E-106 | 2.00E-105 | **** | 3.38E-112 | 6.42E-111 | **** |
| F.C1 | L.P | 1.11E-34 | 2.22E-33 | **** | 4.45E-79 | 7.12E-78 | **** | 4.26E-77 | 5.54E-76 | **** | 7.88E-80 | 1.10E-78 | **** |
| F.C2 | F.C3 | 0.412 | 0.568 | ns | 0.006 | 0.024 | * | 8.65E-52 | 7.79E-51 | **** | 5.21E-33 | 2.08E-32 | **** |
| F.C2 | L.A1 | 0.284 | 0.568 | ns | 1.01E-31 | 1.21E-30 | **** | 2.45E-34 | 1.23E-33 | **** | 1.04E-45 | 7.28E-45 | **** |
| F.C2 | L.R | 0.026 | 0.13 | ns | 1.63E-35 | 2.28E-34 | **** | 1.44E-70 | 1.73E-69 | **** | 8.57E-71 | 9.43E-70 | **** |
| F.C2 | L.B | 1.14E-17 | 1.82E-16 | **** | 3.62E-45 | 5.43E-44 | **** | 1.26E-107 | 2.39E-106 | **** | 4.96E-108 | 8.43E-107 | **** |
| F.C2 | L.P | 1.30E-17 | 1.95E-16 | **** | 1.23E-18 | 1.11E-17 | **** | 3.87E-78 | 5.81E-77 | **** | 4.74E-76 | 5.69E-75 | **** |
| F.C3 | L.A1 | 0.076 | 0.304 | ns | 6.05E-20 | 6.05E-19 | **** | 2.29E-95 | 3.89E-94 | **** | 3.26E-97 | 4.89E-96 | **** |
| F.C3 | L.R | 0.164 | 0.492 | ns | 1.48E-24 | 1.63E-23 | **** | 9.22E-115 | 1.94E-113 | **** | 2.60E-112 | 5.20E-111 | **** |
| F.C3 | L.B | 1.55E-13 | 1.86E-12 | **** | 2.02E-34 | 2.63E-33 | **** | 2.93E-108 | 5.86E-107 | **** | 6.47E-108 | 1.04E-106 | **** |
| F.C3 | L.P | 5.68E-18 | 9.66E-17 | **** | 6.03E-13 | 4.82E-12 | **** | 4.10E-78 | 5.81E-77 | **** | 1.15E-77 | 1.50E-76 | **** |
| L.A1 | L.R | 0.003 | 0.021 | * | 0.002 | 0.01 | ** | 8.59E-05 | 0.0001718 | *** | 0.007 | 0.007 | ** |
| L.A1 | L.B | 9.66E-20 | 1.84E-18 | **** | 9.72E-09 | 6.80E-08 | **** | 1.60E-60 | 1.76E-59 | **** | 1.24E-63 | 1.24E-62 | **** |
| L.A1 | L.P | 1.94E-14 | 2.52E-13 | **** | 0.852 | 0.852 | ns | 1.30E-45 | 7.80E-45 | **** | 9.52E-39 | 4.76E-38 | **** |
| L.R | L.B | 5.15E-08 | 4.64E-07 | **** | 0.022 | 0.066 | ns | 1.74E-56 | 1.74E-55 | **** | 7.93E-61 | 7.14E-60 | **** |
| L.R | L.P | 5.59E-19 | 1.01E-17 | **** | 0.025 | 0.066 | ns | 5.75E-47 | 4.03E-46 | **** | 1.39E-40 | 8.34E-40 | **** |
| L.B | L.P | 3.91E-38 | 8.21E-37 | **** | 6.97E-06 | 4.18E-05 | **** | 0.751 | 0.751 | ns | 0.002 | 0.004 | ** |

Note. gr. = group; *p* = *p value*; *p*.adj = *p value* adjusted; *p*.adj.signif. = *p value* adjusted significance

The results of *Wilcox* test in Table 3 paint a better approximation of how different the value of each characteristic from each collection site and the range of physical characteristics value. It was interesting to note in the SR characteristic from the field collection that only F.C1 that was different significantly in value compared to the rest. On the other hand, the L.R and L.A1 from the laboratory scale was not so different in SR value compared to F.C2 and F.C3 from the field collection. This result may suggest that SR value are stable across most of the collection site that feed on hardwood, the SR value itself is too wide in range that this stable characteristic value might not yield much meaningful information on the fecal pellets formation. Regarding the rest of the characteristics in Table 3, the only other interesting observation was the surface area between L.B (bamboo) and L.P (pine) was not significantly different although the SR characteristics showed a stark difference from each other for these two samples.

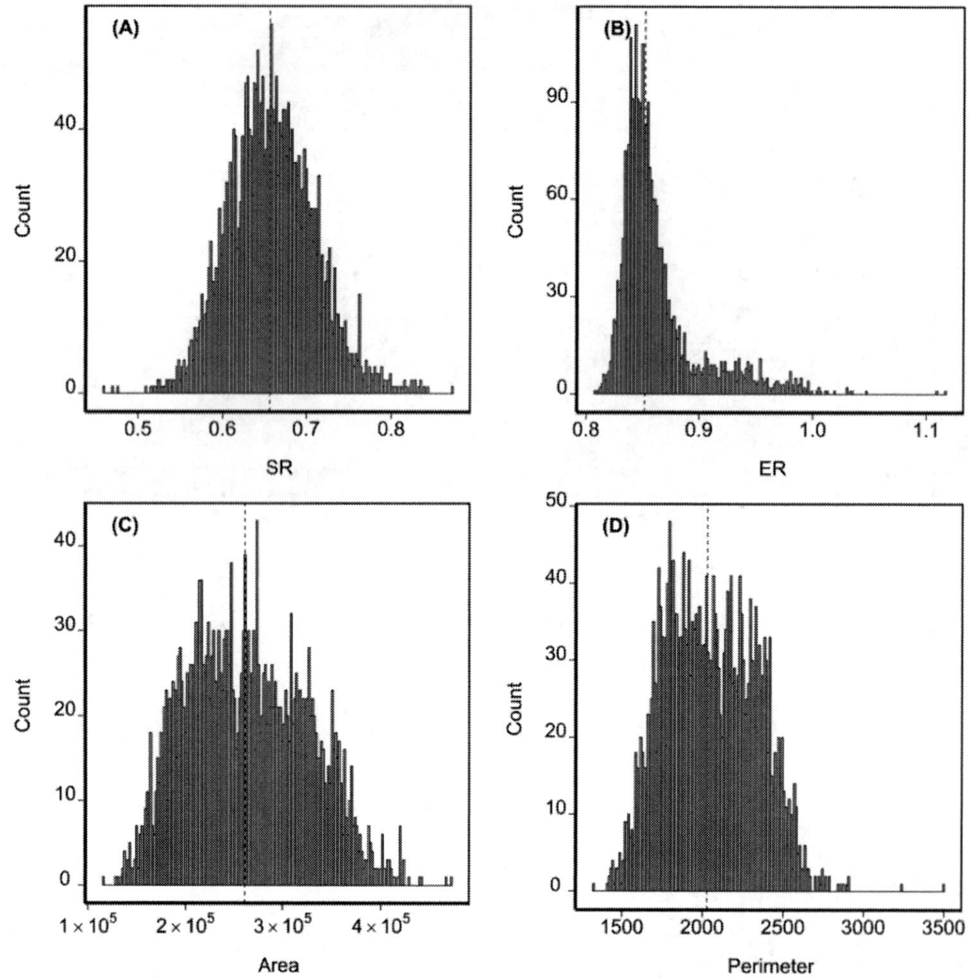


**Figure 2.** Pooled histogram from seven collection site for characteristic (A) Shape Ratio (SR), (B) Enclosure Ratio (ER), (C) Area, and (D) Perimeter. The data was split into 200 bin per characteristic to visualize the break in the data continuity. The vertical dashed line in the middle of each plot was the median marker

Summing up the observation made from Figure 1 together with Table 1 and Table 2 indicated that although various factors comparison showed stark contrast to the rest characteristics value other than in SR, once the individual collection site was directly compared to each other, the results still showed a significant difference between collection site. Therefore, due to each collection site resulted in significantly different physical characteristic's value that differs from one another, we can state that the range of physical characteristics value that belongs to *C. dudleyi* comes from the entire data set almost without repetition other than for SR characteristic.

To get the best approximation of the physical characteristic's value range from each variable, we have built a histogram from each characteristic to visualize the range as presented in Figure 2. The pooled distribution data as presented in figure 2 showed that SR has normal distribution which explains its *Kruskal Wallis* test result. On the other hand, the other characteristics distribution showed unequal distribution with some cases of gradual leaning favoring one side (skewness) or a series of choppy hills across the plot. The observed skewness indicated step-change in fecal pellets features with prominent example in ER histogram plot (Figure 2B) and presence of gap in termite fecal pellet sizes as shown in Figure 2C and 2D.

The result depicted in figure 2 demonstrated that generally fecal pellets of *C. dudleyi* will have these characteristic's value range in which certain smaller range may corresponds to details regarding the species condition whether it was the colony condition or its feed material. The generalized characteristic's means and standard deviation shown below in Table 4.

**Table 4.** Generalized physical characteristics of the observed *C. dudleyi*

| Characteristic | mean | median | SD | IQR | MAD |
|---|---|---|---|---|---|
| SR | 0.6582 | 0.6569 | 0.0512 | 0.0672 | 0.0497 |
| ER | 0.8631 | 0.8523 | 0.0359 | 0.0286 | 0.0193 |
| Area (um2) | 263530.65 | 260374.50 | 61231.52 | 95104.25 | 69960.19 |
| Perimeter (um) | 2046.28 | 2029.00 | 280.92 | 445.25 | 326.17 |

Note. SD = Standard Deviation, IQR = Interquartile Range; MAD = Median Absolute Deviation

## 4. Discussion

The physical characteristics of fecal pellets from *Cryptotermes dudleyi* have been shown to be distributed in large spread of value for each corresponding characteristic (Figure 1 and 2). This range of value was made through data composite by combining each collection site characteristic's value. The need to pool all of the data into one arise due to each collection site individual means and value range data which showed to be significantly different than other sites (Table 3). The key issue remains on what drive the different means for each value range as observed in each collection site.

To address this key issue, we tried group comparison to assess factors that may influence the formation of fecal pellets. While we acknowledge that our data combination was actually shaped by two clumped-factors (colony condition tied with variance in feeding materials) that cannot be separated further cleanly, this data can still give a generalized view of the two factors that was the effect of different colony condition and feeding materials at larger scale once it constituent sub-factors were also analyzed. In our case, the analysis was done by comparing a lot of interrelated factors. These interrelated factors can be seen after the *global factors* in Table 2.

Our analysis showed that there is a clear effect ($p = < 2 \times 10^{-16}$) from the condition of the colony to the fecal pellet formation, in which an orphaned, newly inducted colony will generally yield smaller size fecal pellet than a mature colony. To understand this further, the comparison results in feeding hardwood between two colony condition and direct site-to-site comparison between site in field scale compared to L.R confirmed that they are different in means although the characteristics value range can overlap each other at some range position (Figure 1). This is the first report that documented this condition. There are several biological factors that might answer this observation such as stress reaction after being removed and inducted in new place, and change in general caste composition that favored certain caste [4,19]. Further studies are required to find which exact factor that play into this observed condition.

Another observation also showed that there was a clear effect of different feeding materials. The finding in the orphaned colony found that fecal pellet made of pinewood and grass to be smaller in size and also similar to each other, as indicated by smaller value of surface area and perimeter compared to fecal pellet produced by eating hardwood (Table 2). While this seems to indicate that different feeding material can result in different means value for the characteristics, feeding on the same diet (hardwood) also gave significantly different means with their corresponding value range.

As additional note, while we only restricted our study to physical characteristics measurement, we also observed that some fecal pellet is more fragile compare to other with special note to fecal pellet produce by *C. dudleyi* eating pinewood (L.P). The impression of fragile was indicated by finding a lot of broken fecal pellet that was cut off in the middle or having its cone section cut off which makes them unviable to be measured. The fragileness observed might be related to the type of feed since fecal pellets produced from other source was not as fragile as fecal pellets derived from pinewood (softwood).

To understand the varied sizes of fecal pellets, the mechanism behind the production of fecal pellets would certainly help. As past studies have reported, drywood termite fecal pellets will typically have six-sided/hexagon shape in the middle bounded by a dome in one side and a cone in the other side. This typical shape is most likely generated due to the molding effect of termite anus sphincter muscle that shape the feces into pellets [11]. Based on regular diet of drywood termite, water will be absorbed to the greatest level which could explain the pellet form of the fecal matter as form of water conservation [12,20]. As comparison, past study and our experience in collecting termite feces in subterranean termite found wet and sticky fecal matter since subterranean termite lives in location with water abundance [21]. The pellets also served a function as plug hole to cut off the colony from outside disturbance when needed [1].

After discussing the possibility of factors that influence the varied sizes of fecal pellets, the question is the usage of these physical characteristics as species marker. Our current study clearly indicates that the features of fecal pellets characteristics are too wide and far from being uniform (Table 2). Trying to use them directly as a definitive species marker is possible but due to the value range that is too wide, the filter created through the measured characteristics web will most likely introduced a lot of false positive from other drywood termite species that shared similar measured characteristics to our current data. This will become a hit and miss situation with more miss to be expected.

To find a solution to not-so-sensitive physical value range of *C. dudleyi*, we believe that the solution can be derived by clustering the individual fecal pellets based on its physical characteristics measurement through an axiom with the end goal to reach a data uniformity or a data fingerprint. This approach was noticed once we divided the pooled data back into the group comparison and then to site to site comparison which for each smaller group comparison yielded more detail-oriented information. While this idea is doable to be executed through the existing data, the question remains on how much group that actually exists in our data. Through machine learning, clustering the existing data using clustering algorithm such as k-means clustering is very feasible [22], but we do not have other references, e.g., other drywood termite species fecal pellets, to compare the clustering results. Therefore, we cannot validate our data fingerprint for now.

In the end, our study has demonstrated that the characteristics means value was highly depended on factors such as colony condition and type of dietary materials. We have also given the range of species marker and indicated its usage drawback of possibility to have high rate of false positive. And finally, we come up with an approach to turn existing measurement parameters into more robust species marker for the future works.

## Reference

[1] Himmi SK, Yoshimura T, Yanase Y, Oya M, Torigoe T, Akada M, and Imadzu S 2016 Nest-gallery development and caste composition of isolated foraging groups of the drywood termite, Incisitermes minor (isoptera: Kalotermitidae) *Insects* **7**
[2] Eggleton P and Tayasu I 2001 Feeding groups, lifetypes and the global ecology of termites *Ecological Research* **16** 941-60
[3] Rust MK and Su N-Y 2012 Managing Social Insects of Urban Importance *Annual Review of Entomology* **57** 355-75

[4] Lewis V and Forschler B 2014 Management of drywood termites: past practices, present situation and future prospects *Urban insect pests: sustainable management strategies* (Wallingford: CABI) pp. 130-53
[5] Himmi SK, Yoshimura T, Yanase Y, Torigoe T, Akada M, Ikeda M, and Imazu S 2018 Volume Visualization of Hidden Gallery System of Drywood Termite Using Computed Tomography: A New Approach on Monitoring of Termite Infestation *Sustainable Future for Human Security* vol Vi (Singapore: Springer Singapore) pp. 61-8
[6] Himmi SK, Yoshimura T, Yanase Y, Oya M, Torigoe T, and Imazu S 2014 X-ray tomographic analysis of the initial structure of the royal chamber and the nest-founding behavior of the drywood termite Incisitermes minor *Journal of Wood Science* **60** 453-60
[7] Himmi SK, Yoshimura T, Yanase Y, Mori T, Torigoe T, and Imazu S 2016 Wood anatomical selectivity of drywood termite in the nest-gallery establishment revealed by X-ray tomography *Wood Science and Technology* **50** 631-43
[8] Choi B, Himmi SK, and Yoshimura T 2017 Quantitative observation of the foraging tunnels in Sitka spruce and Japanese cypress caused by the drywood termite Incisitermes minor (Hagen) by 2D and 3D X-ray computer tomography (CT) *Holzforschung* **71** 535-42
[9] Himmi SK, Yoshimura T, Yanase Y, Torigoe T, and Imazu S 2018 Asssesment of initial colony founding by swarming reproductives of the western drywood termite, incisitermes minor *IOP Conference Series: Earth and Environmental Science* **166**
[10] Cabrera BJ and Rust MK 1994 The effect of temperature and relative humidity on the survival and wood consumption of the western drywood termite, Incisitermes minor (Isoptera: Kalotermitidae) *Sociobiology* **24** 95-113
[11] Rust MK, Reierson DA, and Scheffrahn RH 1979 Comparative habits, host utilization and xeric adaptations of the southwestern drywood termites, Incisitermes fruticavus Rust and Incisitermes minor (Hagen)(Isoptera: Kalotermitidae) *Sociobiology* **4** 239-56
[12] Lewis VR, Leighton S, Tabuchi R, Baldwin JA, and Haverty M I 2013 Influence of environmental factors on activity patterns of incisitermes minor (Isoptera: Kalotermitidae) in naturally infested logs *Journal of Economic Entomology* **106** 338-46
[13] Evans TA, Forschler BT, and Grace JK 2013 Biology of invasive termites: a worldwide review *Annual Review of Entomology* **58** 455-74
[14] Scheffrahn RH, Křeček J, Ripa R, and Luppichini P 2009 Endemic origin and vast anthropogenic dispersal of the West Indian drywood termite *Biological Invasions* **11** 787-99
[15] Grace JK, Woodrow RJ, and Oshiro RJ 2009 Expansive gallery systems of one-piece termites (Isoptera: Kalotermitidae) *Sociobiology* **54** 37-44
[16] Haverty MI, Woodrow RJ, Nelson LJ, and Grace JK 2005 Identification of termite species by the hydrocarbons in their feces *Journal of Chemical Ecology* **31** 2119-51
[17] Rojo MJ 2018 Variation in shape and size of fecal pellets as a diagnostic tool for drywood (Kalotermitidae: Blattodea) termite species identification *International Journal of Biosciences* **13** 147-51
[18] Wickham H 2009 *ggplot2: Elegant Graphics for Data Analysis* (New York: Springer New York)
[19] Neoh KB and Lee CY 2011 Developmental stages and caste composition of a mature and incipient colony of the drywood termite, Cryptotermes dudleyi (Isoptera: Kalotermitidae) *Journal of Economic Entomology* **104** 622-8
[20] Korb J and Lenz M 2004 Reproductive decision-making in the termite, Cryptotermes secundus (Kalotermitidae), under variable food conditions *Behavioral Ecology* **15** 390-5
[21] Tarmadi D, Yoshimura T, Tobimatsu Y, Yamamura M, Miyamoto T, Miyagawa Y, and Umezawa T 2017 The effects of various lignocelluloses and lignins on physiological responses of a lower termite, *Coptotermes formosanus Journal of Wood Science* **63** 464-72
[22] Xiong H, Wu J, and Chen J 2009 K-Means Clustering Versus Validation Measures: A Data-Distribution Perspective *IEEE Transactions on Systems, Man, and Cybernetics, Part B (Cybernetics)* **39** 318-31

IOP Conf. Series: Materials Science and Engineering **935** (2020) 012051 doi:10.1088/1757-899X/935/1/012051

# High translucent of polymeric membrane reinforced by nanocellulose from oil palm empty fruit bunches

**N Hastuti[1*], K Kanomata[2], and T Kitaoka[3]**

[1]Forest Products Research and Development Center, Ministry of Environment and Forestry, Bogor, 16610, Indonesia
[2]Graduate School of Pharmaceutical Sciences, Osaka University, 1-1 Yamadaoka, Suita, Osaka, Japan, 565-0871
[3]Graduate School of Bioresource and Bioenvironmental Sciences, Kyushu University, Motooka 744, Nishi-ku, Fukuoka, Japan, 819-0395

*Email: novienov3@yahoo.com

**Abstract.** Polymeric membranes have gained high interest since they provide great potential to be used as functional products in energy, electronics and remediation applications. In this study, cellulose nanocrystals (CNCs) and TEMPO-oxidized cellulose nanofibrils (TOCNs) extracted from oil palm empty fruit bunches (OPEFB) cross-linked with poly(methyl vinyl ether co-maleic acid)/PMVEMA and poly(ethylene glycol)/(PEG) by insitu polymerization to produce polymeric membranes. Characterization of resultant polymeric membranes was employed by microscopic observation to evaluate membrane morphology. UV-Vis analysis and mechanical testing were conducted to observe the effect of nanocellulose addition towards membrane translucent and mechanical properties. Results indicated that the presence of nanocellulose improved membrane translucent and mechanical properties by only 5 weight (wt)% of addition. CNCs addition has higher translucent improvement on polymeric membrane compared to TOCNs. However, TOCNs addition has higher improvement in mechanical properties than CNCs. Polymeric membrane reinforced by nanocellulose has potential utilization for thin film-based composite.

## 1. Introduction

Currently, the fabrication and applications of nanocelluloses (NC) have increasingly gained interest in polymer nanocomposites owing to the growing environmental risks of petroleum based products [1]. Compared to cellulose substrates in larger scale, much smaller NC have higher mechanical strength, better optical transparence, and smoother surface. These properties combined with their low cost, light weight, flexibility, and environmental friendliness, make NC are promising candidates for fabrication of novel composites [2]. Existing literatures have indicated that the addition of NC in lower amount increased mechanical properties, but up to certain level (higher concentration), NC addition causes development of weaker section in polymer/NC composite, increase brittleness and causing fracture [3,4]. The utilization of NC in polymers matrix is not only to improve mechanical properties but also to improve antibiofouling, thermal and high gas barrier properties [5-8]. However, utilization of NC from oil palm empty fruit bunches (OPEFB) as reinforcing agent is still limited to be explored.

Previous study investigated the use of NC as reinforcing fillers in poly(methyl vinyl ether co-maleic acid)/PMVEMA and poly(ethylene glycol)/(PEG) matrix in high concentration (25-75 wt%) [9,10]. In addition, high amount of NC addition drive NC aggregation due to the presence of hydroxy groups on their surface. Meanwhile, the addition of NC did not follow linier pattern of "the more the better". When the higher amount of nanocellulose applied, the thermal stability and elongation at break of the composite films were reduced, and the tensile force remained unchanged or even weakened after reaching maximum value [11]. Therefore, in this study cellulose nanocrystals (CNCs) and TEMPO-oxidized cellulose nanofibrils (TOCNs) extracted from oil palm empty fruit bunches (OPEFB) waste cross-linked with poly(methyl vinyl ether co-maleic acid)/PMVEMA and poly(ethylene glycol)/(PEG) by insitu polymerization to produce polymeric membrane by 5 wt% addition of nanocelluloses. Characteristic of resultant membrane was observed by microscopic, physical and mechanical properties analyses.

## 2. Materials and Methods

### *2.1. Materials*

Poly (methyl vinyl ether maleic acid)/PMVEMA, Mw 1.980.000 (Sigma Aldrich, USA) (Mw 1.980.000) and PEG (Wako Pure Chemical Industries, Osaka, Japan) (Mw 20.000) were used in this study. Hydrochloric acid 0,1 M was used to obtain the solution in pH 2. The water used in this study was purified with an Arium Ultrapure Water System (Sartorius Co., Ltd., Tokyo, Japan). Bleached kraft pulp of oil palm empty fruit bunches (OPEFB) was kindly supplied from Biomaterial Research Institute, Indonesian Institute of Sciences (Bogor, Indonesia) and used as raw materials to produce CNC and TOCNs as described in our previous studies [12,13]. CNCs and TOCNS of OPEFB were denoted as CO and TO, respectively. The CO used in this study has crystallinity index (65%), aspect ratio 23±18. The TO used in this study was TO has crystallinity index 55%, aspect ratio 41±14.

### *2.2. Methods*

A 6.7:1 mass ratio of PMVEMA:PEG (1.5 g) was added in 40 mL of deionized water (DI) preheated at 68°C and acidified with HCl 0.1 M to reach pH 2. At first, PMVEMA was poured into the heated and acidified DI to get it dissolved completely, then followed by PEG and 5 wt% NC of OPEFB addition. The 5 wt% addition of NC in PMVEMA-PEG matrix was chosen because the addition of NC OPEFB below 6% to the PVA considerably results enhancements in tensile strength and modulus as reported by Asad and co-authors [14].

The sample denoted as P100, CO5 and TO5 for sample with PMVEMA and PEG only, PMVEMA and PEG with the addition 5 wt% of CNCs and TOCNs of OPEFB, respectively. Solution was continuously stirred to get all components mixed completely. Then solution was homogenized by electrical homogenizer for 2-3 min, poured into teflon Petri dish, cured at 135°C for 5 minutes and dried at 40°C for 3-4 days. The dried films/membranes were peeled off carefully and characterized.

#### 2.2.1. Membrane characterization

Optical transmittance (T%) was observed using UV-Vis spectrophotometer UH5300 (Hitachi, Tokyo, Japan) at 250-800 nm wavelength. Surface analysis was observed using Atomic Force Microscopy (AFM) in mode Scanning Probe Microscope/Dimension Icon with Scan Asyst function, the results were analyzed by NanoScope Analysis Version 1.80. Color analysis of membranes was measured using handy colorimeter NR-3000 (Nippon Denshoku Industries Co., Ltd., Tokyo, Japan). A white standard color plate (L = 97.75, a = −0.49 and b = 1.96) was used as a background for color measurements. Hunter color (L, a, and b) values were averaged from three readings from each sample. Color values were expressed as L* (whiteness or brightness/darkness), a* (redness/ greenness), and b* (yellowness/blueness) [15]. The total color difference ($\Delta E$) was calculated as follows:

$$\Delta E = \sqrt{(\Delta L)^2 + (\Delta a)^2 + (\Delta b)^2} \quad (1)$$

Tensile tests were observed using Material Testing Instruments (STA-1225, ORIENTEC Co., Ltd., Tokyo, Japan) equipped with a 100 N load cell. Sample specimens were prepared by cutting the film sheets into 40-50 x 10 mm strips. Measurement was conducted at 10% $min^{-1}$ strain rate in 20 mm span length [16]. The sample were kept at 50% humidity for 2-3 days before testing. Sample thickness were measured using Digital Micrometer (Mitutoyo Corp., Kawasaki, Japan); length and width of films was measured using TACKLife Digital Caliper DC01 (Shenzen Take Tools. Co.Ltd., Guangdong, China).

## 3. Results and Discussion

### *3.1. Optical Transmittance*

Optical transmittance (T%) of PMVEMA-PEG membrane only (P100) was 70% at 550 nm. Meanwhile, a light transmittance of membrane with the addition of 5 wt% nanocelluloses (CNCs and TOCNs), denoted as CO5 and TO5 were 90% and 78%, respectively. Polymeric membranes of PMVEMA-PEG by nanocellulose addition had higher light transmittance compared to the control in the order of T% of CO5 was higher than TO5 (Figure 1).

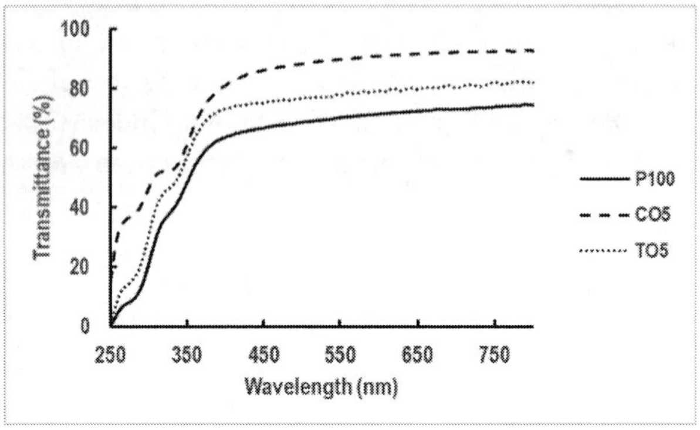


**Figure 1.** Light transmittance of control membrane (PMVEMA-PEG only/P100) and membranes with the 5% wt addition of nanocellulose in PMVEMA-PEG matrices (CO5 and TO5)

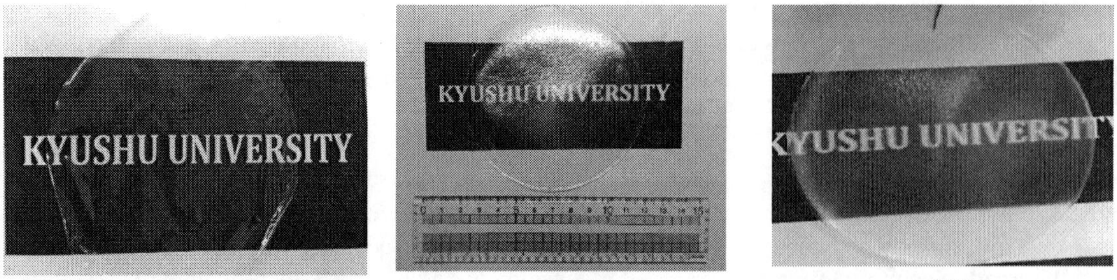


PMVEMA-PEG only (P100) CO5 TO5

**Figure 2.** Optical images of resultant polymeric membranes

The high light transmittance may indicate that nanocellulose had high dispersibility in the water. It was in agreement with the previous study which stated that the rod-like CNC OPEFB which produced by HCl hydrolysis revealed long term nanodispersibility and showed clear birefringence [12]. The interference near infrared region indicates high smoothness on the film surface and the uniformity of film thickness [17]. Nanocellulose fibers orientation in PMVEMA-PEG matrices also influence the optical transmittance. Component rearrangement in the film matrix during the film drying, which defines its internal and surface structure, has an important role in the film optical properties [18]. The optical images of resultant membranes as described in Figure 2.

### *3.2. Surface analysis*

Surface morphology of obtained PMVEMA-PEG-NC membranes were observed using Atomic Force Microscopy (AFM) (Figure 3). Microscopy images of AFM showed that the nanocelluloses have been well dispersed in PMVEMA-PEG matrices. These images were clearly showed the presence of NC in lower amount (5 wt%) of addition compared to the previous study conducted by Goetz and co-authors which used higher amount of NC about 25-75% [9]. The fibrous appearance of TOCNs from OPEFB was easier to be observed than the rod-like appearance of CNC OPEFB.

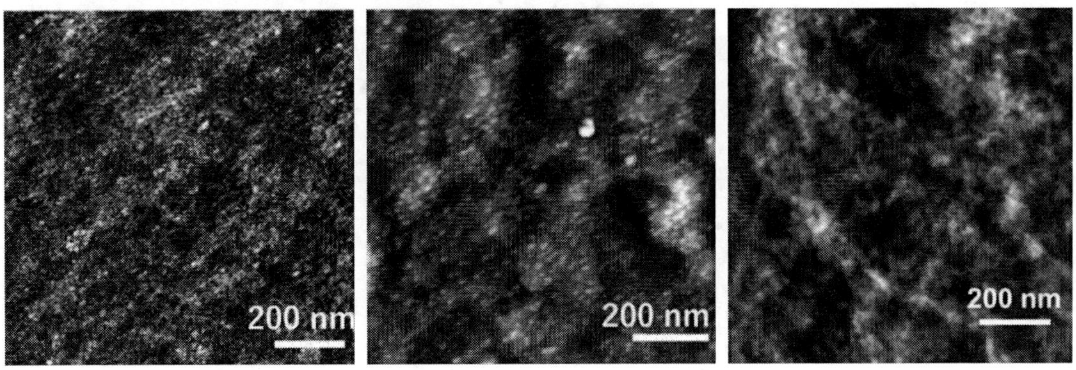


**Figure 3**. Surface analysis of polymeric membranes made from PMVEMA-PEG only (left) and by the presence of 5% wt of CNCs (CO5) (middle) and TOCNs (TO5) (right)

### *3.3. Color analysis*

Color analysis was employed by calculating the color differentiation. It can be calculated by *total color difference* (TCD), denoted as $\triangle E$, using equation (1). In this study, color differentiation referred to the color coordinate system developed by *Commision Internationale de l'Eclairage's* (CIE) L*a*b*. Results of TCD as described on Table 1.

**Table 1**. Color differentiation of obtained films of nanocellulose in PMVEMA-PEG matrices

| Sample | $(\triangle L^*)^2$ | $(\triangle a^*)^2$ | $(\triangle b^*)^2$ | $\triangle E$/TCD (to P100) |
|---|---|---|---|---|
| P100 | 88.86 | -0.55 | 18.61 | n.d |
| CO5 | 90.58 | 0.24 | 4.03 | 14.70 |
| TO5 | 90.59 | 0.56 | 3.77 | 14.99 |

*Remarks: n.d = not determined*

Differences in perceivable color can be classified as very distinct (TCD > 3), distinct (1.5 < TCD < 3), and small differences (TCD < 1.5). Therefore, based on TCD values, the color differences among the obtained membranes were very distinct [15,19]. According to Rheddy and co-author, the TCD increased linierly with increase of nanocelluloses content [20]. However, the addition of nanocellulose increased the brightness and reduced the yellowish of PMVEMA-PEG membrane (Table 1).

### 3.4. *Tensile tests*

Mechanical properties of obtained membranes can be interpreted from tensile stress-strain curves (Figure 4).

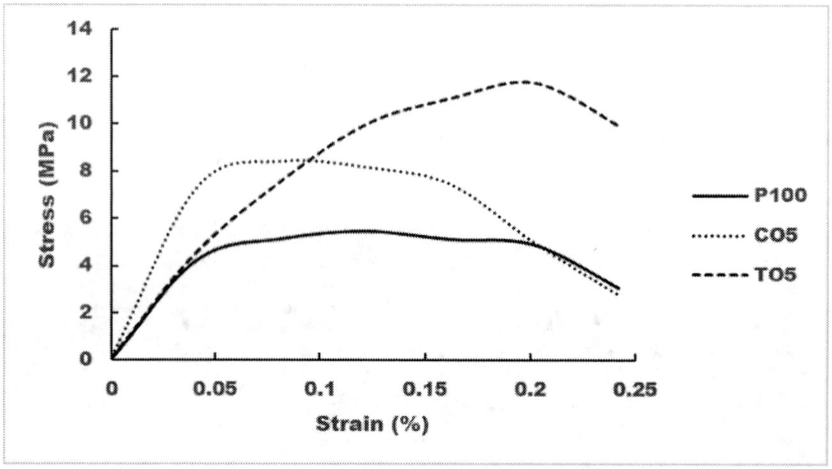


**Figure 4.** Tensile test curves

Tensile strength of PMVEMA-PEG membrane (P100) was lower than the membranes with the nanocellulose addition (CO5 and TO5). However, the strain of membrane reduced by the presence of nanocelluloses. Previous study reported the use of PMVEMA-PEG film in ratio 1:1 with low concentration of PMVEMA (5%) showed relatively low stress about 31 MPa and Young Modulus about 9 MPa [21]. Nanocellulose of OPEFB which was prepared by TEMPO-mediated oxidation (TO) revealed higher reinforcing effect than nanocellulose prepared by acid hydrolysis (CO). Formation of the reactive carboxyl groups on its surface due the mediated oxidation of OH groups is expected to increase the interactions with the polymeric matrix at the interfacial region, therefore TOCN had been examined as interfacial strengthener in continuous fiber-reinforced polymer composite [22].

## 4. Conclusion

The presence of nanocelluloses in PMVEMA-PEG membrane increased the light transmittance and mechanical properties by 5 wt% of addition. Nanocellulose in form of TOCNs gave higher improvement in mechanical properties but lower in optical transmittance improvement compared to CNCs. The presence both of TOCNs and CNCs improved the brightness of polymeric membranes.

## References

[1] Kian LK, Saba N, Jawaid M, and Sultan MTH 2019 A review on processing techniques of bast fibers nanocellulose and its polylactic acid (PLA) nanocomposites *Int. J. Biol. Macromol.* **121** 1314-28

[2] Moon RJ, Martini A, Nairn J, Simonsen J, and Youngblood J 2011 Cellulose nanomaterials review: structure, properties and nanocomposites *Chemical Society Reviews* **40**(7) 3941-94

[3] Butylina S, Geng S, and Oksman K 2016 Properties of as-prepared and freeze-dried hydrogels made from poly(vinyl alcohol) and cellulose nanocrystals using freeze-thaw technique *Eur. Polym. J.* **81** 386-396

[4] Pramanik R, Ganivada B, Ram F, Shanmuganathan K, and Arockiarajan A 2019 Influence of nanocellulose on mechanics and morphology of polyvinyl alcohol *xerogels J. Mech. Behav. Biomed. Mater.* **90** 275-83

[5] Qazanfarzadeh and Kadivar M 2016 Properties of whey protein isolate nanocomposite films reinforced with nanocellulose isolated from oat husk *Int. J. Biol. Macromol.* **91** 1134-40
[6] Hadi P, Yang M, Ma H, Huang X, Walker H, and Hsiao BS 2019 Biofouling-resistant nanocellulose layer in hierarchical polymeric membranes: Synthesis, characterization and performance *J. Memb. Sci.* **579** 162-71
[7] Fukuzumi H, Saito T, Iwamoto S, Kumamoto Y, Ohdaira T, and Suzuki R 2011 Pore size determination of TEMPO-oxidized cellulose nanofibrils films by positron annihilation lifetime spectroscopy *Biomacromolecules* **12** 4057-62
[8] Endo R, Saito T, and Isogai A 2013 TEMPO-oxidized cellulose nanofibril/poly(vinyl alcohol) composite drawn fibers *Polymer (Guildf)* **54** 935-41
[9] Goetz L Mathew A, Oksman K, Gatenholm P, and Ragauskas AJ 2109 A novel nanocomposite film prepared from crosslinked cellulosic whiskers *Carbohydr. Polym.* **75** 85-9
[10] Liang L, Huang C, Hao N, and Ragauskas AJ 2019 Cross-linked poly(methyl vinyl ether-co-maleic acid)/poly(ethylene glycol)/nanocellulosics foams via directional freezing *Carbohydr. Polym.* **213** 346-51
[11] Chen Q, Liu Y, and Chen G 2019 A comparative study on the starch-based biocomposite films reinforced by nanocellulose prepared from different non-wood fibers *Cellulose* **26** 2425-35
[12] Hastuti N, Kanomata K, and Kitaoka T 2018 Hydrochloric acid hydrolysis of pulps from oil palm empty fruit bunches to produce cellulose nanocrystals *J. Polym. Environ.* **26** 3698-709
[13] Hastuti N, Kanomata K, and Kitaoka T 2019 Characteristics of TEMPO-Oxidized cellulose nanofibers from oil palm empty fruit bunches produced by different amounts of oxidant *IOP Conf. Series: Earth and Environmental Science* **35** 012008
[14] Asad M, Saba N, Asiri AM, Jawaid M, Indarti E, and Wanrosli WD 2018 Preparation and characterization of nanocomposite films from oil palm pulp nanocellulose/poly (Vinyl alcohol) by casting method *Carbohydr. Polym.* **191** 103-11
[15] Tiwari BK, Muthulumarappan K, O'Donnell CP, and Cullen PJ 2008 Effects of sonication on the kinetics of orange juice quality parameters *J. Agric. Food Chem.* **56** 2423-28
[16] Tsuboi K, Yokota S, Kondo T 2014 Difference between bamboo and wood derived cellulose nanofibers prepared by the aqueous counter collision method *Nord. Pulp Pap. Res. J.* 29 69-76
[17] Takahashi M, Iyoda K, Miyauchi T, Ohkido S, Tahashi M, and Wakita K 2009 Preparation and characterization of Eu:Ti codoped LiNbO3 films prepared by the sol–gel method *J. Appl. Phys.* **106** 044102
[18] Villalobos R, Chanona J, Hernández P, Gutiérrez G, and Chiralt 2005 A Gloss and transparency of hydroxypropyl methylcellulose films containing surfactants as affected by their microstructure *Food Hydrocoll.* **19** 53-61
[19] Patras A, Brunton NP, Tiwari BK, and Butler F 2011 Stability and degradation kinetics of bioactive compounds and colour in strawberry jam during storage *Food Bioprocess Technol.* **4** 1245-52
[20] Reddy JP and Rhim JW 2014 Characterization of bionanocomposite films prepared with agar and paper-mulberry pulp nanocellulose *Carbohydr. Polym.* **110** 480-8
[21] Raj Singh TR, McCarron PA, Woolfson AD, and Donelly RF 2009 Physochemicql characterization of poly(ethylene glycol) plasticized poly(methyl vinyl ether-co-maleic acid) films *J. Appl. Polym. Sci.* **112** 2792-99
[22] Uribe BEB, Chiromito EMS, Carvalho AJF, Arenal F, and Tarpani JR 2017 TEMPO-oxidized cellulose nanofibers as interfacial strengthener in continuous-fiber reinforced polymer composites *Mater. Des.* **133** 340-8

# Shear strength and subterranean termite resistance of polystyrene impregnated sengon (*Falcataria moluccana*) glulam

**Nurhanifah[1], D Hermawan[1*], Y S Hadi[1], W O M Arsyad[2], and I B Abdillah[1]**

[1]Forest Products Department, Faculty of Forestry, IPB University, 16680, Bogor, Indonesia

[2]Forest Products Research and Development Centre, Ministry of Environment and Forestry, 16610, Bogor, Indonesia

*Email: mr.dede.hermawan@gmail.com

**Abstract.** The purpose of this study is to evaluate shear strength and improve the resistance of glulam from sengon (*Falcataria moluccana*) wood through the preliminary treatment of polystyrene impregnation. Sengon laminas were impregnated with polystyrene and made into two-layer glulam using isocyanate adhesives. The glulam was tested for shear strength according to American Standard for Testing Materials D-905-98, and exposed to subterranean termite attack in the laboratory scale according to Indonesian standard SNI 7207-2014. The untreated glulam and solid wood were used as a comparison purposes, and the tested samples were six replicas. The polymer loading of polystyrene glulam was 28.7%. The results showed that shear strength of solid wood was not significantly different from glulam, it means that impregnation process of the styrene on sengon wood does not affect the gluing process. The polystyrene impregnated glulam had better resistant to termite attack than solid wood, but it was not significantly different from the untreated glulam. To get better resistance it could be further study for higher polymer loading in glulam manufacturing.

## 1. Introduction

Indonesian log supply as raw material for the industry was shifting from natural forests to plantations. The ratio of wood supply from natural forests is less than that of plantations forest [1]. Since wood from plantations dominantly planted with fast-growing wood species that have a short cycle [2]. One type of fast-growing wood that comes from plantations is sengon wood (*Falcataria moluccana*). Sengon wood has a specific gravity of 0.33, strength class IV-V, and wood resistance class IV/V [3]. Sengon wood has sapwood and heartwood, which are usually difficult to distinguish [4]. Also, sengon wood commonly used as furniture, housing, and packaging boxes [5]. According to Basri *et al.* [6] and Fajriani *et al.* [7], fast-growing wood species had inferior properties because they contain juvenile wood. Juvenile wood has a low fiber dimension, so the density and strength of the wood are low [8]. One of several methods that can be done is to make biocomposite products such as glue-laminated lumber (glulam).

Glulam was composed of wood laminate with parallel laminae toward the preparation of fibers and then glued with an adhesive [9]. Glulam has advantages, one of which resulted in a higher strength beam than solid wood in the same dimension [10]. Glulam can be produced from small-diameter wood by utilizing low-density wood [11]. However, glulam products made from fast-growing wood still have low-grade durability [12]. Therefore, preliminary treatment is needed in order to increase service life. Some methods have been developed to improve the quality of wood. One of several methods of wood modification is chemical treatment [13].

One of these chemicals is styrene. Styrene is one of the monomers commonly used for wood modification by filling cells in wood. Research on the impregnation of polystyrene on wood have been carried out on the topic such as physical and mechanical properties [14,15], and also termite and fungi resistance on tropical and subtropical wood [12,16,17]. Devi and Maji [18] were impregnated rubberwood with styrene and glycidyl methacrylate (GMA) showed that the presence of GMA as a crosslinking monomer improved the properties of Wood Plastic Composite (WPC). Impregnation of wood within mixing styrene and methyl methacrylamide (MMA) could increase the dimensional stability and resistance of wood [15]. Darma *et al.* [16] have been researched wood with a mixture of styrene monomers, vinyl acetate, and Copper Chrome Boron (CCB) with tertbutyl hydroperoxide as a catalyst showed produced a high level of resistance to fungus *T. palustris*. Also, Hadi *et al.* [17] and Hadi *et al.* [12] had researched several types of subtropical wood from Poland and tropical wood, resulting in that wood more resistance to dry wood and subterranean termite attacks and in a field test. However, information on impregnated polystyrene of the glulam is still limited. Therefore, the purpose of this research is to improve the quality of glulam from sengon wood through pretreatment with polystyrene impregnation.

## 2. Materials and Methods

### *2.1. Wood preparation*

Glulam was made from sengon wood from Ciampea, Bogor, West Java, Indonesia. Each log had a diameter of around 20 cm and was cut into laminas with dimensions of 2.5 x 5 x 30 cm (thickness, width, and length, respectively). The best laminas were dried in the kiln to a moisture content of approximately 12%. The pretreatment of the lamina was done through immersion processes 24 hours with filling cell voids with a styrene monomer solution. To accelerate the polymerization process, the solution was added with potassium peroxodisulfate 0.5% (w/v). Polymer Loading (PL) was determined through weighing the wood specimen before treatment ($W_1$) and after treatment ($W_2$) at oven-dried condition. The PL was calculated using the following formula (1):

$$PL\ (\%) = (W_2 - W_1)/W_1 \times 100 \qquad (1)$$

### *2.2. Glulam manufacturing*

Glulam samples were made for size (5 x 5 x 30 cm). The sample consists of two-layers of laminas. The laminas were bonded with isocyanate adhesive type PI-3100 with a glue spread of 280 g/m$^2$ [18], followed by cold pressing with specific pressure of 10 kg/cm$^2$. The glulam samples were conditioned at room temperature for ten days before testing. For comparison, solid wood was prepared. Six test specimens were used for each treatment.

The physical and mechanical properties were tested according to JAS 234-2003 standard [19] and ASTM D-905-98 standard [20]. Physical properties were tested based on the Moisture Content (MC) and density, for mechanical properties were tested with a Universal Testing Machine (UTM) Chun Yen to determine the shear strength.

### *2.3. Subterranean termite test methods*

The subterranean termite test using a laboratory scale. The test was based on Indonesian standard SNI 7207-2014 [21] . Before the test, the wood samples were oven-dried at a temperature of 103 ± 3°C until reaching constant weight ($S_1$). Each test specimen was placed in a glass container with 200 g of sterilized sand with enough water to reach a moisture content of 7% under water-holding capacity. Two hundred healthy and active subterranean termite (*Coptotermes curvignathus* Holmgren) workers from a laboratory colony were added to each container ($M_1$). The containers were put in a dark room at a temperature of 25 to 30°C and 80 to 90% relative humidity for four weeks and weighed weekly. If the moisture content of the sand decreased by 2% or more, water was added to achieve moisture content standards. At the end of the 4-week test, the active subterranean termites were counted ($M_2$), and the wood samples were oven-dried at a temperature of 103 ± 3°C until reaching constant weight ($S_2$). Wood percent weight loss and termite mortality were determined using the following formula:

$$\text{Percent weight loss (WL)} = (S_1 - S_2)/S1 \times 100\% \quad (2)$$

$$\text{Termite mortality} = (M_1 - M_2)/M_1 \times 100\% \quad (3)$$

Wood resistance class against subterranean termites was determined by referring to SNI 7207-2014, as shown in Table 1.

**Table 1**. Resistance class against subterranean termite

| Resistance class | Sample condition | Weight loss (%) |
|---|---|---|
| I | Very resistant | < 3.52 |
| II | Resistant | 3.52-7.50 |
| III | Moderately resistant | 7.50-10.96 |
| IV | Poorly resistant | 10.96-18.94 |
| V | Very poorly resistant | > 18.94 |

*2.4. Data analysis*
To analyze the effect of type of materials upon all responses (i.e. moisture content, shear strength, mortality, percent weight loss, and resistance class), a simple completely randomized design was used for data analysis. The data was analyzed using Microsoft Excel and IBM SPSS Statistics (Statistical Product and Service Solution) version 22. The factor was the type of materials (solid wood, untreated glulam, polystyrene glulam). Also, Duncan's multiple range tests were used for further analysis if a factor was significantly different at $p \leq 0.05$.

## 3. Results and Discussion
The densities of solid sengon wood, untreated glulam, and polystyrene glulam were 0.28, 0.34, and 0.28 g/cm$^3$, respectively. The PL of polystyrene glulam is 28.7%. These results are in line with the research of Hadi *et al.* [12], who impregnated with vacuum-pressure. Therefore, the impregnation of styrene monomers using the immersion method in sengon wood in this study is feasible.

**Table 2**. Moisture content, shear strength, mortality, percent weight loss, and resistance class

| Material | Moisture content (%) | Shear strength (kg/cm$^2$) | Mortality (%) | Weight loss (%) | Resistant class | |
|---|---|---|---|---|---|---|
| Untreated glulam | 11.44 (0.38)$^b$ | 43.07 (13.25)$^a$ | 70.0 (46.1)$^b$ | 11.78 (1.56)$^b$ | IV | 3.67 (0.52)$^b$ |
| Polystyrene glulam | 11.60 (0.32)$^b$ | 46.27 (25.21)$^a$ | 71.1 (44.2)$^b$ | 10.79 (2.91)$^b$ | III | 3.33 (0.82)$^b$ |
| Solid wood | 9.79 (0.51)$^a$ | 67.79 (5.80)$^a$ | 8.3 (2.7)$^a$ | 44.48 (9.65)$^a$ | V | 5.00 (0.00)$^a$ |

Note: Values followed by the same letter in a column are not statistically different. The letter in the column is Duncan's multiple range test

*3.1. Mositure content*
Results for MC of samples is shown in Table 2. The average value of the MC of all treatment amounted ranged from 9.79 to 11.60%. The MC of samples met the JAS 234-2003 standard, which requires a maximum value of 15%. Solid wood has a lower moisture content. Based on Duncan's multiple range test of MC in Table 2, moisture content of untreated glulam and polystyrene glulam were same but

significantly different from solid wood. It caused the thin lamina will be easy to adjust to the surrounding environment, other than that the water contained in the adhesive will be trapped in the glulam [22].

**Table 3**. Variance analysis of moisture content, shear strength, mortality, percent weight loss, and resistance class

| Parameter | Sig. |
|---|---|
| Moisture content | $0.000^{**}$ |
| Shear strength | $0.131^{ns}$ |
| Mortality | $0.015^{**}$ |
| Weight loss | $0.000^{**}$ |
| Resistant class | $0.000^{**}$ |

Note: **$p < 0.01$; ns, not significant

*3.2. Shear strength*

The shear strength values of the samples can be seen in Table 2. The average value of the shear strength of all materials ranged from 43.07 to 67.79 kg/cm$^2$. If we compared with the JAS 234-2003 standard, all glulam produced has met the JAS 234-2003 standard, which requires a minimum shear strength of 54 kg/cm$^2$. The results of the variance analysis (Table 3) showed that the type of material did not significantly different. It can be assumed that the pretreatment of polystyrene impregnation on sengon wood lamina does not affect the gluing process that occurs at the level of PL 28.7%. However, solid wood had a higher mean of shear strength.

According to Cahyono *et al.* [23], the value of solid wood shear strength will be higher than the glulam because, during the test, the wood was damaged, not on the glue line. Gluing quality can be affected by the type of wood, thickness of the lamina, and pressing process [24]. According to Komariah *et al.* [18], the lower the moisture content causes the shear strength value will be higher. The value of MC can influence the value of shear strength, if the MC is low, then the cell wall is more compact and dense so that the attractive force between cellulose molecules becomes stronger [25]. The value of sengon glulam shear strength in this study was similar to Lestari *et al.* [26], which made glulam from sengon wood with an isocyanate adhesive that is 47.2 kg/cm$^2$.

*3.3. Resistance to subterranean termite attack*

Results of mortality, percent weight loss, and resistance class of the samples can be seen in Table 2. Termite mortality for all type of material ranged from 8.3 to 71.1%. The type of materials has significantly different on mortality (Table 3). Solid wood has lower mortality than glulam. Based on Duncan's multiple range test of mortality, untreated glulam and polystyrene glulam were same but significantly different from solid wood. This is due to styrene monomers in wood [27], and isocyanate adhesives [28] applied to composite products increase product durability, which causes many termites to die. The criteria for the level of toxicity of an impregnation material can be seen based on its percentage of mortality [29]. Mortality shows the number of termites that die during the testing process [30].

According to Arinana *et al.* [31], several other factors cause termites to die, namely the surrounding temperature, humidity, and light around the termites. According to Sari *et al.*[32], the vacuum-pressure method made styrene deeper into the cell walls, while the immersion method only enters to the lumen level of the wood. In this study, termites that died were most likely due to obstructions by styrene found in the lumen of the wood so that the termites were difficult to eat the wood. According to Hill [13], the impregnation of a monomer in the existing lumen of wood can form a barrier even though the barrier can change with time of use. In this research, the polystyrene impregnation process is said to be useful because it has a mortality value of more than 70%. According to Hadikusumo [33], if the mortality value of a test is a minimum of 70%, then the preservation treatment is effective.

The impact of weight loss (WL) is in line with the three types of materials. The percent WL of glulam ranged from 10.78 to 11.78%. That result was 3.9 times higher than WL from solid wood. The type of materials has significantly difference on WL (Table 2). Based on Duncan's multiple range test of WL, untreated glulam and polystyrene glulam were same but significantly different from solid wood. It could be assumed that the PL of polystyrene into sengon wood should be higher, so that polystyrene glulam is more resistant than untreated glulam. The PL at the level of 28.7% has been able to improve the resistance of sengon wood even though the results are not significantly different from the untreated glulam. These results are in line with research by Fatriasari *et al.* [34], Hadi *et al.* [12], and Wardani *et al.* [27], which showed that wood with low resistance could be improved through product modification and chemical modification. Regarding the Indonesian standard, all samples had average resistance classes ranged from 3.0-5.0, respectively. In other words, impregnation styrene enhanced resistance class to be 1.67 classes higher. Based on the analysis of variance, untreated glulam and polystyrene glulam were same but significantly different from solid wood. The treatment of styrene on wood lamina can increase the class of resistance to termite attack, from class V (solid wood) to class III on polystyrene glulam. We speculate that the styrene and glue line protected it from attack by termites.

## 4. Conclusion

The shear strength value of solid wood and glulams was not significantly different. The polystyrene impregnation process does not affect the adhesion process. Moisture content, mortality, and percent weight loss of polystyrene glulam have the same values as untreated glulam. Impregnation of polystyrene on sengon wood with the addition PL of 28.7% can increase the wood resistance class from class V (solid wood) to class III (polystyrene glulam). In order to get higher resistance class, a higher polymer loading is needed.

## References

[1] [MoEF] Ministry of Environment and Forestry 2019 *Statistik Lingkungan Hidup dan Kehutanan 2018* (Jakarta: Ministry of Environment and Forestry)

[2] Hadi YS, Massijaya MY, Zaini LH, Abdillah IB, and Arsyad WOM 2018 Resistance of methyl methacrylate-impregnated wood to subterranean termite attack *J. Korean. Wood Sci. Technol.* **46**(6) 748-55

[3] Martawijaya A, Kartasujana I, Mandang YI, Prawira SA, and Kadir K 2005 *Atlas Kayu Indonesia Jilid II* (Bogor: Forest Products Research Institute, Forestry Department)

[4] Ogata K, Fujii T, Abe H, and Baas P 2008 *Identification of The Tiimbers of Southeast Asia and The Western Pacific* (Shiga-ken: Kaiseisha Press)

[5] Rahmawati D, Khumaida N, and Siregar UJ 2019 Morphological and phytochemical characterization of susceptible and resistant sengon (*Falcataria moluccana*) tree to gall rust disease *Biodiversitas* **20**(3) 907-13

[6] Basri E, Prayitno TA, and Pari G 2012 Pengaruh umur pohon terhadap sifat dasar dan kualitas pengeringan kayu waru gunung (*Hibiscus macrophyllus* Roxb.) *JPHH* **30**(4) 243-53

[7] Fajriani E, Ruelle J, Dlouha J, Fournier M, Hadi YS, and Darmawan W 2013 Radial variation of wood properties of sengon (*Paraserianthes falcataria*) and jabon (*Anthocephalus cadamba*) *J. Indian. Acad. Wood Sci.* **10**(2) 110-7

[8] Darmawan W, Nandika D, Afaf BDH, Rahayu I, and Lumongga D 2018 Radial variation in selected wood properties of Indonesian merkusii pine *J. Korean Wood Sci. Technol.* **46**(4) 323-37

[9] Moody RC, Hernandez R, and Liu JY 1999 Glued structural members. In: *Wood Handbook, Wood as an Engineering Material* (Madison: USDA Forest Service, Forest Products Laboratory)

[10] Mohamad WHW, Bhkari NM, and Ahmad Z 2019 Bending and bonding properties of mixed-species glue laminated timber from merpauh, jelutong and sesendok *Jurnal Teknologi (Science & Engineering)* **81**(4) 165-70

[11] Sulistyawati I, Nugroho N, Suryokusumo S, and Hadi YS 2008 Kekakuan dan kekuatan lentur maksimum balok glulam dan utuh kayu akasia *Jurnal Teknik Sipil* **15**(3) 113-22

[12] Hadi YS, Massijaya MY, and Arinana A 2016 Subterranean termite resistance of polystyrene-treated wood from three tropical wood species *Insects* **37**(7) 1-6
[13] Hill CAS 2006 *Wood Modification: Chemical, Thermal and Other Processes* (England: John Wiley & Sons Ltd)
[14] Devi RR, and Maji TK 2008 Chemical modification of rubber wood with styrene and glycidyl methacrylate *Polymer Composite* **29**(11) 1258-62
[15] Khalil HPSA, Dungani R, Mohammed IA, Hossain MS, Aprilia NAS, Budiarso E, Rosamah E 2014 Determination of the combined effect of chemical modification and compression of agatis wood on the dimensional stability, termite resistance and morphological structure *BioResources* **9**(4) 6614-26
[16] Darma IGKT, Hadi YS, Atmojo AT 2002 Ketahanan komposit kayu plastik polistirena terhadap serangan jamur pelapuk coklat (*Tyromyces palustris*) *Jurnal Manajemen Hutan Tropika* **8**(1) 31-8
[17] Hadi YS, Nawawi DS, Herliyana EN, and Lawniczak M 1998 Termite attack resistance of four polystyrene-impregnated woods from poland *Forest Products Journal* **48**(9) 60-2
[18] Komariah RN, Hadi YS, Massijaya MY, and Suryana J 2015 Physical-mechanical properties of glued laminated timber made from tropical small-diameter logs grown in Indonesia *J. Korean. Wood Sci. Technol.* **43**(2) 156-67
[19] [JSA] Japanese Standard Association 2003 *Japanese Agricultural Standard for Plywood* JAS SE-1 (Tokyo: Japanese Agricultural Standard)
[20] [ASTM] American Society for Testing and Material 1999 *Standard Test Methods for Strength Properties of Adhesive Bonds in Shear by Compression Loading ASTM D-905-98* (West Conshohocken: ASTM)
[21] [BSN] Badan Standarisasi Nasional 2014 *Uji ketahanan kayu terhadap organisme perusak kayu SNI 7207:2014* (Jakarta: Badan Standarisasi Nasional)
[22] Komariah RN 2014 Karakteristik glued laminated timber dari tiga jenis kayu berdiameter kecil dengan dua ketebalan lamina [Thesis] Pascasarjana Institut Pertanian Bogor
[23] Cahyono TD, Ohorella S, Febrianto F, Priadi T, and Wahyudi I 2014 Physical and mechanical properties of samama wood glulam *JITKT* **12**(2) 186-95
[24] Vick CB 1999 Adhesive bonding of wood material In: *Wood Handbook, Wood as an Engineering Material* (Madison: USDA Forest Service, Forest Products Laboratory)
[25] Tsoumis G 1999 *Science and Technology of Wood Structure, Properties, Utilization* (New York: Van Nostrand Reinhold)
[26] Lestari ASRD, Hadi YS, Hermawan D, and Santoso A 2015 Glulam properties of fast-growing species using mahagony tannin adhesive *BioResources* **10**(4) 7419-33
[27] Wardani L, Risnasari I, Jasni, and Hadi YS 2011 Ketahanan kayu jabon yang dimodifikasi campuran styrene dan MMA terhadap serangan rayap tanah (*Coptotermes curvignathus* Holmgren) *Proceeding of The National Seminar on MAPEKI XIV* pp 534-41
[28] Kusumah SS, Umemura K, Guswenrivo I, Yoshimura T, and Kanayama K 2017 Utilization of weet sorghum bagasse and citric acid for manufacturing of particleboard II: influences of pressing temperature and time on particleboard properties *J. Wood Sci.* **63** 161-72
[29] Sumaryanto A, Hadikusumo SA, and Lukmandaru G 2013 Pengawetan kayu gubal jati secara rendaman dingin dengan pengawet boron untuk mencegah serangan rayap kayu kering (*Cryptotermes cynocephalus*) *Jurnal Ilmu Kehutanan* **7**(2) 95-107
[30] Bignell DE, Roisin Y, and Lo N 2010 *Biology of Termiter: A Modern Synthesis* (London: Springer)
[31] Arinana A, Philipines I, Koesmaryono Y, Nandika D, Rauf A, Harahap IS, Sumertajaya IM, and Bahtiar ET 2016 *Coptotermes curvignathus* Holmgren (Isoptera: Rhinotermitidae) capability to maintain the temperature inside its nets *Journal of Entomology* **13**(5-6) 199-202
[32] Sari RK, Mar'iin M, Jemi R, Jasni J, and Hadi YS 2009 Ketahanan kayu rambutan (*Nephelium* spp. L.) dan kayu pinus (*Pinus merkusii* Jungh, et de Vr.) hasil impregnasi stirena terhadap

rayap *Proceeding of The National Symposium I Forest Products Technology Forum* pp 213-20
[33] Hadikusumo SA 2004 *Pengawetan Kayu* (Yogyakarta: Gadjah Mada University Press)
[34] Fatriasari W, Lestari AS, Prianto AH, and Syamani FA 2012 The resistance of polystyrene treated *Sandoricum koetjape* and *Durio zibethinus* woods against decaying fungi and termites *Proceedings The 2nd International Symposium for Sustainable Humanosphere* pp. 14-22

**Acknowledgments**
The authors also thank to IPB University (Bogor Agricultural University) and Forest Products Research and Development Centre for supporting research facilities.

IOP Conf. Series: Materials Science and Engineering **935** (2020) 012053 doi:10.1088/1757-899X/935/1/012053

# Synthesis of polymeric isocyanate microcapsules via interfacial polymerization and their characterisation using spectroscopy and microscopy techniques

**M A R Lubis[1*], L Suryanegara[1], and M A Pramesti[2]**

[1]Research Center for Biomaterials, Indonesian Institute of Sciences, Cibinong, 16911, Indonesia
[2]Polytechnic STMI, The Ministry of Industry, Jakarta, 10510, Indonesia

*Email: marl@biomaterial.lipi.go.id

**Abstract.** Wood adhesives are holding a key in improving the efficiency of using raw materials in the wood-based panel industry and in durability of the panels during their life in service. Massive progress has been made on research and development of wood adhesives over the past century, but many important challenges are still remaining. In this study, an alternative cross-linker for wood adhesives named microcapsules polymeric 4,4 methane diphenyl isocyanate (M-pMDI) were synthesized at different conditions via interfacial polymerization, and were characterized using spectroscopy and microscopy techniques. Three factors were taken into account to examine the characteristics of M-pMDI, namely isocyanate content, microencapsulation temperature, and microencapsulation agitation rate. Attenuated total reflectance Fourier transform infrared spectroscopy revealed that the isocyanate (−N=C=O) groups of pMDI at 2,250 $cm^{-1}$ disappeared after microencapsulation due to the reaction of pMDI and ethylene glycol to form urethane (−R−NH−C=O−) linkages at 1,650 $cm^{-1}$ as microcapsules shell. Digital microscopy and micro confocal raman imaging hyperspectral spectroscopy confirmed the formation of microcapsules and urethane shell. The results showed that concentration of isocyanate remarkably affected the yield of M-pMDI, while microencapsulation temperature and agitation speed influence the formation of microcapsules itself. Prel-liminary investigation using ATR-FTIR spectroscopy confirmed that free –NCO groups could be released by applying pressure. This study suggested that a combination of 5 mL of pMDI, 60°C of microencapsulation temperature, and 600 rpm of microencapsulation agitation speed could produce M-pMDI with high yield as an alternative cross-linker for wood adhesives in the future.

## 1. Introduction

Isocyanates are one of the major chemicals in the world since found in 1848 due to their versatility [1,2]. As a wood adhesive, isocyanates are commonly used in two ways. The first one is used as a polyurethane pre-polymer that being used in the wood laminating industry, and last is directly used as the isocyanate in the particleboard industry [3]. Toluene diisocyanate (TDI), Isophorone diisocyanate (IPDI), Hexamethylene diisocyanate (HDI), and Methylene diphenyl isocyanate (MDI) are the mostly used isocyanates. The isocyanates are known very sensitive to moisture, which limits their use and lower storage life [4]. To overcome this, a method called blocking has been developed. A blocked isocyanate is formed via a reaction between an –NCO group and a blocking agent which has an active hydrogen atom to block the –NCO [5,6]. However, the use of blocked isocyanates is limited due to their high de-blocking temperature to release the –NCO groups [7]. Another way is to protect the –NCO groups through microencapsulation.

Microencapsulation is a technique of enclosing specific substances with shell materials via chemical and physical processes [8]. Microcapsules are a type of micro containers with an active chemical that prepared through chemical and physical microencapsulation [9]. The first preparation of microcapsules was recorded at around 1950s, when a group of researcher fabricated dyes microcapsules for copying paper [10]. The isocyanate microcapsules are usually prepared by means of interfacial polymerization which produces uniform size of microcapsules [11]. When the microcapsules rupture, it releases the active –NCO groups and reacts with the main polymer. Hence, the isocyanate microcapsules are good options of using as wood adhesive or a cross-linker. This is due to the similarity of protecting and releasing the –NCO groups as blocked isocyanates.

Recent study shows that isocyanate microcapsules derived from polymeric 4,4-methylene diphenyl isocyanate (pMDI) has been prepared via interfacial polymerization using 1,4-butanediol as shell materials and tween 40 as surfactant [12]. The obtained microcapsules show the free –NCO groups after rupture confirmed by attenuated total reflectance Fourier transform infrared (ATR-FTIR). It was observed that microcapsule size as low as 10 microns could be obtained. Another study produced microcapsules with HDI as a core in a polyurethane shell synthesized via interfacial polymerization of pMDI prepolymer and 1,4-butanediol with size as low as 1.7 microns [13]. The resulting isocyanate microcapsules are widely used as self-healing materials and cross-linker in polymer composites [14]. The common polyols used in the preparation of isocyanate microcapsules are 1,4-butanediol, 1,6-hexanediol, and glycerol [8].

Previous studies report the preparation of isocyanate microcapsules for self-healing materials that involved two separate steps [11,15]: (1) interfacial reaction between isocyanates and polyols to build the urethane shell wall, (2) rinsing and filtration of prepared isocyanate microcapsules. Many factors affecting characteristics of isocyanate microcapsules. The isocyanate content, microencapsulation temperature, microencapsulation agitation rate, type of polyol, and type of surfactant are known to be the main parameter on controlling the characteristics of isocyanate microcapsules. It has been reported that the isocyanate droplets are probably cracked by greater shear forces and interfacial tension that has been created from higher agitation rate. Moreover, higher temperature may swell the microcapsules [12,14]. Further research is needed to develop isocyanate microcapsules that can be used as adhesive, additive, and cross-linker.

This present study aimed to synthesis polymeric isocyanates microcapsules derived from polymeric 4,4-methylene diphenyl isocyanate (pMDI) via interfacial polymerization with ethylene glycol that has a potential to be used as cross-linker for wood adhesives. The microcapsules pMDI (M-pMDI) were produce at different isocyanate contents, different microencapsulation temperatures, and different microencapsulation agitation rates. The M-pMDI characteristics were investigated using spectroscopy and microscopy techniques, including attenuated total reflectance Fourier transform infrared, digital microscope, and micro confocal raman hyperspectral imaging spectroscopy.

## 2. Materials and Methods

### *2.1. Materials*
Polymeric 4,4-methylene diphenyl isocyanate (pMDI, Koyobond, Indonesia), acetone, and ethylene glycol were used to prepare microcapsules pMDI (M-pMDI). Tween 80 was used as a surfactant to reduce the surface tension during interfacial polymerization.

### *2.2. Methods*

#### 2.2.1. Preparation of microcapsules pMDI (M-pMDI)
Three parameters were taken into account in the preparation of M-pMDI, namely amount of isocyanate, microencapsulation temperature, and microencapsulation agitation speed. Schematic diagram of microencapsulation of pMDI is presented in Figure 1.

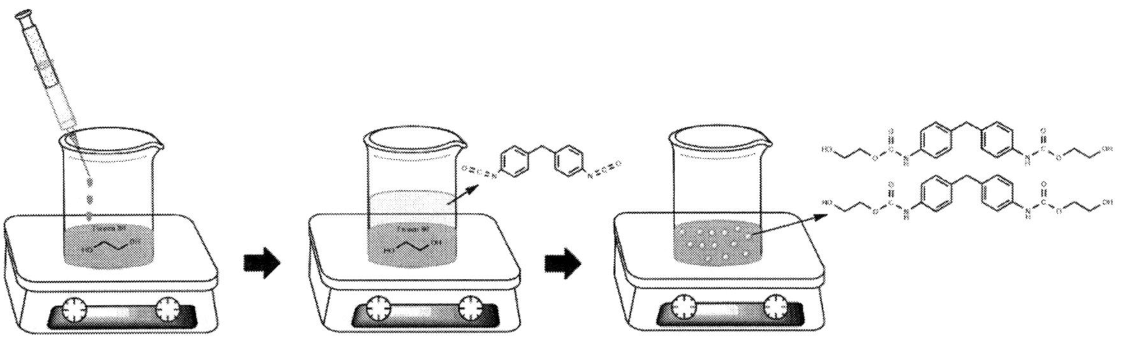


**Figure 1**. Schematic diagram of microencapsulation of pMDI

First, a set of different isocyanate contents such as 3, 5, 7, and 9 mL was used to prepare M-pMDI. The isocyanate was diluted in acetone to make an 80% of concentration prior to microencapsulation. Around 50 g of ethylene glycol was poured into a beaker glass and 0.05 g of tween 80 was added subsequently. Furthermore, the diluted pMDI was dropped into the mixture at a dropping rate of 5.0 mL/min. The microencapsulation was completed after stirring at 600 rpm for 30 min at temperature of 60°C. The mixture was subsequently filtered and then oven dried for 24-h at 60°C to obtain the M-pMDI.

Second, different microencapsulation temperatures such as 25, 40, 60, and 80°C were used to prepare M-pMDI. Around 5 mL of pMDI was diluted in acetone to make an 80% of concentration prior to microencapsulation. Approximately 50 g of ethylene glycol was poured into a beaker glass and 0.05 g of tween 80 was added subsequently. Furthermore, the diluted pMDI was dropped into the mixture at a dropping rate of 5.0 mL/min. The microencapsulation was completed after stirring at 600 rpm for 30 min, and at different temperatures of 25, 40, 60, and 80°C. The mixture was subsequently filtered and then oven dried for 24 hours at 60°C to obtain the M-pMDI.

Third, different microencapsulation agitation rates such as 300, 400, 500, and 600 rpm were also used to prepare M-pMDI. Same as previous, 5 mL of pMDI was diluted in acetone to make an 80% of concentration prior to microencapsulation. Then, 50 g of ethylene glycol was poured into a beaker glass and 0.05 g of tween 80 was added subsequently. The diluted pMDI was then dropped into the mixture at a dropping rate of 5.0 mL/min. The microencapsulation was completed after stirring at different agitation speed of 300, 400, 500, and 600 rpm for 30 min, and at temperatures of 60°C. The mixture was subsequently filtered and then oven dried for 24 hours at 60°C to obtain the M-pMDI.

### 2.2.2. M-pMDI characterization

The yield of M-pMDI was calculated after oven dried by dividing the dried mass of M-pMDI with the mass of pMDI. Functional groups of each M-pMDI were recorded using ATR-FTIR spectroscopy (Spectrum Two, Perkin Elmer, United States). A small amount of sample was placed in ATR-FTIR, and was scanned in the range of 400-4000 $cm^{-1}$ and at room temperature. In addition, pure pMDI was also used as the control. Images of M-pMDI were recorded by putting M-pMDI sample on to a slide glass and covered with a cover glass with a help of digital microscope (VHX 6000, Keyence, Japan) at 500 times of magnification with Dual-Light High-Magnification Zoom Lens VH-Z250T. Micro confocal raman hyperspectral imaging spectrometer (LabRAM HR Evolution, Horiba, Japan) was used to confirm the formation of M-pMDI. M-pMDI sample was put on the slide glass, and the image was capture using an objective lens at 100 times of magnification. The sample image then was bombarded with a laser at a wavelength of 785 nm. The measurement was performed at raman shift of 200-4000 $cm^{-1}$ and at room temperature.

## 3. Results and Discussion

### *3.1. Effect of isocyanate content on characteristics of M-pMDI*

The yield of M-pMDI increased as the amount of isocyanate raised from 3 mL to 5 mL (Table 1). But, as the amount of isocyanate was greater than 5 mL, the yield of M-pMDI decreased remarkably. This is probably due to coagulation of pMDI at 7 and 9 mL of isocyanate. Different phases between oily pMDI and aqueous ethylene glycol are the main reason for the coagulation [16]. Insufficient amount of tween 80 surfactant to reduce the surface tension probably also played some part on coagulation of pMDI that eventually decreased the yield of M-pMDI.

**Table 1**. Yield of M-pMDI at different isocyanate contents prepared at 60°C and 600 rpm

| Isocyanate content (mL) | Mass of pMDI (g) | Mass of M-pMDI | Yield (%) |
|---|---|---|---|
| 3 | 2.40 | 0.16 | 6.87 |
| 5 | 4.09 | 0.83 | 20.22 |
| 7 | 5.91 | 0.95 | 16.13 |
| 9 | 7.88 | 1.09 | 13.85 |

The pure pMDI was dark brown and liquid. The microencapsulation of pMDI resulted M-pMDI with a yellow color (Figure 2a). The morphology of M-pMDI was analyzed using microscopy technique. Digital microscope displays example of M-pMDI produced with 5 mL of isocyanate at 60°C of microencapsulation temperature and 600 rpm of agitation speed (Figure 2b). The spherical isocyanate microcapsules have been formed via interfacial polymerization with observable aggregation. This observation was in agreement with the published works [11,12].

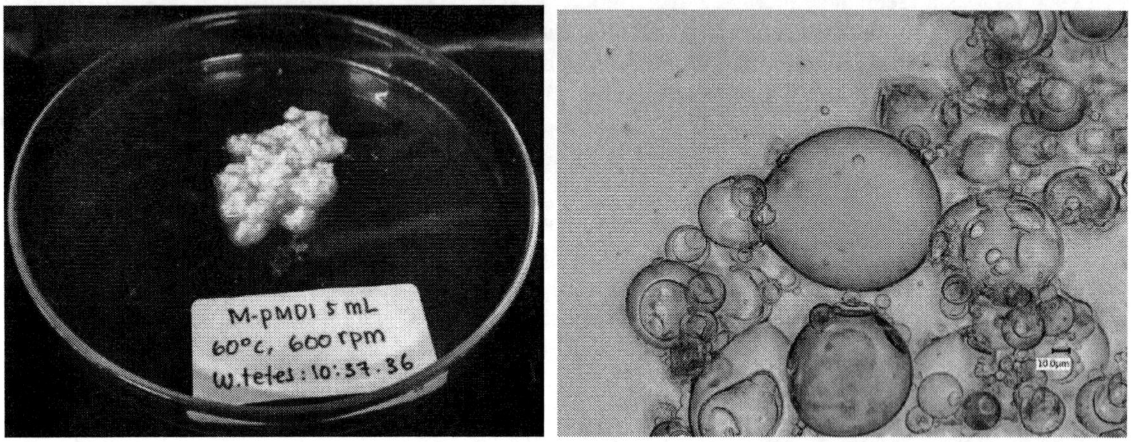


(a) (b)

**Figure 2**. (a) Example of M-pMDI prepared with 5 mL of pMDI at 60°C and 600 rpm, (b) Image of similar M-pMDI sample captured by digital microscope at 500 times magnification

The functional groups of pMDI and M-pMDI were investigated using spectroscopy technique. ATR-FTIR spectroscopy detected the change in functional groups of pMDI after microencapsulation into M-pMDI (Figure 3a). Typical strong peaks at 2250, 1770, and 1525 $cm^{-1}$ were observed only in pure pMDI which is assigned to vibration of –N=C=O groups [9]. Those peaks disappeared after microencapsulation due to the reaction between –NCO groups of pMDI and –OH groups of ethylene glycol to form urethane (−R−NH−C=O−) linkages at 1,650 $cm^{-1}$, which was detected in all M-pMDI. In addition, the vibration of–CH at 2948 and 2880 $cm^{-1}$ were observed in M-pMDI which was attributed to the C–H of urethane linkages. Broad peak at 3300 cm$^{-1}$ was the –OH groups which originated from incorporation ethylene glycol. Micro confocal raman hyperspectral imaging spectroscopy revealed the formation of urethane (–R–NH–C=O–) linkages at 1625 cm$^{-1}$ and the C–H of urethane at 2920 cm$^{-1}$ (figure 3b). The raman spectra was obtained by bombarded a laser with 785 nm of wavelength onto the shell of M-pMDI. This confirmed that the shell of microcapsules was built by urethane linkages from the reaction of –NCO groups of pMDI and –OH groups of ethylene glycol.

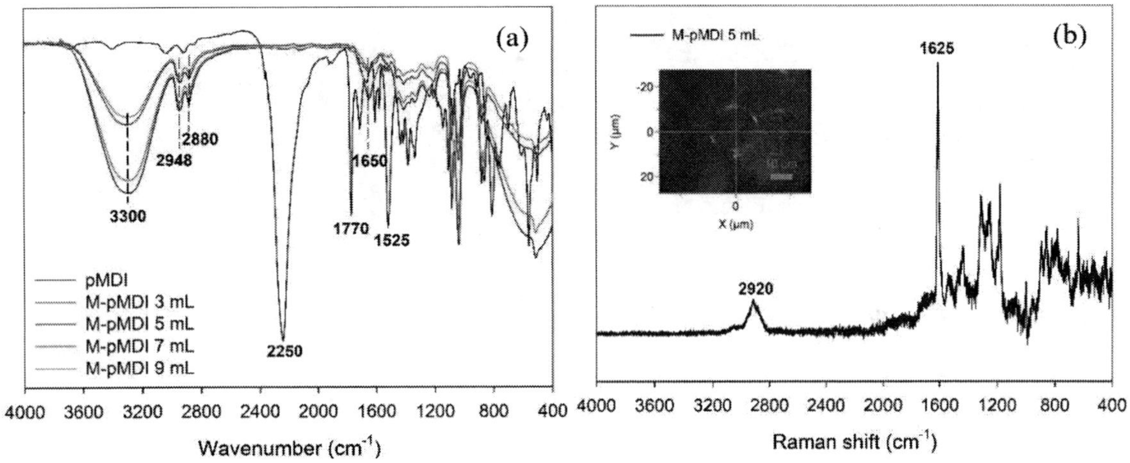


**Figure 3**. (a) Typical ATR-FTIR spectra of pMDI and M-pMDI at different isocyanate contents, (b) formation of microcapsules and raman spectra of M-pMDI 5 mL detected by micro confocal raman hyperspectral imaging spectroscopy at 100 times magnification

### *3.2. Effect of microencapsulation temperature on M-pMDI properties*

Influence of microencapsulation temperature on M-pMDI characteristics was also examined. Table 2 presents the yield of M-pMDI prepared with 5 mL under 600 rpm of agitation speed and at different microencapsulation temperature. The yield of M-pMDI slightly increased by 19.5% as the microencapsulation temperature increased from 25°C to60 °C. This shows that temperature plays a quite significant role on microencapsulation of pMDI.

**Table 2**. Yield of M-pMDI at different temperatures prepared with 5 mL of pMDI and 600 rpm of speed

| Microencapsulation temperature (°C) | Mass of pMDI (g) | Mass of M-pMDI | Yield (%) |
|---|---|---|---|
| 25 | 4.31 | 0.14 | 3.19 |
| 40 | 4.34 | 0.15 | 3.48 |
| 60 | 4.30 | 0.16 | 3.81 |
| 80 | 4.28 | 0.15 | 3.59 |

The M-pMDI morphology was a quite similar to the previous section. Figure 4a displays an example of M-pMDI prepared with 5 mL of pMDI at 40°C and 600 rpm. A yellow color of microcapsules was observed similar to M-pMDI prepared at different isocyanate contents. Figure 4b shows image of M-pMDI produced with 5 mL of isocyanate at 40°C of microencapsulation temperature and 600 rpm of agitation speed provided by digital microscope. Spherical M-pMDI have been formed via interfacial polymerization with observable aggregation similar to M-pMDI prepared at different isocyanate contents. This observation was in agreement with the published works [11,12].

IOP Conf. Series: Materials Science and Engineering **935** (2020) 012053 doi:10.1088/1757-899X/935/1/012053

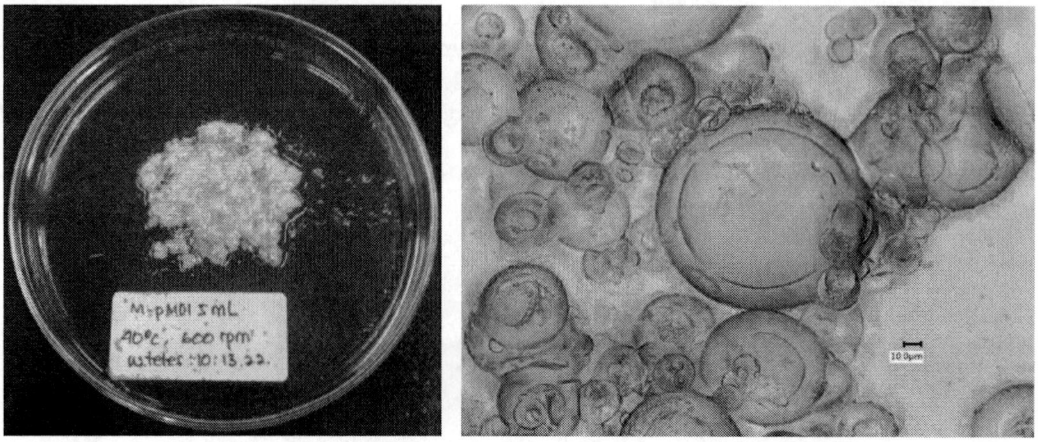


(a) (b)

**Figure 4**. (a) Example of M-pMDI prepared with 5 mL of pMDI at 40°C and 600 rpm. (a) Image of the same M-pMDI sample captured by digital microscope at 500 times magnification

The chemical structures of pMDI and M-pMDI were also investigated using spectroscopy technique. ATR-FTIR spectroscopy detected the change in functional groups of pMDI after microencapsulation into M-pMDI (Figure 5a). Similar to M-pMDI prepared at different isocyanate content, three strong peaks at 2250, 1770, and 1525 cm$^{-1}$ were observed only in pure pMDI, which is assigned to vibration of –N=C=O groups [9]. That peaks disappeared after microencapsulation due to the reaction between –NCO groups of pMDI and –OH groups of ethylene glycol to form urethane (−R−NH−C=O−) linkages at 1650 cm$^{-1}$, which was detected in all M-pMDI. In addition, the vibration of–CH at 2948 and 2880 cm$^{-1}$ were observed in M-pMDI which was attributed to the C–H of urethane linkages. Broad peak at 3300 cm$^{-1}$ was assigned to –OH groups which originated from ethylene glycol. Micro confocal raman hyperspectral imaging spectroscopy revealed the formation of urethane (−R−NH−C=O−) linkages at 1625 cm$^{-1}$ and the C–H of urethane at 2920 cm$^{-1}$ as microcapsules shell (figure 5b). This confirmed that the shell of microcapsules was built by urethane linkages from the reaction of –NCO groups of pMDI and –OH groups of ethylene glycol.

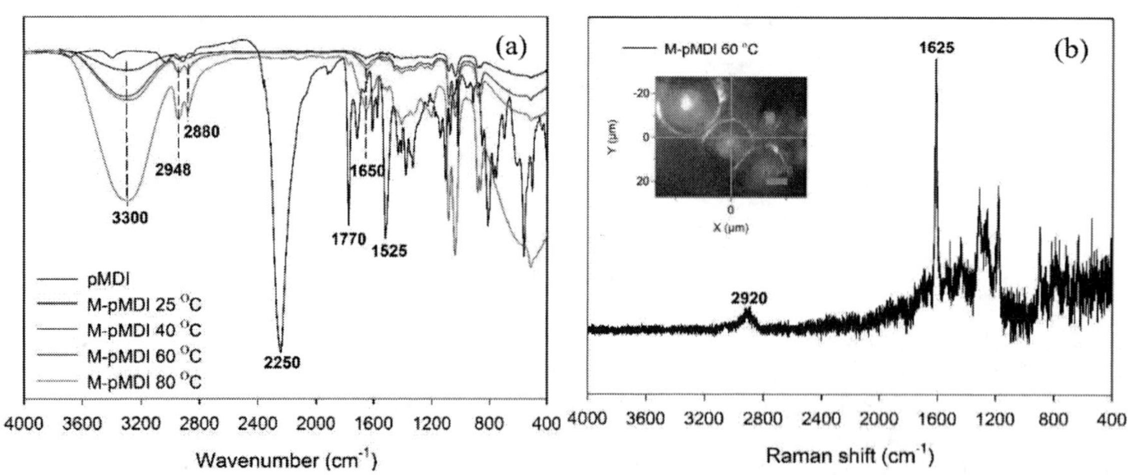


**Figure 5**. (a) Typical ATR-FTIR spectra of pMDI and M-pMDI at different microencapsulation temperatures, (b) formation of microcapsules and raman spectra of M-pMDI 60°C detected by micro confocal raman hyperspectral imaging spectroscopy at 100 times magnification

### *3.3. Effect of microencapsulation agitation rate on M-pMDI characteristics*

Effect of microencapsulation agitation rate on M-pMDI characteristics was also investigated. The yield of M-pMDI prepared with 5 mL at 60 °C and at different microencapsulation agitation speed is presented in Table 3. The yield of M-pMDI remarkably increased by 60.9% as the agitation speed rose from 300 to 600 rpm. By contrast to microencapsulation temperature, agitation speed plays significant role on microencapsulation of pMDI. The isocyanate droplets are possibly broken by greater shear forces and interfacial tension that has been created from higher agitation rate while higher temperature may swell the microcapsules [12,14].

**Table 3**. Yield of M-pMDI at different agitation speed prepared with 5 mL of pMDI and at 60°C

| Microencapsulation agitation rate (rpm) | Mass of pMDI (g) | Mass of M-pMDI | Yield (%) |
|---|---|---|---|
| 300 | 4.30 | 0.40 | 9.47 |
| 400 | 4.31 | 0.41 | 9.56 |
| 500 | 4.30 | 0.51 | 11.93 |
| 600 | 4.30 | 0.65 | 15.23 |

Morphology of M-pMDI prepared at different agitation rates was a quite similar to the M-pMDI produced with different isocyanate contents and at different temperatures. Figure 6a displays an example of M-pMDI prepared with 5 mL of pMDI at 60°C and 300 rpm. A yellow color of microcapsules was observed similar to M-pMDI prepared at different isocyanate contents and at different temperatues. Figure 6b shows image of M-pMDI produced with 5 mL of isocyanate at 60°C of microencapsulation temperature and 300 rpm of agitation speed provided by digital microscope. Spherical M-pMDI have been formed via interfacial polymerization with observable aggregation similar to M-pMDI prepared at different isocyanate contents. This observation was in agreement with the published works [11,12].

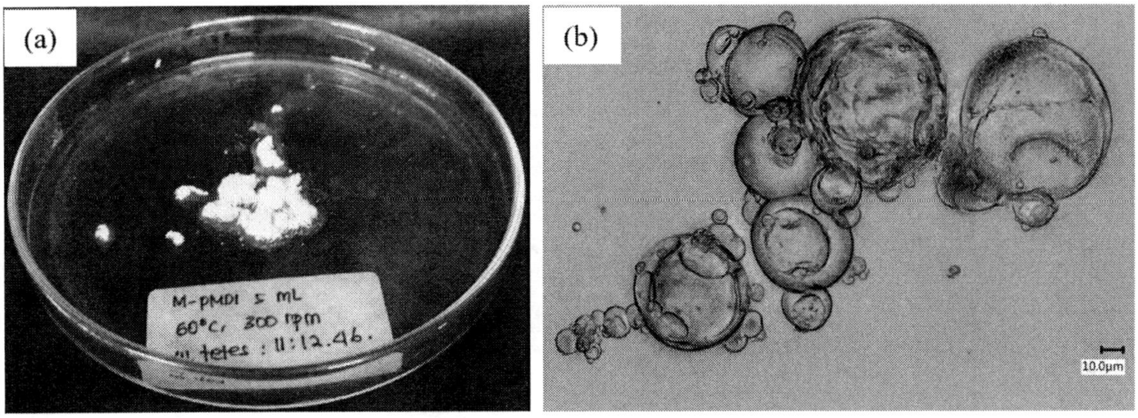


**Figure 6**. (a) Example of M-pMDI prepared with 5 mL of pMDI at 60°C and 300 rpm. (a) Image of the same M-pMDI sample captured by digital microscope at 500 times magnification

ATR-FTIR spectroscopy detected the change in functional groups of pMDI after microencapsulation into M-pMDI (Figure 7a). Similar to M-pMDI prepared at different isocyanate content and different temperatures, three strong peaks at 2250, 1770, and 1525 $cm^{-1}$ were observed only in pure pMDI, which is assigned to vibration of –N=C=O groups [9]. The peaks disappeared after microencapsulation due to the reaction between –NCO groups of pMDI and –OH groups of ethylene glycol to form urethane (−R−NH−C=O−) linkages at 1650 $cm^{-1}$, which was detected in all M-pMDI. In addition, the vibration of–CH at 2948 and 2880 $cm^{-1}$ were observed in M-pMDI which was attributed to the C–H of urethane linkages. Broad peak at 3300 cm$^{-1}$ was assigned to –OH groups which originated from ethylene glycol. The formation of urethane (–R–NH–C=O–) linkages at 1625 cm$^{-1}$ and the C–H of urethane at 2920 cm$^{-1}$ as microcapsules shell was detected by micro confocal raman hyperspectral imaging spectroscopy (figure 7b). This confirmed that the shell of microcapsules was built by urethane linkages from the reaction of –NCO groups of pMDI and –OH groups of ethylene glycol.

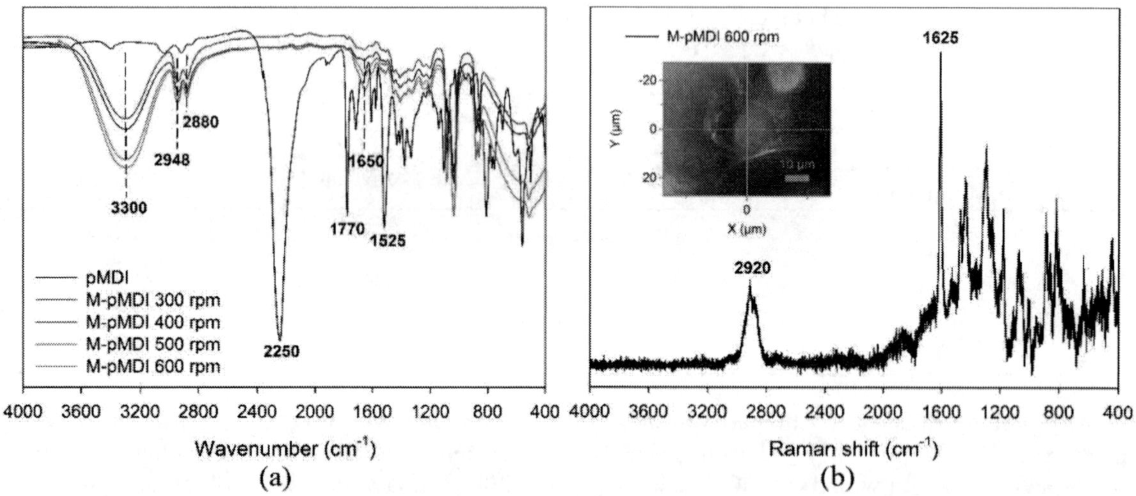


**Figure 7**. (a) Typical ATR-FTIR spectra of pMDI and M-pMDI at different microencapsulation temperatures, (b) formation of microcapsules and raman spectra of M-pMDI 600 rpm detected by micro confocal raman hyperspectral imaging spectroscopy at 100 times magnification

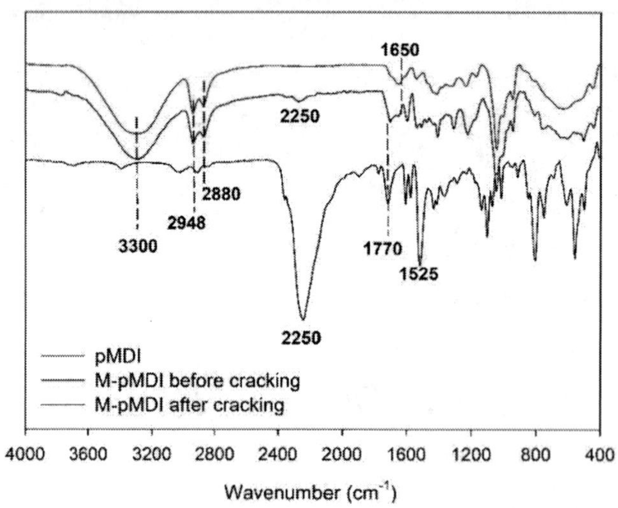


**Figure 8**. Pre-liminary investigation of M-pMDI as cross-linker for wood adhesives using ATR-FTIR

### *3.4. Potential application of M-pMDI*

The M-pMDI were successfully produced via interfacial polymerization between pMDI and ethylene glycol with the help of Tween 80 as surfactant. Figure 8 displays the ATR-FTIR spectra of M-pMDI prepared with 5 mL of isocyanate, at 60°C, and under 600 rpm of agitation rate. The spectra show an interesting result revealing that the core of M-pMDI contained active –NCO groups. This free –NCO groups was detected at 2250 cm$^{-1}$ after cracking the shell of M-pMDI using pressure of ATR-FTIR instrument. The releasing of free –NCO groups was followed by disruption of M-pMDI urethane shell. The ATR-FTIR spectra showed that peak of urethane (−R−NH−C=O−) linkages at 1650 $cm^{-1}$ disappeared after cracking of M-pMDI shell. In addition, the releasing of free –NCO groups also supported by a new peak of C=O at 1,770 $cm^{-1}$ which was previously only detected in pure pMDI. Based on this pre-liminary investigation, M-pMDI have a potential to be used as cross-linker for wood adhesives, such as formaldehyde-based adhesives and bio-based adhesives. Figure 9 depicts the illustration on how M-pMDI can be used as cross-linker for wood adhesives.

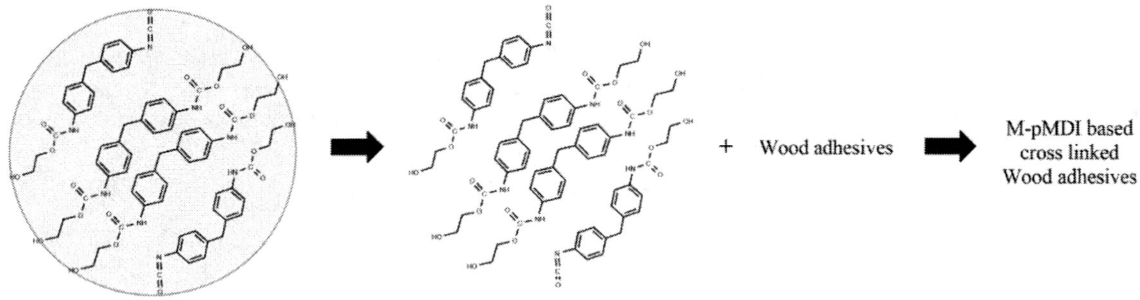


**Figure 9**. Illustration of potential application of M-pMDI as cross-linker for wood adhesives

## 4. Conclusion

Polymeric isocyanate microcapsules derived from Polymeric 4,4-methylene diphenyl isocyanate were prepared via interfacial polymerization with ethylene glycol. The resulting M-pMDI has a yellow color and spherical form with observable aggregation. The yield of M-pMDI remarkably increased as the isocyanate content, microencapsulation temperature, and microencapsulation agitation rate increased to some degrees. ATR-FTIR and micro confocal raman hyperspectral imaging spectroscopy enabled to detect the change of chemical structure of M-pMDI by assessing their functional groups. The –NCO groups of pure pMDI was not detected after microencapsulation due to the covering effect of urethane shell formed from the reaction of –NCO groups of pMDI and –OH groups of ethylene glycol. Images from digital microscope and micro confocal raman hyperspectral imaging spectroscopy revealed the formation of the urethane shell of M-pMDI. The pre-liminary investigration of M-pMDI using ATR-FTIR spectroscopy suggested that M-pMDI prepared with 5 mL isocyanate, at 60°C of microencapsulation temperature, and 600 rpm of microencapsulation agitation rate could be used as cross-linker for wood adhesives.

## References

[1] Ozaki S 1972 Recent advances in isocyanate chemistry *Chem. Rev.* **72** 457-96
[2] Delebecq E, Pascault JP, Boutevin B, and Ganachaud F 2013 On the versatility of urethane/urea bonds: Reversibility, blocked isocyanate, and non-isocyanate polyurethane *Chem. Rev.* **113** 80-118
[3] Nuryawan A and Alamsyah EM 2018 A review of isocyanate wood adhesive: a case study in Indonesia *Applied Adhesive Bonding in Science and Technology* vol i (InTech) p. 13
[4] Wicks D and Wicks ZW 1999 Blocked isocyanates III: Part A. Mechanisms and chemistry *Prog. Org. Coatings* **36** 148-72
[5] Lubis MAR, Park BD, and Lee SM 2019 Performance of hybrid adhesives of blocked-pmdi/melamine-urea-formaldehyde resins for the surface lamination on plywood *J. Korean Wood Sci. Technol.* **47** 200-9
[6] Wicks D and Wicks ZW 2001 Blocked isocyanates III - Part B: Uses and applications of blocked isocyanates *Prog. Org. Coatings* **41** 1-83

[7] Lubis MAR and Park BD 2018 Modification of urea-formaldehyde resin adhesives with oxidized starch using blocked pMDI for plywood *J. Adhes. Sci. Technol.* **32** 2667-81
[8] Wazarkar K, Patil D, Rane A, Balgude D, Kathalewar M, and Sabnis A 2016 Microencapsulation: an emerging technique in the modern coating industry *RSC Adv.* **6** 106964–79
[9] Ma Y, Lu P, Chen W, Zhang Y and Gu J 2019 Preparation of isocyanate microcapsules as functional crosslinking agent by minimalist interfacial polymerization *Adv. Powder Technol.* **30** 1995-2002
[10] Arshady R 1989 Microspheres and microcapsules: A survey of manufacturing techniques. Part 1: Suspension cross linking *Polym. Eng. Sci.* **29** 1746-58
[11] Ma Y, Jiang Y, Tan H, Zhang Y, and Gu J 2017 A rapid and efficient route to preparation of isocyanate microcapsules *Polymers (Basel)* **9** 24-7
[12] Lubis MAR, Park BD, and Lee SM 2020 Microencapsulation of polymeric isocyanate for the modification of urea-formaldehyde resins *Int. J. Adhes. Adhes.* **100** 102599
[13] Huang M and Yang J 2011 Facile microencapsulation of HDI for self-healing anticorrosion coatings *J. Mater. Chem.* **21** 11123-30
[14] Yang J, Keller MW, Moore JS, White SR, and Sottos N R 2008 Microencapsulation of isocyanates for self-healing polymers *Macromolecules* **41** 9650-5
[15] Caruso MM, Blaiszik BJ, Jin H, Schelkopf SR, Stradley DS, Sottos NR, White SR, and Moore JS 2010 Robust, double-walled microcapsules for self-healing polymeric materials *ACS Appl. Mater. Interfaces* **2** 1195-9
[16] Wang W, Xu L, Liu F, Li X, and Xing L 2013 Synthesis of isocyanate microcapsules and micromechanical behavior improvement of microcapsule shells by oxygen plasma treated carbon nanotubes *J. Mater. Chem. A* **1** 776-82

**Acknowledgment**
Authors are grateful for the financial support from PT Startech Multi Invensi and facilities support from Integrated Laboratory of Bioproducts (i-LaB), Research Center for Biomaterials, Indonesian Institute of Sciences.

# Surface modification of eco-friendly particleboard made from sorghum bagasse and citric acid sucrose adhesive

**S S Kusumah[1*], S Y Massijaya[2], K W Prasetyo[1], J Sutiawan[2], M A R Lubis[1], and D Hermawan[2]**

[1]Research center for Biomaterials, Indonesian Institute of Sciences (LIPI), Cibinong, 16911, Indonesia
[2]Division of Bio-Composite, Forest Products Department, Faculty of Forestry, IPB University, Bogor, 16680, Indonesia

[*]Email: sukma.surya@biomaterial.lipi.go.id

**Abstract.** Eco-friendly particleboard made from sweet sorghum bagasse bonded with citric acid and sucrose was developed and met the minimum requirement of JIS standard. The surface roughness of particleboard which is important for coating with thin overlays, however did not measure yet. Therefore, the effect of particle size composition in surface modification of particleboard on the physical and mechanical properties of the panel was investigated. Sorghum bagasse has been used as raw materials in the manufacturing of three-layer particleboard. The composition of coarse particle (core layer) and fine particle (face and back layers) in the fabrication of particleboard was varied into four compositions such as A (0 (face): 100 (core): 0 (back), B (12.75:75:12.75), C (25:50:25), D (37.5:25:37.5)). Particleboard was bonded with 20 wt% of citric acid-sucrose (CAS) adhesive based on dry weight of particle and hot-pressing at 15 MPa and 200°C for 10 minutes. Dimension and target density of particleboards were 30 x 30 x 0.9 and 0.8 g/cm$^3$, respectively. The physical and mechanical properties of the particleboard were evaluated according to JIS A 5908-2003. The Moisture Content (MC), Thickness Swelling (TS), Water Absorbtion (WA), Internal Bonding (IB) strength, Modulus of Elasticity (MOE) and Modulus of Rupture (MOR) values of the particleboard satisfied the requirements for the type 8 JIS A 5908-2003 standard. The particleboard made of composition C had lowest value of surface roughness. The addition of fine particle ratio on the surface layer lowered the WA and TS of the particleboard and slightly increased the IB strength. Further, the MOE and MOR of the particleboard was not affected by increasing fine particles on the surface layer. This study suggests that particle size composition remarkably affects the physical and internal bonding strength of particleboard, while it has less interference to the mechanical properties.

## 1. Introduction

Particleboard production in Europe has increased from 2016 to 2017. Particleboard production in 2016 amounted to 36.61 thousand m$^3$ and increased to 36.80 thousand m$^3$ in 2017 and is predicted to increase to 37.02 thousand m$^3$ in 2018 [1]. However, the increase of particleboard production was not accompanied by the availability of raw materials from natural forests caused by deforestation in the world forests [2]. MoEF data [2] states in Indonesia from 2015 to 2016 there has been deforestation of 0.63 million ha. The future particleboard raw material requirements cannot potentially be [3]. As a result, other lignocellulose materials have an important role as an alternative raw material in the manufacture of particleboards. Another raw material of lignocellulose is sorghum (*Sorghum bicolor* (L.) Moench).

Sorghum has potential to be used as raw material for particleboard. The total area of sorghum of the world in 2020 is 40.81 million ha [4] and according to Pabendon *et al.* [5] the average of sorghum stem biomass is 23.06 tonnes/ha, while the average sorghum bagasse biomass is 4.89 tonnes/ha.

Iswanto *et al.* [6] and Khazaeian *et al.* [7] successfully utilized sorghum bagasse for raw material particleboards using synthetic adhesives. The results of the study showed that the mechanical properties of the particleboard satisfied the requirements of JIS (Japanese Industrial Standard) and EN (European Norm). However, particleboards using synthetic adhesives have poor dimensional stability and commonly release formaldehyde emission. To overcome these problems, particleboard from sorghum bagasse was developed using citric acid (CA) and citric-sucrose (CAS) adhesive [8,9]. CA has been used as a natural cross-linker that promotes ester-type cross-linking [10]. In fact, a combination of CA with polysaccharides can be used as bio-based wood adhesives that produce durable and eco-firendly wood-based panels [11,12]. Particleboards bonded with CAS adhesive has a good quality showing a thickness swelling, modulus of elasticity, modulus of rupture, internal bond, and screw holding that satisfied JIS A 5908-2003 type 18 standards [9].

Sorghum bagasse particleboard bonded with CAS adhesive is still rough on the surface. The roughness of the surface increases the risk of fine defects on the particleboard surface. Fine defects on the board surface can reduced product grade, finishing and bonding qualities [13]. The effort to overcome this limited product particleboard by making a three-layer particle board with a smooth surface [13]. This study aims to evaluate the various layer compositions three-layer sorghum bagasse particleboards (TLSB-particleboard) bonded with CAS adhesives.

## 2. Materials dan Methods

### *2.1. Materials*

The particle sorghum species super 1 were planted in research field at the Biotechnology Research Center on Indonesia Institute of Sciences (LIPI). The particles were milled, then sieved to size of 4-14 mesh for core and -60 mesh for face and back layers. The particle were dried at 80°C for 12 hours to keep moisture content less than 10%. CA and sucrosa of extra purity grade was purchased from MERCK (Jakarta, Indonesia). CA and sucrosa (CAS) was dissolved in distilled water with a final concentration of 59 wt.% and used as adhesive [8].

### *2.2. Methods*

#### *2.2.1.* Particleboard manufactured

Particleboards with size of 30 × 30 × 0.9 cm and target density of 0.8 g/cm$^3$ were manufactured according to Kusumah *et al.* [8], Kusumah *et al.* [3] and Kusumah *et al.* [9]. Around 20 wt.% of adhesive content based on dry weight particle, pressure of 15 MPa, pressing temperature of 200°C, pressing time of 10 min, and CA:sucrose ratios (10/90 wt.%) were used to prepare the panel. Based on the various layer compositions such as surface (face and back) and core layer, the TLSB-particleboard panels were classified into 4 types (Table 1). The particles were sprayed with CAS adhesive using a spray gun (Meiji, Japan) and then were pre-dried at 80°C for 12 hours [8]. Subsequently, the particles were hand-formed into a mat by using a forming box of 30 x 30 cm, followed by hot-pressing into TLSB-particleboard with a distance bar of 9 mm to control the thickness.

**Table 1**. Various layer compositions

| Type of TLSB-particleboard | face:core:back (wt.%) |
|---|---|
| A | 0:100:0 |
| B | 12.5:75:12.5 |
| C | 25:50:25 |
| D | 37.5:25:37.5 |

*2.2.2.* Physical and mechanical testing

The TLSB-particleboard were conditioning for 1 week at a room temperature of 20°C at 60% of relative humidity (RH). The TLSB-particleboard properties were evaluated according to the Japanese Industrial Standard for Particleboards (JIS A 5908-2003) [14]. The Modulus of Rupture (MOR), Modulus of Elasticity (MOE), Internal Bonding (IB) strength, Thickness Swelling (TS), Water Absorption (WA) after water immersion for 24 hours and screw holding (SH) power, were tested. The specimens of 20 x 5 cm were prepared for each board in the bending properties test. The three point bending test was conducted over an effective span of 150 mm at a cross-head speed of 10 mm/min using a universal testing machine (UTM, Shimadzu, Japan). Specimens sized 5 x 5 cm were prepared for each board in IB tests, and two specimens of the same size of each board were prepared for TS and WA tests after 24 hours water immersion at 20°C. The specimens of 10 x 5 cm from each board were prepared for the SH test. Two positions in the tested pieces were vertically screwed, and then the screws pulled out with a cross-head speed of 2 mm/min. The sample 5 x 5 cm were tested for their surface roughness using portable surface roughness tester (Mitutoyo, Japan), which average roughness (Ra) was used to evaluate roughness characteristics of TLSB-particleboard. Each experiment was performed in three replications, and the average values and standard deviations were calculated.

2.3. Data analysis

Data were analyzed using Microsoft Excel and IBM SPSS Statistics (Statistical Product and Service Solution ver. 23, United States). A simple completely randomized design with a single factor was used for data analysis. The factor analysis of variance tests of various layer compositions (face:core:back) (0:100:0, 12.5:75:12.5, 25:50:25 and 37.5:25:37.5) on physical and mechanical properties were evaluated. Duncan's multiple range test was used for further analysis if the significance value of glue spread levels was significantly different at p-value <0.05.

## 3. Results and Discussion

*3.1. Density and moisture content*

The density and MC of TLSB-particleboard was 0.72-0.73 g/cm$^3$ and 6.24%-6.62%, respectively, that was shown in Figure 1. The analysis of variance (ANOVA) shows that the densitiy and MC were not significantly affected by various layer compositions (Table 2). The MC of TLSB-particleboard in this study was lower than the MC of particleboard bonded with synthetic adhesive (2.92-11.06) [6]. This is because the CAS adhesive resistant to water on the bio-composite products [15]. In addition, the formation of ester linkages from the reaction of –COOH of CA and –OH of sucrose or wood could improve the water resistancy of particleboard. The MC of resulting TLSB-particleboard was less than 13% which satisfied JIS A 5908-2003 standard.

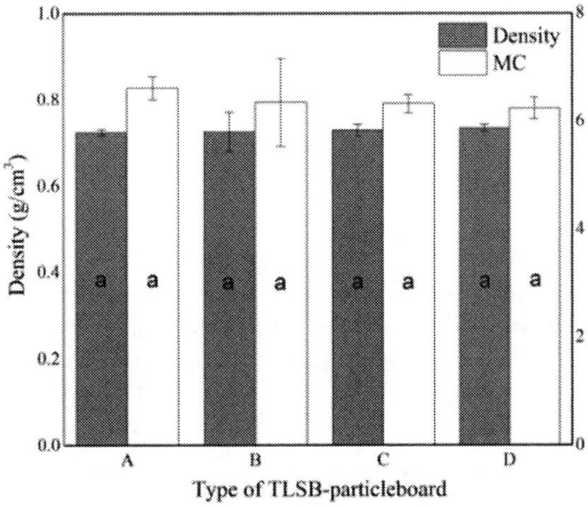


**Figure 1.** Density and MC of TLSB-particleboard

**Table 2.** Analysis of variance (ANOVA) of various layer compositions on the physical properties of TLSB-particleboard

| Properties | Significant |
|---|---|
| Density | $0.959^{ns}$ |
| Moisture content (MC) | $0.751^{ns}$ |
| Water absorption (WA) | $0.020^{**}$ |
| Thickness swelling (TS) | $0.000^{**}$ |

Note: ns: not signifant at 95% confidence interval; **: signifantly different at 95% confidence interval

*3.2. Water absorption and thickness swelling*

Incorporation of fine particles remarkably affected the physical properties of particleboard. The WA and TS values decreased gradually as the fine particles ratio on the surface layer increased. As shown by ANOVA at p-value <0.05 (Table 2), the A type TLSB-particleboard had greater WA and TS than B, C, and D type TLSB-particleboard (Figure 2). These results indicated that the addition of fine particles on the surface layer effectively decreased the WA and TS values. The particleboards made from fine particles has greater water resistance, indicating that the adhesive system has excellent bonding performance. It is probably due to the larger surface contact area of fine particles than that of coarse particles [16]. Tabarsa *et al.* [13] states the combination of particle size and particle cavities on the particleboard will be reduced so the value of WA and TS decreases. Smaller particle size would fill the pores between coarse particles in the core of the particleboard [17]. The WA value of TLSB-particleboard was lower than that of the WA of particleboard bonded with CA [8].The TS value of all the TLSB-particleboard satisfied JIS A 5908 (2003) standard (> 12 %).

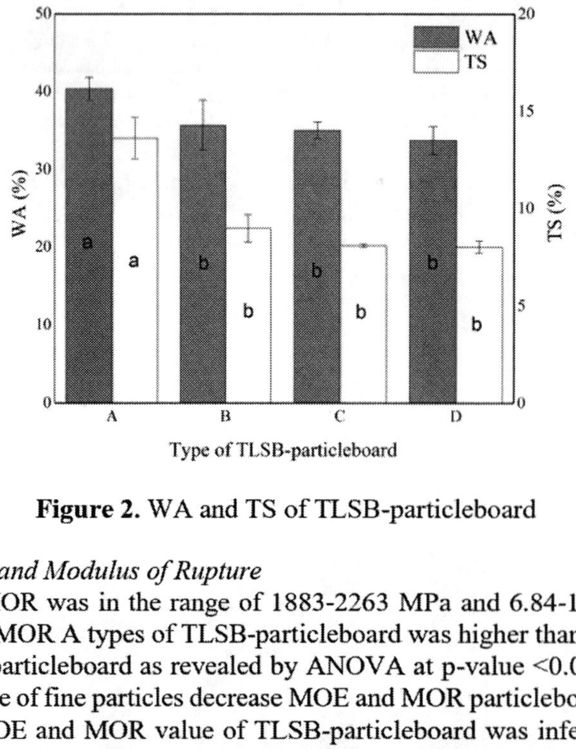


**Figure 2.** WA and TS of TLSB-particleboard

*3.3. Modulus of Elasticity and Modulus of Rupture*

The value of MOE and MOR was in the range of 1883-2263 MPa and 6.84-10.92 MPa, respectively (Figure 3). The MOE and MOR A types of TLSB-particleboard was higher than the MOE and MOR B, C, and D types of TLSB-particleboard as revealed by ANOVA at p-value <0.05 (Table 3). Tabarsa *et al.* [13] reported that the use of fine particles decrease MOE and MOR particleboard due to low adhesion between particles. The MOE and MOR value of TLSB-particleboard was inferior than the MOE and MOR of particleboard teak wood [18]. The type 8 JIS A 5908-2003 met MOE and MOR value of 2000 MPa and 8 MPa, respectively. The A and C particleboard of TLSB-particleboard was satisfied type 8 JIS A 5908-2003, while type B and D of TLSB-particleboard were not.

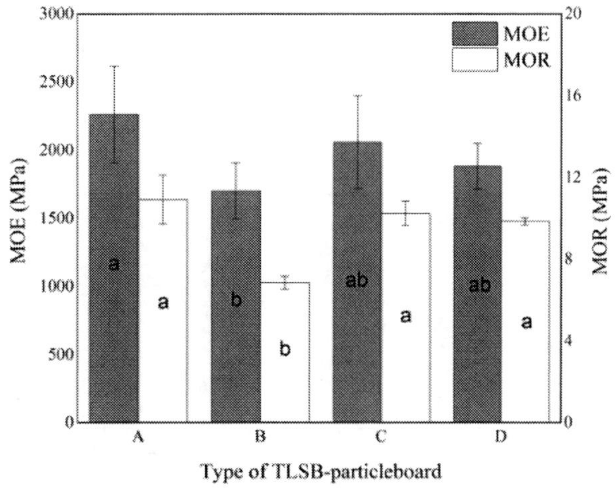


**Figure 3.** MOE and MOR of TLSB-particleboard.

**Table 3.** Analysis of variance (ANOVA) of various layer compositions on the mechanical properties of TLSB-particleboard.

| Properties | Significant |
|---|---|
| Modulus of elasticity (MOE) | $0.046^{**}$ |
| Modulus of rupture (MOR) | $0.000^{**}$ |
| Internal bond (IB) | $0.153^{tn}$ |
| Screw holding (SH) | $0.709^{tn}$ |
| Surface roughness | $0.023^{tn}$ |

Note: ns: not signifant at 95% confidence interval; **: signifantly different at 95% confidence interval

### *3.4. Internal bond and screw holding*

The IB strength values of the TLSB-particleboard gained slightly as the fine particles ratio increased. As depicted in figure 4, the IB strength values of the A, B to C increased, and then decreased for D type TLSB-particleboard. According to ANOVA results, the IB values of the A to D types of TLSB-particleboard were not statistically different at p-value > 0.05 (Table 3). Previous research [15] stated that fine particles might provide a larger contact areas and resulting a stronger bonding. The IB values of TLSB-particleboard were 0.16-0.30 MPa, which complied with the JIS A 5908-2003 standard (≤0.15 MPa). Besides, the IB values of TLSB-particleboard were comparable to particleboards teak wood bonded with CAS (0.2-0.3 MPa) [18].

The SH values of TLSB-particleboard were also investigated. The value was in the range of 129.23-169.86 N (Figure 4). Those SH values were about 50% lower than that of the particleboard bonded with CA (348 N) [3]. In addition, the SH of TLSB-particleboard was not satisfied the type 8 requirement of JIS A 5908-2003 (≥300 N). This result indicates that the addition of fine particle does not affect the SH of TLSB-particleboard as reveals by ANOVA at p-value >0.05 (Table 3). However, the addition fine particles slightly decreased the SH value of TLSB-particleboard.

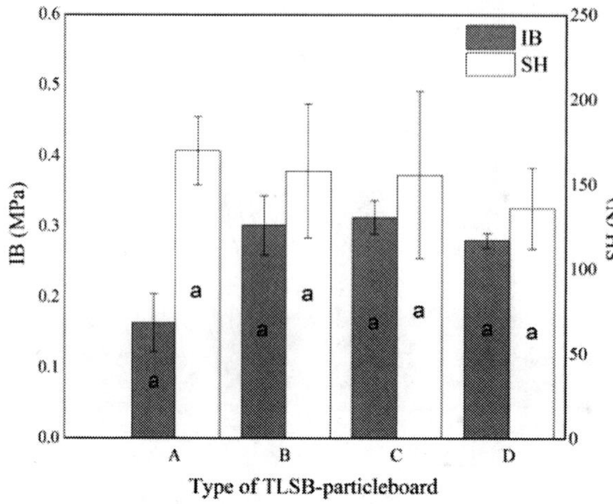


**Figure 4.** IB and SH of TLSB-particleboard

### *3.5. Surface roughness*

The average Ra value for TLSB-particleboard were 1.58-3.01 $\mu$m (Figure 5). The A types of TLSB-particleboard has the highest Ra values than B, C, and D types of TLSB-particleboard as confirmed by ANOVA at p-value <0.05 (Table 2). Increasing the fine particles ratio on the surface layer would cause smoother surface TLSB-particleboard. Particleboard made from fine particles had a smoother surface compared to that of coarse particles [15].

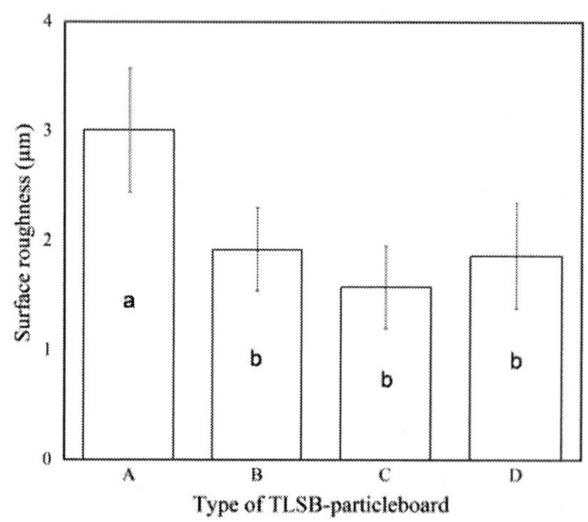


**Figure 5.** The surface roughness of TLSB-particleboard.

## 4. Conclusion

The MC, TS, IB strength, MOE and MOR values of TLSB-particleboard satisfied the requirements for the type 8 JIS A 5908 (2003) standard. The addition of fine particles ratio on surface layer affected the value of WA and TS of TLSB-particleboard and slightly increased IB value of TLSB-particleboard. Meanwhile, the MOE and MOR TLBS-particleboard were not affected by increasing fine particles on the surface layer. The increasing of fine particles composition on surface layer improve the smoothness of the TLSB-particleboard.

**References**

[1] Food and Agriculture Organization 2017 Annual Market Review 2016-2017 Forest Products (Rome: FAO)
[2] [MoEF] Ministry of Environment and Forestry of the Republic of Indonesia 2018 Data Perkembangan Deforestasi Tahun 1990-2017 (Jakarta: Kementrian Lingkungan Hidup dan Kehutanan)
[3] Kusumah SS, Umemura K, Guswenrivo I, Yoshimura T, and Kanayama K 2017 Utilization of sweet sorghum bagasse and citric acid for manufacturing of particleboard II: influences of pressing temperature and time on particleboard properties *J. Wood Sci.* **63** 161-72
[4] Foreign Agricultural Service 2020 World Agricultural Production https://apps.fas.usda.gov/psdonline/circulars/production.pdf
[5] Pabendon MB, Sarungallo R, and Mas'ud S 2012 Pemanfaatan nira batang, bagas, dan biji sorgum manis sebagai bahan baku bioetanol *Penelitian Pertanaman Tanaman Pangan* **31** 180-7
[6] Iswanto AH 2014 Effect of Resin Type, Pressing Temperature and Time on Particleboard Properties made from Sorghum Bagasse *Agric. For. Fish.* **3** 62
[7] Khazaeian A, Ashori A, and Dizaj MY 2015 Suitability of sorghum stalk fibers for production of particleboard *Carbohydr. Polym.* **120** 15-21
[8] Kusumah SS, Umemura K, Yoshioka K, Miyafuji H, and Kanayama K 2016 Utilization of sweet sorghum bagasse and citric acid for manufacturing of particleboard I: Effects of pre-drying treatment and citric acid content on the board properties *Ind. Crops Prod.* **84** 34-42
[9] Kusumah SS, Arinana A, Hadi YS, Guswenrivo I, Yoshimura T, Umemura K, Tanaka S, and Kanayama K 2017 Utilization of sweet sorghum bagasse and citric acid in the manufacturing of particleboard. III: Influence of adding sucrose on the properties of particleboard *BioResources* **12** 7498-514
[10] Hubbe MA, Pizzi A, Zhang H, and Halis R 2018 Critical links governing performance of self-binding and natural binders for hot-pressed reconstituted lignocellulosic board without added formaldehyde: A review *BioResources* **13** 2049-115
[11] Lubis MAR, Park B, and Hong MK 2020 Tuning of Adhesion and disintegration of oxidized starch adhesives for the recycling of medium density fiberboard *BioResources* **15** 5156-78
[12] He, Z ed. 2017 *Bio-based Wood Adhesives: Preparation, Characterization, and Testing*, (Florida, USA: CRC Press)
[13] Tabarsa T, Ashori A, and Gholamzadeh M 2011 Evaluation of surface roughness and mechanical properties of particleboard panels made from bagasse *Compos. Part B Eng.* **42** 1330-5
[14] Japanese Industrial Standard 2003 Particleboard (JIS A 5908-2003)1-20
[15] Umemura K, Putra DR, Prayitno TA, Widyorini R, Awaludin A, and Isnan R 2015 Manufacture and properties of citric acid-bonded particleboard made from bamboo materials *Eur. J. Wood Wood Prod.* **74** 57-65
[16] Okuda N and Sato M 2004 Manufacture and mechanical properties of binderless boards from kenaf core *J. Wood Sci.* **50** 53-61
[17] Nemli G, Aydin I, and Zekoviç E 2007 Evaluation of some of the properties of particleboard as function of manufacturing parameters *Mater. Des.* **28** 1169-76
[18] Widyorini R, Nugraha PA, Rahman MZA, and Prayitno TA 2016 Bonding with citric acid. *BioResources* 11(2) 4526-4535

# Physical mechanical properties of plywood made from lesser known wood veneers bonded with phenol formaldehyde adhesive

**J Sutiawan[1], Nurhanifah[1], S Anggiriani[1], D Hermawan[1*], MY Massijaya[1], IM Sulastiningsih[2], IB Abdillah[1]**

[1]Forest Products Department, Faculty of Forestry, IPB University, Bogor, 16680, Indonesia
[2]Forest Products Research and Development Centre, Ministry of Environment and Forestry, Bogor, 16610, Indonesia

[*]Email: mr.dede.hermawan@gmail.com

**Abstract.** Indonesia plywood production becomes the third-largest after wood particle and pulp production. However, this production is not supported by the continuous availability of wood as raw materials, so the use of less-known wood species such as mempisang wood can be considered as alternative raw materials for plywood. This study aims to develop plywood from mempisang wood and investigate the effect of glue spread on the physical and mechanical properties of plywood. The plywood consisted of three veneers glued with various glue spread of phenol-formaldehyde resin (250 g/m$^2$, 275 g/m$^2$, dan 300 g/m$^2$) using single spread method. The plywood was hot-compressed at 130°C for 5 minutes with specific pressures at 10 kg/cm$^2$. Physical-mechanical properties of plywood were carried out based on JAS 233-2003. The higher of glue spread slightly improves the physical and mechanical properties of the mempisang plywood. The Moisture Content (MC), and Shear Strength (SS) values of the plywood satisfied the requirements for the JAS 233-2003 standard. The best level of glue spread of the plywood glued with 300 g/m$^2$ phenol-formaldehyde.

## 1. Introduction

Indonesia is well known as the biggest plywood producer and exporter. Indonesia plywood production reaches to 3.85 million m$^3$ and becomes the third-largest after wood particle and pulp production [1]. Plywood is one of the most important composite products in Indonesia, also increasingly as a substitute for sawn timber and logs utilization. Meanwhile, wood resources were reduced due to the deforestation of natural forests. MoEF [2] reported that from 2015 to 2016, there had been deforestation of 0.63 million ha in Indonesia. The reduced wood supply can affect the deficiency of wood as raw materials for plywood production. As an alternative, the utilization of lesser-known wood was developed to overcome the problem of wood supply deficiency.

Indonesia is labeled as one of the mega biodiversity countries in the world. The wood species in the tropical rainforest of Indonesia, approximately 4000 potential wood species and only less than 10% that had been utilized [3, 5] According to Martawijaya *et al.* [5], there are about 133 wood species included in lesser-known species. The knowledge of basic properties become essential and needed to determine wood utilization even obtained efficiency the uses of wood. Therefore, it can complement also substitute the uses of commercial wood [6]. One of the lesser-known wood species that can be used mempisang (*Alphonsea spp.*) wood. The results of Supriadi and Abdurachman [6] showed that mempisang wood has poor to excellent class on machining properties. Mempisang wood requires major caution on drilling and turning process on the lathe. Also, information about the utilization that wood as biocomposite products is still limited. Therefore, research on the utilization of mempisang wood as biocomposite products such as plywood should be explored. This study aims to develop plywood from mempisang wood and investigate the effect of glue spread levels on the physical and mechanical properties of the plywood.

## 2. Materials and Methods

### *2.1. Materials*

Veneers mempisang wood with a thickness of 1.5 mm was obtain from Forest Products Research and Development Centre Bogor-Indonesia. The adhesive was used Phenol Formaldehyde (PF) resin from PT Pamolite Adhesive Industry (Jakarta, Indonesia).

### *2.2. Methods*

#### 2.2.1. Plywood manufactured

Mempisang wood veneers were oven-dried at 60˚C for 48 hours to obtained less than 10% of moisture content. Three layers of veneers were manufacture into plywood. The core veneers were glued with PF resin at various glue spread levels (250 g/m$^2$, 275 g/m$^2$, dan 300 g/m$^2$) using a single spread method. After the adhesives were spread and veneer arranged, the plywood was pressed using a hot-pressed machine at 130ºC for 5 minutes with specific pressures at 10 kg/cm$^2$.

#### 2.2.2. Physical and mechanical properties testing

The test of physical and mechanical properties was carried out based on JAS 233-2003 [7] and JIS A 5908-2003 [8]. The physical properties were tested, such as density, Moisture Content (MC), Water Absorption (WA), and Thickness Swelling (TS). The mechanical properties were tested modulus of elasticity (MOE), Modulus of Rupture (MOR), and Shear Strength (SS). The density and MC were measured based on JAS 233-2003 standard using 7.5 x 7.5 x 0.45 cm (in length, width, and thickness, respectively). The WA and TS test were based on JIS A 5908-2003 standard using 5 x 5 x 0.45 cm samples in length, width, and thickness, respectively. The mechanical properties (MOE and MOR) test was carried out according to JAS 233-2003 standard using 20 x 5 x 0.45 cm (in length, width, and thickness, respectively). The test was using UTM Instron (50 kN, USA) with a loading speed of 10 mm/minute for MOE and MOR, also 2 mm/min loading speed for the SS test. The samples for the SS test were 8 x 2.5 x 0.45 cm in length, width, and thickness, respectively.

#### 2.2.3. Data analysis

The data were analyzed using Microsoft Excel and IBM SPSS Statistics (Statistical Product and Service Solution) ver 23. A simple completely randomized design with a single factor was used for data analysis. The factor analysis of variance tests of glue spread levels (250 g/m$^2$, 275 g/m$^2$, dan 300 g/m$^2$) on physical and mechanical properties were evaluated. Duncan's multi-range test was used for further analysis if the significance value of glue spread levels was significantly different $p<0.05$.

## 3. Results and Discussion

### *3.1. Physical properties of plywood*

The density of plywood in this study tends to be same, ranged from 0.53-0.56 g/cm$^3$ (Figure 1). Based on the analysis of variance showed in Table 1. The results showed that the glue spread level did not significantly affect density. There is no requirement of density on the JAS 233-2003 standard. The density of plywood in this study was higher than the density of plywood made from jabon veneers glued with PF resin mixed black liquor (PF-BL) (0.47-0.48 g/cm$^3$) [9]. The results were also higher than 5 layers plywood made from pinus veneers wood (0.48-0.56 g/cm$^3$) [10] and density of 5 layers plywood made from poplar veneers wood about 0.46-0.47 g/cm$^3$ [11].

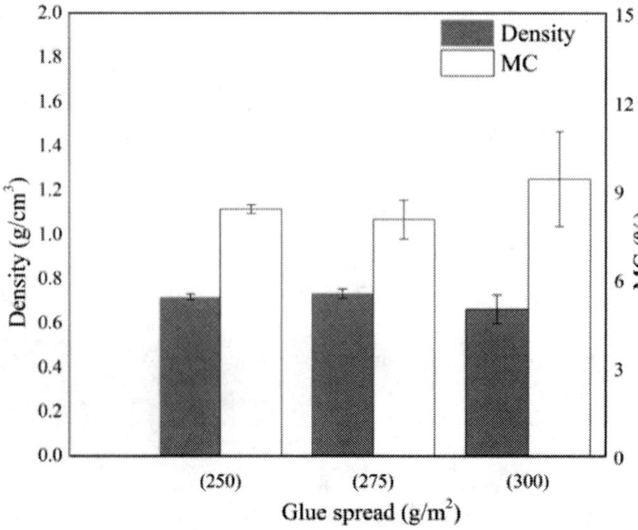


**Figure 1.** Density and MC of mempisang plywood glued with PF resin at various glue spread levels

The average MC of mempisang plywood in this study were 11.72-12.35% (Figure 1). The average MC of all mempisang plywood in this study met JAS 233-2003 standard, which requires a maximum value of 14%. The average MC was lower than three layers plywood made from jabon veneers wood glued with PF-BL (12.61-13.53%) [9]. Meanwhile, the average MC was higher than Suhasman and Agussalim [12], which MC of plywood made from sengon wood veneers reinforced with $H_2O_2$ ranged from 4.92-6.84%. Analysis of variance showed that glue spread levels did not significantly affect the MC of mempisang plywood (Table 1). However, the higher glue spread levels tend to reduce the MC of plywood.

**Table 1.** Analysis of variance (ANOVA) of glue spread levels on the physical properties of mempisang plywood

| Properties | Significant |
|---|---|
| Density | $0.742^{ns}$ |
| Moisture content (MC) | $0.771^{ns}$ |
| Water absorption (WA) | $0.629^{ns}$ |
| Thickness swelling (TS) | $0.260^{ns}$ |

Note: ns: not signifant at 95% confidence interval; **: signifantly different at 95% confidence interval

According to Sulastiningsih *et al.* [13], lignocellulose materials such as wood have hygroscopic properties that can absorb and release water due to temperature, and Relative Humidity (RH) changes. Therefore, WA and TS of plywood are important to determine due to the dimensional stability of the products. The average WA of plywood ranged from 60.00-65.04%. The average WA of mempisang plywood at 300 g/m$^2$ glue spread level has lower than plywood with 250 g/m$^2$ and 275 g/m$^2$ glue spread level. It those due to the higher compatibility between hydrophilic fibers and adhesives of plywood glued with higher glue spread levels. Sreekala *et al.* [14] showed that weak compatibility between the surface of the fibre and the adhesive could cause a weak structure and not optimal. So, it facilitates higher water absorption. The average WA of plywood was lower than WA of plywood made from pine veneers wood glued with PF resin with value 107.6% [15].

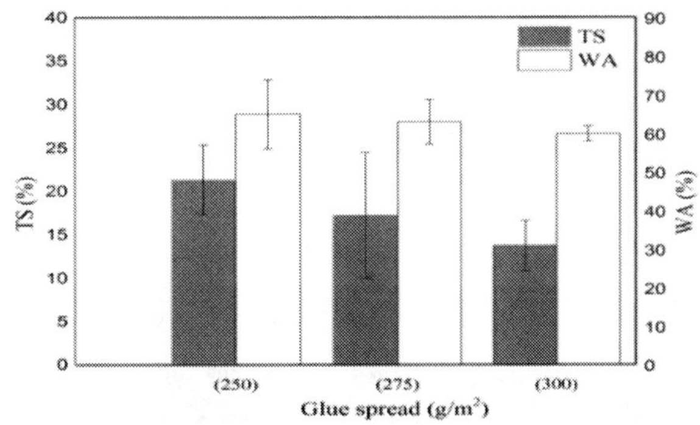


**Figure 2.** WA and TS of mempisang plywood glued with PF resin at various glue spread levels

The average TS value was ranged from 13.72 to 21.35% (Figure 2). The TS value of all mempisang plywood in this study did not meet the requirement of JIS A 5908-2003 standard, which requires a maximum value of 12%. The average TS was higher than plywood made from eucalyptus wood veneers with PF resin, those reported by Almeida *et al.* [16] that the TS value was 7%. Silva *et al.* [15] also reported that the TS value of plywood made from pine wood with PF resin was 8.7%. The results of analysis of variance showed that glue spread levels did not significantly affect WA and TS value of mempisang plywood (Table 1). Therefore, the higher glue spread level tends to reduce the average of WA and TS value of plywood.

### *3.2. Mechanical properties of plywood*

The MOE value of mempisang plywood ranged from 9.6 to 12.9 GPa (Figure 3). The results of the analysis of variance showed that glue spread has no significant effect on the MOE of the plywood (Table 2). However, the higher glue spread produces the higher MOE value of the plywood. It is due to the higher glue spread caused the resin spread and enter the wood pore on the surface and then hardened [17]. The MOE values in this study were higher than 5 layers of poplar plywood (2.1 GPa), and batai plywood (2.4 GPa) glued with UF resin [11,18]. In adition, the results of MOE value was also higher than 5 layers pinewood glued with PF resin that the values were 5.35 GPa [10].

The MOR values of mempisang plywood ranged from 45.35 to 60.39 MPa (Figure 3). The results of ANOVA showed that glue spread levels did not significantly affect on MOR values of plywood (Table 2). However, the higher glue spread level produce showed a higher MOR value. The MOR values of plywood in this study were higher than Bal and Bektas [11] that MOR values of poplar plywood (5.6 MPa) and Batai plywood (20.38 MPa) glued with UF resin [18]. Moreover, the MOR in this study also higher than pine plywood (56.7 MPa) with PF resin [10].

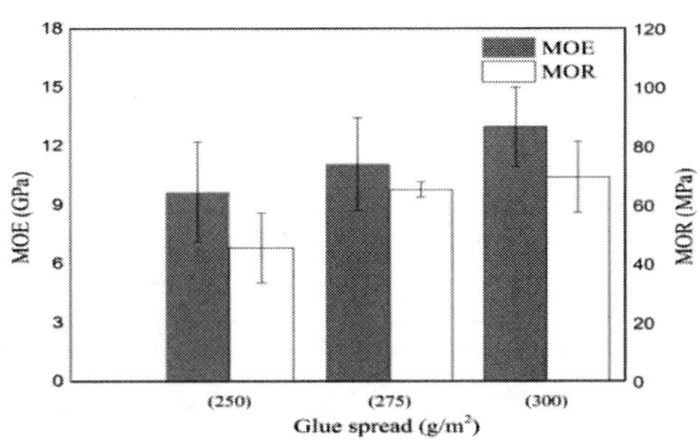


**Figure 3.** MOE and MOR of mempisang plywood glued with PF resin at various glue spread levels

**Table 2.** Analysis of variance (ANOVA) of glue spread levels on mechanical properties of mempisang plywood

| Properties | Significant |
|---|---|
| Modulus of elasticity (MOE) | 0.291$^{ns}$ |
| Modulus of rupture (MOR) | 0.051$^{ns}$ |
| Shear strength (SS) | 0.138$^{ns}$ |

Note: ns: not signifant at 95% confidence interval; **: signifantly different at 95% confidence interval

The SS value in this study ranged from 1.60 to 2.17 MPa, with an average of 1.85 MPa (Figure 4). The SS of plywood in this study was higher than the 3 layers jabon plywood (0.72-1.15 MPa) using PF-BL resin by Karliati *et al.* [9] and sengon plywood reinforced with $H_2O_2$ (0.67-1.32 MPa) by Suhasman and Agussalim [12]. The SS value of mempisang plywood was also higher than 3 layers alder plywood glued with polyethylene adhesive (1.1-1.52 MPa) [19]. The results of the analysis of variance showed that the glue spread level did not significantly affect on the SS values of plywood (Table 2). However, the SS values of mempisang plywood were increased in line with increasing the glue spread (Figure 4). It is following Khalil *et al.* [20], which states that the higher glue spread to a certain level can increase the SS value of plywood. The SS values of mempisang plywood in this study had met the requirements of JAS 233-2003 standard, which require a minimum value of 0.7 MPa.

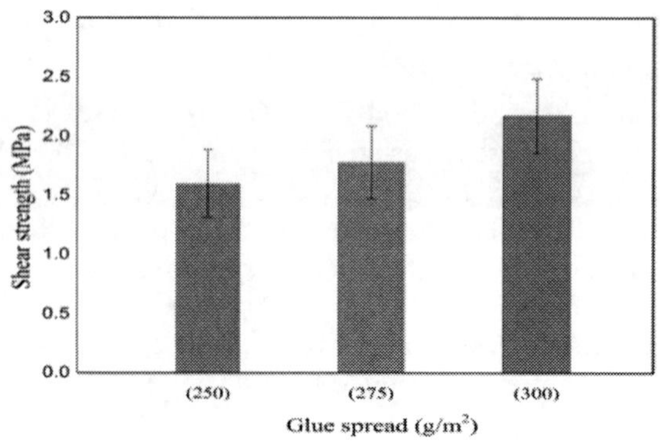


**Figure 4.** SS of mempisang plywood glued with PF resin at various glue spread levels

## 4. Conclusion

The higher glue spread level slightly improves the physical and mechanical properties of the mempisang plywood. The MC, and SS values of the plywood satisfied the requirements for the JAS 233-2003 standard. The best level of glue spread of the plywood glued with 300 g/m$^2$ phenol-formaldehyde.

## References

[1] The Central Bureau of Statistics 2018 S*tatistics of Forestry Production* (Jakarta: BPS)
[2] [MoEF] Ministry of Environment and Forestry of the Republic of Indonesia 2018 *Statistik Lingkungan Hidup dan Kehutanan Tahun 2018* (Jakarta: Kementrian Lingkungan Hidup dan Kehutanan)
[3] Soerianegara I and Lemmens RHMJ 1993 *Plant Resources of South-East Asia:Timber Trees Minor Commercial Timbers* (Wageningen: Pudoc Scientific Publishers)
[4] Mandang YI 2013 *Xylarium Bogoriense and Its Role in Anatomical Research and Introduction of Various Wood Species in Indonesia* (Bogor: Forestry Engineering and Forest Product Processing Research and Development Center-FORDA Ministry of Forestry)

[5] Martawijaya A, Kartasujana I, Kadir K, and Prawira SA 2005 *Indonesian Wooden Atlas Volume III* (Bogor: Forestry Engineering and Forest Product Processing Research and Development Center-FORDA-Ministry of Forestry)
[6] Supriadi A and Abdurachman A 2018 Sifat pemesinan lima jenis kayu asal riau *Jurnal Penelitian Hasil Hutan* **36** 85-100
[7] [JAS] Japanese Agricultural Standard 2008 *Japanese Agricultural Standard for Plywood* (233) 1-78
[8] [JAS] Japanese Industrial Standard 2003 *Particleboard JIS A 5908-2003* 1-20
[9] Karliati T, Sumardi I, Darwis A, and Rumidatul A 2018 Incorporation of phenol formaldehyde based black liquor as an adhesive on the performance of plywood *J. Biol. Sci.* **18** 346-353
[10] Carvalho AG, Zanuncio AJV, Mori FA, Mendes RF, Da Silva MG, and Mendes LM 2014 Tannin adhesive from stryphnodendron adstringens coville in plywood panels *BioResources* **9** 2659-70
[11] Bal BC and Bektaş I 2014 Some mechanical properties of plywood produced from eucalyptus, beech, and poplar veneer *Maderas Cienc. y Tecnol.* **16** 99-108
[12] Suhasman S and Agussalim 2019 Resinless plywood production by using oxidized acacia bark powder as a binder *IOP Conf. Ser. Mater. Sci. Eng.* **593** 1
[13] Sulastiningsih IM, Indrawan DA, Balfas J, Santoso A, and Iskandar MI 2017 Sifat fisis dan mekanis papan untai berarah dari bambu tali (*Gigantochloa apus* (J.A. & J.H. Schultes) Kurz) *Jurnal Penelitian Hasi Hutan* **35** 197-209
[14] Sreekala MS, George J, Kumaran MG, and Thomas S 2002 The mechanical performance of hybrid phenol formaldehyde based composites reinforced with glass and oil palm fibres *Composites Science and Technology* **62**(3) 339-53
[15] da Silva BC, Vieira MC, Oliveira GL, Gonçalves FG, Rodrigues ND, Lelis RCC, and Iwakiri S 2012 Qualidade de compensados fabricados com adesivos à base de tanino-formaldeído de pinus oocarpa e fenol-formaldeído *Floresta e Ambiente.* **19**(4) 511-19
[16] Almeida RR, Júnior GB, and Jankowsky IP 2004 Produção de lâminas a partir da madeira de clones do híbrido *Eucalyptus grandis* x *Eucalyptus urophylla Sci. For. Sci* **65** 49-58
[17] Khalil AHPS, Fazita NMR, Bhat AH, Jawaid M, and Fuad NNA 2010 Development and material properties of new hybrid plywood from oil palm biomass *Mater. Des* **31**(1) 417-24
[18] Fitri MS, Suffian M, Nazri WMWAR, and Yuziah NY 2018 Mechanical properties of plywood from batai (*paraserianthes falcataria*), Eucalyptus (*Eucalyptus Peluta*) and Kelempayan (*neolamarckla cadamba*) with different layer and species arrangement *J. Trop. For. Sci.* **30** (1) 58-66
[19] Öztürk H, Demir A, and Çolak S 2017 Some mechanical properties of nylon composite plywood produced from different wood species *Kastamonu Üniversitesi Orman Fakültesi Derg.* **17**(3) 458-463
[20] Khalil AHPS, Issam AM, Shakri AMT, Suriani R, and Awang AY 2007 Conventional agro-composites from chemically modified fibres *Industrial Crops and Products* **26**(3) 315-23

**Acknowledgment**
The authors also thank to IPB University (Bogor Agricultural University) for supporting research facilities.

# Physical and mechanical properties of lightweight concrete with the addition of crude palm oil production wastes

**I Budiman[1*], Y W Sari[2], and N M Dewi[2]**

[1]Research Center for Biomaterials, Indonesian Institute of Sciences, LIPI, Cibinong, Indonesia
[2]Department of Physics, Faculty of Mathematics and Natural Science, IPB University, Bogor, Indonesia

*Email: budimanismail@gmail.com

**Abstract.** The increased area of oil palm plantations in Indonesia followed by increased production and amount of palm oil waste. One of the waste products of palm oil produced from the process of purifying Crude Palm Oil (CPO) is Spent Bleaching Earth (SBE). This research will utilize SBE waste as sand substitution material in concrete manufacture. The addition of natural fiber from the Empty Fruit Bunches (EFB) in the manufacture of concrete is expected to increase the mechanical strength of concrete that meets the SNI requirements. The process of making concrete begins by mixing all ingredients, stirring, moulding, compacting manual, and curing for 28 days. Concrete making is done by substituting SBE by 10%, 30%, and 50% based on sand weight, and fiber by 2%, 4%, and 6% based on sample volume. The results showed that the substitution of 28.98% SBE for sand weight, and 5.95% EFB based on Light Weight Concrete (LWC) volume, produced concrete with the best properties.

## 1. Introduction

Indonesia is one of the countries with the largest area of oil palm plantations in the world, as well as the largest palm oil-producing country in the world in the form of Crude Palm Oil (CPO). This has consequences for the amount of waste generated from the CPO oil production process which is getting bigger, such as oil palm Empty Fruit Bunches (EFB) and Spent Bleaching Earth (SBE).

EFB fiber is one of the solid wastes produced in the palm oil industry. The amount of EFB fiber is quite large because it is almost the same as the amount of crude palm oil production. The waste has not been used optimally. The data produced in 2015 for EFB amounted to 30.62 million tons [1]. The biggest components of EFB are cellulose (38-65%), besides other smaller components such as hemicellulose (17-35.3%), lignin (13-37%), and ash (1-6%) [2,3]. The great potential of EFB can be utilized to obtain added value and economic value from this material.

Another type of waste, SBE, is an active ingredient used to adsorb the color pigments contained in CPO to produce clearer oil. SBE is a type of clay with the main composition consisting of $SiO_2$, $Al_2O_3$, bound water, and $Ca^{2+}$ ions, magnesium oxide, and iron oxide [4]. SBE bleaching power is due to the presence of $Al^{3+}$ ions on the surface of the absorbing particles so that it can absorb dyes and depends on the ratio of $Al_2O_3$ and $SiO_2$ in SBE. Both types of waste, both EFB and SBE fibers can be utilized as a more useful product, one of which is to be made into the concrete or Light Weight Concrete (LWC).

Several studies using EFB fiber as raw material for making concrete or LWC to improve the mechanical properties of concrete, including conducted by [5-8]. Whereas research on the use of SBE in making concrete as a substitute for cement or sand substitutes was carried out by [9-12].

In this research, LWC manufacture by using SBE as sand substituted with 10%, 30%, and 50%, and the addition of EFB fiber is 2%, 4%, and 6% by cement weight. The purpose of this study was to determine the effect of SBE substitution and the addition of EFB fiber on the physical and mechanical properties of concrete. The optimal value of SBE substitution for sand and EFB fiber addition in producing lightweight concrete also measured using response surface methodology. Besides, seen also the durability of concrete to changing conditions as indicated by the treatment cycles wet/dry, to see the changes in the value of compressive strength.

## 2. Materials and Methods

### *2.1. Materials*

The LWC mixtures were composed of Portland Cement Composites (PCC), sand, water, polycarboxylate based ether superplasticizer from SIKA, SBE, and EFB fibers, as listed in Table 1. Every series of mixtures used to prepare four samples 25 x 5 x 5 cm for bending test, and every four samples 5 x 5 x 5 cm for compressive strength, density, and water absorption test.

**Table 1**. Mortar samples with their code for each composition

| Specimen code | Cement | Sand | SBE | EFB | water | HRWR |
|---|---|---|---|---|---|---|
| Control | 1 | 3.00 | 0.00 | 0.00 | 0.6* | 2 ** |
| S1E0 | 1 | 2.70 | 0.30 | 0.00 | 0.6* | 2 ** |
| S2E0 | 1 | 2.10 | 0.90 | 0.00 | 0.6* | 2 ** |
| S3E0 | 1 | 1.50 | 1.50 | 0.00 | 0.6* | 2 ** |
| S1E1 | 1 | 2.70 | 0.30 | 0.02 | 0.6* | 2 ** |
| S2E1 | 1 | 2.10 | 0.90 | 0.02 | 0.6* | 2 ** |
| S3E1 | 1 | 1.50 | 1.50 | 0.02 | 0.6* | 2 ** |
| S1E2 | 1 | 2.70 | 0.30 | 0.04 | 0.6* | 2 ** |
| S2E2 | 1 | 2.10 | 0.90 | 0.04 | 0.6* | 2 ** |
| S3E2 | 1 | 1.50 | 1.50 | 0.04 | 0.6* | 2 ** |
| S1E3 | 1 | 2.70 | 0.30 | 0.06 | 0.6* | 2 ** |
| S2E3 | 1 | 2.10 | 0.90 | 0.06 | 0.6* | 2 ** |
| S3E3 | 1 | 1.50 | 1.50 | 0.06 | 0.6* | 2 ** |

*) the ratio of water to cement (based on weight)
**) the percentage of superplasticizer to cement (based on weight)

### *2.2. Methods*

#### 2.2.1. EFB pre-treatment

EFB fiber is thought to still contain extractive substances that can reduce compatibility and binding ability with cement, so it requires treatment of EFB fiber, The EFB from the factory is washed thoroughly using water by soaking, beating and draining. Then the fibers are dried by the oven with a temperature of 80°C for 8 hours. After drying, the fiber is mashed up to ± 5cm in size, then scaled down using a refiner disk so that the fiber is between 16 mesh and retained at 40 mesh. To eliminate the extractive substances, the EFB fiber is soaked in clean water for 24 hours, then rinsed with clean water. The treatment is repeated until the extractive substances in the EFB fiber are reduced or lost. Indicators that show a decrease or loss of oil in the fiber, EFB fiber has a pH of 7 or close to 7.

2.2.2. Characterization of SBE, sand, and cement
Density measurements were carried out on SBE, sand, and cement, while the sludge level measurements carried out on the sand. The testing of sludge content also needs to be done, because clay or dirt more than 10% in the sand can inhibit the compacting process and reduce the strength of concrete.

2.2.3. Lightweight concrete manufacture
The first step in the mixing procedure was dry mixing of all the solid components (cement, sand, and SBE) with a mortar mixer. Water was then added to the mixture and further mixed until the mixture was evenly distributed. The superplasticizer was then added and the mixture was again mixed for 3 min. When the mixture was homogenous, the fibers were slowly added and mixing until the mixture was uniform. Molds were filled and the samples were compacted. The samples were removed from the molds after 24 hours and were stored in the water tank for 28 days for the curing process.

2.2.4. Physical testing
The physical parameters investigated in this study were the density and water absorption. The sample size for both tests is 5 x 5 x 5 cm. In the measurement of density, samples that have gone through a period of curing and conditioning are measured by their mass and size. While the measurement of water absorption is done by soaking water samples for 24 hours. Weighing samples before and after soaking. The water absorption test aims to determine the porosity of concrete samples.

2.2.5. Mechanical testing
The mechanical properties investigated in this study were Modulus of Rupture (MOR), Modulus of Elasticity (MOE), and compressive strength. MOR and MOE values were obtained from bending tests, using samples with a size of 25 x 5 x 5 cm, while compressive strength values were obtained from tests using samples with a size of 5 x 5 x 5 cm.

2.2.6. Durability testing with wet/dry cycles treatment
The durability properties investigated in this study was the degradation of compressive strength between LWC with and without wet/dry cycles treatment. Sample with a size of 5 x 5 x 5 cm were treated with wet/dry cycles for 7 times. A wet/dry cycle was defined for this research as 23 hours and 30 min drying in an oven at 60 ± 5°C, air drying at room temperature for 30 min, 23 h and 30 min soaking in water at 20 ± 2°C, and air drying at room temperature for 30 min [13].

2.2.7. Statistical analysis
Multiple averages were compared using an analysis of variance (ANOVA) test with a significance level of 5%. The optimal value of SBE substitution for sand and EFB fiber addition in producing LWC measured using response surface methodology.

## 3. Results and Discussion

### *3.1. Materials characteristics*
Material characterization was carried out to determine the density of SBE, EFB, sand, and cement. Besides, measurements of sludge from sand are also carried out. The results showed that the density of SBE, EFB, sand, and cement respectively of 0.94 g/cm$^3$, 0.49 g/cm$^3$, 2.06 g/cm$^3$, and 2.75 g/cm$^3$. While the sludge content from the sand obtained a value of 9.6%, which means the sand can be directly used in the manufacture of LWC because the sludge content is smaller than 10%.

### *3.2. Physical properties of LWC*

3.2.1 Density of LWC

The LWC density value is shown in Figure 1. The LWC density value with the addition of SBE and or EFB ranges from 1.63-1.91 g/cm3. This density value is lower when compared to the density of the control LWC (without the addition of SBE and EFB), which is 1.93 g/cm$^3$.

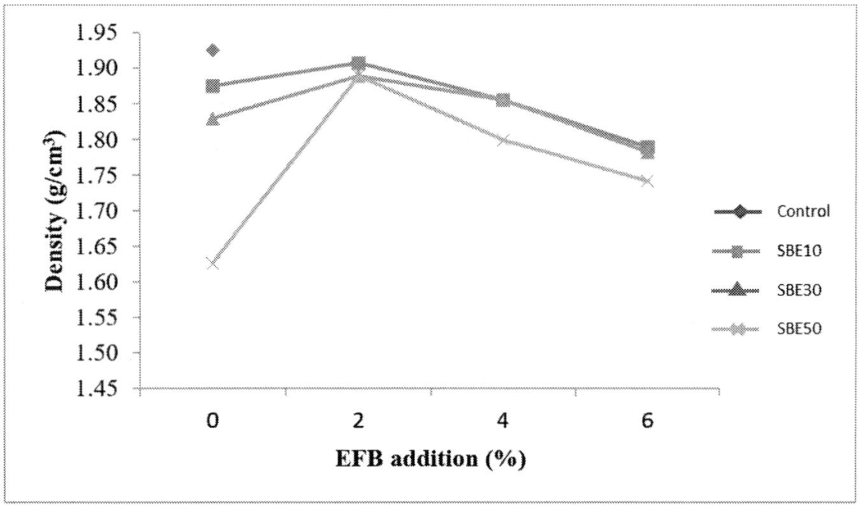


**Figure 1.** The density of lightweight concrete

Based on Figure 1 above, the LWC density value tends to decrease with the addition of SBE. While the addition of EFB tends to increase the density at an additional level of EFB 2% and again decreases at the addition of 4% and 6%. This can occur because the addition of EFB at levels of more than 2% can cause the surface coverage of fibers that can be bound by cement to be reduced so that the bond between the cement and the fiber is not as strong as the LWC with the addition of fiber in low levels. This affects the density of the LWC which decreases when the addition of EFB fiber is more than 2%.

Almost all concrete produced is included in light-weight concrete with a density lower than 1900 kg/m$^3$ (1.90 g/cm$^3$). This shows that the addition of SBE and EFB concrete can be classified as lightweight concrete (LWC) [14]. The surface appearance using SEM from the control LWC and LWC with the addition of SBE and EFB can be seen in Figures 2a and 2b.

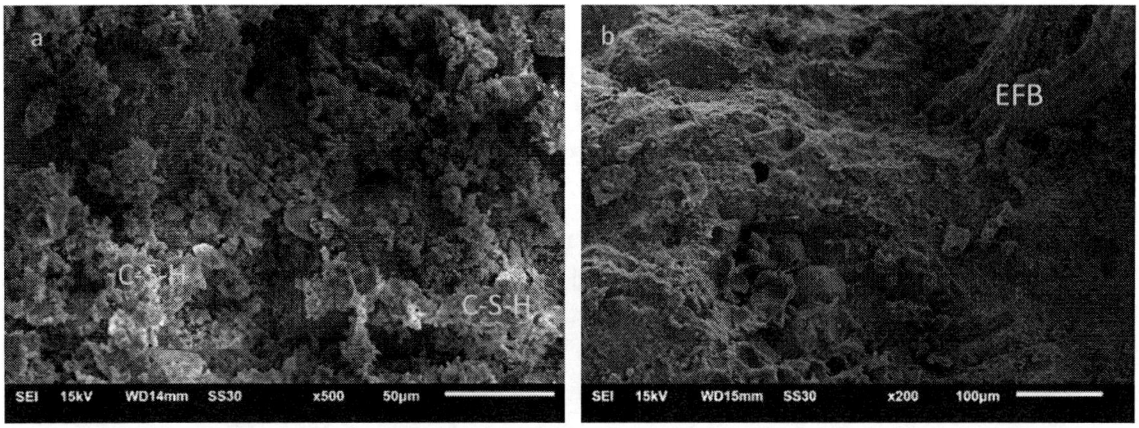


**Figure 2.** SEM micrograph of LWC control (a) with 500x magnification, and LWC with addition of 50% SBE and 6% EFB (b) with 200x magnification

Figure 2a is the result of SEM micrographs in the LWC control sample and Figure 2b is the result of SEM micrographs in the LWC sample with the addition of 50% SBE and 6% EFB. In the SEM micrograph results of the two samples, it can be seen that the calcium silicate hydrate (C-S-H) bond has been formed. The SEM results in Figure 2b show the presence of EFB fibers due to the addition of 6% in the sample.

3.2.2 Water absorption of LWC

The value of water absorption from LWC can be seen in Figure 3. The value of water absorption from LWC with the addition of SBE and or EFB ranges from 0.555-1.316%. This value is greater than the value of LWC control water absorption (0.362%). This happens because EFB as a natural fiber has hygroscopic properties, which means they can absorb water [15].

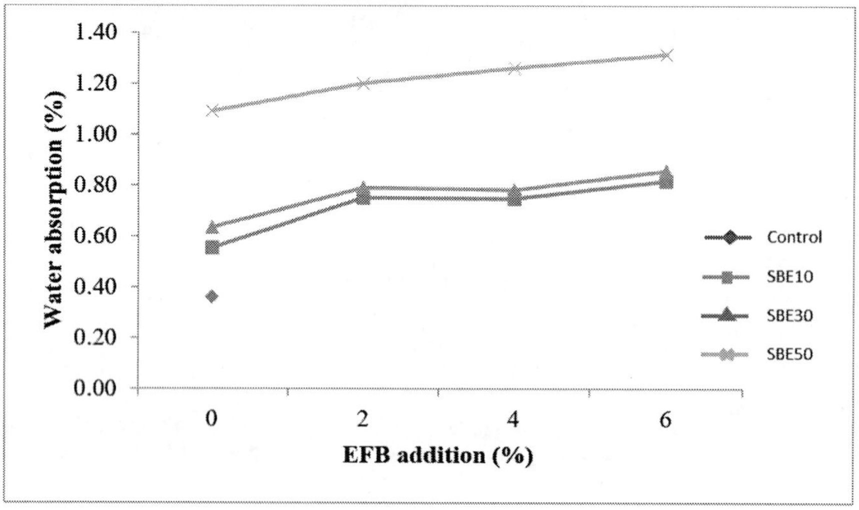


**Figure 3.** Water absorption of LWC

Based on SNI 03-0691-1996, water absorption in concrete does not exceed 10% [16]. So that good concrete based on water absorption value is concrete that has low water absorption capacity. The water absorption value of all concrete formulations meets SNI.

*3.3. Mechanical properties of LWC*

3.3.1 Compressive strength

The compressive strength values of the LWC can be seen in Table 2. The average compressive strength values of the LWC with the addition of SBE and EFB (34.62-149.01 kgf/cm$^2$) are lower when compared to the control concrete (151.20 kgf/cm$^2$). The higher addition of SBE tends to result in a lower of LWC compressive strength value. This happens because there is still oil in SBE [17], and the oil can inhibit the hydration process of cement and reduce mechanical properties, including the compressive strength of concrete [18].

However, when viewed from the level of concrete quality based on Indonesian Standard for walls (SNI 03-0349-1989), most of the LWC is classified at the level I quality or the same as control concrete. The decreasing compressive strength value along with the increasing number of SBE is due to the presence of oil content in SBE, which affects reducing the bond strength between cement and SBE, so the compressive strength decreases. Meanwhile, the effect of the addition of EFB does not show a certain pattern. The optimum value of the addition of EFB tends to increase by 6% based on the LWC volume.

**Table 2.** Distribution of concrete quality based on SNI (03-0349-1989) for walls

| Materials | SBE (%)* | EFB (%)** | Average of compressive strength (kgf/cm$^2$) | Concrete Quality Level |
|---|---|---|---|---|
| Control | 0 | 0 | 151.20 | I |
| S1E0 | 10 | 0 | 130.37 | I |
| S2E0 | 30 | 0 | 99.42 | II |
| S3E0 | 50 | 0 | 34.62 | IV |
| S1E1 | 10 | 2 | 145.01 | I |
| S2E1 | 30 | 2 | 122.70 | I |
| S3E1 | 50 | 2 | 109.17 | I |
| S1E2 | 10 | 4 | 128.00 | I |
| S2E2 | 30 | 4 | 105.16 | I |
| S3E2 | 50 | 4 | 89.78 | II |
| S1E3 | 10 | 6 | 149.01 | I |
| S2E3 | 30 | 6 | 118.62 | I |
| S3E3 | 50 | 6 | 111.76 | I |

*) % weight of sand **) %volume of concrete

3.3.2 Modulus of Rupture (MOR)

The MOR value of the LWC can be seen in Figure 4. The results of the MOR test at the 28 days LWC show that the more SBE additions, the lower the MOR value. The similar reason with the compressive strength, because of the effect of oil, which can inhibit the hydration of concrete and decreasing the mechanical properties of LWC [17,18]. The addition of EFB in the concrete mixture can increase the MOR value, with the optimum value in adding EFB by 4% based on LWC volume.

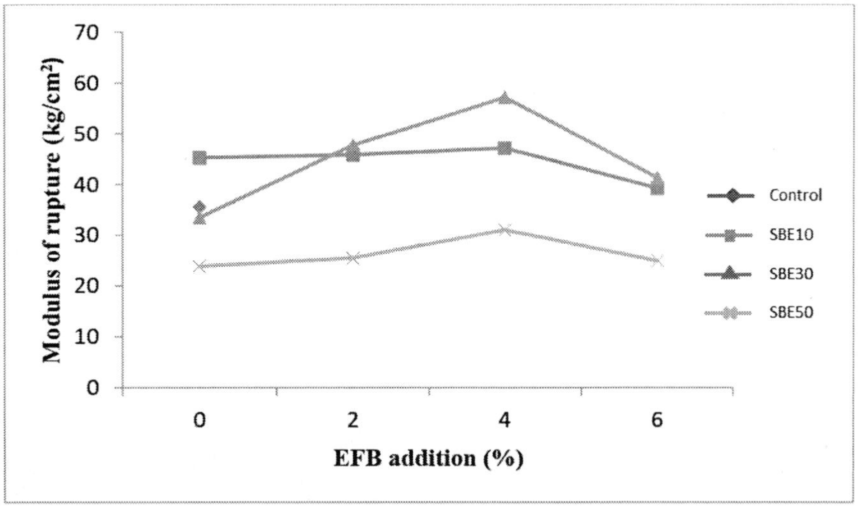


**Figure 4.** Modulus of Rupture of lightweight concrete

Addition of EFB 4% can increase MOR in concrete with SBE 10% by 34.11% (from 35.579 kg/cm$^2$ to 47.716 kg/cm$^2$). Whereas concrete with the addition of SBE 30% increased MOR by 70.77% (from 33.474 kg/cm$^2$ to 57.162 kg/cm$^2$). While concrete with the addition of 50% SBE increased MOR by 7%

(from 23.887 kg/cm$^2$ to 25.554 kg/cm$^2$). Concrete with the addition of 1% recycled fiber can increase concrete MOR by 4% (from 25.194 kg/cm$^2$ to 26.214 kg/cm$^2$).

3.3.3 Modulus of Elasticity (MOE)

The MOE value of the LWC can be seen in Figure 5. MOE of LWC at 28 days showed that the more SBE additions, the lower the MOE value. The addition of EFB in the concrete mix can increase the value of MOE, with the addition of 2% EFB to produce the most optimum MOE value. The addition of EFB fibers at various levels can increase the MOE of the LWC with SBE addition. While the optimum MOE occurs in the addition of 2% EFB fibers.

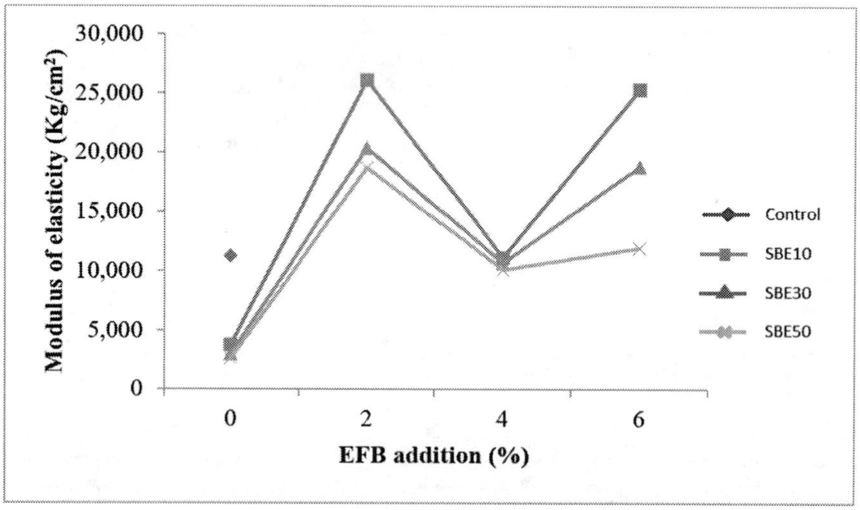


**Figure 5.** Modulus of Elasticity of LWC

3.3.4. Durability of LWC

The results of durability testing in LWC showed that the addition of SBE resulted in decreased LWC's resistance to weather. This is indicated by the percentage decrease in the compressive strength of the concrete increases with the addition of SBE. The addition of EFB to concrete does not significantly reduce the compressive strength of LWC (Figure 6). The decrease in compressive strength in the wet/dry cycle treatment occurs after 2 cycles, where the debonding occurs in the fiber with cement [13]. In addition to the fiber, embrittlement will occur due to mineralization due to this wet/dry cycle treatment [13]. Good concrete is concrete that has high weather resistance so that it can last a long time when applied to walls, roads, parks, and other outdoor applications.

3.3.5. Statistical analysis

Statistical analysis was carried out to determine the effect of different SBE levels and EFB levels on responses in the form of physical properties, mechanical properties, and compressive strength decrease due to the wet/dry cycles process. The results show that there are interactions between SBE factors and EFB factors that affect the response in the form of density, compressive strength, MOR, and MOE. While the response in the form of water absorption is only influenced by the factor of SBE addition, and the durability response is influenced by each factor of SBE addition and EFB addition factor. Based on the analysis of responses surface methodology, the optimum value of the addition of SBE and EFB that can produce LWC with optimum properties is 28.98% SBE and 5.95% of EFB addition.

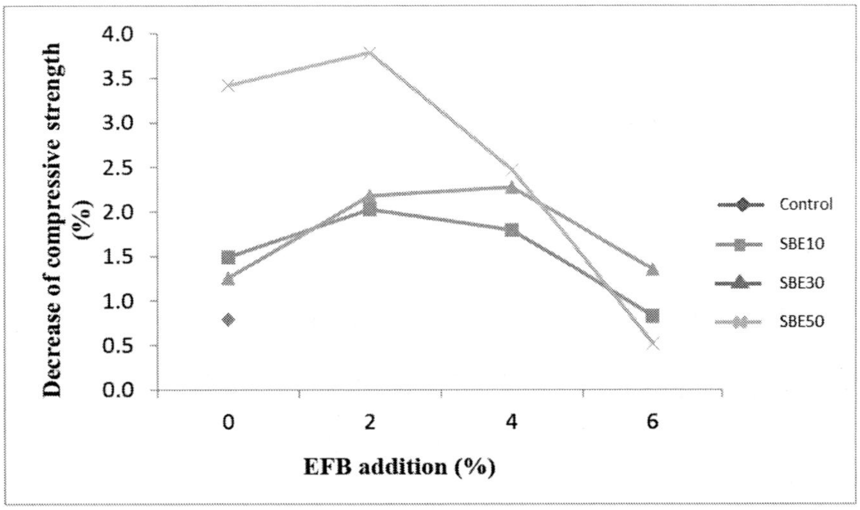


**Figure 6.** Percentage decrease of compressive strength after seven times wet/dry cycles

## 4. Conclusion

The addition of SBE and EFB produces concrete with a density of less than 1.9 g/cm$^3$ and is classified into the LWC. The use of SBE and EFB at various levels has a significant influence on responses. The optimum value of the usage of SBE and EFB that can produce optimum properties of concrete is SBE of 28.98% of the weight of sand, and cement of 5.95% of the volume of concrete.

## References

[1] Hambali E and Rivai M 2017 The Potential of palm oil waste biomass in Indonesia in 2020 and 2030 *IOP Conf. Ser. Earth Environ. Sci.* **65**(1) 012050

[2] Khalil HPSA, Jawaid M, Hassan A, Paridah MT, and Zaidon A 2012 *Oil Palm Biomass Fibers and Recent Advancement in Oil Palm Biomass Fibers Based Hybrid Biocomposites* (*Composites and Their Application*) (London: In Tech Open) chapter 9

[3] Or KH, Putra A, and Selamat MZ 2017 Oil palm empty fruit bunch fibres as sustainable acoustic absorber *Appl. Acoust.* **119** 9-16

[4] Loh SK, James S, Ngatiman M, Cheong KY, Choo YM, and Lim WS 2013 Enhancement of palm oil refinery waste-Spent Bleaching Earth (SBE) into bio organic fertilizer and their effects on crop biomass growth *Ind. Crops Prod.* **49** 775-81

[5] Ling JH, Lim YT, Leong WK, Jusli E, and Sia HT 2019 Properties of cement brick with partial replacement of sand and cement with oil palm empty fruit bunches and silica fume *J. Civ. Eng. Forum* **5**(3) 289

[6] Lim SK, Tiong HY, and Woon KS 2018 Compressive strength and dimensional stability of palm oil empty fruit bunch fibre reinforced foamed concrete *E3S Web Conf.* **65** 1-10

[7] Omoniyi TE 2019 Potential of oil palm (*Elaeis guineensis*) empty fruit bunch fibres cement composites for building applications *Agri. Eng.* **1**(2) 153-63

[8] Musa M, Othuman Mydin MA, and Ghani AN 2019 Thermal properties of foamed concrete with addition of empty fruit bunch (EFB) fiber *Int. J. Innov. Technol. Explor. Eng.* **8**(10) 4662-70.

[9] Tee CK 2010 Performance of spent bleaching earth as cement replacement in concrete [Thesis] University Malaysia Pahang Malaysia

[10] Nalobile P, Wachira JM, Thiong'o JK, and Marangu JM 2019 Pyroprocessing and the optimum mix ratio of rice husks, broken bricks and spent bleaching earth to make pozzolanic cement. *Heliyon* **5**(9) e02443

[11] Wangrakdiskul U, Khonkaew P, and Wongchareonsin T 2015 Use of the spent bleaching earth from palm oil industry in non fired wall tiles *Int. J. Adv Cult Technol.* **3**(2) 15-24

[12] Lalitha Devi G, Srinivasa Pasad K, Rani MS, and Bhanu L 2019 Strength and durability research on concrete with partial replacement of cement by rice husk ash and spent bleaching earth *Int. J. Recent. Technol. Eng.* **8**(2 Special Issue 11) 1035-40
[13] Mohr BJ, Nanko H, and Kurtis KE 2005 Durability of kraft pulp fiber-cement composites to wet/dry cycling *Cem. Concr. Compos.* **27**(4) 435-48
[14] [BSN] Badan Standardisasi Nasional 2002 *Procedures for planning concrete structures for buildings SNI 03-2847-2002* (Jakarta: Standardization National Agency of Indonesia)
[15] Célino A, Fréour S, Jacquemin F, and Casari P 2014 The hygroscopic behavior of plant fibers: A review *Front Chem.* **1** 1-12
[16] [BSN] Badan Standardisasi Nasional 1996 *Paving Block SNI 03-0691-1996* (Jakarta: Standardization National Agency of Indonesia)
[17] Krzyśko-Łupicka T, Cybulska K, Wieczorek A, Możdżer E, and Nowak MJ 2014 The effect of spent bleaching earth ageing process on its physicochemical and microbial composition and its potential use as a source of fatty acids and triterpenes *Environ. Sc.i Pollut. Res.* **21**(18) 10765-74
[18] Almabrok MH, Mclaughlan RG, Vessalas K, and Thomas P 2019 Effect of oil contaminated aggregates on cement hydration *Amer. J. Eng. Res.* **895** 81-9

IOP Conf. Series: Materials Science and Engineering **935** (2020) 012057 doi:10.1088/1757-899X/935/1/012057

# Physical properties of impregnated sengon wood by monoethylen glycol and nano silica from betung bamboo sticks

**I S Rahayu[1*], L H Zaini[1], D Nandika[1], W Darmawan[1], E Prihatini[1], R Agustian[2]**

[1]Department of Forest Product, Faculty of Forestry and Environment, IPB University, Bogor, 16680, Indonesia
[2]Alumni of Department of Forest Product, Faculty of Forestry and Environment, IPB University, Bogor

*Email: istiesr@apps.ipb.ac.id

**Abstract.** Sengon wood (*Falcataria moluccana*) is a fast-growing species that has low dimensional stability and density. The purposes of this study were to analyze the effect of impregnation treatment by using MEG (*monoethylen glycol*) and nano silica from betung bamboo sticks on the dimensional stability and density of sengon wood and to identify the characteristics of impregnated sengon wood. The treatments were consisted of untreated (water treated), 50% MEG (MEG), 0.5% nano silica in 50% MEG (MEGsilica 0.5%), 0.75% nano silica in 50% MEG (MEGsilica 0.75%) and 1% nano silica in 50% MEG (MEGsilica 1%). Weight percent gain, anti-swelling efficiency, water uptake, bulking effect, and density were affected by the treatments. It was due to the coverage of MEG and nano silica on cell walls and vessels of sengon wood (SEM analysis). Also, the presence of nano silica in wood treated (SEM-EDX analysis). The cristalinity of sengon wood decreases because MEG is amourphous. FT-IR showed there was no MEG bonds with sengon wood cell wall components or between nano silica and sengon wood. The optimum treatment for sengon wood was MEGsilika 1%.

## 1. Introduction

Wood is one of the most popular raw materials for various needs in Indonesia. Every year wood production in Indonesia continues to decline with increasing demand. The need for wood in Indonesia was 10,300,872 $m^3$, but natural forests were only able to provide 5,843,179 $m^3$ [1]. Therefore, plantation and community forest are useful in supplying timber for the timber industry in Indonesia. Fast growing species were widely planted in plantation and community forest

Fast growing wood has inferior characteristics i.e. low strength and density and high portion of juvenile wood compared to slow growing wood. Juvenile wood has lower physical, mechanical strength and biodeterioration resistance compared to mature wood [2]. One of the fast-growing woods is sengon . Sengon wood has an average spesific gravity of 0.3-0.5, durabe class IV-V and strong class IV-V [3]. Therefore, wood modification is an alternative to improve wood quality.

Some of the basic properties of wood in the form of durability, dimensional stability, and hardness of wood can be improved by wood modification techniques [4]. Some wood modification treatments include chemical modification, thermal modification, surface modification and impregnation method. Impregnation is performed by introducing chemicals into the wood cell walls in order to increase the characteristics of wood [5].

IOP Conf. Series: Materials Science and Engineering **935** (2020) 012057 doi:10.1088/1757-899X/935/1/012057

The impregnation process is a method to alter wood properties by intervening at cell wall level. The chemicals that are commonly used include styrene [6] and methyl metacrylate (MMA) [7]. Impregnation can keep and precipitate chemicals into cells without damaging the wood so that the wood becomes more dense [8], thus it can overcome the problem of dimensional stability and density of wood. [9] impregnated sengon wood using MEG (Monoethylen glycol) and nano $SiO_2$. The results showed that dimensional stability and density of sengon wood increase because MEG and nano $SiO_2$ can easily enter the pits and wood cell walls.

Silica is a material that usually used for thermal insulators, catalyst supports, adsorbents, drug delivery, glass fillers, glass, and raw materials for making solar cells [10]. Silica sources can be found in inorganic materials such as sand, soil, rocks [11], as well as in organic materials known as bio-silica such as [12] rice husks and betung bamboo [13]. Utilization of silica can be expanded by synthesizing in micro or nano size. Micro-sized silica can improve the dimensional stability of materials, according to [14]. Nano material technology has not been thoroughly explored about aspects of LCA (Life Cycle Assessment), thus in this study, we made nano silica originated from natural materials namely bamboo.

Bamboo is one of Graminae family which is very abundant in tropical countries such as Indonesia with 160 species [15]. There are 9,694,131 bamboo sticks produced throughout Indonesia [16]. One of the most widely cultivated by the community is betung bamboo. Betung bamboo sticks have many benefits, such as wood replacement construction materials, Oriented strand board (OSB), crafts and etc. In addition, the silica content contained in betung bamboo can also be utilized through proper processing. Krisdianto *et al.* [17] stated that there are 0.10-1.78% silica content in betung bamboo.

This study objectives were: 1) to analyze the effect of impregnation treatment using a mixture of MEG and nano silica made from betung bamboo sticks on the dimensional stability and density of sengon wood, and 2) to identify the characteristics of impregnated sengon wood.

## 2. Materials and Methods

### *2.1. Materials*

Three five years old Sengon wood (*Falcataria moluccana*), by the diameter of 25-28 cm and originated from community forests in the Sukabumi region, West Java, Indonesia were used in this study. Other materials were MEG and nano silica from Betung bamboo sticks. Impregnation tubes and ultrasonicator were also used in this study. The logs then were cut into 2 x 2 x 2 $cm^3$ samples [18]. Weight Percent Gain (WPG), Anti Swelling Efficiency (ASE), Water Uptake (WU), Bulking Effect (BE) and density, with 5 replications for each treatment were measured.

### *2.2. Methods*

#### 2.2.1. Nano silica preparation from betung bamboo sticks

Betung bamboo sticks originated from community forest around Bogor. Chipped bamboo sticks, then dried until air-dried moisture content. The chips burned without fuel until they became charcoal [19]. Charcoal were put into a porcelain cup and burned at a temperature of 700°C for 6 hours in a furnace [20] to produce ash. The silica was produced from extracted ash by using the reflux method. Then, silica was mixed with PEG-6000 by using an ultrasonicator for 1 hour. Followed by burning the mixture at a temperature of 600°C for 3 hours in a furnance. The average size of nano silica made from betung bamboo sticks is 502.35 nm.

#### 2.2.2. Impregnation process

The stages of the impregnation process are seen in Figure 1. This stage was adapted from [9].

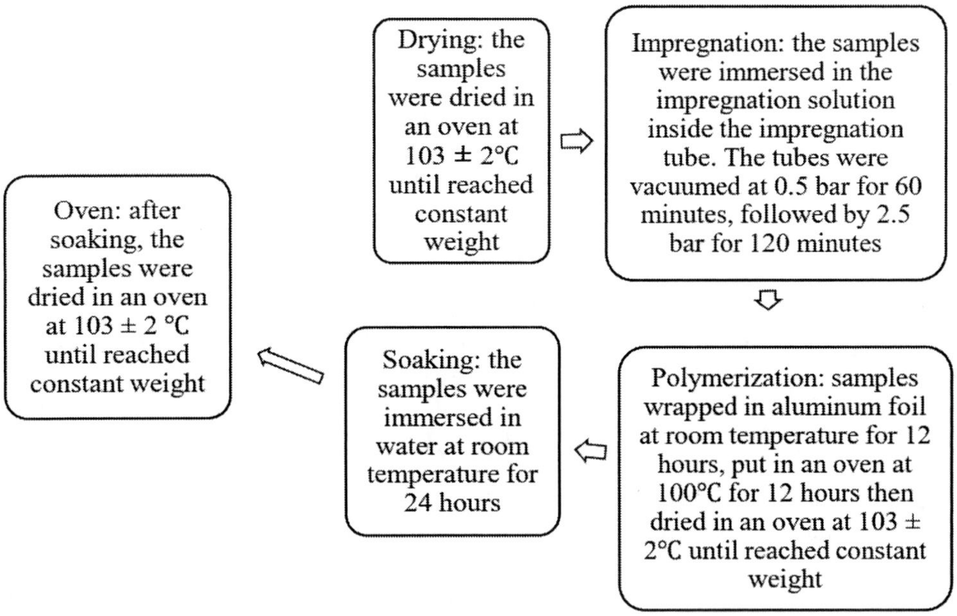


**Figure 1.** The impregnation process using MEG and nano silica solutions

2.2.3. Preparation of impregnation solution
MEG solution and nano silica were mixed using a Cole Parmer ultrasonicator device, with an amplitude of 40% for 1 hour. The composition of the mixture of MEG and nano silica solutions is shown in Table 1.

**Table 1**. Composition of mixed MEG solution and nano silica

| Treatment | MEG (mL) | Water (mL) | Nano Silica (g) |
|---|---|---|---|
| Untreated (water treated) | 0 | 1000 | 0 |
| MEG | 500 | 500 | 0 |
| MEGsilica 0.5% | 500 | 495 | 5 |
| MEGsilica 0.75% | 500 | 492.5 | 7.5 |
| MEGsilica 1% | 500 | 490 | 10 |

2.2.4. Dimensional stability and density
Dimensional stability is measured by using WPG, ASE, WU and BE parameters. The weight percent gain (WPG) measurement formula is as follows:

$$\text{WPG}\ (\%) = \frac{(W_1 - W_0)}{W_0} \text{ X } 100 \quad (1)$$

where, $W_0$= Oven dried weight of the sample before impregnation (g) $W_1$= Oven dried weight of the sample after impregnation (g)

Anti-swelling efficiency (ASE) measurements were carried out using a 24 hours water immersion method with 2 cycles. The ASE formula is as follows:

$$\text{ASE}\ (\%) = \frac{(\text{Su-St})}{\text{Su}} \text{ X } 100 \quad (2)$$

where, Su= Volume shrinkage of the untreated sample (%), St= Volume shrinkage of the treated sample (%)

Measurement of water uptake (WU) after soaking in water for 24 hours is calculated by the formula:

$$WU\ (\%) = \frac{(W_2 - W_1)}{W_1} \times 100 \tag{3}$$

where, $W_1$= Oven-dried weight of the sample after impregnation (g) $W_2$= Weight of the sample after soaking in water (g)

Bulking effect (BE) measurement formula:

$$BE\ (\%) = \frac{(V_1 - V_0)}{V_0} \times 100 \tag{4}$$

where, $V_0$ = Oven-dried volume of the sample before impregnation ($cm^3$) $V_1$ = Oven-dried volume of the sample after impregnation ($cm^3$)

Measurement of density (ρ) is calculated at the time before and after the impregnation treatment, with the formula:

$$\rho = \frac{W_{sample}}{V_{sample}} \tag{5}$$

where, $W_{sample}$= Weight before and after impregnation treatment (g), $V_{sample}$= Volume before and after impregnation treatment ($cm^3$)

*2.2.4.1. Characterization of sengon wood after impregnation*

The characterization of the impregnated Sengon wood was carried out with SEM (Scanning Electron Micrograph), SEM-EDX (Scanning Electron Micrograph-energy X-ray spectroscopy), XRD (X-ray diffraction), and FT-IR (Fourier Transform Infrared Spectrometry) tools.

2.2.5. Data analysis

The experimental design used was a simple randomized complete design (CRD) which was then analyzed by ANOVA. The test was carried out using the IBM SPSS Statistics (Statistical Package for service solutions) version 25.0 calculation program, and continued with the Duncan test at $\alpha = 1\%$ if there were real differences, the equation model used [21] was as follows:

$$Yij = \mu + \tau i + \epsilon ij \tag{6}$$

where:

| | |
|---|---|
| i | = 1, 2,...,t and j = 1, 2, …, r |
| Yij | = response or observation value from the concentration treatment of the i-solution and the j-replication |
| m | = general average |
| τI | = the effect of the i-concentration treatment |
| ∈ij | = the effect of experimental errors from the concentration treatment of the i-impregnation solution and the j-replication |

IOP Conf. Series: Materials Science and Engineering **935** (2020) 012057 doi:10.1088/1757-899X/935/1/012057

## 3. Results and Discussion

### *3.1. Dimensional stability and density*

3.1.1. Weight Percent Gain (WPG)

WPG is a percentage ratio of the weight of impregnated wood to its initial weight. WPG results were shown in Figure 2. Figure 2 shows that the highest level of WPG is the MEG silica 1% treatment (73.55%). The WPG values tend to increase as the increase concentration of nano silica in the impregnation solution. It was due to the pressure in impregnation process caused MEG and nano silica entered the cell cavities and the sengon wood pits. This is in accordance with [5] which states that using a polymer at the time of impregnation causes the penetration of the polymer into the cell wall, because the polymer is able to fill pits in the cell wall. The results of the analysis of variance ($p<0.05$) in Table 2 indicates that treatment had a significant effect on WPG. Duncan's further test (Figure 2) shows that the concentration of each treatment was significantly different from one another.

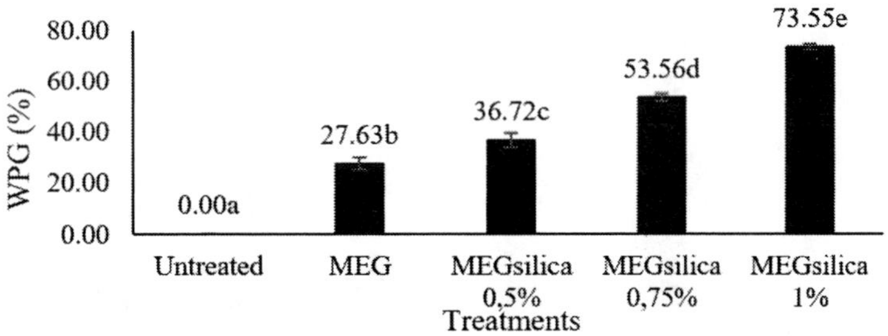


**Figure 2.** Weight percent gain values of various MEG and nano silica impregnation treatments on sengon wood

**Table 2.** F-values of variance analysis

| Source | WPG | ASE | WU | BE | Density |
|---|---|---|---|---|---|
| | F-value | F-value | F-value | F-value | F-value |
| Treatment | 1104.74** | 54.26** | 56.82** | 180.38** | 23.47** |

S*p $\leq$ 0.05 indicates that treatment factor had a significant effect at a 95% confidence level)

3.1.2 Anti-Swelling Efficiency (ASE)

The dimensional stability of wood is the ability of wood not to swell and shrink although there were fluctuating temperature and humidity in the surroundings [22]. Dimensional stability can be measured by anti-swelling efficiency (ASE). ASE test results are shown in Figure 3.

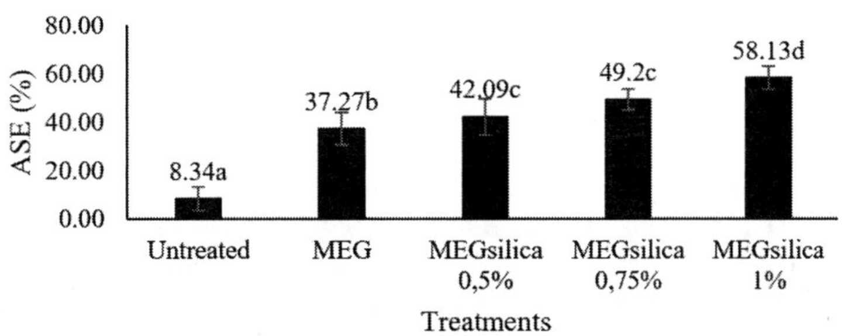


**Figure 3.** Anti-Swelling Efficiency values of various MEG and nano silica impregnation treatments on sengon wood

Figure 3 shows that ASE values increase as the concentration of MEG silica solution increase. The lowest ASE value is untreated (water treated) samples (8.34%) and the highest in MEG silica 1% treatment at 58.13%. The results of the analysis of variance ($p<0.05$) in Table 2 indicates that the ASE value had a significant influence on the concentration level of the treatment. Duncan's further test (Figure 3) states that almost every treatment concentration is significantly different from the others. ASE values are directly proportional to WPG values. The polymer enters the cell wall and covers the empty cavities that are normally occupied by water, so that the polymer coverage in wood could enhance wood dimensional stability [5]. Also, we observed that the higher WPG values were followed by the increase of ASE values. The results of this study are in accordance with [23] who states that the higher the WPG value on rubberwood, the higher the ASE value produced.

3.1.3 Water Uptake (WU)

Water Uptake is a percentage of the ability of wood to absorb water. The lower the WU value, the lower the water uptake in wood. WU results are presented in Figure 4.

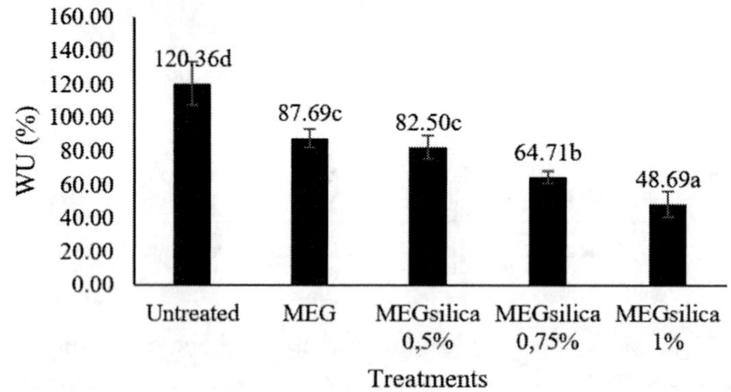


**Figure 4**. Water uptake values of various MEG and nano silica impregnation treatments on sengon wood

Figure 4 shows the highest water uptake value is untreated (water treated) (120.36%), while the lowest is MEGsilica 1% (48.69%). The higher nano silica concentration, the lower WU values. The analysis of variance ($p < 0.05$) in Table 2 shows that the WU value was significantly affected by the concentration level of the treatment. Duncan's further test results (Figure 4) state that almost all treatments are significantly different from other treatments. The WU results in this study are in accordance with the statement of [9] that there is a negative correlation between the value of WPG and WU, an increase in WPG is accompanied by a decrease in WU because the number of MEG and MEGsilica that cover pits in the cell wall, therefor reduces the availability of space for water molecules.

3.1.4 Bulking Effect (BE)

Bulking effect is a percentage ratio of oven-dried wood volume before and after impregnation treatment. The results of BE testing as shown in Figure 5. Figure 5 indicates that the lowest BE value was found in untreated (water treated) sengon wood (1.38%), while the highest was MEG silica 1% (12.90%). The higher nano silica in wood samples resulted the higher BE value. It was due to the coverage of MEG and nano silica on cell cavities and pits. The results of the analysis of variance ($p<0.05$) in Table 2 shows that the sengon wood with the concentration level of treatment significantly affected the value of BE produced. Duncan's further test (Figure 5) shows that almost every treatment concentration gives a significant different to other treatments. The results of BE in this study are in line with [5] who states that dimensional stability is increased due to polymer materials penetrate into cell walls through the diffusion process and also supports by the statement of [24] who states that wood modification using silica material is able to penetrate the cell walls so as to achieve a certain bulking effect.

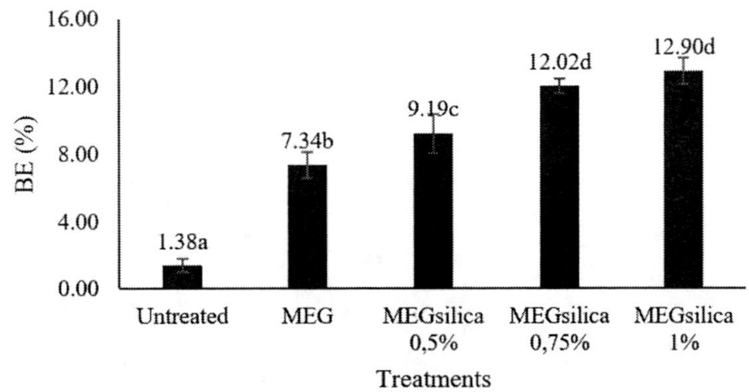


**Figure 5**. Bulking effect values of various MEG and nano silica impregnation treatments on sengon wood

3.1.5 Density

Density is a ratio of weight to volume of wood [25]. The density value obtained is presented in Figure 6.

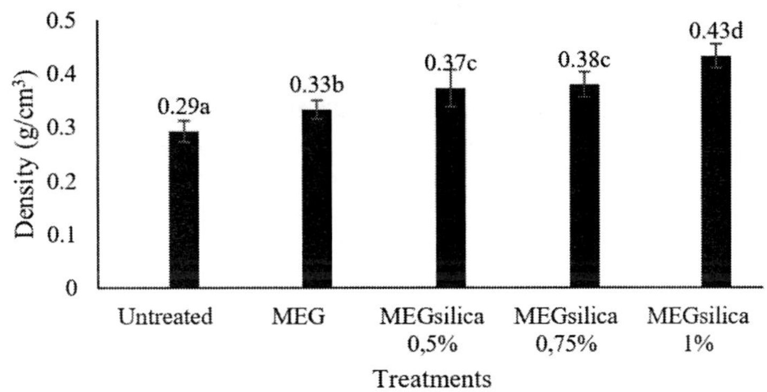


**Figure 6.** Density values for various MEG and nano silica impregnation treatments on sengon wood.

Figure 6 shows that the density of sengon wood tends to increase with increasing MEGsilica concentration. The lowest density value was untreated (water treated) sengon wood (0.29 g/cm$^3$), while the highest was MEGsilica 1% (0.43 g/cm$^3$). Based on the results of the analysis of variance ($p < 0.5$) in Table 2 shows the treatment of sengon wood significantly affected the density value. Duncan's further test (Figure 6) states that almost every treatment is significantly different from one another. The resulting density value increases along with the increase in the concentration of nano silica. It is assumed that MEG and nano silica are able to fill cell cavities and pits in the cell walls of sengon wood. This is consistent with what was stated by [26] that the higher the density of wood, the more wood content in the cell wall which means the thicker the cell wall. The density value is directly proportional to the value of WPG and ASE produced, the higher the value of WPG and ASE, the higher the density value in sengon wood.

### *3.2. Characterization of sengon wood from impregnation results*

3.2.1. Scanning Electron Microscopy analysis (SEM)

SEM analysis (Figure 7a and 7b) shows the morphology of untreated sengon wood with magnification of 100x and 550x. We could observed that many empty pits on cell walls.

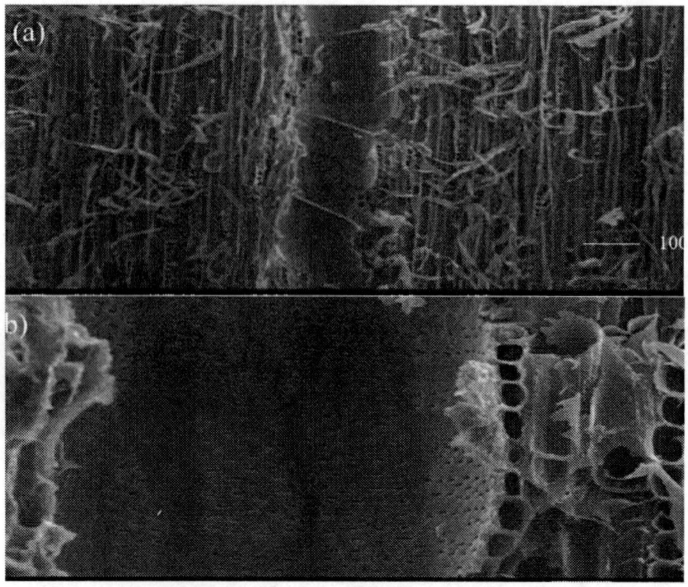


**Figure 7** Scanning Electron Micrograph of untreated sengon wood vessel wall withmagnification of (a) 100x and (b) 500x

Most of the pits in the vessel wall was seen to be filled with MEG, however there were some empty pits were observed due to the uneven distribution of MEG in the sample (Figure 8a and 8b). The 550X magnification in Figure 8 shows that there are several pits that were still not covered by the MEG (Figure 8b - arrowhead). Differences in the characteristics of untreated and MEG treated sengon wood can be clearly seen. The average pit size after being treated by MEG (309.80 μm) were smaller than untreated sengon wood (373.37 μm) [9]. It was due to there were some MEG covered the pits

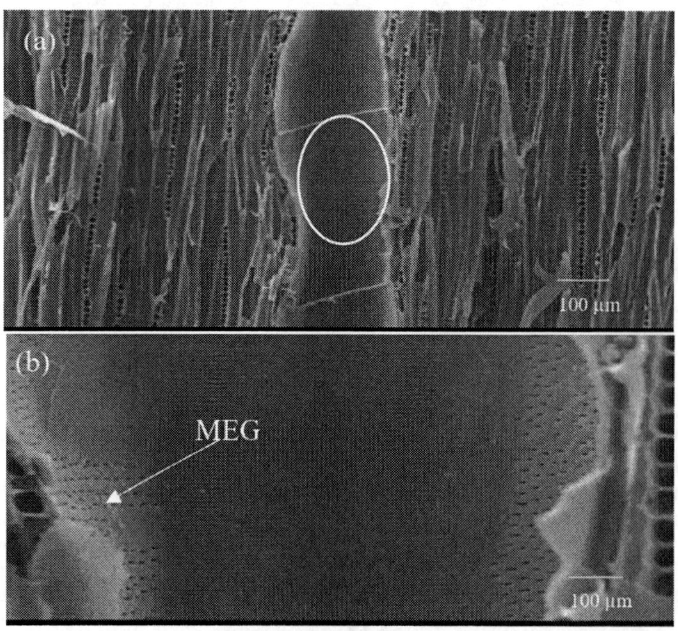


**Figure 8.** Scanning Electron Micrograph of MEG treated sengon wood vessel wallwith magnification of (a) 100x and (b) 500x

Figure 9a and 9b show MEG silica 0.5% treated samples with 100x and 500x magnification, respectively. Nano silica and MEG can cover and fill pits on cell walls. However, we could observe that there were some pits still not covered by nano silica due to small concentration of nano silica used during the impregnation process (Figure 9b-arrow).

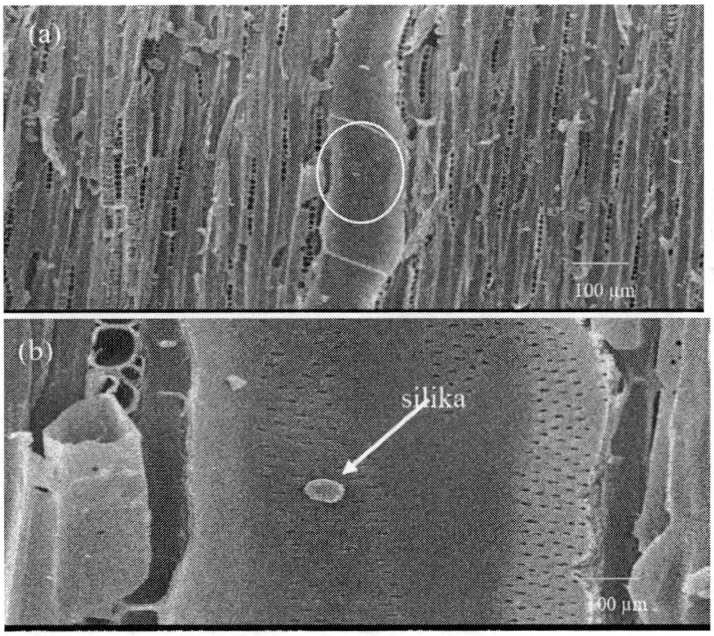


**Figure 9**. Scanning Electron Micrograph of MEGsilica 0.5% treated sengon wood vessels wall with magnification of (a) 100x and (b) 500x

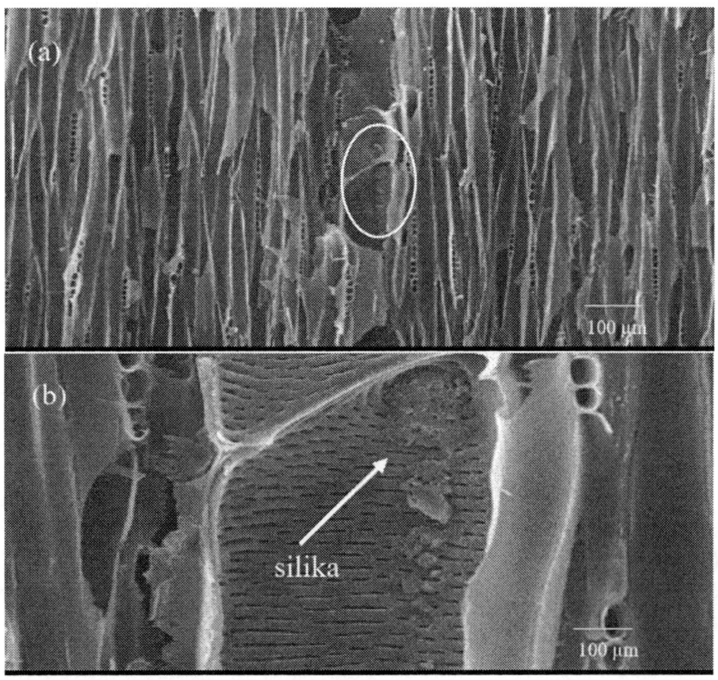


**Figure 10.** Scanning Electron Micrograph of MEGsilica 0.75% treated sengon wood vessel wall with magnification of (a) 100x and (b) 500x

Figures 10 and 11 show MEG and nano silica 0.75% and 1%, respectively. MEG and nano silica can enter and fill the lumen cells and pits of the sengon wood vessel. There are nano silica agglomeration caused by more concentration of nano silica in these treatments. Agglomeration in MEGsilica 0.75% (Figure 10b - arrows) were less than MEGsilica 1% (Figure 11b - arrows).

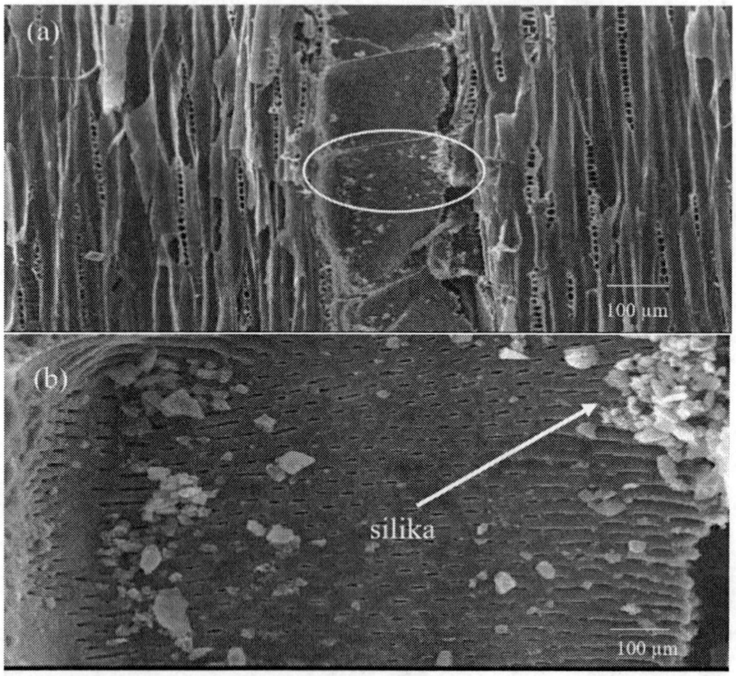


**Figure 11.** Scanning Electron Micrograph of MEGsilica 1% treated sengon wood vessel wall with magnification of (a) 100x and (b) 500x

The SEM analysis results show the reasons why improvement values of WPG, BE, ASE, density and WU occurred in treated sengon wood. According to Xue and Zhao [27] nano particles can enter and penetrate into wood mainly through nano-sized cavities.

### 3.2.2 SEM-EDX Analysis

The SEM-EDX analysis results in Table 3 shows that there was no silica content in untreated (water treated) and MEG treated. In contrast to wood samples impregnated with MEGsilica 0.5%, MEGsilica 0.75% and MEG silica 1%. The silica content increases with increasing nano silica concentrations. It can be concluded that nano silica deposits in wood cells.

**Table 3**. Chemical composition of sengon wood samples using SEM-EDX analysis on various MEG and silica impregnation treatments

| Treatment (wt.%) | Silica (wt.%) |
|---|---|
| Untreated (water treated) | 0 |
| MEG | 0 |
| MEGsilica 0.5% | 1.93 |
| MEGsilica 0.75% | 3.85 |
| MEGsilica 1% | 5.52 |

3.2.3 X-Ray Diffraction (XRD) analysis

XRD testing is carried out at an angle of 2Θ between 5-400. Based on the results of the XRD analysis (Figure 12), there is a peak with an angle of 2Θ = 21.8°-22.87°. According to the JCPDS data, the peak is plane I002 indicating the crystalline phase of cellulose. Then there is a peak with an angle of 2Θ = 16.97° which is the peak of cellulose Iβ in the diffraction plane I101. In accordance with [28] states that cellulose Iβ appears at an angle of 2Θ = 14-160.

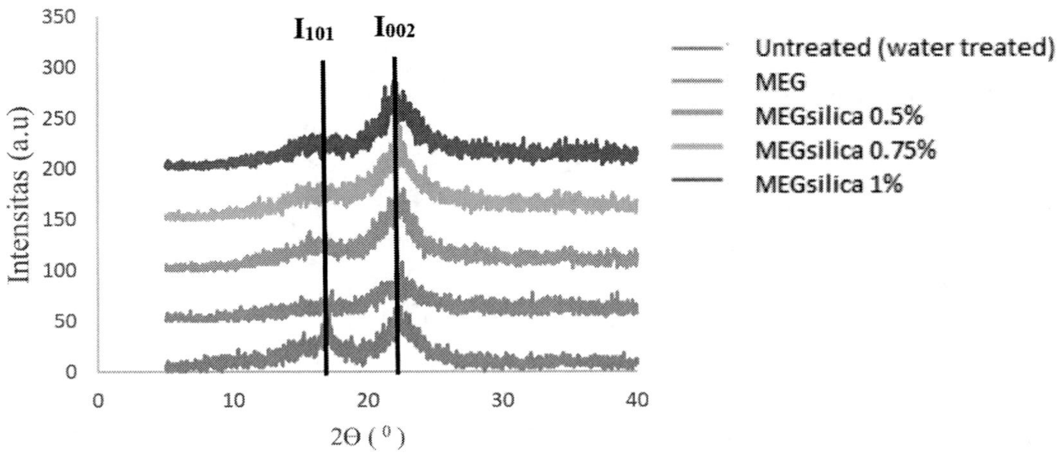


**Figure 12**. Sengon wood XRD spectra of various treatments

Additional information about XRD testing can be seen in Table 4. There was a decrease in the degree of crystallinity in the sengon wood sample impregnated by MEG and nano silica. This is caused by MEG which is amorphous and nano silica which is suspected semi-crystalline because of the sonication treatment that were used. This was in line with Dirna [29] who performed impregnation by using MEG and nano silica from betung bamboo leaves. This result is also supported by Dong *et al.* [30] which states that the crystallinity of poplar wood decreases after the addition of amorphous furfuryl alcohol. XRD analysis shows the effect of nano silica to WPG, ASE, WU, BE and density.

Table 4. Degree of crystallinity of MEG and nano silica impregnation treatments.

| Treatment | Degree of crystallinity (%) |
|---|---|
| Untteated (water treated) | 23.67 |
| MEG | 24.00 |
| MEGsilica 0,5% | 23.31 |
| MEGsilica 0.75%F | 17.98 |
| MEGsilica 1% | 21.39 |

3.2.4. Fourier Transform Infrared Spectrometry (FT-IR) analysis

Fourier Transform Infrared Spectrometery (FT-IR) is an infrared spectroscopy technique that can analyze structural polysaccharides and observe structural changes in biopolymers [31]. Based on Figure 13, it can be seen that the absorption peaks of the infrared spectrum have a significant difference between untreated and treated sengon wood. Absorption bands ranging from 3354.31 to 3375.43 $cm^{-1}$ are hydroxyl groups (O-H) from water in sengon wood. In accordance with the statement of Dachriyanus [32] that in the wave frequency range between 3750 and 3000 $cm^{-1}$ it was detected that there was an O-H group. However, in different sengon wood treated with MEG silica there was no peak at wave 478.30 $cm^{-1}$ which is the bending of the Si-O-Si group, this is because silica does not bond to the cell wall components of the sengon wood . At wave 1057 $cm^{-1}$, C-O groups appearing from MEG which appear new functional groups such as furan rings at waves 1027.67 $cm^{-1}$ and 778.62 $cm^{-1}$ [32,33]. The C = O function group of the sengon wood hemicellulose appeared at wave 1737.32 cm$^{-1}$. Pangau *et al.* [33] state that the C = O group appears in waves ranging from 1723.48 to 1742.53 cm$^{-1}$.

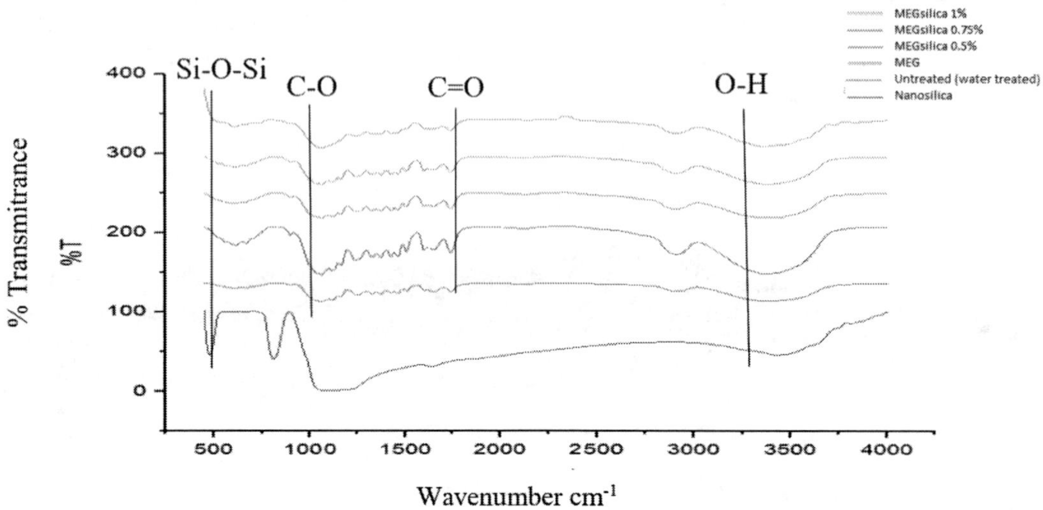


**Figure 13**. FT-IR spectrum of various sengon wood treatments

MEG and nano silica do not bond with the cell wall components of sengon wood which could cause MEG silica to be leachedby water. MEG silica is only able to fill the cell wall cavities which gives a bulking effect on the MEG silica treated sengon wood.

## 4. Conclusion

The MEG and nano silica of betung bamboo sticks impregnation treatment has a significant effect on the dimensional stability and density of the sengon wood. Increasing values of WPG, ASE, BE, and density as well as decreasing the value of WU shows the polymer can cover the lumen cell and pits of sengon wood. The morphological changes of sengon wood after MEG and nano silica impregnation treatment (based on SEM images) showed that the lumen cell and pits were occupied by MEG and nano silica. MEG and nano silica are bulking agents. Based on research, the optimum treatment for five years old sengon wood was MEG silica 1%.

## References

[1] [MoEF] Ministry of Environment and Forestry 2015 *Statistics of the Ministry of Environment and Forestry 2015* (Jakarta)

[2] Hadi YS, Nurhayati T, Jasni, Yamamoto H, and Kamiya N 2012 Resistance of smoked wood to subterranean and dry-wood termite attacks *Inter. Biodeterioration & Biodegradation* **70** 79-81

[3] Pandit IKN 2002 *Guidelines for Identification of Types of Wood in the Field* (Bogor: PROSEA) Foundation

[4] Homan WJ and Jorissen AJM 2004 Wood modification developments *Heron* **49** (4) 361-86

[5] Hill CAS 2006 *Wood modification: Chemical, Thermal and Other Processes. The Atrium, Southern Gate, Chichester* (West Sussex: John Wiley and sons Ltd.)

[6] Darma IGKT, Hadi YS, and Atmojo AT 2002 Resistance of polystyrene plastic wood composites against brown rot fungus (Tyromyces palustris) attacks *J Man Hut Trop* **8(1)** 31-38

[7] Wardani L, Risnasari I, Jasni, and Hadi YS 2012 Resistance of jabon timber modified with styrene and methyl metacrylathe against drywood termites and subterranean termites *Proc. of 9th Pacific Rim Termite Research Group Conference* pp. 27-8

[8] Kollmann FFP, Kuezi WW, and Stamm AJ 1975 *Principles of Wood Science and Technology*, Vol. II (Verlag Berlin: Springer)
[9] Rahayu I, Darmawan W, Zaini LH, and Prihatini E 2019 Characteristic offast growing wood impregnated with nano particles *J. of Forest. Research* **31**(2) 677-85
[10] Gurav JL, Rao AV, Rao AP, Nadargi DY, and Bhagat SD 2009 Physical properties of sodium silicate based silica aerrogel prepared by single step sol-gel process dried at ambient pressure *Alloys and Compounds*. **476** 397-402
[11] Renjanadevi B 2008 Modification of plypropylene and polystyrene using nano silica [Thesis] Cochin University of Science and Technology India
[12] Thuadaij N and Nuntiya A 2008 Preparation of nanosilica powder from rice husk ash by precipitation method *Chiang Mai J. Sci*. **35(1)** 206-11
[13] Ding TP, Zhou JX, Wan DF, Chen ZY, Wang CY, and Zhang F 2009 Silicon isotope fractionation in bamboo and its significance to the biogeochemical cycle of silicon *Geochimica et Cosmochimica Acta* **72**1381-95
[14] Li X 2011 Microencapsulation of Nanoemulsion: Novel trojan particles for bioctive lipid molecule delivery *Inter. J. of Nanomedicine* 1313-25
[15] Widjaya EA 2012 The Utilization of bamboo: At present and for the future *Proc. of International Seminar Strategies and Challenges on Bamboo and Potential Non Timber Forest Products (NTFP) Management and Utilization* pp. 79-85
[16] [BPS] Statistics Indonesia 2016 *Forestry Production Statistics* (Jakarta: Indonesian Central Statistics Agency)
[17] Krisdianto, Sumarni G, and Ismanto, A 2000 Sari hasil penelitian bambu (Bogor: Pusat Penelitian dan Pengembangan Hasil Hutan)
[18] [BS] British Standard 1957 *Methods of Testing Small Clear Specimen of Timber* BS 373:1957 in Annual Book of BS Standard (London: British Standard Institution)
[19] Sa'diyah H, Nurhimawan S, Fatoni SA, Irmansyah, and Irzaman 2016 Ekstraksi silikon dioksida dari daun bambu. *Prosiding Seminar Nasional Fisika* **5** pp. SNF2016-BMP
[20] Thuadaij N and Nuntiya A 2008 Preparation of nano silica powder from rice husk ash by precipitation method *Chiang Mai J. Sci*. **35**(1) 206-11
[21] Sastrosupadi A 2000 *Design of Practical Experiments in Agriculture* (Yogyakarta: Kanisius)
[22] Basri E, Hadjib N, and Saefudin 2014 Effects of heat-pressure treatment on some properties of young-teak wood *Proc.* Jakarta: Forestry Research & Development Agency
[23] Pandey KK, Jayashree V, and Nagaveni HC 2009 Study of dimensional stability, decay resistance, and light stability of phenylisothiosianate modified rubberwood *Bioresources* **4**(1) 257-67
[24] Pries M, Wagner R, Kaesler KH, Militz H, and Mai C 2013 Effect of short-chain silicones bearing different functional groups on theresistance pine (*Pinus sylvestris* L.) and beech (*Fagus sylvatica* L.) against decay fungi *Holzforschung* **67** 447-54
[25] Tsoumis G 1991 *Wood as raw material: Source, Structure, Chemical, Composition, Growth, Degradation and Identification* (Oxford: Pergamon Press)
[26] Bowyer JL, Shmulsky R, and Haygreen JG 2007 *Forest Product and Wood Science an Introduction Fifth Edition* (Iowa: Blackwell Publishing Professional)
[27] Xue F and Zhao G 2008 Optimum preparation technology for Chinese fir wood/Ca montmorillonite (Ca-MMT) composite board *For. Stud. China* **10(3)**199-204
[28] Cheng G, Varanasi P, Li C, Liu H, Melnichenko YB, Simmons BA, Kent MS, and Singh S 2011 Transition of cellulose crystalline structure and surface morphology of biomass as a function of ionic liquid pretreatment and its relation to enzymatic hydrolysis *Biomacromolecules* **12** 933-41
[29] Dirna FC 2019 The characteristics of fast-growing wood results of impregnation of nano silica betung bamboo leaves [Thesis] Bogor Agricultural University
[30] Dong Y, Yan Y, Zhang S, and Li J 2014 Wood/Polymer nanocomposites prepared by impregnation with furfuryl alcohol and nano SiO2 *BioResources* **9(4)** 6028-40

[31] Ahmad A, Anjum FM, Zahoor T, Nawaz H, and Ahmed Z 2010 Extraction and characterization of β-D-glucan from oat for industrial utilization *Inter. J. Bio.Macromolecules* **43** 304-09
[32] Dachriyanus 2004 Structure Analysis of Organic Compounds Spectroscopically (Padang: Institute of Information and Communication Technology Development (LPTIK) Andalas University)
[33] Pangau JR, Sangian HF, and Lumi BM 2017 Characterization of cellulose material by microwave pretreatment irradiation against cempaka wasian wood powder (*Elmerillia ovalis*) in North Sulawesi *J. MIPA Unstrat Online* **6(1)** 53-8

**Acknowledgment**
This work was supported by the Ministry of Research, Technology and Higher Education (RISTEKDIKTI), Republic of Indonesia, under the Leading Research University Competitive Grant with contract 3/E1/KP.PTNBH/2019

IOP Conf. Series: Materials Science and Engineering **935** (2020) 012058 doi:10.1088/1757-899X/935/1/012058

# Combination of citric acid and maltodextrin as bonding agent in sorghum bagasse particleboard

**F A Syamani[1*], D Ramatia[2], Sudarmanto[1], F Akbar[1], S S Kusumah[1], and D Hermawan[2]**

[1]Research Center for Biomaterials, Indonesian Institute of Sciences (LIPI) Cibinong, Jawa Barat 16911, Indonesia
[2]Faculty of Forestry, IPB University, Bogor, Jawa Barat 16680, Indonesia

firda.syamani@biomaterial.lipi.go..id

**Abstract.** The interest of using natural binder in particleboard production is still the primary concern of many researchers. The use of citric acid as an adhesive has been investigated and show satisfying results, especially in particleboard physical properties. However, the particleboard had low bending strength due to high content of citric acid on the particleboard affected the brittleness of the particles. Therefore, the application of citric acid in particleboard production usually combined with other natural adhesives. In this study, a combination of citric acid (CA) and maltodextrin (MD) was used in the production of sorghum bagasse particleboard with a ratio of 1:1. The particleboard was produced with a target density of 0.8 g/cm$^3$, using 15%, 20%, 25% and 30% CA-MD content, hot-pressed at 200°C and 5 MPa for 8 or 10 minutes. The mixture of citric acid and maltodextrin (1:1) was successfully acting as bonding agent in sorghum bagasse particleboard production, mainly when used at 30% adhesive content. The particleboard modulus of rupture and modulus of elasticity, fulfilled the standard of type 8 particleboard, according to JIS 5908. The particleboard physical properties were excellent, proved by the value of particleboard thickness swelling, which was under 12%. This is due to the superb bonding from the combination of citric acid and maltodextrin, as shown by the internal bond values, which exceeded 0.3 N/mm$^2$, the minimum requirement for type 18 particleboards based on JIS A 5908.

## 1. Introduction

Exploring the potential of non-wood lignocellulose as particleboard raw material, still become many researchers concern due to limitation of wood resources. Some considerations to provide sustainable particleboard production were the availability and continuity of the raw material resources. Sweet sorghum bagasse (*Sorghum bicolor* L. Moench) is one of agricultural by products that has been utilized for particleboard raw material, using urea formaldehyde [1] or citric acid [2] as adhesive.

According to Vukusic *et al.* [3], citric acid (CA) is an organic poly-carboxylic acid, which contains three carboxyl groups and has been considered as a cross-linking agent for wood. The use of citric acid as an adhesive in the study by Umemura *et al.* [4,5] showed better results on the physical and mechanical properties of the sweet sorghum bagasse (SSB) particleboard. However, Kusumah *et al.* [6] reported the screw holding power of SSB-CA particleboard was lower than the requirements of JIS A 5908-2003 for type 18 particleboard. According to Jonsson and Martin [7], amorphous polysaccharides from lignocellulose are easily degraded by acid compounds. Amorphous polysaccharide degradation is likely

IOP Conf. Series: Materials Science and Engineering **935** (2020) 012058 doi:10.1088/1757-899X/935/1/012058

to be a major factor affecting fragility in wood [8]. Based on this description, reducing the content of citric acid, is an alternative for resolving the amorphous polysaccharides degradation and fragility in particleboard produced with citric acid as bonding agent.

Maltodextrin is a starch-based polysaccharide produced by the hydrolysis of starch into glucose polymers with an average chain length of 5-10 glucose units. Maltodextrin is soluble in water, which makes it more attractive for industrial applications compared to insoluble starch. Maltodextrin has the potential as a natural binder for the replacement of synthetic adhesive based on petroleum and has a better adhesion to the starch properties [9,10]. The results of Santoso *et al.* [11] research, in the use of sucrose, maltodextrin or citric acid as adhesive in nipa-fronds particleboard production, showed that the particleboards fulfill the JIS A 5908-2003 standard, on mechanical and physical properties. Although it did not work well when pressed at 180°C with maltodextrin. Therefore, in this study, maltodextrin was added to citric acid in the manufacturing of particleboards from sorghum bagasse. The role of maltodextrin is to reduce the amount of citric acid as a major factor affecting particleboard fragility and improve its mechanical and physical properties. This study aims to determine the effect of the pressing time and levels of citric acid and maltodextrin adhesives on the mechanical and physical properties of sorghum bagasse particleboard.

## 2. Materials and Method

This study uses sweet sorghum bagasse which were obtained from the experimental garden at LIPI-Cibinong Science Center. The sorghum bagasse was a by-product from the process of sorghum juice extraction from its stalks. The bonding agents used were citric acid powder and maltodextrin.

### *2.1. Raw material preparations*

Sorghum bagasse was processed by ring flakers to obtain particles, then the particles was screened to obtain particles that passed through a 4-mesh and retained on a 14-mesh screen. All of the particles were dried in oven of 60°C until particles moisture content of around 5%.

Citric acid (anhydrous) of technical grade was used without further purification. Citric acid was dissolved in water at a concentration of 59%, and maltodextrin was added with ratio of 1:1. This solution was used as the binder. No other chemical compounds were used.

### *2.2. Methods*

#### *2.2.1.* Particleboard production

Sorghum bagasse particles were weighed. Then sorghum bagasse particles were put into a drum mixer and sprayed with a mixture of citric acid and maltodextrin (1:1), with adhesive level of 15%, 20%, 25% or 30% using a spray gun. After mixed evenly, the raw material was dried in an oven at 80°C for 12 hours to reduce the moisture content of the material to less than 5%.

The dried sorghum bagasse particles mixed with adhesives were put into the mould with the size (30x30) cm, then hot-pressed with the pressure of 5 MPa, the temperature of 200°C, for 8 or 10 minutes. The particleboard thickness was 1 cm, with the target density of the particleboard was 0.8 g/cm$^3$. After conditioning for one week at a room temperature of about 26°C and relative humidity of approximately 72%, the boards were tested according to the Japanese Industrial Standards for particleboards (JIS A 5908-2003), including physical and mechanical properties.

#### *2.2.2.* Particleboard mechanical properties analysis

The bending properties of the boards, i.e., the modulus of rupture (MOR) and the modulus of elasticity (MOE), were evaluated by conducting a three-point bending testing on 200×30×10 mm of board specimen under dry conditions. The loading speed and effective span were 10 mm/min and 150 mm, respectively. The internal bonding (IB) strength investigated using a 50×50×10 mm board specimen. The screw holding power (SHP) was analysed by pulling out the screws that vertically screwed into the two positions in test pieces vertically. In this test, the pulling-out load speed shall be approximately 2 mm/min.

*2.2.3.* Evaluation of particleboard physical properties after cyclic aging treatment

The thickness swelling (TS) and water absorption (WA) values of each board after water immersion at 20°C for 24 hours, were measured for board specimens of the same size as those used for the IB testing. Particleboard physical properties were thickness swelling and water absorption. Tests carried out using a board specimen with a size of 50×50×10 mm. Analysis of board water absorption was done by measuring the difference in board specimen mass before immersion (B1) and board specimen mass after being immersed for 24 hours (B2). Analysis of board thickness swelling done by measuring the difference in board specimen initial thickness before immersion (T1) and board specimen thickness after being immersed for 24 hours (T2).

To observe particleboards thickness swelling in severe condition, specimens were subjected to a cyclic aging treatment (drying at 105°C for 10 hours, warm-water immersion at 70°C for 6 h, drying at 105°C for 10 hours, immersion in boiling water for 4 hours, and drying at 105°C for 10 hours). The thickness and weight changes of the specimens that occurred throughout the treatment were determined. Each experiment was performed in five replications, and the average values and standard deviations were calculated.

## 3. Results and Discussion

The mechanical properties of particleboards are modulus of rupture, modulus of elasticity, internal bond and screw holding power. Those mechanical properties must fulfil the minimum requirement based on standard JIS A 5908-2003. During particleboards utilization as a shelf in clothes wardrobe or book rack, the force from the book or other materials put on the shelf will give the load to the particleboard. Particleboard type 8, is the lowest standard of particleboard, as a benchmark of the minimum requirement for particleboards utilization.

All modulus of rupture (MOR) of sorghum bagasse particleboards bonded with citric acid and maltodextrin were above 8 N/mm$^2$. Sorghum bagasse bonded with 30% citric acid and maltodextrin (CA-MD) showed the highest MOR, which was 12.66 N/mm$^2$ when hot-pressed for 8 minutes and 12.59 N/mm$^2$ when hot pressed for 10 minutes (Figure 1). The particleboard MOR were higher with the increment of bonding agent content.

Sorghum bagasse particleboards bonded with citric acid and maltodextrin in this study show higher MOR values compare to alang-alang particleboard bonded with 10%, 15% and 20% citric acid [12], which were 10.59 MPa, 7.82 MPa and 6.54 MPa respectively. Still, they show lower MOR values compare to sugarcane bagasse particleboard bonded with 15%, 20%, 25% citric acid [13], which were 17.38 MPa, 18.22 MPa, 21.88 MPa, respectively. Meanwhile, MOR of particleboard from a combination of sorghum bagasse and wood-shaving bonded with urea-formaldehyde were in the range of 7.32-10.47 MPa [1]. Particleboards made from Nipa fronds using citric acid as adhesive showed MOR of 10.41 MPa, and in a variety of 2~ 5 MPa when using maltodextrin [14].

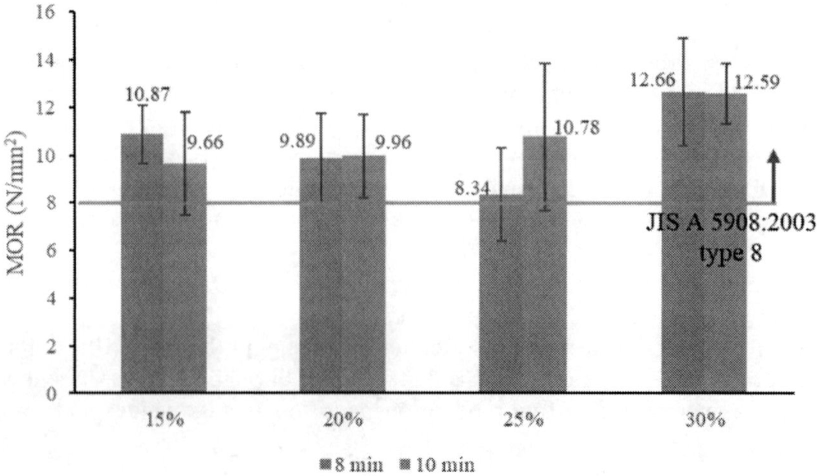


**Figure 1**. Modulus of rupture of particleboards with 15% (A), 20% (B), 25% (C), 30% (D) citric acid+maltodextrin, hot pressed in 8 min or 10 min

The particleboard modulus of elasticity showed the same trend. The highest particleboards MOE was presented by sorghum bagasse particleboards bonded with 30% citric acid and maltodextrin (CA-MD), and it was the only sorghum bagasse particleboards type which fulfilled the requirement of MOE type 8 particleboards (Figure 2). The boards were tough enough to withstand the load. According to Sulastiningsih *et al.* [15], particle length uniformities affected particleboard MOE. In this study, sorghum bagasse particles size was quite uniform that pass 4 mesh and retain 14 mesh screens. So that particleboard MOE values were affected by citric acid and maltodextrin content. Santoso *et a.l* [14] compared the characteristics of nipa fronds particleboard bonded with three different natural binders, which were citric acid, sucrose and maltodextrin. MOE of nipa fronds particleboard bonded with citric acid (3.65 GPa) was higher compared to nipa fronds particleboard bonded with sucrose or maltodextrin. The MOE of sorghum bagasse bonded with 30% CA-MD was 2.66 GPa when hot pressed for 8 min and 2.56 GPa when hot-pressed for 10 min.

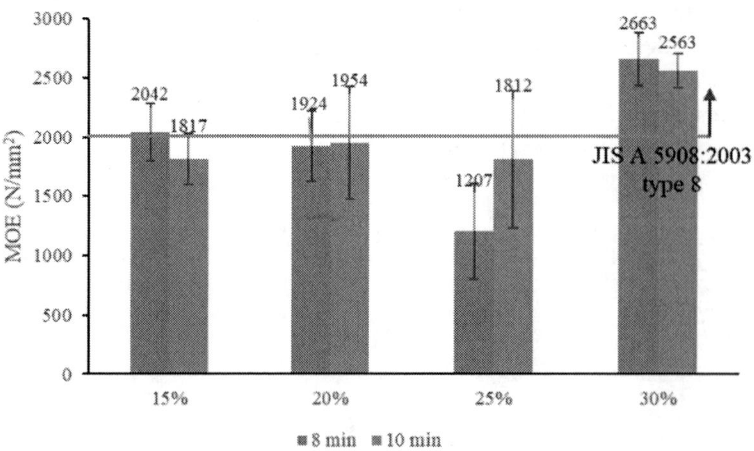


**Figure 2**. Modulus of elasticity of particleboards with 15% (A), 20% (B), 25% (C), 30% (D) citric acid+maltodextrin, hot pressed in 8 min or 10 min

Similar to sugarcane bagasse, sorghum bagasse is also agricultural by products, which collected after extracting sorghum juice. Gnansounou *et al.* [16], stated that about 13% of the sugar will be in the sweet sorghum bagasse. Solihat *et al.* [17], reported that reducing sugar content in sweet sorghum bagasse could reach 45.57%, while sugarcane bagasse containing 61.5% glucose (lowering sugar) [18]. The internal bonds values of all sorghum bagasse particleboards in this study, were higher than 0.22 N/mm$^2$ (Figure 3), that fulfilled the minimum standard for type 13 particleboards (min 0.2 N/mm$^2$) according to JIS A 5908. Moreover, IB values of boards bonded with 30% CA-MD, reached 0.35 N/mm$^2$, that can categorize in type 18 particleboard (min IB 0.3 N/mm$^2$). Therefore, the reducing sugar in sorghum bagasse could act as bonding agent and gave synergetic effect with citric acid and maltodextrin in bonding particles.

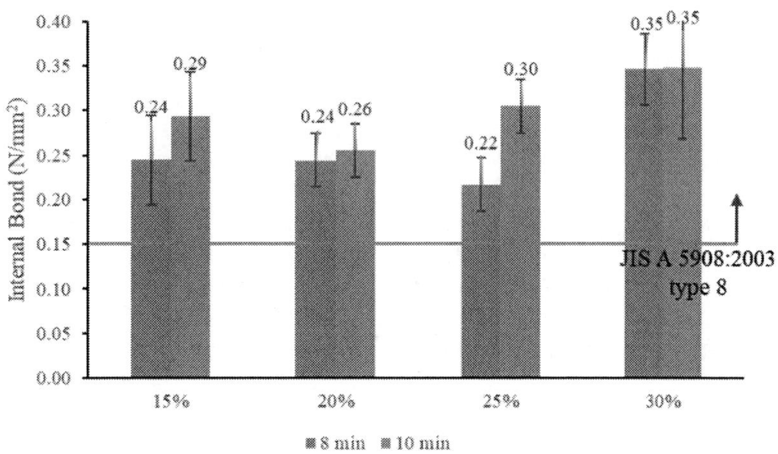


**Figure 3**. Internal Bond of particleboards with 15% (A), 20% (B), 25% (C), 30% (D) citric acid+maltodextrin, hot pressed in 8 min or 10 min

One of the weaknesses of citric acid application as a bonding agent in particleboard production is that the screw holding power of particleboard was low. As presented in Figure 4, the screw holding power (SHP) values of all type sorghum bagasse particleboard in this study were lower than 300 N, which the minimum SHP values of the lowest type of particleboard (type 8). After heat treatment during hot pressing, citric acid reacted with the hydroxyl groups of sorghum bagasse particles to form ester linkages. The formation of ester linkages would result in good adhesive-ness and impart excellent physical properties to the boards [2]. Unfortunately, the particleboard tends to brittle as reported by Kusumah *et al.* [6], then reducing the CA content by adding sucrose was an effective method to decrease particleboard brittleness.

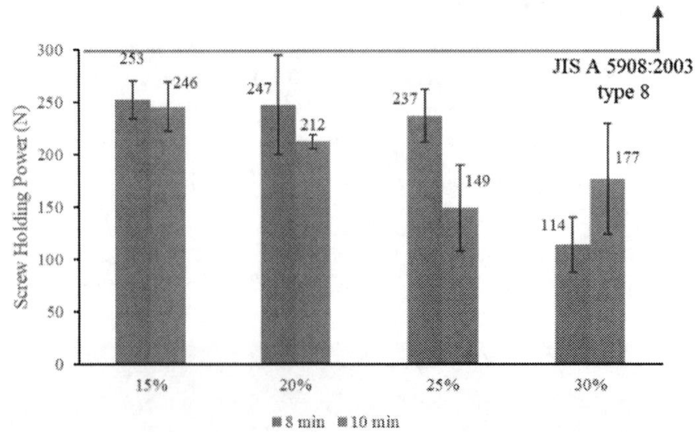


**Figure 4**. Screw holding power of particleboards with 15% (A), 20% (B), 25% (C), 30% (D) citric acid+maltodextrin, hot pressed in 8 min or 10 min

The change in the particleboard thickness and weight with varied CA-MD content and hot-pressed for 8 min, after cyclic aging treatment is presented in Figure 5. The expanded of particleboard thickness and weight change decreased with increasing of citric acid and maltodextrin content. The particleboard thickness and weight change bonded with 20% CA-MD and hot-pressed for 8 min (8B) after boiling treatment, presented better tenacity than the other particleboards. These mean that 20% CA-MD after heat treatment for 8 min could act as a good bonding agent for sorghum bagasse particles and had given good dimensional stability of particleboard even in the severe conditions.

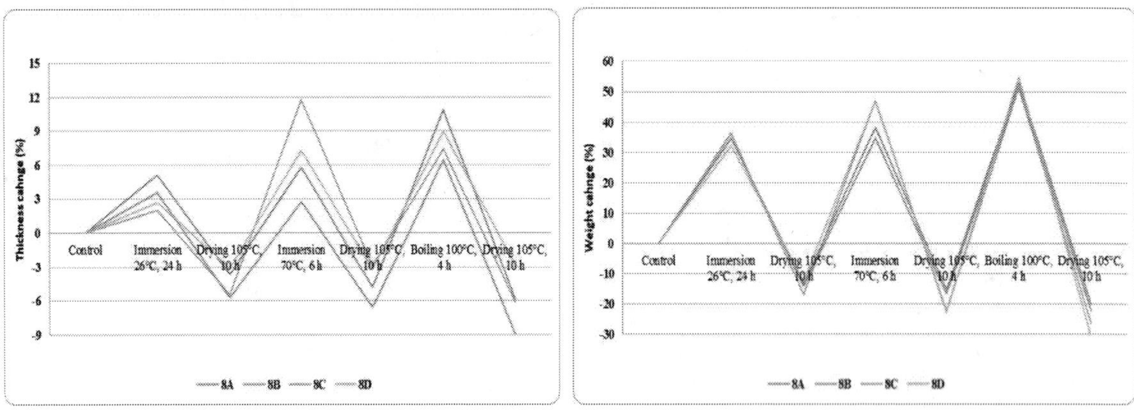


**Figure 5**. Thickness change (%) and weight change (%) of particleboards with 15% (A), 20% (B), 25% (C), 30% (D) citric acid+maltodextrin, hot pressed in 8 min after cyclic aging treatment

The increasing of CA-MD content (30%) require more time to be cured and successful in bond sorghum bagasse particles. The change in the thickness and weight particleboard with varied CA-MD content and hot-pressed for 10 min, after cyclic aging treatment, is presented in Figure 6. The expanded of particleboard thickness and weight change decreased with increasing of citric acid and maltodextrin content. The depth and weight change of the particleboard bonded with 30% CA-MD and hot-pressed for 10 min (10D) after boiling treatment (Figure 6), showed better stability than the other particleboards. These mean that 30% CA-MD after 10 minutes heat treatment before could act as a suitable bonding agent for sorghum bagasse particles. All sorghum bagasse particleboards in this study, show excellent physical properties, even after severe cyclic ageing treatment, proved by the thickness swelling was under 12%.

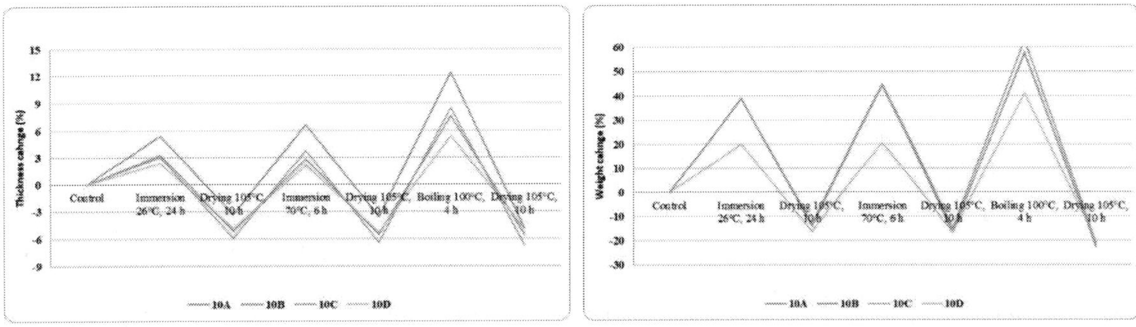


**Figure 6**. Thickness change (%) and weight change (%) of particleboards with 15% (A), 20% (B), 25% (C), 30% (D) citric acid +maltodextrin, hot pressed in 10 min after cyclic aging treatment

## 4. Conclusion

The mixture of citric acid and maltodextrin (1:1) was successfully acting as a bonding agent in sorghum bagasse particleboard production, especially when used at 30% adhesive content. The particleboard modulus of rupture and modulus of elasticity could fulfil the standard of type 8 particleboard, according to JIS A 5908-2003. The particleboard physical properties were excellent, proved by the value of particleboard thickness swelling, which was under 12%. This is due to the excellent bonding from the combination of citric acid and maltodextrin, as shown by the internal bond values, which exceeded 0.3 N/mm$^2$, minimum requirement for type 18 particleboards based on JIS A 5908-2003.

## References

[1] Iswanto AH, Aritonang W, Azhar I, Supriyanto, and Fatriasari W 2016 The physical, mechanical and durability properties of sorghum bagasse particleboard by layering surface treatment *J. Indian Acad. Wood Sci.* **14**(1) 1-8

[2] Kusumah SS, Umemura K, Yoshioka K, Miyafuji H, and Kanayama K 2016 Utilization of sweet sorghum bagasse and citric acid formanufacturing of particleboard I: Effects of pre-drying treatment andcitric acid content on the board properties *Industrial Crops and Products* **84**: 34-42

[3] Vukusic SB, Katovic D, Schramm C, Trajkovic J, and Sefc B 2006 Polycarboxylic acids as non-formaldehyde anti-swelling agents for wood *Holzforschung* **60**(4) 439-44

[4] Umemura K, Sugihara O, and Kawai S 2013. Investigation of a new natural adhesive composed of citric acid and sucrose for particleboard *Journal of Wood Science* **59**(3) 203-8

[5] Umemura K, Sugihara O, and Kawai S 2014 Investigation of a new natural adhesive composed of citric acid and sucrose for particleboard II: Effect of board density and pressing temperature *Journal of Wood Science* **61**(1) 40-44

[6] Kusumah SS, Umemura K, Guswenrivo I, Yoshimura T, and Kanayama K 2017. Utilization of sweet sorghum bagasse and citric acid for manufacturing of particleboard II: Influences of pressing temperature and time on particleboard properties *Journal of Wood Science* **63**(2) 161-172

[7] Jonsson LJ and Martin C 2016 Pretreatment of lignocellulose: Formation of inhibitory by-products and strategies for minimizing their effects *BioResource Technology* **199** 103-12

[8] Phuong LX, Shida S, and Saito Y 2007 Effects of heat treatment on brittleness of Styrax tonkinensis wood *Journal of Wood Science* **53** 181-86

[9] Clare AB, Franich RA, Lomax TD, McDonald AG, and Meder AR 2002 Preparation, characterization, and performance of cross- linkable maltodextrin-based lignocellulosic adhesives *Journal of Adhesion Science and Technology* **16** 317-29

[10] Castro-Cabado M, Casado AL, and Román JS 2016 Bio-based thermosets: effect of the structure of polycarboxylic acids on the thermal crosslinking of maltodextrins *European Polymer Journal* **78** 91-105

[11] Santoso M, Widyorini R, Prayitno TA, Sulistyo J, and Santoso M 2017 bonding performance of maltodextrin and citric acid for particleboard made from nipa fronds *J. Korean Wood Sci. Technol.* **45**(4) 432-43

[12] Syamani FA, Kusumah SS, Astari L, Prasetiyo KW, Wibowo ES, and Subyakto 2018 Effect of pre-drying time and citric acid content on *Imperata cylindrica* particleboards properties *IOP Conf. Series: Earth and Environmental Science* **209** 012034

[13] Syamani FA, Sudarmanto, Subyakto and Subiyanto B 2020. High quality sugarcane bagasse-citric acid particleboards *IOP Conf. Series: Earth and Environmental Science* **415**

[14] Santoso M, Widyorini R, Prayitno TA, and Sulistyo J 2016. the effects of extractives substances for bonding performance of three natural binder on nipa fronds particleboard. *The UGM Annual Scientific Conference Life Sciences* KnE Life Sciences pp. 227-38

[15] Sulastiningsih IM, Indrawan DA, Balfas J, and Santoso, A 2017 Sifat fisis dan mekanis papan untai berarah dari bambu tali (*Gigantochloa apus* (J.A. & J.H. Schultes) Kurz) Bamboo (*Gigantochloa apus* (J.A. & J.H. Schultes) Kurz) **35**(3) 197-209

[16] Gnansounou E, Dauriat A, and Wyman CE 2005 Refining sweet sorghum to ethanol and sugar: economic trade-offs in the context of North China *Bioresource technology* **96** 985-1002

[17] Solihat NN, Fajriutami T, Adi DTN, Fatriasari W, and Hermiati E 2017 Reducing sugar production of sweet sorghum bagasse kraft pulp *AIP Conference Proceedings* **1803** 020012

[18] Zhu G, Xiao Z, Zhu X, Yi F, and Wan X 2013 Reducing sugars production from sugarcane bagasse wastes by hydrolysis in sub-critical water *Clean Techn. Environ. Policy* **15** 55-61

**Acknowledgments**
Authors are wishing to acknowledge financial support from SATREPS project for producing biomass energy and material through revegetation of alang-alang (*Imperata cylindrica*) fields.

IOP Conf. Series: Materials Science and Engineering **935** (2020) 012059 doi:10.1088/1757-899X/935/1/012059

# Microfibrillated cellulose from sweet sorghum (*Shorghum Bicolor* (L.) Moench) fiber by twin screw extruder and its charateristics on polylactic acid biocomposites

**L Suryanegara[1*], W B Kusumaningrum[1], and J Zalesti[2]**

[1]Research Center for Biomaterials, Indonesia Institute of Sciences, Bogor, 16911, Indonesia

[2]Polymeric Chemical Engineering, Polytechnic of STMI, Jakarta, Indonesia

[*]Email: l_suryanegara@yahoo.com

**Abstract.** High tensile strength and modulus could be obtained by microfibrillated cellulose (MFC) addition into PLA matric. MFC production from sweet sorghum fiber with twin screw extruder (TSE) could be an effective mechanical treatment. The objective of this research is to study the effectivity of twin screw extruders for MFC production from sweet sorghum fiber and to analyze its effect on PLA biocomposites properties. Sorghum fiber was firstly delignificated with alkaline treatments and bleached with sodium peroxide solution to produce high cellulose content of fiber. Bleached sorghum fiber was then passed through into TSE for 1, 3, and 5 passes. Afterward, PLA biocomposites with MFC from TSE process were produced by the solvent casting method. Twin screw extruder with co-rotating intermeshing flow is effective for defibrillation and size reduction of cellulose fibers which could achieve 1-10 μm in diameter and millimeters in length.. The tensile strength of PLAM_TSE5 increase 4.54% in 50.60 Mpa and modulus of elasticity increase 40.36% in 2.82 Gpa. Storage modulus of PLAM_TSE5 also increases 10.64% in 3.088 Mpa compare to pure PLA.

## 1. Introduction

Polylactic acid (PLA) is a liniar aliphatic from acetic acid monomer. Commercial PLA produces in amorphous and semicrystalline depend on L-lactide and D-lactic acid content which contributes to different properties [1,2]. PLA is one of the biodegradable polymers which become widely developed in the last decade to consider environmental and green movement issues. Some excellencies of PLA include are high tensile and modulus of elasticity than polypropylene, biodegradable, biocompatible, good clarity, good oil resistance, and good oxygen barrier [1,3,4]. With these properties, PLA could be able for some applications in biomedical, food packaging, and automotive industry [1,5]. However, the weaknesses which are high glass transition temperature (Tg), brittle, low degradation, and low crystallization rate limit the application of PLA [1,2,5]. Modification of PLA whether intramolecular and intermolecular could be as an effective method by the nucleating agent or reinforcing agent addition to overcoming these drawbacks [1,5,6].

Modification via nucleating agent mechanism has been done with various types of organic and inorganic materials. Consider full biodegradability and environmental friendly material, bio-based nucleating agent such as starch, organic plasticizer, and cellulose more preferable than minerals [1, 5] . Cellulose is abundantly available which could be derived from lignocellulosic materials such us stalk, fiber, leaf, and stem [1]. One of potential lignocellulosic fiber is sweet sorghum with high biomass waste because sorghum could be regularly harvested in 3-4 months. In Indonesia, sweet sorghum plantation is cultivated mostly in Java and Nusa Tenggara. Cellulose resources from sweet sorghum could be derived from leaf and stalk part. The cellulose content in leaf part about 64.8% while in stalks is about 65.1% which close to commercial natural fiber such as flax, hemp, and coconut [2].

The preparation of cellulose as a reinforcing agent becomes a crucial role to produce good properties of biocomposites. The disintegration of cellulose into the fiber with a diameter in micro and nano scale hasa modulus of elasticity about 40-70 Gpa and tensile strength 400-700 MPa, while in crystal structure about 130-250 GPa and tensile strength about 800-10,000 MPa [3]. Microfibrillated cellulose (MFC) could produce by mechanical treatments using refiner, high pressure homogenizer, microfluidizer, grinder, and high intensity ultrasonicator [10,11]. Mechanical treatments enable cellulose fiber defibrillated in radially order so could produce high aspect ratio of MFC. Some studies suggest for twin screw extruders as low energy consumption and effective mechanical treatments to produce high solid content of MFC even in a single process [12,13,14] or combination with another process [15,16].

Some studies report for mechanical properties enhancement for PLA and MFC biocomposites. Kneading process of PLA and MFC from kenaf pulp by twin screw extruders could improve tensile strength about 21% and MOE about 34.5% [4]. The addition of fibrillated cellulose up to 1 to 3% enhances tensile strength up to 1.2 fold of PLA composites, but more than 3% result in agglomeration of MFC [5]. In addition, amorphous PLA with 20% of MFC improves tensile strength in 72.1% and modulus 57.6% [6]. Besides tensile strength and modulus enhancement, storage modulus up to 160 MPa under dynamic mechanical analysis resulted in higher value than fully crystallized PLA [7].

As the best know of authors, there is a limited study about PLA and sweet sorghum fiber biocomposites. The only research is PLA biocomposites reinforced with the modified fiber of sweet sorghum in particle form [8]. PLA biocomposites reinforced with MFC from sweet sorghum have not been done, especially MFC which isolated from twin screw extruder. The objective of this research is to study the effectivity of twin screw extruders for MFC production from sweet sorghum fiber and to analyze its effect on PLA biocomposites properties.

## 2. Materials and Methods

### 2.1. Materials

Sweet sorghum (*Shorgum Bicolor* (L.)Moench) was harvested from the Center of Innovation cultivation side, Cibinong. Particle production of sweet sorghum's stem was conducted using hammer mill and disc mill until 40-60 mesh of particle size which could be as raw material for fiber. Acetone 90%, hydrogen peroxide 30%, and sodium hydroxide were used for cellulose isolation. Acetone and hydrogen peroxide ($H_2O_2$) is in technical grade while sodium hydroxide (NaOH) is in pure analysis grade. Amorphous PLA (polylactic acid) 4060 D was purchased from Natureworks, USA. PLA 4060 D has D-lactide content is about 12%, Tg (glass transition temperature) is about 55-60°C, density is 1.24 g/cm$^3$, and MFR (melt flow rate) at 180°C is 3.89 g/10 min.

### 2.2. Methods

#### 2.2.1. Isolation of cellulose from sorghum fiber

Prior to isolate, sorghum fiber was firstly passed through delignification process. As much 100 g of sorghum particle was cooked on 10% (v/v) of NaOH solution at 95°C for 120 min with 250 rpm of magnetic stirring. Particle which transforms into fiber was washed using aquadest until neutral. Subsequently, sorghum fiber was bleached using combination of 30% (v/v) of $H_2O_2$ and 10% (v/v) of NaOH solution at 95°C for 60 min. Cellulose fiber was then washed using aquadest until neutral with pH in 6.2-6.5.

Chemical composition of cellulose derived from sorghum fiber was analyzed using FTIR (Fourier Transform Infra Red). Universal Attenuated total reflectance (UATR)-FTIR Perkin Elmer Spectrum Two was used for qualitative analysis with scanning operation in 4000-400 $cm^{-1}$ of wavelength with 15 times of repetition for each sample.

2.2.2. Preparation of microfibrillated cellulose

Mechanical fibrillation was conducted using TSE Haake Rheomix PTW 16. As much as 10 g of cellulose fibers was feed through sample hooper and then kneaded between two co-rotating screws. Rotational speed was 400 rpm and temperature condition at 26°C for all zone. The fibrillation process was conducted in 1, 3, and 5 passes.

2.2.3. Characterization of microfibrillated cellulose from sorghum fiber

*2.2.3.1. Dispersion stability of microfibrillated cellulose*

Microfibrillated cellulose by TSE process was then dispersibility tested on water. Some samples of MFC which collected on the edge of TSE in 1, 3 , and 5 passes was placed in 50 mL vial bottles then observed the fiber stability after 5 min suspended in water.

*2.2.3.2. Morphology of microfibrillated cellulose*

Morphology of MFC from sorghum fiber was done using FE-SEM (Field Emission Scanning Microscopy) Quattro S Thermo Fisher Scientific. MFC was placed on the step without coating then observed in 1000 times of magnification with scanning electron emission in 2 kV.

2.2.4. Production of polylactic acid and microfibrillated sorghum fiber biocomposite

As much as 54 g of PLA 4060D was dissolved on 600 mL of acetone then stirred for 45 min at room temperature. Treated and untreated MFC in 1% (w/w) was blended into solution for 15 min. The blended compound was cast into 25 x 40 square mold then oven dried at 65°C for 12 hours to vaporize the solvent.

Sample specimens for mechanical properties were molded using a Mini jet Pro to obtain dog bone specimen test according to ASTM D 638 type V. PLA compound was cut into pieces then feed in minijet pro at 155°C of cylinder temperature, 600 bar of molding pressure, and 15 seconds of infection time.

2.2.5. Properties of polylactic acid biocomposites

*2.2.5.1. Mechanical properties*

Mechanical properties were analyzed using Universal Testing Machine (UTM) autograph 10 kN Shimadzu. The tensile test was conducted according to ASTM 638 for plastic composites with load cell 5 kN and crosshead speed 1 mm/min.

*2.2.5.2. Thermal properties*

Thermal analysis was conducted using Differential Scanning Calorymetry (DSC) 4000 Pyris 1 Perkin Elmer equipped with intercooler. 5 mg of PLA pieces were placed into a alumunum pan. The sample was heated from 25°C to 200°C with heating rate 20°C/min, hold the sample at 200°C for 2 min, cooled from 200°C to 25°C with cooling rate 5°C/min, hold at 25°C for 2 min, then heated from 25 to 200°C with heating rate 5°C/min.

*2.2.5.3. Dynamic mechanical properties. Rectangular sample*

Dynamic mechanical properties were conducted using Dynamic Mechanical Analysis (DMA) 8000 Perkin Elmer. The specimens were prepared in rectangular with dimension in 30 x 6 x 2 mm. The analysis was undertaken from 0°C to 100°C with heating rate of 2°C/min and frequency in 1 Hz.

## 3. Results and Discussion

### *3.1. Characterization of cellulose from sorghum fiber*

The main component of sorghum fiber is consist of cellulose, hemicellulose, and lignin. The Visual appearance of sorghum is similar to general natural fiber derived from stalk part of plant such as corn stalk and kenaf which classified as short fiber as seen in figure 1 (a). As a reinforcing agent in composite materials that purposed to enhance mechanical properties as well as thermal and optical properties, natural fibers should be prepared in high cellulose content. Cellulose in the crystalline part considers has better compatibilities to polymer matric than lignin and hemicellulose in amorphous part. Hemicellulose and lignin are interfered with physical bonding with polymer matric and have low ultraviolet degradation resilience [9]. In order to facilitate better properties of the composite, delignification process might be undertaken to isolate cellulose from fibers. Delignification followed with bleaching process result in brighter fibers than the original as seen in figure 1 (b) which could be termed as cellulose fibers. The chemical component of cellulose fibers is consisting of 88% of holocellulose and 3% of lignin. Lignin is still contained after the bleaching process even in small presence and bright appearance.

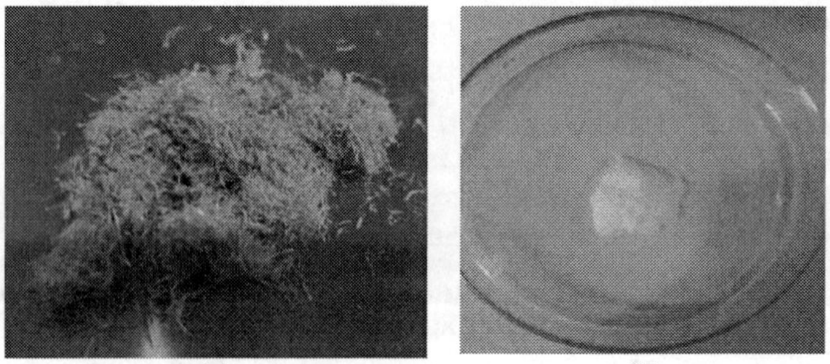

(a) (b)

**Figure 1.** (a) Visual appearance of sorghum particle and (b) sorghum fiber after bleaching process

**Table 1**. Identified functional groups on FTIR spectrogram for sorghum fiber

| Identified functional groups | Wavelength ($cm^{-1}$) | | |
|---|---|---|---|
| | Sorghum particle | Sorghum fiber after delignification process | Sorghum fiber after bleaching process |
| Stretching of –OH | 3330 | 3330 | 3332 |
| Stretching of C-H | 2922 | 2922 | 2922 |
| Stretching of C=O | 1734 | 1734 | - |
| Absorbed O-H conjugated C-O | 1644 | 1644 | 1637 |
| C-C aromatic vibration of lignin | 1510 | 1506 | - |
| Deformation C-H in cellulose and hemicellulose | 1385 | 1377 | 1369 |
| Stretching C-O in cellulose | 1030 | 1027 | 1032 |
| β glicocidic linkage | - | 895 | 895 |

Identified functional groups on FTIR spectrogram for all cellulose isolation steps are represented in Table 1. FTIR Spectrogram identified the specific functional groups of cellulose in 3330 $cm^{-1}$, 2922 $cm^{-1}$ which are identified the stretching of –OH and C-H groups respectively for all fibers. These intensity of functional groups are higher for delignified and bleached fiber than particles as shown in

Figure 2. Consistent with these results, identified functional groups of β-(1-4)-glycosidic bond of cellulose in 895 $cm^{-1}$ is obviously detected in delignified and bleached fiber which correspond to exposed cellulose part. The absence of aromatic ring vibration of lignin in 1515 $cm^{-1}$ for bleached fiber indicate that lignin almost totally dissolved during the bleaching process and result in bright fiber. Furthermore, carbonyl groups that indicates for xylan on hemicellulose in 1734 $cm^{-1}$ is not detected for bleached fibers as seen in Figure 2. In these cases, sorghum fiber that bleached through sodium peroxide solution could be as raw material for MFC with high cellulose content. Alkaline extraction allows for sugar degradation of hemicellulose and disruption of lignin macromolecules then by peroxide oxidation break up chromophore compound on lignin structure [10].

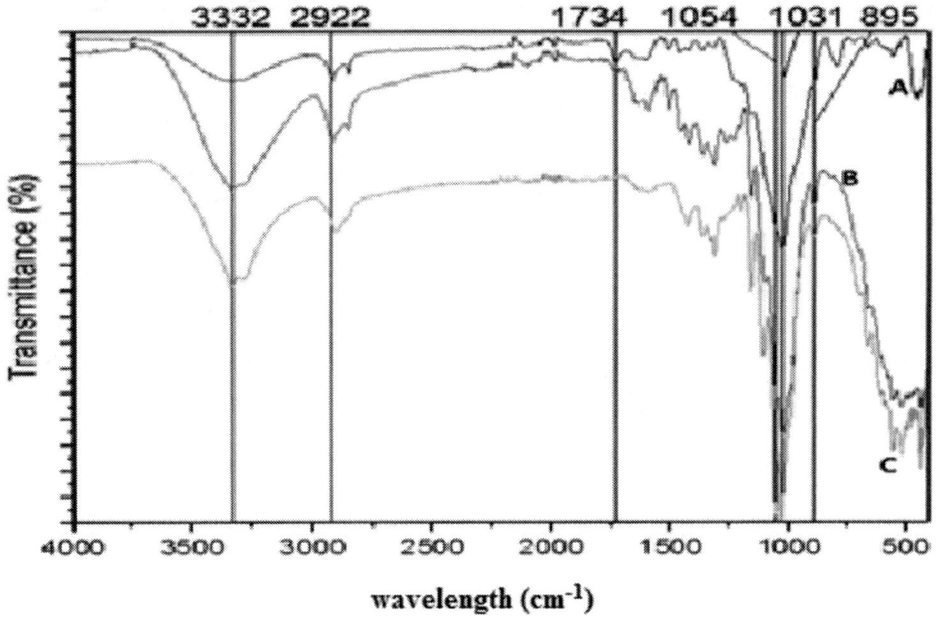


**Figure 2.** (a) FTIR spectrogram from sorghum particle, (b) sorghum fiber after delignification process, and (c) after bleaching process

*3.2. Characterization of microfibrillated cellulose*

3.2.1. Dispersion stability

The dispersion of cellulose in water or a particular solvent could be an indicator of the size reduction and fibrillation effectivity of cellulose. The mechanism of kneading process by pass through the cellulose fibers between two rotors which moving forward in co rotating flow so could defibrillate and reduce the diameter of cellulose simultaneously. The dispersion stability of MFC in 1, 3, and 5 passes are present in Figure 4. The samples start with turbid suspension after the TSE process as shown in figure 3 (a), then after 5 minutes, the MFC with 1 and 3 passes of TSE becomes translucent while MFC with 5 passes remains turbid as shown in Figure 3 (b). These results indicate that smaller particle size is produced by 5 passes of TSE than 1 and 3 passes. Smaller particle size correspond to the wide surface area. The defibrillation process also could expose hydroxyl groups of cellulose which than facilitate hydrogen bonding with water. Therefore, cellulose with smaller particle size and more expose hydroxyl groups is more stable in suspension. MFC build a strong network by hydrogen bonding then well dispersed into the water [11]. MFC which processed minimal in 5 passes more stable in water because result in smaller fiber than 1 and 3 passes of the TSE process [12].

0 min 5 min

**Figure 3.** Dispersion stability of treated MFC with TSE in (A) 1 passes, (B) 3 passes, and (C) 5 passes

3.2.2. Morphology

MFC which resulted by the TSE process performs in micrometers scale of fiber's diameter with millimeters scale of length. Kneading mechanism by the co-rotating flow of TSE effectively produce MFC with high aspect ratio as presented in figure 4 (a, b). Sorghum fiber which has passed through the TSE in 5 passes has smaller fiber's diameter with about 1 up to 10 μm (Figure 4b) than 1 passes with about 10 up to 20 μm (Figure 4a). Figure 4b also show defibrillation and size reduction could be achieved at the same time. Co rotating intermeshing flow of TSE minimize the fibers freely rotate so could hold the fibers remain between on two kneading screw without thrown up into the TSE chamber. Those results confirm for better dispersion stability for MFC with 5 passes than in 1 and 3 passes of TSE. Rol *et al.* [13] report significant size reduction on fibrillated fiber after 7 passes of TSE as high shear in the nip of the kneading part. The shearing force on the fiber creates a rough surface then exposed hydroxyl groups of cellulose so could be stable in water which correlated for stable dispersion stability. Ho *et al.* [12] discover of highly entangled and twisted morphology of MFC with diameters in nanoscale and also suggest for optimum TSE process not more than 6 passes because could encourage for cellulose crystallinity degradation on 10 passes.

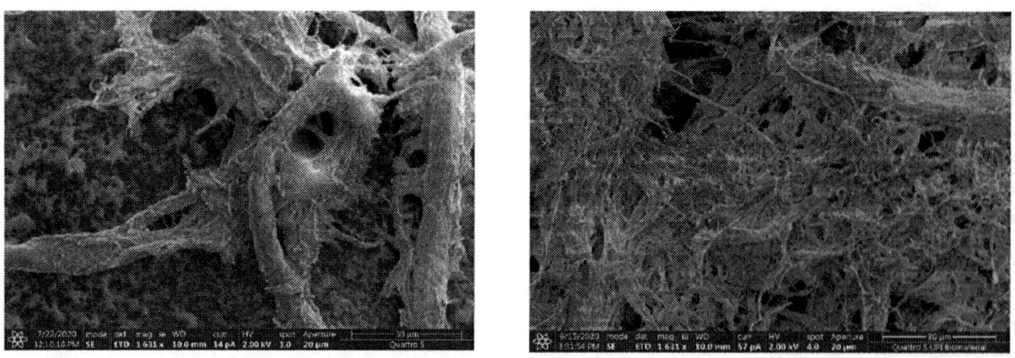

(a) (b)

**Figure 4**. Morphology of treated MFC by TSE in 1 passes (a) and 5 passes (b) with 1000 x of magnification.

IOP Conf. Series: Materials Science and Engineering **935** (2020) 012059 doi:10.1088/1757-899X/935/1/012059

### *3.3. Properties of polylactic acid biocomoposites*

#### 3.3.1. Mechanical properties

PLA biocomposites were produced with MFC without treatment then denoted as PLAM and MFC with 5 passes of TSE then denoted as PLAM_TSE5. Mechanical properties of PLA biocomposites include tensile strength and modulus of elasticity are expressed in figure 5. The addition of MFC improves tensile strength up to 2.5% and 4.54% for PLAM and PLAM_TSE5 respectively. The high aspect ratio of MFC enables better bonding and wettability with PLA matric because of wider surface area. Modulus of elasticity also increase about 38.34% and 40.36% for PLAM and PLAM_TSE5 respectively. MFC addition could enhance the toughness of PLA biocomposites and further with high aspect ratio result in better ductility than pure PLA and PLAM. Overall, MFC with higher aspect ratio, correspond to good interfacial adhesion, could produce PLA biocomposites with higher stiffness and toughness to hold some loads. These result is agree with Joonoobi *et al.* [4] and Suryanegara *et al.* [6], MFC addition improves the interfacial adhesion and physical bond into amorphous PLA and semi-crystalline PLA that contribute to tensile strength and modulus enhancement.

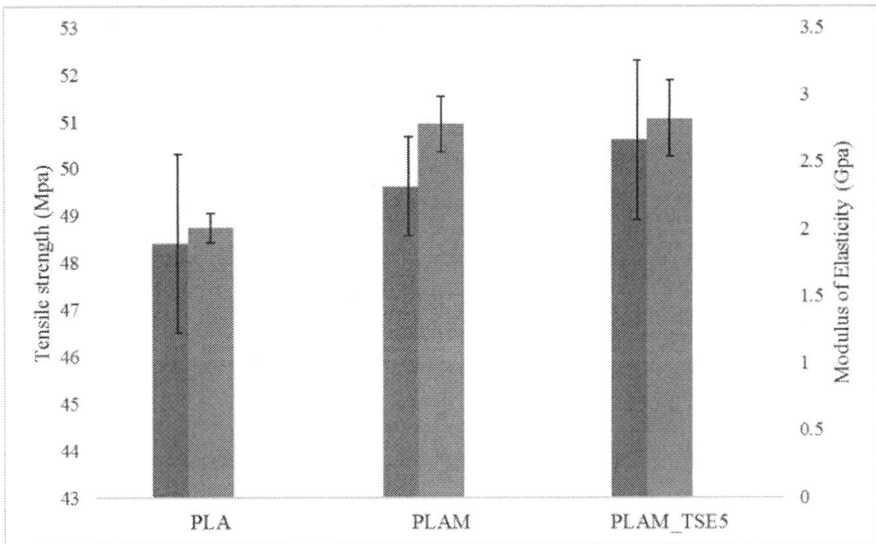


**Figure 5.** Mechanical properties of PLA biocomposites

#### 3.3.2. Thermal properties

Thermal properties of PLA biocomposites are represented with glass transition temperature (Tg) and melting temperature (Tm). Consider that PLA 4060D is an amorphous PLA, there is no crystallization temperature (Tc) and cold crystallization temperature (Tcc) could be detected because of organized crystal structure that could not be formed. The addition of MFC could cause Tg and Tm shift for PLAM and PLAM_TSE5 even in slightly different as presented in Figure 6. These results indicate that MFC has interaction with PLA matric. PLAM_TSE5 has a higher Tg than PLA and PLAM which correspond to the brittleness. These results were similar to Frone *et al.* [14] that higher Tg could be affected by lower chain mobility of PLA because of MFC addition. Different tendencies report by Johari *et al.* [15] that addition MFC together with plasticizer effective to lower the Tg because of accessible chain mobility. Figure 6 reveals only one endothermic peak performed for all biocomposites at around 129 up to 130°C which correlated for one form crystal structure for amorphous PLA. The Tm is affected by the form and perfection of crystal structure [14]. The endothermic peak of PLA with MFC (PLAM and

PLAM_TSE5) has broader peak than PLA which might indicate the change of PLA polymer chain by MFC addition.

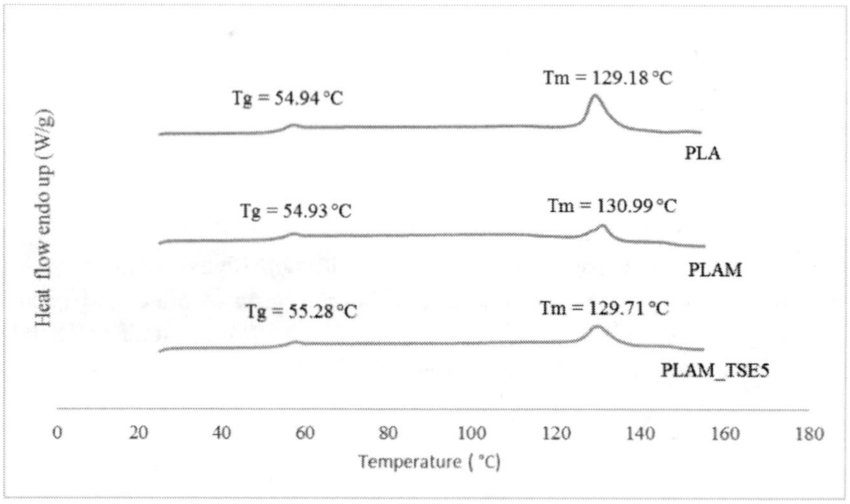


**Figure 6.** Thermal properties of PLA biocomposites

### 3.3.3. Dynamic mechanical properties

Dynamic mechanical analysis allows some tensile force and thermal exposure simultaneously. The addition of MFC increase the storage modulus of PLA biocomposites even in amorphous state. Storage modulus of PLAM_TSE5 is 3.088 Mpa which increase about 10.64% from PLA in amorphous and glassy condition as shown in Figure 7. Contrary, storage modulus tends to be unchanged at the rubbery stage at 75°C. In these cases, MFC might limit the molecular chain mobility of PLA above the Tg [4]. Storage modulus enhancement indicates good interfacial adhesion between MFC and PLA [6]. Tg in dynamic condition tends to increase, similar to thermal analysis, for PLAM and PLAM_TSE5. These results emphasize that lower chain mobility might occur above the Tg than under the Tg with MFC addition.

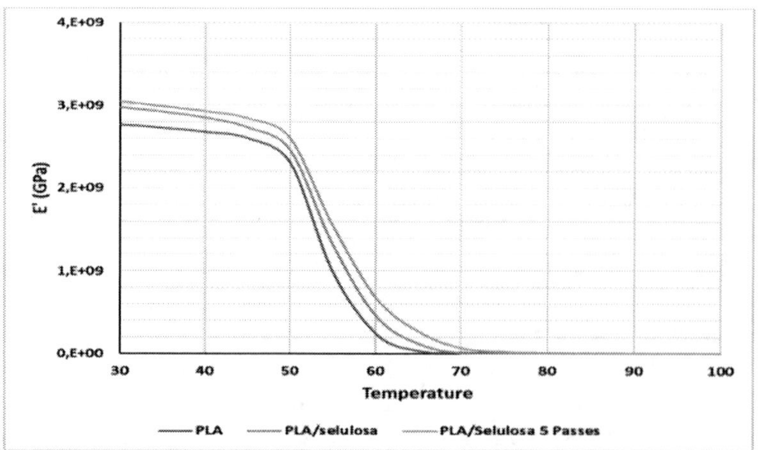


**Figure 7**. Thermomechanical properties of PLA biocomposites

## 4. Conclusion

Sorghum fibers which isolate by sodium peroxide solution could produce high cellulose content fiber. Twin screw extruder with co rotating intermeshing flow is effective for defibrillation and size reduction of cellulose fibers which could achieve 1-10 μm in diameter and millimeters in length. The high aspect ratio of MFC could facilitate bonding and wettability on PLA matric which could improve some properties include mechanical, thermal, and dynamic mechanical. The tensile strength of PLAM_TSE5 increase 4.54% in 50.60 Mpa and modulus of elasticity increase 40.36% in 2.82 Gpa. Storage modulus of PLAM_TSE5 also increases 10.64% in 3.088 Mpa compare to PLA.

## References

[1] Mohanty AK, Misra M, and Drzal LT 2002 Sustainable bio-composites from renewable resources: opportunities and challenges in the green materials world *J. Polym. Environ.* **10**(1) 19-26

[2] Reddy N and Yang YIQI 2007 Structure and properties of natural cellulose fibers obtained from sorghum leaves and stems *J. Agric. Food Chem.* **55**(14) 5569-74

[3] Zimmermann T, Pöhler E, and Geiger T 2004 Cellulose fibrils for polymer reinforcement *Adv. Eng. Mater.* **6**(9) 754-61

[4] Jonoobi M, Harun J, Mathew AP, and Oksman K 2010 Mechanical properties of cellulose nanofiber (CNF) reinforced polylactic acid (PLA) prepared by twin screw extrusion *Compos Sci. Technol.* **70**(12):1742-7

[5] Abdulkhani A, Hoseinzadeh J, Dadashi S, and Mousavi M 2015 A study of morphological, thermal, mechanical and barrier properties of PLA based biocomposites prepared by micro and nano sized cellulosic fibers *Cellul Chem Technol* **49**(7) 597-605

[6] Suryanegara L, Nakagaito AN, and Yano H 2009 The effect of crystallization of PLA on the thermal and mechanical properties of microfibrillated cellulose-reinforced PLA composites *Compos Sci. Technol.* **69**(7–8) 1187-92

[7] Suryanegara L and Norio A 2010 Thermo-mechanical properties of microfibrillated cellulose-reinforced partially crystallized PLA composites *Cellulose* **17** 771-8

[8] Zhong J, Li H, Yu J, and Tan T 2011 Effects of Natural fiber surface modification on mechanical properties of poly(lactic acid) (pla)/sweet sorghum fiber composites *Polym. Plast. Technol. Eng.* **50**(15) 1583-9

[9] Moshiul Alam A M, Beg MDH, Reddy Prasad DM, Khan MR, and Mina MF 2012 Structures and performances of simultaneous ultrasound and alkali treated oil palm empty fruit bunch fiber reinforced poly(lactic acid) composites *Compos Part A Appl. Sci. Manuf.* **43**(11) 1921-9

[10] Liu K, Li H, Zhang J, Zhang Z, and Xu J 2016 The effect of non-structural components and lignin on hemicellulose extraction *Bioresour. Technol.* **214** 755-60

[11] Durand S, Gaillard C, Garnier C, and Doublier JL 2012 Properties of cellulose/pectins composites : Implication for structural and mechanical properties of cell wall *Carbohydr. Polym.* **90**(2) 1081-91

[12] Ho TTT, Abe K, Zimmermann T, and Yano H 2015 Nanofibrillation of pulp fibers by twin-screw extrusion *Cellulose* **22**(1) 421-33

[13] Rol F, Karakashov B, Nechyporchuk O, Terrien M, Meyer V, Dufresne A, Belgacem MN, and Bras J 2017 Pilot-scale twin screw extrusion and chemical pretreatment as an energy-efficient method for the production of nanofibrillated cellulose at high solid content *ACS Sustain Chem. Eng.* **5**(8) 6524-31

[14] Frone AN, Berlioz S, and Panaitescu DM 2013 Morphology and thermal properties of PLA – cellulose nanofibers composites *Carbohydrate* **91**(1) 377-84

[15] Johari AP, Kurmvanshi SK, Mohanty S, and Nayak SK 2016 Influence of surface modified cellulose microfibrils on the improved mechanical properties of poly ( lactic acid ) *Int. J. Biol. Macromol.* **84** 329-39

**Acknowledgements**
The authors acknowledge the facilities, the scientific and technical assistance of the Integrated Laboratory of Bioproducts at the Indonesian Institute of Science. We would like to thanks to Sudarmanto and Sesmi Gutari for preparation and some laboratory works.

IOP Conf. Series: Materials Science and Engineering **935** (2020) 012060 doi:10.1088/1757-899X/935/1/012060

# Properties of fibreboard (FBs) and recycle fibreboard (rFBs) and analysis of their wastage after recycling

**A Nuryawan[1,2*], I Risnasari[1], A P Pohan[1], A U Husna[1], T I Nasution[2,3], R Banurea[2,3], K S Hartini[4]**

[1]Department of Forest Products Technology, Faculty of Forestry, Universitas Sumatera Utara, Medan, 20155, Indonesia
[2]Integrated Research Laboratory, Universitas Sumatera Utara, Medan, 20155, Indonesia
[3]Department of Physics, Faculty of Mathematics and Natural Sciences, Universitas Sumatera Utara, Medan, 20155, Indonesia
[4]Department of Silviculture, Faculty of Forestry, Universitas Sumatera Utara, Medan, 20155, Indonesia

*Email: arif5@usu.ac.id or arif.nury@gmail.com

**Abstract.** Fibreboards (FBs) bonded by urea-formaldehyde (UF) resin derived from industry were intended to undergo cyclist test, a simulation of recycling activity. Objective of this study was to compare quality of FBs before and after recycling including analyse its wastage. Methods of this study were consisted of evaluation physical and mechanical properties of FBs before recycling, cyclist test, remanufacturing FBs, evaluation recycle FBs (rFBs), comparing the properties between FBs and rFBs and analysing the wastage of recycling activity both solid residue and liquid disposal. Results of this study showed physical and mechanical properties of FBs changed after recycling presumably because of the difference in reproduction method. Some physical properties showed better quality, while the mechanical properties performed decrease drastically. Cyclist test predicted that FBs were more resistant in 3 (three) cyclist compared to that of rFBs. There was wastage of recycling consisting of macro solid fibre with length dimension of $1000.51 \pm 339.21 \mu m$ and micro solid fibre with variation in distribution size. Liquid disposal (LD) contained nitrogen (N) with amount 0.05% presumably from sol fraction of UF resin. These findings suggested that remanufacturing of rFBs was feasible even though wastage of recycling activity was inevitable. Fortunately, LD can be utilized further for example for fertilizer because the N content exists.

## 1. Introduction

One of wood-based panel types, fibreboards (FBs), has been substituted solid wood since degradation of natural forest decreased rapidly and plantation forest become trends worldwide. FBs has been used as raw material of furniture: shelving, kitchen worktops, interior signage; furnishing interior: decorative mouldings, cupboard, kitchen set; building complements: door filler, sliding doors, office dividers, laminate flooring, office dividers, walls, ceilings and other wood industrial products [1,2]. FBs are manufactured using low grade quality wood which is converted into fibre by pulping process [3]. Thermosetting resin is used as the binder for dry process production and Urea Formaldehyde (UF) resin is primarily used adhesive for bonding it [4]. Since UF resin is susceptible to hydrolysis, wastage of these products is overabundance. Further, in the tropical countries like Indonesia, where the relative humidity is higher, these products are prone to damage. Various destructions such as creep, cleavage, and crack make these products become garbage and they need to be disposed carefully because of its contaminant. Dry process FBs is consisting of at least two parts namely 90% wood fibre and 10% resin adhesive [3].

Currently, there were at least three recommended choices for dealing with wood litter, namely incinerating, land filling disposal, and recycling [5]. Even though choice either burning or land filling has been conducted for treating this rubbish particularly in the United States [6], both treatments have had negative effects to environment. Incinerating will allow hazardous gases such as CO (Carbon monoxide), NOx (Nitrogen oxides), Sox (Sulphur oxides), PAHs (Polycyclic aromatic hydrocarbons), dioxins, and PCB (Polychlorinated biphenyls) depends on type combustion used [7] while direct disposing as mulch to improve soil condition and prevent erosion caused some disadvantages such as high energy consumption, toxic methane gas/bad odor release and slower reaction kinetics [5]. Therefore, recycling is one of good options for handling FBs wastages.

Here, a simulation for recycling such of materials has been presented. FBs were immersed into water for separating between wood fibre and adhesive components. It is believed that recycled fibre can be utilized as raw a material for remanufacturing FBs. Some scientists have been evaluated quality of FBs made of recycle fibre recently. For instances, Mantanis *et al.* [8] successfully remanufactured FBs made of recycle and virgin fibre with a ratio of 25:75 in industrial scale. Hong *et al.* [9] also examined the quality of FBs made of 10, 20 and 30 parts recycle fibre. They found that a mixture of 20% recycle fibre resulted in better physical and mechanical properties compared to those of 10% and 30%. Comprehensively, Lubis *et al.* [10] examined fibre morphology originated from recycling FBs, nitrogen (N) content within the recycle fibre, and also physical and mechanical characteristics of resulted recycle boards with various mixture of recycle fibre up to 100%. They found only 10% recycle fibre can replace virgin fibre in producing FBs. From this point, activity of recycling of FBs is feasible, and solid wood residue can be utilized as raw material of recycle FBs.

Related to water disposal, it can be used as fertilizer because of rich of N content. UF resin adhesive contained leached chemical component or sol fraction [11]. Grigbsy *et al.* [12] investigated residue and water disposal of FBs bonded UF resin after water soaking. They found that N existed in both fibre and water. Further Lubis *et al.* [13] investigated N content within both solid residue and extract solution after hydrolysis in recycling practice of FBs.

In this contribution, quality of industrial FBs was compared to recycle FBs (rFBs) and wastage after recycling activity was analysed either chemically or morphologically in order to evaluate its feasibility as fertilizer for non-food plants seedling. Thus, the objective of this study was to compare physical and mechanical properties of commercial FBs versus rFBs and to determine impurities occurring in water disposal after recycling activity of FBs.

## 2. Materials and Methods

### *2.1. Materials*

Either commercial FBs purchased from PT Mandiri Jaya, Jakarta or industrial FBs donated by PT Canang Indah, Medan, became main object observation of this research. Both types of FBs were confirmed using UF resin as the binder. UF resin with solid content of 60.52% and ammonium chloride ($NH_4Cl$) as hardener were donated by PT Canang Indah Medan for binding rFBs.

### *2.2. Methods*

#### *2.2.1.* Evaluation physical and mechanical properties of FBs

Both types of FBs were cut and referred into test specimens according to Japanese Industrial Standard/ JIS [14]. The specimens were then exposed to physical examinations such as density and moisture content (MC). The calculation of density used digital balance and veneer caliper, and it was conducted in ambient temperature around 27-31°C. Measurement of thickness swelling and water absorption was carried out both in 2 hours and 24 hours. Investigation of mechanical properties was based on JIS standard with aid of Tensilon Universal Testing Machine. One-point loading was applied for determining Modulus of Elasticity (MOE) and Modulus of Rupture (MOR). An Internal Bonding (IB) test was also determined for evaluating bonding strength of the FBs. Each parameter test was measured 8 (eight) in replications without considering type of the FBs.

*2.2.2.* Remanufacturing FBs and evaluation of their properties
All specimens used in physical and mechanical properties testing were then soaked in water. After the specimens swelled, they underwent hand-stirring for fibres separating completely. A resulted mixture of water and fibre then was filtered using a fabric screen for separating between solid and liquid residue. Solid residues, hereafter recycle fibre, and then underwent sunlight drying. For reaching a target MC of $\pm$ 5%, recycle fibres were oven-dried using laboratory oven.

Remanufacturing rFBs was carried out in a laboratory scale used a column-type hydraulic hot-press. A thickness of 10 mm with dimension of 25 x 25 cm and density of 0.75 g/cm$^3$ were the target size of the FBs. Composition of the adhesive was 8% UF resin based on solid and 3% $NH_4Cl$ as hardener. For adhesive curing, temperature of hot-press was set at 120°C for 15 minutes. Prior to evaluation its properties, each rFBs was conditioned in room temperature for 2 weeks. Physical and mechanical properties examination was done as same procedure of evaluation the FBs referring standard [14].

*2.2.3.* Cyclist test
FBs and rFBs were cut into 5 x 5 cm in size. Each group was consisted of at least five specimens thus totally there were 10 samples. Cyclist test was comprised of cycles of five treatment steps, i.e., first: measurement of initial thickness dimension, second: soaking in 1000 mL water at ambient temperature for 24 hours, third: measurement of thickness swelling dimension, fourth: drying in laboratory oven at (103$\pm$2)°C for 48 hours, and fifth: measurement of oven dry thickness dimension [15]. This cycle was applied until the samples were not possible to be measured their thickness because of sample damage, such as cleavage or crack. Thickness measurement was conducted using a digital veneer caliper. Resulted data were presented, tabulated and compared between FBs and rFBs.

*2.2.4.* Analysis of the recycling wastage
Water disposal originated from recycling activity, hereafter cyclist test, was analyzed its contaminant comprised of individual solid fibre, micro fibre, and N content. Bright-field microscope was used for measuring length and diameter individual of solid fibre with 100 replications. Micro fibre was analysed using particle size analyser (PSA) Malvern Mastersizer 3000 Hydro EV in micron scale with 2 replications. Result of analysis was tabulated and classified according to the diameter class of the particle. N content was measured using Kjeldahl method.

## 3. Results and Discussions
Different from previous work [8-10] which combined recycle fibre and virgin fibre, this study was intended to use all recycle fibre for producing rFBs. However, simulation of recycling of FBs in this study resulted in 75% yield. Commercial/industrial of 8 (eight) FBs specimens were only capable to make 6 (six) rFBs specimens. This condition was influenced by recycling activities such as dissolving of solid fibre residue during board soaking, fibre drying under sunlight as well as under laboratory oven, and trimming sawing for making specimens. All of these made volume of fibre decreased thus reducing the quantity of resulted boards.

### *3.1. Properties of FBs and rFBs*
Table 1 and 2 showed physical and mechanical characteristics of both FBs and rFBs comparing to that of standard, respectively.

**Table 1**. Physical properties of FBs versus rFBs

| Physical properties | FBs | rFBs | JIS |
|---|---|---|---|
| Density (g/cm$^3$) | 0.76 $\pm$ 0.02 | 0.57 $\pm$ 0.04 | > 0.35 |
| MC (%) | 8.52 $\pm$ 0.22 | 2.40 $\pm$ 1.36 | 5-13 |
| Thickness swelling 2 h (%) | 13.66 $\pm$ 1.14 | 10.93 $\pm$ 1.83 | - |
| Thickness swelling 24 h (%) | 18.10 $\pm$ 1.51 | 27.47 $\pm$ 1.07 | 12 |
| Water absorption 2 h (%) | 55.84 $\pm$ 7.25 | 51.28 $\pm$ 21.89 | - |
| Water absorption 24 h (%) | 70.20 $\pm$ 4.77 | 151.15 $\pm$ 19.68 | - |

Remarks: resulted data were measurement of 8 replications for FBs and 6 replications for rFBs, respectively

Density and MC of commercial/industrial FBs were fulfilling the standard however after remanufacturing both properties decreased. Laboratory work with standard equipment presumably affected the quality. For instance, hot-press used in this study was different from in industrial plant. We used a column-type hydraulic hot-press while in industrial scale usually use continuous hot-press [3]. Composition of the final board also influenced the final properties of the boards. Mantanis *et al.* [8] reproduced industrial scale of FBs, the optimum condition was reached at 25% recycle fibre and 75% virgin fibre.

Generally, interior panel products, such as binderless (wet process) FBs or FBs bonded UF resins showed dimensional instability particularly when they contacted or exposed to high moist. Thickness swelling and water absorption examinations on these products proved this phenomenon. Here, either FBs or rFBs was bonded by UF resins. UF resin itself has prone to hydrolytic degradation therefore, dimensional instability was inevitable. Mechanism of degradation of this product included penetration of water molecules into the UF resin within the bond-line either in inter-phase or interface. This situation initiates hydrolysis reaction chain; water attacks the hydroxymethyl ($CH_2OH$) end group, and then splits either C-N-C or C-N-O bonds from the methylene or ether linkages in UF resin [16]. These occurrences resulted in fibre movement, swelling and shrinking thus chemical species degrades into a simpler species [17]. In other words, sol fraction leaches and dissolves in water [18]. Breakage of FBs here enables this board for recycling thus resulted in either simpler species or sol fraction can be used as source of nutrient for plants.

**Table 2**. Mechanical properties of FBs versus rFBs

| Mechanical properties | FBs | rFBs | JIS |
|---|---|---|---|
| MoE (N/mm$^2$) | 1375$\pm$ 166 | 290$\pm$126 | Min 800 for type 5 |
| MoR (N/mm$^2$) | 27.8 $\pm$ 2.7 | 3.4$\pm$ 2.4 | Min 25 for type 25 |
| IB (N/mm$^2$) | 0.25 $\pm$0.13 | 0.02 $\pm$ 0.01 | Min 0.2 for type 5 |

Remarks: resulted data were measurement of 8 replications for FBs and 6 replications for rFBs, respectively

All mechanical properties of industrial/commercial FBs met one of standard types of FBs, however, all mechanical properties belonged to rFBs were fail. The FBs standard is divided into 4 (four) group type of FBs, namely type 5, 15, 25 and 30 according to the bending strength. Poor strength of rFBs was presumably because of difference in reproduction method. Indeed, laboratory scale was different from industrial scale from raw materials, for example in application of gluing and pressing [8]. Because of owing lower strength of industrial/commercial FBs, this product has had also life used shortly. In order to predict and compare the life used of FBs and rFBs, results of cyclist test here were also presented.

### *3.2. Cyclist test*

Cyclist test is an accelerated aging test to predict the life use of the board. It has significant effects on the safety and serviceability of the board over their lifetime [19]. Presented here is a comparison between cyclist tests exposed to FBs versus rFBs (Figure 1). It was clearly showed that Fig 1a exhibits a longer time for FBs undergoing to damage compared to that of belonged to Figure 1b. FBs needed 3 (three) cyclist longer to breakage compared to that of rFBs. FBs were also denser than that of rFBs, therefore FBs were more resistant to water.

Both cyclist tests of FBs and rFBs curves showed similar trends; after one cycle the thickness tend to increase continuously. Previous work on particleboards [20] showed distinct tendency, presumably because wood particles still have greater hysteresis properties comparing to those of wood fibres.

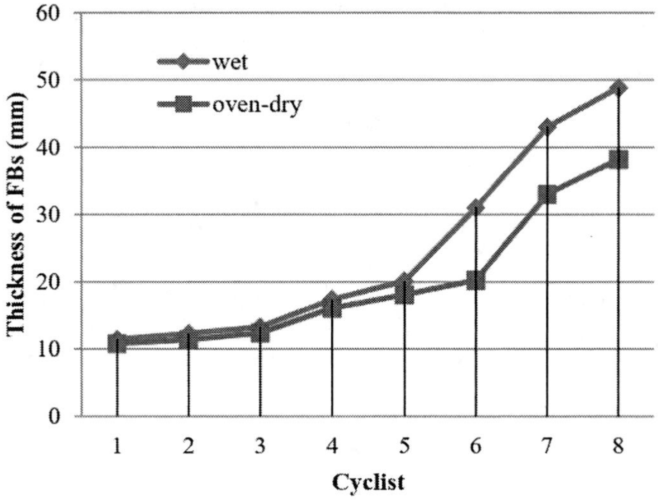


**Figure 1a.** Cyclist test curve measurement of FBs

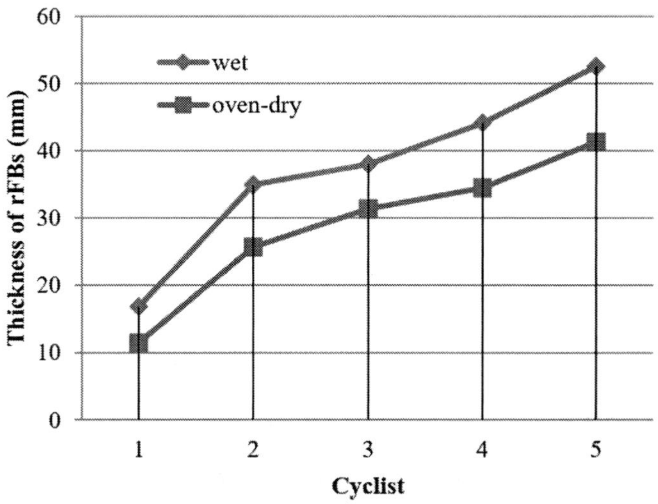


**Figure 1b.** Cyclist test curve measurement of rFBs

Related to wood fibre, there were two kinds of recycle fibre derived from FBs recycling, namely individual fibre and cut fibre or cell fragments, respectively. Both fibres were presented in analysis of recycling wastage section as follow.

3.3. *Analysis of recycling wastage.*

Wastage of recycling activity may be placed in the following categories, namely suspended solid (hereafter fibres and cell fragments), colloidal organic substances originating from the wood material, dissolved solids (soluble organic substances released from the wood material), soluble and insoluble chemical added during the manufacturing of FBs including adhesive resin [3]. As depicted in Figure 2, typical morphology of recycle fibre derived from recycling activity of FBs was shown. Result of measurement of length and diameter fibres were 1000.51±339.21 μm and 25.98±7.93 μm, respectively. Cell fragments and other aforementioned wastage could not be identified except their distribution of size. Figure 3 showed distribution of cell fragments and other aforementioned litter or contaminant found in water disposal of recycling FBs activity in this study. This group was consisted of micro solid waste.

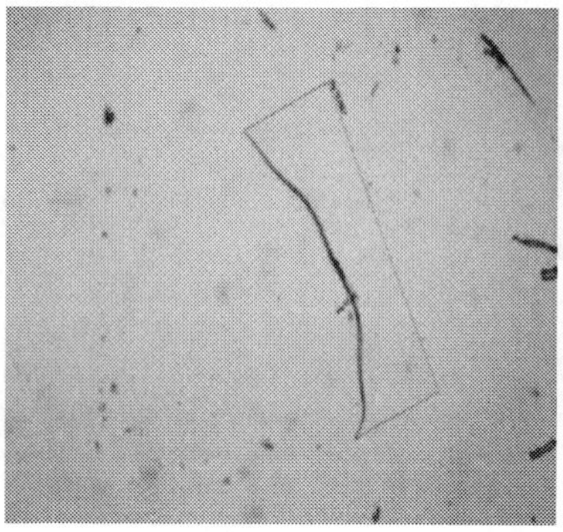

**Figure 2.** Typical individual recycling fibre derived from recycling FBs under microscope measurement

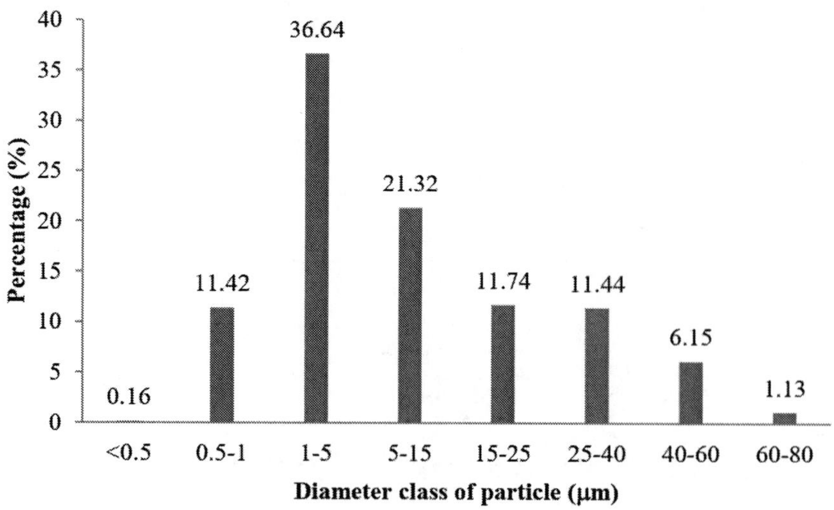


**Figure 3.** Distribution of contaminant within water disposal of FBs recycling

Further, N content as part of sol fraction or chemical found in water disposal was analysed. Using Kjeldahl method, N content amount was 0.05%. This percentage was relatively small compared to that of previous study in particleboard [18] because of amount UF resin applied, type of wood fibre/ particle and amount of water for soaking. All of these have influenced later if the water disposal will be used as liquid fertilizer.

## 4. Conclusion

Remanufacturing of rFBs was feasible even though the resulted quality was poor. Improvement properties of rFBs could be carried out depended on technical production such as application of gluing and pressing. Water or liquid disposal containing impurities can be utilized further for fertilizer because N content exist. Further study was recommended for implementing this finding such as for liquid fertilizer of non-food seedling.

## References

[1] Indrayani Y, Setyawati D, Umemura K, and Yoshimura T 2015 Decay resistance of Medium Density Fibreboard (MDF) made from pineapple leaf fiber *J. Math. Fund. Sci.***47**(1) 76-83

[2] Lee Y-K and Kim H-J 2012 Effect of temperature and bake-out on formaldehyde emission from UF bonded wood composites *J. Korean Wood Sci. Technol.* **40**(2) 91-100

[3] Suchsland O and Woodson GE 1986 *Fiberboard Manufacturing Practices in the United States* (Madison: USDA Forest Service Agriculture Handbook No.640) p. 263

[4] Lee M, Prewitt L, and Mun SP 2014Formaldehyde release from medium density fibreboard in simulated landfills for recycling*J. Korean Wood Sci. Technol.***42**(5) 597-604

[5] Abdel-Shafy HI and Mansour MSM 2018 Solid Waste issue: sources, composition, disposal, recycling, and valorization *Egyptian J. Petroleum* **27** 1275-90

[6] US EPA 2005 *Municipal Solid Waste in The United States: 2003 Facts and Figures* (Washington DC: US Environmental Protection Agency)

[7] Cesprini E, Resente G, Causin V, Urso T, Cavalli R, and Zanetti M 2020 Energy recovery of glued wood waste: a review *Fuel.* **262** 116520

[8] Mantanis G, Athanassiadou E, Nakos P, and Coutinho A 2004 A new process for recycling waste fibreboards *38th International Wood Composites Symposium* pp.119-22

[9] Hong M-K, Lubis MAR, Park B-D, Sohn CH, and Roh J 2018 Effects of surface laminate type and recycled fiber content on properties of three-layer medium density fibreboard *Wood Mater. Sci. Eng.* **15**(3)163-71

[10] Lubis MAR, Hong M-K, Park B-D, and Lee S-M 2018 Effects of recycled fiber content on the properties of medium density fibreboard *Eur J. Wood Wood Prod.***76** 1515-26

[11] Nuryawan A, Singh AP, and Park BD 2015 Swelling Behaviour of cured urea–formaldehyde resin adhesives with different formaldehyde to urea mole ratios *J. Adhes.* **91** 677-700

[12] Grigsby WJ, Carpenter JEP, and Sargent R 2014 Investigating the extent of urea formaldehyde resin cure in medium density fiberboard: resin extractability and fiber effects *J. Wood Chem. Technol.* **34** 225-38

[13] Lubis MAR, Hong MK, and Park BD 2018 Hydrolytic removal of cured urea–formaldehyde resins in medium-density fiberboard for recycling *J. Wood Chem. Technol.* **38**(1) 1-14

[14] [JSA] Japanese Standard Association 2003 *Japanese Industrial Standard for Fibreboard JIS A 5905:2003* (Yokohama: Japanese Standard Association)

[15] McNatt JD and Laufenberg TL 2000 *How Moisture Changes Affect Long-Term Load Performance of Wood-Base Panels in I-Beam and Other Structural Composite Members* (Madison: USDA Forest Service, Forest Products Laboratory)

[16] Lubis MAR and Park B-D 2018 Analysis of the hydrolysates from cured and uncured urea-formaldehyde (UF) resins with two F/U mole ratios *Holzforschung* **72**(9) 759-68

[17] Chuang IS, and Maciel GE 1994 NMR study of the stabilities of urea–formaldehyde resin components toward hydrolytic treatments *J. Appl. Polym. Sci.* **52** 1637-51

[18] Nuryawan A, Rahmawaty, Tambun KDS, Risnasari I, and Masruchin N 2020 Hydrolysis of particleboard bonded with urea-formaldehyde resin for recycling *Heliyon* **6** e03936
[19] Zhou Y, Fushitani M, and Kandem DP 2001 Bending creep behavior of medium density fiberboard and particleboard during cyclic moisture changes *Wood Fiber. Sci.* **33**(4) 609-17
[20] Nuryawan A, Siringo-ringo FJ, Nasution TI, Banurea R, Hartini KS, and Elfiati D 2020 Cyclist test on particleboards and utilization of their liquid disposal for fertilizer*Paper submitted to Iconart2 Seminar Universitas Sumatera Utara Medan*

**Acknowledgment**
This research was funded by Directorate for Research and Community Service, Ministry of Education and Culture, Republic of Indonesia, through scheme of Research Application of Higher Education Competitive (*Penelitian Terapan Unggulan Perguruan Tinggi*) fiscal year of 2020. This research is part of 2$^{nd}$ year a grant research of 2019-2021.

IOP Conf. Series: Materials Science and Engineering **935** (2020) 012061 doi:10.1088/1757-899X/935/1/012061

# Improvement of physical properties of jabon (*Anthocephalus cadamba*) through the impregnation of nano-$SiO_2$ and melamin formaldehyde furfuril alcohol copolymer

**E Prihatini[1], A Maddu[2], I S Rahayu[1*], M Kurniati[2], and W Darmawan[1]**

[1]Department of Forest Product, Faculty of Forestry, IPB University, Bogor, 16680, Indonesia

[2]Department of Physics, Faculty of Mathematics and Natural Sciences, IPB University, Bogor 16680, Indonesia

*Email: istiesr@apps.ipb.ac.id

**Abstract.** Wood polymer nano composites (WPnC) based on nano-$SiO_2$ were prepared by the impregnation of nano-$SiO_2$ and melamine formaldehyde-furfuryl alcohol copolymer. The objectives of this research were to analyze the effect of impregnation of nano-$SiO_2$ and Melamine Formaldehyde Furfuryl Alcohol (MFFA) copolymers on the physical properties of jabon wood and to characterize treated jabon wood. Impregnation method improved the physical properties of jabon wood . Density of jabon become 0.56 g/cm$^3$, weight percent gain (WPG) is 68.34%, Anti-Swelling Efficiency (ASE) is 46.66%, bulking effect (BE) is 7.55%, water uptake (WU) is 75.49%. WPnC composites were characterized by Fourier Transform Infrared Spectroscopy (FTIR), XRD (X-ray Diffraction), SEM (Scanning Electron Microscopy), and SEM-EDX (Energy-dispersive X-ray Spectroscopy). FTIR results showed that the Si-O-Si asymmetric stretching vibration hand slighty shifts toward a higher wave number. It indicated a reaction between nano-$SiO_2$ and wood. XRD studies indicated a decrease in crystallinity of the composites. SEM images observed distribution of nano $SiO_2$ in the composites.

## 1. Introduction

Wood is a hygroscopic and anisotropic biopolymer composite which consists of cellulose, hemicellulose and lignin. Those main components are responsible for most of the physical, mechanical and chemical properties of wood. Wood is an important raw material for buildings and industries in terms of its strength, aesthetics, and low processing costs. However, wood has several disadvantages, such as high water absorption, low durability and changes in dimensions that vary with changes in environmental conditions [1]. Wood is a hygroscopic material because it has a hydroxyl group (-OH) that attracts water molecules through hydrogen bonds in the cell wall polymer [2]. This causes the dimensions of wood keep changing according to moisture content in the cell wall. Anisotropic is wood dimensions alteration due to alteration environmental moisture in three different orientations (longitudinal, radial and tangential). The largest shrinkage percentage is tangential followed by radial and longitudinal directions (the smallest shrinkage percentage). Differences in dimensional changes in these three orientation directions often cause problems in their utilizations.

Another property that can limit the utilization of wood is the presence of juvenile wood. Juvenile wood can experience a longitudinal shrinkage of 10 times that of mature wood. This tends to produce defect such as warping and large checks [3]. Juvenile wood is a xylem mass that are formed in the first few years [4].

Fast-growing timber in Indonesia comes mostly from community forests and community plantations. Fast growing wood species that are widely planted, including sengon (*Falcataria moluccana*), jabon (*Antocephalus cadamba*), africa (*Maesopsis eminii*), mindi (*Melia azedarach*), fruit trees, and others. Fast growing wood has a short cutting cycle (around 6-7 years). At present, the area of community forests in West Java reaches 314.110 ha with a production of 1.3 million $m^3$ [5]. Based on [6], the total potential of forest stands in West Java with a diameter of wood above 20 cm is 27.3% of the overall potential on the island of Java. West Java has an area of production forest of 393 thousand ha or equal to 20% of the total production forest area of Java Island (1.97 million ha). This production area is a huge potential in supporting the supply of wood raw materials for the forestry industry in West Java Province.

Jabon is one of the fast-growing wood species that is widely planted in community forests. Jabon wood has the advantage of being fast growth, easy to adapt to various growing sites and relatively easy in silvicultural treatment. Jabon wood categorize as light wood with several benefits including: raw materials of plywood, lightweight construction, floors, pulp and paper, ceilings, boxes, crates, toys, carvings, lighters, chopsticks and pencils. Jabon has a rather fine wood texture to a bit rough, straight fibrous, less shiny and odorless. Jabon wood has a specific gravity of about 0.42 (0.29–0.56), including Strength Class III-IV, and Durability Class V [7]. Based on Darmawan *et al.* [8], at the age of 7 years, the wood density ranges from 250-580 $kg/m^3$, and has 100% juvenile wood.

Rapid growth will trigger a low density of wood, low wood strength, lots of knots, and a large portion of juvenile wood. Juvenile wood has inferior properties when compared to mature wood [9]. One way to improve the quality of fast-growing wood is to modify wood.

Wood modification is a promising method for improving the quality of fast-growing wood. Efforts to strengthen solid wood with polymers have been carried out in the last few decades. Thermoplastic and thermosetting systems have been used and have succeeded in improving some of the properties of wood, but both show limitations [1,10,11]. Wood modification that has been performed a lot is impregnation with furfuryl alcohol. Furfurylation has succeeded in increasing dimensional stability of wood [12,13].

Nanotechnology-based treatment of wood modification provides new opportunities for improving the quality of fast-growing timber. Wood polymer nano composite (WPnC) that utilizes phenol formaldehyde (PF) and montmorillonite (MMT) impregnation can increase dimensional stability and fire resistance [14]. A mixture of urea formaldehyde (UF) and nano-$SiO_2$ resins can improve the properties of poplar wood, reduce water absorption and increase fire resistance [15]. Addition of nano-$SiO_2$ to melamine formaldehyde-furfuryl alcohol (MFFA) showed that nano-$SiO_2$ can also reduce water absorption, improve mechanical properties, fire resistance, and thermal stability of *Ficus hispida* wood [16]. Nano-$SiO_2$ was treated with FA, experienced increased hardness, water uptake and anti-swelling efficiency in poplar wood [17]. Based on research Rahayu *et al.* [18], nano-$SiO_2$ can effectively increase the dimensional stability and density of sengon wood.

Wood furfurilation can increase the dimensional stability of wood but does not bring changes to the flexural strength and Modulus of Elasticity (MOE) [16]. To fix it, melamine formaldehyde is then used. Melamine formaldehyde resin (MF) is one of the hardest and most rigid polymers that can improve the mechanical and thermal properties of wood [19]. Impregnation of solid wood with water-soluble MF resins has caused a significant increase in surface hardness and MOE [11]. In this research, wood impregnation will be carried out with a mixture of melamine formaldehyde furfuryl alcohol and nano-$SiO_2$.

The objectives of this research were to analyze the effect of impregnation of nano-$SiO_2$ and Melamine Formaldehyde Furfuryl Alcohol (MFFA) copolymers on the physical properties of jabon wood and to characterize treated jabon wood.

## 2. Materials and Methods

### *2.1 Materials*

The sample used in this research was 6 years old jabon wood (*Anthocephalus cadamba*) originated from a community forest in Sukabumi, West Java. The tree used as research material has a 7 meter free branch height and diameters of 28 cm. The chemicals used are melamine, formaldehyde, furfuryl alcohol, NaOH, nano-$SiO_2$ (particle diameter 15 ±5 nm), maleic anhydride, pH paper and distilled water.

The tools used are analytical balance, calipers, ovens, fans, sonicators, impregnation devices, moisture meters and aluminum foil. The characterization of the impregnated jabon wood was carried out with SEM (Scanning Electron Micrograph), SEM-EDX (Scanning Electron Micrograph - energy X-ray spectroscopy), XRD (X-ray diffraction), and FT-IR (Fourier Transform Infrared Spectrometry) tools.

### *2.2. Methods*

#### 2.2.1 Test sample preparation

Jabon wood is sawn without distinguishing heart wood and sapwood. The test sample size used is 2 x 2 x 2 cm and density and dimensioanal stability testing is carried out in form weight percent gain (WPG), bulking effect (BE), water uptake (WU), and anti-swelling efficiency (ASE), with 5 replication for each treatment.

#### 2.2.2 Preparation of impregnant solutions

Preparation of MFFA copolymers. The method created refers to [16] and [20]. Melamine, formaldehyde, furfuryl alcohol (1:3:5 mole ratio) are placed into the flask. The first step in the reaction is the pre-reaction of melamine and formaldehyde to hydroxymethylated melamine. The pre-reaction starts when the pH of the medium is adjusted by dripping 10% NaOH solution until the pH ranges between 9.4-9.6 and the temperature is slowly raised to 98°C. When the solution is clear, furfuryl alcohol is added. After the reaction has been running for 10 minutes, anhydrous maleic catalyst 1.5% is added, and the reaction is continued for 5 minutes. For impregnation, this MFFA copolymer is made with a concentration of 50%.

Nano-$SiO_2$ dispersion in MFFA copolymers. This method refers to [16] by making a sol-gel system. Nano-$SiO_2$ was soaked in FA solution for 24 hours with mechanical stirring, then sonicated for 30 minutes. After that, MFFA copolymers were slowly added to nano $SiO_2$ dispersions under stirring conditions. Mixture of MFFA and nano-$SiO_2$: FA using a volume ratio of 1: 1. Furthermore, this mixture was synchronized for 15 minutes. This mixture is made with three nano-$SiO_2$ compositions (0.5, 1 and 1.5%), this is done to find the most optimal composition for wood impregnation.

#### 2.2.3 Wood polymer nano-composite preparation (WPnC)

This process is a furfurilation process. The furfurilation process consists of three steps, that are impregnation, polymerization and drying. Impregnation is performed by adapting the [16] methods. In this research, wood samples were initially dried, measured and weighed. Wood samples consisting of untreated and treated were put and immersed in impregnation solutions in different containers and put in an impregnation tube. Impregnation was started with a vacuum of 0.5 bar for 1 hour, followed by a pressure of 2 bar for 2 hours. The polymerization process is carried out by cleaning the impregnant residual on wood samples surfaces, then wrapped in aluminum foil, and dried at a temperature of 90°C for 24 hours. For the drying process, after the aluminum foil is opened, the sample is re-dried with a temperature of 103 ± 2°C for 24 hours. After polymerization process completed, samples were measured and weighed to determine density, WPG, BE, WU [2] and ASE [21].

#### 2.2.4 Analysis data

Data analysis was performed using a experimental design that is a simple randomized complete design (CRD) with 1 factor, that is the factor of treatment variation with 4 levels that is MFFA, MFFA nano-$SiO_2$ 0.5%, MFFA nano-$SiO_2$ 1% and MFFA nano-SiO 1.5%. The analysis was carried out by using the

IBM SPSS Statistics (Statistical Package for Service Solutions) version 25.0 calculation program and continued with the Duncan test at $\alpha = 5\%$ if there were significant differences.

2.2.5 Characterization of treated jabon
Determination of the presence of a mixture of MFFA and nano-$SiO_2$ particles in jabon wood was carried out using SEM (Scanning Electron Micrograph), SEM-EDX (Scanning Electron Micrograph - energy X-ray spectroscopy), XRD (X-ray diffraction), and FT-IR (Fourier Transform Infrared Spectrometry).

## 3. Results and Discussion

### *3.1. Effects of MFFA nano-$SiO_2$ impregnation treatment on jabon wood*

Impregnation using MFFA copolymers was proven in increasing jabon physical properties parameters including density, weight percent gain (WPG), bulking effect (BE), water uptake (WU), and anti-swelling efficiency (ASE) (Table 1). Jabon wood density values increase, this is in line with the increasing value of WPG. The higher the WPG value, the higher the density of jabon wood. The increase in WPG value is also directly proportional to the value of BE. But the increase in WPG value is inversely proportional to WU. This is presumably because the impregnant will reduce the ability of wood to absorb water. Hazarika and Maji [16] state that when wood is filled with MFFA, the polymer will fill lumens and empty pits resulting in increased density, WPG, ASE, BE, and decrease in WU. This is in line with the research of Yao *et al.* [20] which showed that impregnation with melamine formaldehyde modified with furfuryl alcohol succeeded in increasing density and WPG of Chinese fir. The addition of nano-$SiO_2$ in MFFA has also been proven to be able to improve its physical properties parameters (when compared to MFFA only). The higher the nano-$SiO_2$ concentration, the higher the physical parameter values [16]. Nano-$SiO_2$ will fill the empty lumens of the wood thereby reducing its water absorption capacity. Jabon wood with the addition of 0.5% and 1% nano-$SiO_2$, increase its physical properties parameters, but at a nano-$SiO_2$ concentration of 1.5% there was a decrease. The decrease in the value of physical properties parameters in the mixture is likely due to the change in the shape of the copolymer mixture into a gel so that it was difficult to penetrate into wood cells. Based on [22] with changes in pH and the distance between close particles due to solvent evaporation, the surface charge will decrease and gelation will take place.

**Table 1**. Density and dimensional stability of untreated and treated jabon wood

| Sample | WPG | BE | Density | WU | ASE |
|---|---|---|---|---|---|
| Untreated | 0.56 (±0.46) | 2.22 (±0.71) | 0.32 (±0.01) | 215.79 (±5.37) | |
| MFFA | 45.83$^a$ (±5.17) | 5.63$^a$ (±2.11) | 0.50$^a$ (±0.04) | 85.60$^b$ (±5.84) | 20.97$^a$ (±5.24) |
| MFFA nano-$SiO_2$ 0.5% | 68.34$^b$ (±5.07) | 7.55$^b$ (±1.00) | 0.56$^b$ (±0.04) | 75.14$^a$ (±5.09) | 46.68$^b$ (±4.29) |
| MFFA nano-$SiO_2$ 1.0% | 69.97$^b$ (±6.02) | 8.15$^b$ (±0.61) | 0.57$^b$ (±0.45) | 72.87$^a$ (±5.92) | 47.99$^b$ (±5.22) |
| MFFA nano-$SiO_2$ 1.5% | 67.33$^b$ (±5.74) | 7.31$^{ab}$ (±1.47) | 0.56$^b$ (±0.03) | 77.69$^a$ (±6.75) | 42.71$^b$ (±5.29) |

Note: The values in parentheses () indicate the standard deviation values. The letter after the value shows Duncan's further test results

The results of statistical analysis showed that the wood impregnated with MFFA copolymer and nano-$SiO_2$ was significantly different compared to MFFA treatment. Jabon wood which experienced the highest increase in physical property parameter values was impregnated with MFFA nano-$SiO_2$ 1%. However, based on the results of statistical analysis with Duncan's further tests (Table 1) shows that there are no significant differences in the three nano-$SiO_2$ concentrations.

### *3.2. FTIR test*

Figure 1 shows the FTIR spectrum of nano-$SiO_2$, a sample of untreated and treated wood. Untreated jabon wood (curve a) is marked with absorption bands at 3425 $cm^{-1}$ (–OH stretching), 2916 and 2843 $cm^{-1}$ (–$CH_2$ asimetric stretching), 1744 $cm^{-1}$ (C = O stretcing), 1666 $cm^{-1}$ for (–OH bending), 1026 $cm^{-1}$ (C – O stretching). The presence of nano-$SiO_2$ in composites causes a decrease in the intensity of hydroxyl groups and carbonyl groups. The peak intensity of the -CH group of treated wood is lower than untreated. The intensity of the carbonyl group also decreases due to the formation of hydrogen bonds with hydroxyl groups on the surface of silica [23]. For Si-O-Si groups, peak forms at different wave numbers. Nano-$SiO_2$ for impregnation has a peak at wave number 810 $cm^{-1}$ (symeric stretching). This value is the same as the research of [24]. In wood, the Si-O-Si group has a peak at wave number 1018 $cm^{-1}$ (asymeric stretching). Peak shifts to higher wave numbers indicate an interaction between nano-$SiO_2$ and wood. A shift towards larger numbers occurs when particle size increases [25]. Hydrogen bonding with the Si-O-Si group causes a reduction in the hydroxyl group. Shifting the wave number in a larger direction indicates the presence of a hydroxyl group replaced by the Si-O-Si group. This has reduced the ability of wood to absorb water. The entry of Si-O-Si also increases the weight of the wood so that WPG and its density will also increase.

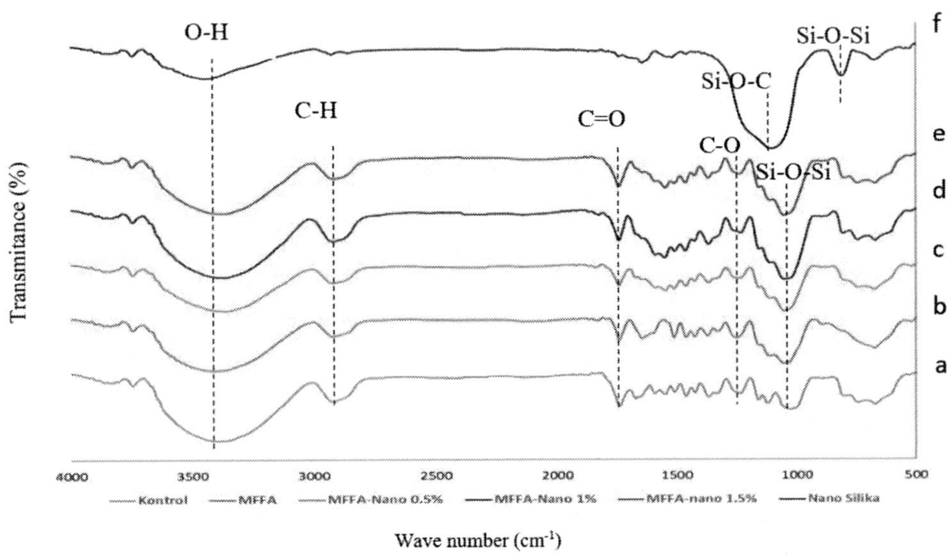


**Figure 1**. FTIR spectrum; (a) Untreated, (b) MFFA, (c) MFFA nano-$SiO_2$ 0.5%, (d) MFFA nano-$SiO_2$ 1%, (e) MFFA nano-$SiO_2$ 1.5% and (f) nano $SiO_2$ powder

### *3.3. XRD test*

From the XRD curve, untreated wood shows a sharper peak so that the crystallite value is high (a). Untreated wood exhibits broad diffraction peaks at 2Θ 22.84° with cellulose crystal fields 012. Two additional small peaks appear at 2Θ 16° and 44.54° with crystal fields 020 and 050 respectively. After being treated with MFFA and MFFA nano-$SiO_2$ decreased peak intensity (b,c,e). In jabon wood with MFFA nano-$SiO_2$ 0.5% treatment, at an angle of 2Θ of $16^{o}$ the graph tends to be flat. This indicates a decrease in the crystallinity of wood cellulose. The graph shows nano $SiO_2$ and MFFA having broad diffraction peaks, this shows both amorph (f and g).

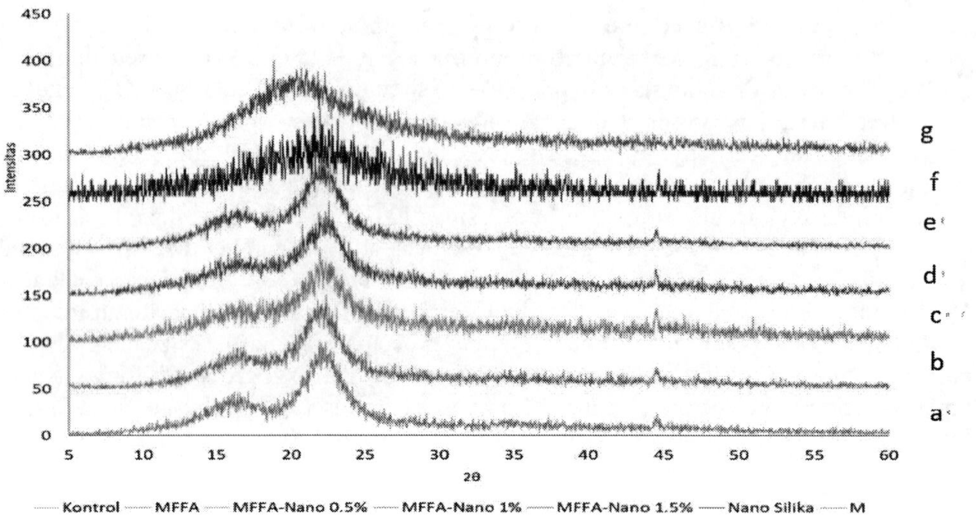


**Figure 2.** XRD results of (a) Untreated, (b) MFFA , (c) MFFA nano-$SiO_2$ 0.5%, (d) MFFA nano-$SiO_2$ 1%, (e) MFFA nano-$SiO_2$ 1.5% treated jabon wood, (f) nano-$SiO_2$ powder, and (g) MFFA copolymer

### *3.4. Results of degree of crystallinity test*

The degree of crystallinity is shown in Table 2. Cellulose from untreated wood shows the highest degree of crystallinity. The lowest degree of crystallinity was shown by wood samples treated with MFFA nano-$SiO_2$ 0.5%. This is consistent with the results of the research of [16] which states that impregnation of nano silica with MFFA copolymers can reduce the degree of crystallinity of cellulose. Rabee *et al.* [24] also stated that coating surface fibers with nano silica reduced the degree of crystalinity by 15%. Tang *et al.* [26] show that the addition of $SiO_2$ increases the formation of hydrogen bonds with the fiber interface. Increased formation of hydrogen bonds with the Si-O-Si group causes the hydroxyl group to decrease. Shiraishi *et al.* [27] reported that chemical graft reactions occur in the amorphous regions of wood cellulose. The polymer reacts on the surface of the crystallite thereby opening several hydrogen-bound cellulose chains, so that more amorphous cellulose is produced. Decreasing the degree of crystallinity indicates the formation of hydrogen bonds between silica and wood, so that the hydroxyl group will decrease. This is in line with FTIR results. Decreasing the crystallinity degree will increase the physical parameter value of the wood. For wood that treated with MFFA nano-$SiO_2$ 1.5% showed a greater degree of crystallinity than those treated with MFFA and MFFA nano-$SiO_2$ (0.5 and 1%). This is probably due to the gelation process which makes it more difficult for the copolymers to penetrate into the wood. Nevertheless, the degree of crystallinity is still smaller when compared to untreated wood.

**Table 2**. Cristalinity degree results of untreated and treated jabon wood

| Treatment | Degree of crystallinity (%) |
|---|---|
| Untreated | 28.99 |
| MFFA | 23.42 |
| MFFA nano-$SiO_2$ 0.5% | 19.79 |
| MFFA nano-$SiO_2$ 1% | 19.99 |
| MFFA nano-$SiO_2$ 1.5% | 23.56 |
| Nano Silica | 21.17 |

### *3.5. SEM test results*

Figure 3 shows SEM images of untreated and treated jabon wood samples. From the picture it can be observed that there are empty lumens and pits on cell walls (Figure 3a and 3b). The impregnation method causes the impregnant material to deposit on the wood cell walls. The presence of polymer deposits in cell walls and pits can be seen in treated wood samples with MFFA (Figure 3c) and MFFA nano-$SiO_2$ (Figure 3d-3f). The pits which previously had empty space, after being treated with MFFA, became evenly closed. Indications for the presence of nano-$SiO_2$ can be detected by several white deposits located in the cell walls and pits of treated jabon wood by MFFA nano-$SiO_2$ (Figure 3d-3f). Nano-$SiO_2$ can enter the pits and cover almost evenly on cell walls. Figure 3d-3f, showed the increase of white deposits as the increase nano-$SiO_2$ concentration.

Additional information obtained from the analysis with SEM is EDX data, which is explained in Table 3. The EDX analysis results from untreated and MFFA treated jabon wood showed no silica content. Jabon wood that was impregnated with MFFA nano-$SiO_2$ showed the presence of silica content. Silica levels in wood impregnated by MFFA nano-$SiO_2$, increased with the increase in the concentration of nano-$SiO_2$.

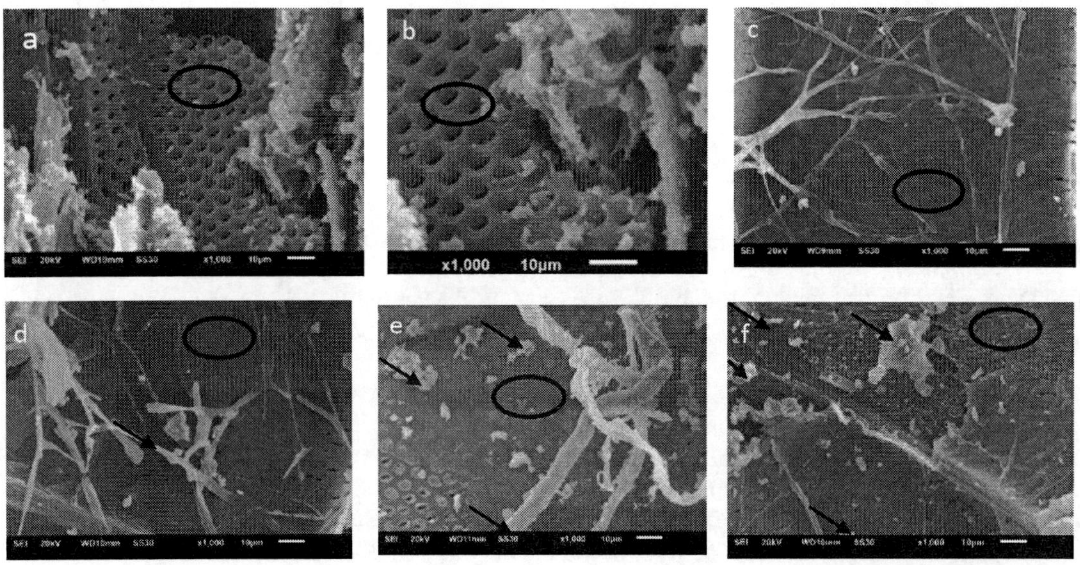


**Figure 3**. Scanning Electron Micrograph images of jabon vessel wall (magnification 1000x) of (a, b) Untreated (c) MFFA, (d) MFFA nano-$SiO_2$ 0.5%, (e) MFFA nano-$SiO_2$ 1%, and (f) MFFA nano-$SiO_2$ 1.5%

**Table 3**. SEM-EDX results of untreated and treated jabon wood

| Treatment | Silica content (mass %) |
|---|---|
| Untreated | 0 |
| MFFA | 0 |
| MFFA nano-$SiO_2$ 0.5% | 0.30 |
| MFFA nano-$SiO_2$ 1% | 0.42 |
| MFFA nano-$SiO_2$ 1.5% | 0.60 |

## 4. Conclusion

Wood samples were impregnated with MFFA and nano-$SiO_2$ copolymers in vacuum and pressure conditions. Impregnation of wood with MFFA and nano-$SiO_2$ copolymers can improve physical properties of the jabon wood parameter, namely increase density, WPG, BE, ASE, and an decrease in WU.

Characterization of MFFA and nano-$SiO_2$ treated samples were carried out by FTIR, XRD, SEM and SEM-EDX. FTIR analysis results shifting wave number in nano-$SiO_2$ peak as an indication of bond between nano-$SiO_2$ and wood. The results of XRD analysis prove the decrease in crystallinity degree of wood cellulose in MFFA nano-$SiO_2$ treated wood. The presence of MFFA and nano-$SiO_2$ polymers in wood cell walls and cell lumen was seen by SEM images. Based on the physical properties parameters and characterization with FTIR, XRD and SEM, MFFA nano-$SiO_2$ 0.5% is an optimum treatment for 6 years old jabon wood.

**Reference**

[1] Kumar S 1994 Chemical modification of wood *Wood and Fib. Sci.* **26**(2)270-80
[2] Hill C 2006 *Wood modification: chemical, thermal and other processes* (London: John Wiley & Sons)
[3] Senft J, Quanci M, and Bendtsen B 1986 Property profile of 60-year-old douglas-fir proc. of *Cooperative Technical Workshop of Juvenile Wood* pp. 17-28
[4] Clark A, Daniels RF, and Jordan L 2006 Juvenile/mature wood transition in loblolly pine as defined by annual ring specific gravity, proportion of latewood and microfibril angle *Wood and Fib. Sci.* **38(2)** 292-299
[5] [Forest Service of West Java Province] 2019 *West Java Forestry Statistics 2018* (Bandung: West Java Provincial Forestry Service)
[6] [KLHK] 2019 *Environmental and Forestry Statistics 2018* (Jakarta: Ministry of Environment and Forestry)
[7] Martawijaya A, Kartasujana I, Mandang YI, Prawira SA, and Kadir K 2005 *Atlas Kayu Indonesia Volume II* (Bogor: Forestry Research and Development Agency, Ministry of Forestry)
[8] Darmawan W, Nandika D, Rahayu I, Fourier M, and Marchal R 2013 Determination of juvenile and mature transition ring for fast growing sengon and jabon wood *J. of Indian Aca. of Wood Sci.* **10**(1)
[9] Zhang S, Qibin Y, and Beaulieu J 2004 Genetic variation in veneer quality and its correlation to growth in white spruce *J. Can. J. For. Res.* **34**(6) 1311-8
[10] Schneider MH, Brebner KI and Hartley ID 1991 Swelling of a cell lumen filled and a cell wall bulked wood polymer composite in water *Wood and Fib. Sci.* **23**(2) 165-72
[11] Deka M and Saikia C 2000 Chemical modification of wood with thermosetting resin :Effect on dimensional stability and strength property *Bioresource Tech.* **73**(2)179-81
[12] Lande S, Westin M, and Schneider M 2004 Properties of furfurylated wood. *Scand J. For Res.* **19**(5) 22-30
[13] Esteves B, Nunies L and Pereira H 2011 Properties of furfurilated wood (*Pinus pinaster*) *Euro. J. Wood Prod.* **69**(4) 521-5
[14] Xue F and Zhao G 2008 Optimum preparation technology for Chinese fir wood/Ca montmorillonite (Ca-MMT) composite board *For. Stud. China* **10**(3)199-204
[15] Shi J, Li J, Zhow W, and Zhang D 2007 Improvement of wood properties by urea formaldehyde and nano-$SiO_2$ *Frontiers of For. in China* **2**(1)104-9
[16] Hazarika and Maji TK 2013 Properties of softwood polymer composites impregnated with nanoparticles and melamine formaldehyde furfuryl alcohol copolymer *J. of Poly. Engine. and Sci.* **54**(5)1019-29
[17] Dong Y, Yan Y, Zhang S, and Li J 2014 Wood/polymer nanocomposites prepared by impregnation with furfuryl alcohol and nano $SiO_2$ *BioResources* **9**(4) 6028-40
[18] Rahayu I, Darmawan W, Zaini LH, and Prihatini E 2019 Characteristic of fast growing wood impregnated with nano particles *J. For. Res.* **31**(2) 677-85

[19] Gindl W, Hansmann C, Gierlinger N, Schwanninger M, Hinterstoisser B, and Jeronimidis G 2004 Using a water- soluble melamine- formaldehyde resin to improve the hardness of Norway spruce wood *J. of Appli. Poly. Sci.* **93**(4) 1900-07
[20] Yao M, Yang Y, Song J, and Yu Y 2017 Melamine formaldehyde modified furfurylation to improve Chinese fir's dimensional stability and mechanical properties *Bioresources* **12**(2) 3057-66
[21] Rowell R and Ellis W. 1978. Determination of dimensional stabilization of wood by water soak method. *Wood Fib. Sci.* **10** 104-11
[22] Milea C, Bogatu C, and Duta A 2011 The influence of parameters in silica sol-gel process *Engine. Sci.* **4**(1) 59-66
[23] Motaung T and Luyt A 2010 Effect of maleic anhydride grafting and the presence of oxidized wax on the thermal and mechanical behaviour of LDPE/silica nanocomposites *Materials Sci. and Engine*er. **527**(3) 761-8
[24] Rabee J, Santos LP, Menezzi CH, and Tonoli GH. 2018. Effect of nano-silica deposition on cellulose fibers on the initial hydration of the Portland Cement *Bioresources* 3525-44
[25] Li M. Yu Y. Li J, Chen B, Wu X, Tian Y, and Chen P 2015 Nanosilica/carbon composite sphere as anodesin li-ion batteries with excellent cycle stability *J. of* Materials *Chem.* **3**(4)1476-82
[26] Tang C, Li X, Li Z, and Hao J 2017 Interfacial hydrogen bonds and their influence mechanism on increasing the thermal stability of nano-$SiO_2$-modified meta-aramid fibres *Polymers* **9**(10) 504
[27] Shiraishi N, Matsunaga T, and Yokota T 1979 Preparation of higher aliphatic acid esters of wood in an N204-DMF cellulose solvent medium. *J. of App. Poly. Sci.* **24**(12) 2347-59

IOP Conf. Series: Materials Science and Engineering **935** (2020) 012062 doi:10.1088/1757-899X/935/1/012062

# Tannin adhesive made from snakeskin fruit and its application for laminated wood

**G F Virginia, S Suhasman*, and A Agussalim**

Laboratory of Forest Product Utilization and Processing, Faculty of Forestry, Hasanuddin University, Makassar, Indonesia

*Email: suhasman@yahoo.com

**Abstract.** Snakeskin fruit called Salak is one type of tropical fruit that is widely grown in Indonesia. Salak fruit peels are known to contain a certain amount of tannin. This research explores the potential of rind salak fruit as the latest alternative adhesive. The extract from the rind had a tannin content of 1.85% and Stiasny number 76.67%. The results showed that the tannin extract of rind salak fruit was reactive to formaldehyde. Characteristics of Tannin Resorsinol Formadehyde (TRF) adhesives with variation in resorcinol concentration have a significant effect on its viscosity, gelatinization, and solid content. Utilizitaion of recorcinol 15% in TRF adhesive formula resulting a highest bonding strength of produced laminated wood. The value was 67.85 kg/cm$^3$ with 53.3% wood failure on laminated wood. Those result indicated that utilization of tannin based adhesive from skin of salak fruit is very potential as an alternative raw material for bio adhesive.

## 1. Introduction

To reduce the utilization of synthetic adhesives, research on the development of natural-based adhesives was considered. Bio adhesive are frequently named as adhesive of natural origin from plants, animals, or another compound [1].·Tannins product from plants are compounds which usually have been developing natural adhesives. This compound was obtained from the extracts of plant body parts in the form of phenol compounds [2]. Tannin can be found on the rind, stems, leaves, fruit, and roots of the plants [3].

The types of plants used as raw materials for tannins are derived from woody plants such as *Acacia mangium, Pinus merkusii, Rhizophora* sp., and *Intsia spp.* Tannin extract was obtained from bark or part of wood species [4,5]. Since the beginning of the development of tannin as a natural adhesive in wood processing technology comes from wood parts. However, the use of wood and bark parts as tannin-producing materials will face several problems. If using wood parts can disrupt the supply of raw materials for the needs of the wood industry. While some tannin raw materials from agricultural products began to research to find out its potential. Pomace grapes which are processed wastes have tannins which have the potential to become raw materials for natural adhesives that are environmentally friendly [6].

The rind of salak fruit is one of the potential plants to be developed as a tannin based adhesive. Research that conducted by Umi Rahman [7] found tannin content in the extraction of salak fruit rind with a blackish color. Based on statistical data, in 2018, salak fruit production in Indonesia amounted to 896,504 tons [8]. These are interest information to produce the right tannin adhesive formulation.

Therefore, this study discusses the use of the rind of salak fruit rind as a source of tannin-producing raw materials to reduce the use of non-renewable adhesives.

## 2. Materials and Methods

### *2.1. Materials*

Salak's rind as a main material for tannin adhesives was obtained from waste originating in Baraka Sub-district, Enrekang District, South Sulawesi Indonesia. The next adhesive was applied to lamina wood from *Pinus merkusii* and *Paraserianthes falcataria* (L.) Nielsen. The two kinds of wood were taken from sawmills located in Tanralili, Maros Regency, South Sulawesi, Indonesia.

### *2.2. Methods*

2.2.1. Tannin extraction

Chip-shaped Salak rind was first extracted in the water using a water bath with a ratio of 1:4 (w:v) at 70°C of temperature for 4 hours. The tannin level was identified by weighing the extraction about 1-2 g and then put into a 100 mL beaker containing 50 mL of water. It was then heated at 40-60°C for 30 minutes. The cold solution is filtered and put into a volumetric flask with addition water until reaching the target. The solution was put into an erlenmeyer and added with an indigo carmine solution to be titrated with 0.1N $KMnO_4$ solution until the color was changed from blue to golden yellow. The second titration using indigo carmine solution was conducted until the color changes from blue to clearer yellow. The tannin level was then calculated using the following formula:

$$\textbf{Tannin Levels} = \frac{\text{d.f} \times (a-b) \times 0{,}006235}{\text{sample weights}} \times 100\% \quad (1)$$

where:
a = mL number of KMnO4 0.1 N sample
b = mL number of KMnO4 0.1 N blank
d.f = dilution factor
1 mL KMnO4 0,1 N ~ 0.006235 g tannin

The tannin reactivity test of formaldehyde uses the Stiansny number test. Salak rind fruit extract around 0.1g was put into a 150 mL beaker containing 10 mL of air. Around 37% formaldehyde and 10 M HCl were added, respectively 2 mL and 1 mL. The formula was then put into the water bath for 30 minutes and then filtered. The filtrate was heated in the oven at 70-80°C and then weighed [9]. The Stiasny number was calculated by using the following formulation:

$$\textbf{Stiasny number (\%)} = \frac{\text{sediment weights}}{\text{sample weight}} \times 100\% \quad (2)$$

2.2.2 Adhesive characteristics

The tannin extract was made into an adhesive formulation by adding formaldehyde (RF) resorcinol solution to improve the adhesive quality. Tannin adhesives made from salak rind extract (T) was added with 50% (R) resorcinol solution, 37% (F) addition of formaldehyde and 40% NaOH. The adhesive ratio (v: v) of resorcinol formaldehyde (TRF) adhesive consists of 3 variations with different percentages of resorcinol, 100:5:10, 100:10:10, 100:15: 0.

*2.2.2.1. Visual appearance*

The sample was poured on the object-glass and spread over to form a thin film layer. Observations were made visually by looking at the color and whether there are solid granules, coarse particles, or other foreign objects.

*2.2.2.2. Viscosity*
Viscosity testing uses a viscometer. A total 100 mL TRF adhesive was put into a 200 mL beaker with a temperature of 25°C, and then the tool was set at a rotor speed of 100 rad/min with a centipoise (cP) measurement unit. Viscosity values were displayed on the scale of the viscometer.

*2.2.2.3 Gelatination*
A total of 10 mL of salak rind tannin adhesive was put into a closed test tube, then it was placed around 2 cm below the water surface in the water bath at 100ºC. Gelatinization time was obtained by noting the time needed by adhesive to get hard.

*2.2.2.4 pH*
pH of the tannin adhesive was measured by using a pH meter at a temperature of 25°C.

*2.2.2.5 Solids content*
Determination of solids content was conducted by weighing the dry weight of an empty Petri dish (B1), then weighing 10 mL of TRF and then placing it in the cup (B2), then the cupped containing TRF is put into the oven at 103 ± 2ºC for 24 hours. The cup was taken out of the oven and put in a desiccator for 15 minutes then was weighed (B3).

$$\mathbf{SC}\ (\%) = \frac{\mathbf{B3-B1}}{\mathbf{B2}}\mathbf{100} \tag{3}$$

2.2.3. Adhesive application
The lamina produced is made from types of *Pinus merkusii* wood and *Paraserianthes falcataria (L,) Nielsen*. The lamina is then dried conventionally to ± 12% moisture content. The weight of the labor for lamina wood is 200 g/m$^2$ with the application of TRF 0.55 g/surface adhesive called the double spread method. The adhesive that has been applied to the lamina wood is then pressed using a cold press for 24 hours at a pressure of 8 kg/cm$^2$. At this stage also observed the percentage of wood failure.

## 3. Results and Discussion

*3.1. Tannin extraction*
The results of the quote from the content of the tannin extract of salak fruit peel obtained 1.42-2.62%. The extraction result does not produce high tannin content, due to there are also other compositions outside of tannin. Other compounds contained from the salak rind fruit were as alkaloid secondary metabolites, polyphenols, flavonoids, quinones, monoterpenes, and sesquiterpenes [10]. Meanwhile, the results of the Stiasny number test on salak rind were at obtained 76.67%. This value is fulfilling the standard Stiasny FAO [11] yields 75-100%. This value shows the extract of salak rind is quite reactive with formaldehyde.

**Table 1**. Salak rind tannin extract

| Parameter | Score of salak rind tannin extract (%) |
|---|---|
| Tannin content | 1.85 ± 0.66 |
| Stiasny numbers | 76.67 ± 5.77 |

*3.2. Characteristic of tannin adhesive*
The variation of resorcinol levels from the adhesive ratio consists of 5%, 10%, and 15%. Like appreciating resorcinol and formaldehyde, each of which has a concentration of 50% and 37% in this TRF ratio, the rind of salak fruit contributes ratio consists are 1.85%. This value was based on the evaluation results of tannin extract obtained from testing tannin levels in salak rind. The results of the physical properties of tannin adhesives made from salak rind extract for the characteristic features of adhesives were presented in Table 2.

**Table 2.** Analysis of characteristics of tannin adhesive

| Physical base | Tannin Resorcinol Formaldehyde (% resorcinol levels) | | | SNI [18] |
|---|---|---|---|---|
| | 5% | 10% | 15% | |
| Visual appearance | Reddish brown | Brownish yellow | Reddish brown | Liquid and free of impurities |
| Viscosity (cP) | 1.79 | 2.27 | 2.57 | 100-150 |
| Gelatinization (minute) | 109 | 62 | 45 | 60 |
| pH (cP) | 7.63 | 7.33 | 7.40 | 7.6 – 8.6 |
| Solid content (%) | 5.38 | 7.70 | 9.83 | 49– 51 |

3.2.1. Visual appearance

Usually, the base color of tannins is a brownish yellow, but in this study, the color varies according to formaldehyde [12]. Another factor for color variation from tannin adhesives made from salak rind is NaOH added from every adhesive. It's known that tannin adhesive colors with a concentration of resorcinol 5% and 15% both have a pH of 10.28. Whiles the resorcinol 10% concentration has a pH of 10.02%. The other visual appearance of the three formulations is free from dirt and foreign matter.

3.2.2. Viscosity

The results of viscosity the tannin adhesives from salak rind using a viscometer continue to increase along with the increase in resorcinol concentration. The addition of resorcinol accompanied requires monomers in adhesive formulations [13]. The condition of liquid adhesive is also the life of the pot period of adhesive of salak rind [5].

3.2.3. Gelatinization

The condition of the adhesive so liquid that it affects the age of tannin adhesive seen from the results of the gelatinization adhesive. The concentration of resorcinol also affects the time of adhesive gelatination. The more resorcinol added faster the tannin adhesive gelatinization time [5].

3.2.4. pH

The rind of salak fruit extract added with 50% resorcinol solution with 37% formaldehyde has a neutral pH containing acid. NaOH as a catalyst is starting to be used to convert the pH of the adhesive into a base for longer service life with the adhesive polymerization process continues [12]. Because of the pH tannin adhesive, the hydrogen bonding reaction is inhibited [14].

3.2.5. Solid content

The result of this study indicated that the solid content of adhesives formulation tends to increase when the resorcinol content increases. Meanwhile, the relationship between feormaldehyde content with solid content of adhesives formulation indicated that the solid content will decrease when formaldehyde content decreases. The trend that found in this study similar with Rachmawati *et al.* [15] that using mangium bark.

*3.3. Adhesive aplication*

The tannin adhesive made from salak rind is applied to lamina *Pinus merkusii* lamina wood and *Paraserianthes falcataria* (L,) Nielsen which uses stickiness and failure of the wood. The value of the constancy check is presented in Figure 1 as follows:

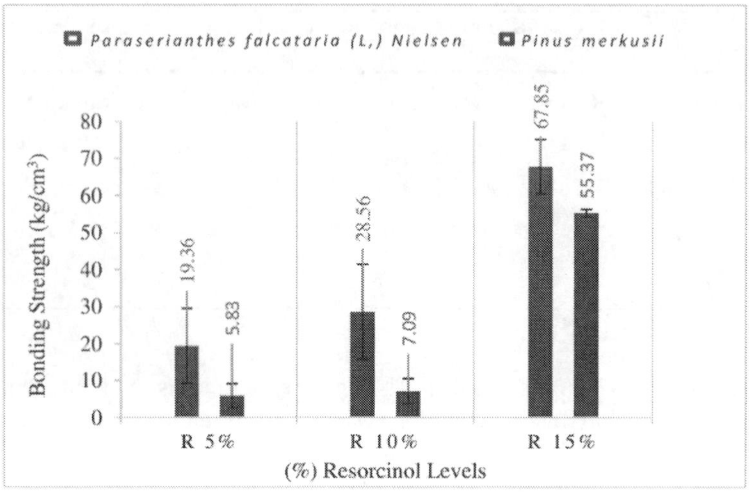


**Figure 1**. Histogram of bonding strength

Three variations of resorcinol levels, namely 5%, 10%, and 15%, were applied to *Pinus merkusii* and *Paraserianthes falcataria* (L.) Nielsen, which have varying average results. The highest bonding strength value is held by *Paraserianthes falcataria* (L.) Nielsen with 15% resorcinol content and then followed by *Pinus merkusii* wood with the same resin. Based on JAS 234-2007 [16], only tannin adhesives with 15% resorcinol content in wood in *Pinus merkusii* and *Paraserianthes falcataria* (L.) Nielsen meets the minimum lamina wood bonding standard of 54 kg/cm$^3$. In general, the addition of resorcinol to adhesives give an effect on its application. Where more concentrations of resorcinol were added, the quality of the adhesive increases [15].

Looking at the results of bonding strength testing of tannin adhesives, the highest percentage of wood failure that reached 53.3% is based on the wood of *Paraserianthes falcataria* (L.) Nielsen from a resorcinol concentration of 15%. The wood density of Pinus merkusii is relatively higher compared to *Paraserianthes falcataria* (L.) Nielsen is also one of the factors that influence the percentage of wood hardness *Paraserianthes falcataria* (L.) Nielsen is higher. Figure 2 shows the visual damage from wood using tannin adhesives made from the rind of salak fruit.

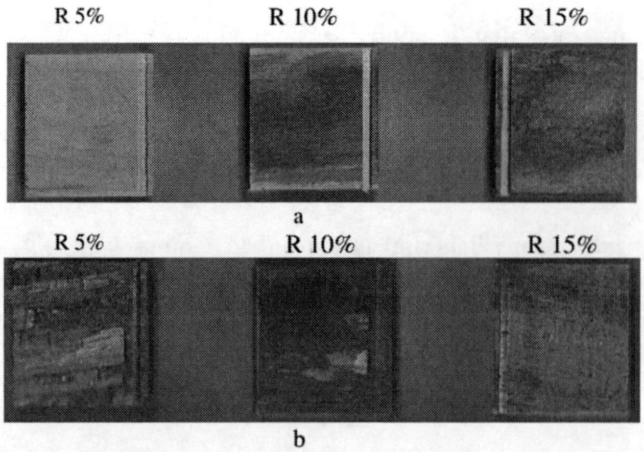


**Figure 2.** Wood damage after wood density testing; a. *Pinus merkusii* and b. *Paraserianthes falcataria* (L.) Nielsen

The results of the study show that tannin extracts up to its application on lamina wood have the potential of salak rind fruit as a tannin adhesive. On a laboratory scale, salak bark as a natural adhesive raw material has relatively high results that meet the standards for the use of adhesives and wood production. This standard then becomes the main requirement for potential adhesive production including its application. By fulfilling the existing standard, the tannin adhesive from salak rind is feasible to be accepted.

The feasibility of further adhesive production for production potential can be looking from the availability of raw material for salak fruit rind. Base on the Badan Pusat Statistik [8] data in catalog No. 5205010 salak products reached 896.504 tons in Indonesia with total rinds of 77.099 tons where the yield of salak rind was 8.6%. From the results of tannin levels from salak rind obtained with the total in Indonesia in 2018, it can be appraiser that the resulting tannin extract is around 2.282 tons in doses. Looking further, the results of tannin extract from salak can produce plywood suitable for production using natural salak rind adhesives as much as 33.064 $m^3$. Indonesia's total plywood production in 2018's according to the BPS [17] 4,213,557 $m^3$ can support 0.7% production using tannin adhesives made from salak leather. Thus, the potential for salak rind as environmentally friendly is quite potential to be used as a natural adhesive. From this good potential, its production strategy does not only prioritize the use of its fruit, including snakeskin.

## References

[1] Ruhendi SDN, Koroh F, Syamani FA, Yanti H, Nurhaida, Saad S, and Sucipto T 2007 *Analysis of Wood Adhesives* (Bogor: Institut Pertanian Bogor)

[2] Pizzi A 1983 *Chemical and Wood Adhesives Technology* (New York: Marcel Dekker. Inc.)

[3] Yusro F 2013 Active levels of tannin extract of Jengkol bark (*Jack Pithecolobium* jiringa) and their reactivity to formaldehyde *Vokasi* **9**(1) 21-6

[4] Santoso A 2005 Utilization of lignin and tannin as alternatives to composite wood adhesive materials *Proceedings of the National Symposium I Polymer* pp. 155-64

[5] Santoso A, Pari G, and Jasni 2015 The quality of lamina board with adhesin resorcinol from merbau wood waste extract *Forest Products Research* **33**(3) 253-69

[6] Ping L, Pizzi A, Guo ZD, and Brossed N 2011 Extraction of condensed tannins from grape pomace: characterization and utilization as wood adhesives for wood particleboard *Industrial Plants and Products* **34**(1) 907-14

[7] Umi R 2016 Effect of salak (*Salacca Zalacca* (Gaertn.) Voss) rind extract on estimation of *Escherichia coli* (Scientific articles) Jambi University

[8] [BPS] Badan Pusat Statistik 2018 *Annual Fruit and Vegetable Plant Statistics* 2018 (Jakarta)

[9] Elbadawi M, Osman Z, Paridah T, Nasroun T, and Canteen T 2015 Particle board properties made from *Acacia seyal* var. seyal using UF - tannin modified adhesives *Cellulose Chem. Technol.* **49** 369-74

[10] Fitrianingsih SPF, Lestari S, and Aminah 2014 Test the antioxidant effect of ethanol extract salakca [*Gaacner Zallacca* (Gaertner) Voss] rind with DPPH reduction method *Science Proceedings Technology and health* **4**(1) 49-54

[11] [FAO] Food and Agriculture Organization 2000 Quantification of tannins in tree leaves Work documents (lab. manual) (Vienna: IAEA) pp. 1-3

[12] Lestari ASRD, Hadi YS, Hermawan D, and Santoso A 2015 Glulam properties of fast-groing species using mahogany tannin adhesive *BioResource* **10**(4) 7419-33

[13] Cowd MA 1991 *Polymer Chemistry* (Bandung: ITB Publisher)

[14] Iskandar MIDA, Prastiwi N, and Wiyantina 2017 Effect of hardener and extender in reorsinol formaldehyde tannin adhesives on plywood formaldehyde emissions *ITEKIMA* **2**(1) 15-26

[15] Rachmawati O, Sugita P, and Santoso A 2015 Synthesis of tannin resinsinol formaldehyde adhesives from mangium tree bark extract to improve palm oil quality *Forest Products Research* **36**(1) 33-46

IOP Conf. Series: Materials Science and Engineering **935** (2020) 012062 doi:10.1088/1757-899X/935/1/012062

[16] [JAS] Japanese Agricultural Standards 2007 *Glued Laminated Wood JIS 234* (Tokyo: Ministry of Agriculture Forestry and fisheries)
[17] [BPS] Badan Pusat Statistik 2018 *Forest Wood Products by Production* ($m^3$) 2002-2018 (Jakarta: Badan Pusat Statistik)
[18] [BSN] Badan Standardisasi Nasional 1998 *Liquid Phenol-Formaldehyde for Plywood Adhesives SNI 06-4567* (Jakarta: Badan Standardisasi Nasional)

IOP Conf. Series: Materials Science and Engineering **935** (2020) 012063 doi:10.1088/1757-899X/935/1/012063

# Effects of compression ratio on physical and mechanical properties of bamboo oriented strand board

**S Maulana[1*], I Sumardi[3], N J Wistara[2], N H Kim[4], and F Febrianto[2]**

[1]Forestry Engineering Program Study, Institut Teknologi Sumatera (ITERA), Lampung Selatan, 35551, Indonesia
[2]Department of Forest Products, Faculty of Forestry and Environment, IPB University (Bogor Agricultural University), Bogor, 16680, Indonesia
[3]School of Life Sciences and Technology, Institut Teknologi Bandung, Bandung 40132, Indonesia
[4]Department of Forest Biomaterials Engineering, College of Forest and Environmental Science, Kangwon National University, Chuncheon 200-701, ReOSBlic of Korea

*Email: sena.maulana@rh.itera.ac.id

**Abstract.** The objective of this study was to evaluate the physical and mechanical properties of a bamboo-oriented strand board (BOSB) under various compression ratios. The raw material of BOSB was betung (*Dendrocalamus asper*) strand that steamed and followed by rinsing with a 1% NaOH solution. BOSB with the dimensions of 30 x 30 x 0.9 $cm^3$ were manufactured with a compression ratio of five levels i.e. 1.15; 1.25; 1.35; 1.44; and 1.54. Three layers of BOSB were made with a shelling ratio of 50:50. Phenol formaldehyde (PF) adhesive (SC=43%) was added with 8% concentration, while paraffin was used as much as 1% of the dry weight of the oven strand. Strand geometry evaluation was carried out on 100 strands randomly. The physical and mechanical properties were evaluated according to JIS A 5908-2003. The results showed that relatively homogeny strand geometry in this research was suitable to manufacture of BOSB. Increasing in BOSB compression ratio up to 1.54 increased mechanical properties. However, the increase of the BOSB compression ratio was decreased the dimensional stability. The optimum BOSB compression ratio was 1.35.

## 1. Introduction

Composite products such as Oriented Strand Board (OSB) are widely developed. Generally, the raw material of OSB is wood. However, wood supply decreases both quality and quantity. Therefore, OSB alternative material search efforts need to be conducted. One of the potential alternative materials for the OSB raw material is bamboo. It's supported by Adrin *et al.* [2] and Chaowana [5] that reported that Bamboo has a match to be used as raw material for biocomposite products, particularly OSB.

Research on bamboo as a raw material of the oriented strand board has been done much. OSB products made of bamboo are called bamboo-oriented strand boards (BOSB). Previous research showed that BOSB had a better mechanical property compared to OSB [1,7]. However, BOSB is still using a relatively expensive, such as MDI. The cheaper adhesives such as PF could be applied, but high concentration was required. Therefore, modifications to the strand were applied to solve this problem [1].

Some research showed that modifications to the strand were successfully carried out to improved BOSB quality. The heat modification in the strand proved to improve BOSB dimensional stability and mechanical properties [1,6,18]. The chemical modifications in the strand after the steam process were done to removed extractive substance on the surface and improved the quality of BOSB [4]. However, the resulting BOSB still has a very high difference in parallel and perpendicular strength. Besides, physical modification i.e. by shelling ratio adjustment successfully carried out to regulated the parallel and perpendicular BOSB strength [7,16].

A series of studies have been conducted indicating that modifications to the raw materials in the form of strand have been done successfully improving the quality of BOSB. Another factor that needs to get attention in the manufacture of BOSB was a comparison between the density of board and raw materials. This comparison value was called the compression ratio. The strength of composite products such as BOSB increased with an increase in compression ratios [15]. However, increased compression ratios require more raw materials. It had been reported that the optimum compression ratio on the particle board was 1.3 [15]. However, there were no scientific reports of the optimum compression ratio value on BOSB. Therefore, the purpose of the study was to evaluate the physical and mechanical properties of BOSB at various compression ratios.

## 2. Materials and Methods

### *2.1. Materials*

The main material used in this research was ±4 years betung bamboo (*Dendrocalamus asper*) 0.52 g/cm$^3$ densities from Sukabumi, West Java, paraffin, PF adhesive (SC=43%). The main equipment used in this research were digital balance, caliper, autoclaves, teflon, rotary blenders, oven, spray guns, hot press, dan Universal Testing Machine (UTM).

### *2.2. Methods*

#### *2.2.1.* Strand treatment

Culms of bamboo with no nodes and bark converted into strand with target dimension 70 x 25 x 0.5 mm$^3$. Strand was given the steam modification treatment at a temperature of 126 °C, pressure 1.4 kg/cm$^2$ for 1 hour and followed by rinsing using a 1% NaOH solution [4]. The modified strand was air-dried to a moisture content of ±15%. Strand was oven dried at 60-80°C for ±36 hours until a moisture content of <5% reached. The strand was measured by 100 samples taken at random to determine strand geometry both the strand aspect ratio (AR) and slenderness ratio (SR) [15].

#### *2.2.2.* BOSB manufacturing

BOSB was made with a dimension 30 x 30 x 0.9 cm$^3$ compression ratio of 1.15; 1.25; 1.35; 1.44; and 1.54. PF adhesive used with a concentration of 8% of the dry weight of the oven strand. BOSB structure was made of three layers with a shelling ratio of 50:50. Paraffin was added as much as 1% of the dry weight of the oven strand. The layers of BOSB was pressed for 9 minutes at a temperature of 135°C with a specific pressure of 25 kg/cm$^2$. The board was conditioned for ±2 weeks to eliminate residual tension.

#### *2.2.3.* Physical and mechanical properties evaluations

Physical and mechanical properties evaluations of BOSB were carried out according to JIS A 5908-2003 standard [13]. Physical properties evaluation include density, Moisture Content (MC), Water Absorption (WA), Thickness Swelling (TS). Mechanical properties evaluation includes Modulus of Elasticity (MOEs) and Modulus of Rupture (MOR) in parallel and perpendicular to the grain, and internal bonding (IB).

#### *2.2.4.* Data analysis

The experimental design used in the study was a simple completely randomized design with a single factor: compression ratio (five levels i.e. 1.15; 1.25; 1.35; 1.44; and 1.54). Each treatment was carried out in four replications. The physical and mechanical properties data obtained in this research were analyzed statistically by using analysis of variance (ANOVA). The value of TS, MOE, MOR, and IB was compared to the OSB commercial standard CSA 0437.0 (Grade O-1) [19].

## 3. Results and Discussions

### *3.1. Strand geometry*

The value of SR and AR was determined in this study to ensure that the strand size used in this study did not become a research factor. The strand geometry distribution is presented in Figure 1. The value of SR and AR obtained from the results of this research in succession was 114.28 and 3.03. This indicated that the strand was suitable for use as a BOSB raw material. The SR value above 60 was classified into a high SR value. To produce a composite with good mechanical properties required a minimum AR value of 3 [14]. High SR and AR values allowed composite boards to had high strength.

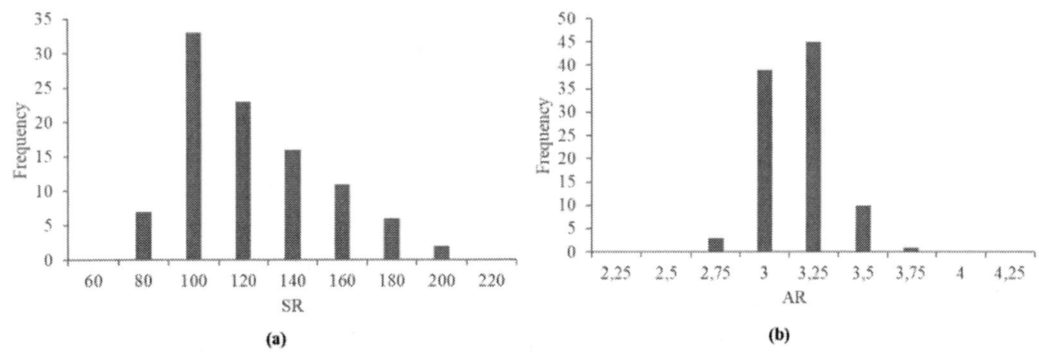


**Figure 1.** (a) Distribution of SR and (b) AR values

### *3.2. Physical properties of BOSB*

The resulting BOSB density values were range from 0.60-0.81 g/cm$^3$ (Figure 2a) and achieved a specified compression ratio. The results of the statistical analysis ($\alpha$=0.05) indicated that the compression ratio provides a significant effect on the OSB density. The compression ratio relies heavily on the board density and the raw material density used. A high-density value of a board with the same raw material resulted in a higher value compression ratio. This caused the compactness of the board to increase. The results showed that the value of a BOSB MC ranged from 11.15-11.92% (Figure 2b). The value indicated that the whole MC of BOSB was uniform. It was supported by statistical analysis results ($\alpha$ = 0.05) which indicated that the compression ratio does not give a significant influence on the MC BOSB.

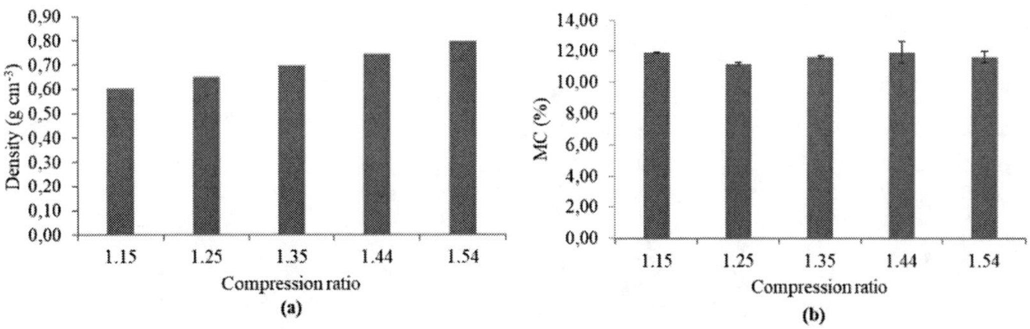


**Figure 2.** (a) The value of density and (b) MC BOSB

IOP Conf. Series: Materials Science and Engineering **935** (2020) 012063 doi:10.1088/1757-899X/935/1/012063

The WA values gained from the study ranged from 22.72-25.73% (Figure 3a). The results of the statistical analysis (α=0.05) indicated that the compression ratio gives a significant effect on the WA of BOSB. Duncan's test results stated that the WA BOSB value of the compression ratio of 1.15 was the lowest WA value and differs from other BOSB. The value of TS obtained was approximately 5.6-9.60% (Figure 3b). The entire BOSB had met to the OSB commercial standard CSA 0437 (Grade O-1) [20]. The results of the statistical analysis (α = 0.05) indicated that the compression ratio provides a significant effect on the value of TS. Duncan's test results stated that the TS BOSB value with the compression ratio of 1.15 differs from other BOSB.

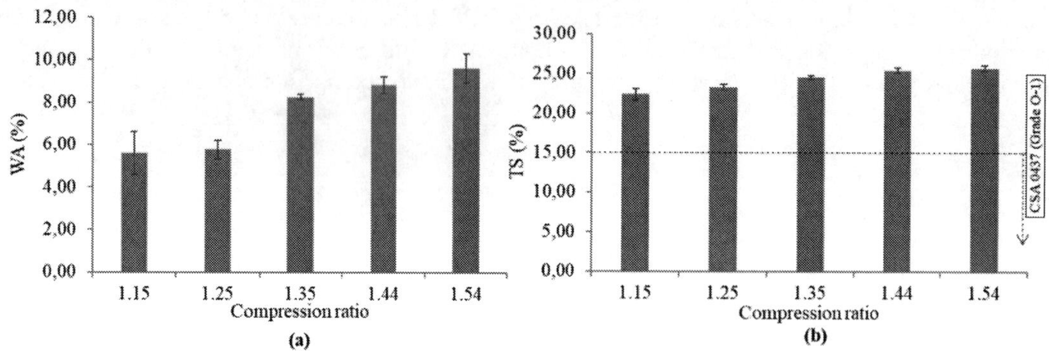


**Figure 3.** The value of WA (a) and TS (b) BOSB

The research showed that increased compression ratios decreased dimensional stability. This was supported by increasing the value of both TS and WA by increasing the compression ratio. Higher compression rate values indicated that in one board there were more cell walls because more materials were needed. This caused BOSB with a higher compression ratio to had lower dimensional stability. This phenomenon in line with some previous reports showing increased compression ratios led to an increased value of TS [8,9,11].

### *3.3. Mechanical properties of BOSB*

The value of parallel and perpendicular MOE BOSB was ranging from 4611-8101 MPa and 1529-2619 MPa, respectively (Figures 4a and 4b). The entire value of MOE BOSB had met the commercial standard CSA 0437 (Grade O-1). The results of the statistical analysis (α=0.05) indicated that the compression ratio gave a significant effect on the value of MOE both parallel and perpendicular to the grain. The highest value of parallel and perpendicular MOE occurred in BOSB with a compression ratio of 1.54. It was also supported by Duncan's advanced test results stating that BOSB with the 1.54 compression ratio was different from other BOSB.

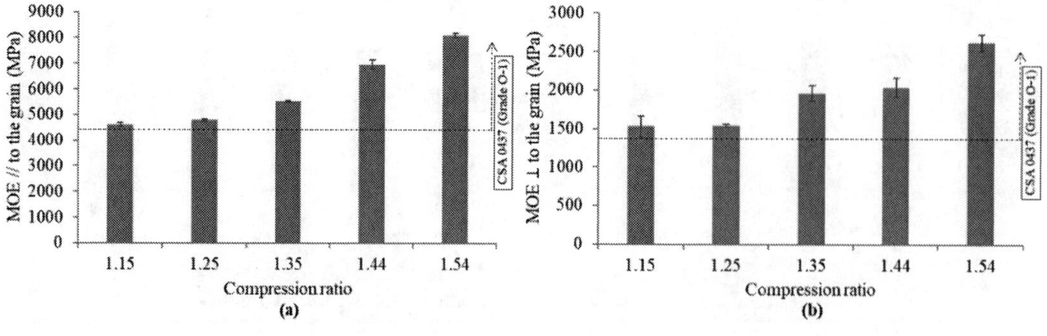


**Figure 4.** (a) The value of MOE BOSB parallel and (b) perpendicular to the grain

The value of parallel and perpendicular MOR BOSB was ranging from 24-46 MPa dan 18-22 MPa (Figures 5a and 5b). The entire MOR values perpendicular and parallel to the grain had met the standard set CSA 0437 (Grade O-1). The results of the statistical analysis ($\alpha$=0.05) indicated that the compression ratio provides a significant effect on both parallel and perpendicular MOR values. Duncan's test results stated that BOSB with the 1.54 compression ratio had a distinct MOR value with other MOR OSB.

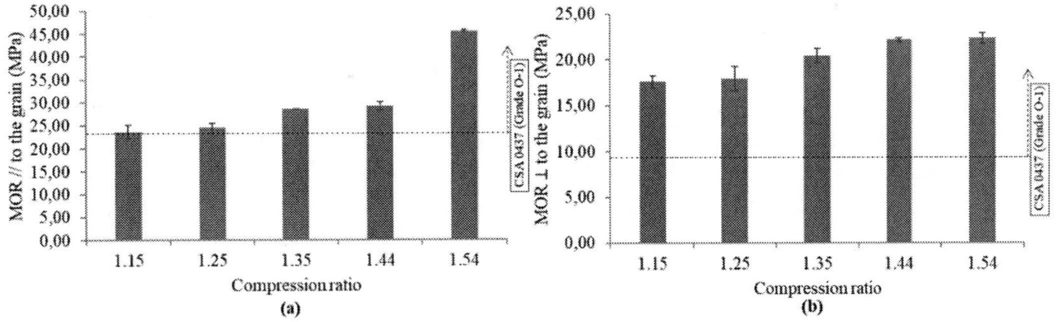


**Figure 5.** (a) The value of MOR BOSB parallel and (b) perpendicular to the grain

The research showed that an increased compression ratio leads to an increased value of MOE and MOR BOSB. This phenomenon was in line with some research suggesting that high compression ratios produced high mechanical properties of composite boards [3,10,12,21,22]. The high compression ratio value indicated that the board was more compact. This was very supportive in the event of press voltage and pull at the time of testing MOE and MOR. Therefore, increasing the compression ratio increased MOE and MOR.

The average IB value of BOSB produced ranges from 0.30-0.34 MPa (Figure 6). The results of the statistical analysis ($\alpha$=0.05) indicated that the compression ratio provides a significant effect on the IB value. Duncan's test results stated that BOSB made with the compression ratio of 1.54 differs significantly from other BOSB. Research showed that an increase in compression ratios increases the IB value. High compression ratio values indicated a high compact board. This greatly affected the BOSB's mechanical properties including the IB value. This phenomenon was similar to previous research that increased IB was found with increasing compression ratios [3,10,12,21].

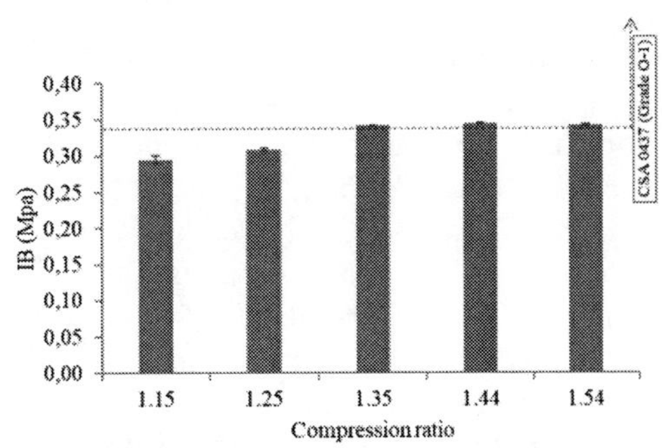


**Figure 6.** The value of IB BOSB

The research showed that IB values that had met the standard were BOSB made with a compression ratio of 1.35; 1.44; and 1.54. In other words, BOSB with a compression ratio of 1.35; 1.44; and 1.54 had met the entire OSB commercial standards. However, it should be noted that the higher the value of the compression ratio, the higher the raw material requirements. Therefore, the optimum compression ratio in BOSB manufacturing from betung was 1.35.

## 4. Conclusions

The results showed that the increase in BOSB compression ratio up to the 1.54 level increased the mechanical properties, but decreased the dimensional stability. Optimum compression ratio in BOSB manufacturing was 1.35.

## References

[1] Adrin, Febrianto F, and Sadiyo S 2013 Properties of oriented strand board prepared from steam treated bamboo strands under various adhesive combinations *JITKT* **11**(2) 109-19

[2] Chaowana P 2013 Bamboo: an alternative raw material for wood and wood-based composites *J. Materials Sci. Research.* **2**(2) 90-102

[3] Chen S, Du C, and Wellwood R 2010 Effect of panel density on major properties of oriented strand board *Wood and Fiber Science* **42**(2) 177-184

[4] Fatrawana A, Maulana S, Nawawi DS, Sari RK, Hidayat W, Park SH, Febrianto F, Lee SH, and Kim NH 2019 Changes in chemical components of steam-treated betung bamboo strands and their effects on the physical and mechanical properties of bamboo-oriented strand boards *European Journal Wood & Wood Products* **77**(5) 731-739

[5] Febrianto F, Sumardi I, Hidayat W, and Maulana S 2017 *Papan Strand Bambu Berarah Material Unggul untuk Komponen Bangunan Struktur* (Bogor: IPB Press)

[6] Febrianto F, Jang JH, Lee SH, Santosa IA, Hidayat W, Kwon JH, and Kim NH. 2015. Effect of bamboo species and resin content on properties of oriented strand board prepared from *steam* treated bamboo strands *BioResources* **10**(2) 2642-55

[7] Febrianto F, Royama LI, Hidayat W, Bakar ES, Kwon JH, and Kim NH 2009 Development of oriented strand board from acacia wood (*Acacia mangium* Willd.): Effect of pretreatment of strand and adhesive content on the physical and mechanical properties of PUB *J. Korean Wood Sci. Technol.* **37**(2) 121-7

[8] Gatchell CJ, Heebink BG, and Hefty FV 1966 Influence of component variables on properties of particleboard for exterior use *For. Prod. J.* **16**(4) 46-59

[9] Halligan AF, and Schniewind AP 1972 Effect of moisture on physical and creep properties of particleboard *For. Prod. J.* **22**(4) 41-8

[10] Hiziroglu S 2009 Properties of strandboard panels manufactured from eastern redcedar *Materials* **2**(3) 926-33

[11] Hse CY 1975 Properties of flakeboards from hardwoods growing on southern pine sites *For. Prod. J.* **25**(3) 48-53

[12] Jin J, Dai C, Hsu WE, and Yu C 2009 Properties of strand boards with uniform and conventional vertical density profiles *Wood Sci. Technol.* **43**(7-8) 559-74

[13] [JIS] Japanese Industrial Standard 2003 *Standards on Particleboard Japanese Standards Association* (Tokyo: Japanese Standards Association)

[14] Kuklewski KM, Blankenhorn PR, and Rishel LE 1985 Comparison of selected physical and mechanical properties of red maple (*Acer rubrum* L.) and aspen (*Populus grandidentata* Michx.) flakeboard *Wood and Fiber Science* **17**(1) 11-21

[15] Maloney TM 1993 *Modern Particleboard and Dry-Process Fibreboard Manufacturing* (Madison, US: Forest Products Society)

[16] Maulana S, Gumelar Y, Fatrawana A, Maulana MI, Hidayat W, Sumardi I, Wistara NJ, Lee SH, Kim NH, and Febrianto F 2019 Destructive and non-destructive tests of bamboo-oriented strand board under various shelling ratios and resin content *J. Korean Wood Sci. Technol.* **47**(4) 519-32

[17] Maulana S, Purusatama BD, Wistara NJ, Sumardi I, and Febrianto F 2016 Pengaruh perlakuan steam pada strand dan *shelling ratio* terhadap sifat fisis dan mekanis *oriented* strand *board* bambu *JITKT* **14**(2) 136-43
[18] Maulana S, Busyra I, Fatrawana A, Hidayat W, Sari RK, Sumardi I, Wistara INJ, Lee SH, Kim NH, and Febrianto F 2017 Effects of steam treatment on phisycal and mechanical properties of bamboo oriented strand board *J. Korean Wood Sci. Technol.* **45**(6) 872-82
[19] [SBA] Structural Board Association 2005 *Oriented* Strand *Board in Wood Frame Construction* (Ontario,CA: Structural Board Association)
[20] [SBA] Stuctural Board Association 2004 *PUB Performance by Design: Oriented Strand Board in Wood Frame Constuction* (Ontario, CA: Structural Board Association)
[21] Sumardi I, Suzuki S, and Ono K 2007 Effect of board density and layer structure on the mechanical properties of bamboo-oriented strand board *J. Wood Sci.* **53** 510-15
[22] Vital BR, Lehmann WF, and Boone RS 1974 How species and board densities affect properties of exotic hardwood particleboards *For. Prod. J.* **24**(12) 37-45

**Acknowledgment**
We sincerely acknowledge the Applied Research (PT) Grant (No. 4168/ IT3.L1/PN/2020) from the Deputy of Strengthening for Research and Development, Ministry of Research and Technology/National Research and Innovation Agency, Republic of Indonesia for the financial support. This research was also supported by Science and Technology Support Program through the National Research Foundation of Korea (NRF) funded by the Ministry of Science and ICT (MSIT) (NRF-2019K1A3A9A01000018) and by Basic Science Research Program through NRF funded by the Ministry of Education (No. NRF-2016R1D1A1B01008339, No. 2018R1A6A1A03025582).

IOP Conf. Series: Materials Science and Engineering **935** (2020) 012064 doi:10.1088/1757-899X/935/1/012064

# Dielectric properties of heat treated teak wood (*Tectona grandis* Linn. F)

**L Karlinasari[1*], U Adzkia[1], R A Tang[1], and Irzaman[2]**

[1]Department of Forest Product, Faculty of Forestry and Environment, IPB University, Bogor, 16680
[2]Department of Physic, Faculty of Mathematics and Natural Science, IPB University, Bogor, 16680

[*]Email: karlinasari@apps.ipb.ac.id

**Abstract.** Heat treatment is a modification method that uses modification of drying temperature or other heat treatment to improve wood quality. Heat treatment with a temperature of 150°C and 180°C for 0, 2, and 6 hours carried out on 8 years old teak wood can improve its quality. The purpose of this study was to evaluate the effect of heat treatment on the dielectric properties of teak wood. The parameters in this study are the capacitance and dielectric constant or relative permittivity. The results showed that heat treatment affected the physical and dielectric properties of wood. Moisture content and wood density decrease after heat treatment. The changes on those physical properties influenced dielectric properties. Heat treated wood have decreased the dielectric wood properties significantly. The capacitance and the dielectric constant decrease on the sample which applied in higher temperature and time. Frequencies below 10 kHz have affected on dielectric properties or both capacitance and dielectric contant Based on the study, the heat treatment was decreasing the ability of electrical insulator properties of wood.

## 1. Introduction

Fast growing teak is a teak species that has a shorter cutting age (20-30 years) [1,2]. The availability of fast growing teak is currently increasing along with the increasing demand for wood. Teak wood is commonly used by the community as raw material for furniture, handicrafts, and construction. However, fast growing teak wood has several disadvantages, such as low dimensional stability, strength, and durability [2]. In addition to these characteristics, [3] stated that 8 years old teak had low machining quality. Therefore, in utilizing teak wood with a short cutting age, modifications need to be made to improve the wood quality. One of the modification methods used is the heat treatment method.

The heat treatment method is an modification method used to improve wood quality through drying or other heat treatments. Hill [4] stated that heat treatment is an environmentally friendly technique because it does not involve chemicals during the process. Heat treatment methods have been carried out on several types of wood such as mahogany (*Swetenia macrophylla*) [5,6] rubberwood (*Hevea brasiliensis*) [7,8], *Eucalyptus urophylla* [9] and some fast-growing wood species such as teak (*Tectona grandis*), mangium (*Acacia mangium*), sengon (*Falcataria moluccana*) and jabon (*Neolamarkcia cadamba*) [3]. Based on these results, heat treatment can improve wood quality as it increases the dimensional stability by reducing the hygroscopicity of wood, and increase the durability of wood against fungi and wood-destroying insects, while also influence on the change of surface roughness, as well as wood machining properties [3,9-11]. Changes in these properties occur due to degradation of chemical components such as cellulose, hemicellulose, and lignin in the wood [4,11]. These changes also affect the electrical properties of wood. The electrical properties of wood undergo large changes with a relatively small change in moisture content below the fiber saturation point [12]. The electrical properties of wood comprise impedance, inductance, dielectric constant, and electrical conductivity [12,13].

The electrical properties can be used to evaluate the material quality both quickly and nondestructively [14]. One of the electrical properties is dielectric properties. The dielectric properties also known as electrical insulator is concerned with the storage and dissipation of electric and magnetic energy in materials. Capacitance and dielectric constant are two variables that illustrate dielectric properties. Capacitance is the amount of charge per unit voltage that two parallel plates can store. Meanwhile, the dielectric constant (often called the relative permittivity) is a measure of the amount of electric potential energy, in the form of induced polarization that is stored in a given volume of material under the action of an electric field. The dielectric properties of material are affected by some paremeter of wood such as moisture content, density, and characteristics wood pore as well as fiber direction [12,14-16]. This properties was also influenced by temperature, frequency, and orientation of electric field with respect to measurements conditions [17]. Changes in wood properties due to heat treatment may affect the dielectric properties of the wood. Therefore, this study was conducted to determine the effect of heat treatment on the capacitance value and the wood dielectric constant. The purpose of this study was to evaluate the effect of heat treatment on the dielectric properties of teak wood.

## 2. Materials and Methods

### *2.1. Materials*

The material used in this study was an 8 years old teak wood (*Tectona grandis*) with diameter about 40 cm from Dramaga Sub District, Bogor. The electrical properties testing was carried out with the 3532-50 LCR Hi-tester (Hioki). The research was conducted from April until October 2018 at some Laboratories in Department of Forest Product, Faculty of Forestry and Environment, IPB University for preparing and treatment of wood samples. The measuring of electrical properties was done in the Spectroscopy Laboratory, Department of Physics, Faculty of Mathematic and Natural Science, IPB University.

### *2.2. Methods*

#### 2.2.1. Heat treatment

The heat treatment was carried out on tangential and radial boards using an oven at a temperature of 150 and 180°C at 0 (control), 2, and 6 hours. After the boards have been treated, they were then conditioned for ± 7 days to release residual stress and stabilize the moisture content of the boards. The boards were then cuts to have sample dimensions of width, length, and thickness 2 x 1 x 0.5 cm and it would have three sections plane of wood namely cross-section, radial, and tangential. The samples number used in this study are summarized in Table 1. The specimens with similar treatment were chosen for the measurement of wood moisture content and density.

**Table 1.** The number of samples electrical properties of heat modified teak wood test

| Temperature | Time variation (jam) | Type and number of sample planes | | |
|---|---|---|---|---|
| | | Cross-section | Radial | Tangential |
| Control | 0 | 3 | 3 | 3 |
| 150°C | 2 | 3 | 3 | 3 |
| 150°C | 6 | 3 | 3 | 3 |
| 180°C | 2 | 3 | 3 | 3 |
| 180°C | 6 | 3 | 3 | 3 |
| Total | | 15 | 15 | 15 |

2.2.2 Physical properties

Testing of the physical properties included the Moisture Content (MC) and wood density (ρ). It was obtained by weighing the sample ($M_m$) before putting it in the oven at a temperature of 103 ± 2°C for 24 hours and followed by measuring the sample using a digital caliper to gain an air-dry volume ($V_m$). Subsequent test samples were removed from the oven and put into the desiccator for ± 15 minutes, then proceeded with weighing the sample after it was in the oven ($M_0$) and measuring the dimensions. The value of moisture content and density were calculated using the following equation.

$$\text{Moisture content (\%)} = \frac{M_m - M_0}{M_0} x\ 100 \quad (1)$$

$$\text{Density (g/cm}^3\text{)} = \frac{M_m}{V_m} \quad (2)$$

where, $M_m$: initial weight (g), $M_0$: oven-dry weight (g), $V_m$: volume of specimen ($cm^3$)

2.2.3. Dielectrical properties

The dielectric properties consisted of the unheated (control) and heated samples. The sample dimension as well as the plate dimension for testing of the electrical properties was 2 x 1 x 0.5 cm. They were attached to a cable connected to the LCR Hi-tester (Hioki) device (Figure 1) with the measurement results on the monitor. The variable which differentiated was the frequency from $10^3$ to $10^6$ Hz. The electrical properties were tested for three sections of wood i.e. cross-section, radial, and tangential with three replications of each. The parameters used in this study consisted of capacitance and dielectric constant. The capacitance and dielectric constant value was calculated with the following equation [12].

$$Cs = \frac{\varepsilon.\varepsilon_0 .A}{d} \quad (3)$$

$$\varepsilon_r = \frac{Cs.d}{\varepsilon_0 .A} \quad (4)$$

where, Cs: capacitance (F), $\varepsilon_r$: dielectric contant or relative permittivity, ε: unitless material parameter, $\varepsilon_0$: permittivity of a vacuum or free space (8.854 x $10^{-12}$ F/m), A: sectional area ($m^2$), d: distance between plate (m)

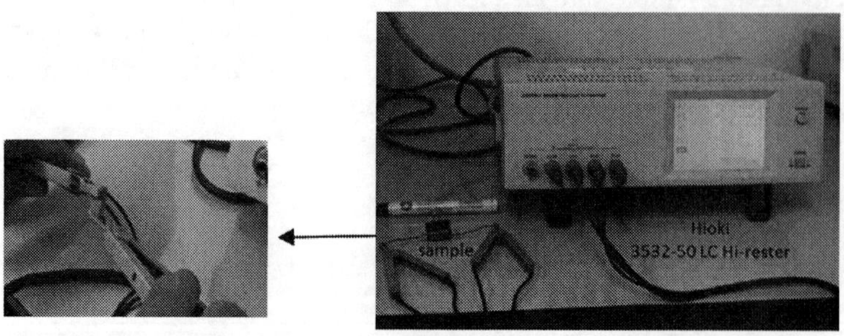


**Figure 1.** Setting sample for dielectric wood properties testing

### *2.3. Data analysis*

Statistical analysis was conducted with a Completely Randomized Factorial Design analysis with two factors used in this study. Factor A is the heating temperature which consists of three levels (control, 2 and 6 hours) and factor B is the heating time which consists of 2 level (150 and 180°C). The test was carried out with three replications. Data was analyzed using Microsoft Excel 2013 application and IBM SPSS ver. 22.0. Duncan's Multi Range Test was used for further analysis if the significance value in the 95% confidence levels.

## 3. Results and Discussion

### *3.1. Physical properties*

#### 3.1.1. Moisture content

Moisture content is the amount of water contained in the wood. The average moisture content decrease with the increase of the temperature and time of heat treatment as shown in Figure 2. The average moisture content of teak without heat treatment was 16%. In treated wood, at 150°C for 2 hours the average moisture content was 10%, while at the same temperature for 6 hours, the average decreased to 9%. Meanwhile, the moisture content at 180°C for 2 and 6 hours were 8% and 7%, respectively. Decreased moisture content due to heat treatment was also conveyed by [3,6,18] who studied on teak (*Tectona grandis*), mangium (*Acacia mangium*), jabon (*Neolamarckia cadamba*), and sengon (*Falcataria moluccana*). The high temperature and longer heating time caused the moisture content to decrease.

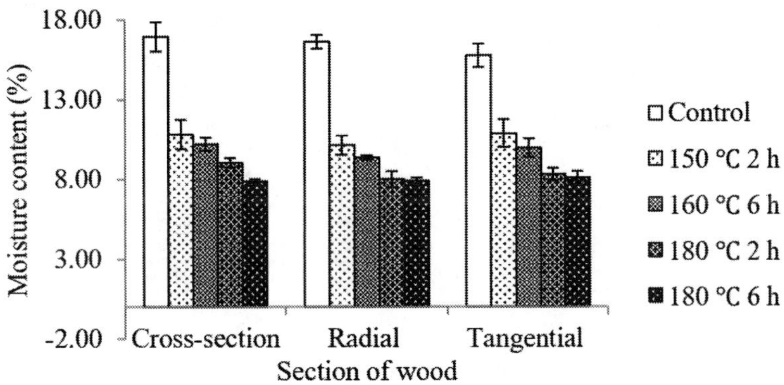


**Figure 2.** Average of moisture content from teak wood with heat modification treatment on the three planes of wood

Statistical analysis at a 95% confidence interval shows that the moisture content in each of the wood was affected by the interaction of temperature factors and heating time (Table 2). Duncan test showed that the high temperature and longer heating time caused the moisture content to decrease. It was due to heat treatment mostly caused degradation in hemicellulose and lead to decrease the ability cell wall to absorb the water [10] and [18]. Degradation of hemicellulose caused a decrease in hydroxyl groups, resulting in a decrease in the moisture content of wood. [19] stated that a significant decrease in the moisture content occurred at 150°C and would continue to do so with additional temperature in the heat treatment.

**Table 2.** Resume of variance analysis (ANOVA) the effect of temperature and heating time on the moisture content of teak wood

| Wood plane | Significant of ANOVA | | |
|---|---|---|---|
| | Temperature | Time | Temperature vs Time |
| Cross-section | 0.382$^{tn}$ | 0.000** | 0.000** |
| Radial | 0.528$^{tn}$ | 0.000** | 0.000** |
| Tangential | 0.360$^{tn}$ | 0.000** | 0.000** |

Note: tn= no significant difference at the confidence level 95%, **= significant difference at the confidence level 95%

3.1.2. Density

The average density of teak before heat treatment was 0.75 g/cm$^3$, whereas after treatment is 0.69 g/cm$^3$, meaning it was reduced about 8% (Figure 2). Study conducted by [20] reported the heat treatment on teak wood lead to a slightly decrease wood density was about 3.58% . While, the study by [21] on alder hardwood species applied heat treatment with temperature at 220°C for 6 hours has led decreasing density wood until was 19.7%. For that reason, wood species and charateristics of heat treatment were factors the decreasing of wood density.

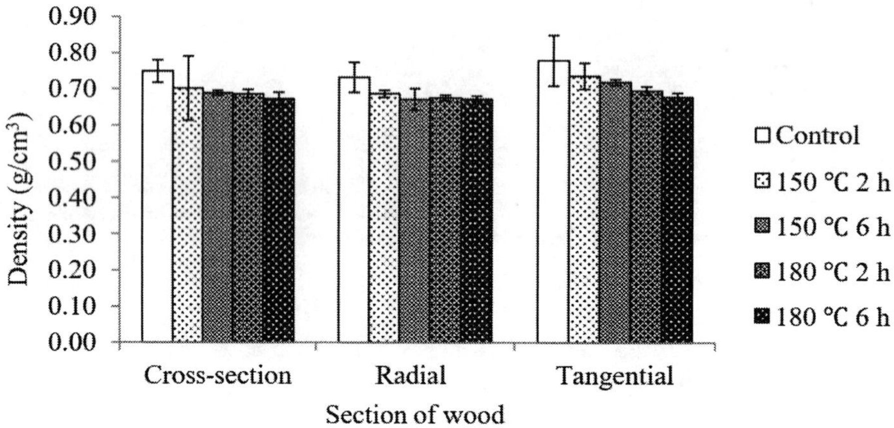


**Figure 2.** The average density of teak wood on the three planes of wood

Statistical analysis at a 95% confidence interval in Table 3 shows that the density of teak in the cross-section and tangential direction is not affected by interaction between temperature and heating time. Meanwhile, the radial plane was influenced by those interaction. The results of the Duncan test analysis showed that the density of teak without heat treatment was significantly higher than that with the treatment. Heat treatment resulted in changes in extractive substances and degradation of hemicellulose, causing the wood density to decrease. Hill [4] and Yildiz [21] mentioned that on alder wood species, the higher heating temperature affected the increase in hemicellulose degradation. Hemicellulose was a component that is more susceptible to heat compared to other wood components.

**Table 3.** Resume of variance analysis (ANOVA) the effect of the temperature and heating time on the density of teak wood

| Wood plane | Significance of ANOVA | | |
|---|---|---|---|
| | Temperature | Time | Temperature#Time |
| Cross-section | 0.605tn | 0.018** | 0.169tn |
| Radial | 0.798tn | 0.001** | 0.031** |
| Tangensial | 0.289tn | 0.020** | 0.090tn |

Note: tn= no significant difference at the confidence level 95%, **= significant difference at the confidence level 95%

### *3.2. Dielectrical Properties*

#### 3.2.1. Capacitance

Capacitance is the ability of a material to maintain an electric charge in Farad unit (F) [22]. The value of teak wood capacitance decreases with increasing frequency (Figure 3). The maximum value of capacitance was at a frequency of $10^3$ Hz. A significant reduction in capacitance occurred at a frequency of $10^3$ to $10^4$ Hz. Meanwhile, the capacitance with frequencies at $10^4$ to $10^6$ Hz tended to be constant. The capacitance value of cross-section differed from the radial and tangential. Below the frequency of $10^4$ Hz, the difference value of capacitance was significant where the cross-section was the highest followed by radial and tangential section in untreated wood (Figure 3a, 3b, 3c). However, the almost similar values in a range 2 to 7 F was found at all section in treated wood in which the lower temperature and shorter time treatment had higher capacitance values. Untreated wood had significantly higher value of capacitance compared with treated wood. It was presumably related to wood porosity characteristcs in term of correlated to wood density as mentioned by [23] In further, they reported wood species differentiantion can be determined based on its dielectric properties.

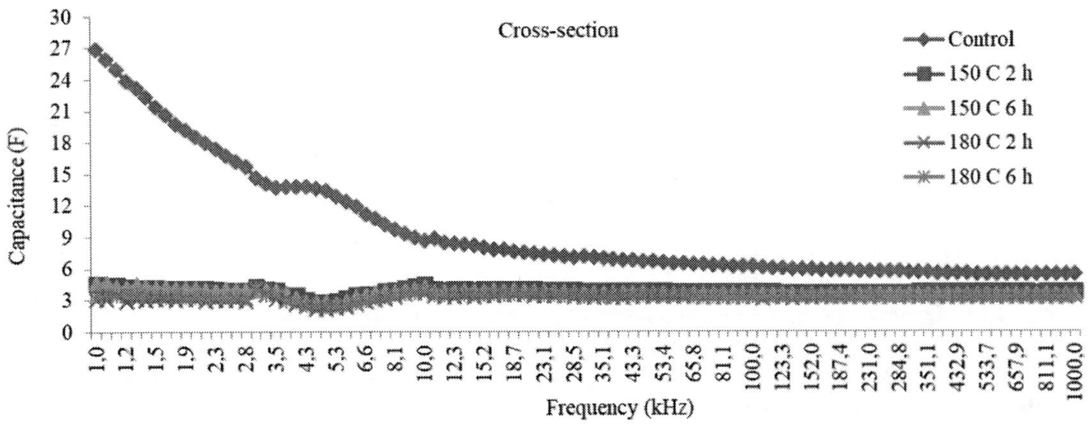


(a)

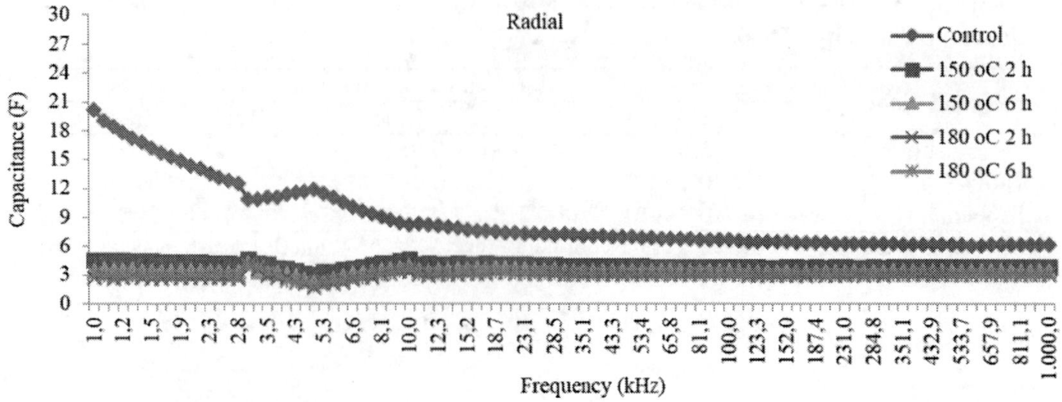


(b)

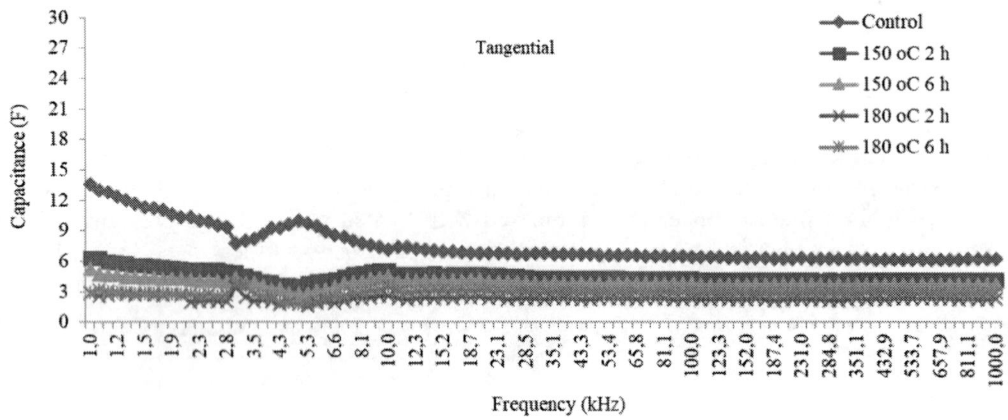


(c)

**Figure 3.** The capacitance of teak wood heat-modification on the three plane (a) cross-section (b) radial (c) tangential

Statistical analysis at a 95% confidence interval shows that the capacitance was influenced by the interaction between the temperature and heating time (Table 4). The higher the temperature and the longer of heating time, the lower the capacitance. The results of Duncan's test showed that the wood capacitance without heat treatment was significantly higher than with heat treatment. Wood capacitance was affected by density, specific gravity, and moisture content [24]. Higher temperature and longer time in heat treatment caused a decreasing of wood moisture content. Higher moisture content made the capacitance lower [14,15,22].

**Table 4.** Resume analysis of variance (ANOVA) the effect of temperature and heating time to the capacitance on the three plane

| Wood plane | Significance ANOVA | | |
|---|---|---|---|
| | Temperature | Time | Temperature vs Time |
| Cross-section | 0.929[tn] | 0.000** | 0.000** |
| Radial | 0.949[tn] | 0.000** | 0.003** |
| Tangential | 0.617** | 0.000** | 0.004** |

Note: tn= no significant difference at the confidence level 95%, **= significant difference at the confidence level 95%

The dielectric constant is an important parameter in the absorption of electromagnetic energy. This value was generated from capacitance. The knowledge of this properties is crucial for a process design, control, and optimazation, as well as simulation of some heating drying technique such as high frequency and microwave drying technique [23].

Figures 4a, 4b, and 4c show the dielectric constant values from three different section of heat-modified teak wood. The highest dielectric constant value of teak was at a cross-secion, followed by radial and tangential section. This phenomena was in line with study by [24-26]. It could be explained that cross-sectional plane could store the electrical energy greater than the radial and tangential. The dielectric constant value was strongly influenced by moisture content, wood density, fiber direction, anatomical structure [27]. Wood porosity have significant affected on dielectric constant value. Higher porosity resulted a lower dielectric contant or relative permittivity [23]. In term of this study, small changes of wood density caused by heat treatment compared with untreated wood have led rélatively large changes in dielectric constant values. The most significant differences appear for lower frequencies. In specific, it was happened on dielectric wood properties below 5 kHz [23,24,28].

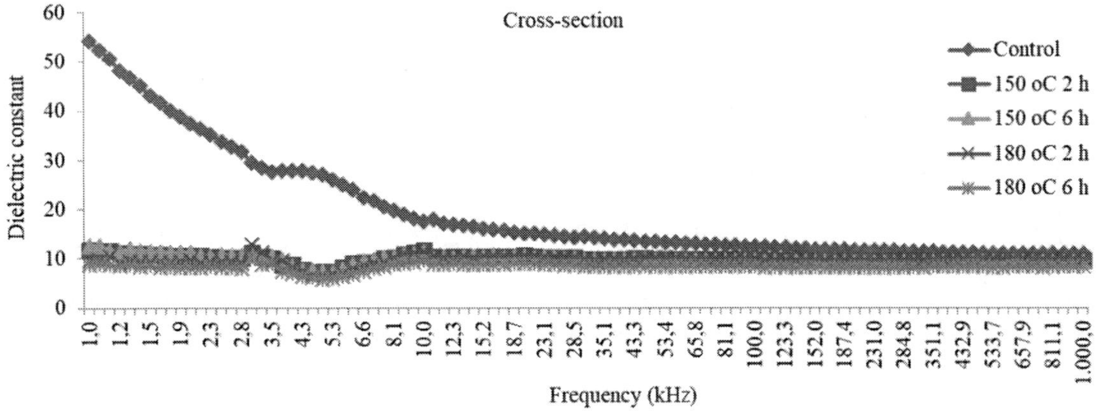


(a)

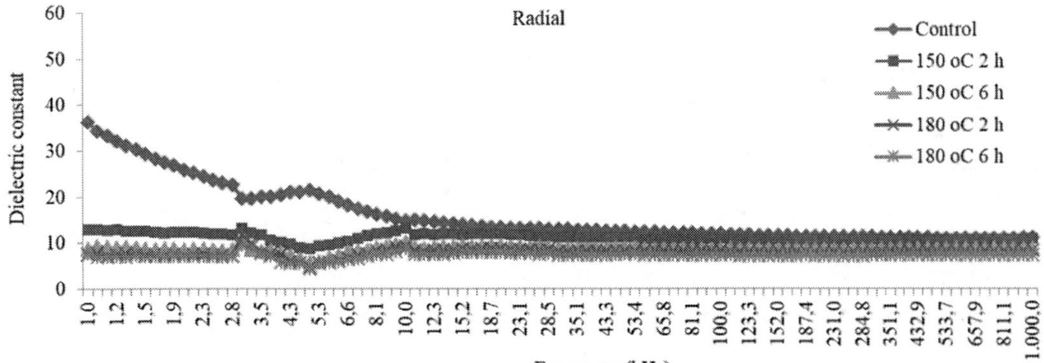


(b)

IOP Conf. Series: Materials Science and Engineering **935** (2020) 012064 doi:10.1088/1757-899X/935/1/012064

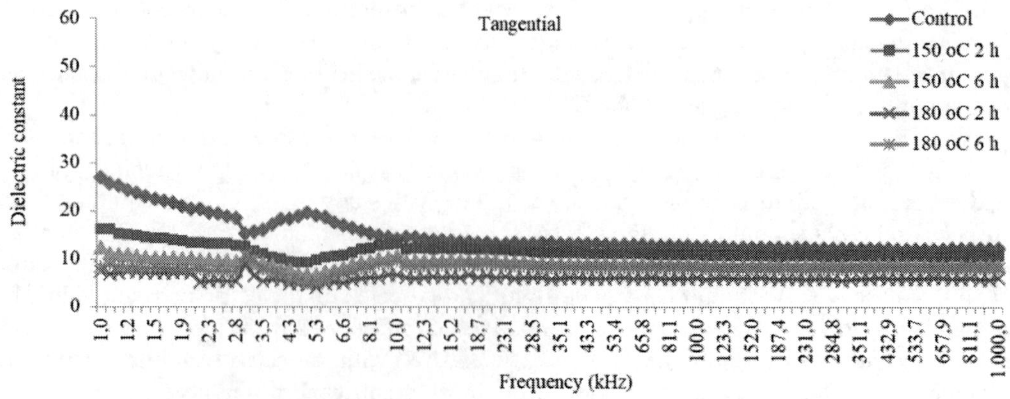

(c)

**Figure 4.** The dielectric constant of teak wood heat modification on the three plane (a) cross-section, (b) radial, (c) tangential

Statistical analysis of variance at a 95% confidence interval shows that the dielectric constant is influenced by the interaction of the temperature and the heating time (Table 5). The higher the temperature and the longer the heating duration, the lower the dielectric constant value. The results of Duncan's analysis showed that the dielectric constant of teak without heat treatment was significantly higher than the heat treated teak wood. This was also conveyed by [17,29,30] that the electrical properties of wood were strongly influenced by moisture content, wood porosity, fiber direction, as well as temperature and frequencyduring testing. Dielectric constant value without heat treatment were higher than with heat treatment. A decrease in the dielectric constant value showed a decrease in the wood's ability to deviate electrical energy. Therefore, wood, which was originally an insulator, it can be modified becaming a semiconductor material.

**Table 5.** Resume of variance analysis(ANOVA) the effect of the temperature and heating time to the dielectric constant on the three planes of wood

| Wood plane | Significance of ANOVA | | |
|---|---|---|---|
| | Temperature | Time | Temperature#Time |
| Cross-section | $0.825^{tn}$ | 0.000** | 0.000** |
| Radial | $0.447^{tn}$ | 0.000** | 0.001** |
| Tangential | 0.207** | 0.000** | 0.004** |

Remarks: tn: not significant at 95% confidence intervals; **: significant difference at 95% confidence intervals

## 4. Conclusion

Wood modification of heat treatment affected on the dielectric properties of teak wood where applied to cross-section, radial and tangential plane. Untreated teak wood possessed had higher dielectric properties of capacitance and dielectric constant compared with the heat treated wood. The higher temperature and the longer time treatment led to decrease dielectric properties. The changes of dielectric properties in term of wood section i.e cross-section, radial, and tangential was lower in treated wood than untreated wood.The most significant decreasing of dielectric properties was at frequency below 10 kHz. Based on this study, the heat treatment decreased the ability of electrical insulator properties of wood.

## References

[1] Bhat KM and Priya PB 2004 Influence of provenance variation on wood properties of teak from the Western Ghat region in India *IAWA J.* **25(3)** 273-82

[2] Wahyudi I, Priadi T, and Rahayu IS 2014 Characteristics and basic properties of 4 and 5 year-old of superioe teakwoods from west java *Indones. J. of Agricultural Sci.* **19(1)** 50-6

[3] Adzkia U, Priadi T, and Karlinasari L 2019 Evaluation of drying and machining defects of four Thermo modified wood of tropical fast growing species *JPIHI* **37(3)** 209-22

[4] Hill CA 2007 *Wood Modification: Chemical, Thermal and Other Processes* (John Wiley & Sons) **5**

[5] Lukmandaru G, Susanti D, and Widyorini R 2018 Chemical properties of modified mahagony wood by heat treatment *Wallacea J. of Forest Research* **7(1)** 37-46

[6] Widyorini R, Khotimah K, and Prayitno TA 2014 The effect temperature and heat-treated method to the physical and finishing of mahoni wood *JIK* **8(2)** 65-74

[7] Taylor P, Srinivas K, and Pandey KK 2013 Effect of heat treatment on color changes, dimensional stability, and mechanical properties of wood *J. of Wood Chem.* 37-41

[8] Bakar BFA, Hiziroglu S, and Tahir P 2013 Properties of some thermally modified wood species *J. Mater.* **43** 348-55

[9] Tu D, Liao L, Yun H, Zhou Q, Cao X, and Huang J 2014 Effects of heat treatment on the machining properties of *Eucalyptus urophylla×E. camaldulensis BioResources* **9(2)** 2847-55

[10] Korkut DS, Korkut S, Bekar I, Budakç M, and Dilik T 2008 The effects of heat treatment on the physical properties and surface roughness of turkish hazel (*Corylus colurna* L .) wood *Int. J. of Molecular Sci.* **9(9)** 1772-83

[11] Piao C, Winandy JE, and Shupe TF 2010 From hydrophilicity to hydrophobicity: a critical review - Part I: wettability and surface behavior *Wood Fiber Science* **42**(4) 490-510

[12] Glass SV and Zelinka SL 2010 *Moisture Relations and Physical Properties of Wood* (Madison: Forest Product Laboratory, University of Wisconsin)

[13] K. Khamoushi 2014 *Characterization and Dielectric Properties of Microwave Rare Earth Ceramics Materials for Telecommunications* (Finlandia: Tampere University of Technology)

[14] Husein I, Agustina A, and Khabibi J 2014 Electrical properties of Indonesian hardwood case study: *Acacia mangium, Swietenia macrophylla* and *Maesopsis eminii Wood Res.* **59**(4) 695-704

[15] Husein I, Sadiyo S, Nugroho N, Cabuy LR, Afif A, Fakhruzy, Zabed M, Indasuary N, Fajriani E, Kabe A, and Sucipto 2011 Electrical properties of Indonesian hardwood *Int. J. of Basic and Apllied Sci.* **11**(6) 161-6

[16] Olmi R, Bini M, Ignesti A, and Riminesi C 2000 Dielectric properties of wood from 2 to 3 GHz *J. Microw. Power Electromagn. Energy* **35**(3) 135-143

[17] Daian G, Taube A, Birnboim A, Daian M, and Shramkov Y 2006 Modeling the dielectric properties of wood *Wood Sci. Technol.* **40**(3) 237-46

[18] Karlinasari L, Lestari AT, and Priadi T 2018 Evaluation of surface roughness and wettability of heat-treated, fast-growing tropical wood species sengon (*Paraserianthes falcataria* (L.) I.C.Nielsen), jabon (*Anthocephalus cadamba* (Roxb.) Miq), and acacia (*Acacia mangium* Willd.) *Int. Wood Prod. J.* **9**(3) 142-8

[19] Zhang Y, Yu W, and Zhang Y 2013 Effect of steam heating on the color and chemical properties of neosinocalamus affinis bamboo *J. Wood Chem. Technol.* **33**(4) 235-46

[20] Priadi T, Suharjo AAC, and Karlinasari L 2019 Dimensional stability and colour change of heat-treated young teak wood *Int. Wood Prod. J.* **10**(3) 119-125

[21] Yildiz S, Yildiz UC, and Tomak ED 2011 The effects of natural weathering on the properties of heat-treated alder wood *BioResources* **6**(3) 2504-21

[22] Voss A, Seppänen A, Siltanen S, Salokangas L, and Baroudi D 2016 Imaging of moisture content in wood using electrical capacitance tomography *World Conf. Timber Eng.* (Austria: Univeristy of Eastern Finland) p. 1621

[23] Pentos K, Luczycka D, and Wysoczanski T 2017 Dielectric properties of selected wood species in poland *Wood Research* **62**(5) 727-736
[24] Kol HS 2009 Thermal and dielectric properties of pine wood in the transverse direction *BioResources* **4**(4) 1663-9
[25] Misato N 1976 Dielectric properties of wood *Bulletin of The Wood Research Institute Kyoto University* **59/60** 106-52
[26] Xu C, Chai H, Cao T, Cai M, Cai Y, and Liu H 2019 Detection of dielectric constant of Pinus sylvestris Var. mongolica and its influencing factors *BioResources* **14**(2) 4532-4542
[27] Torgovnikov GI 1993 *Dielectric Properties of Wood-Based Materials* (Springer-Verlag) p.135-59
[28] Kabir MF, Daud WM, Khalid KB, and Sidek HAA 2001 Temperature dependence of the dielectric properties of rubber wood *Wood Fiber Sci.* **33**(2) 233-238
[29] Sahin H and Ay N 2004 Dielectric properties of hardwood species at microwave frequencies *J. Wood Sci.* **50**(4) 375-80
[30] Li T, Cheng D, Avramidis S, Wålinder MEP, and Zhou D 2017 Response of hygroscopicity to heat treatment and its relation to durability of thermally modified wood *Constr. Build. Mater.* **144** 671-76

**Acknowledgment**
The authors would like to thank the Directorate for Research and Community Service, Ministry of Research Technology and Higher Education of the Republic of Indonesia (RISTEKDIKTI) for research grant 2017-2018.

IOP Conf. Series: Materials Science and Engineering **935** (2020) 012065 doi:10.1088/1757-899X/935/1/012065

# Verification of seismic resistant performance of developed original cross-laminated timber core structure method by shaking table experiment

**I Gunawan[1*], K Moritani[2], H Isoda[3], T Mori[4], M Shinohara[5], T Noda[1], R Hosomi[1], S Kurumada[5], and T Makita[6]**

[1]Structural Design Department 2, Kozo Keikaku Engineering Inc, Nakano-ku Honcho 164-0012 Tokyo, Japan
[2]Life Design KABAYA Inc. Nakasendo, Kita-ku Okayama City Okayama Prefecture 9-11 Japan
[3]Professor, Research Insitute for Sustainable Humanosphere, Kyoto University, Dr. Eng.
[4]Associate Professor, Graduate School of Advanced Science and Engineering., Hiroshima University, Dr. Eng.
[5]Doctor of Agriculture, Structural Design Department 2, Kozo Keikaku Engineering Inc, Nakano-ku Honcho 164-0012 Tokyo, Japan
[5]Meiken Lamwood Corp. Katsuyama, Maniwa City, Okayama Prefecture 1209 Japan
[6]BX Kaneshin Co., Ltd. Okudo, Katsushika City 4-19-12 Tokyo Japan

*Email: indra-gunawan@kke.co.jp

**Abstract.** In recent years, development of wood engineering is gradually increasing. Instead of using many wood columns, cross laminated timber is expected for constructing spacious open space building. Since cross-laminated timber has high rigidity and strength, cross-laminated timber is expected to be used as earthquake resistant wall or floor diaphragm that makes the span of building can be increased and the position of the wall can be adjusted openly. In order to optimize the performance of cross-laminated timber for open space building, original cross laminated timber core structure method was developed. In this paper, the development concept of original cross laminated timber core structure method will be explained. In this method, the joint connection for each element such as joint connection for wall-concrete foundation, wall-beam, and wall to hanging wall was also developed. The experiment to verify the strength and rigidity of each connection has been conducted and the result will be described. The shaking table experiment of 3-story open space building constructed by original cross laminated timber structure using varies earthquake waves was conducted. In this experiment natural period, shear force for each floor, story drift, and building response data is taken. The result shows the structure designed by original CLT core structure method is satisfy the requirement based on Japan cross-laminated panel structure regulation.

## 1. Introduction

In current era, structure engineering needs to consider the environmental problem for designing the structure. Sustainable design engineering is becoming more important, and is expected to be implemented for many engineering designs. Wood is one of the sustainable materials, and its material properties has many benefit for structure. Generally, wood has lighter weight comparing with other structural material such as concrete and steel. Besides that, timber structure also has enough strength to resist earthquake load.

Currently, there are two kinds of engineered wood usually used for structure, glued laminated timber (glulam) and Cross Laminated Timber (CLT). Glulam is made by many layers of wood boards bonded by moisture resistant structural adhesives. Glulam is usually used for structural column and beam. CLT is made by layers of lumber glued, each layers of boards are oriented perpendicular to adjacent layer. As we know that wood has more strength in the fiber parallel direction, CLT makes each layer perpendicular to each other so it has enough strength for parallel and perpendicular direction. [1] Since CLT has comparatively high in-plane rigidity, this engineered wood is usually used for structural wall and floor diaphragm.

Timber structure is commonly used for housing, but recently the demands of timber structure for buildings other than housing such as office, library, and open space buildings are increasing. For housing, many partitions are usually used so that column and structural wall can be put in the partition range. On the other hand, office, library or open space buildings have less partition, so the column is not expected to be put in the middle of the area. In this kind of building CLT is expected to be used.

## 2. Original Cross-Laminated Timber Core Structure Method Outline

Other than housing, currently timber structure is expected to be applied in office or other open space buildings. Frame structure method has many limitations due to less rigidity and strength, so it is very difficult to design large space room. To solve this problem, original cross-laminated timber core structure is proposed in this paper.

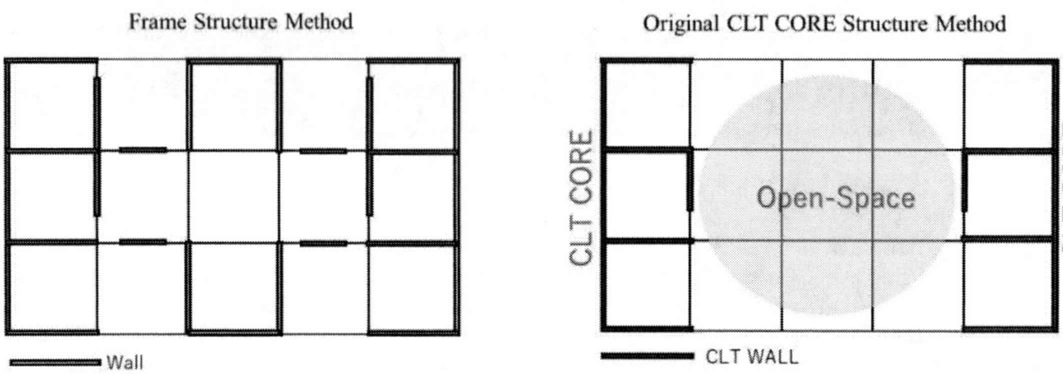


**Figure 1.** Comparison of frame structure method and original CLT core structure method

**Figure 2.** Design of building using original CLT core structure method

The main concept is to design the CLT panel as a core structure that will take horizontal load caused by wind or earthquake. CLT panel will be used as structural wall and floor diaphragm, so the column and beam only take the vertical load. Using this concept, wall to wall span can be increased so the large open space is easy to be designed. [2] The image concept of original CLT core structure method is shown in Figure 3.

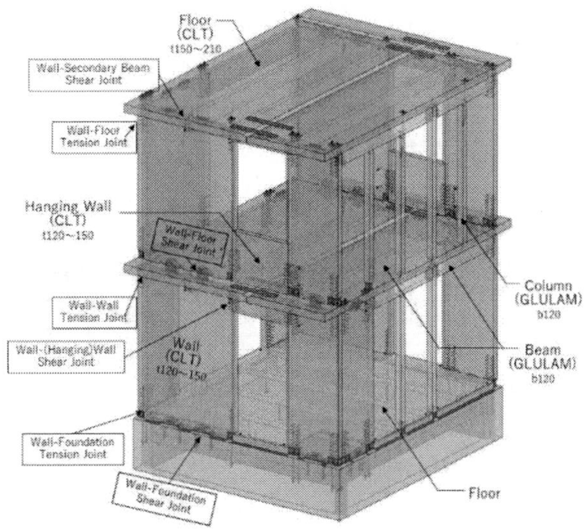


**Figure 3.** 3D image of original CLT core structure method building

## 3. Development of Metal Joint Connection

In original CLT core structure method, CLT wall is the main part of the structure. 100% of horizontal load is expected to be taken by CLT wall. So that the wall will have to transfer shear force to the floor or foundation by the shear joint connection. Due to the horizontal load, CLT wall will also have rocking movement, so tension joint connection is required. In original CLT core structure method, joint connection using metal is developed and the strength is verified by experiment [2].

### *3.1. Metal joint connection composition*

The connection used in original CLT core structure method and each connection composition is mentioned below:

- Wall-foundation tension metal joint connection
  This connection is made by round steel bar hanger with holes that will be inserted to the CLT wall and pinned by drift pin (φ12mm). The round steel bar hanger will be integrated to the box steel plate by high tension bolt (M20), and the box steel plate will be anchored to the foundation by anchor bolt (M20).
- Wall-foundation shear metal joint connection
  This connection is made by steel base plate (t=16) and ellipse-shaped steel plate. The ellipse-shaped steel plate will be inserted to the CLT wall and pinned by drift pin (φ12mm), the base plate will be anchored to the foundation by 4 pieces of anchor bolts (M16).
- Wall-floor shear metal joint connection
  This connection is made by steel plate connected with steel dowel and ellipse-shaped steel plate. The ellipse-shaped steel plate will be inserted to the CLT wall and pinned by drift pin (φ12mm), and steel plate with steel dowel will be inserted to the floor to transfer the shear force.
- Wall-wall (hanging wall) shear metal joint connection

This connection is made by U-shaped steel plate. The steel plate will be inserted to the beam or wall and pinned by drift pin (φ12mm), the plate will be connected to the CLT wall by lag screw bolt (M12) and steel bar.

- Wall-beam shear metal joint connection
  This connection is made by U-shaped steel plate. The steel plate will be inserted to the beam or wall and pinned by drift pin (φ12mm), the plate will be connected to the CLT wall by steel bar.

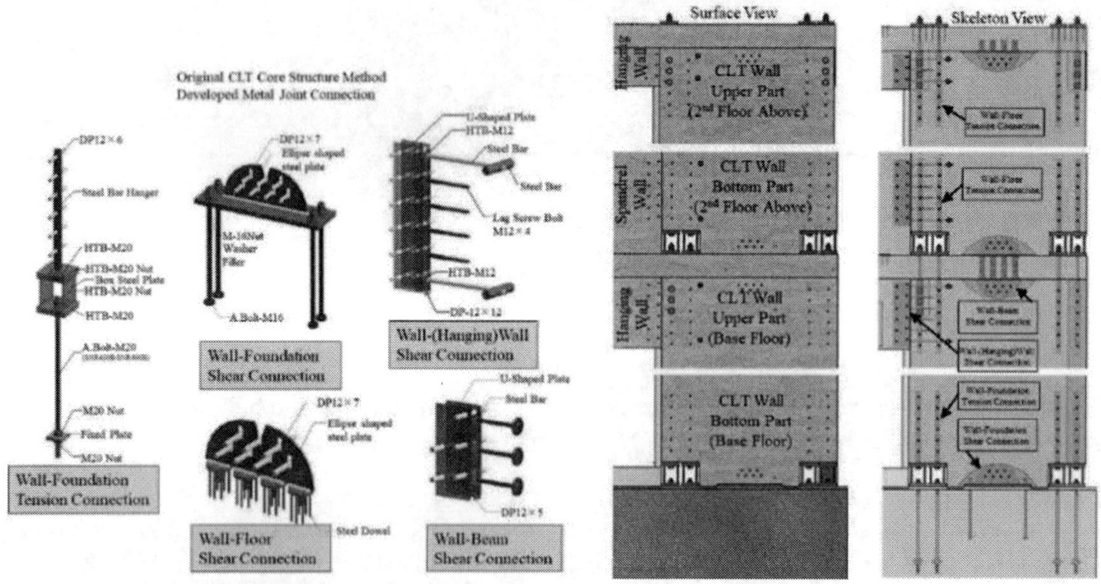


**Figure 4.** Metal joint connection of original CLT core structure method

### *3.2. Metal joint connection experiment results*

To confirm the characteristic and strength of the developed metal joint connection, the experiment of wall-foundation tension joint connection using single and double connection are conducted. In addition, wall-foundation shear connection and wall-floor shear connection are also conducted.

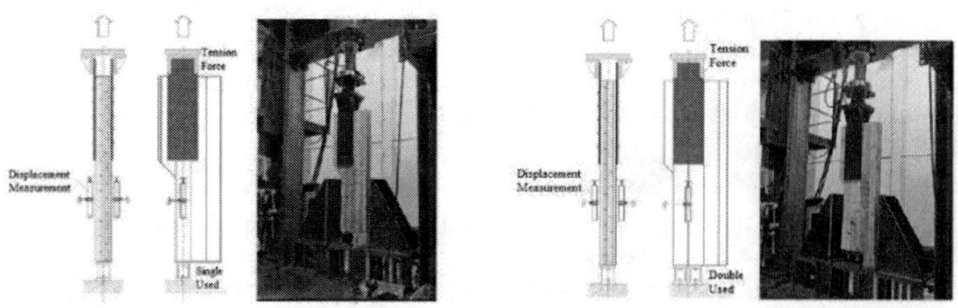

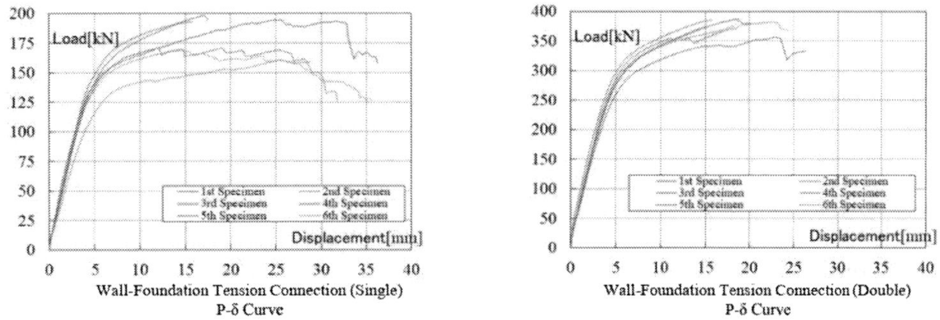


**Figure 5.** Experiment result of wall-foundation for single and double tension connection

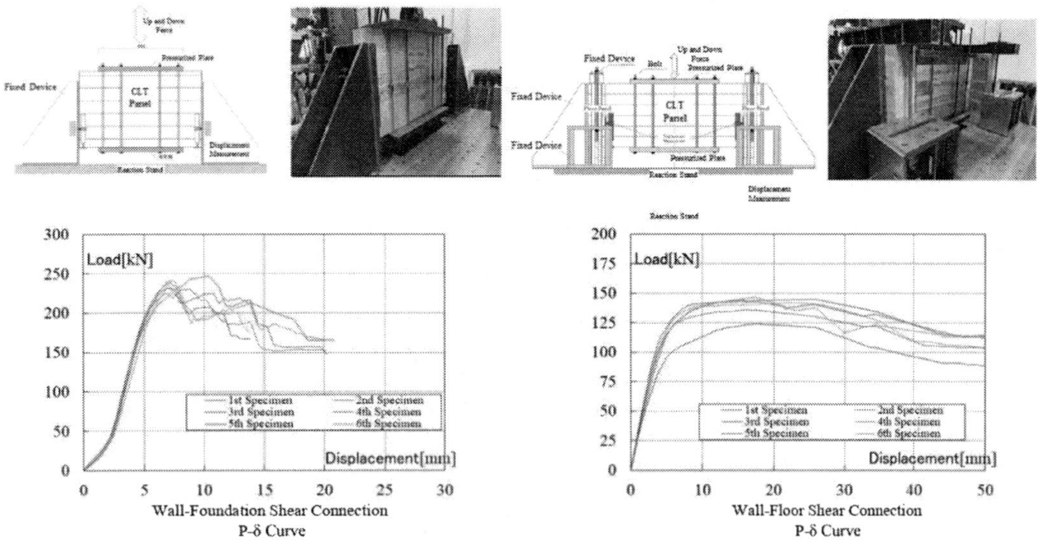


**Figure 6.** Experiment result of wall-foundation and wall-floor shear connection

Using the result of metal joint connection experiment, each metal joint connection rigidity and strength is evaluated below.

- Wall-foundation tension metal joint connection
  For wall-foundation tension metal joint connection, single used metal joint connection and double used metal joint connection are conducted to check whether the strength of using double used metal connection will be linear or nonlinear. For single used, in around 120 kN the drift pin start yielding and drift pin start compressing the CLT panel and finally the outer part of CLT panel is damaged and the test end.
  For double used metal joint connection, the characteristic is almost same with single used, the yield strength is around 2.0-2.1 times of single used, so it can be concluded that the strength of double used tension metal joint connection is increasing linearly.
- Wall-foundation shear metal joint connection
  In this connection experiment, in around 200-250 kN load, CLT panel start cracking and drift pin and steel plate is yielding.
- Wall-floor shear metal joint connection

IOP Conf. Series: Materials Science and Engineering **935** (2020) 012065 doi:10.1088/1757-899X/935/1/012065

In this connection experiment, in around 100-150 kN, steel dowel and nail connected to the floor panel is bending. Steel plate is yielding but no bending is observed at drift pin.

## 4. Shaking Table Experiment
To verify the effectiveness and design performance of original CLT core structure method, the specimen of real scale 3-story building is constructed and shaking table experiment is conducted [2].

### *4.1. Shaking table experiment specimen outline*
The specimen of shaking table test is a building that designed with original CLT core structure method. The building specimen has 3-story with maximum height of 9.588 m, rectangular shape of 7.5×7.6 m plan with 171m$^2$ is adopted. In the middle of the building CLT core structure is used, in the right and left side of CLT core structure, frame structure made by glulam is used. In CLT core structure, CLT floor made by Japan Cedar Mx60-5-7 (JAS) t=210 mm and CLT wall made by Japan Cedar S60-5-5 (JAS) t=210mm are used. In frame structure part, glulam column made by Europe Red Pine E95-F315 (JAS) with size 120 × 120 is used. In this experiment the load equivalent to apartment house in Japan is adopted. For 3$^{rd}$ floor 3.5 kN/m$^2$, for 2$^{nd}$ and 1$^{st}$ floor 5 kN/m$^2$ of load are applied. To give the load equivalent with the design load, steel plate and H-beam are used for loading in the specimen.

In this experiment the response of the structure such as structure natural frequency, story acceleration, and story drift are measured.

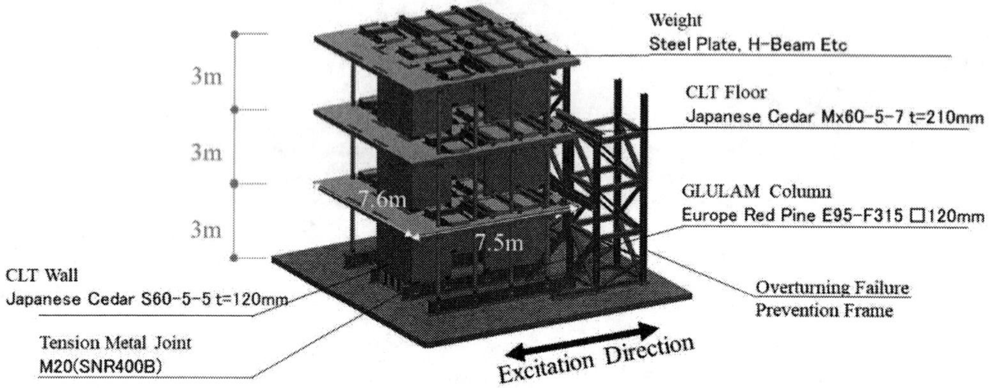


**Figure 7.** 3D model of shaking table experiment specimen

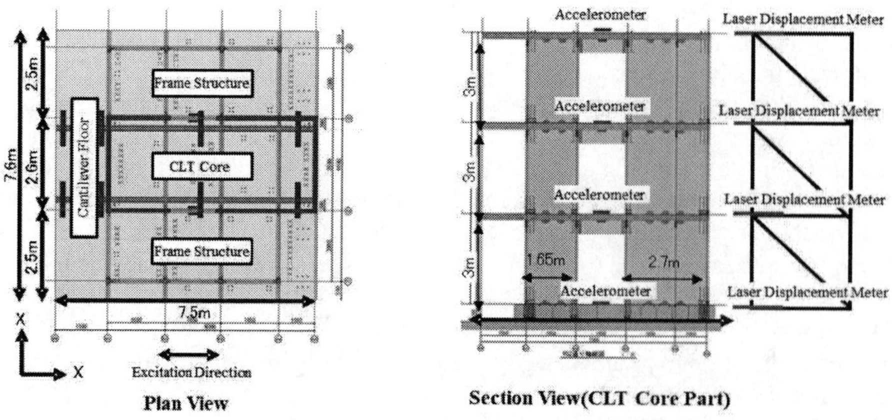


**Figure 8.** Plan view and section view of shaking table experiment specimen

### *4.2. Excitation load plan*

For the excitation load of this shaking table experiment seven earthquake waves are adopted. To check the reliability of the structure, earthquake wave equal to 500 years return period earthquake (BSL18%) and 2500 years return period earthquake (BSL90%) based on Japan Building Standard Law is adopted [3,4]. To understand the performance and behaviour of the structure other waves are also adopted. The list of excitation load plan and each response spectrum are mention in Figure 9.

| Input Earthquake Waves | Max Acceleration (gal) | Remarks |
|---|---|---|
| BSL(18%) | 91 | Return Period 500 years EQ |
| ElCentro NS(L1) | 254 | - |
| BSL(90%) | 455 | Return Period 2500 years EQ |
| BSL(135%) | 682 | 1.5 ×Return Period 2500 years EQ |
| BCJ(L2) | 356 | - |
| JMA Kobe_NS(100%) | 818 | - |
| JMA Kobe_NS(L2) | 451 | - |

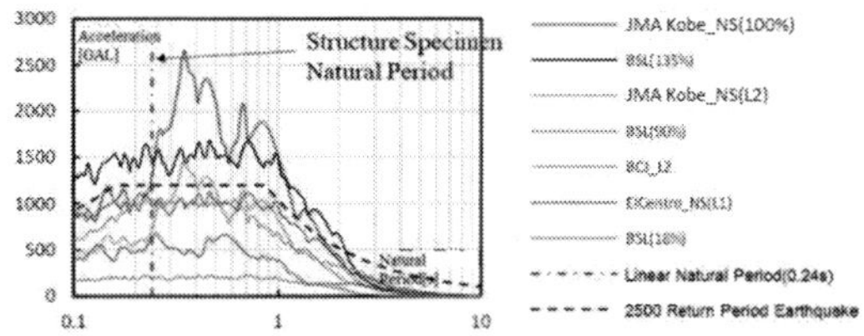


**Figure 9.** Excitation load plan and response spectrum

### *4.3. Experiment result*

The data taken in this experiment are structural natural period, maximum story drift, and story shear force-story drift. As a comparison, the experiment using single used tension joint and double used tension joint is also conducted. The result is presented in 4.3.1 until 4.3.3 subsection.

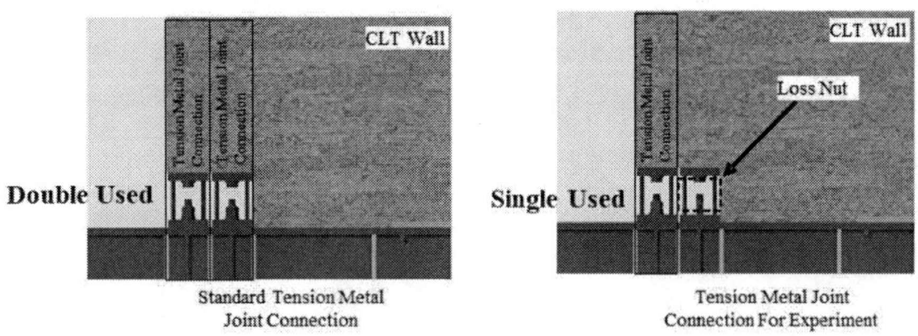


**Figure 10.** Double and single used tension joint specimen

#### *4.3.1.* structure natural period

After each excitation load, the structure specimen is loaded by random wave in small acceleration to check the structure natural frequency. This natural frequency will be compared to the initial natural period. By comparing the natural period, the damage to the structure can be noticed. The measurement result of structure natural period is shown in Figure 11.

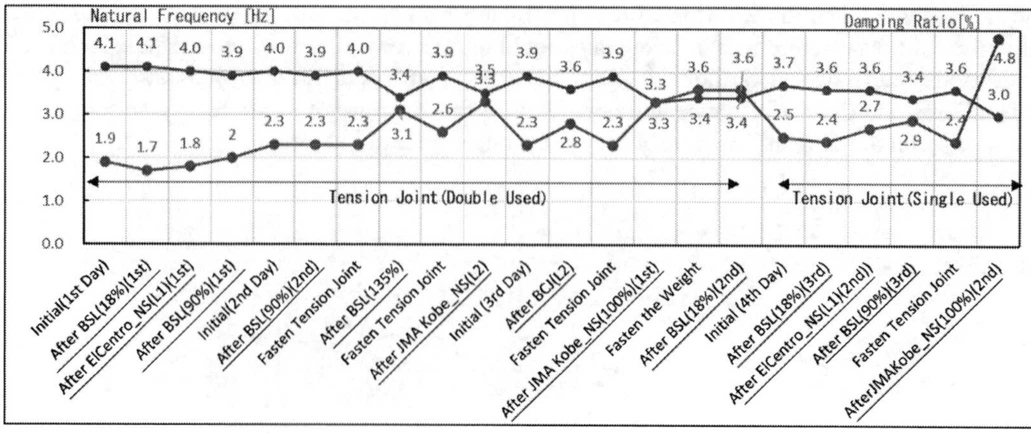


**Figure 11.** Structural natural period result and damping ratio

In the first day until third day specimen using double tension joint is tested. The initial structure natural frequency is about 4.1 Hz, after BSL (18%) and BSL (90%) the natural period is almost having no difference. After BSL (135%), the structure natural period drop to 3.4 Hz, means the rigidity of the structure is decreased, and yielding in some part of the structure is noticed. Same tendency come after JMA Kobe NS L2 and JMA Kobe NS (100%). The condition of the structure wall is almost has no damage, structure rigidity decrement is caused by the loose of the tension connection bolt. In the fourth day, specimen using single tension joint is tested. The initial structure natural frequency is about 3.7 Hz, after BSL (18%) the natural frequency is almost having no difference. After BSL (90%) the structure natural period drop to 3.4 Hz, means the rigidity of the structure is decreased and after JMA Kobe NS (100%) the structure natural frequency is largely drop to 3.0 Hz.

In Figure 11, the damping ratio is also shown. The damping ratio for the specimen is about 1.9%, while the structure natural period decrease, the damping ratio increase, means there is more energy absorption from yielding in some parts of the specimen.

*4.3.2.* Maximum story drift

According to Japan CLT panel structure regulation, the maximum story drift for earthquake with 500 years return period is 1/120 to prevent damage to the building, and for earthquake with 2500 years return period is 1/50 to prevent the collapse of the building [5-7]. The maximum story drift of specimen using double tension joint for BSL (18%) and BSL (90%) is less than 1/120. But for JMA Kobe NS (100%), the story drift is over 1/120 but less than 1/50, means the structure is damaged but still not collapsed. The specimen using single tension joint has bigger story drift, but for BSL (18%) and BSL (90%) the story drift is less than 1/120. For JMA Kobe NS (100%) the story drift is over 1/120 and almost reach 1/50, the structure takes big damage and almost collapse. The maximum story drift measurement result is summarized in Figure 12.

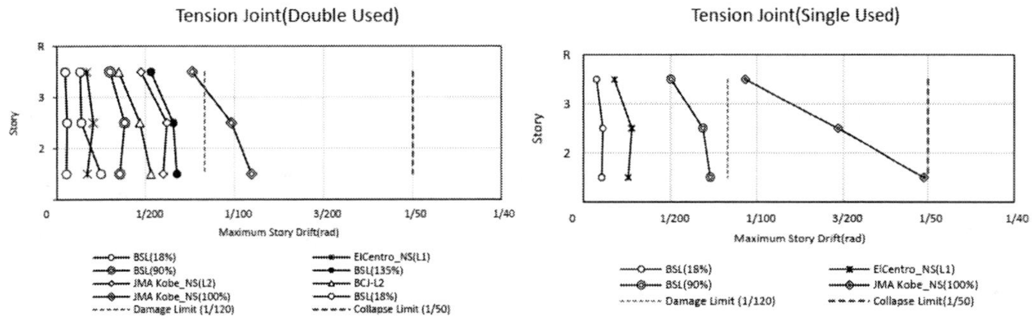


**Figure 12.** Maximum story drift measurement result

*4.3.3.* Shear force-story drift
For comparison according to Japan CLT panel structure regulation, CLT panel structure shall be designed to resist 0.55g of horizontal load [4,8]. The result of shear force and story drift are shown in Figure 13.

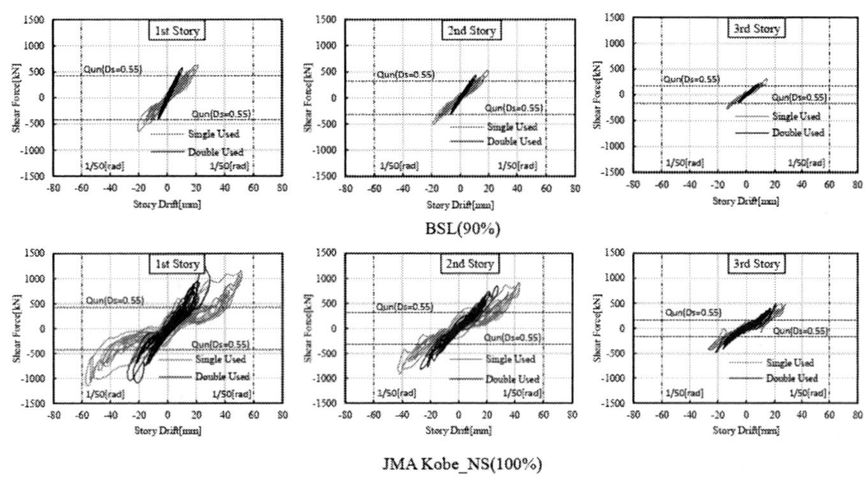


**Figure 13.** Shear force-story drift graph

For specimen using double tension joint, in BSL (90%) shear force-story drift graph, the shear force in structure is about 0.55g of horizontal load range, and the graph show linear line, means in this excitation period there is almost no yielding element or decrement of rigidity. In JMA Kobe NS (100%) the shear force in structure is over 0.55g of horizontal load range and the structure story drift does not exceed 1/50 rad, means the structure is still below collapse limit in this excitation period.

For specimen using single tension joint, almost same tendency can be seen, but the larger displacement for the same shear force is confirmed.

## 5. Conclusion
With original CLT core structure method, larger span and open space timber structure can be more easily designed. The shaking table experiment to test the performance of structure designed by original CLT core structure method shows that the structure is satisfied the Japan CLT panel structure regulation and can resist the earthquake larger than 2500 years return period earthquake.

IOP Conf. Series: Materials Science and Engineering **935** (2020) 012065 doi:10.1088/1757-899X/935/1/012065

**References**

[1] Zmijewki T and Wojtowicz-Jankowska D 2017 Timber material of the future: examples of small wooden architectural structures *IOP Conferences Series: Materials Science and Engineering* **245** 1-9

[2] Moritani K, Isoda H, Mori T, Shinohara M, Hosomi R, Noda T, Kurumada S, and Makita T 2018 Development of original construction method with combination of CLT core wall and wood frame member No.1-4 *Summaries of Technical Papers of Annual Meeting, Architectural Institute of Japan* pp.127-34

[3] Architectural Institute of Japan 2016 *Standard for Structural Design of Timber Structures* (Tokyo: Maruzen Shuppan)

[4] Japan Housing and Wood Technology Center 2016 *Building Design and Construction Manual Using CLT*

[5] Architectural Institute of Japan 2012 *Design Practice for Engineered Timber Joints*

[6] Ministry of Land, Infrastructure, Transport and Tourism National Institute of Land and Infrastructure Policy Research 2017 *Allowable Stress Design of Wooden Frame for Housing Construction*

[7] Ministry of Land, Infrastructure, Transport and Tourism National Institute of Land and Infrastructure Policy Research 2015 *Structural Standard and Technical Standards Manual*

[8] Japan CLT Association and others 2016 *Project Report of Improvement of Wooden Construction Standards Using CLT*

IOP Conf. Series: Materials Science and Engineering **935** (2020) 012066 doi:10.1088/1757-899X/935/1/012066

# Performance of helical router bit in milling of composite boards

**K T P Pangestu[1], W Darmawan[1*], D Nandika[1], I Wahyudi[1], and H Usuki[2]**

[1]Department of Forest Product, IPB University, Bogor, Indonesia
[2]The University of Tokyo, Japan

*Email: wayandar@indo.net.id

**Abstract.** An extreme inclination angle of helical cutting tool edge has been developed in milling of new commercially composite boards. The purpose of this research was to investigate wear resistance, chip formation, surface roughness, and noise level of helical edge router bits in milling composite boards. The composite boards of wood plastic composite, laminated veneer lumber, and strand board were cut by the helical edge router bits (helical angle 15°, 30°, 45°, 60°, and 75°) in a computer numerical control router. The results show that the helical edges provide better wear resistance, smoother surface of machined composite boards, and lower noise emission compared to the conventional edge (0°). Edge wear, surface roughness, and noise level decreased with increasing helix angle of the edge router bits. The differences in structure among the composite boards resulted in the difference in cutting tool wear, chip formation, surface roughness, and noise level phenomenon. The 75° inclination angle of router bit provided the highest wear resistance, better surface of machined composite board, and the lowest noise level.

## 1. Introduction

Composite board is an alternative wood product that manufactured from wood waste or low-quality of wood with combination of an artificial resin and additives of different kinds. Nowadays, composite boards such as Wood Plastic Composite (WPC), Laminated Veneer Lumber (LVL), and Strand Board (SB) have become popular products because of their good in strength, durability, and aesthetic for building construction and decorative purposes. In the composite board manufacturing industry, where the composite boards are machined using a cutting tool, a shape design of cutting tool edge becomes an important economic parameter. The right design of cutting tool edge would optimize and improve the cutting operation resulting better efficiency, stability, accuracy, and tool life during the cutting processes.

Today conventional designs of router bit with straight cutting edge is widely used in the wood working industry for milling purposes. However, the use of straight configuration of cutting tool edge produces several problems during milling. The straight cutting tool edge hits and intermittently engages the surface of the work piece generating a high noise emission, a dull cutting tool edges, and an extreme flight speeds of the formed chips which causes dust emission [1,2]. In addition, this straight configuration also leads to machined surface quality problem due to high splitting, compressing, and damaging the wood cell structure near the surfaces, which causes a costly sanding procedure is needed [2].

An innovation in minimizing the problems of the straight cutting tool edge configuration with a new design of helical edge has been developed. Some research works and investigations were done to find out the effect of inclination angles of the helical cutting tool edge for wood cutting applications. The research on the investigation of dust, chip, and noise behaviors in milling operation using helical cutting tool edge with inclination angles between 0° to 45° were reported [3,4]. It was noted in these studies that inclination angles between 5° and 10° were considered to be useful in lowering dust emissions. Inclination angles larger than 10° lead to chip compression resulting in high axial forces on the cutting tool. The increase in the inclination angle from 0° to 8° leads to a significant decrease in noise emissions. It was found in another study that helical cutting tool edge with inclination angle of 0° to 8° did not provide a significant difference in the chips morphology when milling with shank tools of small diameter

[5]. In addition, the helical edge design of a milling tool improves the quality of machined surface compared to straight cutting edge [6].

It could be considered that when cutting tool edge with the helical configuration engages the surface of the work piece gradually, it is expected that the tools will be always under contact, stability will be improved, and the vibration will be reduced during the milling operation. Due to the unique geometry of cutting tool edge, the work piece chips will be easily deformed and the resulting machined surfaces will be flat and smooth. Considering the fact that the previous researches were limited to the helix angles between 0° and 45°, a new design of helical edge of router bits with extreme inclination angle has been worked and their performance was tested in this research. The aim of this research was to investigate the effect of inclination angles on the edge wear, chip formation, surface roughness, and noise emission of the extreme helical router bit in milling composite boards.

## 2. Materials and Methods

### *2.1. Materials*

Router bits of High Speed Steel (HSS) with helical edges were 80 mm in total length and 12 mm in cutting circle diameter. The router bits consisted of two solid cutting edges with helix angles of 0° (straight edge), 15°, 30°, 45°, 60°, and 75° (Figure 1). Other geometries of the helical router bits are shown in Table 1. The composite boards routed was WPC, LVL, and SB with density of 135 kg/m$^3$, 83 kg/m$^3$, and 59 kg/m$^3$, respectively.

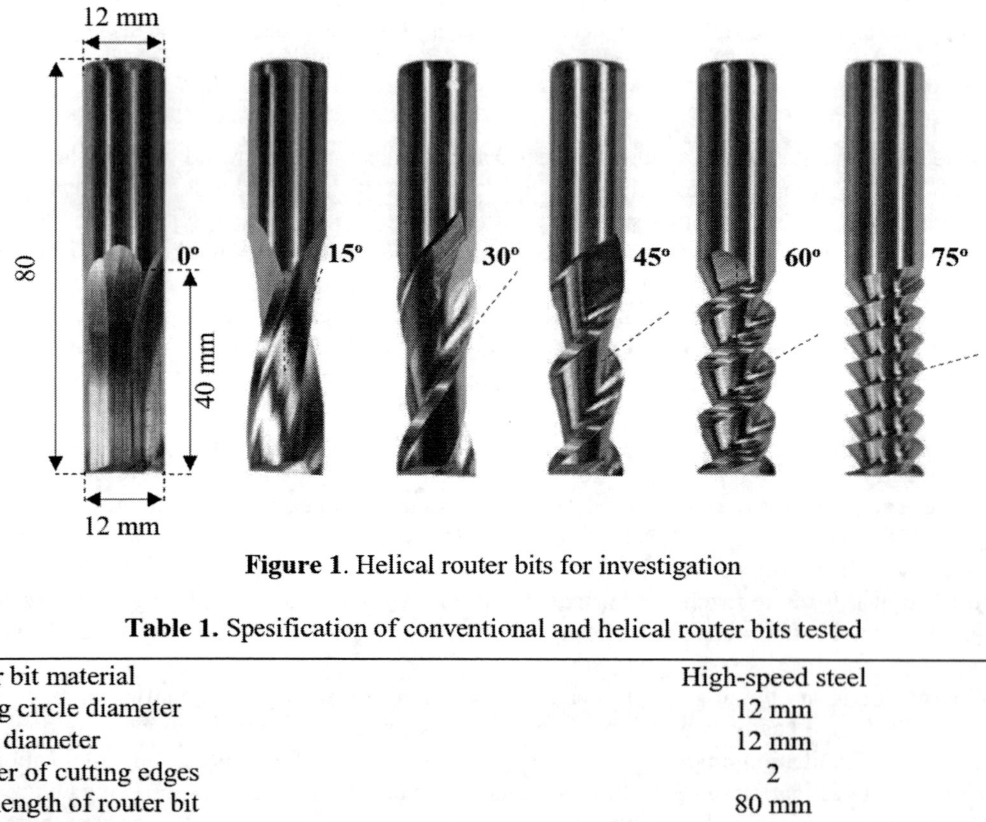


**Figure 1**. Helical router bits for investigation

**Table 1.** Spesification of conventional and helical router bits tested

| | |
|---|---|
| Router bit material | High-speed steel |
| Cutting circle diameter | 12 mm |
| Shank diameter | 12 mm |
| Number of cutting edges | 2 |
| Total length of router bit | 80 mm |
| Length of cut | 40 mm |
| Geometry of the edges | |
| Helix angle | 0° (straight edge), 15°, 30°, 45°, 60°, 75° |
| Orthogonal rake angle | 22° |
| Orthogonal clearance angle | 15° |

### *2.2. Methods*

#### 2.2.1. Milling test

Milling tests were set up on a commercial Computer Numerical Control (CNC) router. Up-milling and down milling processes were performed by setting the rotation of the router spindle in clockwise direction and by feeding the lumber samples in the proper direction with rotation of the spindle. The samples were routed along the length on their side surfaces. Conditions of the milling are shown in Table 2.

**Table 2.** Milling test conditions

| Adjusted parameters | Cutting condition |
|---|---|
| Milling process | Up-milling, down-milling |
| Cutting speed (m/s) | 12.6 |
| Spindle speed (rev/min) | 20,000 |
| Feed speed (mm/min) | 4000 |
| Cutting width (mm) | 1 |
| Cutting depth (mm) | 20 |

#### 2.2.2. *Wear measurements.*

The router bits were inspected before testing to ensure that there are no surface cracks and chippings on cutting edge under an optical video microscope. In this work, the wear resistance was determined by measuring the progress in edge radius at every end of cutting (60 cm). The captured images of the cutting edges were analysed under the ImageJ software for measurements of the radius.

#### 2.2.3. *Chip distribution*

Investigation of the chip distribution was carried out by mesh analysis of the formed chips. The deposited chips on the CNC table were collected and documented. The collected chips were sieved by steel screens of 10 mesh (diameter of holes 2.0 mm), 20 mesh (diameter of holes 0.8 mm), 40 mesh (diameter of holes 0.4 mm), and 80 mesh (diameter of holes 0.2 mm). The distribution of chips was analyzed according to the weight percentage of each screened chip.

#### 2.2.4. *Surface quality measurement*

The machined surface was evaluated in the Ra value of surface roughness. The surface roughness tester SJ-210 was used to measure the roughness on the surface of the machined composite boards. The Ra was measured across the cutting direction of samples with a diamond tip radius of 5 μm. The tracing length was 15 mm and the cut off was 2.5 mm. Five points for roughness measurement were diagonally marked on the routed surface of the samples.

#### 2.2.5. *Noise measurement*

A precision sound level metre Sanfix 356 was used for measurement of the sound level of the audible cutting noise on the A scale, which is usually used for measuring the peak of sound pressure level. The sound level metre was set up at the height of the cutting tool edge (about 1 m above ground level) and at a distance of about 1 m along a straight line extending from the cutting tool edge. The noise level was recorded per 1 second at every milling test, and the values was transferred to a computer for analysis.

## 3. Results and Discussion

### *3.1. Wear resistance*

The results in Figure 2 indicate that the conventional router bits (straight edge/inclination angle 0°) provided lower wear resistance compared to the helical router bit when milling WPC, LVL, and SB. The edge radius wear decreased with increasing inclination angle of router bits. It could be considered that a larger inclination angle would involve a gradual step wise engagement on the surface of machined composite boards. It could eliminate sudden impacts and decrease mechanical loads on the edge surface of router bit.

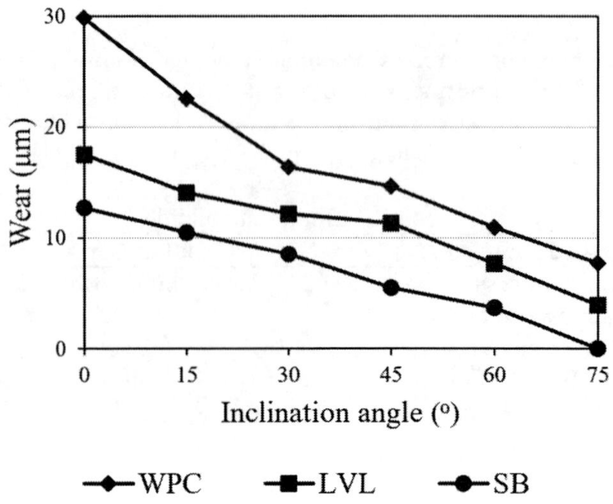


**Figure 2.** Helical router bits wear when cutting composite boards

The amount of edge radius wear resulted from milling the composite boards was different. The average edge radius wear in milling WPC, LVL, and SB were 12.7 µm, 9.6 µm, and 6.3 µm, respectively. The highest density of WPC (135 kg/m$^3$) compared to LVL (83 kg/m$^3$) and SB (59 kg/m$^3$) could be the reason for this phenomenon. In addition, it was noted that the main components of WPC are plastic polymer, lignocellulose materials, fillers (mostly crystalline silicates, titanium dioxide, and other heavy metals), stabilizers and other additives [7,8]. The presences of abrasive materials and additives in the WPC lead to severe mechanical wearing on the cutting tool edge tested [9].

### *3.2. Chip distribution*

The results in Table 3 show that the chip of WPC was more uniform in size compared to the chip of LVL and SB. The chip of WPC mostly netted on 20 mesh sieves. The chip of LVL varied on 20 mesh to pass through 40 mesh sieves. The chip of SB mostly netted on 40 mesh and pass through 40 mesh sieves. It was reported that the chip type netted on 10 mesh, 20 mesh, 40 mesh, and pass through 40 mesh were spiral or splinter, flow, thin, and granule chips, respectively [5,10]. A large amount of granule chips of SB was observed to cause various problems. First, the granule chips tended to adhere and harden on the cutting tools surface which would affect the subsequent cutting. Second, the chips led to produce serious air pollution in the work area.

In Table 3, it was found that the spiral or splinter chip (netted on 10 mesh sieve) generated from cutting composite boards using larger inclination angle of router bits. With increasing of the inclination angle, the spiral or splinter chip was more generated. This chip type was formed from incompletely severed chips through compression on the surface of composite boards. The spiral or splinter chip of WPC was generated only using 75° helical router bit. In milling of LVL, it was generated using 60° and 75° helical router bit. It was generated in milling SB using 45° to 75° helical router bit. This result indicated that the spiral or splinter chip was easier to generate from milling of SB compared to other composite boards. The lowest density and the irregularity in sizes and orientation of the strands in SB could be the reasons for this phenomenon.

**Table 3.** Weight percentage of chips generated during up-milling process

| Composite board/ Incilnation angle (°) | Weight percentage (%) | | | |
|---|---|---|---|---|
| | 10 mesh (> 2 mm) | 20 mesh (0.8-2 mm) | 40 mesh (0.4-0.8 mm) | >40 mesh (< 0.4 mm) |
| WPC | | | | |
| 0 | 0.0 | 94.8 | 4.1 | 1.1 |
| 15 | 0.0 | 91.0 | 7.6 | 1.4 |
| 30 | 0.0 | 74.6 | 19.5 | 5.9 |
| 45 | 0.0 | 93.9 | 4.4 | 1.8 |
| 60 | 0.0 | 97.1 | 2.0 | 0.9 |
| 75 | 5.3 | 90.4 | 3.4 | 0.9 |
| LVL | | | | |
| 0 | 0.0 | 27.7 | 37.2 | 35 |
| 15 | 0.0 | 34.5 | 37.7 | 27.8 |
| 30 | 0.0 | 40.2 | 34.0 | 25.8 |
| 45 | 0.0 | 39.1 | 33.5 | 27.4 |
| 60 | 0.3 | 48.3 | 29.2 | 22.2 |
| 75 | 0.4 | 25.0 | 38.1 | 36.5 |
| SB | | | | |
| 0 | 0.0 | 5.2 | 41.4 | 53.4 |
| 15 | 0.0 | 7.3 | 41.3 | 51.4 |
| 30 | 0.0 | 7.8 | 43.1 | 49.1 |
| 45 | 0.5 | 7.2 | 41.0 | 51.3 |
| 60 | 1.0 | 7.5 | 38.6 | 52.9 |
| 75 | 2.4 | 8.1 | 37.8 | 53.4 |

### *3.3. Surface roughness*

The result in Figure 3 shows that the roughness of machined composite boards (Ra) decreased with increasing inclination angle of router bits tested for both down-milling and up-milling process. This phenomenon could be due to the decrease in router bit wear with increasing the inclination angle. The mechanical abrasion of cutting tools resulted in an irregularity on the cutting edge, which led to irregularities on the surface of wood composites [11]. It was also found in Figure 3 that up-milling process generated rougher surface compared to down-milling process. It was reported in previous another study that the down-milling process is better for workshop application than the up-milling process because the higher periodic cutting force in the latter is considered to be more damaging to the surface [12].

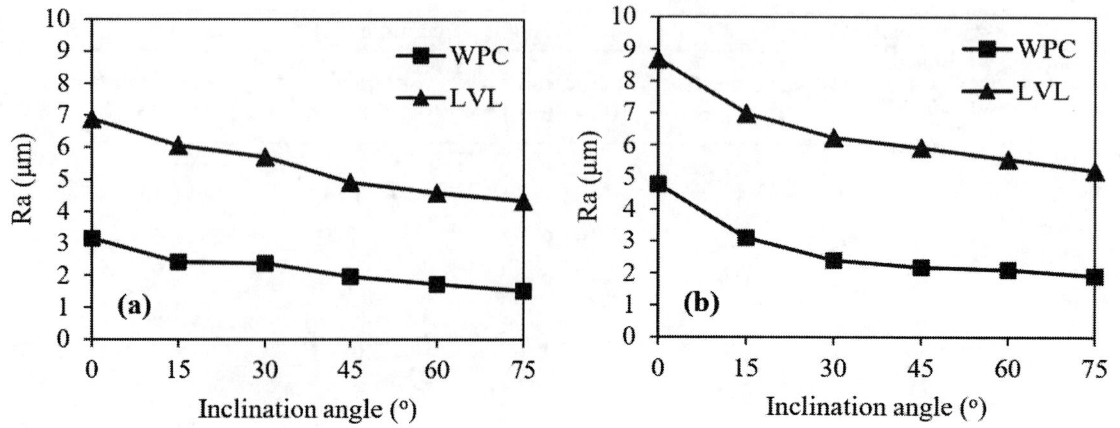


**Figure 3.** Behaviour of surface roughness with respect to the inclination angles of milling cutters at down-milling (a) and up-milling (b) process

The smoothest surface of machined composite board was retained by WPC. The very fine structure of WPC led to generate the smoothest surfaces. However, the surface roughness value of SB was unmeasurable since the Ra of SB much high beyond the maximum range that could be measured by the surface roughness tester. It was caused by the porous in SB structure due to the irregular in sizes and orientation of the strands in the board (Figure 4c).

(a) (b) (c)

**Figure 4.** Appearance of WPC (a), LVL (b), and SB (c) machined using 75° helical router bit in down-milling process

### *3.4. Noise level*

The results in Figure 5 indicate that the conventional router bits generated higher noise levels compared to helical router bits during milling of composite boards. The higher noise level generated by conventional router bit is caused by intermittent hammering and hit of its straight edge into the surface of the composite boards. The noise level decreased with increasing of inclination angle of router bits tested for both down-milling and up-milling process. The noise level remarkable reduced when cutting composite boards using the 45° helical router bit, and then decreased gradually when cutting composite boards using the 75° helical router bit. The helical edge of router bit engaged and penetrated gradually into the surface of the composite boards, which caused the noise to decrease.

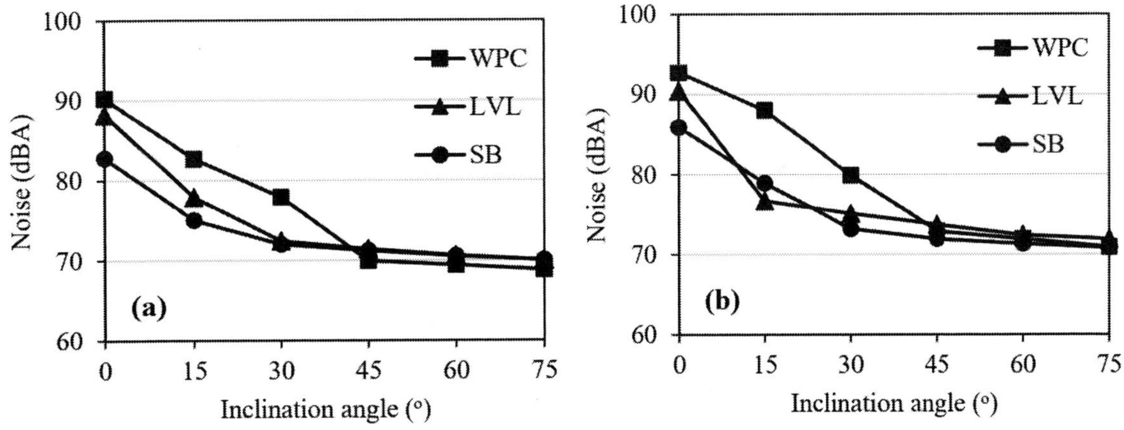


**Figure 5.** Behaviour of noise level with respect to the inclination angles of milling cutters at down-milling (a) and up-milling (b) process

It was also found in Figure 5 that up-milling process (77 dB) tended to generate higher noise level compared to down-milling process (75 dB). It could be considered that up-milling process caused a large impact forces being imposed on the edge router bits [13]. It appears in Figure 5 that the highest and the lowest noise level were generated in cutting WPC (78 dB) and SB (75 dB), respectively. This fact was caused by the highest density of WPC and the lowest density of SB.

In conclusion, both in up-milling and down-milling, the helical router bit provide better wear resistance, better surface quality, and lower noise emission compared to the router bit with conventional edge. The characteristics of composite boards structure would affect the behavior of wear resistance, chip formation, quality of machined surfaces, and noise emission during cutting. The investigations have clearly confirmed that the helical edge is considered to be a valuable design to improve upon the performance of the conventional router bit for composite board milling applications.

## References

[1] Chen WF and Lai HY 2002 *J. Eng. Manuf.* **216** 1493-504
[2] Darmawan W, Gottlober C, Oertel M, Wagenfuhr A, and Fischer R 2011 *Eur. J. Wood Prod.* **69** 565-72
[3] Heisel U and Weiss E 1989 *Holzbearb* **12** 16-21
[4] Heisel U, Niemeyer W, and Weiss E 1993 *Holzbearb* **5** 90-98
[5] Su WC and Wang Y 2002 *J. Wood Sci.* **48** 126-31
[6] Cyra G, Tanaka C, Yoshinobu M, and Nishino Y 1998 *J. Wood Sci.* **44** 169-76
[7] Niska K O and Sain M 2008 *Wood Polymer Composites* (New York: CRC Press)
[8] Masoudifar M, Nosrati B, and Gargari R M 2018 *Int Wood Prod J.* **9** 176-85
[9] Pangestu KTP, Darmawan W, Nandika D, and Usuki H 2019 *Int. Wood Prod. J.* **10** 1-8
[10] Darmawan W, Azhari M, Rahayu I S, Nandika D, Dumasari L, Malela I, and Nishio S 2018 *J. Indian Acad. Wood Sci.* **15** 72-180
[11] Darmawan W, Quesada J, Rossi F, Marchal R, Machi F, and Usuki H 2009 *Eur. J. Wood Prod.* **67** 247-55
[12] Ozoegwu CG, Omenyi SN, Ofochebe SM, and Achebe CH 2013 *Appl. Math.* **3** 1-11
[13] Darmawan W and Tanaka C 2004 *Ann. For. Sci.* **61** 731-6

IOP Conf. Series: Materials Science and Engineering **935** (2020) 012067 doi:10.1088/1757-899X/935/1/012067

# Natural weathering on coated tropical woods

**A T Lestari[1*], W Darmawan [2] and D Nandika[2]**

[1]Forestry Department, Agriculture Faculty, Mataram University, Mataram, Indonesia
[2]Department of Forest Products, Faculty of Forestry, IPB University, Bogor, Indonesia

*Email: atlestari@unram.ac.id

**Abstract**. Teak and pine are common wood species in Indonesia. The purpose of this research is to evaluate the performance of the coating layer in these wood species after six months natural weathering. The durability on natural weathering is evaluated due to microbial disfigurement (fungi or algae), checking and cracking, whereas microbial disfigurement (fungi or algae) is evaluated using ASTM D 3274-95, checking with ASTM D660-93, and cracking with ASTM D 661-93. Wood samples of 30 (l) x 15 (w) x 2 (t) cm are sanded before painted, then preservatives, water repellents, and its combinations are applied on the surface of the samples. The results showed that the coating layer of the combination of paint and preservatives and the combination of paint and water repellents could provide better protection of the surface on long rotation teak, short rotation teak, and pine boards against microbial disfigurement, checking and cracking compared to the other coating layers after 24 weeks of natural weathering. The coating layers on long rotation teak and short rotation teak gave better protection against microbial disfigurement, checking and cracking than pine wood. The coating layer on the radial section gave better protection compared to the tangential section. However, the durability of the coating layer on the radial and tangential section is not significantly different.

## 1. Introduction

Wood as a structural material has not been completely replaced by other materials. Not only because it is strong but also wood has an attractive appearance that cannot be found in other materials. However, wood can be degraded by environmental influences like physically (e.g. changing of water content continuously), biologically (e.g. insects, fungi, bacteria, and marine borer), also chemically (e.g. ultraviolet radiation). One of the solutions to protect the wood from degradation is finishing. Finishing is the final treatment in the woodworking process by applying a coating layer to protect and also to beautify its appearance [1].

One of the most common woods in Indonesia is teak wood (*Tectona grandis* Linn. F.). Demand for teak wood remains high due to wood characteristics that are suitable for various purposes such as materials buildings, furniture, handicrafts, and others [2]. The characteristic is related to surface properties and affect the coating where teak wood can be lacquered and polished well [3] and also affects the absorbency, adhesion, and finishing attributes of wood [4]. On the other hand, the teak wood supply continues to decline from year to year.

The limited of long rotation wood can be overcome by utilizing fast growing species like short rotation teak and pinewood [5]. Short rotation teak and pine wood are common fast growing tropical wood species in plantation forests in Indonesia and likely to be the mainstay raw materials for the wood industry. These short rotation plants are already in demand for use in the light frame structural, woodworking, and wood composite board industries as the main source of raw material and are becoming leading wood product exports [6]. However, fast growing wood species is not recommended for exterior use because it is very susceptible to fungi [7].

Since fast growing wood species are often of inferior quality (e.g. low density, low dimensional stability, low durability, etc.), many modification techniques are used to improve wood properties such as heat treatment [8] and applying a coating layer for exterior use [9]. Exposed outdoors timber needs protection from the effects of sunlight and rain. Surface damage occurs due to oxidation on wood surfaces whose reaction is accelerated by ultraviolet (UV) light from sunlight and accelerated also by other processes such as leaching by rainwater, temperature changes, and abrasion by particles due to wind gusts.

The difference in the coating layer can produce varied quality. However, the coating layer has good quality nor does it necessarily match the structure of certain wood surfaces. The macroscopic structure of wood greatly affects the spread and penetration of the coating layer liquid. The relationship between the macroscopic structure of wood and the ability to drain fluid coating layers has been carried out by several researchers, both on hardwood and softwood [10,11].

Coating techniques with exterior paint have been studied and proven to extend the wood life. Demirci *et al.* [12] uses varnishes on several species of wood to see the effect of temperature and humidity and to determine the sticky strength of varnishes. The results show that varnish applications can protect the wood from changes in temperature and humidity. The main factors that influence sticking strength are moisture content and the level of roughness of wood surfaces.

The most common coating layer is finishing exterior like paints and preservatives. This coating layer can protect the wood from temperature changes. The durability of the paint on wood can decrease due to the influence of the weather. This condition occurs on the surface of all organic materials including wood. Exterior paint used in wood can also experience degradation due to photo oxidation. Exterior paint that forms a thin film can be eroded due to damage to organic polymers because of weathering. In essence, weathering can damage wood and exterior paint layers [13]. Water repelling and permeation on wood surfaces that are only coated with exterior paint are not good. This is because basically giving exterior paint aims to provide wood resistance from weather effects, but exterior paint is not able to withstand the water penetration. To overcome this, it is necessary to make a coating process with water repellents.

Water repellents ingredients can increase wood resistance to the penetration of water particles into the wood or allow the wood to refuse the water penetration. The ability to resist water can be realized through a mixture of wax applications (*beeswax*), oil, or similar substances that can withstand the rate of water. Bees wax is one type of candle that is composed of 71% ester wax, 15% hydrocarbons, fatty acid, and 6% other materials. The ability of water repellents to resist water allows the wood to counteract decay and discoloration. Addition of preservatives and water repellents will increase the effectiveness of coating materials or finishing especially in inhibiting fungal growth. Water repellents can also prevent the occurrence of shrinkage which can cause cracking and warping. These materials can also protect wood that has been given an exterior coat of paint from the sun which can cause the checking of the finishing material.

The information regarding the effectiveness of paint, preservatives, water repellents, and its combination are needed especially those tested for Indonesian tropical wood species that have not widely known. Therefore, this study is conducted to evaluate the effectiveness of the application of coating layer (paint, preservatives, water repellent, and their combination) on long rotation teak and fast-grown wood species (short rotation teak and pine) for exterior use.

## 2. Materials and Methods

### *2.1. Materials*

Samples are prepared according to ASTM D 358-98 [14] from long rotation teak, short rotation teak, and pine. Long rotation teak 45 years old with diameter 45 cm; short rotation teak 10 years old with diameter 25 cm, and Pine 10 years old with diameter 42 cm. Logs are cut using table circular saw machine to produce the samples measuring 30 x 15 x 2 cm. The samples then conditioned at 21°C until reaching dry air content 12 to 15%.

The paint used is an ultran lasur with a solvent based type. Ultran lasur is a deep penetrating protective wood coating specially designed for tropical climate. Water repellents made from beeswax with composition of wax ester 67%, hydrocarbon 14%, acid fat 12%, and other materials 7%. The function of water repellents is to obstructing the water absorption. The preservative used is fungicide substance/material active thiocyanomethylthio benzotiazole (TCMTB). Variations of the coating layer are shown in Table 1.

**Table 1.** Variations of the coating layer.

| Codes | Treatments | Composition of the coating layer |
|---|---|---|
| K | Control (uncoated) | - |
| F | coated with paint | 100% paint |
| P | coated with preservatives | 100% preservatives |
| W | coated with water repellents | 100% water repellents |
| FP | coated with combination of paint and preservatives | 20% preservatives, and 80% paint |
| FW | coated with combination of paint and water repellents | 20% water repellents, and 80% paint |
| FPW | coated with combination of paint, preservatives and water repellents | 10% preservatives, 10% water repellents, and 80% paint |

### *2.2. Methods*

Test samples sanded with sandpaper number 80, 100, and 150 to get a smooth surface. Each variation of the coating layer is applied in test samples. The coating layer is sprinkled with brush 2 times. Before the second layer, the test sample was dried and lightly sandpapered using a 400 sized paper. Then the sample was dried to form a strong coating. The samples are exposed outdoor to test their resistance against weather. This evaluation is to observe the damage that occurs in the coating layer and classify the durability of the coating layer on 11 classes (Table 2). Surface damage due to microbial disfigurement (fungi or algae) was evaluated using ASTM D 3274-95 [15], checking with ASTM D660-93 [16] and, cracking with ASTM D 661-93 [17].

**Table 2.** Classification of the condition of the coating layer after natural weathering [15-17]

| Rating | Percentage of blemished surfaces (%) |
|---|---|
| 0 | 100 |
| 1 | 90 |
| 2 | 80 |
| 3 | 70 |
| 4 | 60 |
| 5 | 50 |
| 6 | 40 |
| 7 | 30 |
| 8 | 20 |
| 9 | 10 |
| 10 | 0 |

## 3. Results and Discussions

The durability of the coating layer is determined based on microbial disfigurement, checking and cracking that occurs in the test sample. Microbial disfigurement is a defect that occurs due to the growth of fungi and algae in the wood and coating layer, thus damaging the appearance of the wood surface (ASTM D 3274-95) [15]. Checking is a defect that occurs in the coating layer and does not penetrate the coating layer (ASTM D 660-93) [16]. Cracking are defects that occur on the surface of the layer that penetrate the coating layer and cause cracks on the wood surface (ASTM D 661-93) [17]. The coating layer has a different level of effectiveness according to weather conditions [13].

Countries that have tropical climate such as Indonesia use Schmidt and Ferguson's classifications based on dry month ratio to wet month. A dry month is a month with total precipitation below 60 mm/month and a wet month is a month with total precipitation above 100 mm/month [18]. The climatic data during the 24 weeks of observation is described in Table 3. The results in Table 3 reveal that the intensity of rainfall in April was 206 mm/month and decrease in June and July then increase in August. This made the surface of the test sample exposed to different conditions during 24 weeks of observation.

**Table 3.** Climate conditions at the test site

| Parameter | Duration of weathering test (month) | | | | | | |
|---|---|---|---|---|---|---|---|
| | Apr | May | Jun | Jul | Aug | Sep | Oct |
| Rainfall (mm/month) | 206.1 | 201.9 | 90.2 | 1.6 | 112.4 | 79.3 | 180 |
| Temperature (°C) | 25.8 | 26.3 | 26.2 | 26.1 | 26.2 | 26.2 | 26.8 |
| Humidity (%) | 86 | 82 | 79 | 74 | 75 | 70 | 75 |
| Solar Radiation Intensity (Cal/cm$^2$) | 313 | 338 | 328 | 353 | 369 | 361 | 375 |

Source: Meteorology and Geophysics Agency Darmaga, Bogor

The intensity of the sun's rays and the fluctuating of the rainfall from April to October made several samples experienced cupping which causes the coating layer to peel and crack (Figure 1). The possibility of splitting and cupping can be reduced by applying the right coating layer [19]. Wood endurance for exterior use is strongly influenced by the type of coating layer used [20].

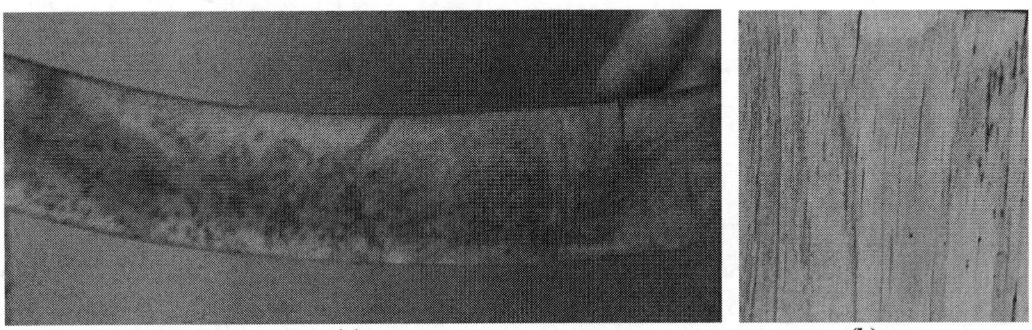

(a) (b)

**Figure 1.** Defects in the tangential section of pine (a) cupping and (b) cracking

Nejad *et al.* [21] report that the life service of the wood can be extended by increasing its dimensional stability. This research is focused on optimizing the application process of coating materials in the form of hot oil in some types of wood that will be used as building floors. The result of the research showed that wood with a coating layer had better resistance to color fading, abrasion, and scratches than the unapplied coating layer wood. The resistance of the coating layer is determined by the large sticking force between paints and wooden surfaces [22]. One of the constraints applying the coating layer is caused by grain raising, the condition of wood surface after sanding, and species of wood [23,24]. On some species of wood, applying exterior paint on a sanded wooden surface can cause the wooden fibers to be lifted. The amount of fiber lifted and the surface roughness are affected by the type of sandpaper [25,26] and the low density of wood [27].

Evaluation of the durability of the coating layer against microbial disfigurement, checking and cracking are presented in Figure 2. From this graph, it can be seen that the durability of the coating layer against microbial disfigurement, checking, and cracking has decreased since the 6$^{th}$ week of exposure. We can assume that this happened because of the high intensity of rainfall in April and May, which then dropped dramatically in July which reached below 60 mm/month. This causes the durability of most of the protective layer to experience damage in the form of checking and cracking.

The durability of the coating layer is affected by sunlight, duration of radiation, and temperature [28-30]. The resistance of the coating layer is influenced by the type of coating layer and the species of wood used [31]. The durability of the coating layer affects the level of water absorption and changes in dimensions [32]. The highest absorption rate was indicated by uncoated wood, and wood with a coating layer with a water solvent, while the lowest absorption rate was shown by wood with a coating layer with oil solvent.

The percentage of the test sample with a coating layer F, FP, and FW have a lower defect compared to the test sample without a coating layer (Figure 2). Without the coating layer, wood can be directly affected by weather changes. Wood which is continuously exposed in the outdoor will experience discoloration and is easily overgrown with fungi [33].

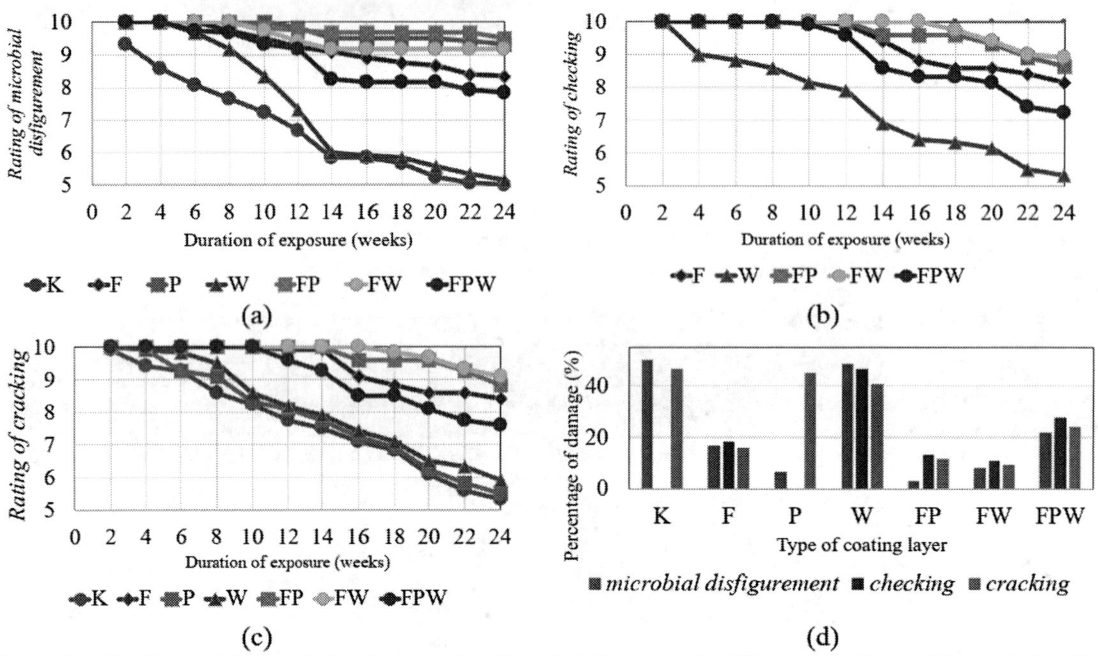


Remarks: K= Control (uncoated); F= coated with paint; P= coated with preservatives; W= coated with water repellents; FP= coated with combination of paint and preservatives; FW= coated with combination of paint and water repellents; FPW= coated with combination of paint, preservatives and water repellents

**Figure 2.** Rating of (a) microbial disfigurement (b) checking and (c) cracking based on the type of coating layer after 24 weeks exposure; (d) Percentage of damage after 24 weeks exposure

Wood with a coating layer had better resistance than wood without a coating layer against microbiological attacks [34]. Wood that is not coated with protective material can be directly affected by weather changes that cause the wood surface to change color due to the washing of extractive substances and fungal attacks. This result confirms that the application of a coating layer can prevent damage to the wood surface so that the life service of the wood will be longer. The coatings layer was able to reduce cracks on the wood surface by 30-40% [35].

Microbial disfigurement can occur on the surface of the coating layer and extends to the surface of the wood when checking and cracking occur more and more. The coating layer first experienced checking and then experienced cracking. After there is a rupture in the coating layer, the wood-destroying organism will easy and fast to damage the wood surface, as shown in Figure 2 (a). This also causes the color change on the surface of all test samples. This indicates that the coating layer is thinning due to photo-degradation of organic polymer components of surface protective materials, especially by ultra-violet light. Most of the surface color of the test sample which was given a coating layer became brighter and seemed yellow-brown after 24 weeks of exposure. This result is due to extractive substances that can evaporate to painted wood surfaces and when yawning will leave a stain of extractive substances that is yellow to brownish red in the paint layer. Rainwater can dissolve extractive substances in the wood so that the wood undergoes genuine color changes [36,37]. Ultra-violet rays penetrate the wood surface and can cause weathering in wood.

### *3.1. The durability of the coating layer based on the wood species*

One of the important factors influencing the durability of the coating layer is the wood species. Wood has different characteristics and properties. Water, dust, and other materials can enter the wood pores. These cause damage to the wood to swell or shrink, checking, cracking or change color. The decrease in the coating layer is presented in Figure 3.

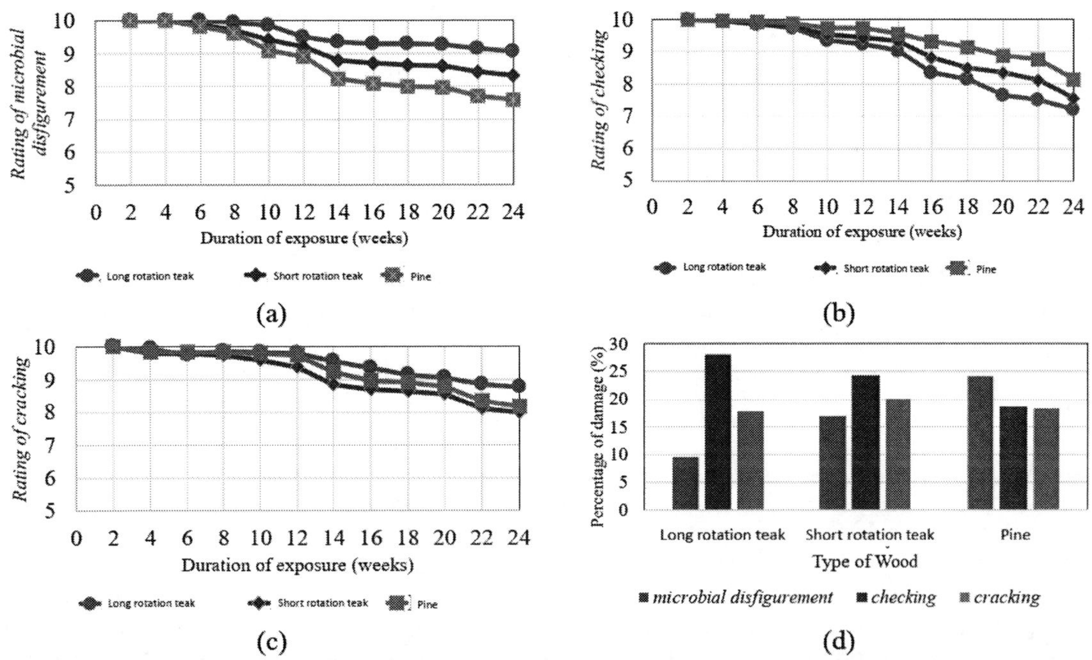


**Figure 3.** Rating of (a) microbial disfigurement (b) checking and (c) cracking based on the type of wood after 24 weeks exposure; (d) Percentage of damage after 24 weeks exposure.

The results in Figure 3 showed that the coating layer on long rotation teak, short rotation teak, and pine begins to degrade due to microbial attack, checking, and cracking after 24 weeks of natural weathering. The results also showed that the durability of the coating layer on pine wood is lower than the coating layer on long rotation teak and short rotation teak. This indicates that long rotation teak and short rotation teak can be used for exterior applications with long life service. Damage to the coating layer due to microbial disfigurement, checking and cracking begins to appear after one month of exposure and develops very quickly so that the class reaches an average Rating 7.

The ability of the adhesion of exterior paint on wood is influenced by the wood species, the direction of the fiber, and the surface texture [34,38-40]. The quality of the finishing results is also influenced by the direction of the fiber, the existence of the knots, and the moisture content of the wood. Meanwhile, the exterior paint resistance can be optimal if the factors that influence the finishing result are fully considered. These factors include wood as a substrate, types of exterior paint, methods for applying exterior paint, and environmental conditions.

Checking and cracking can cause the fungus to enter the coating layer to grow and spread under the coating layer. The result of this study showed that the highest microbial disfigurement occurs was in pine wood. This is because pine wood is less durable and has a greater rate of swelling and shrinking than the other samples. This can be due to pinewood contains extractive substances that slowly disappear when the wood is exposed. This cause water to easily penetrate into the wood and attacked by fungi and other microorganisms [41]. Besides, pinewood contains hydrophobic resin channels [42]. These properties cause the coating layer to bind less strongly to wood. Weather changes can cause degradation in all organic materials including coating layer material consisting of organic polymers, which is caused by changes in humidity and ultraviolet radiation on the wood surface [13].

The percentage of damage after 24 weeks of exposure based on the wood species is presented in Figure 3(d). The results showed that the percentage of checking in long rotation teak and short rotation teak is larger than in pinewood. Differences in anatomical characteristics of both hardwood and softwood affected the difference in the durability of the coating layer against the effects of weather [43]. Wood that has a smoother surface has lower durability than a rough surface. This figure also indicated that the percentage of microbial disfigurement that occur in teak wood is smaller than in pine wood. Based on the results of laboratory experiments and grave experiments on soil fungi and termites, these woods are reported to be resistant to fungal attacks, including *Schizophyllum commune* [3].

### *3.2. The durability of the coating layer based on the wood section*

Wood has anisotropic properties which means that there are differences in properties especially the difference in shrinkage which can affect the durability of the coating layer. The durability of the coating layer against microbial disfigurement, checking, and cracking based on the wood section is presented in Figure 4. The decrease of the coating layer quality begins to occur in the first week because of the high intensity of radiation. This causes the surface of the coating layer is unable to withstand checking and cracking damage. The results in Figure 4 show that the durability of the coating layer against microbial disfigurement, checking, and cracking in the radial section is slightly better than that of the tangential section.

After 14 weeks, the coating layer on the tangential section experienced more checking and cracking compared to the radial section. This is considered because wood with radial section has a lower shrinkage value than a tangential section. The radial section has a lower chance of cracking because of the radial section has lower possibility for swelling and shrinkage than the tangential section in exterior applications [13,44]. The radial section of the same type (camphor) and the same growth level is better in its ability to withstand paint compared to the tangential section [45]. However, the durability of the coating layer to the radial and tangential sections is not significantly different. This can be caused by the radial section and the tangential section of long rotation teak, short rotation teak and pinewood having relatively similar wettability properties. Similar result are also shown in sengon *(Paraserianthes falcataria* (L.) IC Nielsen), jabon *(Anthocephalus cadamba (Roxb.)* Miq), acacia *(Acacia mangium* Willd.) and rajumas wood [8,46].

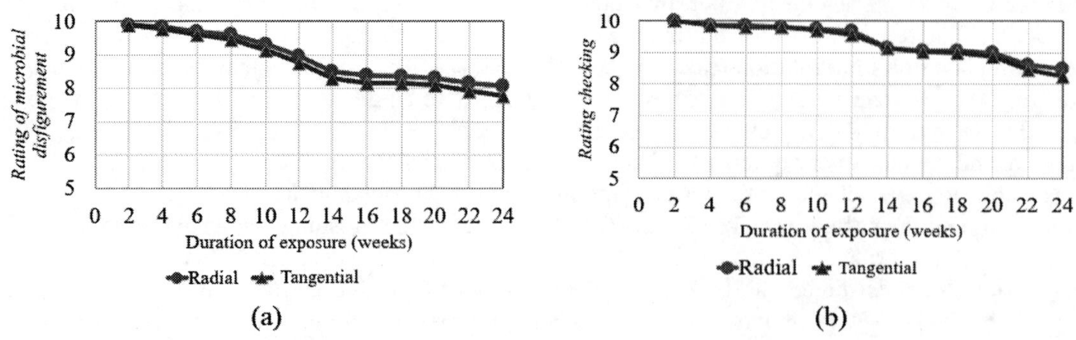

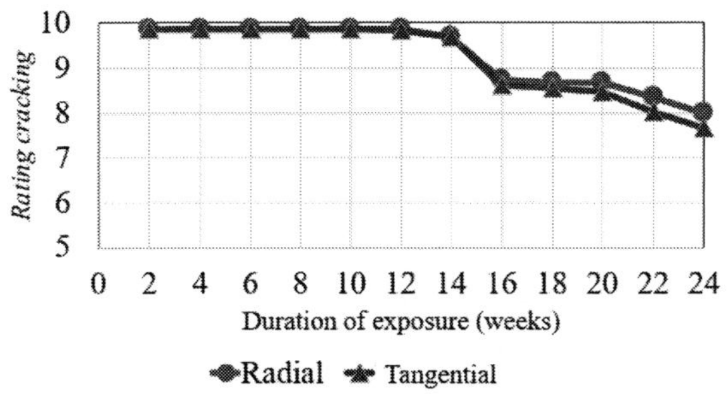


(c)

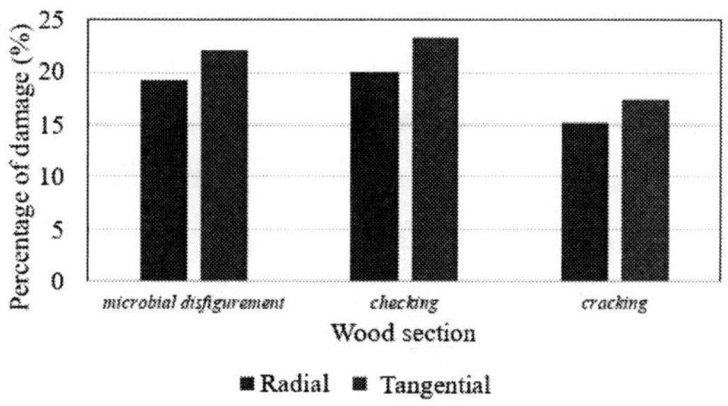


(d)

**Figure 4.** Rating of (a) microbial disfigurement (b) checking and (c) cracking based on the wood section after 24 weeks exposure; (d) Percentage of damage after 24 weeks exposure

## 4. Conclusions

The results of this study indicated that the best coating layers for exterior use are the combination of paint and preservatives and the combination of paint and water repellents. Based on the wood species, it can be concluded that the durability of the coating layer on long rotation teak and short rotation teak are better than the durability of the coating layer on pine wood. The results also showed that the durability of the coating layer in the radial section is slightly better than that of the tangential section. However, the durability of the coating layer on the radial and tangential sections is not significantly different. Further research needs to be conducted is to find out the best concentration of water repellents in a combination of coating layer.

## References

[1] Hoadley RB 2000 *Understanding Wood: A craftsman's guide to wood technology* (United States of America: The Taunton Pr.)

[2] Mahfudz MF, Anis, Yuliah T, Herawan, and Prastyono SH 2007 *Sekilas tentang Jati* (*Tectona grandis*) ed. Parwito D and Donie S (Yogyakarta: Pusat Penelitian dan Pengembangan Bioteknologi dan Pemuliaan Tanaman Hutan)

[3] Martawijaya A, Kartasujana I, Kadir K, and Prawira SA 2005 *Atlas Kayu Indonesia. Jilid I* (Bogor: Departemen Kehutanan)

[4] Esteves BM and Pereira HM 2009 Wood modification by heat treatment: a review *BioResources* **4**(1) 370-404

[5] Liu S, Wu S, and Wang H 2014 Managing planted forests for multiple uses under a changing environment in China *New Zealand Journal of Forestry Science* **44**(S3) 1-10

[6] Abdurachman and Hadjib N 2006 Pemanfaatan hutan rakyat untuk komponen bangunan *Prosiding Seminar Hasil Litbang Hasil Hutan* Bogor pp. 130-48

[7] Delucis RA, Cademartori PHG, Missio AL, and Gatto DA 2016 Decay resistance of four fast-growing Eucalypts wood exposed to three types of fields *Maderas Cienc. Tecnol.* **18**(1)

[8] Karlinasari L, Lestari AT, and Priadi T 2018 Evaluation of surface roughness and wettability of heat-treated, fast-growing tropical wood species sengon (*Paraserianthes falcataria* (L.) IC Nielsen), jabon (*Anthocephalus cadamba* (Roxb.) Miq.), and acacia (*Acacia mangium* Willd.) *Int. Wood Products Journal* 1-7

[9] Evans PD, Haase JG, Shakri A, Seman BM, and Kiguchi M 2015 The search for durable exterior clear coatings for wood *Coatings* **5** 830-64

[10] Yuan Y and Lee TR 2013 *Contact Angle and Wetting Properties* (Surface Science Techniques Springer) pp. 3-34

[11] Bamber RK and Burley J 1983 *The Wood Properties of Radiate Pine* (Oxford: CAB) p. 84

[12] Demirci Z, Somnez A, and Budakci M 2013 Effect of Thermal Ageing on the Gloss and the Adhesion Strength of the Wood Varnish Layers *Bioresources* **8**(2) 1852-67

[13] Williams RS 1999 *Wood Handbook: Wood as an Engineering Material* (Madison: US Department of Agriculture)
[14] [ASTM] American Society for Testing and Materials 2004 *Standard Specification for Wood to Be Used as Panels in Weathering Tests of Coatings ASTM D 358-98* (West Conshohocken: ASTM)
[15] [ASTM] American Society for Testing and Materials 1995 *Standard Test Method for Evaluating Degree of Surface Disfigurement of Paint Films by Microbial (Fungal or Algal) Growth or Soil and Dirt Accumulation ASTM D 3274-95* (West Conshohocken: ASTM)
[16] [ASTM] American Society for Testing and Materials 1993 *Standard Test Method for Evaluating Degree of Checking of Exterior Paints ASTM D 660-93* (West Conshohocken: ASTM)
[17] [ASTM] American Society for Testing and Materials 1993 *Standard Test Method for Evaluating Degree of Cracking of Exterior Paint ASTM D 661-93* (West Conshohocken: ASTM)
[18] Lakitan B 2002 *Dasar-Dasar Klimatologi* (Jakarta: Raja Grafindo Persada)
[19] Dahlen J, Nicholas D, and Schultz T 2008 Water repellency and dimensional stability of southern pine decking treated with waterborne resin acids *Journal of Wood Chemistry and Technology* **28**(1) 47-54
[20] Custodio JEP and Eusebio MI 2006 Waterborne acrylic varnishes durability on wood surfaces for exterior exposure *Progress in Organic Coatings* **56** 59-67
[21] Nejad M, Shafaghi R, Ali H, and Cooper P 2013 Coating performance on oil-heat treated wood for flooring *Bioresources* **8**(2) 888-92
[22] Richter K, Feist WC, and Knaebe MT 1994 The effect of surface roughness on the performance of finishes *Forest Products Journal* **45** 7-8
[23] Landry V, Blanchet P, and Cormier LM 2013 Water-based and solvent-based stains: impact on the grain raising in yellow birch *Bioresources* **8**(2) 1997-2009
[24] Koehler A 1932 Some observation of raised grain *Transaction of the American Society of Mechanical Engineers* **54** 27-30
[25] Marra GG 1943 An analysis of the factors responsible for raised grain on the wood of oak following sanding and staining *Transaction of the American Society of Mechanical Engineers* **65** 177-185
[26] Lestari AT, Darmawan IW, and Nandika D 2016 Pengaruh kondisi permukaan terhadap daya lekat lapisan pelindung *Jurnal Ilmu Teknol. Kayu Tropis* **14**(1) 11-22
[27] Evans P 2009 *Reducing Grain Raising During The Finishing of Wood Water-Based Coatings* (Vancouver: NA) p. 54
[28] Podgorski L Arnold M, and Hora G 2003 Artificial weathering test for wood coatings *Coatings World* **8** 39-48
[29] Ozgenc O, Hiziroglu S, and Yildiz U 2012 Weathering properties of wood species treated with different coating applications *Bioresources* **7**(4) 4875-88
[30] Pastore TCM, Kelly OS and Joel CR 2004 A spectroscopy study on effect of ultraviolet irradiation of four tropical hardwoods. *Bioresource Technology* **93** 37-42
[31] Vitosytė J, Ukvalbergienė K, and Keturakis G 2012 The effect of surface roughness on adhesion strength of coated ash (*Fraxinus excelsior* L.) and birch (*Betula* L.) wood *Materials Science (Medžiagotyra)* **18**(4) 347-51
[32] Meijer M 2001 Review on the durability of exterior wood coatings with reduced VOC-content *Progress in Organic Coatings* **43** 217-25
[33] Hon DNS 1996 *Chemical Modification of Lignocellulosic Materials* (South Carolina: Clemson University Pr.)
[34] Williams RS, Knaebe MT, and Feist WC 1996 *Finishes for Exterior* Wood (Madison: Forest Products Laboratory)
[35] Nejad M and Cooper P 2011 Exterior wood coating. part-1: performance of semitransparent stain on preservative treated wood *Journal of Coating Technology Resources* **8**(4) 449-58

[36] Williams RS 2010 *Finishing of Wood, Wood Handbook: Wood as an Engineering Material* (Madison: United States Department of Agriculture)
[37] Binks BP 1998 *Modern Aspects Of Emulsion Science 1st Edition* London Royal Society of Chemistry **4** 2-4
[38] Amorim MRS, Ribeiro PG, Martins SA, Del Menezzi CHS, and Souza MR 2013 Surface wettability and roughness of 11 amazonian tropical hardwoods *Floresta e Ambiente* **20**(1) 99-109
[39] Ayrilmis N, Candan Z, Akbulut T, and Balqiz OD 2010 Effect of sanding on surface properties of medium density fiberboard *Drvna Industrija* **61**(3) 175-81
[40] Darmawan W, Rahayu IS, Padlinurjaji IM, and Pandit KN 2011 *Pengerjaan Kayu: Ilmu-Ilmu Penunjang dan Teknologi Proses* (Bogor: IPB Press)
[41] Rajkovic VJ, Bogner A, and Mihulja G 2007 Coating adhesion and wettability of aged and preweathered fir wood and pine wood surfaces *Wood Research* **52**(2) 39-48
[42] Kharisma HD 2015 Efektifitas bahan finishing eksterior berpelarut air pada kayu kamper, jati dan pinus karena pengaruh cuaca [Skripsi] Institut Pertanian Bogor
[43] Windt ID, Bulcke JVD, Wuijtens I, Coppens H, and Acker JV 2014 Outdoor weathering performance parameters of exterior wood coating systems on tropical hardwood substrates *European Journal of Wood Products* **72** 261-72
[44] Sandberg D and Sonderstrom O 2006 Crack formation due to weathering of radial and tangential sections of pine and spruce *Wood Material Science Engineering* **1**(1) 12-20
[45] Darmawan W and Purba II 2009 Daya tahan lapisan finishing eksterior beberapa jenis kayu terhadap pengaruh cuaca *JITH* **2**(1) 1-8
[46] Lestari AT 2020 Sifat keterbasahan pada bidang tangensial dan radial kayu rajumas (*Duabanga moluccana* Blume) *Jurnal Perennial* **16**(1) 7-10

# A review of sustainable replanting eucalyptus: higher sustainable productivity

**N Sembiring*, H L Napitupulu, A I Sipahutar, and M T Sembiring**

Department of Industrial Engineering, Faculty of Engineering, Universitas Sumatera Utara, Indonesia.

*Email: nurhayatipandia68@usu.ac.id

**Abstract.** Eucalyptus is a plant that has a very beneficial value to the environment such as can reduce the risk of forest fires, be efficient in water use, nature became more natural, restore land degradation or unproductive land, and the growth of these plants has a sustainable potential for biodiversity. Therefore, the need to replant eucalyptus in order to create a sustainable process for the environment to offers opportunities for wealth and prosperity, while also providing solutions for efficient ways to combat climate change. The purpose of this literature review is to provide information to the public about the benefits of plants that actually have extraordinary functions and characteristics but often get criticism from the community because of community disregard about the benefits It expects in the future eucalyptus can generate more wealth and benefits to the environment. Also, this review literature focuses on eucalyptus plants which provide significant benefits in various sectors, both for the environment and the industrial sector.

## 1. Introduction

The growth process of tree plantations around the world is very fast, it is caused by soil fertility and soil carbon storage [1]. So an important challenge that must be faced in the twenty-first century is the system of replanting sustainable eucalyptus forests to meet the demand for wood production and maintain links with biodiversity [2]. In the coming decades, it is estimated that global eucalyptus demand from plantations will increase. Demand for wood is used for the production of paper and pulp wood, but it can also be used for the production of high-quality multipurpose wood [3]

Initially, eucalyptus plants originated from Tasmania, Australia, and other Indo-Malaysia islands, which numbered around 700 plant species. Eucalyptus plants have great value for the environment, the industrial sector uses 37 of the temporary species for commercial purposes only as many as 15 of the existing species. Eucalyptus are derived from Greek eu (wells) and kalyptus (closed) which can be given protection by an operculum [4].

The development of eucalyptus was more than 200 years ago in Europe, starting from the Iberian Peninsula to Portugal in 1829. Mid-19$^{th}$ century as a result of the flow of migration from Australia and New Zealand this plant was introduced in the United States. Eucalyptus plants are widespread throughout the world because they have a broad ecological level and ability to adapt to the global forest surface area, only about 0.53% of the percentage owned by eucalyptus plants is found in more than 90 countries in the tropics and sub-tropics [4].

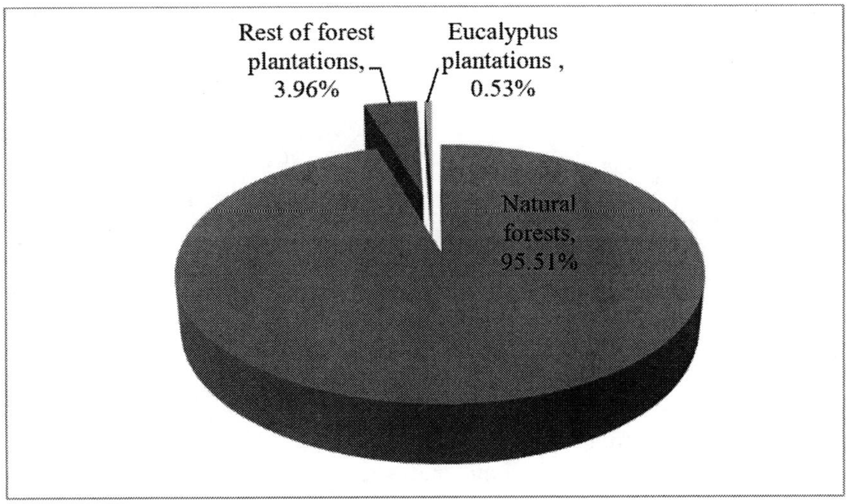


**Figure 1.** Forest masses worldwide (In millions of hectares)

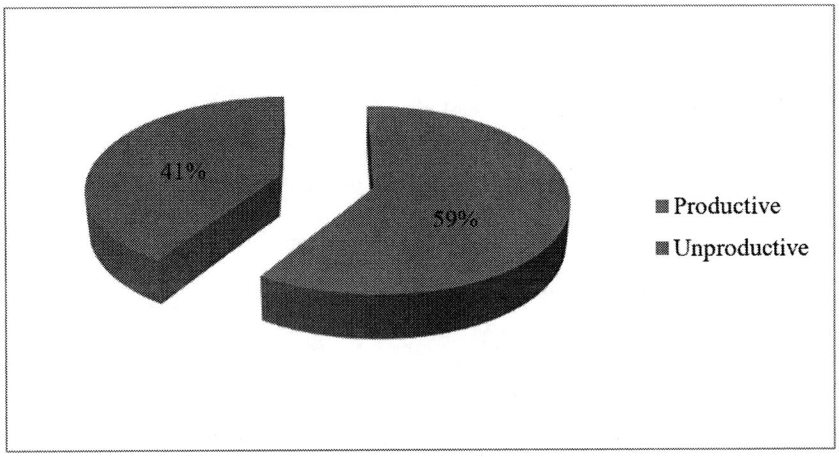


**Figure 2.** Productive eucalyptus forests

For more than 60 years ENCE has conducted various studies on eucalyptus plants. Every year thousands of hectares of forest have been created by ENCE, always developing new research techniques to improve the genetic makeup of eucalyptus plants and focus on species productivity to achieve eucalyptus plants with very diverse conditions and have uses to protect nature and the environment. In addition, the ENCE industrial sector has produced more than 4000 stable jobs that are beneficial to the surrounding area. Because of the many benefits, we can get from replanting eucalyptus plants, this literature review is carried out.

Benefits of eucalyptus plants for all sectors [4]:

*A. Benefits in the environmental sector*

1) Eucalyptus plants are very efficient at capturing CO2, producing oxygen, and improving carbon because these plants have higher growth capacity and dense wood characteristics;
2) Eucalyptus plants act as oxygen storage tanks because of the mass effect in the plants;
3) Eucalyptus plants can improve the amount of carbon $CO_2$ is greater because of rapid renewal and growth every 10 or 15 years;
4) Eucalyptus is more efficient in water consumption than other species;

IOP Conf. Series: Materials Science and Engineering **935** (2020) 012068 doi:10.1088/1757-899X/935/1/012068

5) Forests planted with eucalyptus create new natural space and produce biodiversity;
6) Eucalyptus forests can act like natural forests because of their function, even though these plants are not natural forests;
7) Eucalyptus can increase soil fertility;
8) Eucalyptus can restore land degradation or unproductive land;
9) Eucalyptus leaves and stems can act as a nutrient and not acidic for the soil.

*B. Benefits in the industrial sector*

1) Planting these plants creates employment and wealth in rural areas because using land owned by the community
2) To produce pulp and renewable energy, eucalyptus plants are the right plants to use because they have excellent characteristics
3) Eucalyptus has a quality at a lower cost in making and printing tissue paper.
4) Around the eucalyptus plantations can also be used for many industrial and social purposes raising livestock and hunting
5) There are opportunities for economic and social development in the form of the energy potential offered by eucalyptus biomass

## 2. Material and Methods

### *2.1. Methodological aspects of research*

To conduct a literature review and identification of opportunities to create sustainability for future research is identified replanting of eucalyptus, some literature has been reviewed. From a research literature review can be interpreted from a methodological point of view as systematic approach was taken to evaluate, interpret journal references and identify review literature

2.1.1. Determine the publication period for the literature to be reviewed
Several studies in the period 2018- 020 have been selected as the main reference. The selection of this time period is considered sufficient to represent thoughts about the development of science.

2.1.2. Look for the literature to be reviewed
Therefore, some literature has been identified from four databases: Springer, Scopus, Science Direct and Google Scholar.

2.1.3. Select the literature to review
The main criteria in selecting the literature to be reviewed is to have a clear relationship between the contents of the article and the framework that has been previously defined.

2.1.4. Literature analysis
The next step is to study and examine the articles chosen to find out how the research has progressed over time.

In order to answer to the research questions identified, the research followed the eight steps presented in Figure 3.

**Figure 3.** Research methodology

The range of review of this paper is limited from 2018 to 2020, the search thread used to collect data is Sustainable Replanting Eucalyptus. The search was carried out in an online journal database.

**Table 1.** Literature database search summary (2018-2020)

| Keywords | Journal review | Language |
|---|---|---|
| "Sustainable" "Eucalyptus" and "Replanting" | de Souza *et al.* [5] C. Medauar *et al.* [6] Rezende *et al.* [7] Macana *et al.* [8] | English |
| "Sustainable" and "Eucalyptus" | de Souza et al. [9] Ogunsanwo *et al.* [10] Paper *et al.* [11] Schweier *et al.* [12] Jin *et al.* [13] | English |
| "Replanting" and "Eucalyptus" | Amazonas *et al.* [14] Bandara and Aththanayake [15] Mailard *et al.* [16] Voigtlaender *et al.* [17] | English |
| "Sustainable" and "Replanting" | Kenney-Lazar *et. al* [18] Li *et al.* [19] Vrignan-Brenas *et al.* [20] Faaij [21] | English |

**Table 2.** Number of articles per database

| Database | Quantity |
|---|---|
| Springer | 7 |
| Book | 1 |
| Science Direct | 8 |
| Google Scholar | 19 |
| Total | 35 |

## 3. Results and Discussion

### *3.1. Literature review*

This study uses paper review methods available from books, book reviews, editorials, and brief interpretations not counted in the analysis.

**Table 3.** Literature reviews on various topics in sustainable replanting eucalyptus

| Authors | Focus |
|---|---|
| Cuong *et al.* [22] | The relationship between the government and the State towards the planting of eucalyptus, *Acacia mangium*, and manglietia conifer dandy plants |
| Chu *et al.* [23] | Replanting native eucalyptus plants in the forest is highly recommended. However, it has several influences on the ecological process. This research was conducted in Southern China and focused on 16 years old eucalyptus to assess the effects of replanting eucalyptus trees |
| Nambiar *et al.* [24] | On the islands of Sumatra and Kalimantan, Indonesia and in Sabah, Malaysia, changes in species grown from *Acacia mangium* to eucalyptus were caused by two diseases spread and exacerbated by the destruction of humid tropical environments and diverse fauna |
| Ilseven and Bastas [25] | Various botanists, ecologists, foresters and geographers view eucalyptus plants as plants that are harmful to the environment and can only be considered as assets. However, this has not been proven if eucalyptus plants are planted individually |
| Xu and Zhang [26] | This study focuses on evaluating the impact of eucalyptus replanting on flood control and nutrition management capabilities on the environment |
| Costa *et al.* [27] | This study aims to identify and evaluate the ability to decay naturally occurring fungi on eucalyptus plants |

| | |
|---|---|
| Muthu and Gardetti [28] | Because of the results of industrialization in today's society, industrial development, increasing population etc. have caused great problems for the environment. Negative trends do not seem to stop and instead increase. Therefore, the need to replant forests to overcome these problems |
| Sharma and sethi [29] | Renewable resources can be excessively destroyed as a result of the exploitation of human desire for economic prosperity, material obedience and a desire for pleasure. Therefore, this research focuses on the need for sustainable development planning to solve these problems |
| Subhan [30] | The problem of managing post mining coal through the reclamation of ex-mining land has become a national problem in Indonesia. This research focuses on how to restore environmental sustainability and land reclamation so that environmental problems can be resolved |
| Maghnia *et al.* [31] | This research focuses on providing a description of the main objectives of biotic and abiotic reductions as well as knowledge of the impacts of reduced forest degradation |
| Orewere *et al.* [32] | The need for sustainable planting as a result of land degradation so as not to become a global problem in the 21st century because of its negative impact on the environment, agronomic productivity and its effect on food security and quality of life |
| Yang *et al.* [33] | Yields can increase significantly when replanted and cultivated, but the "benefits of replanting" are unknown as a biological basis. Therefore, to identify candidate genes in eucalyptus plants. So, this research focuses on the use of the RNA-seq technique |
| Shukla *et al.* [34] | In most developing countries, wood from the forests in this study is an attempt to obtain and store information about eucalyptus plants in India |
| Nyakundi *et al.* [35] | To achieve a stable supply of wood for the company, it is necessary to apply forest wood processing through optimal synchronization of plantation forests |

*3.2. Review*

Based on the description in Table 3, it can be concluded that the replanting of native eucalyptus species in the forest is strongly recommended. Many studies have been conducted on eucalyptus, one of which is a study conducted in South China that focuses on 16 years old eucalyptus to assess the effects of replanting eucalyptus trees. While on the islands of Sumatra and Kalimantan, Indonesia and, Sabah, Malaysia, changes were also made to the plants that originally grew *Acacia mangium* to be converted into eucalyptus. In point 3 also explained the importance of sustainable planting as a result of land degradation so as not to become a global problem in the $21^{st}$ century due to its negative impact on the environment, agronomic productivity, and its effect on food security and quality of human life.

## 4. Conclusion

The conclusions that can be drawn after a literature review are as follows:

a. Eucalyptus was first discovered in Australia, but in the Iberian Peninsula, especially in Galicia and in the Cantabria region, it has ideal conditions especially in the soil and climate for the development of these plants so as to produce higher sustainability productivity.
b. There are about 700 species of eucalyptus, all of which have good value for the environment, but the industrial sector uses 37 of these species while only 15 are used for commercial purposes.

c. Even though only 13 million hectares have productivity and are attractive from an industry perspective, eucalyptus plants are now present in more than 90 countries and are spread out over more than 22 million hectares worldwide.
d. Replanting eucalyptus plants has been proven to have so many characteristics and benefits from both the environmental and industrial sectors. These benefits have been explained in detail in this review literature. Therefore, it is very necessary to replant the eucalyptus plants in order to create a process of sustainability into the future.

**Reference**

[1] López DM, Arturi MF, Goya JF, Pérez CA, and Frangi JL 2020 *Eucalyptus grandis* plantations: effects of management on soil carbon, nutrient contents and yields *Journal of Forestry Research* **31**(2) 601-11
[2] Cardoso BN, de Moraes Gonçalves JL, de Carvalho Balieiro F, and Franco AA 2020 *Mixed plantations of eucalyptus and leguminous trees* (Springer International Publishing)
[3] de Oliveira RK, Higa AR, Silva LD, Silva IC, and Gonçalves MD 2018 Emergy based sustainability assessment of a loblolly pine (*Pinus taeda*) production system in southern Brazil *Ecological Indicators* **93** 481-9
[4] Grossberg SP 2009 *Forest Management* (Nova Science Publishers)
[5] de Souza HP, Schumacher MV, Ludvichak AA, Momolli DR, do Couto Guimarães C, and Malheiros AC 2019 Effects of fertilization on biomass and macronutrient content of eucalyptus urophylla ST Blake in Arenized soil of pampa biome *Journal of Experimental Agriculture International* **13** 1-0
[6] Medauar CC, Silva SD, Carvalho LC, Tibúrcio RA, Lima JS, and Medauar PA 2018 Monitoring of eucalyptus sprouts control using digital images obtained by unmanned aerial vehicle *Journal of Sustainable Forestry* **37**(7) 739-52
[7] Rezende GD, Lima JL, da Costa Dias D, de Lima BM, Aguiar AM, Bertolucci FD, and Ramalho MA 2019 Clonal composites: An alternative to improve the sustainability of production in eucalypt forests *Forest Ecology and Management* **449** 117445
[8] Macana YA, de Toledo FH, de Vicente Ferraz A, de Moraes Gonçalves JL, López SM, Cubillos CA, Rojas ÁN, and Bolognani HA 2020 Soil fertility and fine root distribution after gypsum application in Eucalyptus plantations with different tolerance to water deficit *New Forests* **22** 1-6
[9] de Souza Nascimento E, de São Pedro Filho F, Piacentini AL, Tadeu Simões Piacentini M, and Ramos EG. Directives for Sustainability Management in the Amazon Forest Economy. *International Journal of Advanced Engineering Research and Science* **5**(4) 237452
[10] Ogunsanwo OY, Attah VI, Adenaiya AO, and Umar M 2014 Sustainable utilization of firewood as a form of energy in Nigeria. In Sudano-sahelian landscapes and renewable natural resources development in Nigeria. *Proceedings of the 37th Annual Conference of the Forestry Association of Nigeria* pp. 816-830
[11] Paper C, Mani A, and Czech T 2018 Forestry potential, short rotation and agroforestry for climate change mitigation and sustainability *Agrofor. Clim. Resil. Rural Livelihood*
[12] Schweier J, Spinelli R, Magagnotti N, Wolfslehner B, and Lexer MJ 2018 Sustainability assessment of alternative thinning operations in Mediterranean softwood plantations *Forests* **9**(7) 375
[13] Jin E, Mendis GP, and Sutherland JW 2019 Integrated sustainability assessment for a bioenergy system: A system dynamics model of switch grass for cellulosic ethanol production in the US mid-west *Journal of Cleaner Production* **234** 503-20
[14] Amazonas NT, Forrester DI, Silva CC, Almeida DR, Rodrigues RR, and Brancalion PH 2018 High diversity mixed plantations of Eucalyptus and native trees: An interface between production and restoration for the tropics *Forest Ecology and Management* **417** 247-56

[15] Bandara WA and Aththanayake AM 2018 Generic assumption vs. empirical model in stem carbon prediction of *Eucalyptus grandis* Wall ex. Maiden *Journal of the National Science Foundation of Sri Lanka* **46**(3)

[16] Maillard F, Leduc V, Bach C, de Moraes Gonçalves JL, Androte FD, Saint-André L, Laclau JP, Buée M, and Robin A 2019 Microbial enzymatic activities and community-level physiological profiles (CLPP) in subsoil layers are altered by harvest residue management practices in a tropical *Eucalyptus grandis* plantation *Microbial ecology* **78**(2) 528-33

[17] Voigtlaender M, Laclau JP, de Moraes Gonçalves JL, de Cássia Piccolo M, Moreira MZ, Nouvellon Y, Ranger J, and Bouillet JP 2012 Introducing *Acacia mangium* trees in *Eucalyptus grandis* plantations: consequences for soil organic matter stocks and nitrogen mineralization. *Plant and Soil* **352**(1-2) 99-111

[18] Kenney-Lazar M, Wong G, Baral H, and Russell AJ 2018 Greening rubber? Political ecologies of plantation sustainability in Laos and Myanmar *Geoforum* **92** 96-105

[19] Li ZR, Amist N, and Bai LY 2019 Allelopathy in sustainable weeds management *Allelopathy Journal* **48**(2) 109-38

[20] Vrignon-Brenas S, Gay F, Ricard S, Snoeck D, Perron T, Mareschal L, Laclau JP, Gohet É, and Malagoli P 2019 Nutrient management of immature rubber plantations: A review *Agronomy for sustainable development* **39**(1) 11

[21] Faaij AP 2018 Securing sustainable resource availability of biomass for energy applications in europe (Review of Recent Literature) Hg. v. University of Groningen

[22] Cuong T, Chinh TT, Zhang Y, and Xie Y 2020 Economic Performance of Forest Plantations in Vietnam*: Eucalyptus*, *Acacia mangium*, and *Manglietia conifera Forests* **11**(3) 284

[23] Chu S, Ouyang J, Liao D, Zhou Y, Liu S, Shen D, Wei X, and Zeng S 2019 Effects of enriched planting of native tree species on surface water flow, sediment, and nutrient losses in a Eucalyptus plantation forest in southern China *Science of The Total Environment* **675** 224-34

[24] Nambiar EK, Harwood CE, and Mendham DS 2018 Paths to sustainable wood supply to the pulp and paper industry in Indonesia after diseases have forced a change of species from acacia to eucalyptus *Australian Forestry* **81**(3) 148-61

[25] İlseven S and Baştaş M 2018 The place of eucalyptus within the vegetation of mesaoria plain (Cyprus) and the views of vegetation geography lecturers *EURASIA Journal of Mathematics, Science and Technology Education* **14**(7) 3381-8

[26] Xu X and Zhang Q 2019 Sustainable configuration of bioretention systems for nutrient management through life-cycle assessment and cost analysis *Journal of Environmental Engineering* **145**(5) 04019016

[27] Costa LG, Paes JB, WC JJ, Brocco VF, and Pinho DB 2020 Isolation and identification of fungi with potential for biological stump removal of *Eucalyptus spp*. *Journal of Tropical Forest Science* **32**(2)

[28] Muthu SS, and Gardetti MA 2020 *Sustainability in the Textile and Apparel Industries: Sourcing Synthetic and Novel Alternative Raw Materials* (Springer Nature)

[29] Sharma AB and Sethi G 2018 Media and challenges to sustainable development *Research Guru* **12**(1) 83-96

[30] Subhan E 2020 Preliminary study of coal mining reclamation using forest plants in Barito East District, Central Kalimantan Province *Journal of Advanced Research in Engineering and Technology* **11**(1)

[31] Maghnia FZ, Abbas Y, Mahé F, Prin Y, El Ghachtouli N, Duponnois R, and Sanguin H 2019 The rhizosphere microbiome: A key component of sustainable cork oak forests in trouble *Forest Ecology and Management* **434** 29-39

[32] Orewere E, Hassan B, Faiza M, Owonubi A, and Ogunrayewa MO 2020 Curbing land degradation through sustainable landscaping and building resilient cities *International Journal of Geography and Environmental Management* **6**(2) 1-21

[33] Yang YH, Li MJ, Yi YJ, Li RF, Dong C, and Zhang ZY 2018 The root transcriptome of *Achyranthes bidentata* and the identification of the genes involved in the replanting benefit. *Plant cell reports* **37**(4) 611-25
[34] Shukla G, Kumari A, and Chakravarty S 2018 Not targeting trees and forests to extract firewood by indigenous, rural and urban communities: case study from Jharkhand *India Journal* **2**(3)
[35] Nyakundi F, Mulwa R, and Kabubo-Mariara J 2018 Determination of optimal rotation period for management of lumbering forests in Kenya *Journal of Sustainable Forestry* **37**(7) 645-60

**Acknowledgements**
This Research was funded by Directorate of Research and Community Service, Directorate General for Research and Development at the Ministry of Research, Technology and Higher Education of Indonesia

IOP Conf. Series: Materials Science and Engineering **935** (2020) 012069 doi:10.1088/1757-899X/935/1/012069

# Wood decay diagnostic of *Joannesia princeps* Vellozo at Bali Botanical Garden using arborsonic acoustic 3D tomograph

**S F Hanum[1]*, R Iryadi[1], A Rahayu[1], T M Bangun[1], I D P Darma[1] and Sutomo[1]**

"Eka Karya" Bali Botanic Garden, Research Center for Plant Conservation and Botanic Garden - Indonesian Institute of Sciences, Bali, 82191, Indonesia

*Email: siti030@lipi.go.id

**Abstract.** Living collection such as tree in botanical garden has hazard risk. Tree hazard risk must be evaluated to give safety for visitor and employee. However, standing tree with decay wood inside is often the cause of tree failure. Internal tree defect such as wood decay cannot be seen only from visual investigation but must be done with ArborSonic acoustic tomograph to get accurate result. The objective of this research was to diagnostics wood decay in the stand of *Joannesia princeps* Vellozo tree using visual investigation and ArborSonic measurement technology. The result shows *Joannesia princeps* has severe damage cause of wood decay especially in the base of the trunk. Wood decay percentage range from 21-49%. The alleged caused of this damage is stem borer and termites. This information give advantage for Bali Botanical Garden manager to minimize this tree hazard risk by doing activity such as crown pruning, give hazard risk tree sign to limit activity around this tree and do pest management.

## 1. Introduction

Public safety is the important point for visitor to enjoy botanic garden. Tree failure often leads to personal injury or property damage on private or public land. However, standing tree with decay wood inside is often the cause of tree failure [1]. Moreover, non- destructive testing methods to understand the locating and quantifying wood decay and defect are needed to promote public safety. Stress wave method has been commonly used to assess internal tree condition [2]. Stress wave velocity measured in trees has also been found effective to detect moderate to severe decay in Cibodas and Bogor botanical gardens [3,4,5]. ArborSonic 3D acoustic tomograph is one example of non-destructive commercial equipment [2].

Stress wave is a sound wave produced manually, in a complex mixture of frequencies from hammer tapping on one pin to each other pin, that are inserted into the bark. Detecting decay using stress wave method is based on the observation that stress wave travels slower in decayed wood than in sound wood. By using multiple transducers, stress wave computed tomography technology has been developed, where an increase in the number of installed transducer and high frequency can give more accurate information on decays [1].

*Joannesia princeps* Vellozo (Euphorbiaceae) is a tree widespread in Brazil where it grows up to 50 m high [6]. This tree is one of Bali Botanical Garden (BBG) collection which has vulnerable status [7]. Vulnerable status means this tree is not include in endangered and critically endangered status but in future, this species facing risk of extinction in the nature [8]. This tree has several functions such as the leaves have shown antibacterial, antioxidant and anti-inflammatory activities [9], meanwhile the seed exhibits laxative effects and has been used in many regions of Brazil as laxative [10]. *Joannesia princeps* also a pioneer tree which has potency for carbon offset plantations in degraded area and has percentage of survival rate more than 50% [11]. However, this tree collection comes from Brazil and only one planted in XIIIM at BBG [12]. This tree also located on highly visited area. Therefore, the aim of this research was to assess wood decay of *Joannesia princeps* standing tree by using acoustic tomograph of ArborSonic

## 2. Materials and Methods

This study was conducted in June 2020 on standing tree of *Joannesia princeps* at Bali Botanic Garden. The tree has 22 m height, 12 m diameter of crown, 73,6 cm of DBH (diameter at breast height) and three main stems branch. The equipment used is a set of ArborSonic equipment (sensors, hammers, Amplifier boxes and computers connected to ArborSonic), tape diameter, roll meter 30 meters, digital cameras. The structural tree stability was evaluated based on visual observation and acoustic tomograph of ArborSonic 3D Acoustic Tomograph (Fakopp Enterprise, Agfalva, Hungary. Visual observation of the condition of tree trunks was also done by looking at the outermost trunk of the sample tree, especially to see whether there were any symptoms or signs of deterioration on root, trunk and crown. To get accurate information about tree health conditions, visual observation was done by a minimum of four person.

Further assessment of internal trunk defect using a tomograph was conducted with ArborSonic 3D. The step to use ArborSonic 3D were: (1) drive eight sensors perpendicular to the trunk with the equal distance in counterclockwise order; (2) connect the sensors to the amplifier boxes then connect the amplifiers in a line. Connect the battery box on any end of line then connect to PC; (3) each sensor was tapped with a steel hammer to generate sound waves; (4) the software was calculated and displayed the internal sound-velocity distribution of the tree. To get accurate information about internal trunk condition, it is important to assess the tree in several layers. Tomogram result were showed in graphics with green color showed intact, red color showed decayed and blue color showed hollow tree.

The information of several layers proportion area was used to analysis the volumetric content with the cross-sectional area multiplied by based height as the calculation cylinder volume, so there are three layers can be calculated by:

$$\mathbf{V} = \mathbf{\Delta h_{1\text{-}2}} \left( \frac{A1 + A2}{2} \right) + \mathbf{\Delta h_{2\text{-}3}} \left( \frac{A2 + A3}{2} \right) \tag{1}$$

where: V = Volume; Δh = interval high, A= area several layer class, with knowing delineation area of the segmen (hollow, decayed and intact), this math can calculate the several volume portion of class.

ArborSonic measurement results determine the type of tree handling. Trees with a high priority will get immediate treatment and special treatment (regular monitoring) compared to trees with low priority. Trees with high priority are trees located in areas that are densely populated, have large trunk diameters, are aged, or are seen to be attacked by pests and diseases [3]. The level of tree risk is obtained from the maximum percentage of decay level. A decayed area is the percentage of an area of decay compared to the total area of the selected layer

## 3. Results and Discussion

The results of the visual investigation to detect wood decay at *Joannesia princeps* shows that tree has three branches with cavity on the base trunk, the crown and the root intact. One branch already fallen several years ago (Figure 1a). The cavity already filling with cement to avoid fallen tree, but unfortunately one of the branches still fallen in June 2020 (Figure 1b).

IOP Conf. Series: Materials Science and Engineering **935** (2020) 012069 doi:10.1088/1757-899X/935/1/012069

a b

**Figure 1**. Tree hazard risk on *Joannesia princeps*. Picture taken in December 2019 (a) failure tree leads to property damage in June 2020 (b).

ArborSonic 3D is used for advance wood decay detection. To get accurate information about internal tree condition several layers with different height already done. The blue colour mean hollow, the red color mean decay and the green color mean intact. Hollow still found from height 107 cm until 258 cm, only the defect size decreases as the height increase (Figure 2). Table 1 show the percentage of wood decay decrease as the height increase. Since ArborSonic only determine decay percentage per layers (Figure 2,3), volumetric calculation will support information about tree volume. This tree has the percentage hollow area 12.7% and the decaying area 32.3%. The severe defect found in the base of the trunk because this already make cavity. This can be derived that tree has high risk potentially.

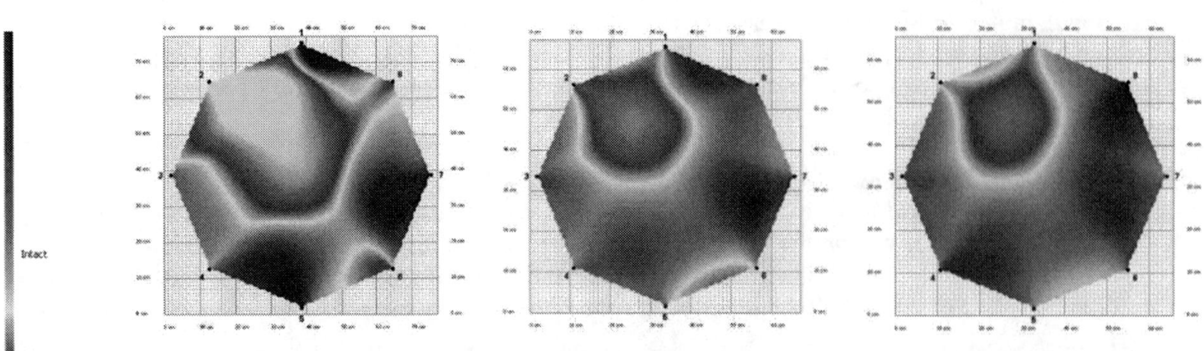


(a) (b) (c)

**Figure 2**. Tomogram of *Joannesia princeps* on different height (a) 107 cm, (b) 182 cm and (c) 258 cm.

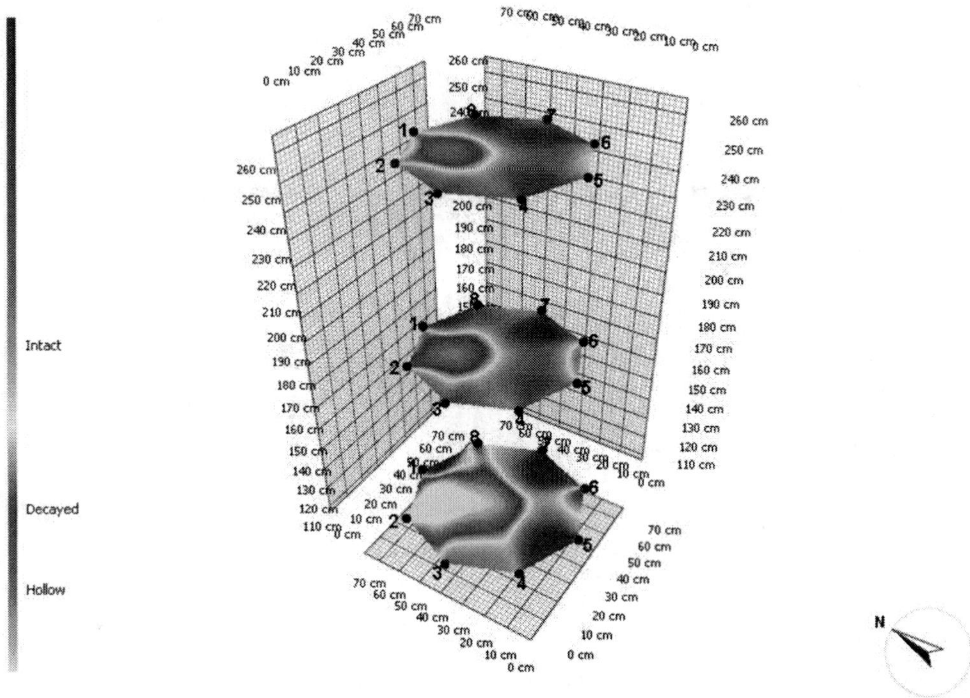


**Figure 3.** Tomogram vertical layers of *Joannesia princeps*

**Table 1**. Decay percentage with ArborSonic measurement on *Joannesia princeps*

| Layer | Height (cm) | Decay (%) |
|---|---|---|
| 1 | 107 | 49 |
| 2 | 182 | 27 |
| 3 | 258 | 21 |

### *3.1. Identification of* Joannesia princeps *stem damage*

Based on observations, the most severe tree damage occurred at the base of the trunk, as shown in Figure 4a. From that picture, it appears that almost half of the base of the stem has been damaged. The damage to the base of the stem of *Joannesia princeps* is thought to be caused by a pest, that is stem borer and termite. Although at the time the data was collected, no pests were found, but the alleged cause of damage to the base of the stem was obtained from the presence or trace of the existence of stem borer and termites. Symptoms of attacks caused by stem borer are the presence of small holes in the main stem section as shown in Figure 4b and 4c. Stem borer larvae attack cambium and xylem areas, which can cause plants to wither, dry and die [13]. While the symptoms of termite attacks are evidenced by the discovery of the termite house attached to the bark. Termites use wood as nesting sites and food sources, thus causing damage to wood [14]. Furthermore, termite attacks generally occur at the root, neck, or stem [15]. The attack on the trunk is marked by the existence of a tunnel that runs along the gap of the bark to the end of the trunk, as shown in Figure 4d.

**Figure 4.** Severe tree damage at the base of the trunk (a), the presence of small holes in the main trunk (b), the location of small holes (c), the tunnel a tunnel that runs along the gap of the bark to the end of the trunk (d).

Furthermore, to prevent the negative effect of tree hazard risk, BBG management give sign tree hazard risk on the trunk and do crown pruning to prevent the visitor and the employee from tree hazard. Moreover, the existence of stem borer and termites must be minimized by doing pest management

## 4. Conclusion

Visual investigation and ArborSonic acoustic tomograph shows *Joannesia princeps* has serious wood decay. Internal decay in standing tree will affect structural safety. The percentage of wood decay range from 21%-49%. It shows the severe damage found on the base of the trunk. This information help BBG management to determine tree's treatment. The alleged cause of damage was stem borer and termites which need further research.

## References

[1] Goh CL, Rahim RA, Rahiman MH, Talib MT, and Tee ZC 2018 Sensing wood decay in standing trees: A review *Sensors and Actuators A: Physical* **269** 276-82

[2] Li G, Weng X, Du X, Wang X, and Feng H 2016 Stress wave velocity patterns in the longitudinal–radial plane of trees for defect diagnosis *Computers and Electronics in Agriculture* **124** 23-8

[3] Helmanto H, Kristiati E, Wardhani FF, Zulkarnaen RN, and Rachmadiyanto AN 2018 Tree health assessment of *Agathis borneensis* Warb. in Bogor Botanical Garden using arborsonic *IOP Conference Series: Earth and Environmental Science* **203**(1) 012032
[4] Rachmadiyanto A, Helmato H, Rinandio D, and Abdurachman 2019 Tree Health Analyzing of Malvaceae Collections in Bogor Botanical Gardens Tree Health Analyzing of Malvaceae Collections in Bogor Botanical Gardens *IOP Conference Series: Earth and Environmental Science* **298**(1) 012017
[5] Zuhri M, Sunandar D, Rustandi U, Nadhifah A, Kurniawati F, and Iskandar E 2018 The detection of wood decay of trees collection *Agathis borneensis* and *Castanopsis argentea* at the public area in Cibodas Botanical Garden *IOP Conference Series: Earth and Environmental Science* **203**(1) 012034
[6] Waibel R, Benirschke G, Benirschke M, and Achenbach H 2003 Sesquineolignans and other constituents from the seeds of *Joannesia princeps Phytochemistry* **62**(5) 805-11
[7] IUCN 2020 retrieved from https://www.iucnredlist.org/ on 14 July 2020
[8] Mogea J P, Gandawidjaja D, Wiriadinata H, and dan Irawati REN 2001 *Tumbuhan Langka Indonesia* (Bogor: Pusat Penelitian dan Pengembangan Biologi-LIPI Balai Penelitian Botani Herbarium Bogoriense) p. 86
[9] Camero CM, Temraz A, Braca A, and De Leo M 2017 Phytochemical study of *Joannesia princeps* Vell (Euphorbiaceae) leaves *Biochemical Systematics and Ecology* **70** 69-72
[10] Araújo AC, Guiguer EL, Barbalho SM, Bueno PC, Lopes JA, da Silva BF, Girotto LC, de Paula MG, Zeber PV, and de Alvares Goulart R 2016 Phytochemical characteristics of seeds and its effects on the intestinal motility and toxicity of *Joannesia princeps Journal of medicinal food* **19**(1) 68-72
[11] de Morais Junior VT, Jacovine LA, Torres CM, Alves EB, de Paiva HN, Alcántara-de la Cruz R, and Zanuncio JC 2019 Early assessment of tree species with potential for carbon offset plantations in degraded area from the southeastern Brazil *Ecological Indicators* **98** 854-60
[12] Plant Registration Unit 2020 *The Number of Plant Living Collection and Herbarium in January 2020 (Monthly Report, unpublished)* (Tabanan: Bali Botanic Garden)
[13] Pujiastuti,Y and Adam T 2010 Keanekaragaman serangga penggerek batang (Coleoptera: Cerambycidae) pada tanaman mangga dan nangka *Prosiding Semirata Bidang Ilmu-ilmu Pertanian BKS-PTN wilayah Barat Tahun 2010* Pp. 83-6
[14] Pratiwi, Lely S and Rahmat S 2018 Penilaian Vitalitas Pohon Jati dengan Forest Health Monitoring di KPH Balapulang *Ecogreen* **4**(1) 9-14
[15] Hadi, Teguh TW, and Mahfudz 2012 *Hama Hutan Indonesia: Catatan 20 Tahun Peneliti* (Manado: Balai Penelitian Kehutanan Manado)

**Acknowledgement**
The author would thank to the Head of Bali Botanical Garden for facilitate and funded this research through internally foundation

IOP Conf. Series: Materials Science and Engineering **935** (2020) 012070 doi:10.1088/1757-899X/935/1/012070

# The suitability of forest tree species for enrichment planting at PT Semitau Region, West Kalimantan

**A A Vinton[1], A Yudana[2], S Puspoyo[2], A Wijaya[2], and I Z Siregar[1]**

[1]Department of Silviculture, Faculty of Forestry and Environment, IPB University, Bogor, 16680, Indonesia
[2]Sustainability Implementation Division PT Sinar Mas Agro Resources and Technology (SMART Tbk), Jakarta, 10350, Indonesia

E-mail: aureliaaranti@gmail.com

**Abstract.** Forest tree enrichment can improve biodiversity and ecosystem functions. Tree enrichment planting in the riverside rehabilitation area of palm oil plantation needs to be monitored to prevent the risk of erosion and sedimentation. However, regular monitoring of enrichment planting is not yet optimally carried out including the benefits and identified suitable species through field observations. This study was aimed to determine the three most suitable tree species to be recommended for enrichment planting in the riverside rehabilitation areas at PT Semitau Region oil palm plantation. Samples were determined through a combination of linear systematic sampling methods. The measured variables included the survival percentage, health, diameter, height, light intensity, and soil analysis. The results showed that the three species had the best adaptability, namely kawi (*Shorea balangeran*), pulai (*Alstonia spp.*), and ubah (*Syzygium darifolium*) with 21%, 15%, and 10% dominance frequency, respectively. Total dominance frequency of 45 % for all three species are expected to be useful information for increasing successfully riverside rehabilitation area of oil palm plantation.

## 1. Introduction

Nowadays, oil palm has the most rapid growth in the Southeast Asia region [1]. Overall oil palm plantation areas in 2017 were estimated at approximately 18.7 Mio ha [2]. The expansion of oil palm has resulted in significant impact on biodiversity and ecosystem functions. Therefore, the policy rules in the form of sustainable certification have been created to reduce the environmental impact caused by palm oil, such as the Roundtable on Sustainable Palm Oil (RSPO), Indonesian Sustainable Palm Oil (ISPO), and Malaysian Sustainable Palm Oil (MSPO) [3]. The regulation guidelines relate to soil fertility and erosion, water management and riparian zones, integrated pest management, and protected areas of high conservation value [4]. The benefit of riparian zones on the environment is about maintaining water quality and hydrological patterns, protecting biodiversity, and increasing the potential for carbon storage in agricultural landscapes [5]. Management of riparian zones in oil palm plantations is carried out through the rehabilitation program with a distance of 50 m to the left and right from the buffer area. However, the literature discussed riparian zone rehabilitation strategies in oil palm plantations remains limited [3]. PT Semitau Region is one of the palm oil plantation companies that has implemented a rehabilitation strategy through enrichment planting on its riversides. Enrichment planting is one way to carry out a rehabilitation program through the insertion of new plants in the gap planting between existing plants [6]. The condition of the land prior to the implementation of PT Semitau Region's operational activities were rubber plantations, former shifting cultivation of vacant land and shrubs, peat, and secondary forests of former timber companies.

PT Semitau Region oil palm plantation permit area was carried out in 2008 while [7] the planting rehabilitation area was begun in 2016. PT SR area located near the Kapuas Hulu river, whose flow forms a dendritic pattern and characterized by spreading flow. Overall, the indigenous tribe who inhabit the village area around the plantation area is Dayak Kantu with 90% composition [8]. The enrichment strategy that undertaken by PT Semitau Region was to insert forest species seedling with 4 m spacing from the main oil palm. However, the rehabilitation efforts did not completely well-run, many things that hampered success including internal and external factors. Internal factors include the suitability of plant species, while external factors include outside circumstances such as soil conditions, nutrient availability, competition between plants, and the presence of pests and diseases [9].

Monitoring the success of a rehabilitation project is very important to evaluate the efforts over the last four years. The scope of the study is adjusting the suitable forest tree species as enrichment plants. The result of the research is appropriate forest tree species information in order to minimize company losses due to rehabilitation failure. This study was aimed to determine the three most suitable forest species to be recommended as enrichment plants in the riverside rehabilitation area of PT Semitau Region oil palm plantation. Based on the last four years (2016-2019) climate data obtained from the Pangsuma Meteorological Station, Kapuas Hulu (Figure 1), it is known that Kapuas Hulu area which is the location of PT SR had quite high rainfall and temperatures in April and October, while the lowest in the month of July. Kapuas Hulu had an average temperature of 28°C with a range temperature of 23.1-33.3°C. This data could help to determine the best season for nurseries, acclimatization, and direct planting.

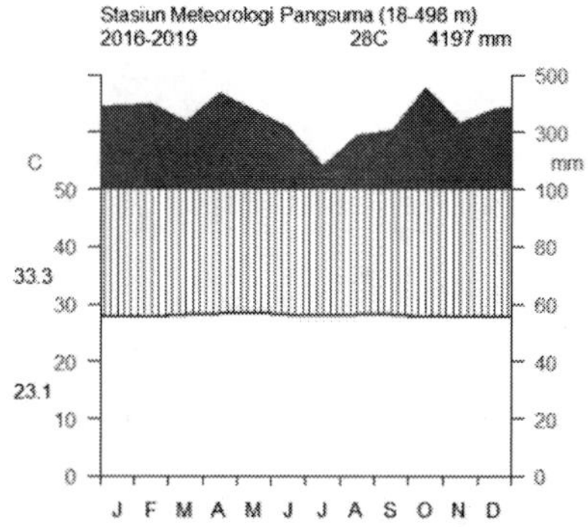


**Figure 1.** Climate graph of Kapuas Hulu

## 2. Materials and Methods

### *2.1 Materials*

Data collection measurement tools consisted of digital calipers, walking sticks, GPS, lux meters, stationery, and camera. Soil sampling equipment consisted of the ground drill, plastic samples, and labels. The application used for data processing was Ms. Visio, Ms. Excel, Ms. Word, MiniTab16, RStudio, Avenza Map, and ArcMap 10.5. The primary data materials are rehabilitation enrichment plants and soil samples in the PT SR riverside area. The secondary data was obtained from year 2016-2019 climate data of

Meteorology and Geophysics Agency (BMKG) and secondary soil analysis data was obtained from Sucofindo Pontianak Laboratory in 2020 and SMART Research Institute-Analytical Laboratory in 2015.

*2.2 Methods*

2.2.1 Sampling technique

The main data included survival percentage and health of forest tree species; while supporting data was climate data, light intensity, and soil properties. The natural tillers were also observed at 2 m distance around the rehabilitation points (Figure 2). The dominance of forest tree species in rehabilitation area was mostly at the seedling and sapling level. Vegetation data was collected based on linear systematic sampling method which representatives adjusting the rehabilitation plant population per hectare. [10] Systematic sampling is a method using random numbers in the initial selection and then followed by a pattern of multiple figures. [11] In general, linear systematic sampling (LLS) is recommended to choose a representative value (n) that can be determined with certainty. The minimum sample number from all locations of 13 628 individual population were adjusted by Slovin formula [12]:

$$S = \frac{N}{1 + N\ x\ e^2} \tag{1}$$

where: S = sample, N = population, e = degrees of accuracy (0.05%)

In order to get a more significant number of representatives and fit the given deadline, the linear systematic sampling method was adjusted to the sampling intensity and number of individual rehabilitation plants in block concerned as listed in Table 1.

**Table 1.** Determination of the representative value of individual rehabilitation plants per planting block

| Rehabilitation tree population | Selected sample |
|---|---|
| $\leq 4$ | Census [13] |
| $> 4$ and $\leq 31$ | 4 sample [13] |
| $> 31$ and $\leq 401$ | Sampling intensity 10% [14] |
| $> 401$ | Sampling intensity 5% [14] |

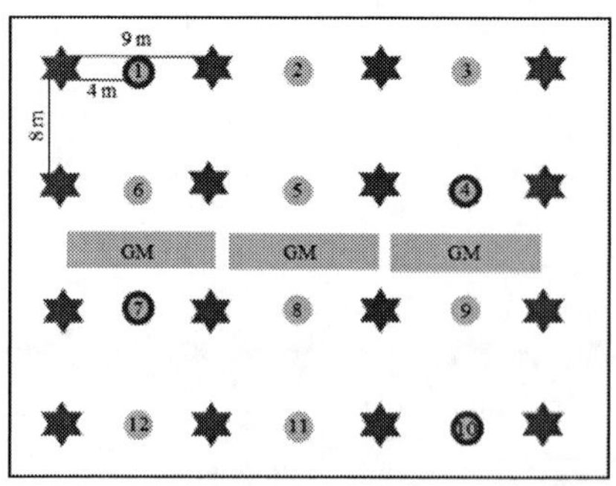


**Figure 2.** Sketch of oil palm and rehabilitation plants patterns and selected sample (0.12 ha)
remarks:

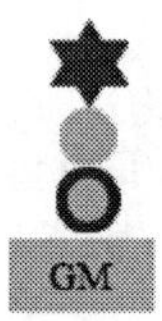


Oil palm, planting distance 9 x 8 m
Rehabilitation plants, planting distance 4 m from oil palm
Selected rehabilitation plant samples by linear systematic sampling method
"Gawangan mati", heaps of oil palm fronds

2.2.2 Measurement of diameter and height
Measurements were done on living plants by calculating the average diameter and height of plants.

$$d = \frac{\Sigma\, di}{\Sigma\, Ni} \times 100 \qquad t = \frac{\Sigma\, ti}{\Sigma\, Ni} \times 100 \quad (2)$$

where:
d = average diameter (cm)
t = average height (m)
at, ti = diameter and height of the i- plant
Ni = number of plants measured

2.2.3 Survival percentage
Data analysis was performed through the results of the calculation of the percentage of living plants with the following formula:

$$H = \frac{\Sigma\, hi}{\Sigma\, Ni} \times 100\ \% \quad (3)$$

where:
H = percent of plant life (%)
hi = number of i- living plants
Ni = number of i- plants planted

2.2.4 Health
Health of plants was determined by calculating the area and intensity of pest and disease attacks with the following formula:

$$LS = \frac{\Sigma\, n}{\Sigma\, N} \times 100\ \% \qquad IS = \frac{\Sigma\, (ni \times vi)}{\Sigma\, N \times Z} \times 100\ \% \quad (4)$$

where:
LS = attack area (%)
IS = attack intensity (%)
n = number of plants attacked
ni = number of plants attacked by i score
vi = i score
N = total number of plants observed
Z = highest score

2.2.5 Measurement of light intensity
This action was determined the amount of sun light that could enter between oil palm canopy gaps. Measurements were done using lux meter which has symbol as lux (Lx) under with and without shade with 3x replications. The value obtained could help the selection of more suitable species based on whether or not resistant to shade (tolerant, intolerant, and semi-tolerant). Light intensity was measured in the morning (07.00-10.00) or afternoon (14.00-16.00) to obtain more significant irradiation value. However, there are also several conditions that affect as drizzle or cloudy weather. Irradiation conditions when raining have lower intensity compared to not raining conditions.

2.2.6 Soil sampling
Soil sampling was taken to determine the chemical properties of the selected soil from several locations through composite technique and quadrant method with 20 cm depth [15] that taken from 5 points of the planting block. Soil samples were analyzed at Sucofindo Pontianak Laboratory, West Kalimantan.

IOP Conf. Series: Materials Science and Engineering **935** (2020) 012070 doi:10.1088/1757-899X/935/1/012070

2.2.7 Data analysis

Survival percentage data was completed using the Mini Tab16 application. The height, diameter, and health data were presented in bar charts, lines, box plots, and tabulations using the Microsoft Excel application.

## 3. Results and Discussion

### *3.1 Survival percentage*

Measurement of the percentage of survival was done based on an assessment of the species adaptability in each location. Based on representative sampling results at 846 rehabilitation points spread across 19 river border locations, 24 plant species were found with total of 627 individuals.

**Table 2.** Species variants in PT Semitau Region

| No | Local name | Scientific name | Family | Individual total | Dominance frequency (%) |
|---|---|---|---|---|---|
| 1 | Ampening | - | - | 13 | 2.1 |
| 2 | Cempedak | *Artocarpus champedens* | Moraceae | 14 | 2.2 |
| 3 | Durian | *Durio zibethinus* | Malvaceae | 9 | 1.4 |
| 4 | Jambu Monyet | *Melastoma marginata* | Melastomaceae | 72 | 11.5 |
| 5 | Jengger | *Tristaniopsis bornensis* | Myrtaceae | 12 | 1.9 |
| 6 | Jengkol | *Archidendron jiringa* | Fabaceae | 1 | 0.2 |
| 7 | Karet | *Hevea braziliensis* | Euphorbiaceae | 1 | 0.2 |
| 8 | Kawi | *Shorea balangeran* | Dipterocarpaceae | 132 | 21.1 |
| 9 | Keladan | *Dipterocarpus gracillis* | Dipterocarpaceae | 1 | 0.2 |
| 10 | Kopi | *Theobroma cacao* | Rubiaceae | 1 | 0.2 |
| 11 | Laban | *Vitex pubescens* | Lamiaceae | 25 | 4.0 |
| 12 | Manaung | - | - | 20 | 3.2 |
| 13 | Markubung | *Macaranga gigantea* | Euphorbiaceae | 20 | 3.2 |
| 14 | Mengadai | - | - | 7 | 1.1 |
| 15 | Pulai | *Alstonia spp.* | Apocynaceae | 96 | 15.3 |
| 16 | Pulai Pipit | *Alstonia angustiloba* | Apocynaceae | 70 | 11.2 |
| 17 | Purang | *Macaranga triloba* | Euphorbiaceae | 34 | 5.4 |
| 18 | Rambutan | *Nephelium lappaceum* | Sapindaceae | 1 | 0.2 |
| 19 | Rengas | *Gluta renghas* | Anacardiaceae | 1 | 0.2 |
| 20 | Simpur | *Dillenia maxima* | Dilleniaceae | 2 | 0.3 |
| 21 | Tengkawang | *Shorea stenoptera* | Dipterocarpaceae | 13 | 2.1 |
| 22 | Tepung | - | - | 19 | 3.0 |
| 23 | Trembesi | *Samanea saman* | Fabaceae | 2 | 0.3 |
| 24 | Ubah | *Syzygium darifolium* | Myrtaceae | 61 | 9.7 |
| Total | | | | 627 | 100.0 |

The most dominant species was kawi (*Shorea balangeran*) with 132 individuals (21%). The next dominant species comes from the Apocynaceae family named pulai rawa (*Alstonia spp.*) with 96 individuals (15%) and pipit pulai (*Alstonia angustiloba*) with 70 individuals (11%). Furthermore, from the Melastomaceae family, it was dominated by understorey, but for this species is able to grow high and part of woody species namely jambu monyet (*Melastoma marginata*) with 72 individuals (11%). Then, from the Myrtaceae family was ubah (*Syzygium darifolium*) with 61 individuals (10%). The average survival percentage of rehabilitation plants and natural tillers at PT Semitau Region was 74.11%.

In addition to internal factors, external factors such as disruption of human activities (the continuous harvesting activity in the rehabilitation area because the oil palm still produces fruit even though there was no special treatment has been given such as fertilization) are things that also affect the success of plant growth. The impact of oil palm harvesting activities such as the frond falling that hit the rehabilitation plant resulted from a plant in the failure to grow and die. The highest survival percentage was in Besar and Tekalong River with 100% total of living plants, while the lowest was in Rukam Hilir River with a 25% total percentage (Figure 3). The species that were known to be the most adaptive and had the highest population of each location were kawi (*Shorea balangeran*) from 6 locations, pulai rawa (*Alstonia spp.*) and ubah (*Syzygium darifolium*) from 4 locations, respectively (Figure 4).

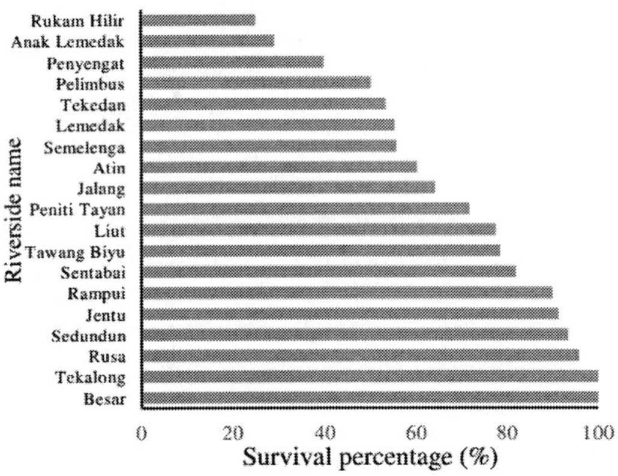


**Figure 3.** Percentage of plant life per location

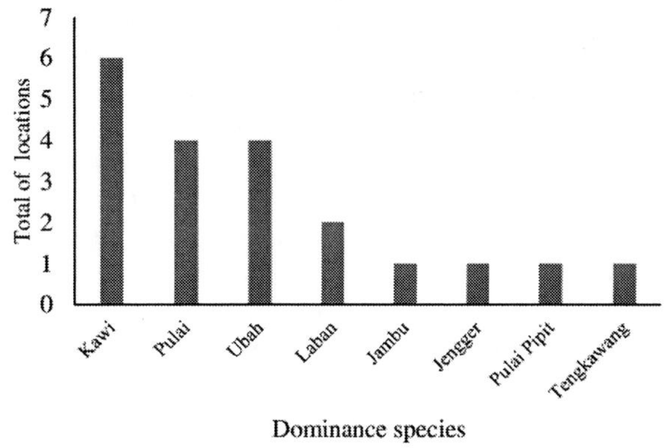


**Figure 4.** The dominance species in each location

### *3.2 Plant height and diameter*

The diameter of the three most adaptive plants, namely kawi, pulai, and ubah, had diversities that were not too significantly different. There were kawi and pulai species individuals which had quite different diameters that were shown by outliers. The median of the three plants showed in the range of 1 cm value indicating that the plant was in the seedling or sapling level (Figure 5). Figure 6 showed that the three species had quite different diversity values. Pulai had outlier at the smallest and largest which indicates that this species also had diversity in the seedling and sapling strata.

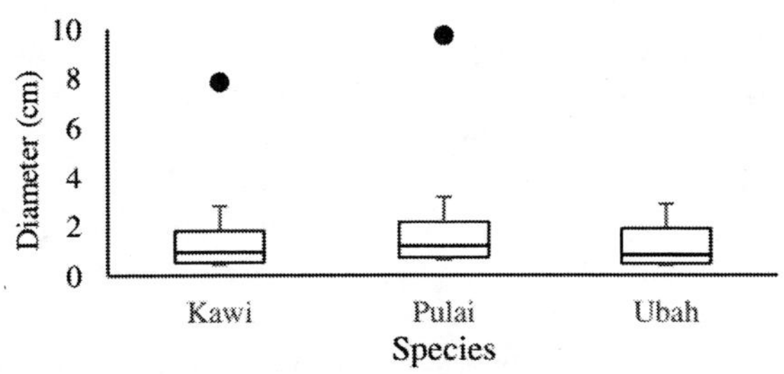


**Figure 5.** Box plot of diameter of adapted plants

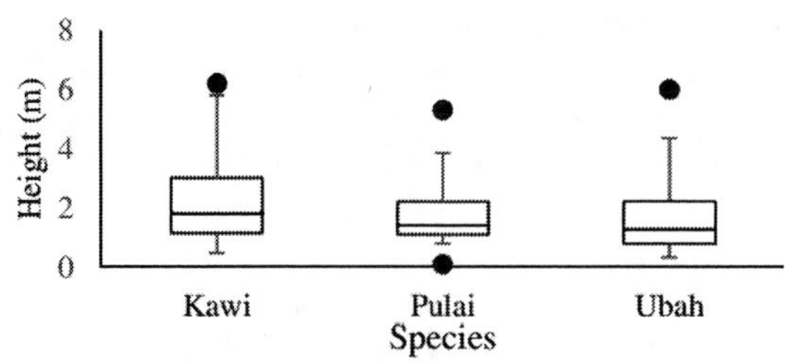


**Figure 6.** Box plot of height of adapted plants

**Table 3.** Total data of height and diameter values

| Data | Height | Diameter |
|---|---|---|
| Maximal Value | 6.50 m | 11.94 cm |
| Minimal Value | 0.10 m | 0.07 cm |
| Average Value (X) | 1.90 m | 1.61 cm |
| Standard Deviation (S) | 1.30 m | 1.54 cm |
| Coefficient Variance (CV) | 67.24 % | 95.50 % |

The coefficient variance, both height and diameter showed medium-high value, 67.24% for height and 95.50% for diameter (Table 3). The higher coefficient of variance indicates that the spread of species was increasingly heterogeneous. This was evidenced by a total of 24 plant species were found. Based on the value of diameter distribution, it could be seen that the dominant species of rehabilitation area were in the diameter range of 0.06 - 2.05 cm, which indicates that the dominance value was at the seedling or sapling level of 470 individual. Based on the height distribution value, it was known that the dominant species were in the seedling stage as evidenced by the highest total in the range of 0.1-1.5 m, followed by dominance at the next interval of the sapling stage (>1.5 m). The highest value location was located on the Tekalong River with 2.56 cm of average diameter and 2.75 m of height. The worst location was in the Atin River with 0.77 cm of diameter and 0.65 m of height. The increasing of diameter and height plants were greatly influenced by the availability of water and nutrient balance in the soil, if there is a lack of that then the plants will experience deficiency [16].

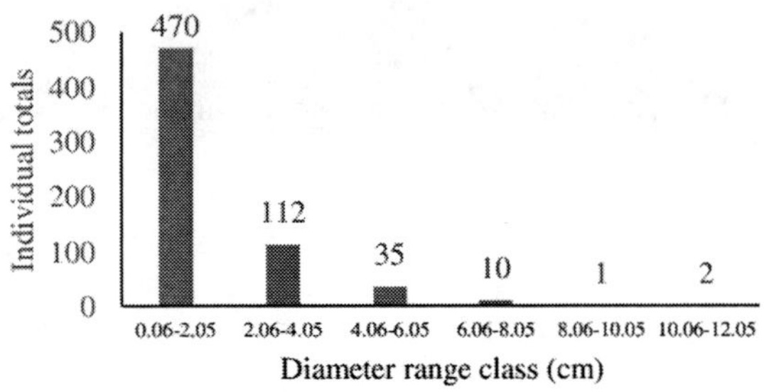


**Figure 7.** Diameter frequency distribution

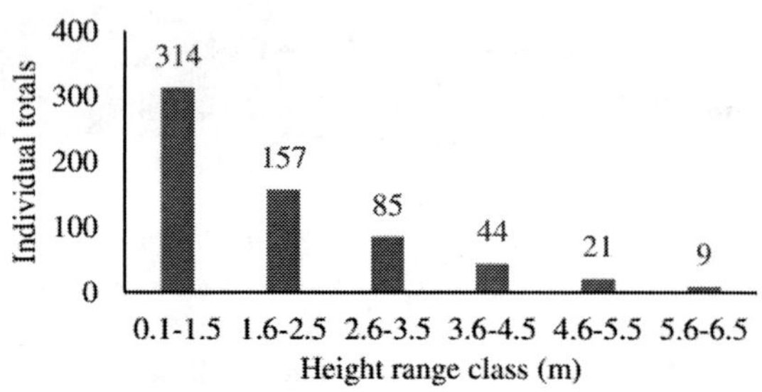


**Figure 8.** Height frequency distribution

### *3.3 Health*

The health percentage of plants was measured by scoring the area of attack (LS) and the intensity of attack (IS) from suspected pests and diseases occurred. The total average value of PT SR rehabilitation area was 76.36% for pest attack area and the disease attack area was 32.19%, while for pest attack intensity was 42.36% and disease attack intensity was 18.60%. Based on the values obtained, it was known that pests had a greater role in causing damage compared to disease factors. Figure 9 presents the different symptoms of plant diseases due to pest and disease attacks. The information shows that plants that were attacked by pests might not necessarily be attacked by diseases. The pest attack was dominated by score 1 which showed that the percentage of attacks ranged between 26-50%. Although dominated by healthy plants, the highest disease attack was at score 1 which was mostly marked by the presence of sooty mold on the leaves. The sooty mold was caused by the *Capnodium sp.* fungi. which has the black mycelium scattered characteristics that covered the leaf surface [17]. Sooty mold fungus was associated with plant-eating insects so that black-colored hyphae were formed [18]. Species with the highest pest and disease resistance were pulai rawa (*Alstonia spp.*), kawi (*Shorea balangeran*), and laban (*Vitex pubescens*). The dominance of pest resistance for pulai, kawi, and laban was 42%, 21%, and 16%, respectively. Meanwhile, the disease resistances were 21%, 26%, and 16%, respectively. The more resistance of these plants compared to another was due to the fact that they contain secondary metabolite compounds that function as defense mechanisms against pests and diseases [19-21]. Secondary metabolite of alkaloid base compounds causes pulai's resistance of herbivorous pests [19]. Laban contains secondary metabolites consisting of alkaloids, flavonoids, terpenoids, and steroids which cause this plant to function as an antimicrobial and cure infection [20]. Kawi's content of phytochemical compound functions as antioxidant and decreasing proliferation of disease [21]. The highest integration value between attack area and pest attack intensity was located in the Rukam Hilir River with 100% and 66.67% of the total, respectively. While the highest disease was in the Liut River with the 65.00% and 26.67% of acquisition value (Figure 10).

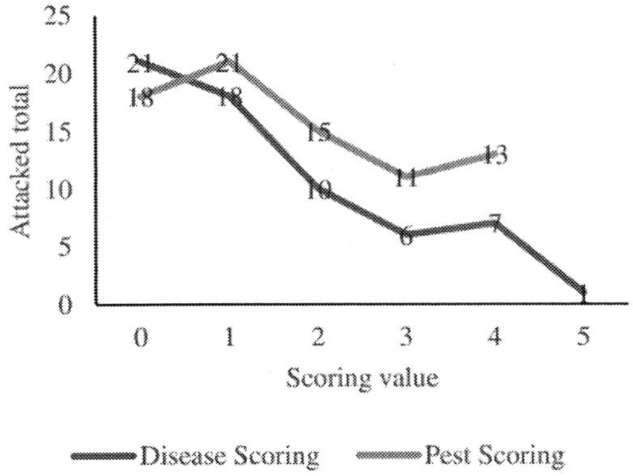


**Figure 9.** Frequency scoring of pests and disease

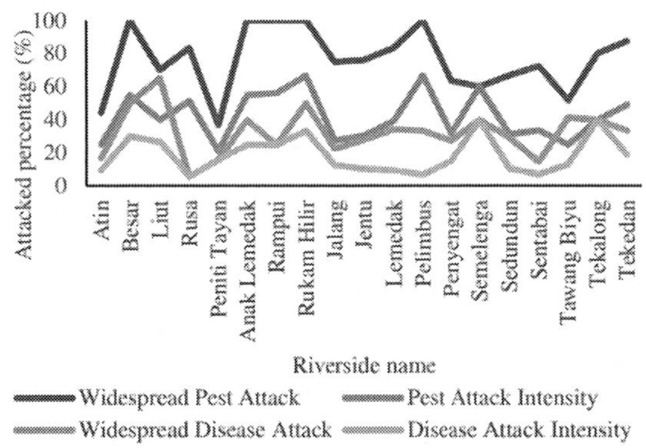


**Figure 10.** Health graph per locations

### *3.4 Soil chemical analysis*

PT Semitau Region has peat and mineral soil types, but the composition which is fully spread in the rehabilitation area is mineral soil of latosol and podsolic. Besides considering the physical and chemical properties of soil, the calculation of light intensity is also needed as additional data to find out the species that are suitable for planting in each location. The average value of light intensity obtained from all locations of rehabilitation areas, with canopy was 542.88 Lx while without canopy was 2483.88 Lx.

**Table 4.** Soil chemical properties in selected locations

| Location | Analysis Results | | | | |
|---|---|---|---|---|---|
| | pH | C-Organic (%) | N-Total (%) | C/N Ratio | KTK |
| Tekalong[1] | 5.42 | 3.08 | 1.69 | 1.82 | 4.90 |
| Atin[1] | 5.49 | 3.65 | 0.84 | 4.35 | 2.40 |
| Besar[2] | 3.84 | 1.41 | 0.12 | 11.75 | - |
| Rukam Hilir[2] | 4.88 | 1.55 | 0.13 | 11.92 | - |
| Peniti Tayan[2] | 4.90 | 1.57 | 0.12 | 13.08 | - |
| Liut[2] | 4.89 | 1.57 | 0.13 | 12.07 | - |
| Rusa[2] | 4.89 | 1.54 | 0.11 | 14.00 | - |

[1]Sucofindo (2020) and [2]SMART Research-Institute Analytical Laboratory (2015)

The value of light intensity was useful for determining the species suitability of tolerant, intolerant, or semi-tolerant. Based on Table 4 it is known that the C/N ratio at the location was quite varied. The year difference in the analysis of soil chemical properties greatly influenced the value of the C/N ratio. Tekalong and Atin Rivers which were analyzed directly at Sucofindo Laboratory in 2020 had a lower value of C/N ratio, 1.82, and 4.35, respectively. Meanwhile, Besar, Rukam Hilir, Penayan Tayan, Liut, and Rusa Rivers which were analyzed by secondary data from the SMART Research Institute-Analytical Laboratory in 2015 showed higher C/N ratio with a lower value of N-total ranged between 0.11-0.13. The difference significant in the C/N ratio showed that there has been a decomposition of soil organic matter within 5 years. This was also certainly influenced by the presence of rehabilitation plants that have been planted since 2016, it could be seen that N-total increasingly significant with the 0.11-0.13 range to the 0.84-1.69 range. Generally, there are similarities of symptoms caused by disease and nitrogen deficiency, the leaves show pale and yellowish color. The symptoms of N deficiency are first seen on older leaves, namely pale green leaves then chlorosis appears into bright yellow or pale and subsequently experience necrosis [22].

### *3.5 Implications for recommendation species*

Consideration for selecting suitable forest tree species was done based on the analysis of the variables that have been discussed. These species were kawi (*Shorea balangeran*), pulai rawa (*Alstonia spp.*), and ubah (*Syzygium darifolium*). Kawi or was known as red meranti or balangeran is a local species which only spread naturally on the Sumatra and Kalimantan islands, Indonesia. Kawi could grow well in areas of peat swamp forest reaching 20-25 m of height, 50 cm of diameter, and producing damar (resin). The field reality showed that kawi could still grow even in submerged areas. Pulai has natural distribution in almost all regions of Indonesia such as Java, Sumatra, Bali, Nusa Tenggara, Maluku, South Kalimantan, West Kalimantan, and Irian Jaya. Pulai is scattered in peat swamp forests, tidal areas, and dry areas with varieties of heights. The field reality showed that this species is able to re-grow through lateral branches even though it has been affected by the falling frond of oil palm. Ubah is a species that grows mostly in peat swamp. This plant comes from the Myrtaceae family with Syzygium genus. This species was also found quite a lot in the rehabilitation area as natural tillers.

## 4. Conclusion

The best survival rate of three plants species for further development is kawi (*Shorea balangeran*), pulai rawa (*Alstonia spp.*), and ubah (*Syzygium darifolium*) which is 45% of total individual plants that have survived. Information on the type of adaptation potential is expected to be useful information for increasing successfully riversides through replication technique of rehabilitation area of oil palm plantation.

As a recommendation, further research is needed based on the best adaptability of the three forest tree species so that they can be developed for future replication of mass planting in the field.

## References

[1] FAO 2020 *http://www.fao.org/faostat/en*

[2] Meijaard E, Garcia-Ulloa J, Sheil D, Wich SA, Carlson KM, Juffe-Bignoli D, and Brooks TM 2018 *Oil Palm and Biodiversity: A Situation Analysis by the IUCN Oil PALM Task Force* (Gland: International Union for Conservation of Nature)

[3] Luke SH, Advento AD, Aryawan AAK, Adhy DN, Ashton-Butt A, Barclay H, Dewi JP, Drewer J, Dumbrell AJ, Edi and *et al.* 2020 Managing oil palm plantations more sustainably: large-scale experiments within the Biodiversity and Ecosystem Function in Tropical Agriculture (BEFTA) programme *Front. For. Glob. Change* **2** 75

[4] [RSPO] Roundtable on Sustainable Palm Oil 2018 *https://ga.rspo.org/ga15/Resolutions/RSPO_P&C_ 2018.pdf*

[5] Barclay H, Gray CL, Luke SH, Nainar A, Snaddon JL, and Turner EC 2017 *RSPO Manual on Best Management Practices (BMPs) for the Management and Rehabilitation of Riparian Reserves* (Kuala Lumpur: Roundtable on Sustainable Palm Oil)
[6] Camirand R 2002 *Guidelines for Forest Plantation Establishment and Management in Jamaica* (Quebec: Tecsult International)
[7] *Progress Report of River Sides Rehabilitation Semitau Region (2016-2018)* 2019 (Jakarta: PT SMART Tbk) Unpublished
[8] *Main Report on Environmental Impact Analysis (AMDAL)* 2006 (Putussibau: PT Semitau Region) Unpublished
[9] Buntoro BH, Rogomulyo R and Trisnowati S 2014 The effect of manure dose and light intensity on growth and yield of white intersection (*Curcuma zedoaria* L.) *J. Vegetalika* **3**(4) 29-39
[10] Srivastava AK 2011 *Regional Workshop on the Use of Sampling in Agricultural Surveys* (Montevideo: Food and Agriculture Organization of the United Nations)
[11] Subramani J 2018 On circular systematic sampling in the presence of linear trend *Biometrics and Biostatistics International Journal* **7**(4) 286-92
[12] Yusuf M 2014 *Research Methods: Quantitative, Qualitative, and Combined Research First Edition* (Jakarta: Kencana)
[13] Supranto J 2000 *Sampling Technique for Survey and Experiment* (Jakarta: PT Rineka Cipta)
[14] Soerianegara I and Indrawan A 2005 *Indonesia Forest Ecology* (Bogor: IPB)
[15] Arifin M 2010 Study of soil physical properties and various land uses in relation to estimating soil erosion *Agriculture Journal MAPETA* **12**(2) 72-144
[16] Harjanti RA, Tohari , and Utami SNH 2014 Effect of nitrogen and silica fertilizer doses on initial growth (*Saccharum officinarum* L.) to inceptisol *Vegetalika* **3**(2) 35-44
[17] Labib MA, Yuliani, Ratnasari E, and Dwiastuti ME 2015 Application of celery herb extract (*Apium graveolens*) to the distribution of *Capnodium citri* fungus which causes soot dew in various citrus plants *LenteraBio Journal* **4**(1) 93-98
[18] Nelson S 2008 *Sooty Mold Plant Disease* (Manoa: College of Tropical Agriculture and Human Resources University of Hawaii)
[19] Silalahi M 2019 Botanical and bioactivity of pulai (*Alstonia scholaris*) *Prolife Journal* **6**(2) 136-147
[20] Sirait EU, Khotimah S, and Turnip M 2014 Laban fruit extract (*Vitex pubescens* Vahl) as an inhibitor growth of *Salmonella thypi* and *Staphylococcus aureus Journal of Protobiont* **3**(3) 40-45
[21] Wardani M and Susilo A 2016 Description of grow place, morphological diversity, and content of phytochemical compounds *Shorea balangeran* (Burck) at Bangka Belitung Forest *Bul. Plasma Nutfah* **22**(2) 81-92
[22] Matana YR and Mashud N 2015 Response of N, P, K and Mg fertilization on soil and leaf nutrient content in young oil palm plants *Journal of Palma* **16**(1) 23-31

# Sustainability of masoyi (*Cryptocarya massoy* (Oken) Kosterm) for essential oil industry materials

**I Yeny*, Darwo, H S Nuroniah**

Forestry Research and Development Centre, Ministry of Environment and Forestry, Bogor, 16610, Indonesia

*Email: irmayeny.kemenhut@yahoo.com

**Abstract**. Excessive exploitation without cultivation caused harvest of masoyi (*Cryptocarya massoy* (Oken) Kosterm) from forest to be more difficult. Cultivation outside of natural habitat becomes important to ensure sustainable masoyi oil production. This research presents information on the prospects for cultivation and growth rates of masoyi plants outside natural habitats. The research method used a plot design conducted at three locations, namely (1) Haurbentes Research Forest, Bogor Regency, West Java Province; (2) Boalemo Protection Forest Management Unit (PFMU), Gorontalo Province; and (3) community land in East Lombok Regency, West Nusa Tenggara Province. The results showed that the soil characteristics in the Haurbentes Research Forest were more suitable for masoyi plant growth. This is because the Haurbentes area has a texture of clay dust, climate classification tipe A. However, the characteristics of growing areas in Boalemo and East Lombok showed that masoyi plants were still able to grow. Thus, masoyi plants have the potential to be cultivated outside their natural habitat.

## 1. Introduction

Essential oil is the result of distillation from plant parts (roots, stems, bark, rhizomes, leaves, fruits, seeds, and flowers) that are liquid and volatile [1]. This oil is classified as vegetable oil and has a distinctive aroma. Indonesia has around 500 types of essential oil-producing plants. At least 40 out of 80 types of essential oil-producing plants traded in the world grow in Indonesia [2]. Based on Minister of Forestry Regulation No. P. 35 of 2007 concerning Non-Timber Forest Products (NTFPs), masoyi (*Cryptocarya massoy* (Oken) Kosterm) is one of 20 types of NTFPs producing essential oils under the jurisdiction of the Ministry of Forestry. Masoyi oil produces a distinctive aroma (such as the aroma of coconut, caramel, milk) and high levels of lactone.

Masoyi is a tree species with a height above 30 m, the upright trunk is straight and untwisted. It has greenish-gray bark with 5-15 mm thickness [3]. Topul *et al.* [4] mentions that the hydro distillation from bark, wood core, and fruit parts of masoyi produces pale yellow oil, with the main components of bark and wood core oil being massoia lactone C-10 and massoia lactone C-12, while the main fruit oil constituents is benzyl benzoate (68.3%). Core wood also contains C-14 massoia lactone (1.4%) and saturated d-decalactone C-10 derivatives (2.5%). Massoia lactone is needed in the chemical industry, especially in the food industry, cosmetics as a flavoring agent and raw material for drugs. Massoia lactone (synonym 5 hydroxy 2 decenoic acid delta lactone cas No. 54814-64-1) meets regulation 872/2012/EC as a safe flavoring agent. Some research results mention that masoyi bark essential oil is a potential in vitro immunomodulator, anti-bio-film and antimicrobial (Hertiani et al., 2018; Permanasari et al., 2017).

Masoyi oil ranks 8th in the production of essential oils in Indonesia. The amount of masoyi oil production in 2016 has an average production value of 15-20 tons/year [2]. The selling price of masoyi oil is divided based on its lactone compound content, which is 50% lactone content (IDR 2.5 million/kg), 70% (IDR 4 million/kg) and 90% (IDR 9 million/kg) (personal communication, 2019). The high selling value has an impact on the high demand for raw materials in masoyi oil production.

Until now, the raw material for making masoyi oil is in the form of tree bark obtained through collection activities from nature in the masoyi endemic region, Papua Island. Exploitation was carried out by cutting down trees and directly skinning the trunk has accelerated the reduction of masoyi stands at the tree level. If the oil yield is assumed to be 2%, 750-1000 tons/year masoyi bark is needed to produce 15-20 tons/year of masoyi oil. If a 30 cm diameter masoyi tree produces 100 kg of masoyi bark [7], then in one year 7500-10000 masoyi trees will be felled. This condition threatens the continued availability of raw material for masoyi oil in Indonesia. This is evidenced from the research of Hutapea *et al.* [8] which shows the presence of masoyi at the pole and tree level is very low due to harvesting activities. To ensure the sustainability of the raw material for masoyi oil, the Research and Innovation Agency of the Ministry of Environment and Forestry has carried out masoyi development outside its natural habitat, namely Sulawesi Island (Boalemo), the West Nusa Tenggara Islands (Lombok Timur), and Java Island (Bogor). This paper presents information on the prospects for cultivation and growth rates of masoyi plants outside natural habitats to support the continuity of supply of essential oils. It is expected that research results can provide information on the potential of masoyi plants to be developed outside their natural habitat.

## 2. Materials and Methods

### *2.1. Materials*

Data collection was carried out from three planting trials of masoyi in: 1) Haurbentes Research Forest, Bogor Regency, West Java Province, 2) Boalemo Production Forest Management Unit (PFMU), Boalemo Regency, Gorontalo Province, and 3) community land in Timbenuh Village, Pringgasela District, Lombok Timur Regency, West Nusa Tenggara Province.

**Figure 1**. Location of masoyi planting trials (location is shown on red dots)

### *2.2. Method*

#### 2.2.1. Characteristics of growth places

Characteristics of masoyi planting trials was collected through desk study from literature and reports from relevant agencies.

2.2.2 Trial planting of masoyi outside the natural habitat
Planting trials were carried out through three stages: (a) provision of planting material (seedlings), (2) planting, and (3) maintenance.

*2.2.2.1. Seedlings*
Fruit was collected from the masoyi population in the natural forests of Fak-Fak Regency, West Papua Province. The seeds were sown 2 weeks after harvesting. Before sowing, fruit flesh was discarded. Furthermore the seeds were soaked 24 hours with water at room temperature. Seeds were then soaked in Dithane solution for 2 hours to kill the fungus carried by seeds. The seeds were sowed in a media mixture of soil+charcoal+cocopeat in a greenhouse with 50% paranet shade. After the seeds produced seedlings with a number of leaves > 2 strands, seedlings were transplanted to polybags containing soil+sand+charcoal husks. Seedlings used for planting were 6 months old with a height of over 45 cm.

*2.2.2.2. Planting*
Masoyi planting is carried out in three locations as follows:

a. Haurbentes Research Forest, Bogor Regency, West Java Province.
   The plot was established in 2016. The initial condition land was shrubs. The observation form was in a path with a distance between plants in 2 m lane and a distance between path in 10 m lane (2 m x 10 m).
b. Boalemo Production Forest Management Unit (PFMU), Boalemo Regency, Gorontalo Province.
   The initial condition was unproductive natural forest production and dominated by shrubs. The research plot was built in 2015 with a spacing of 4 m x 5 m.
c. The community land of Timba Nuh Village, in East Lombok Regency, West Nusa Tenggara Province.
   The plot was built in 2018 which was planted along the boundaries of community land with a distance between plants of 5 m.

*2.2.2.3. Maintenance*
Maintenance carried out in the three research plots in the form of:

a. Basic fertilization with 2 kg of manure per planting hole 30 x 30 x 30 cm.
b. Weeding by removing the weeds around the 1 m diameter plant. Weeding was done 2x/year.
c. No fertilization was added.

*2.2.2.4. Growth measurement*
The observation plot was made by a line plot containing 20 plants. The number of plot in each location was as follows: Haurbentes Research Forest was 12 plots, Boalemo PFMU was 5 plots, and East Lombok was 19 plots. Variables of growth parameters included survival rate, height, and diameter. The diameter of the stem was measured at the base of the stem as high as 10 cm from the ground. Masoyi stand at Boalemo and East Lombok were measured once at 2 years old. Masoyi stand at Haurbentes was measured regularly each six months since plantation until 3 years old (6 measurements).

*2.2.2.5. Data analysis*
Data were analyzed to obtain the mean and standard deviation of height, diameter, survival rate, and MAI. Specifically for growth data at the Haurbentes Research Forest site, a regression analysis was performed to determine the relationship between diameter and height to plant age.

## 3. Results and Discussion

### *3.1. Results*
3.1.1. Characteristics of the masoyi planting area
Based on the literature study, it showed that the masoyi planting trial area has a medium range of topography, but has different climate with wide variations and varied soil textures (Table 1).

**Table 1.** Land characteristics of masoyi planting trials

| Variables | Location | | |
|---|---|---|---|
| | Haurbentes[1] | Boalemo[2] | Lombok Timur[3] |
| Elevation (masl) | 200-250 | 400-600 | 300-400 |
| Slope (%) | 25-45 | 15-25 | 15-25 |
| Soil type | Red yellow podsolik | Red yellow podsolik | Reddish brown latosol |
| Soil texture | Clay sandy dust | Dusty clay sand | Clay dust sand |
| Precipitation) (mm/years) | 4276 | 1475 | 1100 |
| Wet season (months) | 11 | 7 | 6 |
| Dry season (months) | 1 | 5 | 6 |
| Minimum temperature (°C) | 23 | 21 | 20 |
| Maximum temperature (°C) | 28 | 34 | 30 |

Sources: [1][9]; [2][10]; [3]KPHL Rinjani Timur

The site condition provide a varied growth response. In general, the site condition in Haurbentes Research Forest has a sandy clay dust, while in Boalemo and East Lombok the texture of the land is dusty sand and clay dust sand. The soil texture in Boalemo and East Lombok tends to be higher in sand content than in Haurbentes Research Forest. Thus, soil conditions at the Boalemo and East Lombok research sites are more porous compared to at the Haurbentes Research Forest site.

Based on the Schmidt and Ferguson classification, the rainfall type in Haurbentes Research Forest area is type A and does not have a dry month (rain fall less than 60 mm/month) [9]. The climate type in Boalemo PFMU is type D, which is temperate and is influenced by the rainforest season [10]. The climate type in the Rinjani Timur KPH area is type D (moderate) and E (slightly dry), which is the condition of the region with rainfall between semi-dry to moderate.

3.1.2. Trial planting of masoyi outside the natural habitat

*3.1.2.1. Seedlings*

Masoyi seeds are semi recalcitrant, so the seeds were sown directly (2 weeks after harvesting). Single leaf emerged after 30-32 days. Rate germination was 70%. Seedlings were raised in the nursery for 6 months to reach a standard height of 45 cm for planting.

*3.1.2.2. Growth*

The growth of 2 years old masoyi in the three sites showed that the highest to lowest masoyi plant successively was in the Haurbentes Research Forest, East Lombok, and Boalemo. The average diameter of the largest to the smallest is in the Haurbentes Research Forest, Boalemo, and East Lombok. Planting trials in the three sites showed that the growth of the masoyi plants in the Haurbentes Research Forest was the best. Data recapitulation is presented in Table 2. Growth curves of height and diameter and picture of 2 years old masoyi can be seen in Figures 2, 3, and 4.

**Table 2.** The growth of 2 years old masoyi in three sites

| No. | Location | Diameter | | | Height | | | Survival (%) | |
|---|---|---|---|---|---|---|---|---|---|
| | | Mean (cm) | SD (cm) | MAI (cm/years) | Mean (m) | SD (m) | MAI (m/years) | Mean | SD |
| 1. | Lombok Timur, Nusa Tenggara Barat | 0.34 | 0.07 | 0.17 | 1.29 | 0.24 | 0.69 | 75.8 | 22.6 |
| 2. | Boalemo, Gorontalo | 0.83 | 0.80 | 0.42 | 0.82 | 0.27 | 0.41 | na | - |
| 3. | Bogor, Jawa Barat | 1.12 | 0.20 | 0.56 | 1.65 | 0.23 | 0.86 | 81.0 | 7.5 |

Remarks: SD = Standard of deviation; MAI = Mean Annual Increment, na= not available

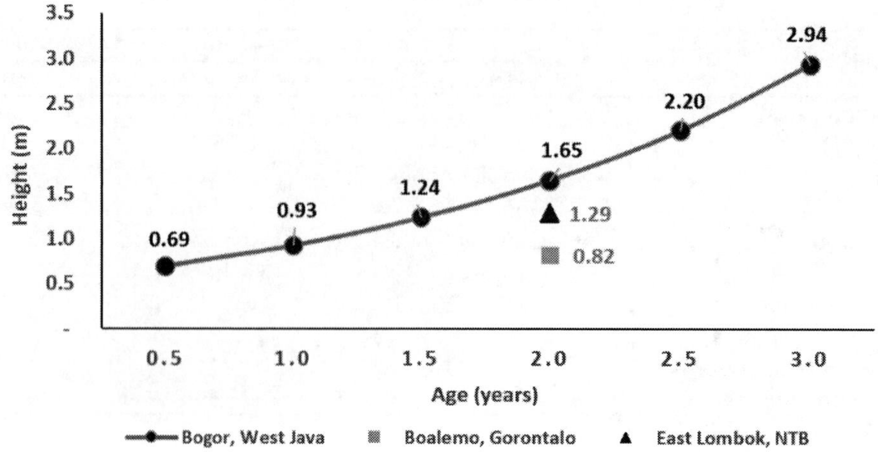


**Figure 2**. Height growth of masoyi at three planting sites

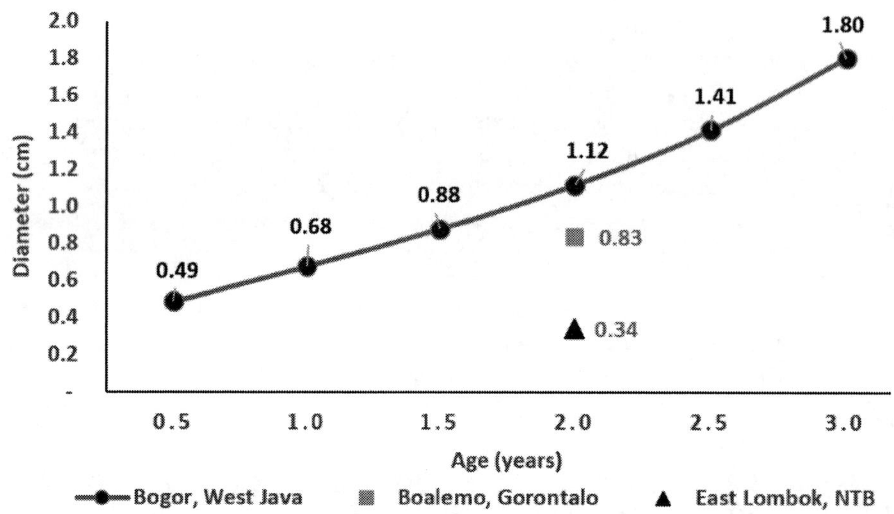


**Figure 3.** Diameter growth of masoyi at three planting sites

Bogor Boalemo Lombok Timur

**Figure 4.** Growth performance of 2 years old masoyi at three sites

### *3.2. Discussion*

3.2.1. Characteristics of planting sites

Planting a species outside of natural habitat requires taking consideration the similarity of the site with its natural habitat. The parameters observed are the physical properties of the site (altitude and soil type) and climate (rainfall, temperature, humidity, and canopy cover).

The natural distribution of masoyi is spread throughout the provinces of West Papua and Papua (Figure 5). In West Papua province, it is found in Manokwari Regency, Ransiki District, Windesi, Teluk Wondama District (Wasior), Teluk Bintuni Regency, Fak-Fak Regency, Kaimana Regency, Sorong Regency (Sorong District, Moraid District), and Teminabuan Regency; whereas in Papua Province it is spread in Nabire Regency, Yapen Waropen Regency (South Yapen District), Biak Nunfor Regency (East Biak District, North Biak, West Biak, South Supriori), parts of Jayapura Regency and parts of Merauke Regency [11]. Hutapea *et al.* [8] mentioned the potential of masoyi in Teluk Bintuni Regency was 1593 individuals/ha, higher than the Kaimana District which was 871 individuals/ha. However, this potential was dominated by seedling (80-90%) and sapling (5-13%), and only 1 individual/ha at tree level. The low natural distribution and plant at tree level seem to be caused by the need for habitus and high exploitation of masoyi at tree level. Therefore the biophysical information of the masoyi natural habitat becomes important to be considerd on the suitability of the edaphic factor for masoyi development.

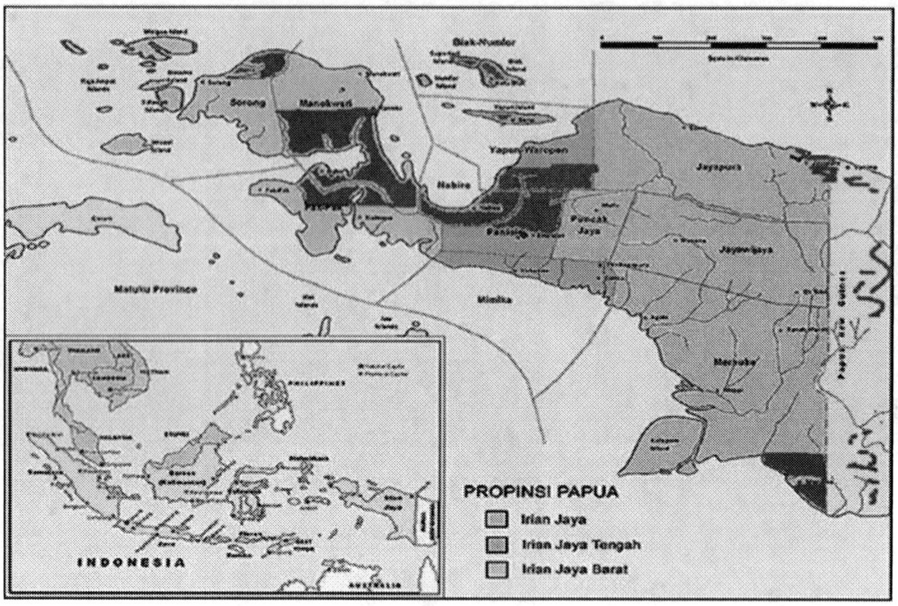


**Figure 5.** The natural distribution of masoyi in Papua (marked in red)

Based on natural distribution of masoyi in three districts namely Nabire, Bintuni, and Manokwari, Yeny *et al.* [11] concluded that the masoyi natural habitat is on a fairly wide slope range of 2-60% and rainfall 1300-2500 mm/year, but has a narrow range in terms of elevation, soil type, soil texture, and effective depth. Furthermore Hutapea *et al.* [8] conducted a survey of the natural distribution of masoyi in Bintuni and Kaimana and stated that masoyi grew at an altitude of between 50-1100 masl, and grew scattered (not clustered). Based on the study of Yeny *et al.*[12], Hutapea *et al.* [8] and this study (Table 1) can be concluded that masoyi can adapt to climate types A (very wet regions), D (temperate regions), and E (slightly arid regions) according to the Schmidt & Fergusson classification, and elevation 200-475 masl, Masoyi can grow in areas with temperatures of 21-34°C. The results of masoyi planting trials until 3 years old indicate that masoyi are able to grow in the elevation range of 250-600 masl with rainfall between 1100-4276 mm/year (Table 1).

Based on the characteristics of three planting sites, the Haurbentes Research Forest area is most suitable for masoyi plantation compared to the two sites of Boalemo and East Lombok. This is presumably because the Haurbentes Research Forest has climate type A with wet months throughout the year, and texture conditions of sandy clay dust. However, in Boalemo and East Lombok the texture of the clay is dusty clay sand and clay dust sand with low rainfall. Soil and climate characteristics in Boalemo and East Lombok have higher sand content, the soil is more porous so it dries faster. More porous soil conditions with low rainfall result in slower growth of masoyi plants (Table 1). However, with the characteristics of growing places in Boalemo and East Lombok it turns out that masoyi plants still have a high survival rate, which is above 75%. Thus, masoyi plants have the potential to be developed outside their habitat on the terms of agro-climate and edaphis such as in the Haurbentes Research Forest, Boalemo, and East Lombok.

3.2.2. Trial planting of masoyi outside of natural habitat

Propagation of masoyi plants can be done by generative and vegetative. Vegetative propagation can be obtained through seeds and wildlings. Germination rate of seeds reach 70%. Since masoyi seeds are recalcitrant and seed sources/parent trees are rare, vegetative propagation should be forced. It can be done by cuttings and the rooting percentage reached 72% using mix media soil+sand (2:1, v/v) [13].

The study showed that Masoyi had better growth in Bogor than two sites. This is indicated by the MAI diameter and height of 2 years old masoyi in the Haurbentes Research Forest which are 0.65

cm/year and 0.86 m/year, respectively. Compared with other forest tree, the growth of masoyi is similar to the growth of *Vatica sumatrana, Hopea mangarawan, Dryobalanops lancelolata, Dipterocarpus* sp., with MAI diameter and plant height at 2 years old less than 1 cm/year and 1 m/year [14,15]

Planting trials showed that masoyi is a semi-tolerant plant. Therefor, masoyi is able to grow in open areas by gap planting. Masoyi also can be used as an enrichment plant in unproductive Log Over Area (LOA) forest by applying a line planting or gap planting.

Using vegetative propagation method, the production of masoyi seedlings can be performed in any places. Then, it allows masoyi as a noteworthy NTFPs in forest areas and community forests. Local people in Papua Island cut the masoyi tree with minimum diameter 15 cm, which is generally reached by trees aged 10 years. Based on masoyi growth up to 3 years old in Haurbentes Research Forest, the harvest age will be achieved at the age of 15 years. If one masoyi tree with a diameter of 15 cm produces 7.5 kg of masoyi bark, then to produce 20 tons of masoyi oil/year (same as production value in 2016) takes 133,333 trees/year. If the planting distance is 5 x 5 m, then planting and logging are needed with an area of 333 ha/year.

Masoyi is slow-growing tree species, so it requires a long observation time for predicting growth and yield of stands up to 15 cm in diameter. Therefore, observations will continue to predict the economic cycle of the masoyi stand. Masoyi economic cycle is the time length to produce an economic diameter of the masoyi tree for harvesting. Economic cycle information data is important because it deals with the prospects of masoyi plantation on a business scale. Other effort to be conducted is to develop improved masoyi by breeding to find faster growth of masoyi.

## 4. Conclusion

Masoyi showed better growth at Haurbentes site, Bogor Regency, than at the Boalemo and East Lombok sites. Site characteristics in the Haurbentes Bogor Research Forest are clay sandy dust texture with climate classification tipe A. However, masoyi is able to grow in the site characteristics of Boalemo and East Lombok. Thus, masoyi has the potential to be cultivated outside their habitats as long as the agro-climatic and edaphic conditions are the same type as natural habitat.

## References

[1] Bangkit TPS, Sirait R, and Iriany 2012 Penentuan kondisi keseimbangan unit leaching pada produksi eugenol dari daun cengkeh *J. Tek. Kim. USU* **1**(1) 10-4
[2] Atsiri Council 2018 *Minyak Atsiri Hasil Hutan Bukan Kayu*
[3] Rostiwati T and Efendi R 2013 *Mendulang Uang Tanpa Tebang, Lima Jenis HHBK Unggulan* (Bogor: Forda Press)
[4] Topul R, Wossa SW, and Leach DN 2007 Comparative chemical analysis of the essential oil constituents in the bark, heartwood and fruits of *Cryptocaryamassoy* (Oken) Kosterm. (Lauraceae) from Papua New Guinea *J. Mol.* **12** 149-54
[5] Hertiani T, Yuswanto A, Pratiwi SUT, and Mashar MH 2018 Effect of massoia (*Massoia aromatica* Becc.) bark on he phagocytic activity ofwistar rat macrophages *Journals Sci. Pharm.* **86**(2) 19
[6] Permanasari P, Triana H, Agustinus, and Yuswanto 2017 In vitro evaluation of massoia bark essential oil and c-10 massoialactone potency as immunomodulator *J. Essent. oil brearing plants* **20** 2
[7] Heyne K 1987 *Tumbuhan Berguna Indonesia III* (Jakarta: Departemen Kehutanan)
[8] Hutapea F J, Kuswandi R, and Asmoro J P 2020 Potensi dan sebaran masoi (*Cryptocarya massoy* (Oken) Kosterm) di Kabupaten Teluk Bintuni dan Kabupaten Kaimana *Jurnal Penelitian Kehutanan FALOAK* **4** 1
[9] Pusprohut 2015 Rencana pengelolaan KHDTK Haurbetes Bogor
[10] KPHP Boalemo 2015 Rencana Pengelolaan Hutan Jangka Panjang Tahun 2015-2024 Kesatuan Pengelolaan Hutan Produksi (KPHP) Model Boalemo Boalemo

[11] Yeny I and Minarningsih 2018 Potensi dan peluang pengembangan *Cryptocarya massoy* Oken Konstrem untuk kesejahteraan masyarakat dan kelestarian keragaman hayati *Prosiding Seminar Nasional Biodiversitas untuk kehidupan* p. 372-87
[12] Yeny I, Narendra B, and Nuroniah S 2018 Wilayah pengembangan masoyi berdasarkan kesiapan masyarakat dan kesesuaian lahan di UPTD KPH Unit V Boalemo *Jurnal Penelitian Hutan Tanaman* **15**(2) 67-145
[13] Darwo and Yeny I 2018 Penggunaan media dan zat pengatur tumbuh terhadap keberhasilan stek masoyi *Cryptocarya massoy* (Oken) Kosterm) *J. Penelitian Hutan Tanaman* **15**(1) 43-55
[14] Darwo 2020 Peningkatan produktivitas hutan tidak produktif dengan menerapkan TPTJ-Silin (Policy Brief) Pusat Pengembangan Perubahan Iklim dan Kebijakan Bogor
[15] Mardhatillah R, Pamoengkas P, and Darwo, 2019 The growth of kapur tanduk (*Dryobalanops lanceolata* Burck.) on different levels of canopy opening and fertilization *IOP Conf. Series: Earth and Environmental Science* **394** *01206* p. 1-10

**Acknowledgements**
We thank to all parties who helped with the data collection in three sites (Bogor, Boalemo, and East Lombok). Our gratitudes go to the Head of FRDC, Ir. Djimlan (Head of Unit V Boalemo PFMU 2014-2018), the community of Timba Nuh Village, Pringgasela District for the data and information provided during the research. We thank Faturahman, S.Hut, Ari S. Hut, Irwan, and Gafar for their assistance throughout the research.

IOP Conf. Series: Materials Science and Engineering **935** (2020) 012072 doi:10.1088/1757-899X/935/1/012072

# Health asessment of large and old trees in Ragunan Zoo, Jakarta

**D Nandika[1*], H Kusuma[1], D T Kusumawardhani[1], E Rumiyati[2], Tata[2], L Karlinasari[1], and I Z Siregar[3]**

[1]Department of Forest Product, Faculty of Forestry and Environment, IPB University, Bogor, 16680, Indonesia
[2]Taman Margasatwa Ragunan (TMR)- Ragunan Zoo, Jakarta Selatan 12550, Indonesia
[3]Department of Silviculture, Faculty of Forestry and Environment, IPB University, Bogor, 16680, Indonesia

[*]Email: nandikadodi@gmail.com

**Abstract.** Taman Margasatwa Ragunan (TMR) or Ragunan Zoo in Jakarta is the largest zoo in Indonesia. The zoo has around 52,700 trees that serve as part of the landscape and shades for animals and visitors. However, the presence of the trees in zoo also carries some risks, such as broken branches or even falling trees, that may threaten the safety of the animals, visitors, and infrastructures. Because most of the trees are large and old with high historical and ecological values, their health needs to be evaluated regularly. A study was conducted to evaluate it with around fifty selected trees, particularly the ones that were old and located closest to the targets (i.e. animals, visitors, infrastructures). The evaluation used visual observation as well as integrated with sonic tomography technology. The result showed that five trees (10%) were healthy and safe, 31 trees (62%) were healthy and fairly safe, 12 trees (24%) were less healthy and fairly dangerous, and the rest trees were unhealthy or dangerous and very dangerous. The most common defects found in them are poor tree architecture (62%), root problems (58%), and codominant trunks or branches (40%). Therefore, it is vital to take care of the tree health in Ragunan Zoo. Trees with the risk of endangering its surroundings need to be cut off immediately. Biomechanics on sloping trees with codominant trunks need to be strengthened with the help of dynamic bracing sling system and support.

## 1. Introduction

Taman Margasatwa Ragunan (TMR) or Ragunan Zoo in Jakarta is the biggest zoo in Indonesia. This zoo covers 140 hectares with an animal shelter capacity of 2000 animals. Based on the tree inventory in 2014, TMR has 92 tribes of trees with 252 genera and 426 species. If they are divided into two utilization groups, at least 222 species belonging to 150 genera and 51 tribes are classified into ornamental plants, while, 223 species from 159 genera and 51 tribes with a total of 52,733 trees are grouped as shade trees [1].

The presence of trees in TMR is vital in supporting the zoo's role and function. Shade trees in the middle of a metropolitan not only serve as an identity but also to absorb air pollution, reduce noise, and support the conservation of the city's soil and beauty. Trees in urban areas also provide important resources to other species. Any damages in them will reduce their functions and benefits. Early identification of risks can reduce tree fractures or fall and increase the efficiency of the management [2].

Monitoring tree health can be done through visual observation of its physical condition. However, this method cannot reveal the overall health condition of the tree, considering that some parts of it cannot be seen visually (ex. inside the trunk). Meanwhile, various technologies to determine the inside of the trunk have now developed nondestructively, including a technology based on the velocity of sound wave propagation. This tool can determine the internal condition of the tree [3,4]. It is spot on to observe the internal of the trunks because it doesn't require over-damage actions on the tree [5]. Study by Nicolotti *et al.* [6] reported that the most effective technology for detecting tree decay is sonic tomography. It can determine the position of anomalies and estimate the size and shape of the tree trunks. The technique starts with creating a database of trees in TMR then conducting their comprehensive health evaluation. This is important considering if the tree is unhealthy, they can be prone to falling and threatening the safety of their surrounding in TMR. This study aims to evaluate tree health in TMR based on sonic tomography technology as a scientific basis for the management of the trees in the zoo.

## 2. Materials and Methods

### *2.1. Identfication of targeted tree*

The research was conducted from October to December 2019 at Taman Margasatwa Ragunan, Jakarta. Fifty trees which diameter more than 30 cm, age more than 30 years, and closed to "target" i.e. animal, visitor, property targetedwere then identified by location, species, and dendrometry, such as diameter, height, canopy proportionality, and special markings. Each of them was labeled with a mica plastic of 6 x 5 cm then numbered using colored paint and tied using a rope around the trunk (Figure 1)

**Figure 1.** Labelling of target tree

### *2.2. Visual tree assessment*

Visual Tree Assessment (VTA) was done by observing the condition and characteristics of all parts of the targeted tree, in all 360° directions around it [7]. It is important to find signs and symptoms of defects in each targeted tree, especially the roots, stems, branches, canopies, leaves, and micro-environmental conditions. Visual inspection was conducted following the modified VTA form from the International Society of Arboriculture (ISA). This form provides information on (1) tree characteristics, (2) tree defects or damages, (3) environmental conditions, and (4) the level of risk of failure. The intended target was people, animals, and infrastructures around trees that were potentially affected if there was any failure on the trees (e.g. falling trees).

IOP Conf. Series: Materials Science and Engineering **935** (2020) 012072 doi:10.1088/1757-899X/935/1/012072

### *2.3. Tree assessment using sonic tomography*

Sonic Tomography (SOT) was developed as a nondestructive test tool to detect internal damages in tree trunks by producing velocity of sound wave propagation that was transformed into tomogram image forms (two-dimensional images) [8]. SOT became a sufficient tool to identify the presence of intial deterioration in the tree trunks or branches, because inspection using this technology was very effective [9-11].

Testing using Picus®*Sonic* was carried out by installing 10-12 Picus®*Sonic Tomography* transducers on the surface of the targeted tree trunk with a height of ±130 cm from the ground. Sound wave generation was done by hitting a nail near the transducer using an electronic hammer. Each transducer was connected each other so that the propagation of sound waves could spread from one point to another (Figure 2). All of these were connected to computer software that processed the velocity of sound wave data. Data recorded were the speed of sound waves propagation, then turned into image gradations of color following the speed of sound waves reflected in the internal trunk of the tree. The results of the tomogram analysis of each tree could complement the results of visual observations that had been made, especially increasing accuracy in trials that prevented the failure of the trunks that could be done on defective trees [12].

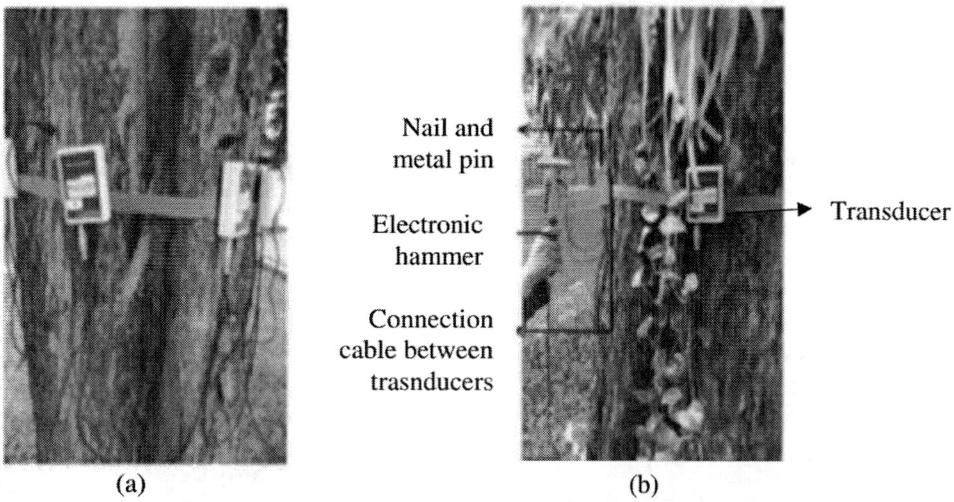


**Figure 2.** (a) Transducers in targeted tree trunk (b) Testing activity by tapping the electronic hammer into the metal pin

### *2.4. Interpretation of tree's condition based on sonic tomographye*

The health of targeted trees in TMR was determined based on the color image produced from the processing of the sound velocity matrix with the sonic tomography tool. Based on the results of the tomogram on the device, it was known if the trees contain solid or damaged wood. Table 1 shows the criteria developed for the inner-part trees reunk condition namely solid, moderate, and deteriorated based on the percentage (5) of solid wood conditions [13].

**Table 1.** Inner-pat tree condition based on solid wood volume percentage (%) and recorded volume of the tree trunk in the tomogram (SOT) [13]

| Solid wood percentage (%) | Innermost condition of target tree |
|---|---|
| 80-100 | Solid |
| 50-79 | Moderate |
| <50 | Deteriorated |

### 2.5. *Determination of targeted tree vigor*
The determination of the targeted tree vigor was based on the proportion ratio of solid wood or wood thickness with a cross-section radius (t/r ratio). This study found that when most trees reached linear decay measurements greater than 70% or the thickness was less than 30% (2/3 of cross-section), the tree had a high probability of failure [7]. The wood thickness percentage was a result of the tomogram image, while the cross-section radius was measured directly through those results. Furthermore, the ratio calculation was assisted with the ImageJ application.

### 2.6. *Integration of visual and sonic tomography assessment*
In determining the risk of tree failure, visual observations data were integrated with sonic tomography observations. It was presented in a table and analyzed with Microsoft Office Excel 2010. Table 2 shows the category of tree failure risk ranking [13].

**Table 2.** The category of targeted tree failure risk ranking based on visual observations and *sonic tomography* (SOT) [13]

| Visual assessment result (risk category) | Sonic tomography result (based on the percentage of solid wood) | | |
|---|---|---|---|
| | >80% | 50% -79% | <50% |
| Low | Safe | Fairly safe | Dangerous |
| Moderate | Fairly safe | Fairly dangerous | Dangerous |
| High | Fairly dangerous | Fairly dangerous | Very dangerous |
| Extreme | Fairly dangerous | Dangerous | Very dangerous |

## 3. Results and Discussion

### 3.1. *Characterisation of targeted trees*
The result showed that the 50 targeted trees consisted of 12 species, namely Sengon (*Paraserianthes falcataria*) 3 trees, Trembesi (*Samanea saman*) 12 trees, Khaya (*Khaya anthotheca*) 11 trees, Kemenyan (*Styrax benzoin*) 10 trees, Beringin (*Ficus benjamina*) 1 tree, Kedawung (*Parkia timoriana*) 2 trees, Cempaka (*Magnolia champaca*) 4 trees, Petir (*Parkia javanica*) 1 tree, Kapuk (*Ceiba pentandra*) 2 trees, Nyatoh (*Palaquium rostratum*) 1 tree, Karet (*Hevea brasiliensis*) 2 trees, and Mahoni (*Swietenia mahagoni*) 1 tree (Figure 3).

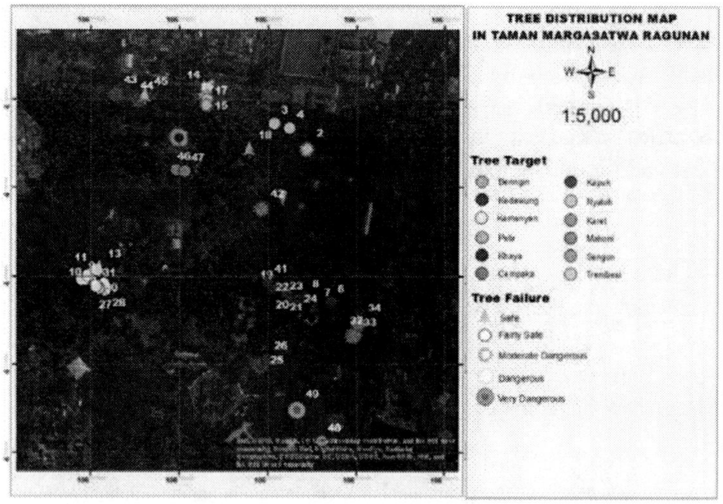


**Figure 3.** Spatial distribution of targeted tree in TMR

The fifty targeted trees have various diameters, ranging from 31-150 cm. Most of them have 61-90 cm diameter (20 trees), followed by 121-150 cm (13 trees), then trees with 91-120 cm diameter (10 trees), and lastly 31-60 cm (7 trees). Meanwhile, the height ranges from 11-40 m. Most of them are 21-30 m of height (27 trees), followed by 11-20 m of height (13 trees), and 31-40 m (10 trees) (Figure 4).

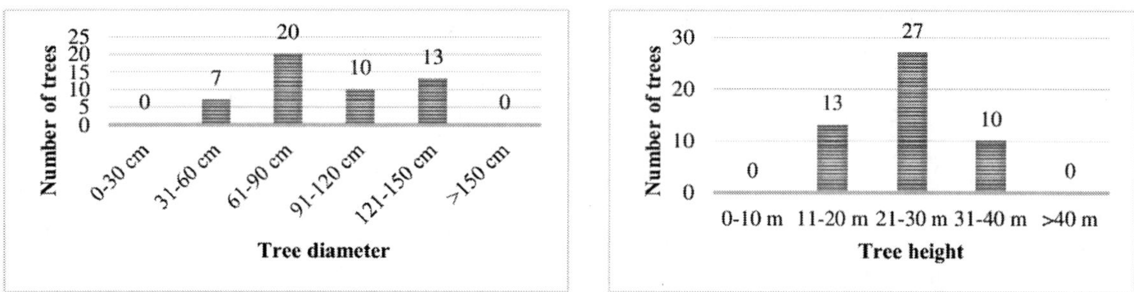


(a) (b)

**Figure 4.** (a) Tree diameter distribution and (b) Tree height distribution of targeted tree

The results of the target trees diameter and height distribution indicate that most of them are old. Old trees are defined as the main ecological structure because they provide resources for other species. They also have high cultural values because of their aesthetic and spiritual qualities [14,15]. Old trees are characterized by large volume appearance, peeled bark, spiral grains on the trunks, and some dead branches [16,17]. Large and old trees are most likely to face removal in urban landscapes because they have more safety risks (fallen branches or trees) posed to the public and infrastructures. Their removal can cause serious ecological implications. Therefore, the regular health evaluations at TMR are necessary to monitor the condition of the tree to maintain critical habitats for biodiversity in the long run and prevent any safety risks.

IOP Conf. Series: Materials Science and Engineering **935** (2020) 012072 doi:10.1088/1757-899X/935/1/012072

*3.2. Visual tree asessment result*

This result shows several case findings found in the targeted trees. Most of them are bad tree architecture (31 cases), followed by root problems (29 cases), codominant branches (14 cases), termite attacks (10 cases), other pests attacks (8 cases), cracks/ruptures (4 cases), cancer (2 cases), and dead tree tips (1 case) (Figure 5). The root problem that usually arises is the disruption of root development by infrastructures such as roads and sidewalks, causing the roots to be cut off or mechanically compressed.

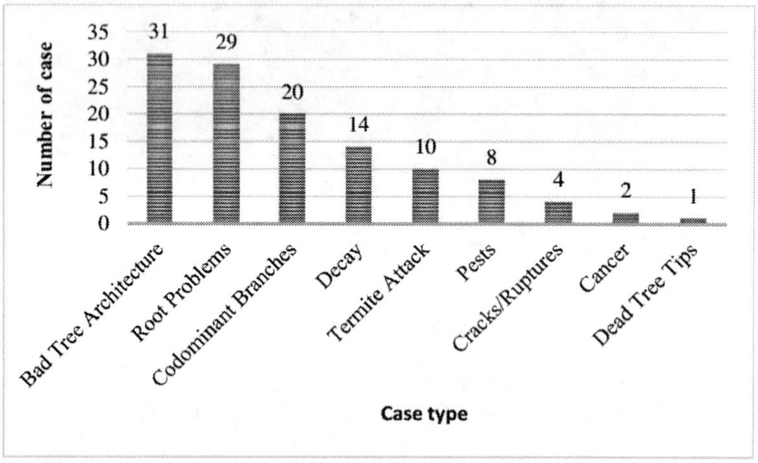


**Figure 5.** Case distribution on targeted tree

The measurement of tree architecture was conducted to find out the symmetry of the tree canopy and Live Crown Ratio (LCR). LCR is calculated based on the ratio of the height of the canopy to the total height of the tree expressed in percentage (%). LCR is a characteristic of canopy condition which correlates with tree growth [18,19]. Whereas according to Zhao *et al.* [20], LCR indirectly becomes an indicator of photosynthetic capacity and stands density. That is why it becomes a general indicator of the tree fitness. Tree fitness and normal diameter growth are not less than 30% of LCR. LCR itself can be categorized as high (>50%), moderate (30-50%), and low (<30%) [21].

The Live Crown Ratio (LCR) measurements showed that 22 trees had LCRs ranging from 30-59%, followed by 16 trees with LCR $\geq$ 60%, and 12 trees have $\leq$ 30% of LCR. This shows that some targeted trees need special care for their canopy symmetry, considering it is not only related to the fitness and photosynthesis of the tree, but also the beauty and safety of its surrounding environment. The canopy width is positively correlated with how the roots obtain minerals in the soil, which means photosynthesis will automatically affect the growth of many parts of the tree [22]. From photosynthesis, the tree crown provides carbohydrates for the roots, while the roots are absorbing water and nutrients from the soil to meet the canopy needs. The distribution of LCR percentage and the shape of the tree crown is shown in Figure 6.

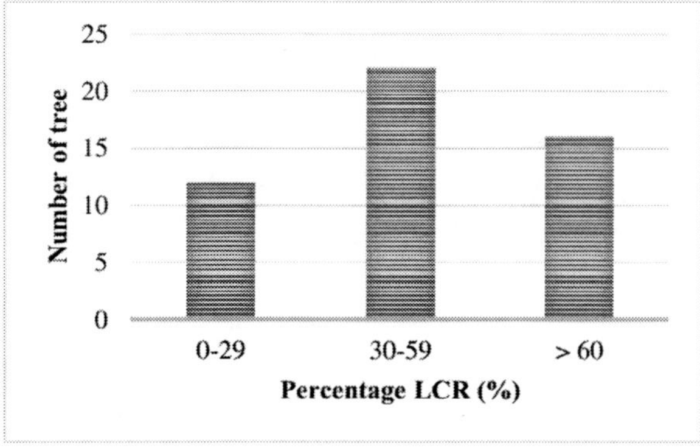


(a)

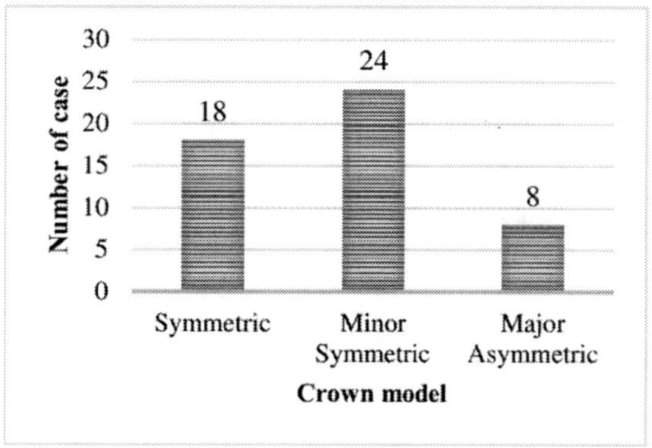


(b)

**Figure 6.** (a) Distribution of percentage LCR and (b) Crown model distribution of targeted tree

Every damage to every part of the targeted tree was observed (Figure 7). The most common defect on the roots was mechanical damage, such as cut off roots, wounds, and others caused by the surrounding's construction and buildings. Weathering and decay that occurs in trees can be caused by many factors, such as termite attack, fungus, or water infiltration. According to Tainter and Baker [23], decay usually starts from the middle (heart decay), but it can also start from the outside (tree bark). If decay occurs from the middle, it will reduce more than double the wood volume. Economically, trees that experience decay will have low value [24]. In terms of security, trees with large diameter and are located in human-accessible locations are considered dangerous, especially if the stems are weathered and porous. Therefore, they need regular maintenance so as not to endanger the tree structure.

The sympthom defects, such as subterranean termite tunnels, and the fungus fruiting bodies on the targeted tree must be anticipated because they can endanger the condition of the tree. The presence of subterranean termites was marked by soil traces in the tree. According to Zulkaidhah *et al.* [25], subterranean termites always come in contact with the soil because they live there. Meanwhile, the pests absorb food sources from the host tree so its health will be affected. The higher the attack frequency, the higher the aggressiveness, which means more damages to the host tree [26]. The wound, cavity, and cracks of the stem need to be anticipated so that they will not interfere with the tree structure. If not, they will trigger infection from various pathogens (fungi, bacteria, viruses, microplasma, nematodes) into plant tissue [27].

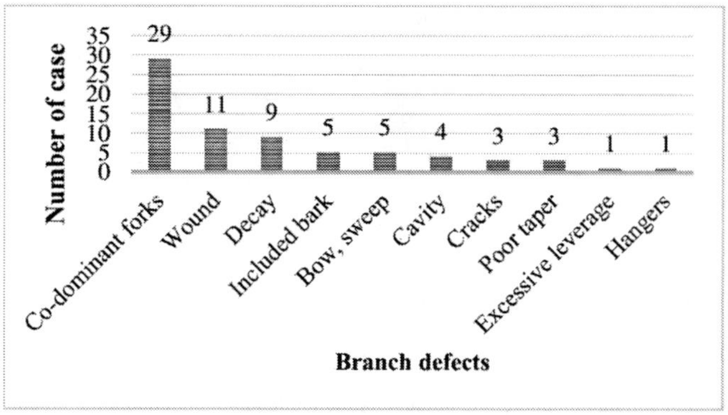


(a)

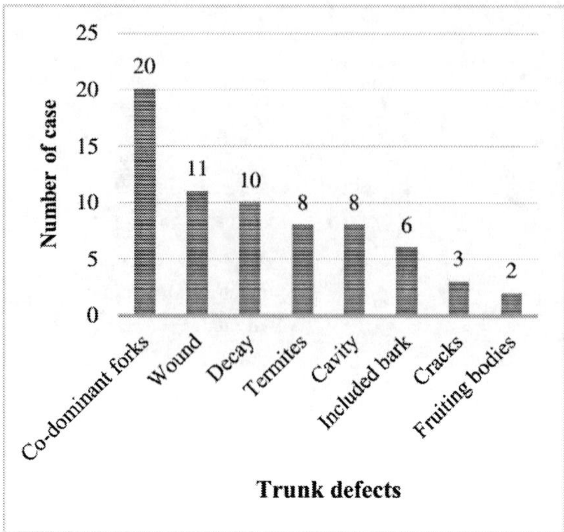


(b)

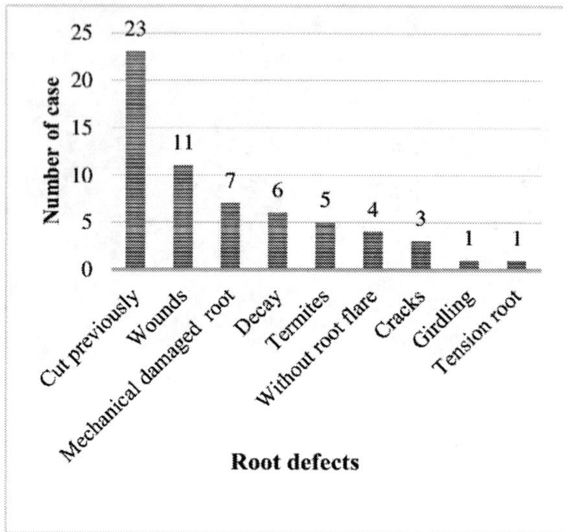


(c)

**Figure 7.** (a) Branch defects distribution and (b) Trunk defects distribution of targeted tree (c) Root defects distribution of targeted tree

The recapitulation result of visual assessment showed that 41 trees were at moderate risk of failure, 6 trees at high risk of failure, and 3 trees at low risk of failure. Six trees at high risk of failure were Trembesi (no 2), Trembesi (no 15), Kedawung (no 25), Kemenyan (no 28), Cempaka (no 32), and Sengon (no 49) (Figure 8).

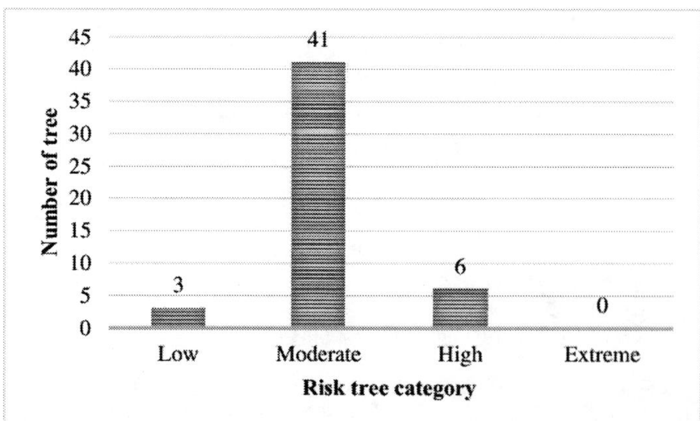


**Figure 8.** Frequency of risk tree category based on visual assessment

### 3.3. *Tomogram of trunk*

The *sonic tomography* result showed there were 41 trees in solid condition, 7 trees in moderate condition, and 2 trees in a deteriorated condition. This refers to the percentage of solid wood on the tomogram results of each tree (Figure 9). Trees with moderate conditions ranged from 70-79% of solid wood found in number

17 (Trembesi), number 33 (Cempaka), number 34 (Cempaka), number 36 (Petir), number 42 (Cempaka), number 43 (Kapuk), and number 44 (Kapuk). Meanwhile, the deteriorated trees are shown by number 37 (Sengon) and number 49 (Sengon).

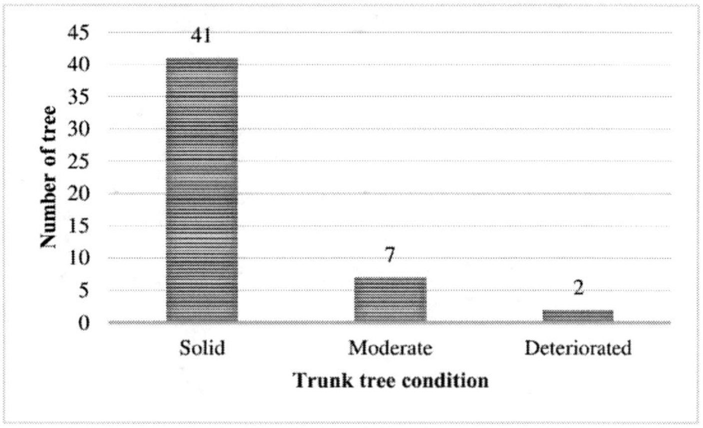


**Figure 9.** Frequency of trunk tree condition based on *sonic tomography*

### *3.4. Integration of visual and sonic tomography assessment results*

The result showed 31 trees were only fairly safe, while 5 of them were classified as safe. The dangerous and very dangerous trees have one tree each (Figure 10). Number 37, which is classified as dangerous, is a Sengon tree whose trunks are damaged due to a subterranean termite attack. Meanwhile, number 49 (Sengon) is categorized as very dangerous because it also has some heavily damaged trunks and was located on a puddle that causes heavy weathering. The LCR of both trees is at <45%.

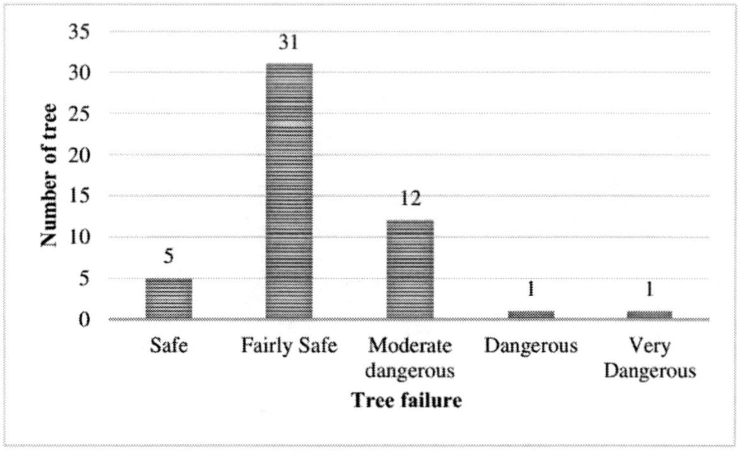


**Figure 10.** Tree failure risk based on visual dan *sonic tomography* observation

(a) (b)

**Figure 11.** Targeted tree number 37 (Sengon, a) and number 49 (Sengon, b)

*3.5. Tree fitness*

The percentage of solid wood is related to the integrity and fitness of the inner-part of the tree trunk. This percentage can also be used to determine the value t/R ratio, which is the percentage of solid wood mass to the diameter of the tree trunk. This method was adopted from [28] that requires trees to have a t/R ratio of more than 0.3. A value of t/R ratio that is less than 0.3 indicates that the percentage of the mass is too small compared to the diameter of the tree trunk, so the action needed is to cut down the tree.. The result showed that 48 trees have a t/R ratio of more than 3.0. Meanwhile, trees with a ratio of less than 0.3 only amounts to 2 trees (number 37 and 49) (Figure 12).

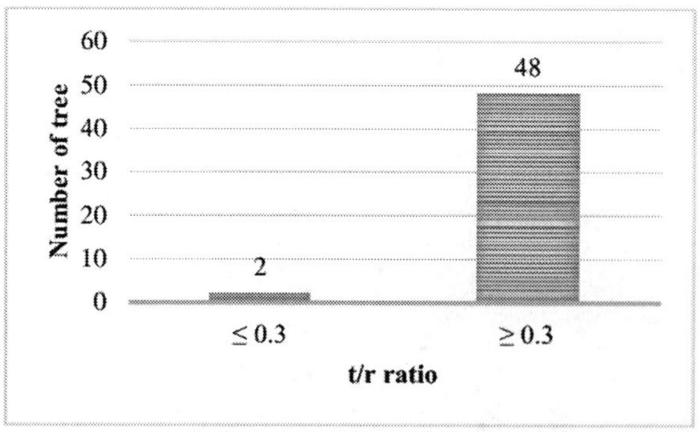


**Figure 12.** Frequency of t/R ratio of targeted tree

*3.6. Recommendation*

Several target trees have a tree trunk with a slope of >20° that requires treatment with the support of a certain height using reinforced construction made from two intersected galvanized iron trunks (10 cm in diameter). For some of the trees with codominant branches, dynamic slings can be made (Figure 13). Trees with a risk of failure are fairly dangerous to make similar observations no later than 6 months after the previous one. Meanwhile, trees with a risk of very dangerous failure are recommended to be cut down. Most of them are Sengon a species which are not historical tree.

(a) (b)

**Figure 13.** Simple lines for (a) Application technique of dynamic "sling" (b) Tree stand for leaning targeted tree

## 4. Conclusion

All targeted trees in Taman Margasatwa Ragunan (TMR) are old. Most of them have symmetrical and minor asymmetrical crowns. However, 68% of them have Live Crown Ratio (LCR) of less than 60%, which indicates the disproportionation of the canopy. The most common defects found in targeted trees are poor architecture, roots disturbance, decayed, and codominant branches. Regarding the risk of failure, most of the trees are in a reasonable and safe condition. Some of them are in dangerous and very dangerous conditions in urgent need of intensive treatments. The evaluation results of the tree fitnees through the t/r ratio state that in general, the trees are still in a solid condition.

## Reference

[1] Tim TMR 2014 *Inventarisasi Pohon di Taman Margasatwa Ragunan* (Jakarta: PT Airmas ntercontilnental)
[2] Ellison MJ 2005 Quantified tree risk assessment used in the management of amenity trees *J. Arboric.* **31**(2) 57-65
[3] Team Argus Electronic Gmbh 2003 *PICUS® Sonic Tomograph* (Germany: Argus Electronic Gmbh)
[4] Bethge K, Mattheck C, and Hunger E 1996 Equipment for detection and evaluation of incipient decay in trees *J. Arboric.* **20**(1) 13-37

[5] Mattheck CG and Bethge KA 1993 Detection of decay in trees with the Metriguard stress wave timer *J. Arboric.* **19**(6) 374-78

[6] Nicolotti G, Socco LV, Martinis R, Godio A, and Sambuelli L 2003 Application and comparison of three tomographic techniques for detection of decay in trees *J. Arboric.* **29**(2) 66-78

[7] Mattheck C and Breloer H 2003 *The Body Language of Trees: A Handbook for Failure Analysis* (London, UK: The Stationary Office)

[8] Schubert S, Gsell D, Dual J, Motavalli M, and Niemz P 2009 Acoustic wood tomography on trees and the challenge of wood heterogeneity *Holzforschung* **63**(1) 107-12

[9] Wunder J, Manusch C, Queloz V, Brang P, Ringwald V, and Bugman H 2012 Does increment coring enhance tree decay? New insights from tomography assessments *Can J. For. Res.* **43** 711-8

[10] Nicoletti G and Miglietta P 1998 Using high-technology instruments to detect decay in trees *J. Arboric* **24(6)** 297-302

[11] Wang X, Allison RB, Wang L, and Ross RJ 2007 *Acoustic Tomography for Decay Detection in Red Oak Trees Research Paper PFL-RP-642* (United States: Forest Laboratory, Department of Agriculture)

[12] Indahsuary N, Nandika D, Karlinasari L, and Santoso E 2014 Reliability of sonic tomography to detect agarwood in *Aquilaria microcarpa* Baill. *J Indian Acad Wood Sci.* **11**(1) 65-71

[13] Karlinasari L, Nandika D, Siregar IZ, Kusumawardhani DT, Rahman M, and A Hermadi 2018 *Laporan Akhir Pemeriksaan Kesehatan Pohon di Lanskap PT Bio Farma* (Bogor: Institut Pertanian Bogor)

[14] Pederson N 2010 External characteristics of old trees in the eastern deciduous forest *Natural Areas J.* **30**(4) 396-407

[15] Goodburn JM and Lorimer CG 1998 Cavity trees and course woody debris in oldgrowth and managed northern hardwood forests in Wiscondisn and Michigan *Canadian J. of Forest Research.* **28** 427-38

[16] Lindenmayer DB and Wood JT 2010 Long-term patterns in the decay, collapse, and abundance of trees with hollows in the mountain ash (*Eucalyptus regnans*) forests of Victoria, southeastern Australia *Canadian J. of Forest Research* **40** 48-54

[17] Stahle DW and Chaney PL 1994 A predictive model for the location of ancient forests *Natural Areas J.* **14** 151-58

[18] Adeyemii AA, Jimoh SO, and Adesoye PO 2013 Crown ratio models for tropical rainforests species in Oban Division of the cross River National Park, Nigeria *J. of Agricultre and Social Research* **13**(1) 63-76

[19] Putri KP, Supriyanto, and Sayufina L 2016 The health assessment of *Shorea* spp . seed source in FASP haurbentes using *JPHT* **13**(1) 37-48

[20] Zhao D, Kane M, and Borders BE 2012 Crown ratio and relative spacing relationships for loblolly pine plantations *Open J. of Forest* **2**(3) 110-5

[21] Djamhuri E, Supriyanto, Siregar IZ, Siregar UY, Sukendro A, Wilarso S, Pamungkas P, and Safei R 2006 *Petunjuk Teknis Seleksi Pohon Induk* (Bogor: Laboratorium Silvikultur, Fakultas Kehutanan, Institut Pertanian Bogor)

[22] Raharjo JT and Sadono R 2008 Crown model on several family of 9 years old of teak (*Tectona grandis* L.f) progeny trial *JIK* **2**(2) 89-95

[23] Tainter EH and Baker EA 1996 *Principles of Forest Pathology* (New York: John Wiley and Sons Inc.)

[24] Liang S, Wang X, Wiedenbeck J, Cai Z, and Fu F 2008 Evaluation of acoustic tomography for tree decay detection In *Proc. of the 15th Int. Symp. on Nondestructive Testing of Wood* (Duluth: Forest Product Society) **1** pp. 49-54

[25] Zulkaidhah, Djumali M, and Chandradewana B 2007 Identification of Termite Attacking Plants at Different Function Zones in Kebun Raya Unmul Samarinda (KRUS) *Forest J. Unmul.* **3**(2) 201-13
[26] Sunaryo, Rochman E, and Uji T 2007 Identification of Identification of Plant Damage in the Bali Botanic Garden by parasites *Environmental Engineering J.* **8**(2) 172-180
[27] Widyastuti, Sumardi, and Harjono 2005 *Patologi Hutan* (Yogyakarta: Gadjah Mada University Press)
[28] Mattheck, Claus, and Breloer H 1994 *The Body Language of Trees: A Handbook for Failure Analysis* (London: HMSO)

**Acknowledgement**
The authors are indebted to the management of Taman Margasatwa Ragunan Jakarta for research access and collaboration. In particular, we are grateful to Bapak Tata and his technical staff for guiding the field evaluation

IOP Conf. Series: Materials Science and Engineering **935** (2020) 012073 doi:10.1088/1757-899X/935/1/012073

# Knowledge transfer on sustainable bamboo forest management through social capital approach in Ngada Regency, Indonesia

**B D Prasetyo[1*], D Ekawati[1], D Djaenudin[1], E Y Suryandari[1], Indartik[1], Handoyo[1], G K Sari[1], and D Pamungkas[2]**

[1]Center for Research and Development of Social Economic Policies and Climate Change, Bogor, 16610, Indonesia

[2]Environment and Forestry Research and Development Institute of Kupang, East Nusa Tenggara, Indonesia

*Email: budiyanto.dwiprasetyo@gmail.com

**Abstract**. Bamboo is known as multi-purposes plants and currently, has potentially used as wood substitution products. The demand for bamboo from industrial sector is even higher. Although many countries have practiced bamboo cultivation, in Indonesia Bamboo tends to be allowed to grow naturally and still lack of treatment. The threat of unsustainable exploitation can cause the decreasing of bamboo productivity and lead to its scarcity. The sustainable bamboo forest management system then emerged as a solution. But the knowledge of such system has not been transmitted massively among bamboo farmers and owners. This paper will discuss the transfer of knowledge on sustainable bamboo forest management using social capital as an approach. This paper uses data from research project conducted in 2018 to 2019, which is located in Ngada Regency, East Nusa Tenggara, Indonesia. The results indicated that social capitals such as trust, organizations, social networks, and norms or rules are embedded with social institutions that exist among community. This research shows that *Sa'o* and BUMDes could be the most potential media as a means on transferring knowledge about sustainable bamboo forest systems. Both can be used as an entry point for any actors to run a small-scale bamboo industry development program. However, there are some potential obstacles could be occurred during the process of knowledge transfer, such as in-group feeling among indigenous community, the assumption that bamboo is a social good, not an economic good, do not concerned with the commercialization of bamboo, the complexity of inheritance law in the customary (*adat*) system, and the involvement of *adat* elites in political practices related local elections.

## 1. Introduction

It is undeniable that bamboo known as a multi-purpose plant and currently has potential used as a wood substitution product. Through the modern utilization of bamboo with higher value-added opportunities, it will increase the community income. This has been proven in China where bamboo plays an important role in the development of industry in rural areas in Anji City, Zhejiang Province [1]. While globally, the demand for bamboo from industry players is increasing [2]. This phenomenon has attracted industrial actors who want to boost the bamboo supply for the market. The bamboo engineering for industrial needs has been implemented in several countries such as in China, India, Vietnam, and the US [3,4]. Specifically, for example, there are industries that use bamboo as raw materials for constructions [2], furniture [5], foods [6], and even medicines [6-8].

Many countries have practiced bamboo cultivation such as in India dan China [9,10]. Despite this, bamboo cultivation is still rarely implemented in many regions in Indonesia. In this country, bamboo tends to be allowed to grow naturally and still lack of treatment. However, it does not mean that people in Indonesia are not familiar with bamboo. Studies show that in Java and Bali, for example, people know about bamboo and can differentiate bamboo species through traditional name [12]. They use bamboo as raw material for building houses, creating tools and appliances, as well as musical instruments [11,12]. Some parts of bamboo also could be eaten as an alternative food source [6]. Currently, as the tourism sector grows rapidly in Indonesia, many people create souvenirs using bamboo as its material for selling it to tourists [14].

Start in 2012, the Environmental Bamboo Foundation (EBF) introduced sustainable bamboo clumps management known as "hutan bambu lestari" (HBL) (sustainable bamboo forestry/SBF) system. Nevertheless, the practice of sustainable bamboo forestry (SBF) system has not much yet been applied among rural communities in Indonesia, even though there were supports from the government and private sectors to introduce these techniques. For example, such has been introduced to bamboo farmers in Ngada District where bamboo (*Dendrocalamanus asper)* forest areas are massive. However, the knowledge of sustainable forest management techniques is still limited distributed and has not been transferred adequately to the bamboo farmers and owners. At the end, it is likely that the problem would have appeared if the demand of bamboo for industry grows up but not simultaneously followed by the massive process of knowledge transfer among bamboo farmers and owners. This could potentially cause the decreasing of bamboo stocks. It also can lead to its scarcity on community bamboo forest areas. The extinction of bamboo is more likely to happen if this problem cannot be solved.

Knowledge transfer on SBFM is crucial to be conducted. However, it needs an appropriate strategy to do so. Social capital among the community, such as trust, social networks, organizations, and norms [15], that is embedded in social institutions [16] then could be used as an alternative approach to begin the process of knowledge transfer. Although there is still lack of literature regarding the use of social capital in introducing SBFM systems, such approach has commonly been used in, for example, transforming knowledge on agroforestry systems among farmers in many countries [16,17], particularly in Indonesia [19]. Therefore, this paper aims to discuss how the knowledge transfer on sustainable bamboo forest management would be processed if social capital were taken as an approach.

## 2. Materials and Methods

### *2.1. Location*

The study was conducted in 2018 to 2019 and it was located in Golewa Sub-District, Ngada Regency, East Nusa Tenggara Province, Indonesia. Ngada Regency is a district located at the Center of Flores Island. It has Bajawa as its district capital. It has 1,776.72 $km^2$ of land area and 708.64 $km^2$ of the waters area. The length of the beach is 102.318 km with the width of the North Coast waters 381.58 $km^2$ with the coast length 58.168 km, the area of the South Coast waters 327.06 $km^2$ with the coast length 44.15 km [20].

Geographically, Ngada Regency located in between 8°20'24.28"-8°57'28.39" South Latitude and 120°48"-121°11" East Longitude coordinates. It has geographical boundaries such as in the eastern area bordering with Nagekeo Regency, while in the western part with East Manggarai Regency, in the Northern area bordering the Flores Sea and in the Southern part bordering the Savu Sea. The population in Ngada Regency is 162,299 and it has three major tribes, such as Nagekeo, Bajawa, and Riung.

In terms of bamboo, Ngada Regency is highly potential to be developed on it. Bamboo in Ngada is dominated by *bambu betung* species (*Dendrocalamus asper*). This species grows well in Ngada as it is suitable for local climatic and geographical conditions. There is also strong connection between bamboo and local culture. Ngada people have customary laws to protect bamboo from any threat. It is called as *waja* and *ri'i*. Since 2012, there has been a laminated bamboo industry in Ngada Regency, which is still operating up to now. It has bamboo supply from the bamboo forest surround Golewa Sub-District. The bamboo industry cooperates with NGOs in assisting the bamboo farmers and owners in the supply of their industrial materials.

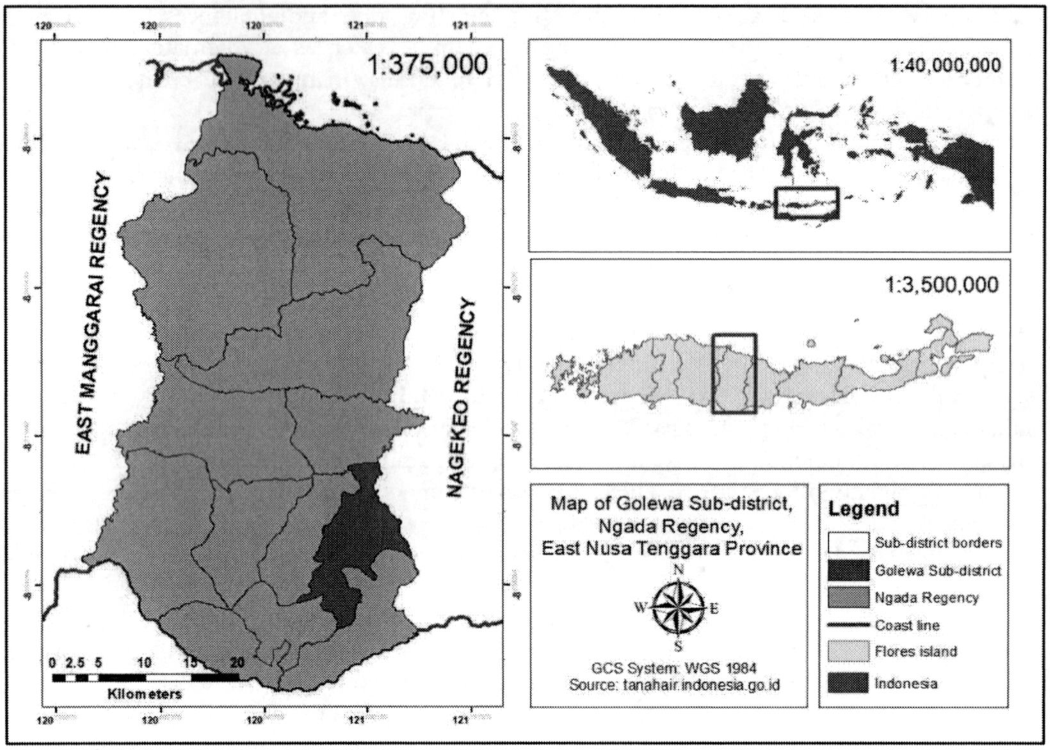


**Figure 1.** Map of research location in Ngada Regency

### 2.2. Tools and materials

This study used tools such as voice recorders and notebooks for recording interviews; visual aids tools such as flipchart papers, pens and markers for assisting participative rural appraisals and focused group discussions; as well as socio-economic tools such as questionnaires and other documentation equipment e.g. digital cameras, smartphones, and computers. Meanwhile, materials that was used for this study were informants, respondents, and other objects e.g. artifacts, momentums, and socio-cultural events that fit with the study objectives and appeared the during data collection.

### 2.3. Procedures

This study combined quantitative and qualitative approaches in order to target the research objective. There four methods of data collection processed as follow.

First was socio-economic survey. This method was conducted in 2018 and 2019 towards ten purposed villages in Golewa Sub-District, Ngada Regency, they were Village Were I, Were IV, Dadawea, Radabata, Ratogesa, Ulubelu, Wajamala, Langagedha, Rakateda II, dan Sarasedu I. The 111 respondents (N=111) were interviewed as samplings using semi-structured questions written on questionnaire sheets. However, of those questions that were asked to the respondents only some relevant topics would be discussed within the paper, such as the asset of social institutions among communities.

Second was ethnographic interview. This method was used to obtain detailed and in-depth information related to social and cultural realities [21]. In this case, this tool was employed to generate the information related to the construction of social capital on SBFM in Ngada. The interviews were conducted in July to August 2019 towards 8 key informants. The informants were selected with some criteria such as mastering knowledge of bamboo and its social, culture, and philosophical dimensions. The snowball technique was employed to determining the key informants. The open interview techniques were used with focusing on the study of customary dominance in bamboo forest management.

Third was focused group discussion (FGD). This method invited stakeholders to brainstorm about bamboo forest management, its problems as well as the possible solutions [21,22]. Two FGDs were done in 2018. The first was in Village Were I, in which focusing on the traditional bamboo management system. About 20 people attended this FGD with an equal gender proportion. The second FGD was in Village Ratogesa, which attended by 18 young people due to its stressing on the role of young people and its prospects on bamboo management. Meanwhile, another FGD was conducted in 2019 where by more than 30 villagers attended this meeting. Village Radabata was selected purposively as the host for the meeting. This FGD was focusing on identifying the role of BUMDes (village owned enterprises) and its potentials on supporting market of bamboo.

Fourth was participatory rural appraisal (PRA) [24]. This method was done towards two Sa'o (the smallest unit of traditional family system in Ngada) namely Sa'o Susuteme in Village Dadawea and Sa'o Gedhe Ana in Village Waia. About 25 people attended the PRA for each Sa'o. Those PRA were held in the first and at the end of March 2019. This method was aimed to identify the traditional bamboo forest management system among local communities.

### *2.4. Data analysis*
Quantitative data from socio-economic survey were statistically analyzed using frequency tables and presented in the narrative manner. Qualitative data from ethnographic interviews, FGD and PRA were analyzed using thematic analysis technique which relied on the needs of this study, such as identification of social capital and social institutions, as well as potential barriers to knowledge transfer through social capital among communities. Those primary data were triangulated by dialogue with the literature gathered from scientific journals and popular writings. The final results of the overall data analysis were presented in a descriptive narrative form.

## 3. Social Capital: An Invisible Power Inside the Social Institutions

Lyda Judson Hanifan was the person who first used the term social capital. In his writing, The Rural School Community Center, the words social capital appeared when there was an explanation of the community's ability to overcome various problems independently [25]. The word capital refers to something that is owned or an asset that can make the community grow properly. Social capital can be in the form of good will, friendship, mutual sympathy, and social relations and cooperation between individuals and families that form a social group [25].

Social capital can be an overall effort related to the mastery of relationships between groups in an institutional network that is based on knowing and recognizing each other [26]. Social capital is determined by social structure and access to that structure. The social structure consists of organizations and rules that govern its members [15]. There are three pillars supporting social capital, namely trust, access of information, and norms [15]. Moreover, social capital is also can be seen as a feature of social organizations consisting of networks, norms, trusts that facilitate coordination and collaboration among social groups' members in order to reach common benefits [27]. Therefore, social capital is basically a potential asset that owned by the social groups, such as organizations, institutions, communities, or societies, which is commonly used for overcoming common issues and reaching common goals. There are important elements that facilitate social groups activities, such as trust, networks, and norms.

In Indonesia, the discourse on social capital as a hidden power possessed by social groups has made a hot topic discussed among social scientists. For instance, there was a finding from social capital studies in Indonesia, which said that the strong social capital in the community is determined by effective communication and is colored by the similarity of concepts, competencies, connections, credibility as well as care among the group members [28]. While in the forestry sector, it was proven that social capital in Jambi and West Sumatera could encourage forest sustainability. This finding was contrast to the economic benefits of forests that lead to forest destruction [29]. Furthermore, there was a study about social capital bonding, which is an innovation based on traditional concept of social capital, that can be a power in encouraging the adaptation capacity of rural communities when carrying out infrastructure development in East Java [30].

Meanwhile, in the institutional perspective, social capital is perceived as an entity that is integrated with the organization or groups in society. Social capital has a connection with the capacity of the state in natural resource management [16]. Figure 2 shows that the private institutional model is the weakest in terms of social capital and state capacity. It is normally occurred with private companies. Whereas state management institutions have a strong state capacity but do not put social capital out as its priority. It makes its position then becomes weak. This is commonly happened in government institutions or BUMDes. On the other hand, strong social capital and weak state domination lie in the community-based management institutional model. This model is commonly found in natural resource management systems in Indonesia. Lastly, the most ideal institutional model is so called collaborative management. In this model, social capital and state capacity are equally strong in sustaining the institutional superstructure and infrastructure. Partnerships in social forestry schemes in Indonesia may be a good example for this model.

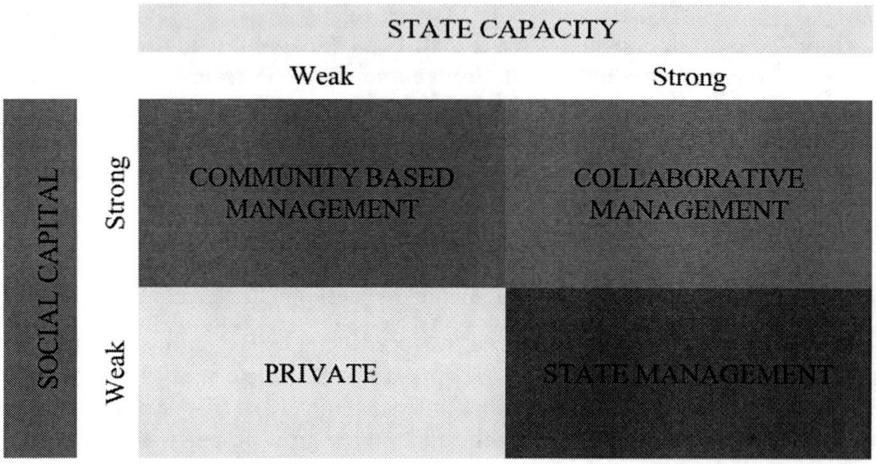


**Figure 2.** Natural resource management options based on state capacity and social capital [16]

## 4. Results and Discussions

### *4.1. Identifying social capital in Ngada*

The existence of social capital in a society cannot be seen explicitly. However, it will be more easily perceived through the institutions that exist in society, whether formal and informal [16]. Identification of social capital in Ngada Regency was carried out through the existing social institutions that exist in the community.

**Table 1.** Social institutions in Golewa, Ngada

| No. | Institutions | Answers | HBL-patrilineal | HBL-matrilineal | Non-SBFM | Total | % |
|---|---|---|---|---|---|---|---|
| Desa | | | Were 1, Were 4 | Dadawea, Radabata, Ratogesa, Ulubelu, Wajamala | Langagedha, Rakateda II, Sarasedu I | | |
| 1 | Adat | Yes | 27 | 47 | 17 | 91 | 81.98 |
| | | No | 2 | 10 | 8 | 20 | 18.02 |
| | | Total | 29 | 57 | 25 | 111 | |
| 2 | Farmers | Yes | 13 | 35 | 16 | 64 | 57.66 |
| | | No | 16 | 22 | 9 | 47 | 42.34 |
| | | Total | 29 | 57 | 25 | 111 | |
| 3 | Fisheries | Yes | 0 | 0 | 0 | 0 | - |
| | | No | 29 | 57 | 25 | 111 | 100.00 |
| | | Total | 29 | 57 | 25 | 111 | |
| 4 | Water users | Yes | 11 | 10 | 0 | 21 | 18.92 |
| | | No | 18 | 47 | 25 | 90 | 81.08 |
| | | Total | 29 | 57 | 25 | 111 | |
| 5 | Religious | Yes | 29 | 57 | 25 | 111 | 100.00 |
| | | No | 0 | 0 | 0 | 0 | - |
| | | Total | 29 | 57 | 25 | 111 | |
| 6 | Ethnicities | Yes | 1 | 3 | 1 | 5 | 4.50 |
| | (Place of origin) | No | 28 | 54 | 24 | 106 | 95.50 |
| | | Total | 29 | 57 | 25 | 111 | |
| 7 | Women/Arisan | Yes | 24 | 51 | 20 | 95 | 85.59 |
| | | No | 5 | 6 | 5 | 16 | 14.41 |
| | | Total | 29 | 57 | 25 | 111 | |
| 8 | Young people | Yes | 11 | 20 | 12 | 43 | 38.74 |
| | | No | 18 | 37 | 13 | 68 | 61.26 |
| | | Total | 29 | 57 | 25 | 111 | |
| 9 | Cooperation | Yes | 19 | 25 | 14 | 58 | 52.25 |
| | (gotong royong) | No | 10 | 32 | 11 | 53 | 47.75 |
| | | Total | 29 | 57 | 25 | 111 | |

Source. Primary data of socio-economic survey 2018-2019

Table 1 shows that of the nine social institutions that were asked to respondents, only fisherman groups were not available. It made sense as the location of the Golewa Sub-District was in mountainous areas, which was fertile. Another reason was because the sample villages were also located in the mountainous region. Of the nine social institutions identified on survey, religious groups and indigenous (adat) groups played a very dominant role in people's lives. Meanwhile, the results of in-depth interviews, FGDs, and PRA showed that, generally, social institutions in the community consisted of formal and informal institutions as seen in table 2.

**Table 2**. Formal and informal institutions in Golewa, Ngada

| Formal institutions | Informal institutions |
|---|---|
| BUMDes | Indigenous groups/Adat |
| Koperasi | Religious Groups |
| Farmer Groups | Young People |
| Village Government | Women/Arisan |

Source: Primary data of ethnographic interviews, FGD, and PRA, 2018-2019

The forms of formal institutions that have been identified in Ngada District and have relevance to sustainable bamboo management were BUMDes, Koperasi, Farmer Groups, and Village Governments. BUMDes is a village government-owned business institution that aims as a source of income for the village government. This institution is established based on government regulations.

IOP Conf. Series: Materials Science and Engineering **935** (2020) 012073 doi:10.1088/1757-899X/935/1/012073

BUMDes venture capital comes from the government as well. While, Koperasi is alternative institutions besides banks and moneylenders, which is operating in the countryside. The main function of Koperasi is generally as a savings and loan institution. Communities borrow money from Koperasi mostly to have business capital, finance children's schooling, and traditional party events purposes. Communities re-pay Koperasi loans by paying in installments from the profits of their business or harvests. On the other hand, farmer groups were apparently less popular in the Golewa. Many farmer groups have been formed but they were not working. The coffee farmer group was the most active group in the study site. Farmer groups are a place for its members to share knowledge, information, and control the harvest price by middlemen. Finally, the last institution is the village government. This is the most formal institution that exists at the village level. This institution is an extension of the central government. In addition to carrying out administrative functions of government, the village government is the front line for succeeding government development programs. The effectiveness of village government depends on the quality of the leaders and apparatus in it.

While informal institutions that have been identified and have a relevance to sustainable bamboo management were adat, religious groups, young people, and women/arisan. The adat institution was the most influential informal organizations in Ngada Regency. The results of the ethnographic interview explained, since the initial phase of its formation, the Ngada community has been using a tribal social system to regulate their living governance. The smallest unit of the tribe as a social system was so called as Sa'o. Sa'o managed several family heads who were members in one tribe. The regulation included a matter of kinship law. This kinship law would ultimately determine inheritance law and land tenure law as well as economic regulation.

In Village Were, most of the tribes applied a patrilineal system. Men played an important role in making decisions and especially about accepting inheritance. Whereas in other regions, such as Villages Radabata, Dadawea, Ratogesa and other lower regions, they had tendency to adhere to matrilineal kinship. In the matrilineal system mothers had the right to make decisions in a negotiation in Sa'o. Women also had the right to occupy Sa'o's traditional house as a place to live. However, in reality these women still sought the opinions and considerations of adult men in Sa'o before deciding on a matter. For example, it occurred when women called for decision in terms of determining the use of bamboo, land use for planting, types of commodities to be planted.

Sa'o operates a prohibition system to avoid ecosystem damage and crop failure. *Waja* and *rii* customary law are clear examples for this prohibition system. *Waja* is basically a prohibition for not doing anything in a location (garden) in a certain period of time. This purposed to restore the condition of the damaged ecosystem so that it can be reused in the future. While *rii* is a traditional ritual to operate a ban on taking the harvest of certain commodities where the Sa'o will use the commodity for certain purposes. The *rii* application also has a certain period of time to protect the yield to be harvested. Therefore, the harvest can be maximized as expected.

Relations in between Sa'o as well as in between tribes could be a powerful power for indigenous people in Ngada. The social relations can be seen clearly, for example when the traditional houses of Sa'o are being repaired. The depth of the relationship is shown by the presence of each tribe representatives and other Sa'o groups to the traditional ceremony. They usually bring rice, moke (palm wine), and livestocks (buffalo, pigs, chickens) to give to the Sa'o family who are repairing their house. All those gifts then were cooked and then to eat together.

Next is the religious group. The majority of Ngada people embrace Catholicism so that the existing religious social groups are generally organized by and have affiliation with Catholic Church. For example, the Catholic Base Group (Kelompok Umat Basis-KUB) whose members are also members of the Catholic Churches. In addition to the religious activities, KUB also carries out social activities such as choir training, village cleansing, and early childhood education. The influence of the church can also be seen in youth. The role of youth is organized by the Catholic Church institution, which is named as the Catholic Youth (Orang Muda Katolik - OMK). Almost all young people in the Catholic Church in Golewa are OMK members. OMK activities are generally in the fields of youth and sports, such as organizing volleyball competitions, football, choirs, and performing arts. OMK also facilitates training and sharing knowledge about entrepreneurship independently. Lastly, another informal group is arisan. This group is dominated by mothers and commonly formed in the neighborhood level. Arisan is not just a matter of who gets money for social gathering, but rather it has substantial matters on the communality and communication between members regarding the actual social problems being faced together.

*4.2. The potential institutions with powerful social capital*

Based on the results of social capital identification, it is known that there are two social institutions that have the potential to be utilized in order to support the transfer of knowledge on SBFM, namely Sa'o, which is an informal institution and BUMDes originating from formal institutions. The characteristics of those institutions are explained in Table 3.

**Table 3**. The potential of social capital to support SBFM

| Community based management (Sa'o) | State management (BUMDes) |
|---|---|
| 1. Trust is high<br>2. Bottom up<br>3. Obedience to leaders/patrimonial is high<br>4. Have a management system at the Sa'o level (extended family). It has impacts on regulating roles and functions in smaller level (nuclear family).<br>5. A high work ethic based on communal interest<br>6. Minimum of vested/political interest<br>7. Traditional (adat) legal base (waja/rii)<br>8. Traditional lobby approach (socio-cultural trust building) through ritual, traditional ceremony, live-in, and sustainable accompaniment. | 1. Trust is low (if leader were charismatic, trust could be high)<br>2. Top down<br>3. Obedience to leaders/patrimonial is low (If leader were charismatic, this obedience could be high)<br>4. Bureaucratic management system (formal organizational structure)<br>5. Project based work ethic<br>6. Strong vested/political interest<br>7. Government legal based (village decrees)<br>8. Political lobby approach and bureaucratic regulations. |
| e.g. Sa'o Negu Wulla in Village Dadawea | e.g. BUMDes in Village Radabata |

4.2.1. Sa'o

The most powerful and potential social capital in informal institution was reflected in Sa'o. This institution undoubtedly could be categorized as the community-based management model because it has strong social capital and weak state domination [16]. Sa'o is the smallest unit of tribal social system adopted by the Ngada people. Social cohesion in Sa'o is constructed through the elements of social capital, such as high trust. Trust grows from the intensive interactions that last for a very long time [28]. This certainly has been carried out since the ancestors of the Ngada people traveled to the foot of Mount Inerie in Flores from Rear Yunnan in China, as explained in ethnographic interview. The similarity of fate, history, ancestors and forged through the long journey of living together as a group makes trust naturally flourish among Sa'o members. It leads to create a patrimonial leadership in Sa'o, which is hereditary based on the obedience of its members. The leader of Sa'o will be customarily determined according to his descendant line.

Initiations at Sa'o appeared in a bottom up manner. The close relationship and the high trust among members made up each individual problem become a common problem. Therefore, to determine a solution it is often to held a deliberated meeting between fellow members of Sa'o. Patrimonialism in the Sa'o makes leadership run stable. The decision was taken after listening to input and information from traditional deliberated meeting. That decision was always firmly determined, undoubted, and implemented consistently by the leader of Sa'o and all members.

Sa'o also has a fairly good household management system. Sa'o is an extended family [31] and has many communal assets, which were jointly managed such as land, houses, gardens, or human

IOP Conf. Series: Materials Science and Engineering **935** (2020) 012073 doi:10.1088/1757-899X/935/1/012073

resources. The amount of the assets is regulated in the division of tasks that is appointed to each family leader in Sa'o (nuclear family) [31]. The division of tasks did not conduct authoritatively by the leader Sa'o but it was based on deliberated meeting. It aimed to avoid misunderstandings and to ensure that each Sa'o member accepted the decisions. Due to those processes, the predetermined work was usually carried out happily and fostered a positive work ethic that encouraged Sa'o members.

Sa'o is an informal social institution based on adat (traditional system) and it is non-political. Therefore, the adat elites must not put their political interests into Sa'o. It aims to maintain the trust, norms and rules of adat, as well as social cohesion that have long been developed. On the other hand, Sa'o is referring itself towards customary law. For example, there were customary law of *waja* and *rii*, which are applied for protecting natural resource. However, according to the PRA, currently, *waja* and *rii* need to be formally strengthened if they want to be implemented in today's society. Strengthening the role of customary law in village level could be processed throughout the village regulations [32]. For example, making village regulations (Perdes) regarding *waja* or *rii* in villages that has wide scale of bamboo forest

Adat lobbying is the approach that is importantly needed when put Sa'o as an entry point for the knowledge transfer activities, such as promoting the thousand-bamboo village program or SBFM knowledge. Building social and cultural trust between program implementers and the Sa'o community could be a part of lobbying. The lobbying can be also conducted through traditional rituals, traditional ceremonies, staying periodically in a time that is not short (live-in), and building a network with Sa'o members or its representative in order to carry out ongoing assistance. The possible output of the succeeded lobbying could produce the best model of institution, which is a collaborative management model [16].

4.2.2. BUMDes

In contrast, BUMDes could be categorized as part of state management institutions [16]. It is because BUMDes has a strong state capacity but it does not put social capital out as its priority. However, it is still possible to select BUMDes as formal institution, which is a potential media to support knowledge transfer on SBFM. This institution is often touted as the most possible entrance to transfer knowledge about SBFM to the community. It is because BUMDes has a strong legal basis and financial resources, which is certainty. Human resources at BUMDes can also be appointed at any time to work based on a decree or bylaws. Financially, the BUMDes funding source is already available, for example there is village funds that could be used as financial support for BUMDes.

The distribution of tasks is carried out based on formal documents. The work approach is BUMDes environment is top-down. It is often that the head of BUMDes takes decisions unilaterally without any deliberation processes. This makes BUMD bureaucratic and causes low obedience to leaders. The work ethic that grows is also low. Different situations will occur if the BUMDes is led by person who has social influence, whether inside or outside the organization. That influence can certainly change the work ethic, obedience, and trust in its members to be a positive direction.

BUMDes is also vulnerable to vested interests and political interests because of its existence that cannot be separated from village government institutions. The local democratic events such as regional general election to choose local leader e.g. governor, regent, major, or village head, known as pilkada, will strongly affect the performance of BUMDes. The will to power instinct of each village head or local political elites will make it possible to threaten the existence of BUMDes in carrying out their duties and functions.

An example for previous explanation was occurred in BUMDes Radabata. Although the foundation of the formal legal superstructure was strong and the infrastructure of funding and human resources were adequate, in practice, the implementation was not able to last long and not sustainable. Out of the six BUMDes institutions, only one was running with not optimal condition. The BUMDes of the brick plant was a clear example of which the state capacity did not always guarantee the sustainability of programs or projects. Conversely, potential social capital that is built from the bottom is very important needed in moving the community in the business sector.

However, there is still possible and necessary approach if want to use BUMDes as the entry point for applying knowledge transfer on SBFM towards community. It needs a strong political lobbying supported by formal and bureaucratic regulations in order to ensure that all plans would be run as expected.

### *4.3. Potential issues in knowledge transfer on SBFM*

Although the two institutions are very potential to be an option for knowledge transfer processes on SBFM, it does not mean that there will be no obstacles. There are some potential obstacles that might be encountered in the process of knowledge transfer on SBFM.

#### 4.3.1. In-group attitude

Although the survey shows that public is open to outsiders, nevertheless the attitude of in-group feeling still cannot be eliminated. In-group feeling is an attitude that is inherent in traditional society [33]. By neglecting this attitude, it will potentially ignite resistance local people to foreign people. The process of transferring knowledge about SBFM will be difficult and can even fail if there are problems in the community.

#### 4.3.2. Bamboo is a social goods, not economic goods

For local people, Bamboo is not a priority commodity for meeting economic needs. The fertile and cold in Golewa region makes local people choose the plantation sector (coffee, cloves, corn, sugar palm) and agriculture (ginger, pumpkin, and vegetables) as the main commodity for their income. Instead, bamboo is used in subsistent manner by almost all of its owners, both by the Sa'o community and individuals. Although there are some people who claim to have sold it, but the buyer usually is their own neighbor. And they sold it not for a large-scale industry that consistently sustainable. This will be a limiting factor for the knowledge transfer process on SBFM.

#### 4.3.3. Do not care about the commercialization of bamboo

The player of bamboo industry in Ngada is only PT Indo Bambu. Although they have implemented SBFM, the SBFM's knowledge and practice remains in the company workers. The bamboo owners only signing contract letter in order to give permit for company to manage their bamboos using SBFM system. Bamboo owners also sell their bamboo pieces. Meanwhile, through their workers, the company carries out the SBFM in the bamboo forest areas, which belong to bamboo owners. The attitude, which merely wanting to earn money easily (without work), will certainly hamper the process of knowledge transfer on SBFM.

#### 4.3.4. The complexity of inheritance law in customary systems

Most of the young people reluctant to be involved in managing bamboo with their family members. The reason was because they realized that they would not benefit the rights to sell bamboo. The youths (teenagers) who have participated in bamboo management in their family (both extended and nuclear family) claimed that they were only seconded without receiving a share of bamboo sales profits. Youth understanding of communal and private land rules and inheritance law is also poor. Therefore, young people tend to have a negative perception of bamboo management system, which operated in their family, such as bamboo is the business of the elderly, does not make money for youth, and is complicated to divide the land (inheritance) that has bamboo in it.

#### 4.3.5. The traditional elite involvement in political practices

The rise of political processes such as general election results in fragmented village communities. This political fragmentation has the potential to cause social and cultural fragmentation in society. Local elites who need votes will use the strategy to enter Sa'o and influence citizens to vote themselves. This is a significant latent challenge for facilitators of knowledge transfer on SBFM who will do their job. Furthermore, it is also potentially disrupting and even frustrating the efforts to transfer knowledge on SBFM. This socio-cultural fragmentation has potential for conflict outbrake.

## 5. Conclusion

This study concludes that social capital in the Ngada community is inherent in the existence of social organizations. Sa'o and BUMDes are two potential organizations to become strategic options as media for knowledge transfer on SBFM. However, some obstacles are still likely to arise when the two options are chosen, such as in-group feeling, bamboo is a social goods and not an economic goods, do not concerned with the commercialization of bamboo, the complexity of inheritance law in customary systems, and traditional elite involvement in political practices.

## References

[1] Ruiz-Pérez M, Maoyi F, Xiaosheng Y, and Belcher B 2001 Bamboo forestry in China: toward environmentally friendly expansion *J. For.* **99**(7) 14-20

[2] Walling M, and Puro N 2018 Bamboo Diversity and Utilization in Mokokchung District, Nagaland. EPH-International *J. Agric. Environ. Res.***4**(9) 14-25

[3] Tamang DK, Dhakal D, Gurung S, Sharma NP, and Shrestha DG 2013 Bamboo diversity, distribution pattern and its uses in Sikkim (India) Himalaya *Int. J. Sci. Res. Publ.* **3**(2) 1-6

[4] Khalil HPSA, Bhat IUH, Jawaid M, Zaidon A, Hermawan D, and Hadi YS 2012 Bamboo fibre reinforced biocomposites: A review *Mater. Des.* **42** 353-68

[5] Boran S, Ayfer DÇ, and Barbu MC 2013 Evaluation of bamboo as furniture material and its furniture designs *Pro. Ligno.* **9**(4)

[6] Choudhury D, Sahu JK, and Sharma GD 2012 Value addition to bamboo shoots: A review *J. Food Sci. Technol.* **49**(4) 407-14

[7] Singhal P, Bal LM, Satya S, Sudhakar P, and Naik SN 2013 Bamboo shoots: A novel source of nutrition and medicine *Crit. Rev. Food Sci. Nutr.* **53**(5) 517-34

[8] Hossain MF, Islam MA, and Numan SM 2015 Multipurpose uses of bamboo plants: A review *Int Res J Biol Sci* **4**(12) 57-60

[9] Das M 2019 Bamboo: Inherent source of nutrition and medicine *J. Pharmacogn Phytochem.* **8**(2) 1338-44

[10] Lu H-F, Cai C-J, Zeng X-S, Campbell DE, Fan S-H, and Liu G-L 2018 Bamboo vs. crops: an integrated emergy and economic evaluation of using bamboo to replace crops in south Sichuan Province, China *J. Clean Prod.* **177** 464-73

[11] Aseri GK, Sinha BKP, Agarwal D, Sharma A, Jain N, and Babu V 2012 New and improved method of bamboo cultivation in semi arid areas of Indian Thar desert *African J. Biotechnol.* **11**(84) 15002-11

[12] Arinasa IBK and Bagus I 2010 Bamboo Diversity and Utilization in Balinese Rituals at Angsri Village- Bali, Indonesia *J. Am. Bamboo Soc.* **23**(1) 29-37

[13] Setiawati T, Mutaqin AZ, Irawan B, Ana A, and Iskandar J 2018 Species diversity and utilization of bamboo to support life's the community of Karangwangi Village, Cidaun Sub-District of Cianjur, Indonesia *Biodiversitas J. Biol. Divers.* **18**(1)

[14] Walansendow A and Lagarense BES 2017 *Monitoring Growths and Product Quality of Local Souvenirs to Enhance Tourist Destination Image of North Sulawesi, Indonesia: Balancing Development and Sustainability in Tourism Destinations* (Springer) pp. 277-90

[15] Coleman JS 1988 Social capital in the creation of human capital *Am J. Sociol.* **94** 95-120

[16] Nurrochmat DR, Darusman D, and Ekayani M 2016 *Kebijakan Pembangunan Kehutanan Dan Lingkungan* (Bogor: Bogor Agricultural University)

[17] Partey ST, Sarfo DA, Frith O, Kwaku M, and Thevathasan NV 2017 Potentials of bamboo-based agroforestry for sustainable development in Sub-Saharan Africa: A review *Agric. Res.* **6**(1) 22-32

[18] Kittur BH, Sudhakara K, Kumar BM, Kunhamu TK, and Sureshkumar P 2016 Bamboo based agroforestry systems in Kerala, India: performance of turmeric (*Curcuma longa* L.) in the subcanopy of differentially spaced seven year-old bamboo stand *Agrofor Syst* **90**(2) 237-50

[19] Utami NWF, Arifin HS, Nurhayati HSA, and Wijaya S 2018 Community-based bamboo stands management in the Kali Bekasi Watershed, Indonesia *Environ. Nat. Resour. Res.* **8**(1)
[20] Kabupaten Ngada 2017 Review Rencana Terpadu dan Program Investasi Jangka Mengengah Kabupaten Ngada Tahun 2017-2021: Profil Kabupaten Ngada [Internet] p. II-8 Available from: http://sippa.ciptakarya.pu.go.id/sippa_online/ws_file/dokumen/rpi2jm/DOCRPIJM_1501387908Bab_2_PROFIL.pdf
[21] Reeves S, Kuper A, and Hodges BD 2008 Qualitative research methodologies: ethnography. *Bmj* **337** 1020
[22] Parker A and Tritter J 2006 Focus group method and methodology: current practice and recent debate *Int J Res Method Educ* **29**(1) 23-37
[23] Baral H, Jaung W, Bhatta LD, Phuntsho S, Sharma S, Paudyal K, Zarandian A, Sears R, Sharma R, Dorji T, and Artati Y 2017 *Approaches And Tools For Assessing Mountain Forest Ecosystem Services* (Center for International Forestry Research)
[24] Chilisia B 2014 Indigenous research methodologies *Can. J. Progr. Eval.* **29**(1)
[25] Hanifan LJ 2000 The rural school community centre (Reprinted in R. Putnam. Bowl Alone) p. 130
[26] Bourdieu P 1986 The forms of capital **249**
[27] Putnam RD, Leonardi R, and Nanetti RY 1994 *Making Democracy Work: Civic Traditions in Modern Italy* (Princeton university press)
[28] Ancok D 2003 Modal sosial dan kualitas masyarakat *Psikologika J Pemikir dan Penelit Psikol.* **8**(15) 4-14
[29] Ekawati S and Nurrochmat DR 2014 Hubungan modal sosial dengan pemanfaatan dan kelestarian hutan lindung *Jurnal Analisis Kebijakan Kehutanan* 29269
[30] Kusumastuti A 2015 Modal sosial dan mekanisme adaptasi masyarakat pedesaan dalam pengelolaan dan pembangunan infrastruktur *Masy. J. Sosiol.* 81-97
[31] Georgas J, Mylonas K, Bafiti T, Poortinga YH, Christakopoulou S, Kagitcibasi C, Kwak K, Ataca B, Berry J, Orung S, and Sunar D 2001 Functional relationships in the nuclear and extended family: A 16- culture study *Int. J. Psychol.* **36**(5) 289-300
[32] Prasetyo B 2017 Ketika Masyarakat Adat Menolak Punah: Peluang dan hambatan pencatatan hukum adat pengelolaan sumberdaya alam di Nusa Tenggara. Seri Studi Kebudayaan I: Pluralisme, Multikulturalisme, dan Batas-batas Toleransi Program Studi Antropologi Fakultas Ilmu Budaya, Universitas Brawijaya pp. 312-323
[33] Wright SC, Aron A, McLaughlin-Volpe T, and Ropp SA 1997 The extended contact effect: Knowledge of cross-group friendships and prejudice *J. Pers. Soc. Psychol.* **73**(1) 73

**Acknowledgements**
Acknowledgments were addressed to the institutions that supported all the research project activities, including: Forestry and Research, Development and Innovation Agency (FORDIA), Research and Development Center for Socio-Economic Policies and Climate Change (P3SEKPI), Environment dan Forestry Research and Development Institute of Kupang (BP2LHK-Kupang), the Environmental Bamboo Foundation (EBF)/*Yayasan Bambu Lestari* (YBL), ICRAF-ACIAR, PT Indobambu, and all institutional leaders who were involved in Kanoppi 2 project activities. Gratitude is also given to the communities in ten sample villages in Golewa Sub-District, The Ngada Regency Government, and ethnography key informants and all respondents of socio-economic surveys who have supported the research processes.

# AUTHOR INDEX